1
2
11 p 265 282
11 p 393 - 398
3
4
5
12
14
622 - 656
11 265 - 306

Plant Pathology

Plant Pathology

George N. Agrios

Professor and Chairman
Department of Plant Pathology
University of Florida, Gainesville

Third Edition

ACADEMIC PRESS, INC.
Harcourt Brace Jovanovich, Publishers
San Diego New York Berkeley Boston
London Sydney Tokyo Toronto

Dedicated with thanks to:
The 4-H Foundation,
its International Farm Youth Exchange Program,
and the Fulbright Grant Program,
for helping me get started in plant pathology.

ACADEMIC PRESS, INC.
San Diego, California 92101

United Kingdom Edition published by
ACADEMIC PRESS LIMITED
24-28 Oval Road, London NW1 7DX

ISBN 0-12-044563-8

PRINTED IN THE UNITED STATES OF AMERICA
89 90 91 9 8 7 6 5 4 3 2

CONTENTS

6 Genetics of Plant Disease 116

7 Environmental Effects on Infectious Plant Disease Development 147

8 Plant Disease Epidemiology 156

PREFACE

Since the second edition of *Plant Pathology* appeared, numerous new developments have taken place in plant pathology and in the more general fields of biology and agriculture. Biology has been revolutionized by developments in molecular biology and genetic engineering. Agriculture, which has already made considerable progress toward more plentiful food production, is now rapidly embracing the new biology and is looking toward greater gains from biotechnologies employing tissue culture and genetic engineering. Plant pathology is at the center of developments in plant and microbial molecular biology and is a main focal point of their biotechnological applications to agriculture. Plant pathology provides and benefits from information on gene location and action, gene vectors and transformation mechanisms, and also from information on plant tissue culture systems used for biological studies or for agricultural biotechnologies.

Plant pathology has also responded to the need for reduced useage of pesticides for plant disease control by increased research on mechanisms of disease induction and resistance and on breeding for resistant varieties, by developing basic epidemiological information on pathogen survival, infection, and spread, by increased research on and applications in biological control of plant diseases, and by developing systemic fungicides effective in much smaller amounts and aimed only at the specific pathogen. Furthermore, some plant pathogenic bacteria and viruses have received extraordinary attention recently for their role and potential as vehicles for moving useful genetic material among plants, and additional plant pathogens may soon be shown to have similar potential. In the meantime, a myriad of new information items have been published on the specific pathogens and diseases discussed in this book.

The third edition of *Plant Pathology* incorporates all the important new developments in the field. Although it follows closely the organization and format of the last edition, it also includes two new chapters, "Plant Disease Epidemiology" and "Applications of Biotechnology in Plant Pathology." Extensively updated new information has also been added about the history of plant pathology, the stages in the development of disease, the chemical weapons of attack by pathogens, and the genetics of plant disease. The chapter on control of plant diseases has been reorganized and expanded to include recent developments on all aspects of integrated pest management and of new chemicals and biological agents used for plant disease control.

The chapters dealing with specific plant diseases have also been revised to include all pertinent developments that have occurred since the second edition. Special attention has been given to the recent taxonomy of plant pathogenic fungi, prokaryotes, viruses, and nematodes. The information on distribution, biology, epidemiology, and control of each pathogen and disease has been updated. Information on viral diseases of plants has been reorganized and expanded to reflect the increasing importance of these diseases among plants. Finally, recent knowledge on the effects of environmental factors such as low or high temperature, air pollution, and acid rain on plants is discussed.

Several colleagues provided comments and suggestions for improvement of the second edition, or reviewed and commented on the manuscript of the third edition. I am much obliged to them all for their constructive criticisms and suggestions, which I made every effort to incorporate into the third edition.

I thank Drs. Richard Berger, Laurence Madden, and Gail Schumann for reviewing the chapter on "Epidemiology of Plant Disease." I also thank all individuals and organizations who provided me with photographs; they are acknowledged individually in the text. Finally, I want to publically thank my wife Annette, who most enthusiastically did all the typing, and some retyping of the entire manuscript of the third edition, and had the even more tedious job of cutting and pasting the old and the new manuscript into a unified and, hopefully, cohesive work.

George N. Agrios

I

General Aspects

1 INTRODUCTION TO PLANT PATHOLOGY

The welfare of plants is of particular interest to those most directly concerned with the growth of plants and the manufacture and distribution of plant products. Such persons include not only farmers and workers in industries that process agricultural products but also innumerable workers in supporting industries whose livelihood depends on making equipment or products used in processing plant products—for example, machinery for textile and canning industries—or on distributing the raw or manufactured agricultural products. Most importantly, however, the welfare of plants should be of concern to every one of us as growers of plants for food or pleasure, as individuals concerned with the beauty and safety of our natural environment, and particularly, as consumers of plants and of the endless series of products derived from plants.

The growth and yield of plants depend on the availability of nutrients and water in the soil where they grow and on the maintenance within certain ranges of such environmental factors as temperature, moisture, and light. Anything that affects the health of plants is likely to affect their growth and yield and may seriously reduce their usefulness to humans. Plant pathogens, unfavorable weather, weeds, and insect pests are the most common causes of reduction or destruction of plant growth and production. Plants suffer from diseases whose causes are similar to those affecting animals and humans. Although we have no evidence that plants feel pain or discomfort, the development of disease follows the same steps and is usually as complex in plants as it is in animals.

Plant pathology is the study of (1) the living entities and the environmental conditions that cause disease in plants; (2) the mechanisms by which these factors produce disease in plants; (3) the interactions between the disease-causing agents and the diseased plant; and (4) the methods of preventing or controlling disease and alleviating the damage it causes.

Plant pathologists study diseases caused by fungi, bacteria, mycoplasmas, parasitic higher plants, viruses, viroids, nematodes, and protozoa. They also study plant disorders caused by the excess, imbalance, or lack of such physical or chemical factors as moisture, temperature, and nutrients. Plant damage caused by insects, humans, or other animals is not ordinarily included in the study of plant pathology.

Plant pathology uses the basic knowledge and techniques of botany, mycology, bacteriology, virology, nematology, plant anatomy, plant physiology, genetics, molecular biology, genetic engineering, biochemistry, horticulture, tissue culture, soil science, forestry, chemistry, physics, meteorology, and many other branches of science. Plant pathology profits from advances in any one of these sciences, and many advances in other sciences have been made in an attempt to solve plant pathological problems. A knowledge of at least the basic facts of the related sciences is indispensable for the efficient performance of any plant pathologist.

As a science, plant pathology attempts to increase our knowledge of the causes and the development of plant diseases. It is also a science with a more practical goal: to develop controls for all plant diseases in order to save the produce that today is destroyed by plant diseases and to make it available to the hungry and ill-clothed millions of our increasingly overpopulated world.

The Concept of Disease in Plants

A plant is healthy, or normal, when it can carry out its physiological functions to the best of its genetic potential. These functions include normal cell division, differentiation, and development; absorption of water and minerals from the soil and translocation of these substances throughout the plant; photosynthesis and translocation of the photosynthetic products to areas of utilization or storage; metabolism of synthesized compounds; reproduction; and storage of food supplies for overwintering or reproduction.

Whenever plants are disturbed by pathogens or by certain environmental conditions and one or more of these functions is interfered with beyond a certain deviation from the normal, then the plants become diseased. The primary causes of disease are either pathogenic living organisms **(pathogens)** or factors in the physical environment. The specific mechanisms by which diseases are produced vary considerably with the causal agent and sometimes with the plant. At first the plant reacts to the disease-causing agent at the site of affliction. The reaction, of a chemical nature, is invisible. Soon, however, the reaction becomes more widespread, and histological changes take place that manifest themselves macroscopically and constitute the symptoms of the disease.

Affected cells and tissues of diseased plants are usually weakened or destroyed by disease-causing agents. The ability of such cells and tissues to perform their normal physiological functions is reduced or completely eliminated; as a result, plant growth is reduced or the plant dies. The kinds of cells and tissues that become infected determine the type of physiological function that will be affected first. Thus, infection of the root (for example, root rots) interferes with absorption of water and nutrients from the soil; infection of the xylem vessels (vascular wilts, certain cankers) interferes with translocation of water and minerals to the crown of the plant; infection of the foliage (leaf spots, blights, mosaics) interferes with photosynthesis; infection of the cortex (cortical canker, viral and mycoplasmal infections of phloem) interferes with the downward translocation of photosynthetic products; flower infections (bacterial and fungal blights, viral, mycoplasmal, and fungal infections of flowers) interfere with reproduction; and infections of fruit (fruit rots) interfere with reproduction or storage of reserve foods for the new plant (Figure I-1).

In contrast, there is another group of diseases in which the affected cells, instead of being weakened or destroyed, are stimulated to divide much faster (hyperplasia) or to enlarge a great deal more (hypertrophy) than normal cells.

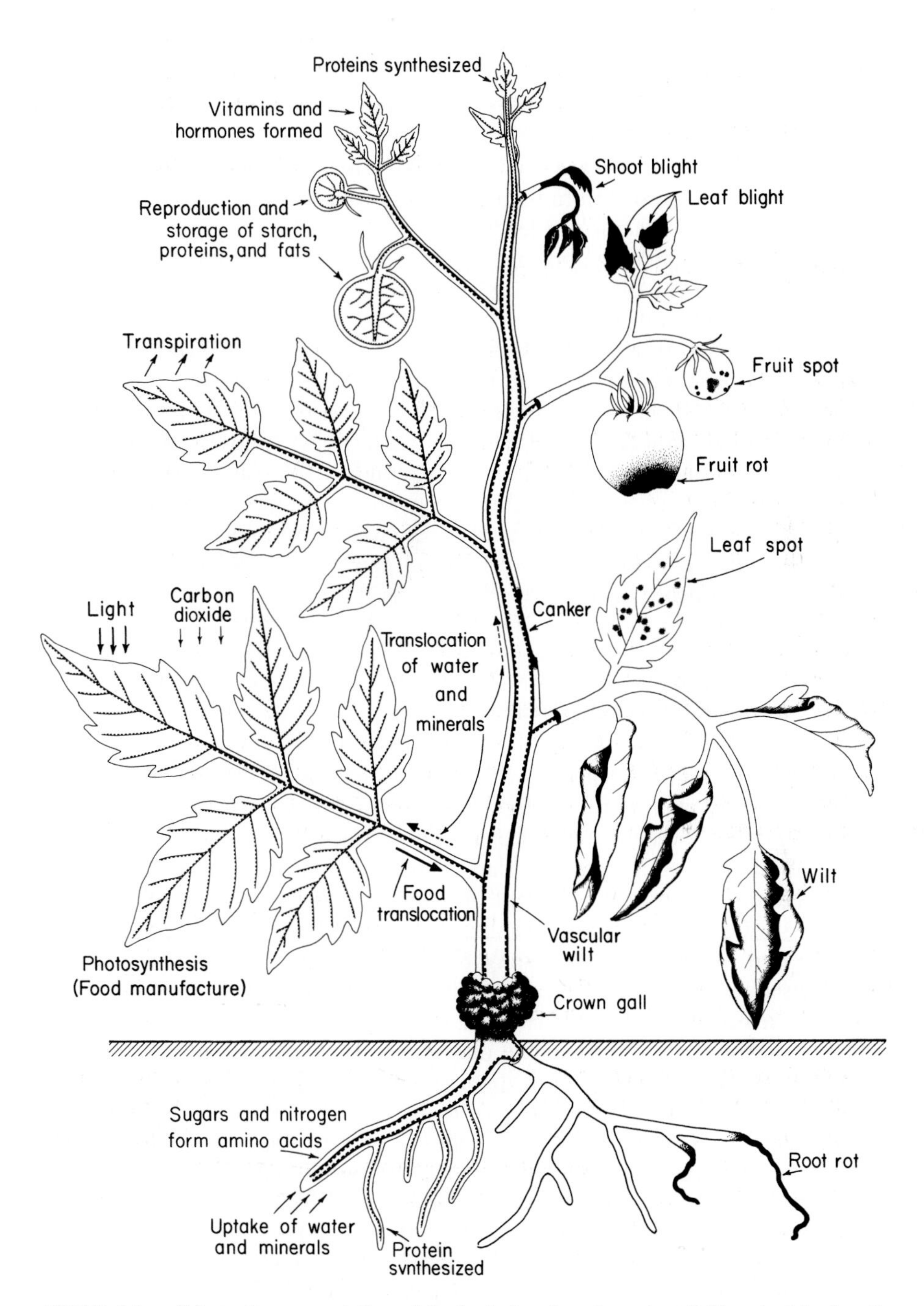

FIGURE 1-1 Schematic representation of the basic functions in a plant (left) and of the interference with these functions (right) caused by some common types of plant diseases.

Such hyperplastic or hypertrophied cells result in the development of usually nonfunctioning, abnormally large, or abnormally proliferating organs or in the production of amorphous overgrowths on normal-looking organs. Overstimulated cells and tissues not only divert much of the available foodstuffs to themselves and away from the normal tissues, but frequently, by their excessive growth, crush adjacent normal tissues and interfere with the physiological functions of the plant.

In plants, then, **disease** can be defined as the malfunctioning of host cells and tissues that results from their continuous irritation by a pathogenic agent or environmental factor and leads to the development of symptoms. Disease is a condition involving abnormal changes in the form, physiology, integrity, or behavior of the plant. Such changes may result in partial impairment or death of the plant or its parts.

Pathogens may cause disease in plants by (1) weakening the host by continually absorbing food from the host cells for their own use; (2) killing or disturbing the metabolism of host cells through toxins, enzymes, or growth-regulating substances they secrete; (3) blocking the transportation of food, mineral nutrients, and water through the conductive tissues; and (4) consuming the contents of the host cells upon contact. Diseases caused by environmental factors result from extremes in the conditions supporting life (temperature, moisture, light, and so on) and in excessive or deficient amounts of chemicals absorbed or required by plants.

Classification of Plant Diseases

Tens of thousands of diseases affect cultivated plants. On the average, each kind of crop plant can be affected by one hundred or more plant diseases. Each kind of pathogen may affect anywhere from one variety to several dozen or even hundreds of species of plants. To facilitate the study of plant diseases, therefore, they must be grouped in some orderly fashion. This is necessary also for the identification and subsequent control of any given plant disease. Any one of several criteria may be used to classify plant diseases. Plant diseases are sometimes classified according to the symptoms they cause (root rots, cankers, wilts, leaf spots, scabs, blights, anthracnoses, rusts, smuts, mosaics, yellows), according to the plant organ they affect (root diseases, stem diseases, foliage diseases, fruit diseases), or according to the types of plants affected (field crop diseases, vegetable diseases, fruit tree diseases, forest diseases, turf diseases, diseases of ornamental plants). The most useful criterion for classification of a disease, however, is the type of pathogen that causes the disease (see Figures 1-2 and 1-3). The advantage of such a classification is that it indicates the cause of the disease, knowledge of which suggests the probable development and spread of the disease and also possible control measures. On this basis, plant diseases are classified as follows:

I. Infectious, or biotic, plant diseases
   1. Diseases caused by fungi
   2. Diseases caused by prokaryotes (bacteria and mycoplasmas)
   3. Diseases caused by parasitic higher plants
   4. Diseases caused by viruses and viroids

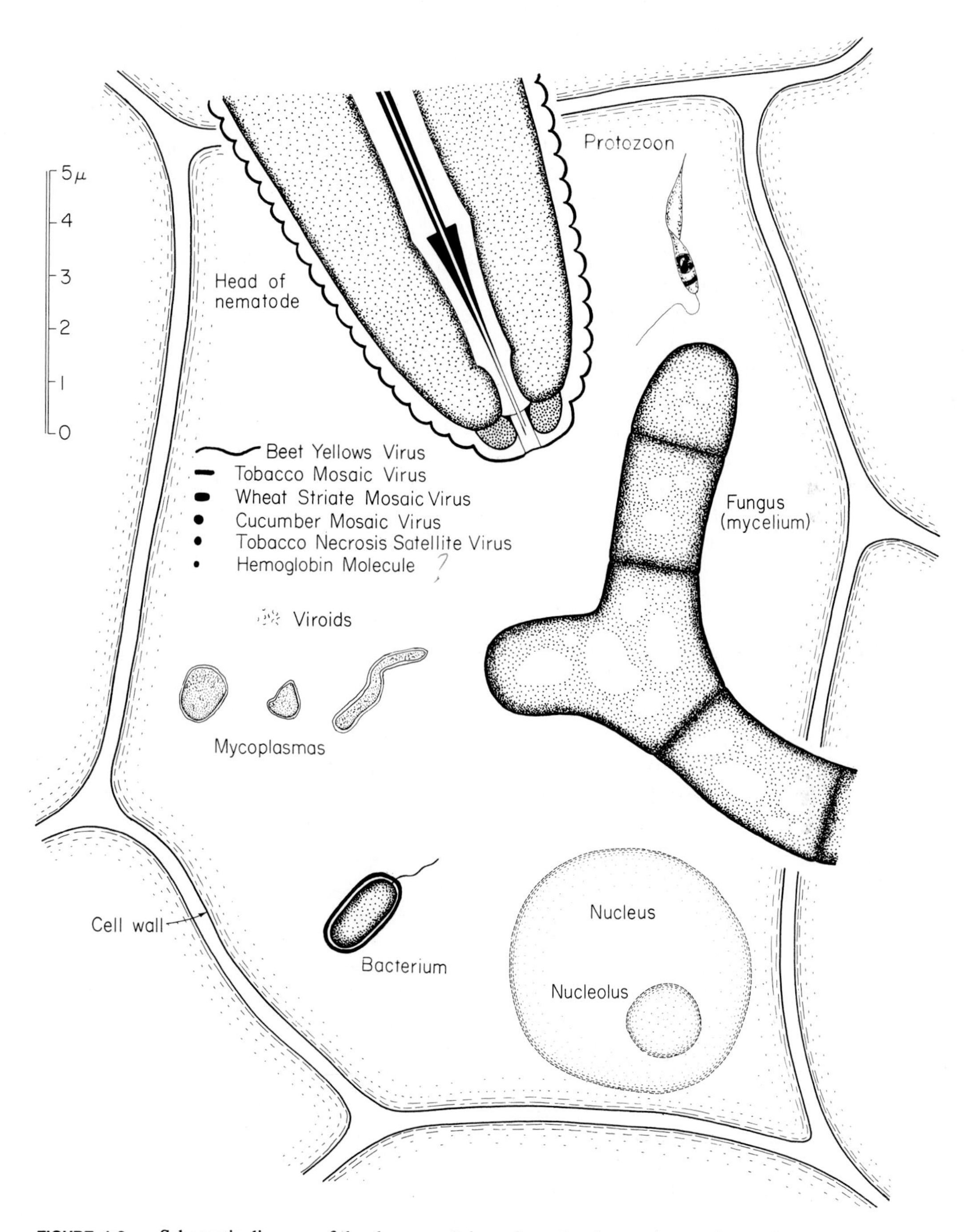

FIGURE 1-2 Schematic diagram of the shapes and sizes of certain plant pathogens in relation to a plant cell.

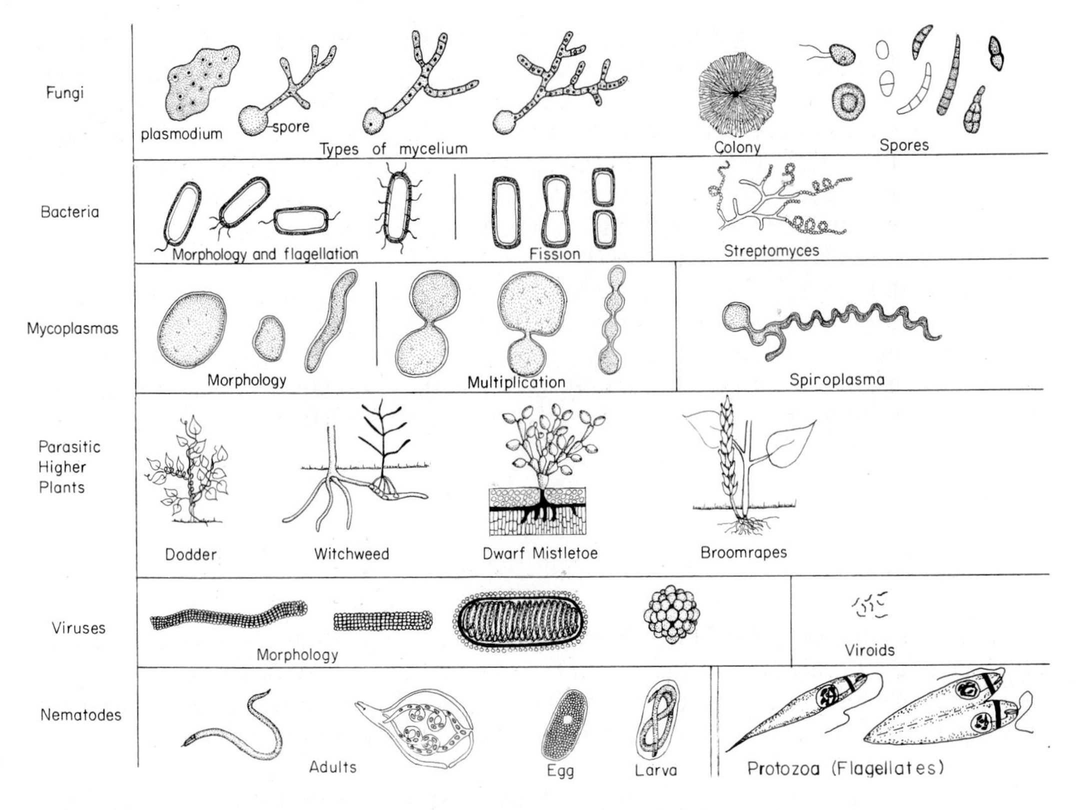

FIGURE 1-3 Morphology and multiplication of some of the groups of plant pathogens.

5. Diseases caused by nematodes
6. Diseases caused by protozoa

II. Noninfectious, or abiotic, diseases are diseases caused by:

1. Too low or too high a temperature
2. Lack or excess of soil moisture
3. Lack or excess of light
4. Lack of oxygen
5. Air pollution
6. Nutrient deficiencies
7. Mineral toxicities
8. Soil acidity or alkalinity (pH)
9. Toxicity of pesticides
10. Improper cultural practices

History of Plant Pathology

That human beings became painfully aware of plant diseases in antiquity is evidenced by the inclusion in the Old Testament of blasting and mildew, along with human diseases and war, among the great scourges of mankind. The Greek philosopher Theophrastus (370–286 B.C.) was the first actually to study and write about diseases of trees, cereals, and legumes. Among other things, he observed that plant diseases generally were more severe in lowlands than on the hillsides and that rust was more common on cereals than on legumes. His approach, however, was observational and speculative rather than experimental. The Romans became so painfully aware of the devastating effects of the rusts of grain crops that they created a special rust god, Robigo, to protect them from these diseases. In the spring of each year, just before the rusts appeared, they celebrated the Robigalia, a special holiday that involved sacrifices of red dogs and sheep in an attempt to appease Robigo.

During the following 2000 years, little was added to the knowledge of plant pathology, although references to the ravages of plant diseases appeared in the writings of several contemporary historians. The invention of the compound microscope around the middle of the seventeenth century opened a new era in the life sciences. The anatomy of plants was studied and described, and the fungi, bacteria, and many other microorganisms were discovered.

Discovery of the Role of Fungi

In 1729, Micheli observed that dust particles taken from a fungus and placed on freshly cut slices of melon generally reproduced the same kind of fungus. He concluded that the dust particles were the seeds (spores) of the fungus and that the occasional different fungi that appeared were produced from spores carried through the air.

In 1755, Tillet added the black dust from bunted wheat to seed from healthy wheat and observed that bunt was much more prevalent in plants produced from such seed than from nondusted seed. He thus showed that bunt, or stinking smut, of wheat is a contagious plant disease. He also showed that its occurrence can be reduced by seed treatments. Tillet, however, believed that it was a poisonous substance contained in the dust, rather than living microorganisms, that caused the disease.

In 1807, Prévost proved conclusively that bunt is caused by a fungus; he studied the spores, their production, and germination. He was able to control the disease by dipping the seed in a copper sulfate solution, and he pointed out the importance of the environment in induction and development of the disease. Prévost's findings, however, were ahead of his time and were rejected by almost all his contemporaries, who believed in spontaneous generation of microorganisms and of disease and that the microorganisms and their spores were the result rather than the cause of disease.

The devastating epidemics of late blight of potato in Northern Europe, particularly Ireland, in the 1840s tragically dramatized the effect of plant diseases on humans and greatly stimulated interest in their causes. The destruction of the potato crop in Ireland in 1845 and 1846 caused widespread famine, which resulted in the death of hundreds of thousands of people and the emigration of more than one and a half million people from Ireland to the United States. Several investigators described various aspects of the disease and of the pathogen and suggested that a fungus was the cause of late blight. In 1861 deBary finally proved experimentally that the fungus *Phytophthora infestans* was the cause of the disease.

DeBary (1853), working at first with smut and rust fungi, established conclusively that fungi are causes, not results, of plant disease. He described the microscopic structure and development of many smut and rust fungi and the relationships of these fungi to the tissues of the diseased plants. DeBary also made great contributions with his studies of the Peronosporales and the diseases they incite (downy mildews), especially the late blight of potato, his discovery of the occurrence of two alternate hosts in the rusts, and his studies of the physiology of the *Sclerotinia* rot diseases of carrots and other vegetables.

Brefeld (1875, 1883, 1912) contributed greatly to plant pathology by introducing and developing modern techniques for growing microorganisms in pure culture. He was greatly assisted by the methods and refinements developed by Koch, Petri, and others. Brefeld also studied and illustrated the complete life cycles of the smut fungi and diseases of cereal crops.

Control of Plant Diseases

In 1878, the downy mildew of grape was introduced into Europe from the United States. It spread rapidly and threatened to ruin the vineyards of Europe. In 1882, Millardet noticed that vines that had been sprayed with a bluish-white mixture of copper sulfate and lime to deter pilferers retained their leaves throughout the season, whereas the leaves of untreated vines were killed by the disease and fell to the ground. After numerous spraying experiments, Millardet concluded in 1885 that a mixture of copper sulfate and hydrated lime could effectively control the downy mildew of grape. This mixture became known as "Bordeaux mixture," and its success in controlling downy mildews and many other foliage diseases was spectacular. Even today, Bordeaux mixture is one of the most widely used fungicides around the world. The discovery of Bordeaux mixture gave great encouragement and stimulus to the study of the nature and control of plant diseases.

In 1882, Ward, investigating the coffee rust in Ceylon, observed that the disease was much more destructive in the uniform stands of coffee tree plantations in Ceylon than in Brazil, where coffee trees are often surrounded by other kinds of trees, and he warned against monoculture of crop plants in large areas to avoid such destructive outbreaks. In 1913, Riehm introduced seed treatment with organic mercury compounds, and such treatments became routine until the 1960s when, because of their general toxicity, all mercury-containing pesticides were removed from the market. In the meantime, in 1934, Tisdale discovered the first dithiocarbamate fungicide (thiram), which led to the development in the subsequent ten years of a series of effective and widely used fungicides including ferbam, zineb, and maneb (Tisdale and Williams, 1934). Many other important protective fungicides followed, and in 1965 von Schmeling and Kulka introduced the first systemic fungicide, carboxin (von Schmeling and Kulka, 1966). Antibiotics were first used to control plant diseases in 1950. In 1954, a few strains of bacterial plant pathogens were noticed that were resistant to certain antibiotics and, in 1963, strains of fungal plant pathogens were found that were resistant to certain protectant fungicides. It was in the 1970s, however, when the use of systemic fungicides became widespread, that, in numerous diseases, new races of their pathogens appeared that were resistant to a fungicide that had previously been effective. In 1972, New and Kerr introduced the first commercial biological control of a bacterial plant disease (crown gall) by use of another bacterium.

Physiology of Plant Disease

Once it became apparent, through the work of Tillet, Prévost, and deBary, that fungi are the causes rather than the results of plant disease, efforts began to elucidate the mechanisms by which fungi cause disease. In 1886, deBary, working with the Sclerotinia rot diseases of carrots and other vegetables, noted that host cells were killed in advance of the invading hyphae of the fungus and that juice from rotted tissue could break down healthy host tissue. Boiled juice from rotted tissue had no effect on healthy tissue. DeBary concluded that the pathogen produces enzymes and toxins that degrade and kill plant cells from which the fungus can then obtain its nutrients. In 1905 Jones reported the involvement of cytolytic enzymes produced by bacteria in several soft rot diseases of vegetables. In 1915, Brown recognized the significance of pectic enzymes in pathogenesis of some fungi on plants. It was not until the 1940s, however, that cellulases were implicated in plant disease development. After deBary, many attempted to show that most plant diseases, particularly vascular wilts and leaf spots, were caused by toxins secreted by the pathogens, but those claims could not be confirmed. The first involvement of a toxin in a plant disease that was later confirmed was suggested in 1933 by Tanaka, who noted that culture filtrates of a species of the fungus *Alternaria,* when applied on the fruit of susceptible pear varieties, could cause black spots that were identical to those caused by the fungus itself. Tanaka's work, however, went unnoticed until the early 1950s. In the meantime, Clayton in 1934 confirmed an earlier suggestion by Johnson and Murwin (1925) that the bacterium *Pseudomonas tabaci,* which causes the wildfire disease of tobacco, produces a toxin that is responsible for the bacteria-free chlorotic zone ("halo") surrounding the necrotic leaf spots that contain the bacteria. The wildfire toxin was the first toxin to be isolated in pure form by Woolley *et al.* in 1952. In 1947, Meehan and Murphy showed that a species of the fungus *Helminthosporium,* that attacked and caused blight on oats of the variety Victoria and its derivatives only, produced a toxin, named victorin, that could induce the symptoms of the disease only on the susceptible varieties. Several other fungal and bacterial toxins were subsequently identified, as were several distinctive mechanisms of action affecting specific sites on mitochondria, chloroplasts, plasma membrane, specific enzymes, or specific cells such as guard cells.

The observation that in many diseases the affected plant exhibits stunting, excessive growth, tumors, and other growth abnormalities led many investigators to look for imbalances in growth regulator levels. In 1926, Kurosawa showed that the excessively tall, slender seedlings of the rice foolish seedling disease caused by the fungus *Gibberella (Fusarium)* could also be reproduced by treating healthy seedlings with sterile culture filtrates of the fungus. In 1939, Yabuta and Hayashi showed that the active component of the filtrate responsible for the excessive elongation was a growth regulator named gibberellin. By the late 1950s, numerous plant pathogenic fungi and bacteria were shown to produce the plant hormone indoleacetic acid (IAA).

In 1966, Klambt *et al.* and Helgeson and Leonard showed that the bacterium *Corynebacterium fascians,* which causes the fasciation disease of peas, produces a cytokinin and that the fasciation symptoms could be induced on healthy plants after treatment with kinetin, which is an animal-derived cytokinin. Although IAA and cytokinins were shown to accumulate in plants affected by diseases that produce overgrowths, wilts, and various malformations, and to be present in lower amounts in plants stunted by disease, it has not yet been made clear whether, or to what extent, the hormonal imbalance is the cause or the result of the infection. Cytokinins and IAA have often been found at higher than normal concentrations in tumors of the crown gall disease caused by *Agrobacterium tumefaciens.* In 1977, Chilton *et al.* showed that in crown gall disease the bacterium introduces part of the DNA (T-DNA) of its plasmid (Ti-plasmid) into a plant cell and the T-DNA is then incorporated into plant chromosome DNA. Also in 1977, Drummond *et al.* showed that the incorporated T-DNA is transcribed by the plant cell.

Genetics of Resistance to Disease

In the early 1900s, the first studies on the inheritance of disease resistance appeared. In 1905, Biffen described the Mendelian inheritance of resistance of two wheat varieties and their progeny to a rust fungus. In 1909, Orton, working with the Fusarium wilts of cotton, watermelon, and cowpea distinguished among disease resistance, disease escape, and disease endurance. In 1911, Barrus demonstrated that there is genetic variability within a pathogen species; that is, there are different pathogen races, which are restricted to certain varieties of a host species. Soon after that, Stakman (1914) discovered that within a pathogen species physiological races of the pathogen exist that are morphologically indistinguishable but differ in their ability to infect a set of host differential varieties. This helped explain why a variety that was resistant in one geographic area was susceptible in another, why resistance changed from year to year, and why resistant varieties suddenly became susceptible—that is, because a different physiological race was involved!

At first (Walker, 1923) it was thought that resistance to a pathogen was a result of the presence of a toxic substance in the plant. In 1946, however, Flor, working with the rust disease of flax, showed that for each gene for resistance in the host there was a corresponding gene for avirulence in the pathogen (a gene-for-gene relationship). In 1963, Vanderplank suggested that there are two kinds of resistance: one, controlled by few "major" genes, is strong but race specific (vertical resistance); the other, determined by many "minor" genes, is weaker but is effective against all races of a pathogen species (horizontal resistance). Several types of plant cell structures and substances have been shown to provide the means by which such genes are believed to confer disease resistance in plants. In 1946, Gäumann proposed that, in many host–pathogen combinations, plants remain resistant through hypersensitivity; that is, the attacked cells are so sensitive to the pathogen that they and some adjacent cells die immediately, thus isolating or causing the death of the pathogen. Muller proposed in 1961 and Cruickshank confirmed and expanded in 1963 that disease resistance is often brought about by phytoalexins, that is, antimicrobial plant metabolites that are absent, or present at nondetectable levels, in healthy plants but accumulate to high levels as a result of some pathological stimulus.

Bacteria in Plant Disease

In 1876, Pasteur and Koch proved that the animal disease anthrax is incited by a bacterium. In 1878, Burrill showed that fire blight of pear and apple is also caused by a bacterium. Shortly thereafter, several other plant diseases were shown to be caused by bacteria; E. F. Smith's numerous and excellent contributions from 1895 on to the study of bacterial diseases of plants, particularly of the bacterial wilts of cucurbits, of solanaceous crops, and of crucifers, established beyond any doubt the role of bacteria as phytopathogens. As with fungal plant pathogens, however, acceptance of bacteria as causes of disease in plants was slow. Alfred Fischer, a prominent German botanist who had studied under deBary, rejected the results of Smith and others who claimed to have seen bacteria in plant cells. A heated controversy developed between Smith and Fischer, and their replies to each other in the scientific literature stand out in plant pathology as one of the best documented cases of scientific disagreement of historical interest. Smith also was among the first to notice (1893–1894) and to study the crown gall disease. He showed it to be caused by bacteria, studied its anatomy and development, and considered it to resemble cancerous tumors of humans and animals. In 1977, Chilton *et al.* showed that the crown gall bacterium transforms normal plant cells into tumor cells by introducing into them a plasmid, part of which becomes inserted into the plant cell chromosome DNA.

In 1972, Windsor and Black observed rickettsialike organisms in the phloem of clover plants infected with the club leaf disease. The following year similar organisms were observed in grape infected with Pierce's disease, in peach infected with phony peach, and others. These pathogens are fastidious bacteria transmitted by leafhoppers and are present exclusively or primarily in the phloem or xylem elements of plants. They are apparently a new kind of bacterium, but so far relatively little is known about their nature and properties.

Nematodes in Plant Disease

The first report of plant parasitic nematodes was in 1743 by Needham who observed nematodes within wheat galls (kernels). It was not until the 1850s, however, that other nematodes such as the root knot, the bulb and stem, and the cyst nematodes were observed. From 1913 to 1932, Cobb made a series of studies on plant parasitic nematodes, and these studies contributed greatly to nematode taxonomy, morphology, and methodology.

Viruses in Plant Disease

In 1886, Mayer reproduced the "tobacco mosaic" disease by injecting juice from infected tobacco plants into healthy plants. The juice of diseased plants remained infective even after continual heating at 60°C, although it lost its infectivity after several hours of heating at 80°C. Since no fungi were present on the diseased plant or the filtered juice, Mayer concluded that tobacco mosaic was probably caused by a bacterium. In 1892, Ivanowski showed that the causal agent of tobacco mosaic could pass through a filter that retains bacteria. This led him to believe that the disease was caused by a toxin secreted by bacteria or by small bacteria that passed through the pores of the filter. Beijerinck (1898) finally concluded that tobacco mosaic was caused not by a microorganism but by a *contagium vivum fluidum,* which he called a **virus.**

In 1929, F. O. Holmes provided a tool by which the virus could be measured by showing that the amount of virus present in a plant sap preparation is proportional to the number of local lesions produced on appropriate host plant leaves rubbed with that sap. It was not until 1935, however, that Stanley precipitated an infectious crystalline protein by treating juice from infected tobacco plants with ammonium sulfate and concluded that the virus was an autocatalytic protein that could multiply within living cells. Although his results and conclusions were later proved incorrect, for this discovery Stanley received the Nobel Prize. In 1936, Bawden and his colleagues demonstrated that the crystalline preparations of the virus actually consisted of protein and ribonucleic acid. In 1939 Kausche and his colleagues saw virus particles for the first time with the electron microscope. In 1956, Gierer and Schramm showed that the protein could be removed from the virus and that the nucleic acid carried the genetic information so that inoculation with the nucleic acid alone could cause infection and could reproduce the complete virus.

Numerous advances in instrumentation and methodology in the 1940s and 1950s, particularly the widespread use of the ultracentrifuge, the electron microscope, electrophoresis, and serological techniques, contributed immensely to the understanding of plant virus structure, chemistry, replication, and genetics. The agar double diffusion serological test (1962) and, later, the enzyme-linked immunosorbent assay (ELISA) (1977) and production of monoclonal antibodies (1975) made possible the quick and accurate detection and identification of plant viruses, thus allowing extensive and detailed studies of virus taxonomy and epidemiology. While the nucleic acid of all plant viruses studied from 1936 to 1966 was a single-stranded RNA, some of the subsequently studied viruses were found to have double-stranded RNA (1963), some double-stranded DNA (1968) and some single-stranded DNA (1977).

In 1971, Diener determined that the potato spindle tuber disease was caused by a small (250–400 bases long), single-stranded, circular molecule of infectious RNA, which he called a **viroid.** Since then about a dozen other viroids that cause disease in a variety of plants have been isolated. In 1982, a circular, single-stranded, viroidlike RNA 300–400 bases long was found encapsidated together with the single-stranded linear RNA (about 4,500 bases long) of velvet tobacco mottle virus. The small circular RNA was called a **virusoid.** Virusoids seem to form an obligatory association with the viral RNA and have since been found in several other plant viruses. Viroids seem to be the smallest nucleic acid molecules to infect plants, but no viroids have so far been found in animals. On the other hand, an even smaller type of infectious agent, composed of a small proteinaceous infectious particle, has been shown to cause the scrapie disease of sheep and goats. Such agents, which Prusiner in 1982 called "prions," are believed to cause several chronic degenerative diseases of animals and humans. Although prions have so far been found only in animals, it is not unreasonable to expect that they will soon be shown to affect plants also.

In 1980, cauliflower mosaic virus, whose genome is a circular double-stranded DNA chromosome, was the first plant virus for which the exact sequence of all its 8,000 base pairs was determined. In 1982, the complete sequence of the bases in the single-stranded tobacco mosaic virus RNA was determined, as were those of smaller viral RNAs and of viroids.

Protozoa in Plant Disease

In 1909, Lafont observed flagellate protozoa in the latex-bearing cells of laticiferous plants of the family Euphorbiaceae, but protozoa found in laticiferous plants were thought to be parasitizing the latex without causing disease to the host plants. In 1931, Stahel found flagellates infecting the phloem of coffee trees and causing abnormal phloem formation and wilting of the trees (Stahel, 1931, 1933). In 1963, Vermeulen presented additional and more convincing evidence of the pathogenicity of flagellates to coffee trees, and in 1976 flagellates were also reported to be associated with several diseases of coconut and oil palm trees in South America and in Africa.

Mycoplasmas in Plant Disease

In 1967, Doi and his colleagues in Japan observed mycoplasmalike bodies in the phloem of plants infected with several leafhopper-transmitted diseases. The same year Ishiie and his colleagues showed that the mycoplasmalike bodies and the symptoms disappeared temporarily when the plants were treated with tetracycline antibiotics. Similar bodies have since been found in many other yellows- or witches'-broom-type diseases previously thought to be caused by viruses. In 1972, Davis and his colleagues observed a motile, helical microorganism associated with corn stunt disease; they called it **spiroplasma.** Spiroplasmas also lack a cell wall; therefore, they are helical mycoplasmas. It has since been shown that spiroplasmas are the cause of corn stunt and citrus stubborn diseases.

Genetic Engineering and Plant Pathology

In 1944, Avery and his colleagues showed for the first time that DNA carries genetic information. They concluded this from experiments in which DNA from heat-killed pathogenic pneumonia bacteria *(Diplococcus pneumoniae)* could transform nonpathogenic bacteria into pathogenic ones. In 1952, Hershey and Chase showed that in bacterial viruses (phages), only the DNA of the phages entered the host bacteria and carried the genetic information into the bacteria, while the viral protein coat remained outside and thus could be ruled out as potential genetic material. In 1953, Watson and Crick showed that DNA consists of a double helix of intertwined polynucleotides with complementary base pairing that allows self-replication. By 1960 it was shown that messenger RNA carries information from DNA to the cytoplasm, where proteins are synthesized. In 1961 it was shown that the genetic code is read in stepwise groups (codons) consisting of three base pairs each. That same year it was also shown that three forms of RNA (messenger RNA, transfer RNA, and ribosomal RNA) are involved in protein synthesis. In 1964, it was shown that when mutations (base changes) appear in a gene, the order of mutations in the gene is the same as the order of amino acid changes in the gene's protein product. By 1966, the genetic code was fully deciphered, assigning specific base triplets (codons) for each kind of amino acid, and vice-versa, although most amino acids could be coded by more than one codon.

DNA is self-replicating and exists in cells as long molecules consisting of millions of nucleotide pairs (chromosomes) or as smaller molecules consisting of thousands or hundreds of thousands of nucleotide pairs **(plasmids).** Some plasmids can move into and become part of chromosomes of the same or other hosts. Viruses have DNA or RNA "chromosomes." The expression (functioning) of genes is controlled by special nucleotide sequences on the DNA and the mRNA of each gene. In 1970 Hamilton Smith discovered the first **restriction nuclease,** an enzyme that cuts DNA at a certain point of a specific nucleotide sequence. In 1973, Boyer and Cohen inserted foreign genes (DNA) into a plasmid and then introduced the plasmid into bacteria, which multiplied **(cloned)** the plasmid and the foreign DNA in them. In 1975, DNA fragments of a mammalian gene were cloned in a bacterial virus. In 1975, also, a hybrid cell was obtained by fusing an antibody-producing spleen cell with an autonomously dividing myeloma tumor cell. The tumorous hybrid cell **(hybridoma)** had the potential to produce identical **(monoclonal) antibodies** indefinitely. Restriction enzymes allowed relatively rapid determination of the nucleotide sequence in DNA. By 1977, Sanger *et al.* determined the complete sequence of the 5386 base-pairs of the DNA phage ϕX174. In 1977 also, it was discovered that most genes are split, that is, they do not consist of a continuous stretch of DNA but are split into several segments by non-gene DNA. In the same year the first sequencing of a complete mammalian (globin) gene was accomplished. In 1979, foreign DNA was cloned in eukaryotic yeast cells. In 1980, foreign genes were introduced into and expressed in mammalian (mouse) cells.

Along with the tremendous, rapid advances of genetic engineering in microorganisms and mammals, somewhat slower but no less important developments were taking place in the plant sciences. Plant tissue culture began in 1934, when White demonstrated that excised tomato root tips had a potentially unlimited growth in a liquid medium. In 1939, White and others obtained the first long-term cultures of tobacco and carrot callus. In 1946, Ball regenerated whole plants in vitro from isolated explants of the shoot apex and, in 1951, it was shown that whole plants could be regenerated from 100–250 μm-long isolated shoot apices bearing only one to two leaf primordia. In 1953, Muir obtained single cells from callus and showed that they could be propagated by subculture. In 1958, Stewart demonstrated that carrot suspension cultures could be induced to form embryos, that is, embryogenesis from somatic cells. In 1960, Cocking obtained isolated p enzymatic digestion of cell walls, and in 1971, Takebe *et al.* plants from isolated mesophyll protoplasts of tobacco. In th 1966, Guha and Maheshwari obtained the first haploid plants pollen.

One of the earliest applications of plant tissue culture was in plant tumors caused by *Agrobacterium tumefaciens.* The first c autonomously growing bacteria-free crown gall tissue was obtained b White and Braun in 1942. Tissue culture techniques have also been used to produce pathogen-free plants via apical meristem cultures. Plant protoplasts have been used to study virus infection and replication, the action of toxins, and to regenerate plants or obtain new somatic hybrids by protoplast fusion that exhibit different degrees of resistance to various pathogens. Genetic engineering techniques have also made possible the elucidation of the nature of the tumor-inducing principle in crown gall and of the genetic make-up of plant viruses and bacteria. The already accomplished use of *Agrobacterium* and of certain plant viruses as vectors of foreign genetic material into plants is expected to open a whole new era of genetic transformations of plants, the promise and potential of which cannot be imagined today.

During the twentieth century, plant pathology has matured as a science. Thousands of diseases have been described, pathogens have been identified, new kinds of plant pathogens have been discovered, and control measures have been developed. The studies of genetics and of the physiology of diseases have been expanded greatly, and new chemical compounds are being developed continually to combat plant diseases. Still, this is probably just the beginning of plant pathology and of the hope that it holds for the future. The huge losses in plants and plant products that occur annually are the single best reminder of how much is yet to be learned about plant diseases and their control.

Significance of Plant Diseases

Kinds and Amounts of Losses

Plant diseases are significant to humans because they cause damage to plants and plant products. For millions of people all over the world who still depend on their own plant produce for their existence, plant diseases can make the difference between a happy life and a life haunted by hunger or even death from starvation. The death from starvation of a quarter million Irish people in 1845 and much of the hunger of the underfed millions living in the underdeveloped countries today are examples of the consequences of plant diseases. For countries where food is plentiful, plant diseases are significant because they cause economic losses to growers, they result in increased prices of products to consumers, and they destroy the beauty of the environment by damaging plants around homes, along streets, in parks, and in forests.

Plant diseases may limit the kinds of plants that can grow in a large geographic area by destroying all plants of certain species that are extremely susceptible to a particular disease. For example, the American chestnut was annihilated in North America as a timber tree by the chestnut blight disease, and the American elm is being eliminated as a shade tree by Dutch elm disease. Plant diseases may also determine the kinds of agricultural industries and the level of employment in an area by affecting the amount and kind of produce available for local canning or processing. On the other hand, plant diseases are responsible also for the creation of new industries that develop chemicals, machinery, and methods to control plant diseases; the annual expenditures to this end amount to billions of dollars in the United States alone.

The kinds and amounts of losses caused by plant diseases vary with the plant or plant product, the pathogen, the locality, the environment, the control measures practiced, and combinations of these factors. The quantity of loss may range from slight to 100 percent loss. Plants or plant products may be reduced in quantity by disease in the field, as indeed is the case with most plant diseases, or by disease during storage, as is the case of the rots of stored fruits, vegetables, grains, and fibers. Sometimes, destruction by disease of some plants or fruits is compensated by greater growth and yield of the remaining plants or fruits as a result of reduced competition. Frequently, severe losses may be incurred by reduction in the quality of plant products. For instance, while spots, scabs, blemishes, and blotches on fruit, vegetables, or ornamental plants may have little effect on the quantity produced, the inferior quality of the product may reduce the market value so much that production is unprofitable or a total loss. With some produce—for example, apples infected with apple scab—even as little as 5 percent disease may cut the price in half, while with others—for example, potatoes infected with potato scab—there may be no effect on price in a market with slight scarcity, but there may be a considerable price reduction in years of even minor gluts of produce. Some diseases, such as ergot of rye, make plant products unfit for human or animal consumption by contaminating them with poisonous fruiting structures.

Plant diseases may cause financial losses in the following ways: Farmers may have to plant varieties or species of plants that are resistant to disease but are less productive, more costly, or commercially less profitable than other varieties. They may have to spray or otherwise control a disease, thus incurring expenses for chemicals, machinery, storage space, and labor. Shippers may have to provide refrigerated warehouses and transportation vehicles, thereby increasing expenses. Plant diseases may limit the time during which products can be kept fresh and healthy, thus forcing growers to sell during a short period of time when products are abundant and prices are low. Healthy and diseased plant products may need to be separated from each other, thus increasing handling costs.

Some plant diseases can be controlled almost entirely by one or another method, thus resulting in financial losses only to the amount of the cost of the control. Sometimes, however, this cost may be almost as high as, or even higher than, the return expected from the crop, as in the case of certain diseases of small grains. For other diseases, no effective control measures are yet known, and only a combination of cultural practices and somewhat resistant varieties makes it possible to raise a crop. For most plant diseases, however, practical controls are available, although some losses may be incurred in spite of the control measures taken. In these cases, the benefits from the control applied are generally much greater than the combined direct losses from the disease and the indirect losses due to expenses for control.

In spite of the variety of types and size of financial losses that may be caused by plant diseases, well-informed farmers who use the best combinations of available resistant varieties and proper cultural, biological, and chemical control practices not only manage to produce a good crop in years of severe disease outbreaks but may also obtain much greater economic benefits from increased prices after other farmers suffer severe crop losses.

Some Historical and Present Examples of Losses Caused by Plant Diseases

For thousands of years, mankind has depended on a few crop plants for its sustenance and survival. Wheat, rice, corn, a few other cereals, potatoes, and some legumes have provided the staple food for humans in different parts of the world. The same or related plants are used as feed for all domesticated animals, which humans then use for food, as energy sources, or for pleasure. As societies developed, the needs of people for fiber plants for more and better clothing increased. Cotton was and still is the main fiber crop, but flax, hemp, jute, and sisal have been important in some parts of the world. At first, wood and wood products filled the needs for tools, shelter, and furniture, but recently, the industrial use of paper, plastics, and so on, have increased their demand tremendously. Industry has also been reaching more and more for plants as raw materials, for example, rubber, synthetic fibers, drugs, and a large variety of organic compounds. Improved living conditions also created and increased the demand for more and better fruits, vegetables, sugar, and oil crops, which are part of a normal, healthful diet, as well as the demand for luxury or pleasure crops such as tobacco, coffee, tea, and cacao. Finally, plants have always been a necessary part of the human environment for aesthetic reasons and also because they provide a moderating force in balancing the concentration of carbon dioxide in the atmosphere and in preventing floods and erosion of the soil and because they improve the physical properties and fertility of the soil by adding organic matter to it.

Plant diseases have affected the existence, adequate growth, and productivity of each of the above kinds of plants and thereby one or more of the basic prerequisites for a healthy, safe life for humans since the time they gave up their dependence on wild game and wild fruit for their existence, became more stationary and domesticated, and began to practice agriculture more than 6000 years ago. Destruction of food and feed resources by diseases has been an all too common occurrence in the past and has resulted in malnutrition, starvation, migration, and death of people and animals on numerous occasions, several of which are well documented in history. Similar effects are observed annually in underdeveloped, agrarian societies, in which families and nations are dependent for their sustenance on their own produce. In more developed societies, losses from diseases in food and feed produce result primarily in financial losses and higher prices. It should be kept in mind, however, that lost food or feed produce means less such available in the world economy and, considering the perennially inadequate amounts of food avail-

Table 1-1. Examples of Severe Losses Caused by Plant Diseases

Disease	Location	Comments
FUNGAL DISEASES		
1. Cereal rusts	Worldwide	Frequent severe epidemics. Huge annual losses.
2. Cereal smuts	Worldwide	Continuous losses on all grains.
3. Ergot of rye and wheat	Worldwide	Poisonous to humans and animals.
4. Late blight of potato	Cool, humid climates	Epidemics—Irish famine (1845–46)
5. Brown spot of rice	Asia	Epidemics—the great Bengal famine (1943)
6. Southern corn leaf blight	U.S.	Epidemic 1970, $1 billion lost.
7. Powdery mildew of grapes	Worldwide	European epidemics (1840–1850s)
8. Downy mildew of grapes	U.S., Europe	European epidemic (1870s–1880s)
9. Downy mildew of tobacco	U.S., Europe	European epidemic (1950s–1960s). Epidemic in North America (1979)
10. Chestnut blight	U.S.	Destroyed all American chestnut trees (1904–1940)
11. Dutch elm disease	U.S., Europe	Destroying all American elm trees (1930 to date)
12. Coffee rust		Destroyed all coffee in Southeast Asia (1870s–1880s). Since 1970 present in Brazil.
13. Banana leaf spot or Sigatoka disease	Worldwide	Great annual losses.
14. Rubber leaf blight	S. America	Destroys rubber tree plantations.

able, many poor people somewhere in the world will be worse off for these losses and will go hungry.

Some examples of plant diseases that have caused severe losses in the past are shown in Tables 1-1 and 1-2.

Plant Diseases and World Crop Production

World population today is about 5.0 billion, and at the present rate of 2.14 percent annual growth, it is expected to be 5.4 billion by 1990 and 6.7 billion by the year 2000 (see Figure 1-4). Paradoxically, the developing countries, in which 57.6 percent of the population is engaged in agriculture, have the lowest agricultural output, their people are living on a substandard diet, and they have the highest population growth rates (2.64 percent). Because of the current distribution of usable land and population, educational and technical levels for food production, and of general world economics, it is estimated

Table 1-1. Examples of Severe Losses Caused by Plant Diseases *(continued)*

Disease	Location	Comments
VIRAL DISEASES		
15. Sugarcane mosaic	Worldwide	Great losses on sugarcane and corn.
16. Sugarbeet yellows	Worldwide	Great losses every year.
17. Citrus quick decline (tristeza)	Africa, Americas	Millions of trees being killed.
18. Swollen shoot of cacao	Africa	Continuous heavy losses.
19. Plum pox or sharka	Europe	Spreading severe epidemic on plums, peaches, apricots.
20. Barley yellow dwarf	Worldwide	Important on small grains worldwide.
BACTERIAL DISEASES		
21. Citrus canker	Asia, Africa, Brazil, United States	Killed millions of trees in Florida, 1910s and again in the 1980s.
22. Fire blight of pome fruits	North America, Europe	Kills numerous trees annually.
MYCOPLASMAL DISEASES		
23. Peach yellows	Eastern U.S., Russia	10 million peach trees killed.
24. Pear decline	Pacific coast states and Canada (1960s), Europe	Millions of pear trees killed.
NEMATODE DISEASES		
25. Root knot	Worldwide	Continuous losses on vegetables and most other plants.
26. Sugarbeet cyst nematode	Severe in Northern Europe and the Western U.S.	

that even today some 800 million people are undernourished and 2.0 billion suffer from hunger or malnutrition or both. To feed these people and the additional millions to come in the next few years, all possible methods of increasing the world food supply are currently being pursued, including: (1) expansion of crop acreages, (2) improved methods of cultivation, (3) increased fertilization, (4) use of improved varieties of crops, (5) increased irrigation, and (6) improved crop protection.

There is no doubt that the first five of the above measures must provide the larger amounts of food. Crop protection from pests and diseases can only reduce the amount lost after the potential for increased food production has been attained by proper utilization of the other parameters. Crop protection, of course, has been important in the past and is important now. For example, it is estimated that in the United States alone, each year, crops worth $9.1 billion are lost to diseases, $7.7 billion to insects, and $6.2 billion to weeds. But crop protection becomes even more important in an intensive agriculture, where increased fertilization, genetically uniform high-yielding varieties,

Table 1-2. Additional Diseases Likely to Cause Severe Losses in the Future

Disease	Comments
FUNGAL DISEASES	
1. Downy mildew of corn and sorghum	Just spreading out of Southeast Asia.
2. Soybean rust	Also spreading from Southeast Asia and from Russia.
3. Monilia pod rot of cocoa	Very destructive in South America; spreading elsewhere.
4. Sugarcane rust	Destructive in the Americas and elsewhere.
VIRAL DISEASES	
5. African cassava mosaic	Destructive in Africa; threatening Asia and the Americas.
6. Streak disease of maize (corn)	Spread throughout Africa on sugarcane, corn, wheat, etc.
7. *Hoja blanca* (white tip) of rice	Destructive in the Americas so far.
8. Bunchy top of banana	Destructive in Asia, Australia, Egypt, Pacific islands.
BACTERIAL DISEASES	
9. Bacterial leaf blight of rice	Destructive in Japan and India; spreading.
10. Bacterial wilt of banana	Destructive in the Americas; spreading elsewhere.
MYCOPLASMAL DISEASES	
11. Lethal yellowing of coconut palms	Destructive in Central America; spreading into U.S.
VIROID DISEASES	
12. Cadang-cadang disease of coconut	Killed more than 15 million trees in the Philippines to date.
NEMATODE DISEASES	
13. Burrowing nematode	Severe on citrus in Florida and on banana in many areas.

increased irrigation, and other methods are used. Crop losses to diseases and pests not only affect national and world food supplies and economies but affect even more individual farmers, whether they grow the crop for direct consumption or for sale. Since operating expenditures for the production of the crop remain the same, harvests lost to disease and pests directly lower the net return.

Table 1-3 provides an estimate of the actual 1982 world crop production and of the preharvest losses to diseases, insects, and weeds. As the table shows, huge quantities of produce are either never produced because of direct interference by pests or are destroyed by pests. The amount of each crop lost to pests varies with the crop (for example, 23.4 percent for fruits, 34.5 percent for cereals, 55.0 percent for sugarcane). Table 1-4 shows that crop loss varies with the degree of development of the country in which the crop is produced. Also, the importance of each kind of pest (diseases, insects, weeds) varies with the

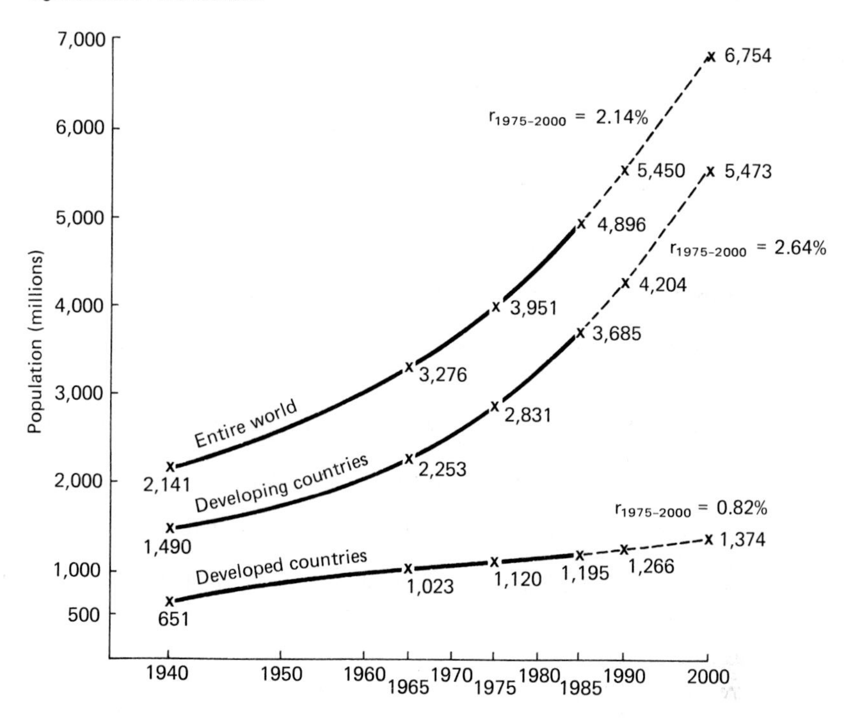

FIGURE 1-4 Real and projected population changes from 1937 to 1985 to the year 2000. The rates of population growth were estimated for the years 1975–1982 and, for this graph, were assumed unchanged to the year 2000. The agricultural population in 1982 was 11.6 percent of total in developed countries and 57.6 percent in developing countries.

crop, but in general diseases and insects each destroy quite similar percentages (12 percent) of the crop before harvest and weeds about 10 percent of the crop, for a total of 33.7 percent, or one-third of the potential production of the world (Table 1-3). To these losses should be added 9–20 percent postharvest losses to pests, which brings the total (preharvest and postharvest) food losses to pests in the United States to about 40 percent and for the entire world to about 48 percent of all food crops—in spite of all types of pest controls used. This is indeed a huge loss of needed food. It is apparent that losses are much greater in underdeveloped areas than they are in the more developed ones. Another point that can be made is that, although diseases and insects cause approximately equal losses to world crops overall, insects are much more easily controlled in developed countries than in the underdeveloped ones, especially Asia, while losses caused by diseases seem to be as great in developed as they are in underdeveloped countries.

The crop losses caused by diseases, insects, and weeds become particularly striking and alarming when one considers their distribution among countries of varying degrees of development. In developed countries (Europe, North America, Australia, New Zealand, Japan, Israel, and South Africa), in which only 11.6 percent of the population is engaged in agriculture, the estimated losses and percentages of losses are considerably lower than those in developing countries (the rest of the world) (Table 1-4), in which 57.6 percent of the population is engaged in agriculture. The situation becomes particularly painful if one compares Fig. 1-4 and Table 1-3, which points to the fact that the developing countries, which have much greater population than the

Table 1-3. Estimated 1982 World Crop Production[a] and Preharvest Losses (in Millions of Tons) and Percent[b] of World Production Lost to Diseases, Insects and Weeds

	Actual production	Estimated losses to dis-ins-w	Potential production	Estimated losses to diseases	% of crop lost to			Total % of crop lost
					Diseases	Insects	Weeds	
		Millions of Tons			%	%	%	%
Cereals	1,695	893	2,588	238	9.2	13.9	11.4	34.5
Potatoes	255	121	376	82	21.8	6.5	4.0	32.3
Other root crops	556	420	976	163	16.7	13.6	12.7	43.0
Sugarbeets	319	104	423	44	10.4	8.3	5.8	24.5
Sugarcane	811	991	1,802	346	19.2	20.1	15.7	55.0
Legumes	45	22	67	8	11.3	13.3	8.7	33.3
Vegetables	368	141	509	51	10.1	8.7	8.9	27.7
Fruits	302	92	394	50	12.6	7.8	3.0	23.4
Coffee-cocoa-tea	8	7	15	3	17.7	12.1	13.2	42.4
Oil crops	240	106	346	34	9.8	10.5	10.4	30.7
Fiber crops	40	18	58	6	11.0	12.9	6.9	30.8
Tobacco	6	3	9	1	12.3	10.4	8.1	30.8
Natural rubber	4	1	5	0.6	15.0	5.0	5.0	25.0
				Average percentages lost	11.8	12.2	9.7	33.7

[a] Production data from 1982 FAO Production Yearbook (Ford and Agriculture Organization, 1982).
[b] Percentages of losses taken from Cramer (1967).

Table 1-4. Estimated 1982 Crop Production[a] and Preharvest Losses (in Millions of Tons) and Percent[b] Lost to Diseases and Other Pests (Insects, Weeds) in Developed and Developing Countries.

	Developed countries				Developing countries			
	Population (millions): 1,186	% in Agriculture: 11.6			Population (millions): 3,405	% in Agriculture: 57.6		
	Arable land (millions of hectares): 672,335				Arable land (millions of hectares): 796,264			
	Actual Production	Estim. loss to Dis-Ins-W	% lost to Dis-Ins-W	MT Lost to Dis	Actual Production	Estim. loss to Dis-Ins-W	% lost to Dis-Ins-W	MT Lost to Dis
Cereals	874	190	17.8	51	822	703	46.1	187
Potatoes	204	59	22.4	40	51	23	32.0	21.6
Other Root Crops	205	64	23.8	25	350	356	50.4	138.0
Sugarbeets	290	89	23.5	38	29	15	34.5	6.4
Sugarcane	77	40	34.0	14	734	951	56.4	332
Legumes (Dry)	12	2.5	17.2	0.8	33	19.5	37.1	6.4
Vegetables	145	43	23.0	16	223	98	30.5	36
Fruits	138	32	19	17	164	60	26.8	32
Coffee-Cocoa-Tea	0	—	—	—	8	5.9	42.4	2.5
Oil Crops	99	35	26	11	141	71	33.5	23
Fiber Crops	13	4.6	26	1.6	27	13.4	33.2	4.8
Tobacco	2	0.8	29	0.3	4	2.2	35.5	0.9
Natural Rubber	0	—	—	—	4	1.3	25	0.8

[a]Production data from 1982 FAO Production Yearbook.
[b]Percentages of losses taken from Cramer (1967).

Table 1-5. Percentage of All Produce Lost to Diseases, Insects and Weeds by Continent or Region[a]

Continent or Region	Percent of Produce Lost to Diseases, Insects, and Weeds
Europe	25
Oceania	28
North and Central America	29
USSR and China	30
South America	33
Africa	42
Asia	43

[a]From H. H. Cramer, "Plant Protection and Crop Production" (translated from German by J. H. Edwards), *Pflantzenschutz Nachrichten,* 20/1967.1 Farbenfabriken Baver AG. Leverkusin.

developed countries, produce relatively less food and fiber and suffer much greater losses to diseases and to the other pests. Taking into account the kinds of crops grown in temperate climates, where most developed countries are, and in the tropics, where developing countries are located, the total percentage losses differ considerably with the continent, as shown in Table 1-5.

Effects of Changes in Agricultural Methods and in Human Society on the Development and Spread of Plant Diseases

The last several decades have seen a rapid increase in world population with resultant food deficits, greater mobility of people and products over the earth, a rapid increase in knowledge in every field of endeavor, industrialization, and the increased cooperation of scientists and governments in solving problems common to several parts of the world. As a result, new agricultural methods have been developed to meet the food and economic needs of growers, nations, and the world. All these changes in human society and agricultural methods, however, have had an effect on the kinds, severity of development, and rates of spread of the diseases that attack crop plants.

Improving crop plants by breeding high-yielding varieties has been and continues to be one of the better and cheaper ways of increasing crop production. This is being done with every single kind of cultivated crop plant. However, it has achieved its greatest success and was responsible for a tremendous upsurge in food production, the so-called "green revolution," in the case of the high-yielding dwarf and semidwarf wheat varieties that were also resistant, at least for some years, to the stem rust disease. These varieties, produced and distributed at first by the International Maize and Wheat Improvement Center in Mexico (CIMMYT), increased wheat production in Mexico in the mid-1960s 6.5 times that in 1945, thus changing Mexico, for a while at least, from a wheat-importing to a wheat-exporting country. They also behaved similarly and were just as productive in Africa and Asia. To produce high yields with these varieties, many agronomic practices had to be altered drastically. Plant density per acre was increased, the date of planting had to be changed, higher levels of fertilizers and heavier and more frequent irrigation had to be employed. Soon, enormous areas were sown with a few high-yielding, genetically uniform dwarf varieties, and in many areas local pathogens or local strains of common pathogens attacked the dwarf wheats. For example, monocultures of these wheats in India, West Pakistan, Afghanistan, and Turkey increased from about 23,000 acres in 1966 to 30 million acres in 1971, replacing hundreds of local varieties and coming in contact with numerous new pathogens or new pathogen races. When virulent pathogens or new virulent races that may arise come in contact with such huge expanses of genetically uniform crops, devastating epidemics may develop within a short time. Already, new races and biotypes of stem rust (*Puccinia graminis* f. sp. *tritici*), leaf rust (*P. recondita*), and stripe rust (*P. striiformis*) have been identified and have caused severe epidemics in some areas that reduced yields of dwarf varieties by as much as 55 percent. In the same or other areas *Septoria* leaf blotch and glume blotch also caused severe losses on some dwarf varieties. For example, in the Mediterranean countries, *Septoria* almost completely destroyed one dwarf variety causing yield losses between 80 and 87 percent. Many of the dwarf wheats are susceptible to powdery mildew, while others are more susceptible than the older local varieties to seedling blights, to ergot, to smuts, or to certain local bacterial, viral, or nematode diseases. Local breeding programs are now an essential part of the CIMMYT program; after severe losses occurred to original releases, the local programs have performed the adaptation breeding to incorporate resistance to local pathogens.

A similar "green revolution" with respect to improvement of rice varieties has been carried out by the International Rice Research Institute in the Philippines. New nonlodging dwarf rice varieties that respond favorably to high nitrogen fertilization and produce high yields were developed and distributed widely in southeast Asia and elsewhere. Soon, however, many of these varieties became susceptible to diseases, such as bacterial blight caused by *Xanthomonas oryzae* and bacterial leaf streak caused by *X. oryzicola*, that were either unknown or unimportant when old local varieties were planted, but which now, due to high nitrogen fertilization and double cropping of large expanses of genetically homogeneous varieties, reached catastrophic proportions. In some countries, rice blast, caused by the fungus *Pyricularia oryzae* also became severe on the new high-nitrogen fertilized rice varieties.

The need to reduce costs in the production of high-yielding hybrid corn seed led to the search for and development of the male-sterile plants that would not need detasseling. This, however, led to hybrids that were genetically uniform in carrying in the cytoplasm the trait for male sterility. The same cytoplasm, however, also carried a gene that made these hybrids susceptible to a previously unimportant race of the fungus *Helminthosporium maydis*, and as a result the southern corn leaf blight destroyed more than a billion dollars worth of corn in the United States in just one year.

In Venezuela, expansion of irrigation made possible the production of two rice crops per year where only one had been grown before. As a result, a serious outbreak of the virus disease *hoja blanca* occurred because the new conditions favored the multiplication and spread of the insect vector of the virus from the one rice crop to the other. Irrigation by most techniques (flooding, furrows, or sprinkling) also increases the population and distribution of many fungal, bacterial, and nematode pathogens that affect the roots and lower parts of the stem. Trickle irrigation, however, which allows great economy in water use, use of somewhat saline water, use of shallow soil with concurrent fertilization, irrigation under plastic mulches, and irrigation of slopes, reduces or practically eliminates many diseases such as anthracnose, powdery mildews, leaf spots and blights. The scarcity and increased costs of labor and fuel, and the concern over soil moisture and soil erosion, as well as successful weed control through herbicides have resulted in a tremendous increase in acreage of crops grown under some kind of reduced or no-tillage system. These systems sometimes involve crop rotation, ecofallow, multiple cropping, or intercropping. Many foliar pathogens of crops survive better and cause more severe diseases in minimal tillage than when the plant debris is buried by plowing, but other diseases, for example, stalk rot of grain sorghum and corn caused by *Fusarium moniliforme,* are decreased dramatically, especially when two different crops are used in the reduced tillage system.

The grafting of varieties on different rootstocks, either to secure trueness of the variety or to provide resistance to a factor to which the variety is susceptible, often leads to complications and heavy losses. In addition to the cases of true horticultural incompatibility between rootstock and scion, rootstocks often carry latent viruses or mycoplasmas that may be injurious to the scion, for example, East Malling clonal stocks used in apple tree propagation. In some cases, new pathogens attack the rootstocks through the soil, for example, *Fusarium javanicum* var. *ensiforme* attacking the cucurbit rootstock on which greenhouse cucumbers were grafted because the rootstock was resistant to *F. oxysporum.* Finally, hypersensitive resistant rootstocks may appear susceptible to viruses and mycoplasmas that are transmitted by insect vectors to tolerant scions, as it happened with the citrus tristeza virus causing the decline of sweet orange trees grafted on sour orange rootstocks and with the mycoplasma causing the decline of pear varieties grafted on oriental pear. In these cases, necrosis of the phloem just below the graft union as a result of the hypersensitive resistance of the rootstock causes the decline and death of the trees. On the other hand, apple and stone fruit varieties hypersensitive resistant to tomato ringspot virus become necrotic at the graft union, decline, and die when grafted on certain virus-tolerant rootstocks, and the virus is carried to the rootstock by its nematode vector.

Mechanization of agricultural practices often creates a number of plant disease problems. This is usually the result of unnoticed and more indiscriminate contamination of cultivators, harvesters, conveyors, and farm equipment with pathogens upon contact with diseased plants or infested soil, and of the more widespread dissemination of these pathogens to other products, other fields, or other parts of the same, large field.

The increased use and amounts of fertilizers, particularly nitrogen, for production of greater yields is generally considered to increase the severity of such diseases as powdery mildew, rusts, and fire blight, caused by pathogens that prefer young succulent tissues, and to decrease the diseases caused by pathogens that attack primarily mature or senescent tissues. It is now known, however, that it is generally the form of nitrogen (nitrate or ammonium) available to the host or pathogen that affects disease severity or resistance rather than the amount of nitrogen. In either case, increased fertilization does affect the susceptibility of plants to diseases, and this must be taken into account in the efforts to increase productivity through fertilization.

The weed killers that are increasingly used in cultivated fields may cause injury to cultivated crop plants directly, but they also influence several soil pathogens and soil microorganisms antagonistic to pathogens. Other chemicals, too, such as fertilizers, insecticides, and fungicides, alter the types of microorganisms that survive and thrive in the soil, and this sometimes leads to reduction in the numbers of useful predators and antagonistic microorganisms of pathogens or their vectors. The use of fungicides and other pesticides specific against a particular pathogen often leads to increased populations and disease severity caused by other pathogens not affected by the specific pesticide. This occurs even with some rather broad spectrum systemic fungicides, for example, benomyl, which control most but not all pathogens. Where such fungicides are used regularly and widely, some fungi, such as *Pythium*, that are not affected by them, may become more important as pathogens than when other more general fungicides were used.

The use of pesticides to control plant diseases and other pests has been increasing steadily at an annual rate of about 14 percent since the mid-1950s. By 1977 approximately 2.3 billion kg (5 billion pounds) of pesticides, including fungicides, insecticides, and herbicides, were applied to crops throughout the world each year, and it is estimated that by the year 2000 more than 3 billion kg of pesticides will be used annually worldwide. So far, about 35 percent of all pesticides are applied in the United States and Canada, 45 percent in Europe, and the remaining 20 percent in the rest of the world. It is likely that as more of the developing countries become developed, the use of pesticides will increase sharply. There is little doubt that pesticide use has increased yields of crops in most cases in which they have been applied. The cost of production, distribution, and application of pesticides is, of course, another form of economic loss caused by plant diseases and pests (Table 1-4). Furthermore, such huge amounts of poisonous substances damage our environment and food as they are spread over our crop plants several times each year.

The public awareness of the direct, indirect, and cumulative effects of pesticides on organisms other than the pests they are intended to control has led to increased emphasis on the protection of the environment. As a result, many pesticides have been abandoned or their use has been restricted and their functions have been taken over by other less effective or more specific pesticides, or by more costly or less efficient methods of control. The effort to control diseases and other pests by biological and cultural methods is still growing while at the same time more restrictions are being imposed in the testing, licensing, and application of pesticides. The pesticide producers must provide more detailed data on the effectiveness, toxicity, and persistence of each pesticide, and the application of each pesticide must be licensed for each crop on which it is going to be applied. Furthermore, in some countries each prospective commercial applicator of pesticides must pass an examination and be licensed to apply pesticides on crop plants.

The desirability of using fewer and safer pesticides, however, is counteracted by the increasing demand of consumers over the last several decades for high-quality produce, especially fruits and vegetables free of any kind of blemishes caused by diseases or insects. A change in the attitude of consumers to demand less extravagant esthetic quality of produce could reduce the use of pesticides and the waste of perfectly wholesome foodstuffs, but such a change in attitude may not come for some time yet.

The economics of agricultural production continue to lead to consolidation of smaller farms into increasingly larger ones, often devoted to monoculture of a single profitable crop or a single stage of it. Monoculture is made more imperative by mechanization, since different crops would require additional expenditures for the specialized equipment needed for sowing, cultivation, spraying, harvesting, storage, and handling of the various crops. The concentration into a continuous area of many fields and many plants of the same species and variety, however, carries many special risks, particularly the appearance or introduction and rapid spread of a destructive pathogen.

The tendency in recent years of farm enterprises to specialize in the year-round production of young seedlings or cuttings, for example, tomato, strawberry, chrysanthemum, which they subsequently sell to commercial growers throughout the world, carries with it the danger not only of a destructive disease spreading rapidly within that farm but, much more importantly, of a destructive disease being carried on the propagative material to the rest of the country and the world. This has already happened with a chrysanthemum rust that spread from Japan to South Africa and from there throughout the globe.

The increased mobility of all kinds of nursery stock and produce throughout the world has been a major factor in the spread or appearance of many new diseases in many parts of the world. In agriculturally advanced countries, plant quarantine inspectors at each port of entry intercept most of the pathogens and other pests. But many pathogens do get through nevertheless, and if they happen to be carried in an area where the environment is favorable and there are susceptible host plants, a new disease may appear. The chances for new diseases to appear are much greater in underdeveloped countries, where new improved varieties are constantly imported from developed countries. Many times the imported propagative material carries pathogens that may be serious not only to this same variety but, more importantly, to some or all of the local varieties of the same and related species. Moreover, even when the imported propagative material is disease free, once it has been planted extensively in the new area or country, it may be attacked by one of the locally existing pathogens or races of pathogens, and this may lead to an unexpected epidemic and the failure of the new variety.

Increased tourist and business travel has undoubtedly contributed to the introduction of some plant pathogens to new areas, but no specific cases are known.

Industrialization and increased travel harm plants in more direct ways. Air pollutants produced by factories, automobiles, and airplanes cause direct injury to most plants and reduce their growth and productivity. Also, much productive land is continually being turned into residential areas, huge industrial complexes, shopping centers, parking lots, highways, and lesser roads. It is estimated that in the United States and Canada highway building alone takes a quarter of a million acres of arable land and that much more pasture land out of production in a single year! In the United States, two million acres of land each year are converted from agricultural to nonagricultural uses, including 420,000 acres for urban development, an equal amount for reservoirs and flood control, and nearly one million acres for parks, wilderness, and wildlife areas. The amount of cropland is decreasing at an annual rate of

3 percent. How long can this continue before we run out of food-producing land?

Diagnosis of Plant Diseases

Pathogen or Environment?

To diagnose a plant disease it is necessary to first determine whether the disease is caused by a pathogen or an environmental factor. In some cases, in which typical symptoms of a disease or signs of the pathogen are present, it is fairly easy for an experienced person to determine not only whether the disease is caused by a pathogen or an environmental factor but by which one. Frequently, comparing the symptoms with those listed in "host indices,"—books that list the known diseases and their causes for specific plant hosts—or in books like those of the compendia series of the American Phytopathological Society help narrow the field of likely causes and often help identify the cause of the disease. In most cases, however, a detailed examination of the symptoms and an inquiry into characteristics beyond the obvious symptoms are necessary for a correct diagnosis.

Infectious Diseases

Diseases caused by pathogens (fungi, bacteria, parasitic higher plants, nematodes, viruses, mycoplasmas, and protozoa) are characterized by the presence of these pathogens on the surface of the plants (some fungi, bacteria, parasitic higher plants, and nematodes) or inside the plants (most pathogens). The presence of such pathogens in an active state on the surface of a plant indicates that they are probably the cause of the disease. Their detection and identification can, in some cases, be determined with the experienced naked eye or with a magnifying lens (some fungi, all parasitic higher plants, some nematodes) or, more frequently, by microscopic examination (fungi, bacteria, and nematodes) (see Figure 1-3). If no such pathogens are present on the surface of a diseased plant, then it will be necessary to look for additional symptoms and, especially, for pathogens inside the diseased plant. Such pathogens are usually at the margins of the affected tissues, at the vascular tissues, or at the base of the plant, and on or in its roots.

Diseases Caused by Parasitic Higher Plants

The presence of a parasitic higher plant (for example, dodder, mistletoe, witchweed, or broomrape) growing on a plant is sufficient for diagnosis of the disease.

Diseases Caused by Nematodes

The presence on or in a plant or in its rhizosphere of a species of plant parasitic nematodes, which can be distinguished from the nonparasitic ones by the stylet (spear) they possess, indicates that the nematode is probably the pathogen that caused the disease, or at least is involved in the production of the disease. If the nematode can be identified as belonging to a species or genus known to cause such a disease, then the diagnosis of the disease can be made with a degree of certainty.

Diseases Caused by Fungi and Bacteria

When fungal mycelium and spores or bacteria are present on the affected area of a diseased plant, two possibilities must be considered: (1) The fungus or bacterium may be the actual cause of the disease; or (2) they may be one of the many saprophytic fungi or bacteria that can grow on dead plant tissue once the latter has been killed by some other cause—even other fungi or bacteria.

Fungi To determine whether the observed fungus is a pathogen or a saprophyte the morphology of its mycelium, fruiting structures, and spores are first studied under a microscope. The fungus can then be identified and checked in an appropriate book of mycology or plant pathology to see whether it has been reported to be pathogenic or not, especially on the plant on which it was found. If the symptoms of the plant correspond to those listed in the book as caused by that particular fungus, then the diagnosis of the disease is in most cases considered complete. If no such fungus is known to cause a disease on plants, especially one with symptoms similar to the ones under study, then the fungus found should be considered a saprophyte or, possibly, a previously unreported plant pathogen, and the search for or the proof of the cause of the disease must continue. In many cases, neither fruiting structures nor spores are initially present on diseased plant tissue, and therefore no identification of the fungus is possible. For some fungi, special nutrient media are available for selective isolation, identification, or to promote sporulation. Others need to be incubated under certain temperature, aeration, or light conditions to produce spores. With most fungi, however, fruiting structures and spores are produced in the diseased tissue if the tissue is placed in a glass or plastic "moisture chamber," that is, a container in which wet paper towels are added to increase the humidity in the air of the container.

Bacteria The diagnosis of a bacterial disease and the identification of the causal bacterium is based primarily on the symptoms of the disease, the constant presence of large numbers of bacteria in the affected area, and the absence of any other pathogens. Bacteria are small ($0.8 \times 1-2\ \mu m$), however, and although they can be seen with the compound microscope, they are all tiny rods and have no distinguishing morphological characteristics for identification. Care must be taken, therefore, to exclude the possibility that the observed bacteria are secondary saprophytes, that is, bacteria that are growing in tissue killed by some other cause. Selective media are available for the selective cultivation of almost all plant pathogenic bacteria free of common saprophytes so that the genus and even some species can be identified. The easiest and surest way to prove that the observed bacterium is the pathogen is through isolation and growth of the bacterium in pure culture and, using a single colony for reinoculation of a susceptible host plant, reproducing the symptoms of the disease and comparing them with those produced by known species of bacteria. More recently, immunodiagnostic techniques, including agglutination and precipitation, fluorescent antibody staining, and enzyme-linked immunosorbent assay have been used to detect and identify plant pathogenic bacteria. Such techniques are quite sensitive, specific, rapid, and easy to perform, and it is expected that soon standardized, reliable antisera will be available for serodiagnostic assays of plant pathogenic bacteria.

Identifying the fastidious vascular bacteria (the so-called rickettsialike bacteria) and diagnosing the plant vascular diseases they cause are more difficult. Most fastidious bacteria are quite small and difficult or impossible to see under the light microscope, although they are visible by phase-contrast and, of course, electron microscopy. Besides, they are present in very small numbers only in xylem or phloem elements of their hosts and so are difficult to find. In addition, none of these bacteria can grow on the usual bacteriological media, although a few can be grown on specialized media. Antisera have been developed for most of them, and therefore serodiagnostic techniques, particularly fluorescent antibody staining, are quite useful for their detection and diagnosis. Also, some of these bacteria are sensitive to penicillin and most are sensitive to tetracyclines and to heat therapy. Therefore, these agents, plus symptomatology, graft transmissibility, and transmission by certain insect vectors are often used to diagnose a disease as being caused by a fastidious bacterium, although it does not always identify the particular bacterium involved.

Diseases Caused by Mycoplasmas

Diseases caused by mycoplasmas appear as stunting of plants, yellowing or reddening of leaves, proliferation of shoots and roots, abnormal flowers, and eventual decline and death of the plant. The mycoplasmas are small, polymorphic, wall-less bacteria that live in young phloem cells of their hosts, are generally visible only under the electron microscope, and except for the genus *Spiroplasma,* cannot be cultured on nutrient media. The diagnosis of such diseases, therefore, is based on symptomatology, graft transmissibility, transmission by certain insect vectors, electron microscopy, sensitivity to tetracycline antibiotics but not to penicillin, sensitivity to moderately high (32–35°C) temperatures, and in a few cases in which specific antisera have been prepared, on serodiagnostic tests.

Diseases Caused by Viruses and Viroids

Many viruses (and viroids) cause distinctive symptoms on their hosts and so the disease and the virus (or viroid) can be quickly identified by the symptoms. In the many other cases in which this is not possible, the diseases are diagnosed and the viruses are identified primarily through: (1) virus transmission tests to specific host plants by sap inoculation or by grafting, and sometimes by certain insect, nematode, fungus, or mite vectors; (2) for viruses for which specific antisera are available, by using serodiagnostic tests, primarily the enzyme-linked immunosorbent assay (ELISA), gel-difusion test, microprecipitin test, and fluorescent antibody staining; (3) by electron microscopy techniques such as negative staining of virus particles in leaf dip or purified preparations, or immune specific electron microscopy (a combination of serodiagnosis and electron microscopy); (4) by microscopic examination of infected cells for specific crystalline or amorphous inclusions, which sometimes are diagnostic of the group to which the virus belongs; (5) electrophoretic tests, useful primarily for detection and diagnosis of viroids and of nucleic acids of viruses; (6) hybridization of commercially available radioactive DNA complementary to a certain viroid RNA, with the viroid RNA present in plant sap and attached to a membrane filter.

Diseases Caused by More Than One Pathogen

Quite frequently a plant may be attacked by two or more pathogens of the same or different kinds and may develop one or more types of disease symptoms. It is important to recognize the presence of the additional pathogen(s). Once this is ascertained, the diagnosis of the disease(s) and the identification of the pathogen(s) proceeds as described above for each kind of pathogen.

Noninfectious Diseases

If no pathogen can be found, cultured from, or transmitted from a diseased plant, then it must be assumed that the disease is caused by an environmental factor. The number of environmental factors that can cause disease in plants is almost unlimited, but most of them affect plants by interfering with normal physiological processes. Such interference may be a result of an excess of a toxic substance in the soil or in the air or a lack of an essential substance (water, oxygen, or mineral nutrients) or a result of an extreme in the conditions supporting plant life (temperature, humidity, oxygen, CO_2, or light). Some of these effects may be the result of normal conditions (for example, low temperatures) occurring at the wrong time or of abnormal conditions brought about naturally (flooding or drought) or by the activities of people and their machines (air pollutants, soil compaction, and weed killers).

The specific environmental factor that has caused a disease might be determined by observing a change in the environment, for example, a flood or an unseasonable frost. Some environmental factors cause specific symptoms on plants that help determine the cause of the malady, but most of them cause nonspecific symptoms that, unless the history of the environmental conditions is known, make it difficult to accurately diagnose the cause.

Identification of a Previously Unknown Disease—Koch's Rules

When a pathogen is found on a diseased plant, the pathogen is identified by reference to special manuals; if the pathogen is known to cause such a disease, and the diagnostician is confident that no other causal agents are involved, then the diagnosis of the disease may be considered completed. If, however, the pathogen found seems to be the cause of the disease but no previous reports exist to support this, then the following steps are taken to verify the hypothesis that the isolated pathogen is the cause of the disease:

1. The pathogen must be found associated with the disease in all the diseased plants examined.
2. The pathogen must be isolated and grown in pure culture on nutrient media, and its characteristics described (nonobligate parasites), or on a susceptible host plant (obligate parasites), and its appearance and effects recorded.
3. The pathogen from pure culture must be inoculated on healthy plants of the same species or variety on which the disease appears, and it must produce the same disease on the inoculated plants.
4. The pathogen must be isolated in pure culture again, and its characteristics must be exactly like those observed in step 2.

If all the above steps, usually known as Koch's rules, have been followed and proved true, then the isolated pathogen is identified as the organism responsible for the disease.

Koch's rules are possible to implement, although not always easy to carry out, with such pathogens as fungi, bacteria, parasitic higher plants, nematodes, some viruses, some viroids, and the spiroplasmas. These organisms can be isolated and cultured, or can be purified, and can then be introduced into the plant to see if they cause the disease. With the other pathogens, however, such as some viruses, mycoplasmas, fastidious vascular bacteria, and protozoa, culture or purification of the pathogen is not yet possible, and the pathogen often cannot be reintroduced into the plant to reproduce the disease. Thus, with these pathogens, Koch's rules cannot be carried out and their acceptance as *the* pathogens of the diseases with which they are associated is more or less tentative. In most cases, however, the circumstantial evidence is overwhelming, and it is assumed that further improvement of techniques of isolation, culture, and inoculation of pathogens will someday prove that today's assumptions are justified. On the other hand, in the absence of the proof demanded by Koch's rules and as a result of insufficient information, all plant diseases caused by mycoplasmas (for example, aster yellows) and fastidious vascular bacteria (for example, Pierce's disease of grape) were for years thought to be caused by viruses.

REFERENCES

Avery, O. T., MacLeod, C. M., and MacCarty, M. (1944). Studies on the chemical nature of the substance inducing transformation of pneumococcal types. *J. Exp. Med.* **79,** 137–158.

Ball, E. (1946). Development in sterile culture of stem tips and subjacent regions of *Tropaeolum majus* and *Lupinus albus* L. *Am. J. Bot.* **33,** 101–118.

Barger, G. (1931). "Ergot and Ergotism." Gurnev & Jackson, London.

Barrus, M. F. (1911). Variation in varieties of beans in their susceptibility to anthracnose. *Phytopathology* **1,** 190–195.

Bawden, F. C., Pirie, N. W., Bernal, J. D., and Fankuchen, I. (1936). Liquid crystalline substances from virus-infected plants. *Nature (London)* **138,** 1051–1052.

Beijerinck, M. W. (1898). Ueber ein contagium vivum fluidum als Ursache der Fleckenkrankheit der Tabaksblatter. *Verh. K. Akad. Wet. Amsterdam* **65**(2), 3–21; Engl. transl. by J. Johnson in *Phytopathol. Classics* No. 7 (1942).

Berkeley, M. J. (1845). Disease in potatoes. *Gard. Chron.,* 593.

Biffen, R. H. (1905). Mendel's laws of inheritance and wheat breeding. *J. Agric. Sci.* **1,** 4–48.

Brefield, O. (1875). Methoden zur Untersuch der Pilze. *Landwirtsch. Jahrb.* **4,** 151–175.

Brown, W. (1915). Studies in the physiology of parasitism. I. The action of *Botrytis cinerea. Ann. Bot. (London)* **29,** 313–348.

only a few of them are serious parasites of crop plants. Recently, several fastidious vascular bacteria and viroids have also been shown to cause diseases in plants. A single crop, the tomato, is attacked by more than 80 species of fungi, 11 bacteria, 16 viruses, several mycoplasmas, and several nematodes. This number of diseases is average since corn has 100, wheat 80, and apple and potato each are susceptible to about 200 diseases.

Host Range of Pathogens

Pathogens differ with respect to the kinds of plants that they can attack, with respect to the organs and tissues that they can infect, and with respect to the age of the organ or tissue of the plant on which they can grow. Some pathogens are restricted to a single species, others to one genus of plants, while others have a wide range of hosts, including many taxonomic groups of higher plants. Some pathogens grow especially on roots, others on stems, some mainly on the leaves or on fleshy fruit or vegetables. Some pathogens, for example, vascular parasites, attack specifically certain kinds of tissues, such as vascular tissues. Others may produce different effects on different parts of the same plant. With regard to the age of plants, some pathogens attack seedlings or the young tender parts of plants, while others attack only mature tissues.

Most obligate parasites are usually quite specific as to the kind of host they attack, possibly because they have evolved in parallel with their host and require certain nutrients that are produced or become available to the pathogen only in these hosts. Nonobligate parasites usually attack many different plants and plant parts of varying age, possible because they depend for their attack on nonspecific toxins or enzymes that affect substances or processes found commonly among plants. Some nonobligate parasites, however, produce disease on only one or a few plant species. In any case, the number of plant species currently known to be susceptible to a single pathogen is smaller than the actual number in nature since only a few species out of thousands have been studied for their susceptibility to each pathogen. Furthermore, because of genetic changes, a pathogen may be able to attack hosts previously immune to it. It should be noted, however, that each plant species is susceptible to attack by only a relatively small number of all known plant pathogens.

Development of Disease in Plants

A plant becomes diseased when it is attacked by a pathogen or affected by an abiotic agent. Therefore, for a plant disease to occur, at least two components (plant and pathogen) must come in contact and must interact. If at the time of contact and for some time afterwards conditions are too cold, too hot, too dry, or some other extreme, the pathogen may be unable to attack or the plant may be able to resist the attack, and in spite of the two being in contact, no disease develops. Apparently, then, a third component—a set of environmental conditions within a favorable range—must also occur for disease to develop. Each of the three components, however, can display considerable

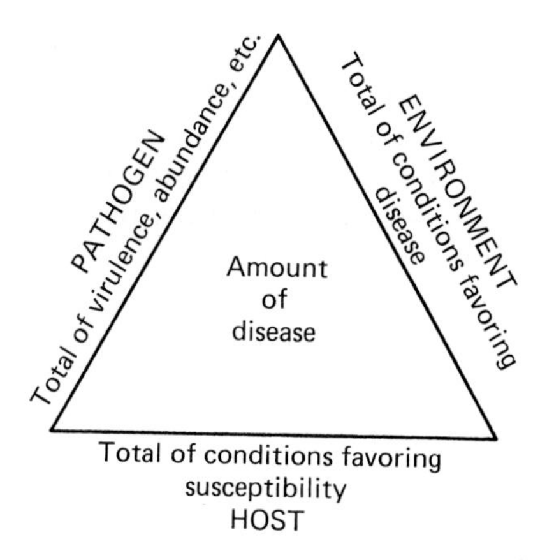

FIGURE 2-1 The disease triangle.

variability, and as one component changes it affects the degree of disease severity within an individual plant and within a plant population. For example, the plant may be of a species or variety that may be more or less resistant to the pathogen, or it may be too young or too old, or plants over a large area may show genetic uniformity, all of which can either reduce or increase the rate of disease development by a particular pathogen. The pathogen may be of a more or less virulent race, may be present in small or extremely large numbers, it may be in a dormant state, or it may require a film of water or a specific vector. The environment may affect both the growth and resistance of the host plant and also the rate of growth or multiplication and degree of virulence of the pathogen, as well as its dispersal by wind, water, vector, and so on.

The interactions of the three components of disease have often been visualized as a triangle (Figure 2-1), generally referred to as the "disease triangle." Each side of the triangle represents one of the three components. The size of each side is proportional to the sum total of the characteristics of each component that favor disease. For example, if the plants are resistant, the wrong age, or widely spaced, the host side—and the amount of disease—would be small or zero, while if the plants are susceptible, at a susceptible stage of growth, or densely planted, the host side would be long and the potential amount of disease could be great. Similarly, the more virulent, abundant, and active the pathogen, the longer the pathogen side would be and the greater the potential amount of disease. Also, the more favorable the environmental conditions that help the pathogen (for example, temperature, moisture, and wind) or that reduce host resistance, the longer the environment side would be and the greater the potential amount of disease. If the three components of the disease triangle could be quantified, the area of the triangle would represent the amount of disease in a plant or in a plant population.

Stages in the Development of Disease: The Disease Cycle

In every infectious disease a series of more or less distinct events occurs in succession and leads to the development and perpetuation of the disease and the pathogen. This chain of events is called a **disease cycle.** A disease cycle sometimes corresponds fairly closely to the life cycle of the pathogen, but it refers primarily to the appearance, development, and perpetuation of the disease as the pathogen relates to it rather than to the pathogen itself. The disease cycle involves the changes in the plant and the plant's symptoms as well as those in the pathogen and spans periods within a growing season and from one growing season to the next. The primary events in a disease cycle are inoculation, penetration, establishment of infection, colonization (invasion), growth and reproduction of the pathogen, dissemination of the pathogen, and survival of the pathogen in the absence of the host, that is, overwintering or oversummering of the pathogen.

Inoculation

Inoculation is the coming in contact of a pathogen with a plant. The pathogen or pathogens, that land on, or are otherwise brought into contact with the plant are called the **inoculum.** Inoculum is any part of the pathogen that can initiate infection. Thus, in fungi inoculum may be spores, sclerotia (a compact mass of mycelium) or fragments of mycelium. In bacteria, mycoplasmas, viruses, and viroids, inoculum is always whole individuals of bacteria, mycoplasmas, viruses, and viroids, respectively. In nematodes, inoculum may be adult nematodes, nematode larvae, or eggs. In parasitic higher plants inoculum may be plant fragments or seeds. Inoculum may consist of a single individual of a pathogen, for example, one spore or one multicellular sclerotium, or of millions of individuals of a pathogen, for example, bacteria carried in a drop of water. One unit of inoculum of any pathogen is called a **propagule.**

Types of Inoculum

Inoculum that survives the winter or summer and causes the original infections in the spring or in the autumn is called **primary inoculum** and the infections it causes are called **primary infections.** Inoculum produced from primary infections is called **secondary inoculum** and that, in turn, causes **secondary infections.** Generally, the more abundant the primary inoculum and the closer it is to the crop, the more severe the disease and the losses that result.

Sources of Inoculum

Inoculum sometimes is present right in the plant debris or soil in the field where the crop is grown; other times it comes into the field with the seed, transplants, tubers, or other propagative organs, or it may come from sources outside the field. Outside sources of inoculum may be nearby plants or fields or fields many miles away. In many plant diseases, especially those of annual crops, the inoculum survives in perennial weeds or alternate hosts, and every season it is carried from them to the annual and other plants. Fungi, bacteria, parasitic higher plants, and nematodes either produce their inoculum on the surface of infected plants or their inoculum reaches the plant surface when the infected tissue breaks down. Viruses, viroids, mycoplasmas, and fastidious bacteria produce their inoculum within the plants; such inoculum almost never reaches the plant surface in nature and, therefore, cannot by itself escape from one plant and spread to another.

Landing or Arrival of Inoculum

The inoculum of most pathogens is carried to host plants passively by wind, water, and insects. Air-borne inoculum usually gets "out of the air" and onto the plant surface not just by gravity but by being "washed out" by rain. Only a tiny fraction of the inoculum produced actually lands on susceptible host plants; the bulk of the inoculum lands on things that cannot become infected. Some types of inoculum in the soil, for example, zoospores and nematodes, may be attracted to the host plant by such substances as sugars and amino acids diffusing out of the plant roots. Vector-transmitted pathogens are usually carried to their host plants with an extremely high efficiency.

Prepenetration Phenomena

Germination of Spores and Seeds

All pathogens in their vegetative state are capable of initiating infection immediately. Fungal spores and seeds of parasitic higher plants, however, must first germinate; for that they require favorable temperature and also moisture in the form of rain, dew, or a film of water on the plant surface or at least high relative humidity. The moist conditions must last long enough for the pathogen to penetrate; otherwise it desiccates and dies. Most spores can germinate immediately after their maturation and release, but others (the so-called resting spores) require a dormancy period of varying duration before they can germinate. When a spore germinates it produces a germ tube, that is, the first part of the mycelium, that can penetrate the host plant. Some fungal spores germinate by producing other spores, such as zoospores or basidiospores.

Spore germination is often favored by nutrients diffusing from the plant surface; the more nutrients (sugars and amino acids) exuded from the plant, the more spores germinate and the faster they germinate. In some cases, spore germination of a certain pathogen is stimulated only by exudates of plants susceptible to that particular pathogen. In other cases, spore germination may be inhibited to a lesser or greater extent by materials released into the surrounding water by the plant, by substances contained within the spores themselves, especially when the spores are highly concentrated, and by saprophytic microflora present on or near the plant surface.

Fungi in soil coexist with a variety of antagonistic microorganisms that cause an environment of starvation and of toxic metabolities. As a result, spores of many soil-borne fungi are often unable to germinate in some soils, and this phenomenon is called **fungistasis,** or their germ tubes lyse rapidly. Soils in which such events occur are known as **suppressive soils**. Fungistasis, however, is generally counteracted by root exudates of host plants growing nearby, and the spores are then able to germinate and infect.

After spores germinate, the resulting germ tube must grow, or the motile secondary spore (zoospore) must move, toward a site on the plant surface at which successful penetration can take place. The number, length, and rate of growth of germ tubes, or the number of motile spores, may be affected by physical conditions, such as temperature and moisture, and also by the kind and amount of exudates the plant produces at its surface and by the saprophytic microflora. The growth of germ tubes in the direction of successful penetration sites seems to be regulated by several factors, including: greater humidity or chemical stimuli associated with such openings as wounds, stomata, and lenticels; thigmotropic (contact) responses to the topography of the leaf surface resulting in germ tubes growing at right angles to cuticular ridges that generally surround stomata and thus eventually reaching a stoma; and nutritional responses of germ tubes toward greater concentrations of sugars and amino acids present along roots. The direction of movement of motile spores (zoospores) is also regulated by similar factors, that is, chemical stimuli emanating from stomata, wounds, or the zone of elongation of roots, physical stimuli related to the structure of open stomata, and the nutrient gradient present in wound and root exudates.

Seeds germinate by producing a radicle, which either penetrates the host plant or produces a small plant that penetrates the host plant by means of specialized feeding organs called **haustoria.** Most conditions described above as affecting spore germination and the direction of growth of germ tubes also apply to seeds.

Hatching of Nematode Eggs

Nematode eggs also require conditions of favorable temperature and moisture to become activated and hatch. In most nematodes, the egg contains the first larval stage before or soon after the egg is laid. This larva immediately undergoes a molt and gives rise to the second larval stage that may remain dormant in the egg for various periods of time. Thus, when the egg finally hatches it is the second stage larva that emerges, and it either finds and penetrates a host plant or undergoes additional molts that produce further larval stages and the adults.

Once nematodes are in close proximity to plant roots, they are attracted to roots by certain chemical factors associated with root growth, particularly carbon dioxide and some amino acids. These factors may diffuse and may have an attractant effect on nematodes present several centimeters away from the root. Nematodes are generally attracted to roots of both host and nonhost plants, although there may be some cases in which nematodes are attracted more to the roots of host plants.

Attachment of Pathogen to Host

Pathogens such as viruses, mycoplasmas, fastidious bacteria, and protozoa are placed directly into cells of plants by their vectors, and they are probably immediately surrounded by cytoplasm and cytoplasmic membranes. On the other hand, almost all fungi, bacteria, and parasitic higher plants are first brought into contact with the external surface of plant organs. Before they can penetrate and colonize the host, they must first become attached to the host surface. The propagules of these pathogens have on their surface mucilaginous sheaths consisting of mixtures of polysaccharides, glycoproteins, polymers of hexosamines, and fibrillar materials, which, when moistened, become sticky and help the pathogen adhere to the plant. After hydration, as spores and seeds germinate, the germ tubes also produce mucilaginous materials that allow them to adhere to the cuticular surface of the host, either along their entire length or only at the tip of the germ tube. In regions of contact with the germ tube, the structure of the host cuticle and cell walls often appears altered, presumably as a result of degradative enzymes contained in the mucilaginous sheath.

Recognition between Host and Pathogen

It is still unclear how pathogens recognize their hosts and vice-versa. It is assumed that when a pathogen comes in contact with a host cell, an early event takes place that triggers a fairly rapid response in each organism that either allows or impedes further growth of the pathogen and development of disease. The nature of the "early event" is not known with certainty in any host–parasite combination, but it may be one of many biochemical substances, structures, and pathways. These substances, structures, and pathways may include specific elicitor molecules produced by one organism that induce formation of specific products by the other organism; production of host-specific toxins that react with specific receptor sites of the host; release of hormones that affect growth and development of the other organism; enzymes that may alter the cell wall or membrane; substances that trigger activation of latent enzymatic pathways; chelating substances that alter membrane permeability and affect the balance of ions; polysaccharides that interfere with the passage of water and nutrients; and proteins or glycoproteins that react with cell wall or membrane polysaccarides.

When the initial recognition signal received by the pathogen favors growth and development, disease may be induced; if the signal suppresses pathogen growth and activity, disease may be aborted. On the other hand, if the initial recognition signal received by the host triggers a defense reaction, pathogen growth and activity may be slowed or stopped and disease may not develop; if the signal either suppresses or bypasses the defense reaction of the host, disease may develop.

Penetration

Pathogens penetrate plant surfaces by direct penetration, through natural openings, or through wounds (Figure 2-2). Some fungi penetrate tissues in one way only, others in more than one. Bacteria enter plants mostly through wounds, less frequently through natural openings, and never directly (Figure 2-3). Viruses, viroids, mycoplasmas, and fastidious bacteria enter through wounds made by vectors, although some viruses and viroids may also enter through wounds made by tools and other means. Parasitic higher plants enter their hosts by direct penetration. Nematodes enter plants by direct penetration and, sometimes, through natural openings.

Penetration does not always lead to infection. Many organisms actually penetrate cells of plants that are not susceptible to these organisms and that

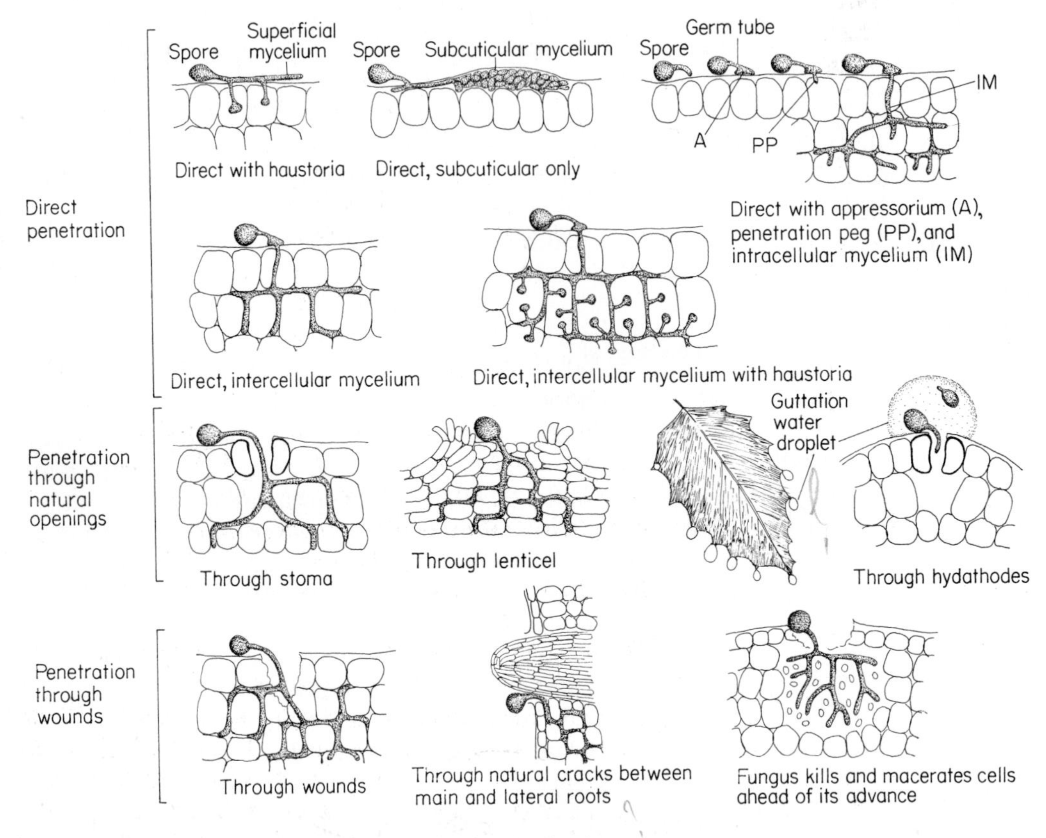

FIGURE 2-2 Methods of penetration and invasion by fungi.

do not become diseased; these organisms cannot proceed beyond the stage of penetration and die without producing disease.

Direct Penetration through Intact Plant Surfaces

Direct penetration through intact plant surfaces is probably the most common type of penetration in fungi and nematodes and the only type of penetration in parasitic higher plants. None of the other pathogens can enter plants by direct penetration.

Fungi that penetrate their host plants directly do so through a fine hypha or an **appressorium** (Figures 2-2, 2-4, and 2-5). The fine hypha or appressoria are formed at the point of contact of the germ tube or mycelium with a plant surface. The fine hypha grows toward the plant surface and pierces the cuticle and the cell wall through mechanical force and enzymatic softening of the cell

FIGURE 2-3 Methods of penetration and invasion by bacteria.

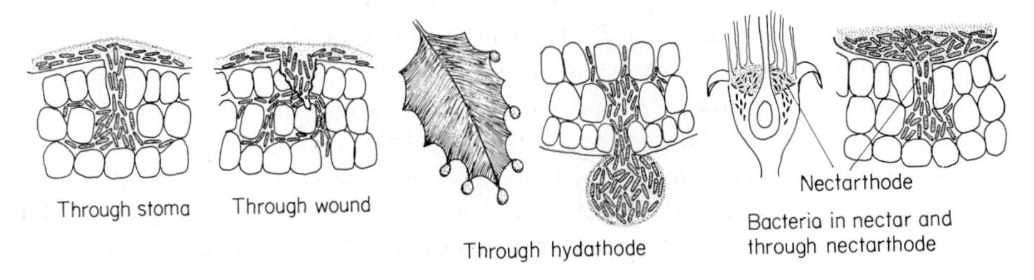

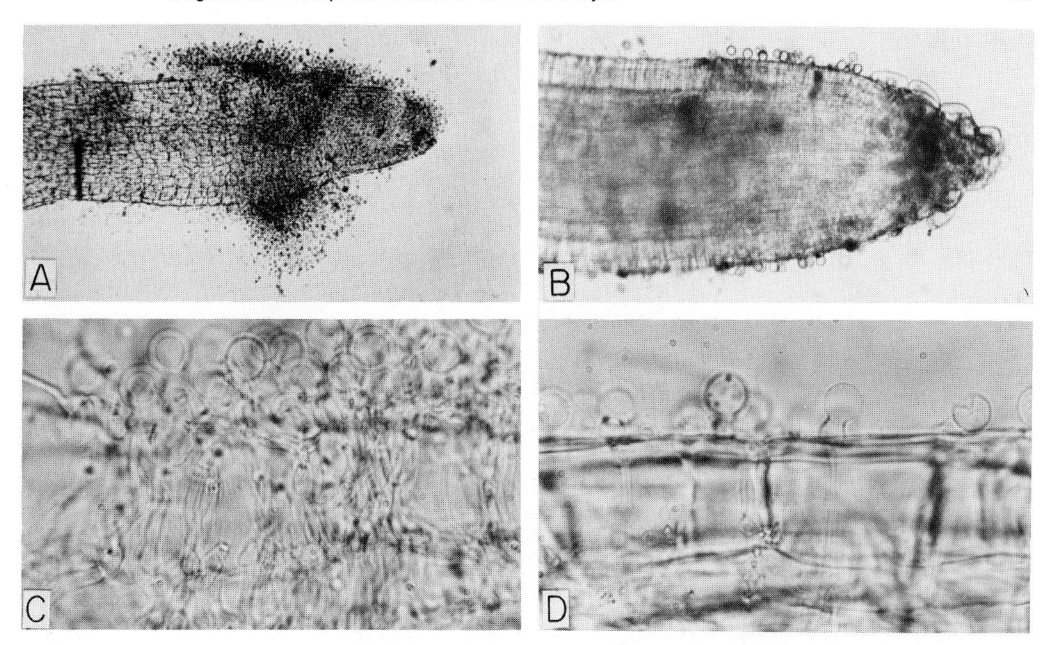

FIGURE 2-4 Attraction of zoospores of *Phytophthora cinnamomi* to roots of two types of blueberry (A and B) and infection of the roots by the zoospores (C and D). Attraction of zoospores to roots one hour after inoculation (A and B) and infection and colonization of the root after 24 hours (C and D) are greater in the susceptible highbush blueberry (A, C) than in the more resistant rabbiteye blueberry (B, D). (Photos courtesy R. D. Milholland, from *Phytopathology* **65,** 789–793).

wall substances. Most fungi, however, form an appressorium at the end of the germ tube, the appressorium being usually bulbous or cylindrical with a flat surface in contact with the host plant's surface. Then, a fine hypha, usually called a **penetration peg,** grows from the flat surface of the appressorium toward the host and pierces the cuticle and the cell wall. The penetration peg is generally much smaller in diameter than a normal hypha of the fungus, but it regains its normal diameter once inside the cell lumen. In most fungal diseases the fungus penetrates the plant cuticle and the cell wall, but in some, such as apple scab, the fungus penetrates only the cuticle and stays between the cuticle and the cell wall.

Parasitic higher plants also form an appressorium and penetration peg at the point of contact of the radicle with the host plant, and penetration is similar to that in fungi.

Direct penetration in nematodes is accomplished by repeated back-and-forth thrusts of their stylets. Such thrusts finally create a small opening in the cell wall, and the nematode sends its stylet into the cell or the entire nematode enters the cell (Figure 2-6.)

Penetration through Wounds

All bacteria, most fungi, some viruses, and all viroids can enter plants through various types of wounds (Figures 2-2, 2-3), and viruses, mycoplasmas, and fastidious vascular bacteria enter plants through wounds made by their vectors. The wounds utilized by bacteria and fungi may be fresh or old and may consist of lacerated or killed tissue. These pathogens may grow briefly on such

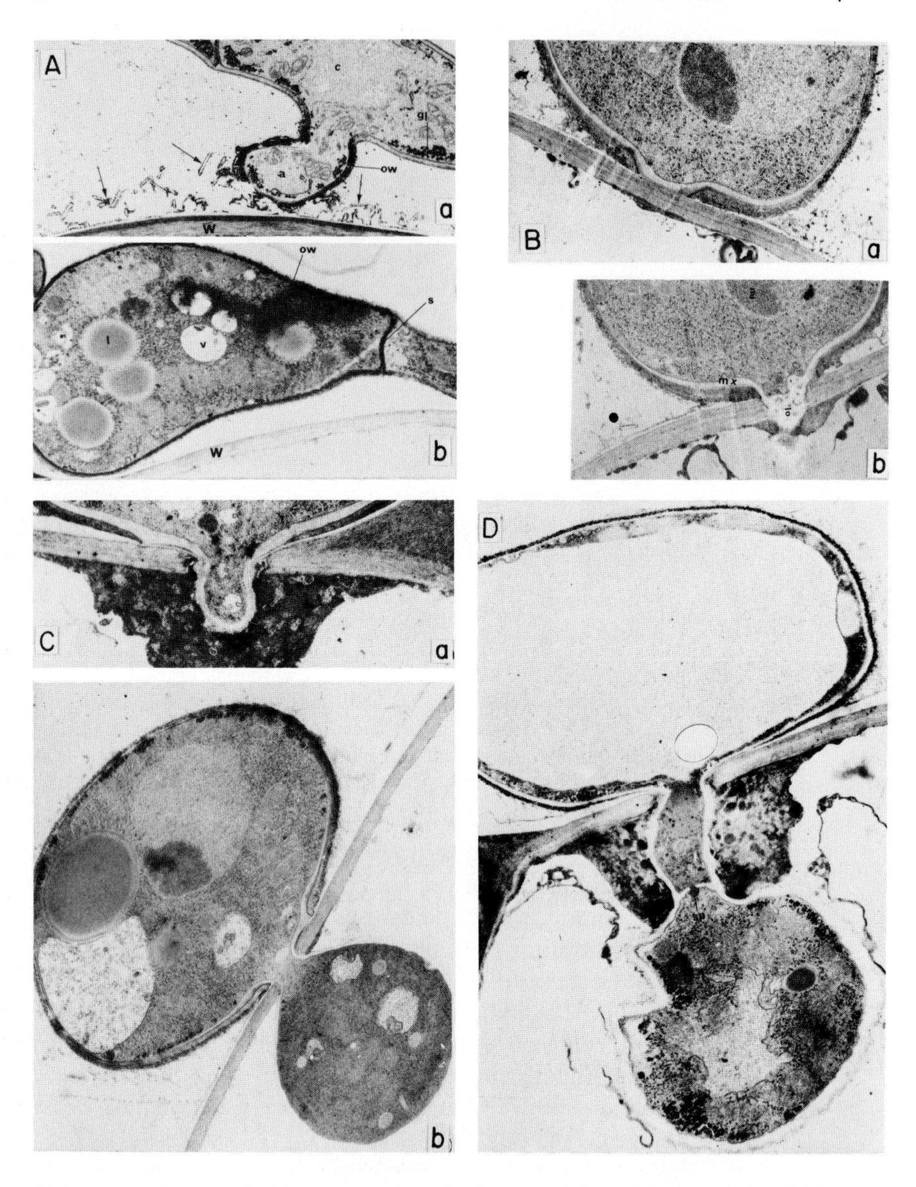

FIGURE 2-5 Electron micrographs of direct penetration of a fungus *(Colletotrichum graminicola)* into an epidermal leaf cell. (A) (a) Developing appressorium from a conidium. Note wax rods (arrows) on leaf surface. (b) A mature appressorium separated by a septum from the germination tube. (B) (a) Formation of a penetration peg at central point of contact of appressorium with cell wall. (b) Lomasomelike structures in the infection peg which has already penetrated the cell wall and the papilla produced by the invaded cell. (C) Development of infection hyphae. (a) Infection peg penetrating the papilla. (b) Appressorium and swollen infection hypha after penetration. (D) Upon completion of penetration and establishment of infection, the appressorium consists mostly of a large vacuole and is cut off from the infection hypha by a septum. (Photos courtesy D. J. Politis and H. Wheeler, from *Physiol. Plant Pathol.* **3,** 465–471.)

tissue before they advance into healthy tissue. Laceration or death of tissues may be the result of: environmental factors, such as wind breakage or rubbing, sand blasting, hail, frost, heat scorching, and fire; animal feeding, for example, insects, nemotodes, worms, and large animals; cultural practices of humans, such as cultivation, weeding, pruning, grafting, transplanting, spray-

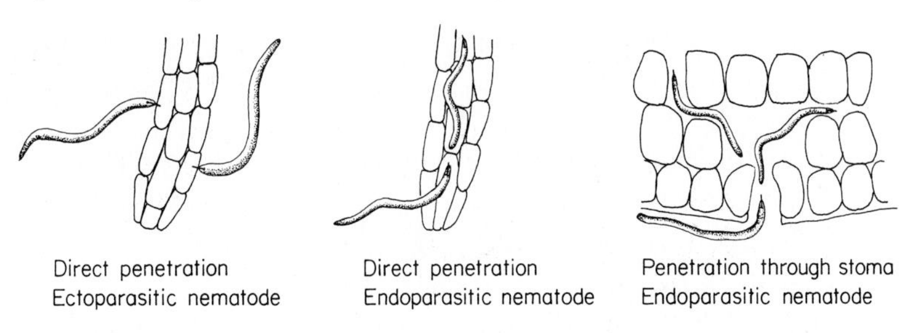

FIGURE 2-6 Methods of penetration and invasion by nematodes.

ing, and harvesting; self-inflicted injuries, such as leaf scars, and root cracks; and, finally, wounds or lesions caused by other pathogens. Bacteria and fungi penetrating through wounds apparently germinate or multiply in the sap present in fresh wounds or in a film of rain or dew water present on the wound. Subsequently, the pathogen invades adjacent plant cells directly or through haustoria, or it secretes enzymes and toxins that kill and macerate the nearby cells.

Penetration of viruses, mycoplasmas, and fastidious bacteria through wounds depends on the deposition of these pathogens by their vectors (insects for all three pathogens and also nematodes, mites, and fungi for viruses, and human hands and tools for some viruses and viroids) in fresh wounds created at the time of inoculation. In most cases, these pathogens are carried by one or a few kinds of specific vectors and can be successfully inoculated only when they are brought to the plant by these particular vectors.

Penetration through Natural Openings

Many fungi and bacteria enter plants through stomata and some enter through hydathodes, nectarthodes, and lenticels (Figures 2-2, 2-3). Most stomata are present in large numbers on the lower side of leaves, measure about $10-20 \times 5-8$ μm, and are open in the daytime but more or less closed at night. Bacteria present in a film of water over a stoma can easily swim through the stoma and into the substomatal cavity where they can multiply and start infection. Fungal spores generally germinate on the plant surface, and the germ tube may then grow through the stoma. Frequently, however, the germ tube forms an appressorium that fits tightly over the stoma, and usually one fine hypha grows from it into the stoma (Figure 2-7). In the substomatal cavity the hypha enlarges, and from it grow one or several small hyphae that actually invade the cells of the host plant directly or through haustoria. Although some fungi can apparently penetrate even closed stomata, others penetrate stomata only while they are open, and some, for example, the powdery mildew fungi, may grow over open stomata without entering them.

Hydathodes are more or less permanently open pores at the margin and tip of leaves; they are connected to the veins and secrete droplets of liquid containing various nutrients. Some bacteria use these pores as a means of entry into leaves, but few fungi seem to enter plants through hydathodes. Some bacteria also enter blossoms through the nectarthodes or nectaries, which are similar to hydathodes.

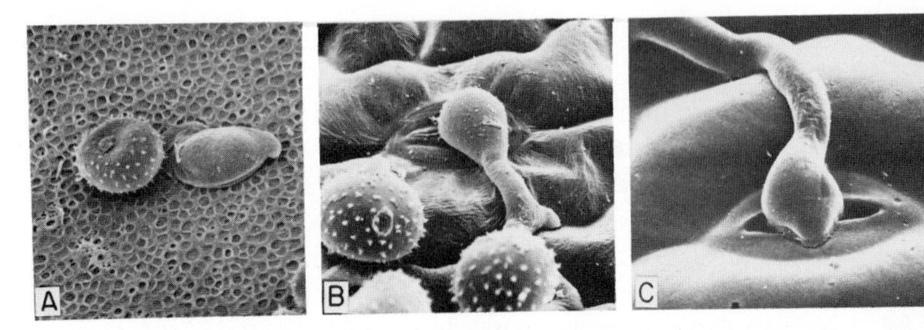

FIGURE 2-7 Scanning electron micrographs of appressorium formation and penetration through a stoma by the bean rust fungus *Uromyces phaseoli.* (A) Uredospore, short germ tube, and large, flattened appressorium forming on a membrane. (B) Uredospore, germ tube, and appressorium formed after 6-hour germination over closed stoma on bean leaf. (C) Young appressorium over open stoma of bean leaf. (Photos courtesy W. K. Wynn, from *Phytopathology* **66,** 136–146).

Lenticels are openings on fruit, stems, and tubers that are filled with loosely connected cells to allow passage of air. During the growing season, lenticels are open, but even so relatively few fungi and bacteria penetrate tissues through them, growing and advancing mostly between the cells. Most pathogens that penetrate through lenticels can also enter through wounds, lenticel penetration being apparently a less efficient, secondary pathway.

Infection

Infection is the process by which pathogens establish contact with the susceptible cells or tissues of the host and procure nutrients from them. During infection pathogens grow or multiply, or both, within the plant tissues and invade and colonize the plant to a lesser or greater extent. Thus, invasion of the plant tissues by the pathogen, and growth and reproduction (colonization) of the pathogen in or on infected tissues are actually two concurrent substages of disease development within the stage of infection.

Successful infections result in the appearance of discolored, malformed, or necrotic areas on the host plant, which are called **symptoms.** Some infections, however, remain **latent,** that is, they do not produce symptoms right away but at a later time when the environmental conditions may be more favorable, or at a different stage of maturity of the plant.

All the visible changes in the appearance of infected plants make up the symptoms of the disease. Symptoms may change continuously from the moment of their appearance until the entire plant dies, or they may develop up to a point and then remain more or less unchanged for the rest of the growing season. Symptoms may appear as soon as 2 to 4 days after inoculation, as happens in some localized virus diseases of herbaceous plants, or as late as 2 to 3 years after inoculation, in the case of some viral, mycoplasmal, and other diseases of trees. In most plant diseases, however, symptoms appear from a few days to a few weeks after inoculation.

The time interval between inoculation and the appearance of disease symptoms is called the **incubation period.** The length of the incubation period of various diseases varies with the particular pathogen-host combination, with the stage of development of the host, and with the temperature in the environment of the infected plant.

During infection, some pathogens obtain nutrients from living cells, often without killing the cells or at least not for a long time; others kill cells and utilize their contents as they invade them; still others kill cells and disorganize surrounding tissues. During infection pathogens release in the host a number of biologically active substances (such as enzymes, toxins, and growth regulators), which may affect the structural integrity of the host cells or their physiological processes. In response, the host reacts with a variety of defense mechanisms, which result in varying degrees of protection of the plant from the pathogen.

For a successful infection to occur it is not sufficient that a pathogen come in contact with its host, but several other conditions must also be satisfied. First of all, the plant variety must be susceptible to the particular race of the pathogen—in which case that race of the pathogen is said to be **virulent** on the variety of the host plant. The host plant must be in a susceptible stage, since some pathogens attack only young seedlings, others only mature or senescing plants, some only the leaves, others only the flowers or the fruit, or only ripe fruit. The pathogen must be in a pathogenic stage, for example, fungal mycelium or spores and seeds that can germinate and infect immediately without requiring a resting (dormancy) period first, or infective larval stages or adults of nematodes. Finally, the temperature and moisture conditions in the environment of the plant must favor the growth and multiplication of the pathogen. When these conditions occur at an optimum, the pathogen can invade the host plant up to the maximum of its potential, even in the presence of plant defenses, and as a consequence disease develops.

Invasion

Various pathogens invade hosts in different ways and to different extents (Figures 2-2, 2-3, and 2-6). Some fungi, such as those causing apple scab and black spot of rose, produce mycelium, which grows only in the area between the cuticle and the epidermis (subcuticular); others, such as those causing the powdery mildews, produce mycelium only on the surface of the plant but send haustoria into the epidermal cells. Most fungi spread into all the tissues of the plant organs (leaves, stems, roots) they infect, either by growing directly through the cells—**intracellular mycelium**—or by growing between the cells—**intercellular mycelium**—. The fungi that cause vascular wilts invade the xylem vessels of plants.

Bacteria invade tissues intercellulary, although when parts of the cell walls dissolve, bacteria also grow intracellularly. Bacteria causing vascular wilts, like the vascular wilt fungi, invade the xylem vessels. Most nematodes invade tissues intercellularly, but some can invade intracellularly as well. Many nematodes do not invade cells or tissues at all but feed by piercing epidermal cells with their stylet.

Viruses, viroids, mycoplasmas, and fastidious bacteria invade tissues by moving from cell to cell intracellularly. Viruses and viroids invade all types of living plant cells, mycoplasmas invade phloem sieve tubes and perhaps a few adjacent phloem parenchyma cells, while fastidious bacteria invade either xylem vessels or phloem sieve tubes.

Many infections caused by fungi, bacteria, nematodes, viruses, and parasitic higher plants are **local,** that is, they involve a single cell, a few cells, or a small area of the plant. These infections may remain localized throughout the growing season or they may enlarge slightly or very slowly. Other infections enlarge more or less rapidly and may involve an entire plant organ (flower, fruit, leaf), a large part of the plant (a branch), or the entire plant.

All infections caused by mycoplasmas and fastidious bacteria and all natural infections caused by viruses and viroids are **systemic,** that is, the pathogen, from one initial point, spreads and invades most or all susceptible cells and tissues throughout the plant. Vascular wilt fungi and bacteria invade xylem vessels internally, but they are usually confined to a few vessels in the roots, the stem or the top of infected plants, and only in the final stages of the disease do they invade most or all xylem vessels of the plant. Some fungi, primarily among those causing downy mildews, smuts, and rusts, also invade their hosts systemically, although in most cases the older mycelium degenerates and disappears and only younger mycelium survives in actively growing plant tissues.

Growth and Reproduction (Colonization) of the Pathogen

Individual fungi and parasitic higher plants generally invade and infect tissues growing into them from one initial point of inoculation. Most of these pathogens, whether producing a small spot, a large infected area, or a general necrosis of the plant, continue to grow and branch out within the infected host indefinitely so that the same pathogen individual spreads into more and more plant tissues until the spread of the infection is stopped or the plant is dead. In some fungal infections, however, while the younger hyphae continue to grow into new healthy tissues, the older ones in the already infected areas die out and disappear, so that an infected plant may have several points where separate units of mycelium are active. Also, fungi causing vascular wilts often invade plants by producing and releasing spores within the vessels, and as the spores are carried in the sap stream, they invade vessels far away from the mycelium, germinate there, and produce mycelium, which invades more vessels.

All other pathogens,—bacteria, mycoplasmas, viruses, viroids, nematodes, and protozoa—do not increase much, if at all, in size with time, since their size and shape remain relatively unchanged throughout their existence. These pathogens invade and infect new tissues within the plant by reproducing at a rapid rate and increasing their numbers tremendously in the infected tissues; the progeny then are either carried passively into new cells and tissues through plasmodesmata (viruses and viroids only), phloem (viruses, viroids, mycoplasmas, fastidious bacteria, protozoa), xylem (some bacteria), or as happens somewhat with bacteria, and more with protozoa and nematodes, they may move through cells on their own power.

Plant pathogens reproduce in a variety of ways (see Figure 1-3 in Chapter 1). Fungi reproduce by means of spores, which may be either asexual (equivalent to the buds on a twig or the tubers of a potato plant), or sexual (equivalent to the seeds of plants). Parasitic higher plants reproduce just like all plants, that is, by seeds. Bacteria, mycoplasmas, and protozoa reproduce by fission—one mature individual splits into two equal, smaller individuals. Viruses and viroids are replicated by the cell, just as a page placed on a photocopying machine is replicated by the machine as long as the machine is operating and paper supplies last. Nematodes reproduce by means of eggs.

The great majority of plant pathogenic fungi produce mycelium only within the plants they infect. Relatively few fungi produce mycelium on the surface of their host plants, and only the powdery mildew fungi produce mycelium on the surface of, and not within, their hosts. The great majority of fungi produce spores on, or just below, the surface of the infected area of the host, and the spores are released outward into the environment. Some of the lower fungi, however, such as the clubroot pathogen, and the fungi causing vascular wilts produce spores within the host tissues, and these spores are not released outward until the host dies and disintegrates. Parasitic higher plants produce their seeds on aerial branches, and some nematodes lay their eggs at or near the surface of the host plant. Bacteria reproduce between or within host cells, generally inside the host plant, and come to the host surface only through wounds, cracks, and so on. Viruses, viroids, plant mycoplasmas, and fastidious bacteria reproduce only inside cells and apparently do not reach or exist on the surface of the host plant.

The rate of reproduction varies considerably among the various kinds of pathogens, but in all of them, one or a few pathogens can produce tremendous numbers of individuals within one growing season. Some fungi produce spores more or less continuously while others produce them in successive crops. In either case, several thousand to several hundreds of thousands of spores may be produced per square centimeter of infected tissue. Even small specialized sporophores can produce millions of spores, and the number of spores produced per infected plant is often in the billions or trillions. The numbers of spores produced in an acre of heavily infected plants, therefore, are generally astronomical, and enough spores are released to land on and inoculate every conceivable surface in the field and the surrounding areas.

Bacteria reproduce rapidly within infected tissues. Under optimum nutritional and environmental conditions (in culture) bacteria divide (double their numbers) every 20 to 30 minutes, and presumably, bacteria multiply just as fast in a susceptible plant as long as the temperature is favorable. Millions of bacteria may be present in a single drop of infected plant sap, so the number of bacteria per plant must be astronomical. Fastidious bacteria and mycoplasmas appear to reproduce more slowly than typical bacteria, and although they spread systemically throughout the vascular system of the plant, they are present in relatively few xylem or phloem vessels, and the total number of these pathogens in infected plants is relatively small.

Viruses and viroids reproduce within living host cells—the first new virus particles can be detected several hours after infection. Soon after that, however, virus particles accumulate within the infected living cell until as many as 100,000 to 10,000,000 virus particles may be present in a single cell. Viruses and viroids infect and multiply in most or all living cells of their hosts, and it is apparent that each plant may contain innumerable particles of these pathogens.

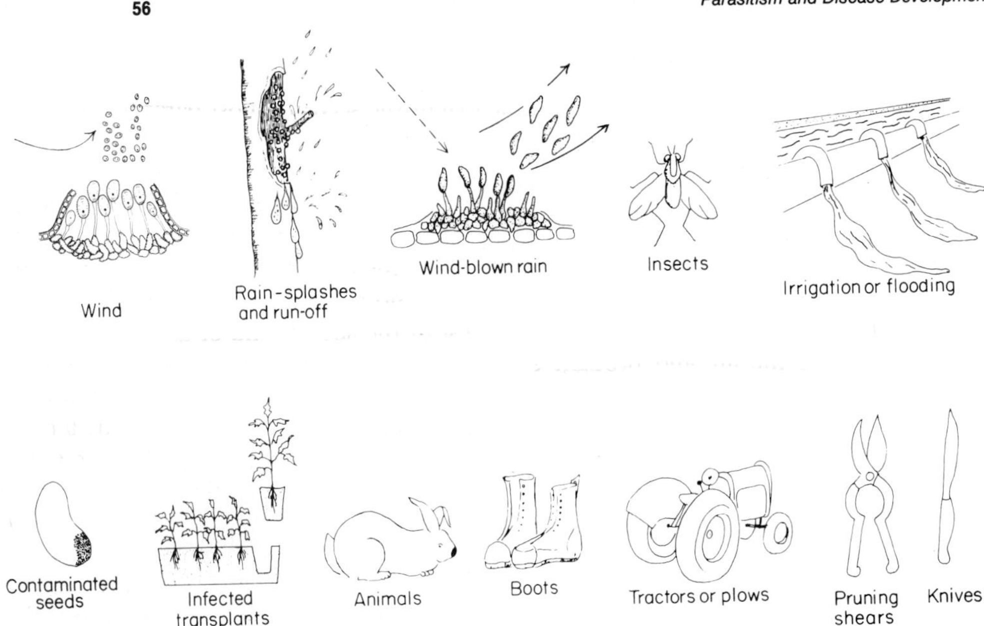

FIGURE 2-8 Means of dissemination of fungi and bacteria.

Nematode females lay about 300 to 600 eggs, about half of which produce females and these again lay 300 to 600 eggs each. Depending on the climate, the availablity of hosts, and the duration of each life cycle of the particular nematode, the nematode may have from two to more than a dozen generations per year. If even half of the females survived and reproduced, each generation time would increase the number of nematodes in the soil by more than 100-fold. Thus the buildup of nematode populations within a growing season and in successive seasons is often quite dramatic.

Dissemination of the Pathogen

A few pathogens, such as nematodes, fungal zoospores, and bacteria, can move short distances on their own power and thus can move from one host to another one very close to it. Fungal hyphae and rhizomorphs can grow between tissues in contact and sometimes through the soil toward nearby roots. Both of these means of dissemination, however, are quite limited, especially in the cases of zoospores and bacteria.

The spores of some fungi are expelled forcibly from the sporophore or sporocarp by a squirting or puffing action that results in successive or simultaneous discharge of spores up to a centimeter or so above the sporophore. The seeds of some parasitic plants are also expelled forcibly and may arch over distances of several meters.

Almost all dissemination of pathogens that is responsible for plant disease outbreaks, and even for disease occurrences of minor economic importance, is carried out passively by such agents as air, water, insects, certain other animals, and humans (Figure 2-8).

Dissemination by Air

Most fungal spores and to some extent the seeds of most parasitic plants are disseminated by air currents that carry them as inert particles to various distances. Air currents pick up spores and seeds off the sporophores, or while they are being forcibly expelled or are falling at maturity, and depending on the air turbulence and velocity, may carry the spores upward or horizontally in a way similar to that of particles contained in smoke. While airborne, some of the spores may touch wet surfaces and get trapped, and when air movement stops or when it rains, the rest of the spores land or are "washed out" from the air and brought down by the raindrops. Most of the spores, of course, land on anything but a susceptible host plant. Also, the spores of many fungi are actually too delicate to survive a long trip through the air and are therefore successfully disseminated for only a few hundred or a few thousand meters. The spores of other fungi, however, particularly those of the cereal rusts, are very hardy and occur commonly at all levels and at high altitudes (several thousand meters) above infected fields. Spores of these fungi are often carried over distances of several kilometers, even hundreds of kilometers, and in favorable weather may cause widespread epidemics.

Air dissemination of other pathogens occurs rather infrequently and only under special conditions, or indirectly. Thus, the bacteria causing fire blight of apple and pear produce fine strands of dried bacterial exudate containing bacteria, and these strands may be broken off and disseminated by wind. Bacteria and nematodes present in the soil may be blown away along with soil particles in the dust. Wind also helps in the dissemination of bacteria, fungal spores, and nematodes by blowing away rain splash droplets containing these pathogens, and wind carries away insects that may contain or are smeared with viruses, bacteria, or fungal spores. Finally, wind causes adjacent plants or plant parts to rub against each other, and this may help the spread by contact of bacteria, fungi, some viruses and viroids, and possibly of some nematodes.

Dissemination by Water

Water is important in disseminating pathogens in three ways: (1) Bacteria, nematodes, and spores, sclerotia and mycelial fragments of fungi present in the soil are disseminated by rain or irrigation water that moves on the surface or through the soil. (2) All bacteria and the spores of many fungi are exuded in a sticky liquid and depend for their dissemination on rain or (overhead) irrigation water, which either washes them downward or splashes them in all directions. (3) Raindrops or drops from overhead irrigation pick up the fungal spores and any bacteria present in the air and wash them downward where some of them may land on susceptible plants. Although water is less important than air in long-distance transport of pathogens, water dissemination of pathogens is more efficient in that the pathogens land on an already wet surface and can move or germinate immediately.

Dissemination by Insects, Mites, Nematodes, and Other Vectors

Insects, particularly aphids and leafhoppers, are by far the most important vectors of viruses, while leafhoppers are the main vectors of mycoplasmas and fastidious bacteria. Each one of these pathogens is transmitted, internally, by only one or a few species of insects during feeding and movement of the insect vectors from plant to plant. Specific insects also transmit certain fungal and bacterial pathogens, such as those causing Dutch elm disease and the bacterial wilt of cucurbits. In all diseases in which the pathogen is carried internally or externally by one or a few specific vectors, the dissemination of the pathogen depends to a large extent, or entirely, on that vector. In many diseases, however, such as bacterial soft rots, anthracnoses, and ergot, insects become smeared with various kinds of bacteria or sticky fungal spores as they move among plants. The insects carry these pathogens externally from plant to plant and deposit them on the plant surface or in the wounds they make on the plants during feeding. In such diseases, dissemination of the pathogen is facilitated by but is not dependent on the vector. Insects may disseminate pathogens over short or long distances, depending on the kind of insect, the insect-pathogen association, and the prevailing weather conditions, particularly wind.

A few species of mites and nematodes can transmit internally several viruses from plant to plant. In addition, mites and nematodes probably carry externally bacteria and sticky fungal spores with which they become smeared as they move on infected plant surfaces.

Almost all animals, small and large, that move among plants and touch the plants along the way can disseminate pathogens such as fungal spores, bacteria, seeds of parasitic plants, nematodes, and perhaps some viruses and viroids. Most of these pathogens adhere to the feet or the body of the animals, but some may be carried in contaminated mouthparts.

Finally, some plant pathogens, for example, the zoospores of some fungi and certain parasitic plants, can transmit viruses as they move from one plant to another (zoospores), or as they grow and form a bridge between two plants (dodder).

Dissemination by Humans

Human beings disseminate all kinds of pathogens over short and long distances in a variety of ways. Within a field, humans disseminate some pathogens, such as tobacco mosaic virus, through successive handling of diseased and healthy plants. Other pathogens are disseminated through tools, such as pruning shears, contaminated when used on diseased plants (for example, pear infected with fire blight bacteria) and then carried to healthy plants. Humans also disseminate pathogens by transporting contaminated soil on their feet or equipment and on infected transplants, seed, nursery stock, and budwood and by using contaminated containers. Finally, humans disseminate pathogens by importing into an area new varieties that may carry pathogens that go undetected and by travels throughout the world and importing food or other items that may carry harmful plant pathogens. Examples of the human role as a vector of pathogens can be seen in the introduction into the United States of the fungi causing Dutch elm disease and white pine blister rust and of the citrus canker bacterium and in the introduction in Europe of the powdery and downy mildews of grape.

Overwintering and/or Oversummering of Pathogens

Pathogens that infect perennial plants can survive in them during the low winter temperatures or during the hot, dry weather of the summer, or both, regardless of whether the host plants are actively growing or are dormant at the time.

Annual plants, however, die at the end of the growing season, as do the leaves and fruits of deciduous perennial plants and even the stems of some perennial plants. In colder climates, annual plants and the tops of some perennial plants die with the advent of the low winter temperatures, and their pathogens are left without a host for the several months of cold weather. On the other hand, in hot, dry climates, annual plants die during the summer and their pathogens must be able to survive such periods in the absence of their hosts. Thus, pathogens that attack annual plants and renewable parts of perennial plants have evolved mechanisms by which they can survive the cold winters or dry summers that may intervene between crops or growing seasons (Figure 2-9).

Fungi have evolved a great variety of mechanisms for persisting between crops. On perennial plants, fungi overwinter as mycelium in infected tissues, for example cankers, and as spores at or near the infected surface of the plant or on the bud scales. Fungi affecting leaves or fruits of deciduous trees usually overwinter as mycelium or spores on fallen, infected leaves or fruits, or on the bud scales. Fungi affecting annual plants usually survive the winter or summer as mycelium in infected plant debris, as resting or other spores and as

FIGURE 2-9 Forms and locations of survival of fungi and bacteria between crops.

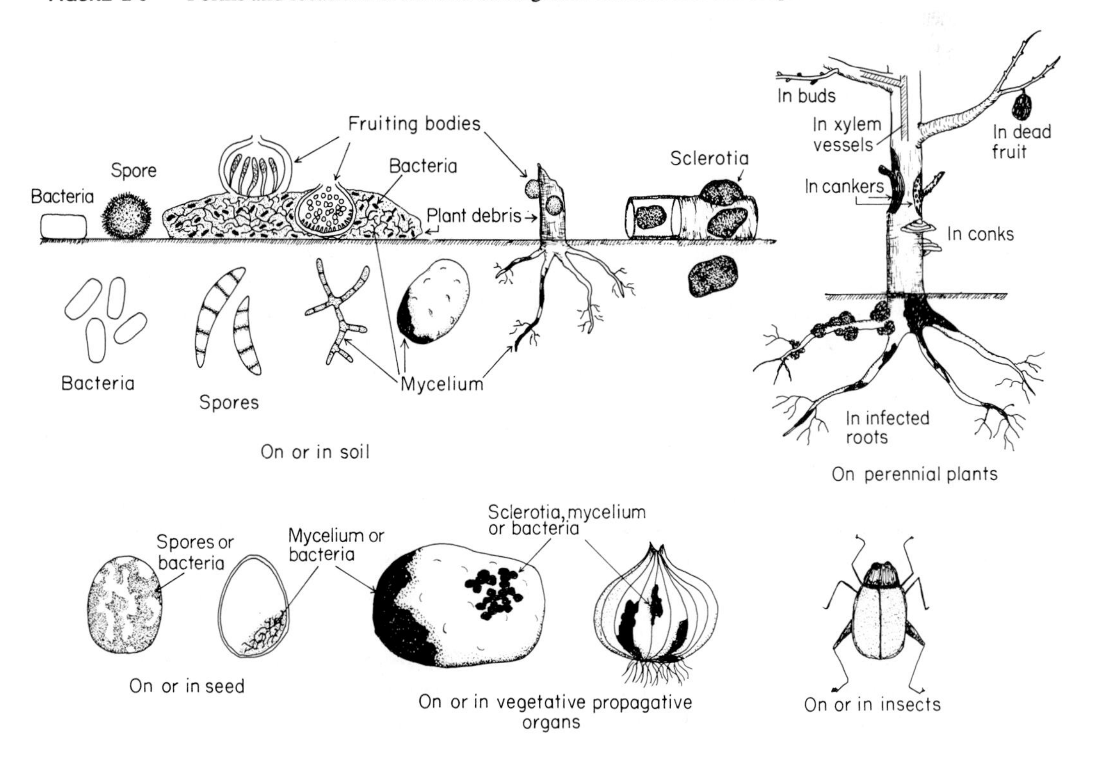

sclerotia in infected plant debris or in the soil, and as mycelium, spores, or sclerotia in or on seeds and other propagative organs, such as tubers. Some plant pathogenic fungi are **soil inhabitants**; that is, they are able to survive indefinitely as saprophytes (for example, *Pythium, Fusarium, Rhizoctonia*). Soil inhabitants are generally unspecialized parasites that have a wide host range. Others are **soil transients;** that is, they are rather specialized parasites that generally live in close association with their host but may survive in the soil for relatively short periods of time as hardy spores or as saprophytes. In some areas, fungi survive by continuous infection of host plants grown outdoors throughout the year, such as cabbage, or of plants in the greenhouse in the winter and outdoors in the summer. Similarly, some rust and other fungi overwinter on winter crops grown in warmer climates and move from them to the same hosts grown as spring crops in colder climates. Also, some fungi infect cultivated or wild perennial, as well as annual, plants and move from the perennial to the annual ones each growth season. Some rust fungi infect alternately an annual and a perennial host, and the fungus goes from the one to the other host and overwinters in the perennial host.

Bacteria overwinter and oversummer as bacteria in essentially the same ways as described for fungi, that is, in infected plants, seeds, and tubers, in infected plant debris and, some of them, in the soil. Bacteria survive poorly when present in small numbers and free in the soil but survive well when masses of them are embedded in the hardened, slimy polysaccharides that usually surround them. Some bacteria also overwinter within the bodies of their insect vectors.

Viruses, viroids, mycoplasmas, fastidious bacteria, and protozoa survive only in living plant tissues such as the tops and roots of perennial plants, the roots of perennial plants that die to the soil line in the winter or summer, in vegetative propagating organs, and in the seeds of some hosts. A few viruses survive within their insect vectors, and some viruses and viroids may survive on contaminated tools and in infected plant debris.

Nematodes usually overwinter or oversummer as eggs in the soil and as eggs or nematodes in plant roots or in plant debris. Some nematodes produce larval stages or adults that can remain dormant in seeds or on bulbs for many months or years.

Parasitic higher plants survive either as seeds, usually in the soil, or as their infective vegetative form on their host.

Relationships between Disease Cycles and Epidemics

Some pathogens complete only one, or even part of one, disease cycle in one year and are called **monocyclic,** or single-cycle, pathogens. Diseases with monocyclic pathogens include the smuts in which the fungus produces spores at the end of the season (these spores serve as primary—and the only—inoculum for the following year), many tree rusts, which require two alternate hosts and at least one year to complete one disease cycle, and many soil-borne diseases, for example, root rots and vascular wilts. In root rots and vascular wilts, the pathogens survive the winter or summer in decaying stems and roots or in the soil, infect plants during the growth season, and at the end of the growth season, produce new spores in the infected stems and roots. These spores remain in the soil and serve as primary inoculum the following growth season. In monocyclic pathogens the primary inoculum is the only inoculum available for the entire season, since there is no secondary inoculum and no secondary infection. The amount of inoculum produced at the end of the season, however, is greater than that present at the start of the season, and so in monocyclic diseases the amount of inoculum may increase steadily from year to year.

In most diseases, however, the pathogen goes through more than one generation per growth season, and such pathogens are called **polycyclic,** or multi-cycle, pathogens. Polycyclic pathogens can complete many (2–30) disease cycles per year, and with each cycle the amount of inoculum is multiplied manyfold. Polycyclic pathogens are disseminated primarily by air or air-borne vectors (insects) and are responsible for the kinds of diseases that cause most of the explosive epidemics on most crops, for example, the downy mildews, late blight of potato, powdery mildews, leaf spots and blights, grain rusts, and aphid-borne viruses. In polycyclic fungal pathogens, the primary inoculum generally consists of the sexual (perfect) spore or, in fungi that lack the sexual stage, some other hardy structure of the fungus such as sclerotia, pseudosclerotia, or mycelium in infected tissue. The number of sexual spores or other hardy structures that survive and cause infection is usually small, but once primary infection takes place, large numbers of asexual spores (secondary inoculum) are produced at each infectious site, and these spores can themselves cause new (secondary) infections that produce more asexual spores for more infections.

In some diseases of trees, for example, fungal vascular wilts, mycoplasmal yellows, and viral infections, the infecting pathogen may not complete a disease cycle within one year but instead may take several years before the inoculum it produces can be disseminated and initiate new infections. Such pathogens are called **polyetic** (multi-year). Although polyetic pathogens may not cause many new infections over a given area within a single year, and their amount of inoculum does not increase greatly within a year, because they survive in perennial hosts they have the advantage that, at the start of each year, they have almost as much inoculum as they had at the end of the previous year. Therefore, the inoculum may increase steadily (exponentially) from year to year and may cause severe epidemics when considered over several years. Examples of such diseases are Dutch elm disease, pear decline, and citrus tristeza.

Whether the pathogen involved in a particular disease is monocyclic, polycyclic, or polyetic has great epidemiological consequences since it affects the amount of disease caused by the specific pathogen within a given period of time. The rate of inoculum or disease increase *(r)* has been calculated for many diseases and varies from 0.1 to 0.5 *per day* for polycylcic foliar diseases such as Southern corn leaf blight, potato late blight, grain rusts, and tobacco mosaic to 0.02 to 2.3 *per year* for polyetic diseases of trees such as dwarf mistletoe of conifers, Dutch elm disease, chestnut blight, and peach mosaic. These values of r signify an increase in the amount of inoculum or disease (number of plants infected, amount of plant tissue infected, and so on) from 10 to 50 percent per day for foliar diseases and from 2 to 230 percent per year for diseases of trees like those listed above.

SELECTED REFERENCES

Andrews, J. H. (1984). Life History strategies of plant parasites. *Adv. Plant Pathol.* **2,** 105–130.

Asada, Y., Bushnell, W. R., Ouchi, S., and Vance, P., eds. (1982). "Plant Infection: The physiological and Biochemical Basis". Springer-Verlag, Berlin and New York.

Asahi, T., M., Kojima M., and Kosuge, T. (1979). The energetics of parasitism, pathogenesis and resistence in plant disease. *In* "Plant Disease" (J. G. Horsfall and E. B. Cowling, eds.), Vol. 4, pp. 47–74, Academic Press, New York.

Chong, J., Harder, D. E., Rohringer, R. (1981). Ontogeny of mono- and dikaryotic rust haustoria: Cytochemical and ultrastractural studies. *Phytopathology* **71,** 975–983.

Daly, J. M. (1984). The role of recognition in plant disease. Annu. *Rev. Phytopathol.* **22,** 273–307.

Dodman, R. L. (1979). How the defenses are breached. *In* "Plant Disease" (J. G. Horsfall and E. B. Cowling, eds.), Vol. 4, pp. 137–153. Academic Press, New York.

Dunkle, L. D. (1984). Factors in pathogenesis. *In* "Plant-Microbe Interactions: Molecular and Genetic Perspectives" (T. Kosuge and E. W. Nester, eds.), Vol. 1, pp. 19–41. Macmillan, New York.

Ellingboe, A. H. (1968). Inoculum production and infection by foliage pathogens. *Annu. Rev. Phytopathol.* **6,** 317–330.

Emmett, R. W., and Parbery, D. G. (1975). Appressoria. *Annu. Rev. Phytopathol.* **13,** 147–167.

Hancock, J. G., and Huisman, O. C. (1981). Nutrient movement in host-pathogen systems. *Annu. Rev. Phytopathol.* **19,** 309–331.

Hornby, D. (1983). Suppressive soils. *Annu. Rev. Phytopathol.* **21,** 65–85.

Horsfall, J. G., and Cowling E. B., eds. (1978). "Plant Disease," Vol. 2. Academic Press, New York.

Horsfall, J. G., and Cowling, E. B., eds. (1979). "Plant Disease." Vol. 4, Academic Press, New York.

Lippincott, J. A., and Lippincott, B. B. (1984). Concepts and experimental approaches in host-microbe recognition. *In* "Plant-Microbe Interactions: Molecular and Genetic Perspectives" (T. Kosuge and E. W. Nester, eds.), Vol. 1, pp. 195–214. Macmillan, New York.

Littlefield, L. J., and Heath, M. C. (1979). "Ultrastructure of Rust Fungi." Academic Press, New York.

McKeen, W. E., and Svircev, A. M. (1981). Early development of *Peronospora tabacina* in the *Nicotiana tabacum* leaf. *Can. J. Plant Pathol.* **3,** 145–158.

Meredith, D. S. (1973). Significance of spore release and dispersal mechanisms in plant disease epidemiology. *Annu. Rev. Phytopathol.* **11,** 313–342.

Mount, M. S., and Lacey, G. H., eds. (1982). "Phytopathogenic Prokaryotes," Vols, 1 and 2. Academic Press, New York.

Nelson, R. R. (1979). The evolution of parasitic fitness. *In* "Plant Disease" (J. G. Horfall and E. B. Cowling, eds.), Vol. 4, pp. 23–46. Academic Press, New York.

Rotem, J., and Palti, J. (1969). Irrigation and plant disease. *Annu. Rev. Phytopathol.* **7,** 267–288.

Royle, D. J., and Thomas, G. G. (1973). Factors affecting zoospore responses towards stomata in hop downy mildew *(Pseudoperonospora humuli)* including some comparisons with grapevine downy mildew *(Plasmopara viticola). Physiol. Plant Pathol.* **3,** 405–417.

Schneider, R. W., ed. (1982). "Suppressive Soils and Plant Disease." Phytopathol. Soc., St. Paul, Minnesota.

Schuster, M. L., and Coyne, D. P. (1974). Survival mechanisms of phytopathogenic bacteria. *Annu. Rev. Phytopathol.* **12,** 199–221.

Tarr, S. A. J. (1972). "The Principles of Plant Pathology." Winchester Press, New York.

Vanderplank, J. E. (1975). "Principles of Plant Infections." Academic Press, New York.

Wood, R. K. S., and Graniti, A., eds. (1976). "Specificity in Plant Diseases." Plenum, New York.

Wynn, W. K. (1981). Tropic and taxic responses of pathogens to plants. *Annu. Phytopathol.* **19,** 237–255.

3 HOW PATHOGENS ATTACK PLANTS

The intact, healthy plant is a community of cells built in a fortresslike fashion. The plant surfaces that come in contact with the environment either consist of cellulose as in the epidermal cells of roots and in the intercellular spaces of leaf parenchyma cells, or they consist of a cuticle which covers the epidermal cell walls, as is the case in the aerial parts of plants. Often an additional layer, consisting of waxes, is deposited outside the cuticle, especially on younger parts of plants (Figure 3-1).

Pathogens attack plants because during their evolutionary development they have acquired the ability to live off the substances manufactured by the host plants, and some of the pathogens depend on these substances for survival. Many substances, however, are contained in the protoplast of the plant cells, and, if pathogens are to gain access to them, they must first penetrate the outer barriers formed by the cuticle and/or cell walls. Even after the outer cell walls have been penetrated, further invasion of the plant by the pathogen necessitates penetration of more cell walls. Furthermore, the plant cell contents are not always found in forms immediately utilizable by the pathogen and must be transformed to units which the pathogen can absorb and assimilate. Moreover, the plant, reacting to the presence and activities of the pathogen, produces structures and chemical substances that interfere with the advance or the existence of the pathogen; if the pathogen is to survive and to continue living off the plant, it must be able to overcome such obstacles.

Therefore, for a pathogen to infect a plant it must be able to make its way into and through the plant, obtain nutrients from the plant, and neutralize the defense reactions of the plant. Pathogens accomplish these activities mostly through secretions of chemical substances that affect certain components or metabolic mechanisms of their hosts. Penetration and invasion, however, seem to be aided by, or in some cases be entirely the result of, mechanical force exerted by certain pathogens on the cell walls of the plant.

Mechanical Forces Exerted by Pathogens on Host Tissues

Plant pathogens are, generally, tiny microorganisms that cannot apply a "voluntary" force to a plant surface. Only some fungi, parasitic higher plants, and the nematodes appear to apply mechanical pressure to the plant surface they are about to penetrate. The amount of pressure, however, may vary greatly with the degree of "presoftening" of a plant surface by enzymatic secretions of the pathogen.

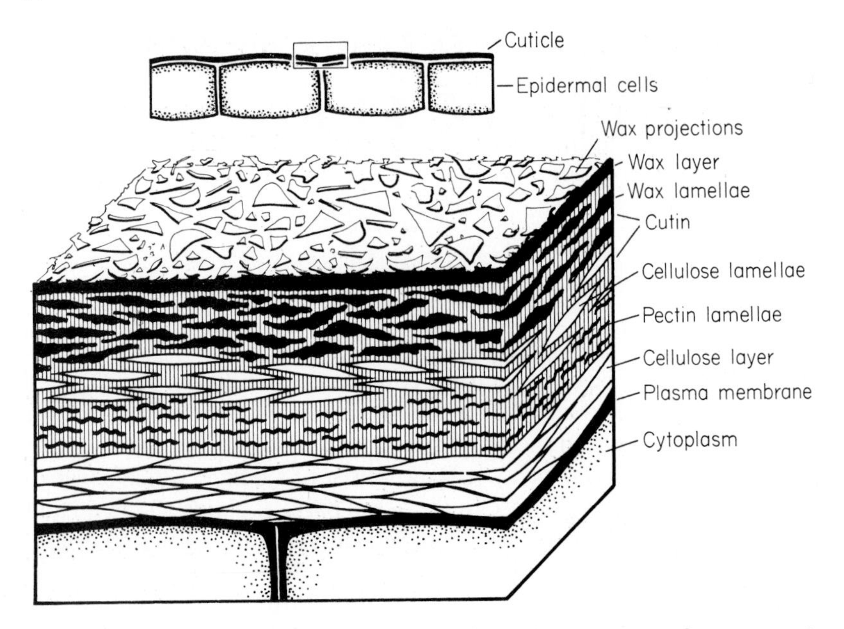

FIGURE 3-1 Schematic representation of the structure and composition of the cuticle and cell wall of foliar epidermal cells. Adapted from Goodman, R.N., Kiraly, Z., and Zaitlin, M. (1967). "The Biochemistry and Physiology of Infectious Plant Disease." Van Nostrand, Princeton, New Jersey.

For fungi and parasitic higher plants to penetrate a plant surface, they must, generally, first adhere to it. Although hyphae and radicles are usually surrounded by mucilaginous substances, their adhesion to the plant seems to be brought about primarily by the intermoleculer forces developing between the surfaces of plant and pathogen upon close contact with each other. After contact is established, the diameter of the part of hypha or radicle in contact with the host increases and forms a flattened, bulblike structure called the **appressorium.** This increases the area of adherence between the two organisms and securely fastens the pathogen to the plant. From the appressorium, a fine growing point, called the **"penetration peg,"** arises and advances into and through the cuticle and cell wall. If the underlying host wall is soft, penetration occurs easily. When the underlying wall is hard, however, the force of the growing point may be greater than the adhesion force of the two surfaces and may cause the separation of the appressorial and host walls, thus averting infection. Penetration of plant barriers by fungi and parasitic higher plants is almost always assisted by the presence of enzymes secreted by the pathogen at the penetration site, resulting in the softening or dissolution of the barrier.

While the penetration tube is passing through the cuticle, it usually attains its smallest diameter and appears threadlike. After penetration of the cuticle, the hyphal tube diameter often increases considerably. The penetration tube attains the diameter normal for the hyphae of the particular fungus only after it has passed through the cell wall (see Figures 2-2 and 2-5 in Chapter 2).

Nematodes penetrate plant surfaces by means of the stylet, which is thrust back and forth and exerts mechanical pressure on the cell wall. The nematode first adheres to the plant surface by suction, which it develops by bringing its fused lips in contact with the plant. After adhesion is accomplished, the nematode brings its body, or at least the forward portion of its body, to a position vertical to the cell wall. With its head stationary and fixed to the cell wall, the nematode then thrusts its stylet forward while the rear part of its body sways or rotates slowly round and round. After several consecutive thrusts of the stylet, the cell wall is pierced and the stylet or the entire nematode enters the cell.

Once a fungus or nematode has entered a cell it generally secretes increased amounts of enzymes, which, presumably, soften or dissolve the opposite cell wall and make its penetration easier. Mechanical force, however, probably is brought to bear in most such penetrations, although to a lesser extent.

Considerable mechanical force is also exerted on host tissues by some pathogenic fungi upon formation of their fructfications in the tissues beneath the plant surface. Through increased pressure, the sporophore hyphae as well as fruiting bodies, such as pycnidia and perithecia, push outward and cause the cell walls and the cuticle to expand, become raised in the form of blister-like protuberances, and finally break.

Chemical Weapons of Pathogens

Although some pathogens may use mechanical force to penetrate plant tissues, the activities of pathogens in plants are largely chemical in nature. Therefore, the effects caused by pathogens on plants are almost entirely the result of biochemical reactions taking place between substances secreted by the pathogen and those present in, or produced by, the plant.

The main groups of substances secreted by pathogens in plants that seem to be involved in production of disease, either directly or indirectly, are enzymes, toxins, growth regulators, and polysaccharides. These groups vary greatly as to their importance in pathogenicity, and their relative importance may be different from one disease to another. Thus in some diseases, such as soft rots, enzymes seem to be by far the most important, whereas in diseases like crown gall, growth regulators are apparently the main substances involved, and in the *Helminthosporium* blight of Victoria oats, the disease is primarily the result of a toxin secreted in the plant by the pathogen. Enzymes, toxins, and growth regulators, probably in that order, are considerably more common and more important in plant disease development than are polysaccharides.

Among the plant pathogens, all except viruses and viroids can probably produce enzymes, growth regulators, polysaccharides, and possibly toxins. Plant viruses and viroids are not known to produce any substances themselves, but they induce the host cell to produce either excessive amounts of certain substances already found in healthy host cells or substances completely new to the host. Some of these substances are enzymes, others may belong to one of the other groups mentioned above.

Pathogens produce these substances either in the normal course of their activities or upon growth on certain substrates. Undoubtedly, natural

selection has favored the survival of pathogens that are assisted in their parasitism through the production of such substances. The presence or the amount of any such substance produced, however, is not always a measure of the ability of the pathogen to cause disease. As a matter of fact, many substances, identical to those produced by pathogens, are also produced by the healthy host plant.

In general, plant pathogenic enzymes disintegrate the structural components of host cells, break down inert food substances in the cell, or affect the protoplast directly and interfere with its functioning systems. Toxins seem to act directly on the protoplast and interfere with the permeability of its membranes and with its function. Growth regulators exert a hormonal effect on the cells and either increase or decrease their ability to divide and enlarge. Polysaccharides seem to play a role only in the vascular diseases in which they passively interfere with the translocation of water in the plants, or they may also be toxic.

Enzymes

Enzymes are large protein molecules that catalyze all the interrelated reactions in a living cell. For each kind of chemical reaction that occurs in a cell, there is a different enzyme that catalyzes that reaction.

Enzymatic Degradation of Cell Wall Substances

Most plant pathogens secrete enzymes throughout their existence or upon contact with a substrate. Usually, the first contact of pathogens with their host plants occurs at a plant surface. Such a surface may consist primarily of cellulose, which makes up the epidermal cell walls, or (on the aerial plant parts) of cellulose plus cuticle. Cuticle consists primarily of cutin, more or less impregnated with wax and frequently covered with a layer of wax. Protein and lignin may also be found in epidermal cell walls. Penetration of pathogens into parenchymatous tissues is brought about by the breakdown of the cell walls, which consist of cellulose, pectins, and hemicelluloses, and of the middle lamella, which consists primarily of pectins. In addition, complete plant tissue disintegration involves breakdown of lignin. The degradation of each of these substances is brought about by the action of one or more sets of enzymes secreted by the pathogen.

Cuticular Wax Plant waxes are found as granular or rodlike projections or as continuous layers outside or within the cuticle of many aerial plant parts. To date, no pathogens are known that produce enzymes that can degrade waxes. Fungi and parasitic higher plants apparently penetrate wax layers by means of mechanical force alone.

Cutin Cutin is the main component of the cuticular layer. The upper part of the layer is admixed with waxes, while its lower part, in the region where it merges into the outer walls of epidermal cells, is admixed with pectin and cellulose (see Figure 3-1). Cutin is an insoluble polyester of mostly unbranched derivatives of c_{16} and C_{18} hydroxy fatty acids, which may also have one or more internal reactive groups that may result in additional ester crosslinking.

Many fungi and at least one bacterium *(Streptomyces scabies)* have been shown to produce **cutinases,** that is, enzymes that can degrade cutin. Cutinases are esterases—they break the ester linkages between cutin molecules and release monomers (single molecules) as well as oligomers (small groups of molecules) of the component fatty acid derivatives from the insoluble cutin polymer.

$$-R-\overset{O}{\overset{\|}{C}}-OCH_2-(\overset{R_1}{\overset{|}{C}}H_2)n-\overset{O}{\overset{\|}{C}}-OR \quad \left\{\begin{array}{l} R = \text{cutin polymer} \\ R_1 = OH, -COOH, \text{ or epoxide} \\ n = 14 \text{ or } 16 \end{array}\right.$$

$$\downarrow \text{Cutinase}$$

$$-R-\overset{O}{\overset{\|}{C}}-OH + HOCH_2(\overset{R_1}{\overset{|}{C}}H_2)n-\overset{O}{\overset{\|}{C}}-OH + HOR$$

Fungi seem to constantly produce low levels of cutinase, which upon contact with cutin, releases small amounts of monomers that subsequently enter the pathogen cell and stimulate it to produce almost a thousand times more cutinase than before. Cutinase production by the pathogen, however, may also be initiated by some of the fatty acids present in the wax normally associated with cutin in plant cuticle. It is not certain yet whether pathogens that produce higher levels of cutinase are more pathogenic than others. At least one study has shown, however, that the germinating spores of a fungus *(Fusarium)* produced much more cutinase than those of an avirulent isolate of the same fungus, and that the avirulent isolate could be turned into a virulent one if purified cutinase was added to its spores. This observation suggests that production of different levels of cutinase may affect the ability of a pathogen to produce disease.

Pectic Substances Pectic substances constitute the main components of the middle lamella, that is, the intercellular cement that holds in place the cells of plant tissues (Figure 3-2). Pectic substances also make up a large portion of the primary cell wall, in which they form an amorphous gel filling the spaces between the cellulose microfibrils (Figure 3-3).

Pectic substances are polysaccharides consisting mostly of chains of galacturonan molecules interspersed with a much smaller number of rhamnose molecules and small side chains of galacturonan and some other sugars. Several enzymes known as **pectinases** or **pectolytic enzymes** degrade pectic substances (Figure 3-4). Some (**pectin methyl esterases**) remove small branches off the pectin chains and have no effect on the length of the pectin chains, but they alter their solubility and affect the rate at which they can be attacked by the chain-splitting pectinases. The latter cleave the pectic chain and release shorter-chain portions containing one or a few molecules of galacturonan. Some chain-splitting pectinases (**polygalacturonases**) split the pectic chain by adding a molecule of water and breaking (hydrolyzing) the linkage between two galalcturonan molecules; others (**pectin lyases or trans-eliminases**) split the chain by removing a molecule of water from the linkage and thereby breaking it and releasing products with an unsaturated double bond (Figure 3-4). Each of these enzymes occurs in types that can either break the pectin chain at random sites (endo-pectinases) and release shorter chains or can break only the terminal linkage (exo-pectinases) of the chain and

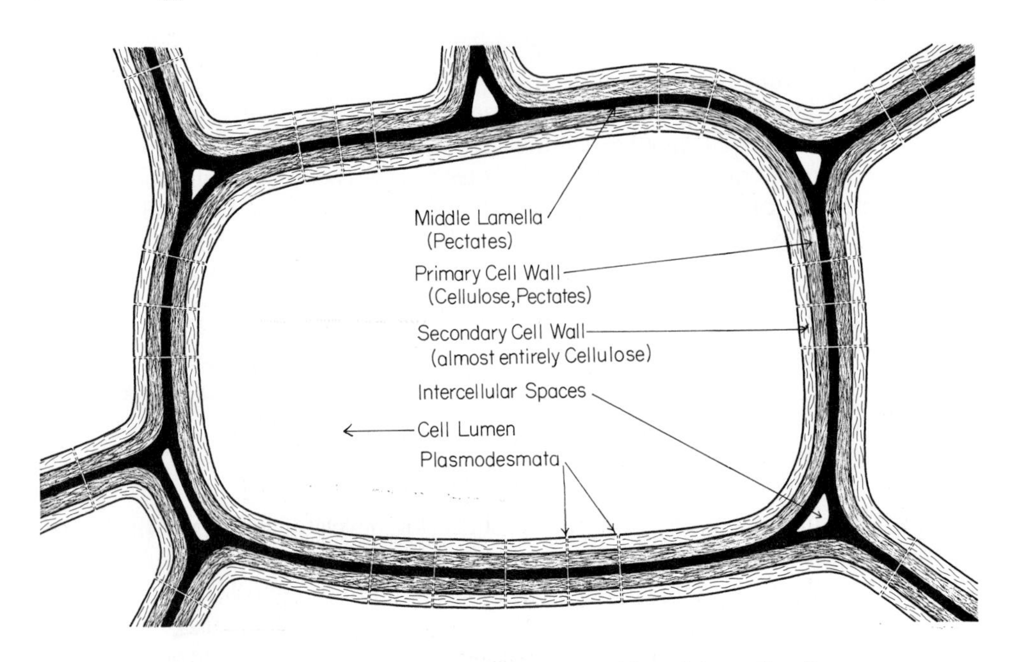

FIGURE 3-2 Schematic representation of structure and composition of plant cell walls.

release single units of galacturonan. The rhamnose and other sugars that may be forming part or branches of the pectin chain are hydrolyzed by other enzymes that recognize these molecules.

As with cutinases, and with other enzymes involved in degradation of cell wall substances, production of extracellular pectolytic enzymes by pathogens is regulated by the availability of the pectin polymer and the released galacturonan units. The pathogen seems to produce at all times small, base-level amounts of pectolytic enzymes which, in the presence of pectin, release from it a small number of galacturonan monomers or oligomers. These molecules, when absorbed by the pathogen, serve as inducers for enhanced synthesis and release of pectolytic enzymes **(autocatalytic induction),** which further increase the amount of galacturonan monomers. The latter are readily assimilated by the pathogen, but at higher concentrations they act to repress the synthesis of the same enzymes **(catabolite repression),** and so production of the enzymes and subsequent release of galacturonan monomers are reduced. The production of pectolytic enzymes is also repressed when the pathogen is grown in the presence of glucose and some other sugars, but repression can be abolished by addition of cyclic AMP (cAMP), that is, adenosine-3′, 5′-cyclic monophosphate.

The pectin-degrading enzymes have been shown to be involved in the production of many diseases, particularly those characterized by softrotting of tissues. Pectic enzymes are produced by germinating spores and apparently, acting together with other pathogen metabolites (cutinases and cellulases), assist in the penetration of the host. Pectin degradation results in liquifaction of the pectic substances that hold plant cells together and in weakening of cell walls, leading to tissue **maceration** (softening and loss of coherence of plant tissues and separation of individual cells, which eventually die). The weakening of cell walls and tissue maceration undoubtedly facilitate the inter- or

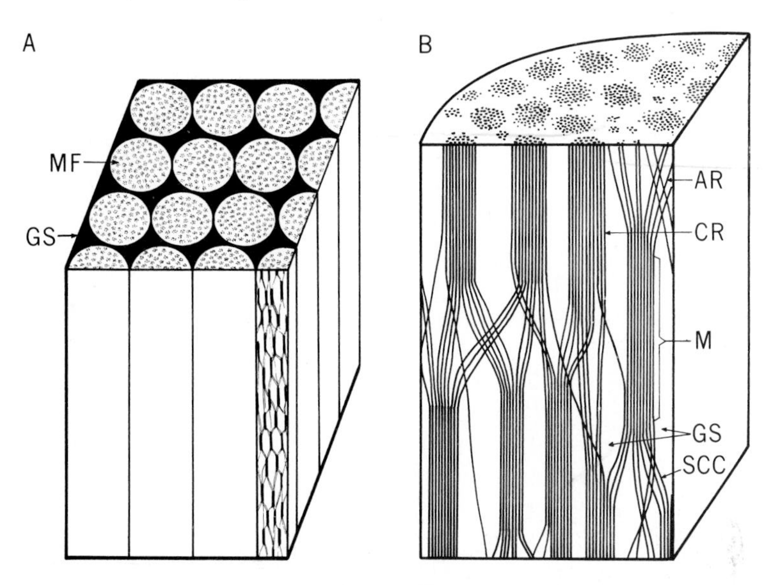

FIGURE 3-3 Schematic diagram of the gross structure of cellulose and microfibrils (A), and of the arrangement of cellulose molecules within a microfibril (B). MF = microfibril; GS = ground substance (pectin, hemicelluloses, or lignin); AR = amorphous region of cellulose; CR = crystalline region; M = micelle; SCC = single cellulose chain (molecule). Adapted from H. P. Brown, A. J. Panshing, and C. C. Forsaith (1949). "Textbook of Wood Technology," Vol. 1. McGraw-Hill, New York.

intracellular invasion of the tissues by the pathogen. Pectic enzymes also provide nutrients for the pathogen in infected tissues. Pectic enzymes, by the debris they create, seem to be involved in the induction of vascular plugs and occlusions in the vascular wilt diseases. Although cells are usually quickly killed in tissues macerated by pectic enzymes, how these enzymes kill cells is not yet clear. It is thought that cell death results from the weakening by the pectolytic enzymes of the primary cell wall, which then cannot support the osmotically fragile protoplast, and the protoplast bursts.

Cellulose Cellulose is also a polysaccharide, but it consists of chains of glucose molecules. The glucose chains are held to each other by a large number of hydrogen bonds. Cellulose occurs in all higher plants as the skeletal substance of cell walls in the form of microfibrils (see Figures 3-2 and 3-3). Microfibrils, which can be perceived as bundles of iron bars in a reinforced concrete building, are the basic structural units (matrix) of the wall, even though they account for less than 20 percent of the wall volume in most meristematic cells. The cellulose content of tissues varies from about 12 percent in the nonwoody tissues of grasses to about 50 percent in mature wood tissues to more than 90 percent in the cotton fibers. The spaces between microfibrils and between micelles or cellulose chains within the microfibrils may be filled with pectins and hemicellulose and probably some lignin at maturation.

The enzymatic breakdown of cellulose results in the final production of glucose molecules. The glucose is produced by a series of enzymatic reactions carried out by several **cellulases** and other enzymes. One cellulase (C_1) attacks native cellulose by cleaving cross-linkages between chains. A second cellulase

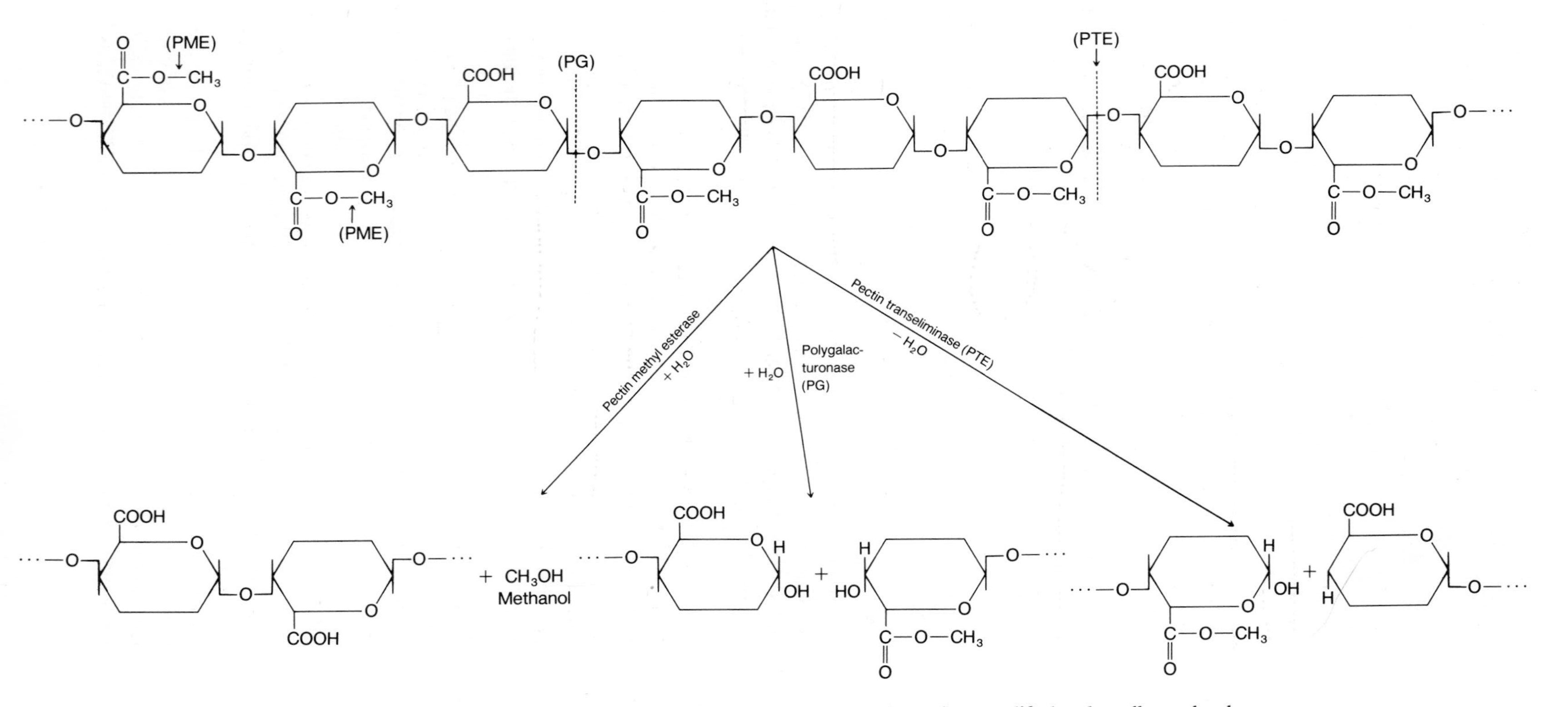

FIGURE 3-4 Degradation of a pectin chain by the three types of pectinases into modified and smaller molecules.

(C_2) also attacks native cellulose and breaks it into shorter chains, which are then attacked by a third group of cellulases (Cx), which degrade them to the disaccharide cellobiose. Finally, cellobiose is degraded by the enzyme β-glucosidase into glucose.

Cellulose-degrading enzymes (cellulases) have been shown to be produced by several phytopathogenic fungi, bacteria, and nematodes, and they are undoubtedly produced by parasitic higher plants. Saprophytic fungi, mainly certain groups of basidiomycetes, and to a smaller degree saprophytic bacteria cause the breakdown of most of the cellulose decomposed in nature. In living plant tissues, however, cellulolytic enzymes secreted by pathogens play a role in the softening and disintegration of cell-wall material. They facilitate the penetration and spread of the pathogen in the host and cause the collapse and disintegration of the cellular structure, thereby aiding the pathogen in the production of disease. Cellulolytic enzymes may further participate indirectly in disease development by releasing, from cellulose chains, soluble sugars, which serve as food for the pathogen and, in the vascular diseases, by liberating into the transpiration stream large molecules from cellulose, which interfere with the normal movement of water.

Hemicelluloses Hemicelluloses are complex mixtures of polysaccharide polymers, the composition and frequency of which seem to vary among plant tissues, plant species, and with the developmental stage of the plant. Hemicelluloses are a major constituent of the primary cell wall and may also make up a varying proportion of middle lamella and the secondary cell wall. The hemicellulosic polymers include primarily xyloglucan, but also glucomannans, galactomannans, arabinogalactans, and others. Xyloglucan, for example, is made of glucose chains with terminal branches of smaller xylose chains and lesser amounts of galactose, arabinose, and fucose. Hemicelluloses link the ends of pectic polysaccharides and various points of the cellulose fibrils.

Enzymatic breakdown of hemicelluloses appears to require the activity of many enzymes. Several **hemicellulases** seem to be produced by many plant pathogenic fungi. Depending on the monomer released from the polymer on which they act, the particular enzymes are called xylanase, galactanase, glucanase, arabinase, mannase, and so on. In spite of the fact that fungal pathogens produce these enzymes, it is still not clear how much they contribute to cell wall breakdown or to the ability of the pathogen to cause disease.

Lignin Lignin is found in the middle lamella, in the cell wall of xylem vessels, and in the fibers that strengthen plants. It is also found in epidermal and occasionally hypodermal cell walls of some plants. The lignin content of mature woody plants varies from 15 to 38 percent and is second only to cellulose.

Lignin is an amorphous, three-dimentional polymer that is different from both carbohydrates and proteins in composition and properties. The most common basic structural unit of lignin is a phenylpropanoid:

—C—C—C

with one or more of its carbons having a —OH, $—OCH_3$, or =O group.

Lignin forms by oxidative condensation (C—C and C—O bond formation) between such substituted phenylpropanoid units. The lignin polymer is perhaps more resistant to enzymatic degradation than any other plant substance.

It is obvious that enormous amounts of lignin are degraded by microorganisms in nature, as is evidenced by the decomposition of all annual plants and a large portion of perennial plants that disintegrate annually. It is generally accepted, however, that only a small group of microorganisms is capable of degrading lignin. Actually, only about 500 species of fungi, almost all of them basidiomycetes, have been reported so far as being capable of decomposing wood. About one-fourth of these fungi (the brown-rot fungi) seem to cause some degradation of lignin but cannot utilize it. Most of the lignin in the world is degraded and utilized by a group of fungi called white-rot fungi. It appears that the white-rot fungi secrete one or more enzymes (**ligninases**), which enable them to utilize lignin.

In addition to the wood-rotting fungi, several other pathogens, primarily several Ascomycetes and Imperfect fungi and even some bacteria, apparently produce small amounts of lignin-degrading enzymes. However, it is not known to what extent, if any, the diseases they cause are dependent on the presence of such enzymes.

Enzymatic Degradation of Substances Contained in Plant Cells

Most kinds of pathogens live all or part of their lives in association with or inside the living protoplast. These pathogens obviously derive nutrients from the protoplast. All the other pathogens—the great majority of fungi and bacteria—obtain nutrients from protoplasts after the latter have been killed. Some of the nutrients, for example, sugars and amino acids, are probably sufficiently small molecules to be absorbed by the pathogen directly. Some of the other plant cell constituents, however, such as starch, proteins, and fats, can be utilized only after degradation by the enzymes secreted by the pathogen.

Proteins Plant cells contain innumerable different proteins, which play diverse roles as catalysts of cellular reactions (enzymes) or as structural material (membranes). Proteins are formed by the joining together of numerous molecules of about twenty different kinds of amino acids.

$$\underset{\displaystyle NH_2}{R-\underset{|}{CH}}-CO\boxed{OH+H}-\overset{\displaystyle R-CH-COOH}{\overset{|}{NH}} \longrightarrow \underset{\displaystyle NH_2}{R-\underset{|}{CH}}-CO-\overset{\displaystyle R-CH-COOH}{\overset{|}{NH}} + H_2O$$

All pathogens seem to be capable of degrading many kinds of protein molecules. The plant pathogenic enzymes involved in protein degradation are similar to those present in higher plants and animals and are called **proteinases.**

Considering the paramount importance of proteins as enzymes, constituents of cell membranes, and structural components of plant cells, degradation of host proteins by proteolytic enzymes secreted by pathogens can profoundly affect the organization and function of the host cells. The nature and extent of such effects, however, has been investigated little so far, and the significance in disease development is not known.

Starch Starch is the main reserve polysaccharide found in plant cells. Starch is synthesized in the chloroplasts and, in nonphotosynthetic organs, in the amyloplasts. Starch is a glucose polymer and exists in two forms: amylose, an essentially linear molecule, and amylopectin, a highly branched molecule of various chain lengths.

Most pathogens utilize starch, and other reserve polysaccharides, in their metabolic activities. The degradation of starch is brought about by the action of enzymes called **amylases.** The end product of starch breakdown is glucose, and it is used by the pathogens directly.

Lipids Various types of lipids occur in all plant cells, the most important being: the oils and fats found in many cells, especially in seeds where they function as energy storage compounds; the wax lipids, found on most aerial epidermal cells; and the phospholipids and the glycolipids, both of which, along with protein, are the main constituents of all plant cell membranes. The common characteristic of all lipids is that they contain fatty acids that may be saturated or unsaturated.

Several fungi, bacteria, and nematodes are known to be capable of degrading lipids. Lipolytic enzymes, called **lipases, phospholipidases,** and so on, hydrolyze the liberation of the fatty acids from the lipid molecule. The fatty acids are presumably utilized by the pathogen directly.

Microbial Toxins in Plant Disease

Living plant cells are complex systems in which many interdependent biochemical reactions are taking place concurrently or in a well-defined succession. These reactions result in the intricate and well-organized processes essential for life. Disturbance of any of these metabolic reactions causes disruption of the physiological processes that sustain the plant and leads to development of disease. Among the factors inducing such disturbances are substances that are produced by plant pathogenic microorganisms and are called toxins. Toxins act directly on living host protoplasts, seriously damaging or killing the cells of the plant. Some toxins act as general protoplasmic poisons and affect many species of plants representing different families. Others are toxic to only a few plant species or varieties and completely harmless to others.

Fungi and bacteria may produce toxins in infected plants as well as in culture medium. Toxins, however, are extremely poisonous substances and are effective in very low concentrations. Some are unstable or react quickly and are tightly bound to specific sites within the plant cell.

Toxins injure host cells either by affecting the permeability of the cell membrane or by inactivating or inhibiting enzymes and subsequently interrupting the corresponding enzymatic reactions. Certain toxins act as antimetabolities inducing a deficiency for an essential growth factor.

Toxins That Affect a Wide Range of Host Plants

Several toxic substances produced by phytopathogenic microorganisms have been shown to produce all or part of the disease syndrome not only on the host plant but also on other species of plants that are not normally attacked by the pathogen in nature. Such toxins are called **non-host-specific toxins** and increase the extent of disease caused by a pathogen but are not essential for the pathogen to cause disease. Several of these toxins, for example, tabtoxin and phaseolotoxin, affect the cellular transport system, especially the H^+/K^+ exchange at the cell membrane. Others, for example, cercosporin, act as photosensitizing agents causing peroxidation of membrane lipids.

Tabtoxin Tabtoxin was first reported to be produced by the bacterium *Pseudomonas syringae* pv. *tabaci,* which causes the wildfire disease of tobacco, but it is now known to be produced by other strains of pv. *tabaci* occurring on other hosts such as bean and soybean, and by other pathovars (sub-species) of *Ps. syringae,* such as those occurring on oats, maize, and coffee. Toxin-producing strains cause necrotic spots on leaves, each spot surrounded by a yellow halo. Sterile culture filtrates of the organism, as well as purified toxin, produce symptoms identical to those characteristic of wildfire of tobacco not only on tobacco but in a large number of plant species belonging to many different families (non-host-specific toxin!). Strains of *P. syringae* pv. *tabaci* sometimes produce mutants that have lost the ability to produce the toxin (Tox^-). Tox^- strains show reduced virulence and cause necrotic leaf spots without the yellow halo. Tox^- strains are indistinguishable from *Ps. angulata,* the cause of angular leaf spot of tobacco, which is now thought to be a nontoxicogenic form of pv. *tabaci.*

HN—┐ (β-lactam ring with =O and OH)

$$\begin{array}{ccc} CH_2 & & CH_3 \\ | & & | \\ CH_2 & & CHOH \\ | & & | \\ H_2N—CH—CO & \vdots & NH—CH—CO_2H \end{array}$$

(Tabtoxinine) *(Threonine)*

Tabtoxin

Tabtoxin is a dipeptide composed of the common amino acid threonine and the previously unknown amino acid tabtoxinin. Tabtoxin as such is not toxic, but in the cell it becomes hydrolyzed and releases tabtoxinine, which is the active toxin. Tabtoxin, through tabtoxinine, is toxic to cells because it inactivates the enzyme glutamine synthetase, which leads to depleted glutamine levels and a reduced ability of the plant to respond actively to the bacterium. As a consequence of inactivation of glutamine synthetase, ammonia accumulates during photorespiration, thus inhibiting photosynthesis and photorespiration and causing chlorosis and eventually necrosis.

Phaseolotoxin Phaseolotoxin is produced by the bacterium *Pseudomonas syringae* pv. *phaseolicola,* the cause of halo blight of bean and some other legumes. The localized and systemic chlorotic symptoms produced in infected plants are identical to those produced on plants treated with the toxin alone, so they are apparently the results of the toxin produced by the bacteria. Infected plants and plants treated with purified toxin also show reduced growth of newly expanding leaves, disruption of apical dominance, and accumulation of the amino acid ornithine.

```
                O                                       NH
                ||                                      ||
           NH—P(OH)—O—SO2NH2                      NH—C—NH2
           |                                       |
          (CH2)3               CH3                (CH2)4
           |                   |                   |
    H2N—CH—CO ┼ NH—CH—CO ┼ NH—CH—COOH
```

Phaseolotoxin

Phaseolotoxin is an ornithine-alanine-arginine tripeptide carrying a phosphosulfamyl group. Soon after the tripeptide is excreted by the bacterium into the plant, plant enzymes cleave the peptide bonds and release alanine, arginine, and phosphosulfamylornithine. The latter is the biologically functional moiety of phaseolotoxin. The toxin affects cells by binding to the active site of and inactivating the enzyme ornithine carbamoyltransferase, which normally converts ornithine to citrulline, a precursor of arginine. By its action on the enzyme, the toxin thus causes accumulation of ornithine and depleted levels of arginine. The tripeptide form of the toxin, while still in the bacterial cell, seems to both protect the bacterium's own enzyme from the more toxic phosphosulfamylornithine moiety and to facilitate export of the toxin from the bacterial cell to its surroundings. The mechanism by which the toxin induces chlorosis in plants has not yet been elucidated, and it is not known whether the inactivated enzyme is the primary and only site of toxin action or whether the toxin also affects other sites leading to more irregular biochemical reactions and chlorosis.

Tentoxin Tentoxin is produced by the fungus *Alternaria tenuis,* which causes chlorosis in seedling plants of many species. Seedlings with more than one-third of their leaf area chlorotic die, and those with less chlorosis are much less vigorous than healthy plants.

```
                                  H3C   CH3
                                     \ /
                                      |
    H3C—N—CH—CO ┼ NH—CH—C=O
        --┼--                 --┼--
      O=C—CH2—NH ┼ CO—C—N—CH3
                          ||
                          CH
                          |
                          C
                        //  \
                      HC     CH
                      ||     |
                      HC     CH
                        \  //
                          C
                          H
```

Tentoxin is a cyclic tetrapeptide that binds to and inactivates a protein (chloroplast-coupling factor) involved in energy transfer into chloroplasts and also inhibits the light-dependent phosphorylation of ADP to ATP. Both the inactivation of the protein and the inhibition of photophosphorylation are much greater in plant species susceptible to chlorosis after tentoxin treatment than in species not sensitive to the toxin. In sensitive species, tentoxin interferes with normal chloroplast development and results in chlorosis by interfering with chlorophyll synthesis, but it is not certain that these effects are solely related to tentoxin binding to the chloroplast-coupling factor protein. An additional but apparently unrelated effect of tentoxin on sensitive plants is that it inhibits the activity of polyphenyloxidases, enzymes involved in several resistance mechanisms of plants. Both effects of the toxin—stressing the host plant with events that lead to chlorosis and suppressing the host's resistance mechanisms—tend to enhance the virulence of the pathogen. The molecular site of action of tentoxin, however, and the exact mechanism by which it brings about these effects are still unknown.

Fusicoccin Fusicoccin is produced by fungus *Fusicoccum amygdali,* which causes twig blight on almond and peach trees. The toxin, when introduced into the xylem of these and many other plants, also induces similar symptoms.

Fusicoccin is a complex molecule, a carbotricyclic terpene glycoside, and is one of many similar derivatives produced by *Fusicoccum.* Fusicoccin appears to affect the cellular transport systems, particularly the H^+/K^+ pumps, through which it increases the uptake by affected cells of several anions, sugars, and amino acids. Fusicoccin also stimulates stomatal opening, respiration, and plant cell enlargement. Fusicoccin brings about these effects by binding to receptor sites in the membrane of affected cells and activating a membrane-based enzyme, ATPase, which not only accelerates the use of ATP bonds but also mediates the exchange of outward-moving H^+ with inward-moving k^+.

Other Non-Host-Specific Toxins Several other non-host-specific toxic substances have been isolated from cultures of pathogenic fungi and bacteria and have been implicated as contributing factors in the development of the disease caused by the pathogen. Among such toxins produced by fungi are fumaric acid, produced by *Rhizopus* spp. in the almond hull rot disease; oxalic acid, produced by *Sclerotium* and *Sclerotinia* spp. in various plants they infect, and by *Endothia parasitica,* the cause of chestnut blight; alternaric acid, alternariol, and zinniol produced by *Alternaria* spp. in leaf spot diseases of various plants, cerato-ulmin, produced by *Ceratocystis ulmi* in Dutch elm disease; ophiobolins, produced by several *Helminthosporium* spp. in diseases of grain crops; pyricularin, produced by *Pyricularia oryzae* in the rice blast disease; cercosporin, produced by *Cercospora* spp. in leaf spots of beet and other plants; fusaric acid and lycomarasmin, produced by *Fusarium oxysporum* in tomato wilt, and many others. Other non-host-specific toxins produced by bacteria are coronatine, produced by *Pseudomonas syringae* pv. *atropurpurea* and other forms infecting grasses and soybean; rhizobitoxine, produced by *Rhizobium japonicum* in soybean root nodules; syringomycin, produced by *Pseudomonas syringae* pv. *syringae* in leaf spots of many plants; syringotoxin, produced by *P. s.* pv. *syringae* in citrus plants; and tagetitoxin, produced by *P. s.* pv. *tagetis* in marigold leaf spot disease.

Host-Specific Toxins

A **host-specific toxin** is a substance produced by a pathogenic microorganism that, at physiologic concentrations, is toxic only to the hosts of that pathogen and shows little or no toxicity against nonsusceptible plants. Most host-specific toxins must be present for the producing microorganism to be able to cause disease. So far, host-specific toxins have been shown to be produced only by certain fungi (*Helminthosporium, Alternaria, Periconia, Phyllosticta, Corynespora,* and *Hypoxylon),* although certain bacterial polysaccharides from *Pseudomonas* and *Xanthomonas* have been reported to be host-specific.

Victorin, or HV, Toxin Victorin, or HV, toxin is produced by the fungus *Helminthosporium victoriae,* which appeared in 1945 after the introduction and widespread use of the oat variety Victoria and its derivatives, all of which contained the gene V_b for resistance to crown rust disease. *H. victoriae* infects the basal portions of susceptible oat plants and produces a toxin that is carried to the leaves, causes a leaf blight, and destroys the entire plant. All other oats and other plant species tested were either immune to the fungus and to the toxin or their sensitivity to the toxin was proportional to their susceptibility to the fungus. Resistance and susceptibility to the fungus, as well as tolerance and sensitivity to the toxin, are controlled by the same pair of alleles, although different sets of these alleles may be involved in cases of intermediate resistance. The toxin not only produces all the external symptoms of the disease induced by the pathogen, but also produces similar histochemical and biochemical changes in the host, such as changes in cell wall structure, loss of electrolytes from cells, increased respiration, decreased growth, and protein synthesis. Moreover, only fungus isolates that produce the toxin in culture are pathogenic to oats, while those that do not produce toxin are nonpathogenic.

Victorin has been purified but its chemical structure has not yet been determined. Also, although victorin was the first host-specific toxin to be studied, we still do not know either the identity or the cellular location of the molecule (or molecules) that is the primary target of the toxin.

T-Toxin (*Helminthosporium maydis* race T toxin). T-toxin is produced by race T of *Helminthosporium maydis,* the cause of southern corn leaf blight. Race T, indistinguishable from all other *H. maydis* races except for its ability to produce the T-toxin, appeared in the corn belt in 1968. By 1970, it had spread throughout the corn belt, attacking only corn that had the Texas male sterile (Tms) cytoplasm. Corn with normal cytoplasm was resistant to the fungus and the toxin. Resistance and susceptibility to *H. maydis* T and its toxin are inherited maternally (in cytoplasmic genes). The ability of *H. maydis* T to produce T-toxin and its virulence to corn with Tms cytoplasm are controlled by one and the same gene. T-toxin does not seem to be necessary for pathogencity of *H. maydis* race T, but it increases the virulence of the pathogen.

T-toxin is a mixture of linear, long (35 to 45 carbon) polyketols, the most prevalent having the formula:

T-toxin apparently acts specifically on mitochondria of susceptible cells, in which it causes early loss of matrix density, renders them nonfunctional, and inhibits ATP synthesis. Although no specific receptor molecule has been identified within sensitive mitochondria, the toxin-sensitive site seems to be located on the inner mitochondrial membrane, which is thereby made highly permeable to small molecules and ions. In addition to, or as a result of, its effects on mitochondria, T-toxin also causes selective uptake of certain ions, inhibition of root growth in seedlings, and closure of stomata.

AK-toxin

AK-Toxin Ak-toxin is produced by a distinct pathotype of *Alternaria alternata,* previously known as *A. kikuchiana,* the cause of black leaf spot disease of Japanese pears *(Pyrus serotina)*. Pears of susceptible varieties sprayed with the toxin or fungus culture filtrates are damaged, while those of resistant varieties are unharmed. The AK-toxin causes sensitive leaf cells to instantaneously lose K^+ and phosphate, while the plasma membrane develops pronounced invaginations and the cell walls show conspicuous degradation. It is thought that the initial toxic effect of the toxin occurs at the interface between the cell wall and the plasma membrane, but no specific receptor molecule has been identified.

AM-Toxin AM-Toxin is produced by the apple pathotype of *Alternaria alternata,* previously known as *A. mali,* the cause of Alternaria blotch of apple. The toxin molecule is a cyclic depsipeptide, and it usually exists as a mixture of three forms. The toxin is extremely selective for susceptible apple varieties, while resistant varieties can tolerate more than ten thousand times as much toxin without showing symptoms. The site and mechanism of action of AM-toxin resemble those of AK-toxin, but AM-toxin also causes rapid loss of chlorophyll, suggesting that this toxin has more than one site of action.

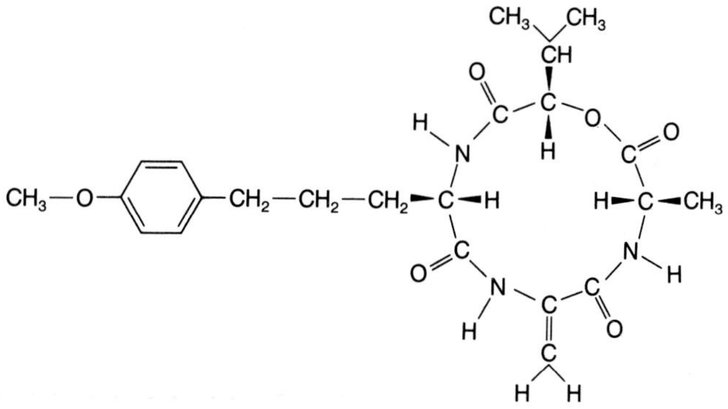

AM-toxin

Other Host-Specific Toxins At least a dozen additional host-specific toxins are known, and many more will undoubtedly be discovered in the future. Several of the additional toxins are also produced by species of the fungi *Helminthosporium* or *Alternaria:* HC-Toxin, produced by *H. carbonum* race I, affects corn; HS-toxin, produced by *H. sacchari,* affects sugarcane; ACL-toxin, produced by *Alternaria citri* (lemon race), affects rough lemon; ACT-toxin, produced by *A. citri* (tangerine race) affects Dancy tangerine; AL-toxin, produced by *A. alternata f. lycopersici,* affects tomato; other *Alternaria* species also produce AF-toxin on strawberry and Japanese pear and AT-toxin on tobacco.

At least two other fungi produce well-known host-specific toxins: *Periconia circinata* produces the PC-toxin in sorghum, and *Phyllosticta maydis* produces the PM-toxin in corn that has Texas male sterile cytoplasm.

Another fungus, *Corynespora cassiicola,* produces the CC-toxin in tomato. Toxins produced by some other fungi, for example, *Hypoxylon mammatum* on poplar and *Perenophora teres* on barley, seem to be species-selective rather than host-specific.

Growth Regulators in Plant Disease

Plant growth is regulated by a small number of groups of naturally occurring compounds that act as hormones and are generally called growth regulators. The most important growth regulators are auxins, gibberellins, and cytokinins, but other compounds, such as ethylene and growth inhibitors, play important regulatory roles in the life of the plant. Growth regulators act in very small concentrations, and even slight deviations from the normal concentration may bring about strikingly different plant growth patterns. The concentration of a specific growth regulator in the plant is not constant, but it usually rises quickly to a peak and then quickly declines as a result of the action of hormone-inhibitory systems present in the plant. Growth regulators appear to act, at least in some cases, by promoting synthesis of messenger-RNA molecules, which leads to the formation of specific enzymes, which in turn control the biochemistry and the physiology of the plant.

Plant pathogens may produce more of the same growth regulators as those produced by the plant or more of the same inhibitors of the growth regulators as those produced by the plant; they may produce new and different growth regulators or inhibitors of growth regulators; or they may produce substances that stimulate or retard the production of growth regulators or growth inhibitors by the plant.

Whatever the mechanism of action involved, pathogens often cause an imbalance in the hormonal system of the plant and bring about abnormal growth responses incompatible with the healthy development of a plant. That pathogens can cause disease through secretion of growth regulators in the infected plant or through their effects on the growth-regulatory systems of the infected plant is made evident by the variety of abnormal plant growth responses they cause, such as stunting, overgrowths, rosetting, excessive root branching, stem malformation, leaf epinasty, defoliation, and suppression of bud growth. The most important groups of plant growth regulators, their function in the plant, and their role in disease development, where known, are discussed below.

Auxins

The auxin naturally occurring in plants is indole-3-acetic acid (IAA). Continually produced in growing plant tissues, IAA moves rapidly from the young green tissues to older tissues but is constantly being destroyed by the enzyme indole-3-acetic acid oxidase, which explains the low concentration of the auxin.

CH_2COOH
N
H

Indole-3-acetic acid

The effects of IAA on the plant are numerous. Required for cell elongation and differentiation, absorption of IAA to the cell membrane also affects the permeability of the membrane; IAA causes a general increase in respiration of plant tissues and promotes the synthesis of messenger RNA and, subsequently, of proteins—enzymes as well as structural proteins.

Increased auxin (IAA) levels occur in many plants infected by fungi, bacteria, viruses, mycoplasmas, and nematodes, although some pathogens seem to lower the auxin level of the host. Thus, the fungi causing clubroot of cabbage (*Plasmodiophora brassicae*), late blight of potato (*Phytophthora infestans*), corn smut (*Ustilago maydis*), cedar apple rust (*Gymnosporangium juniperi-virginianae*), banana wilt (*Fusarium oxysporum* f. *cubense*), the root-knot nematode (*Meloidogyne* sp.), and others not only induce increased levels of IAA in their respective hosts but are themselves capable of producing IAA. In some diseases, however, increased levels of IAA are wholly or partly due to the decreased degradation of IAA through inhibition of IAA oxidase, as has been shown to be the case in several diseases, including corn smut and stem rust of wheat.

The production and role of auxin in plant disease have been studied more extensively in some bacterial diseases of plants. *Pseudomonas solanacearum,* the cause of bacterial wilt of solanaceous plants, induces a 100-fold increase in the IAA level of diseased plants compared with that of healthy plants. How the increased levels of IAA contribute to the development of wilt of plants is not yet clear, but the increased plasticity of cell walls as a result of high IAA levels renders the pectin, cellulose and protein components of the cell wall more accessible to, and may facilitate their degradation by, the respective enzymes secreted by the pathogen. Increase in IAA levels seems to inhibit lignification of tissues and may thus prolong the period of exposure of the nonlignified tissues to the cell-wall degrading enzymes of the pathogen. Increased respiratory rates in the infected tissues may also be due to high IAA levels, and since auxin affects cell permeability, it may be responsible for the increased transpiration of the infected plants.

In crown gall, a disease caused by the bacterium *Agrobacterium tumefaciens* on more than one hundred plant species, galls or tumors develop on the roots, stems, and petioles of host plants. Crown gall tumors develop when crown gall bacteria enter fresh wounds on a susceptible host. Immediately after wounding, cells around the wound are activated to divide. *Agrobacterium* bacteria do not invade cells but attach to cell walls. During the intense cell division of the second and third days after wounding, the cells are somehow conditioned and made receptive to a piece of bacterial plasmid DNA (called T-DNA, for tumor DNA), which becomes integrated into the nuclear plant DNA and transforms normal plant cells into tumor cells. Tumor cells subsequently grow and divide independently of the bacteria, and their organization, rate of growth, and rate of division can no longer be controlled by the host plant.

Tumor cells contain higher than normal amounts of IAA and also of cytokinin. The crown gall bacteria, of course, produce IAA, and genes coding for IAA as well as cytokinin production have been found on Ti plasmids. Since even tumors free of bacteria contain increased levels of IAA, it is certain that the tumor cells themselves have been rendered capable of generating the abnormal levels of IAA and cytokinins they contain. Although the increased levels of IAA and cytokinins of tumor cells are sufficient to cause the autonomous enlargement and division of these cells once they have been transformed to tumor cells, high IAA and cytokinin levels alone cannot cause the transformation of healthy cells into tumor cells. What other substances are involved in the transformation of healthy cells into tumor cells is not known.

In another hyperplastic disease, the knot disease of olive, oleander, and privet caused by the bacterium *Pseudomonas savastanoi,* the pathogen produces IAA, which induces infected plants to produce galls. The more IAA a strain produces, the more severe the symptoms it causes. Strains that do not produce IAA fail to induce production of galls. The bacterial genes for IAA production are in a plasmid carried in the bacterium, but some IAA synthesis is also carried out by a gene in the chromosome of the bacterium.

On the other hand, it is known that certain toxins, for example, fusicoccin, produced by the fungus *Fusicoccum amygdali,* and coronatin, produced by the bacterium *Pseudomonas syringae* pv. *atropurpurea,* exhibit auxinlike activity, while the toxin victorin, produced by the fungus *Helminthosporium victoriae,* inhibits cell elongation induced by IAA.

Gibberellins

Gibberellins are normal constituents of green plants and are also produced by several microorganisms. Gibberellins were first isolated from the fungus *Gibberella fujikuroi,* the cause of the "foolish seedling disease" of rice. The best known gibberellin is gibberellic acid. Compounds such as vitamin E and helminthosporol also have gibberellinlike activity.

Gibberellic acid

Gibberellins have striking growth-promoting effects. They speed elongation of dwarf varieties to normal sizes, promote flowering, stem and root elongation, and growth of fruit. Such elongation resembles in some respects that caused by IAA, and gibberellin also induces IAA formation. Auxin and gibberellin may also act synergistically. Gibberellins seem to activate genes that have been previously "turned off."

The foolish seedling disease of rice, in which rice seedlings infected with fungus *Gibberella fujikuroi* grow rapidly and become much taller than healthy plants, is apparently the result, to a considerable extent at least, of the gibberellin secreted by the pathogen.

Although no difference has been reported so far in the gibberellin content of healthy and virus- or mycoplasma-infected plants, spraying of diseased plants with gibberellin overcomes some of the symptoms caused by these pathogens. Thus, stunting of corn plants infected with corn stunt mycoplasma and of tobacco plants infected with severe etch virus was reversed after treatment with gibberellin. Axillary bud suppression, caused by sour cherry yellows virus (SCYV) on cherry and by leaf curl virus on tobacco, was also overcome by gibberellin sprays. The same treatment also increased fruit production in SCYV-infected cherries. In most of these treatments the pathogen itself does not seem to be affected, and the symptoms reappear on the plants after gibberellin applications are stopped. It is not known, however, whether or not the pathogen-caused stunting of plants is actually due to reduced gibberellin concentration in the deseased plant, especially since the growth of even healthy plants is equally increased after gibberellin treatments.

Cytokinins

Cytokinins are potent growth factors necessary for cell growth and differentiation. In addition, they inhibit the breakdown of proteins and nucleic acids, thereby causing inhibition of senescence, and they have the capacity to direct the flow of amino acids and other nutrients through the plant toward the point of high cytokinin concentration. Cytokinins occur in very small concentrations in green plants, their seeds, and in the sap stream.

Kinetin

The first compound with cytokinin activity to be identified was kinetin which, however, was isolated from herring sperm DNA and does not occur naturally in plants. Several cytokinins, for example, zeatin and isopentenyl adenosine (IPA), have since been isolated from plants.

Cytokinins act by preventing genes from being "turned off" and by activating genes that have been previously "turned off."

Zeatin

The role of cytokinins in plant disease has just begun to be studied. Cytokinin activity increases in clubroot galls, in crown galls, in smut and rust galls, and in rust-infected bean and broad bean leaves. In the latter, cytokinin activity seems to be related to both the juvenile feature of the green islands around the infection centers and to the senescence outside the green island. On the other hand, cytokinin activity is lower in the sap and in tissue extracts of cotton plants infected with Verticillium wilt and in plants suffering from drought. In the *Helminthosporium* blight disease of Victoria oats, cytokinins increase the quantity of toxin absorbed by the cells, but tobacco leaves injected with the wildfire toxin and treated with kinetin fail to develop the typical toxin-induced chlorosis. A cytokinin is partly responsible for the "leafy" gall disease caused by the bacterium *Corynebacterium fascians,* and it has been suggested that cytokinins may be responsible for the witches'-broom diseases caused by fungi and mycoplasmas.

Treating plants with kinetin before or shortly after inoculation with a virus seems to reduce the number of infections in local-lesion hosts and to reduce virus multiplication in systemically infected hosts.

Ethylene: $CH_2{=}CH_2$

Naturally produced by plants, ethylene exerts a variety of effects on plants, including chlorosis, leaf abscission, epinasty, stimulation of adventitious roots, and fruit ripening. Ethylene also causes increased permeability of cell membranes, which is a common effect of infections. On the other hand, ethylene induces phytoalexin formation in some tissues and stimulates synthesis or activity of several enzymes that may play a role in increasing plant resistance to infection. Ethylene is produced by several plant pathogenic fungi and bacteria. In the fruit of banana infected with *Pseudomonas solanacearum,* the ethylene content increases proportionately with the (premature) yellowing of the fruit, while no ethylene can be detected in healthy fruits. Ethylene has also been implicated in the leaf epinasty symptom of the vascular wilt syndromes and in the premature defoliation observed in several types of plant diseases.

Abscisic Acid

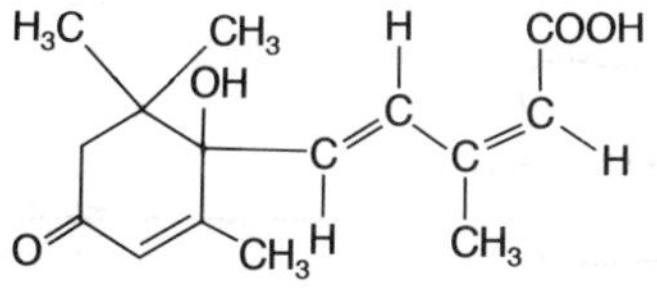

Abscisic acid is one of several growth inhibitors produced by plants and by at least some plant pathogenic fungi. It has several growth-inhibiting and hormonal functions, such as induction of dormancy, inhibition of seed germination, inhibition of growth, closure of stomata, and stimulation of germination of fungal spores. Several diseases are known in which infected plants show varying degrees of stunting, for example, tobacco mosaic and cucumber mosaic, both caused by viruses, southern bacterial wilt of tobacco, and Verticillium (fungus) wilt of tomato. In all these cases, diseased plants contain higher than normal levels of abscisic acid, and it is thought that abscisic acid is one of the factors responsible for the observed stunting of infected plants.

Polysaccharides

Fungi, bacteria, nematodes, and possibly other pathogens constantly release varying amounts of mucilaginous substances that coat their bodies and provide the interface between the outer surface of the microorganism and its environment.

The role of slimy polysaccharides in plant disease appears to be limited primarily to the wilt diseases caused by pathogens that invade the vascular system of the plant. In the vascular wilts, large polysaccharide molecules released by the pathogen in the xylem may be sufficient to cause a mechanical blockage of vascular bundles and thus initiate wilting. Although such an effect by the polysaccharides alone may occur rarely in nature, when it is considered together with the effect caused by the macromolecular substances released in the vessels through the breakdown of host substances by pathogen enzymes, the possibility of polysaccharide involvement in blockage of vessels during vascular wilts becomes obvious.

SELECTED REFERENCES

Bateman, D. F., and Basham, H. G. (1976) Degradation of plant cell walls and membranes by microbial enzymes. *Encycl. Plant Physiol. New Ser.* **4,** 316–355.

Blanchette, R. A. (1984). Selective delignification of eastern hemlock by *Ganoderma tsugae. Phytopathology* **74,** 153–160.

Collmer, A., Berman, P., and Mount, M. S. (1982). Pectate lyase regulation and bacterial soft rot pathogenesis. *In* "Phytopathogenic Prokaryotes" (M. S. Mount and G. H. Lacy, eds.), Vol. 1, pp. 395–422. Academic Press, New York.

Comai, L., Surico, G., and Kosuge, T. (1982). Relation of plasmid DNA to indoloacetic acid production in different strains of *Pseudomonas syringae* pv. *savastanoi*. *J. Gen. Microbiol.* **128,** 2157–2163.

Coty, P. J., Misaghi, I. J., Hine, R. B. (1983). Production of zinniol by *Alternaria tagetica* and its phytotoxic effect on *Tagetes erecta*. *Phytopathology* **73,** 1326–1328.

Cutler, D. F., Alvin, K. L., and Price, C. E., eds. (1982). "The Plant Cuticle," Linn. Soc. Symp. Ser. No. 10. Academic Press, London.

Daly, J. M., and Deverall, B. J., eds. (1983). "Toxins in Plant Pathogenesis." Academic Press, New York.

Daly, J. M., and Knoche, H. W. (1982). The chemistry and biology of pathotoxins exhibiting host-selectivity. *Adv. Plant Pathol.* **1,** 83–138.

Darvill, A., McNeil, M., Albersheim, P., and Delmer, D. P. (1980). The primary cell walls of flowering plants. *In* "The Biochemistry of Plants" (P. K. Stumpf, ed.), Vol. 4, pp. 91–162. Academic Press, New York.

Daub, M. E. (1982). Cercosporin, a photosensitizing toxin from *Cercospora* species. *Phytopathology* **72,** 370–374.

Dekhuijzen, H. M., and Overeem, J. C. (1971). The role of cytokinins in clubroot formation. *Physiol. Plant Pathol.* **1,** 151–161.

Durbin, R. D. (1979). How the beachhead is widened. *In* "Plant Disease" (J. G. Horsfall and E. B. Cowling, eds.), Vol. 4, pp. 155–162. Academic Press, New York.

Durbin, R. D., ed. (1981). "Toxins in Plant Disease." Academic Press, New York.

Durbin, R. D. (1982). Toxins and pathogenesis. *In* "Phytopathogenic Prokaryotes" (M. S. Mount and G. H. Lacy, eds.), Vol. 1, pp. 423–441. Academic Press, New York.

Gelvin, S. B. (1984). Plant tumorigeneses. *In* "Plant-Microbe Interactions: Molecular and Genetic Perspectives" (T. Kosuge and E. W. Nester, eds.), Vol. 1, pp. 343–377. Macmillan, New York.

Goodman, R. N., Kiraly Z., and Wood, K. R. (1986). "The Biochemistry and Physiology of Plant Disease." Univ. of Missouri Press, Columbia.

Goto, M., Yaguchi, Y., and Hyodo, H. (1980). Ethylene production in citrus leaves infected with *Xanthomonas citri* and its relation to defoliation. *Physiol. Plant Pathol.* **16,** 343–350.

Hancock, J. G., and Huisman, O. C. (1981). Nutrient movement in host-pathogen systems. *Annu. Rev. Phytopathol.* **19,** 309–331.

Heitefuss, R., and Williams, P. H., eds. (1976). "Physiological Plant Pathology," Encycl. Plant Physiol. New Ser. Vol. 4. Springer-Verlag, Berlin and New York.

Helgeson, J. P. (1978) Alteration of growth by disease. *In* "Plant Disease" (J. G. Horsfall and E. B. Cowling, eds.), Vol. 3, pp. 183–200. Academic Press, New York.

Horsfall, J. G., and Cowling, E. B., eds. (1977-1980). "Plant Disease," Vols. 1–5. Academic Press, New York (see especially Volume 4).

Kiraly, Z., El Hammady, M., and Pozsar, B. I. (1967). Increased cytokinin activity of rust-infected bean and broad bean leaves. *Phytopathology* **57,** 93–94.

Kirk, T. K. (1971). Effects of microorganisms on lignin. *Annu. Rev. Phytopathol.* **9,** 185–210.

Kolattukudy, P. E. (1981). Structure, biosynthesis and degradation of cutin and suberin. *Annu. Rev. Plant Physiol.* **32,** 539–567.

Kolattukudy, P. E. (1985). Enzymatic penetration of the plant cuticle by fungal pathogens. *Annu. Rev. Phytopathol.* **23,** 223–250.

Koller, W., Allan, C. R., and Kolattukudy, P. E. (1982). Role of cutinase and cell wall degrading enzymes in infection of *Pisum sativum* by *Fusarium solani* f. sp. *pisi*. *Physiol. Plant Pathol.* **20,** 47–60.

Kosuge, T., and Nester E. W., eds. (1984). "Plant-Microbe Interactions: Molecular and Genetic Perspectives," Vol. 1. Macmillan, New York.

Kuriger, W. E., and Agrios, G. N. (1977). Cytokinin levels and kinetin-virus interactions in tobacco ringspot virus-infected cowpea plants. *Phytopathology* **67,** 604–609.

Lee, R. F., Raju, B. C., Nyland, G., and Goheen, A. C. (1982). Phytotoxin(s) produced in culture by the Pierce's disease bacterium. *Phytopathology* **72,** 886–888.

Liu, S.-T., Perry, K. L., Schardl, C. L., and Kado, C. I. (1982). *Agrobacterium* Ti plasmid indoleacetic acid gene is required for crown gall oncogenesis. *Proc. Natl. Acad. Sci. U.S.A.* **79,** 2812–2816.

Maramorosch, K. (1957). Reversal of virus-caused stunting in plants by gibberellic acid. *Science* **126,** 651–652.

Mills, L. J., and Van Staden, J. (1978). Extraction of cytokinins from maize, smut tumorss of maize and *Ustilago maydis* cultures. *Physiol. Plant Pathol.* **13,** 73–80.
Misaghi, I. J. (1982). "Physiology and Biochemistry of Plant-Pathogen Interactions." Plenum, New York.
Misaghi, I. J., DeVay, E., and Kosuge, T. (1972). Changes in cytokinin activity associated with the development of Verticillium wilt and water stress in cotton plants. *Physiol. Plant Pathol.* **2,** 187–196.
Mitchel, R. E. (1984). The relevance of non-host-specific toxins in the expression of virulence by pathogens. *Annu. Rev. Phytopathol.* **22,** 215–245.
Mount, M. S. (1978). Tissue is disintegrated. *In* "Plant Disease" (J. G. Horsfall and E. B. Cowling, eds.), Vol. 3, pp 279–297. Academic Press, New York.
Nester, E. W., and Kosuge, T. (1981). Plasmids specifying plant hyperplasias. *Annu. Rev. Microbiol.* **35,** 531–565.
Nishimura, S., and Kohmoto, K. (1983). Host-specific toxins and chemical structures from *Alternaria* species. *Annu. Rev. Phytopathol.* **21,** 87–116.
Pozsar, B. I., and Kiraly, Z. (1966). Phloem-transport in rust-infected plants and the cytokinin directed long-distance movement of nutrients. *Phytopathol. Z.* **56,** 297–309.
Rowan, S. J. (1970). Fusiform rust gall formation and cytokinin of Loblolly pine. *Phytopathology* **60,** 1225–1226.
Scheffer, R. P. (1983). Toxins as chemical determinants of plant disease. *In* "Toxins and Plant Pathogenesis" (J. M. Daly and B. J. Deverall, eds.), pp. 1–40. Academic Press, New York.
Scheffer, R. P., and Livingston, R. S. (1984). Host-selective toxins and their role in plant disease. *Science* **223,** 17–21.
Sequeira, L. (1973). Hormone metabolism in diseased plants. *Annu. Rev. Plant Physiol.* **24,** 353–380.
Stall, R. E., and Hall, C. B. (1984). Chlorosis and ethylene production in pepper leaves infected by *Xanthomonas campestris* pv. *vesicatoria. Phytopathology* **74,** 373–375.
Stermer, B. A., Scheffer, R. P., and Hart, J. H. (1984). Isolation of toxins of *Hypoxylon mammatum* and demonstration of some toxin effects on selected clones of *Populus tremuloides. Phytopathology* **74,** 654–658.
Tsuyumu, S., and Chatterjee, A. K. (1984). Pectin lyase production in *Erwinia chrysanthemi* and other soft-rot *Erwinia* species. *Physiol. Plant Pathol.* **24,** 291–302.
VanEtten, H. D., and Kistler, H. C. (1984). Microbial enzyme regulation and its importance for pathogenicity. *In* "Plant-Microbe Interactions: Molecular and Genetic Perspectives" (T. Kosuge and E. W. Nester, eds.), Vol. 1, pp. 42–68. Macmillan, New York.
Wheeler, H (1975). "Plant Pathogenesis." Springer-Verlag, Berlin and New York.
Yoder, O. C. (1980). Toxins in pathogenesis. *Annu. Rev. Phytopathol.* **18,** 103–129.

4 EFFECTS OF PATHOGENS ON PLANT PHYSIOLOGICAL FUNCTIONS

Effects of Pathogens on Photosynthesis

Photosynthesis is the basic function of green plants that enables them to transform light energy into chemical energy, which they can then utilize in their cell activities. Photosynthesis is the ultimate source of all energy used in plant or animal cells, since in a living cell all activities except photosynthesis expend the energy provided by photosynthesis.

In photosynthesis, carbon dioxide from the atmosphere and water from the soil are brought together in the chloroplasts of the green parts of plants and, in the presence of light, react to form glucose with concurrent release of oxygen:

$$6\ CO_2 + 6\ H_2O \xrightarrow[\text{chlorophyll}]{\text{light}} C_6H_{12}O_6 + 6\ O_2$$

In view of the fundamental position of photosynthesis in the life of plants, it is apparent that any interference by pathogens with photosynthesis results in a diseased condition in the plant. That pathogens do interfere with photosynthesis is obvious from the chlorosis they cause on many infected plants, from the necrotic lesions or large necrotic areas they produce on green plant parts, and from the reduced growth and amounts of fruits produced by many infected plants.

In leaf spot, blight, and other kinds of diseases in which there is destruction of leaf tissue or defoliation, photosynthesis is reduced because the photosynthetic surface of the plant is lessened. Even in other diseases, however, plant pathogens reduce photosynthesis, especially in the late stages of diseases, by affecting the chloroplasts and causing their degeneration. The overall chlorophyll content of leaves in many fungal and bacterial diseases is reduced, but the photosynthetic activity of the remaining chlorophyll seems to remain unaffected. In some fungal and bacterial diseases, photosynthesis is reduced because the toxins, such as tentoxin and tabtoxin, produced by these pathogens inhibit some of the enzymes that are directly or indirectly involved in photosynthesis. In plants infected by many vascular pathogens, stomata remain partially closed, chlorophyll is reduced, and photosynthesis stops even before the plant eventually wilts. Most virus, mycoplasma, and nematode diseases induce varying degrees of chlorosis. In the majority of such diseases, photosynthesis of infected plants is reduced greatly. In advanced stages of disease, the rate of photosynthesis is no more than one-fourth the normal rate.

Effect of Pathogens on Translocation of Water and Nutrients in the Host Plant

All living plant cells require an abundance of water and an adequate amount of organic and inorganic nutrients in order to live and to carry out their physiological functions. Plants absorb water and inorganic (mineral) nutrients from the soil through their root system. These substances are generally translocated upward through the xylem vessels of the stem and into the vascular bundles of the petioles and leaf veins, from which they enter the leaf cells. The minerals and part of the water are utilized by the leaf and other cells for synthesis of the various plant substances, but most of the water evaporates out of the leaf cells into the intercellular spaces and from these diffuses into the atmosphere through the stomata. On the other hand, nearly all organic nutrients of plants are produced in the leaf cells, following photosynthesis, and are translocated downward and distributed to all the living plant cells by passing for the most part through the phloem tissues. When a pathogen interferes with the upward movement of inorganic nutrients and water or with the downward movement of organic substances, diseased conditions will result in the parts of the plant denied these materials. The diseased parts, in turn, will be unable to carry out their own functions and will deny the rest of the plant their services or their products, thus causing disease of the entire plant. For example, if water movement to the leaves is inhibited, the leaves cannot function properly, photosynthesis is reduced or stopped, and few or no nutrients are available to move to the roots, which in turn become starved, diseased, and may die.

Interference with Upward Translocation of Water and Inorganic Nutrients

Many plant pathogens interfere in one or more ways with the translocation of water and inorganic nutrients through plants. Some pathogens affect the integrity or function of the roots, causing them to absorb less water; other pathogens, by growing in the xylem vessels or by other means, interfere with the translocation of water through the stem; and, in some diseases, pathogens interfere with the water economy of the plant by causing excessive transpiration through their effects on leaves and stomata.

Effect on Absorption of Water by Roots

Many pathogens, such as the damping-off fungi, the root-rotting fungi and bacteria, most nematodes, and some viruses cause an extensive destruction of the roots before any symptoms appear on the aboveground parts of the plant. Root injury directly affects the amount of functioning roots and decreases proportionately the amount of water absorbed by the roots. Some vascular parasites, along with their other effects, seem to inhibit root hair production, which reduces water absorption. These and other pathogens also alter the permeability of root cells, an effect that further interferes with the normal absorption of water by roots.

Effect on Translocation of Water through the Xylem

Fungal and bacterial pathogens that cause damping-off, stem rots, and cankers may reach the xylem vessels in the area of the infection and, if the affected plants are young, may cause their destruction and collapse. Affected vessels may also be filled with the bodies of the pathogen and with substances secreted by the pathogen or by the host in response to the pathogen, and may become clogged. Whether destroyed or clogged, the affected vessels cease to function properly and allow little or no water to pass through them. Certain pathogens, such as the crown gall bacterium *(Agrobacterium tumefaciens),* the clubroot fungus *(Plasmodiophora brassicae),* and the root-knot nematode (*Meloidogyne* sp.) induce gall formation in the stem or roots or both. The enlarged and proliferating cells near or around the xylem exert pressure on the xylem vessels, which may be crushed and dislocated, thereby becoming less efficient in transporting water.

The most typical and complete dysfunction of xylem in translocating water, however, is observed in the vascular wilts caused by the fungi *Ceratocystis, Fusarium,* and *Verticillium,* and bacteria like *Pseudomonas* and *Erwinia.* These pathogens invade the xylem of roots and stems and produce diseases primarily by interfering with the upward movement of water through the xylem. In many plants infected by these pathogens the water flow through the stem xylem is reduced to a mere 2 to 4 percent of that flowing through the stems of healthy plants. In general, the rate of flow through infected stems seems to be inversely proportional to the number of vessels blocked by the pathogen and by the substances resulting from the infection. Evidently, more than one factor is usually responsible for vascular dysfunction in the wilt diseases. Although the pathogen is the single cause of the disease, some of the factors responsible for the disease syndrome originate directly from the pathogen, while others originate from the host in response to the pathogen. The pathogen can reduce the flow of water through its physical presence in the xylem as mycelium, spores, or bacterial cells and by production of large molecules (polysaccharides) in the vessels. The infected host may reduce the flow of water through reduction in the size or collapse of vessels due to infection, development of tyloses in the vessels, release of large-molecule compounds in the vessels as a result of cell wall breakdown by pathogenic enzymes, and reduced water tension in the vessels due to pathogen-induced alterations in foliar transpiration.

Effect on Transpiration

In plant diseases in which the pathogen infects the leaves, transpiration is usually increased. This is the result of destruction of at least part of the protection afforded the leaf by the cuticle, an increase in permeability of leaf cells, and the dysfunction of stomata. Diseases like the rusts, mildews, and apple scab destroy a considerable portion of the cuticle and epidermis, and this results in uncontrolled loss of water from the affected areas. If water absorption and translocation cannot keep up with the excessive loss of water, loss of turgor and wilting of leaves follows. The suction force of excessively transpiring leaves is abnormally increased and may lead to collapse or dysfunction of underlying vessels through production of tyloses and gums.

Interference with the Translocation of Organic Nutrients through the Phloem

Organic nutrients produced in leaf cells through photosynthesis move through plasmodesmata into adjoining phloem elements. From there they move down the phloem sieve tubes and eventually, again through plasmodesmata, into the protoplasm of living nonphotosynthetic cells, where they are utilized, or into storage organs, where they are stored. Thus, in both cases, they are removed from "circulation." Plant pathogens may interfere with the movement of organic nutrients from the leaf cells to the phloem or with their translocation through the phloem elements and, possibly, with their movement from the phloem into the cells that will utilize them.

Obligate fungal parasites, such as the rust and mildew fungi, cause an accumulation of photosynthesis products, as well as inorganic nutrients, in the areas invaded by the pathogen. In these diseases, the infected areas are characterized by reduced photosynthesis and increased respiration. However, synthesis of starch and of other compounds as well as dry weight are temporarily increased in the infected areas, indicating translocation of organic nutrients from uninfected areas of the leaves or from healthy leaves toward the infected areas.

In some virus diseases, particularly the leaf-curling type and some yellows diseases, starch accumulation in the leaves is a common phenomenon. In most of these diseases, starch accumulation in the leaves is mainly the result of degeneration (necrosis) of the phloem of infected plants, which is one of the first symptoms. It is also possible, however, at least in some virus diseases, that the interference with translocation of starch stems from inhibition by the virus of the enzymes that break down starch into smaller translocatable molecules. This is suggested by the observation that in some mosaic diseases, in which there is no phloem necrosis, infected, discolored areas of leaves contain less starch than "healthy," greener areas at the end of the day, a period favorable for photosynthesis; but the same leaf areas contain more starch than the "healthy" areas after a period in the dark, which favors starch hydrolysis and translocation. This suggests not only that virus-infected areas synthesize less than healthy ones, but also that starch is not easily degraded and translocated from virus-infected areas, although no damage to the phloem is present.

Effect of Pathogens on Host Plant Respiration

Respiration is the process by which cells, through enzymatically controlled oxidation (burning) of the energy-rich carbohydrates and fatty acids, liberate energy in a form that can be utilized for the performance of various cellular processes. Plant cells carry out respiration in, basically, two steps. The first step involves the degradation of glucose to pyruvate and is carried out, either in the presence or in the absence of oxygen, by enzymes found in the ground cytoplasm of the cells. The production of pyruvate from glucose follows either the **glycolytic pathway,** otherwise known as **glycolysis,** or to a lesser extent, the **pentose pathway.** The second step involves the degradation of pyruvate, however produced, to CO_2 and water. This is accomplished by a series of reactions known as the **Krebs cycle,** which is accompanied by the so-called **terminal oxidation** and is carried out in the mitochondria only in the presence of oxygen. Under normal (aerobic) conditions, that is, in the presence of oxygen, both steps are carried out, and one molecule of glucose yields, as final products, six molecules of CO_2 and six molecules of water,

$$C_6H_{12}O_6 + 6\ O_2 \rightarrow 6\ CO_2 + 6\ H_2O$$

with concomitant release of energy (678,000 calories). Some of this energy is lost, but almost half is converted to 20–30 reusable high-energy bonds of adenosine triphosphate (ATP). The first step of respiration contributes two ATP molecules per mole of glucose, and the second step contributes the rest. Under anaerobic conditions, however—that is, in the absence of oxygen—pyruvate cannot be oxidized, but it instead undergoes **fermentation** and yields lactic acid or alcohol. Since the main process of energy generation is cut off, for the cell to secure the necessary energy a much greater rate of glucose utilization by glycolysis is required in the absence of oxygen than is in its presence.

The energy-storing bonds of ATP are formed by the attachment of a phosphate (PO_4) group to adenosine diphosphate (ADP), at the expense of energy released from the oxidation of sugars. The coupling of oxidation of glucose with the addition of phosphate to ADP to produce ATP is called **oxidative phosphorylation.** Any cell activity that requires energy utilizes the energy stored in ATP by simultaneously breaking down ATP to ADP and inorganic phosphate. The presence of ADP and phosphate in the cell, in turn, stimulates the rate of respiration. If, on the other hand, ATP is not utilized sufficiently by the cell for some reason, there is little or no regeneration of ADP and respiration is slowed down. The amount of ADP (and phosphate) in the cell is determined, therefore, by the rate of energy utilization; this rate, in turn, determines the rate of respiration in plant tissues.

The energy produced through respiration is utilized by the plant for all types of cellular work, such as accumulation and mobilization of compounds, synthesis of proteins, activation of enzymes, cell growth and division, defense reactions, and a host of other processes. The complexity of respiration, the number of enzymes involved in respiration, its occurrence in every single cell, and its far-reaching effects on the functions and existence of the cell make it easy to understand why respiration of plant tissues is one of the first functions to be affected when plants are infected by pathogens.

Respiration of Diseased Plants

When plants are infected by pathogens, the rate of respiration generally increases. This means that affected tissues use up their reserve carbohydrates faster than healthy tissues would. The increased rate of respiration appears shortly after infection—certainly by the time of appearance of visible symptoms—and continues to rise during the multiplication and sporulation of the pathogen. After that, respiration declines to normal levels or to levels even lower than those of healthy plants. Respiration increases more rapidly in infections of resistant varieties, in which large amounts of energy are needed and used for rapid production or mobilization of the defense mechanisms of the cells. In resistant varieties, however, respiration also declines quickly after it reaches its maximum. In susceptible varieties, in which no defense mechanisms can be mobilized quickly against a particular pathogen, respiration increases slowly after inoculation, but it continues to rise and remains at a high level for much longer periods.

Several changes in the metabolism of the diseased plant accompany the increase in respiration after infection. Thus, the activity or concentration of several enzymes of the respiratory pathways seem to be increased. The accumulation and oxidation of phenolic compounds, many of which are associated with defense mechanisms in plants, are also greater during increased respiration. Increased respiration in diseased plants is also accompanied by an increased activation of the pentose pathway, which is the main source of phenolic compounds. Increased respiration is sometimes accompanied by considerably more fermentation than observed in healthy plants, probably as a result of an accelerated need for energy in the diseased plant under conditions in which normal aerobic respiration cannot provide sufficient energy.

The increased respiration in diseased plants is apparently brought about, at least in part, by uncoupling of the oxidative phosphorylation. In that case, no utilizable energy (ATP) is produced through normal respiration in spite of the use of existing ATP and accumulation of ADP, which stimulates respiration. The energy required by the cell for its vital processes is then produced through other, less efficient ways including the pentose pathway and fermentation.

The increased respiration of diseased plants can also be explained as the result of increased metabolism. In many plant diseases, growth is at first stimulated, protoplasmic streaming increases, and materials are synthesized, translocated, and accumulated in the diseased area. The energy required for these activities derives from ATP produced through respiration. The more ATP is utilized, the more ADP is produced and further stimulates respiration. It is also possible that the plant, because of the infection, utilizes ATP energy less efficiently than a healthy plant. Because of the waste of part of the energy, an increase in respiration is induced and the resulting greater amount of energy enables the plant cells to utilize sufficient energy to carry out their accelerated processes.

Although oxidation of glucose via the glycolytic pathway is by far the most common way through which plant cells obtain their energy, part of the energy is produced via the pentose pathway. The latter seems to be an alternate pathway of energy production to which plants resort under conditions of stress. Thus, the pentose pathway tends to replace the glycolytic pathway as the plants grow older and differentiate, and to increase upon treatment of the plants with hormones, toxins, wounding, starvation, and so on. Infection of plants with pathogens also tends, in general, to activate the pentose pathway over the level at which it operates in the healthy plant. Since the pentose pathway is not directly linked to ATP production, the increased respiration through this pathway fails to produce as much utilizable energy as the glycolytic pathway and is, therefore, a less efficient source of energy for the functions of the diseased plant. On the other hand, the pentose pathway is the main source of phenolic compounds, which play important roles in the defense mechanisms of the plant against infection.

Effect of Pathogens on Permeability of Cell Membranes

Cell membranes consist of a double layer of lipid molecules in which are embedded many kinds of protein molecules, parts of which usually protrude on both sides of the lipid bilayer. Membranes function as permeability barriers that allow passage into a cell only of substances the cell needs and inhibit passage out of the cell of substances needed by the cell. The lipid bilayer is impermeable to most biological molecules. Small water-soluble molecules such as ions (charged atoms or electrolytes), sugars, and amino acids flow through or are pumped through special membrane channels made of proteins. In plant cells, because of the cell wall, only small molecules reach the cell membrane. In animal cells, however, and in artificially prepared plant protoplasts, large molecules or particles may also reach the cell membrane and enter the cell by **endocytosis:** A patch of the membrane surrounds and forms a vesicle around the material to be taken in, brings it in and releases it inside the cell. Disruption or disturbance of the cell membrane by chemical or physical factors alters (usually increases) the permeability of the membrane with subsequent uncontrollable loss of useful substances as well as inability to inhibit the inflow of undesirable substances or excessive amounts of any substances.

Changes in cell membrane permeability are often the first detectable responses of cells to infection by pathogens, to most host-specific and several nonspecific toxins, to certain pathogen enzymes, and to certain toxic chemicals, such as air pollutants. The most commonly observed effect of changes in cell membrane permeability is the loss of **electrolytes,** that is, of small water-soluble ions and molecules from the cell. It is not certain, however, whether the cell membrane is the initial target of pathogen toxins and enzymes and that the accompanying loss of electrolytes is the initial effect of changes in cell membrane permeability, or whether the pathogen products actually affect other organelles or reactions in the cell, in which case cell permeability changes and loss of electrolytes are secondary effects of the initial events. If pathogens do, indeed, affect cell membrane permeability directly, it is likely that they bring this about by stimulating certain membrane-bound enzymes, such as ATPase, which are involved in the pumping of H^+ in and K^+ out through the cell membrane, by interfering with processes required for maintenance and repair of the fluid film making up the membrane, or by degrading the lipid or protein components of the membrane by pathogen-produced enzymes.

Effects of Pathogens on Transcription and Translation

Transcription of cellular DNA into messenger RNA and translation of messenger RNA to produce proteins are two of the most basic, general, and

Unwin, N., and Henderson, R. (1984). The structure of proteins in biological membranes. *Sci. Am.* **250**(2), 78–94.
Wheeler, H. (1975). "Plant Pathogenesis." Springer-Verlag, Berlin and New York.
Wheeler, H. (1978). Disease alterations in permeability and membranes. *In* "Plant Disease" (J. G. Horsfall and E. B. Cowling, eds.), Vol. 3, pp. 327–347. Academic Press, New York.
Zimmerman, M. H., and McDonough, J. (1978). Dysfunction in the flow of food. *In* "Plant Disease" (J. G. Horsfall and E. B. Cowling, eds.), Vol. 3, pp. 117–140. Academic Press, New York.

5 HOW PLANTS DEFEND THEMSELVES AGAINST PATHOGENS

Each plant species is affected by approximately one hundred different kinds of fungi, bacteria, mycoplasmas, viruses, and nemotodes. Frequently, a single plant is attacked by hundreds, thousands, and in the leafspot diseases of large trees, probably hundreds of thousands of individuals of a single kind of pathogen. Although such plants may suffer damage to a lesser or greater extent, many survive all these attacks and, not uncommonly, manage to grow well and to provide appreciable yields.

In general, plants defend themselves against pathogens by a combination of weapons from two arsenals: (1) structural characteristics that act as physical barriers and inhibit the pathogen from gaining entrance and spreading through the plant, and (2) biochemical reactions that take place in the cells and tissues of the plant and produce substances that either are toxic to the pathogen or create conditions that inhibit growth of the pathogen in the plant. The combinations of structural characteristics and biochemical reactions employed in the defense of plants are different in different host–pathogen systems. In addition, even within the same host and pathogen, the combinations vary with the age of the plant, the kind of plant organ and tissue attacked, the nutritional condition of the plant, and with the weather conditions.

Structural Defense

Preexisting Defense Structures

A plant's first line of defense against pathogens is its surface, which the pathogen must penetrate if it is to cause infection. Some structural defenses are present in the plant even before the pathogen comes in contact with the plant. Such structures include the amount and quality of wax and cuticle that cover the epidermal cells, the structure of the epidermal cell walls, the size, location, and shapes of stomata and lenticels, and the presence on the plant of tissues made of thick-walled cells that hinder the advance of the pathogen.

Waxes on leaf and fruit surfaces form a water-repellent surface and thereby prevent the formation of a film of water on which pathogens might be deposited and germinate (fungi) or multiply (bacteria). A thick mat of hairs on a plant surface may also exert a similar water-repelling effect and may reduce infection.

A thick cuticle may increase resistance to infection in diseases in which the pathogen enters its host only through direct penetration. Cuticle thickness, however, is not always correlated with resistance, and many plant varieties with cuticle of considerable thickness are easily invaded by directly penetrating pathogens.

The thickness and toughness of the outer wall of epidermal cells are apparently important factors in the resistance of some plants to certain pathogens. Thick, tough walls of epidermal cells make direct penetration by fungal pathogens difficult or impossible. Plants with such walls are often resistant, although if the pathogen is introduced beyond the epidermis of the same plants by means of a wound, the inner tissues of the plant are easily invaded by the pathogen.

Many pathogenic fungi and bacteria enter plants only through stomata. Although the majority of pathogens can force their way through closed stomata, some, like the stem rust of wheat, can enter only when stomata are open. Thus, some wheat varieties, in which the stomata open late in the day, are resistant because the germ tubes of spores germinating in the night dew desiccate owing to evaporation of the dew before the stomata begin to open. The structure of the stomata—for example, a very narrow entrance and broad, elevated guard cells—may also confer resistance to some varieties against certain of their pathogens.

The cell walls of the tissues being invaded vary in thickness and toughness and may sometimes inhibit the advance of the pathogen. The presence, in particular, of bundles or extended areas of sclerenchyma cells, such as are found in the stems of many cereal crops, may stop the further spread of pathogens like the stem rust fungi. Also, the xylem, bundle sheath, and sclerenchyma cells of the leaf veins effectively block the spread of some fungal, bacterial, and nematode pathogens that cause the various "angular" leaf spots because of their spread only into areas between, but not across, veins.

Defense Structures Formed in Response to Infection by the Pathogen

In spite of the preformed superficial or internal defense structures of host plants, most pathogens manage to penetrate their hosts and to produce various degrees of infection. Even after the pathogen has penetrated the preformed defense structures, however, plants usually respond by forming one or more types of structures that are more or less successful in defending the plant from further pathogen invasion. Some of the defense structures formed involve tissues ahead of the pathogen (deeper into the plant) and are called **histological defense structures**; others involve the walls of invaded cells and are called **cellular defense structures**; still others involve the cytoplasm of the cells under attack, and the process is called **cytoplasmic defense reaction.** Finally, the death of the invaded cell may protect the plant from further invasion, and this is called **necrotic** or **hypersensitive defense reaction.**

Histological Defense Structures

Formation of Cork Layers Infection by fungi or bacteria and even by some viruses and nematodes frequently induces plants to form several layers of

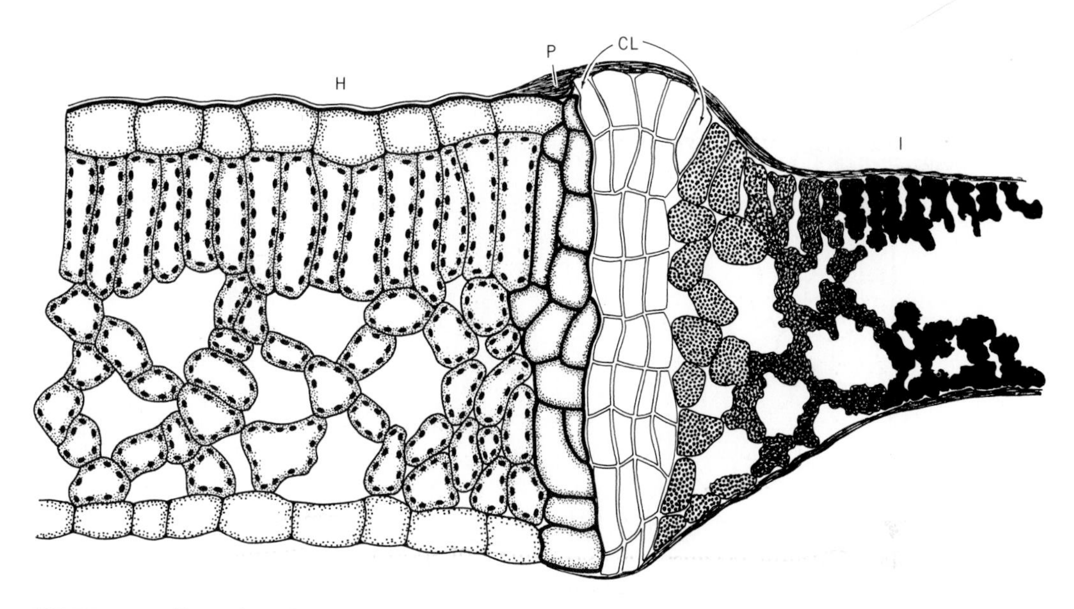

FIGURE 5-1 Formation of cork layer between infected and healthy areas of leaf. CL = cork layer; H = healthy leaf area; I = infected; P = phellogen. After Cunningham (1928).

cork cells beyond the point of infection (Figures 5-1 and 5-2), apparently as a result of stimulation of the host cells by substances secreted by the pathogen. The cork layers inhibit the further invasion by the pathogen beyond the initial lesion and also block the spread of any toxic substances that the pathogen may secrete. Furthermore, cork layers stop the flow of nutrients and water from the healthy to the infected area and deprive the pathogen of

FIGURE 5-2 Formation of cork layer on potato tuber following infection with *Rhizoctonia*. After G. E. Ramsey, *I. Agric. Res.* **9**, 421–426 (1917).

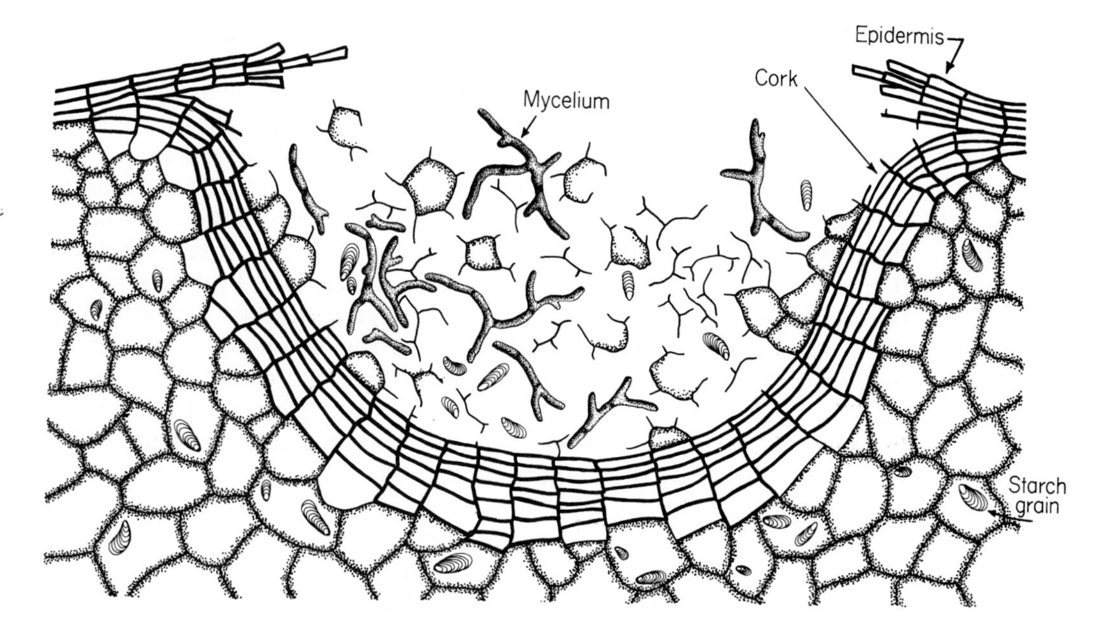

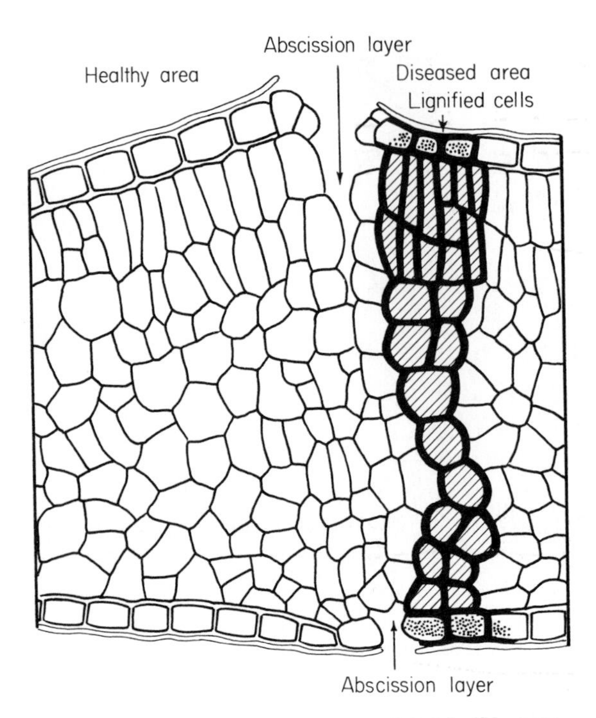

FIGURE 5-3 Formation of abscission layer around a diseased spot of a *Prunus* leaf. After Samuel (1927).

nourishment. The dead tissues, including the pathogen, are thus delimited by the cork layers and either remain in place forming necrotic lesions (spots) that are remarkably uniform in size and shape for a particular host–pathogen combination (see Figures 11-39, 11-43, 11-46C, 11-48 to 11-53, 11-86, and 14-10) or are pushed outward by the underlying healthy tissues and form scabs that may be sloughed off, thus removing the pathogen from the host completely (see Figures 11-65, 11-103, 12-10, 12-19B, 12-31).

Formation of Abscission Layers Abscission layers are formed on young, active leaves of stone fruit trees after infection by any of several fungi, bacteria, or viruses (see Figure 12-11). An abscission layer consists of a gap between two circular layers of cells of a leaf surrounding the locus of infection. Upon infection, the middle lamella between these two layers of cells is dissolved throughout the thickness of the leaf, completely cutting off the central area from the rest of the leaf (Figure 5-3). Gradually this area shrivels, dies, and sloughs off, carrying with it the pathogen. Thus, the plant, by discarding the infected area along with a few yet uninfected cells, protects the rest of the leaf tissue from being invaded by the pathogen and from becoming affected by the toxic secretions of the pathogen.

Formation of Tyloses Tyloses form in xylem vessels of most plants under various conditions of stress and during invasion by most of the vascular pathogens. Tyloses are overgrowths of the protoplast of adjacent living parenchymatous cells, which protrude into xylem vessels through pits (Figure 5-4). Tyloses have cellulosic walls and may, by their size and numbers, clog the vessel completely. In some varieties, tyloses form abundantly and quickly

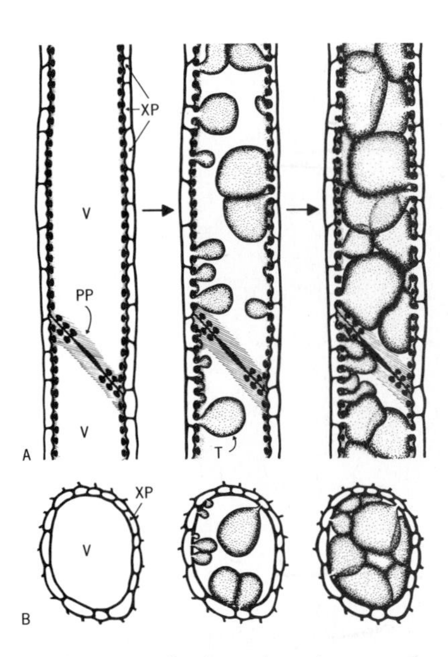

FIGURE 5-4 Development of tyloses in xylem vessels. Longitudinal (A) and cross-section (B) views of healthy vessels (left), and of vessels with tyloses. Vessels on right are completely clogged with tyloses. PP = perforation plate; V = xylem vessel; XP = xylem parenchyma cell; T = tylosis.

ahead of the pathogen, while the pathogen is still in the young roots, and block the further advance of the pathogen. The plants of these varieties remain free of, and therefore resistant to this pathogen. Varieties in which few, if any, tyloses form ahead of the pathogen are susceptible to disease.

Deposition of Gums Various types of gums are produced by many plants around lesions after infection by pathogens or injury. Gum secretion is most common in stone fruit trees but occurs in most plants. The defensive role of gums stems from the fact that they are quickly deposited in the intercellular spaces and within the cells surrounding the locus of infection, thus forming an impenetrable barrier that completely encloses the pathogen (Figure 5-5). The pathogen then becomes isolated, starved, and sooner or later dies.

Cellular Defense Structures

Cellular defense structures involve morphological changes in the cell wall or changes derived from the cell wall of the cell being invaded by the pathogen. The effectiveness of these structures as defense mechanisms seems to be rather limited, however. Three main types of such structures have been observed in plant diseases: (a) The outer layer of the cell wall of parenchyma cells coming in contact with imcompatible bacteria swells, and this swelling is accompanied by production of an amorphous, fibrillar material that surrounds and traps the bacteria and prevents them from multiplying; (b) Cell walls thicken in response to some viral and fungal pathogens, for example, in the local lesion response to virus infections and in the cucumber response to

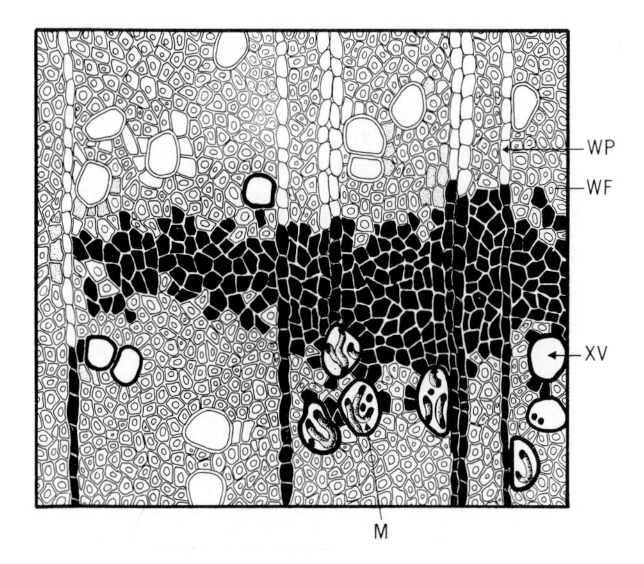

FIGURE 5-5 Gum barrier in apple twig infected with *Physalospora cydoniae.* M = mycelium in vessels; XV = xylem vessel; WF = wood fiber; WP = wood parenchyma. After Hesler (1916).

the scab fungus *Cladosporium cucumerinum.* The thickening material appears to be cellulosic but is often infused with phenolic substances that further increase its resistance to penetration. (c) Callose **papillae** are deposited on the inner side of cell walls in response to invasion by fungal pathogens (see Figure 2-5). Papillae seem to be produced by cells within minutes after wounding and within 2–3 hours after inoculation within microorganisms. Although the main function of papillae seems to be to repair cellular damage, sometimes —especially if papillae are present before inoculation—they also seem to prevent the pathogen from subsequently penetrating the cell. In some cases,

FIGURE 5-6 Formation of sheath around hypha penetrating a cell wall. CW = cell wall; H = hypha; A = appressorium; AH = advancing hypha still enclosed in sheath; HC = hypha in cytoplasm; S = sheath.

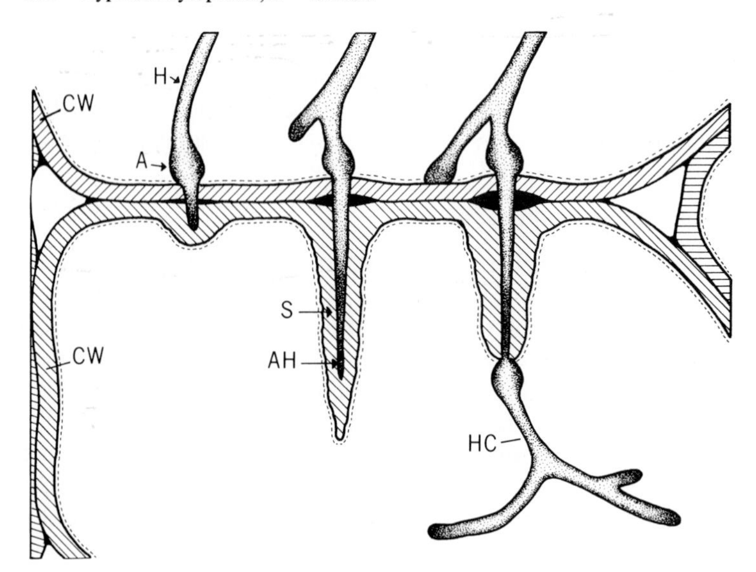

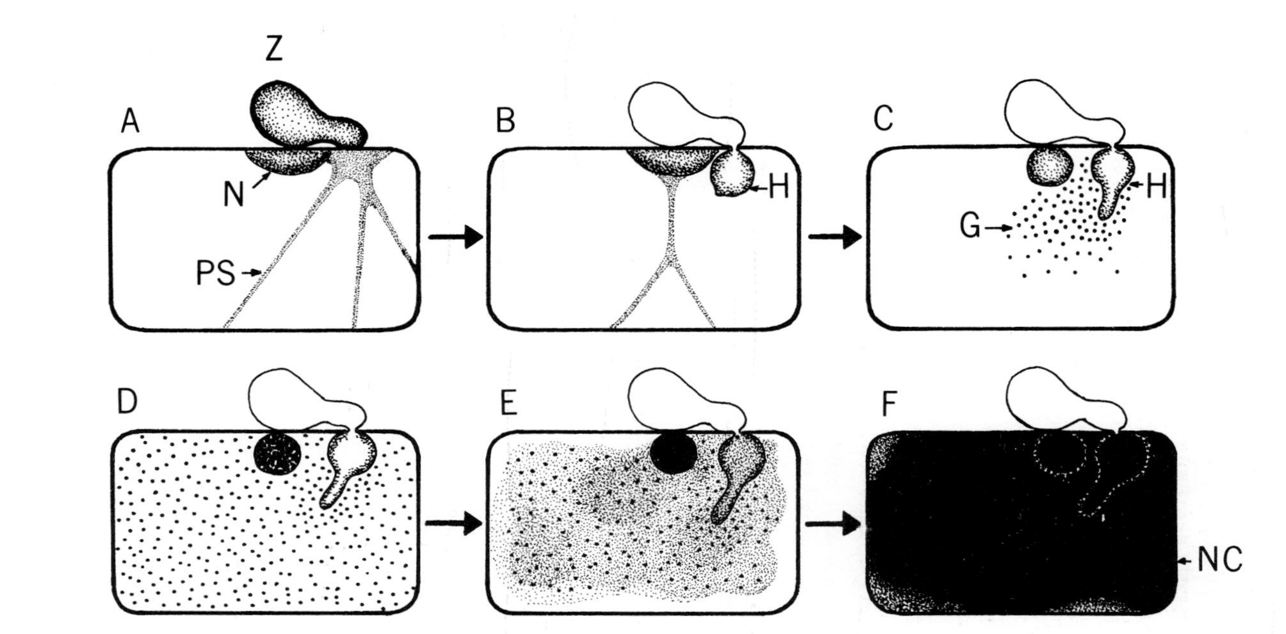

FIGURE 5-7 Stages in the development of necrotic defense reaction in a cell of a very resistant potato variety infected by *Phytophthora infestans*. N = nucleus; PS = protoplasmic strands; Z = zoospore; H = hypha; G = granular material; NC = necrotic cell. After K. Tomiyama, *Ann. Phytopathol. Soc. Jpn.* **21,** 54–62 (1956).

hyphal tips of fungi penetrating a cell wall and growing into the cell lumen are enveloped by cellulosic (callose) materials that later become infused with phenolic substances and form a **sheath or "lignituber"** around the hypha (Figure 5-6).

Cytoplasmic Defense Reaction

In a few cases of slowly growing, weakly pathogenic fungi that induce chronic diseases or nearly symbiotic conditions, the cytoplasm surrounds the clump of hyphae, and the nucleus is stretched to the point where it breaks in two. In some cells, the cytoplasmic reaction is overcome and the protoplast disappears while fungal growth increases. In some of the invaded cells, however, the cytoplasm and nucleus enlarge. The cytoplasm becomes granular and dense, and various particles or structures appear in it. Finally, the mycelium of the pathogen disintegrates and the invasion stops.

Necrotic Defense Reaction: Defense through Hypersensitivity

In many host-pathogen combinations, the pathogen may penetrate the cell wall, but as soon as it establishes contact with the protoplast of the cell, the nucleus moves toward the invading pathogen and soon disintegrates, and brown, resinlike granules form in the cytoplasm, first around the pathogen and then throughout the cytoplasm. As the browning discoloration of the cytoplasm of the plant cell continues and death sets in, the invading hypha begins to degenerate (Figure 5-7). In most cases the hypha does not grow out of such cells and further invasion is stopped. In bacterial infections of leaves, the hypersensitive reaction results in destruction of all cellular membranes of cells in contact with bacteria, and that is followed by desiccation and necrosis of the leaf tissues invaded by the bacteria.

The necrotic or hypersensitive type of defense is quite common, particularly in diseases caused by obligate fungal parasites and by viruses and nematodes. Apparently, the necrotic tissue isolates the obligate parasite from the living substance on which it depends absolutely for its nutrition and, therefore, results in its starvation and death. The faster the host cell dies after invasion, the more resistant to infection the plant seems to be.

SELECTED REFERENCES

Agrios, G. N. (1980). Escape from disease. *In* "Plant Disease" (J. G. Horsfall and E. B. Cowling, eds.), Vol. 5, pp. 17–37. Academic Press, New York.

Aist, J. R. (1976). Papillae and related wound plugs of plant cells. *Annu. Rev. Phytopathol.* **14,** 145–163.

Aist, J. R. (1983). Structural responses as resistance mechanisms. *In* "The Dynamics of Host Defense" (J. A. Bailey and B. J. Deverall, eds.), pp. 33–68. Academic Press, Orlando, Florida.

Akai, S. (1959). Histology of defense in plants. *In* "Plant Pathology" (J. G. Horsfall and A. E. Dimond, eds.), Vol. 1, pp. 391–434. Academic Press, New York.

Akai S., and Fukutomi, M. (1980). Preformed internal physical defenses. *In* "Plant Disease" (J. G. Horsfall and E. B. Cowling, eds.), Vol. 5, pp. 139–159. Academic Press, New York.

Beckman, C. H. (1980). Defenses triggered by the invader: Physical defenses. *In* "Plant Disease" (J. G. Horsfall and E. B. Cowling, eds.), Vol. 5, pp. 225–245. Academic Press, New York.

Beckman, C. H. (1987). "The Nature of Wilt Diseases of Plants." APS Press, St. Paul, Minnesota.

Beckman, C. H., and Talboys, P. W. (1981). Anatomy of resistance. *In* "Fungal Wilt Diseases of Plants" (M. E. Mace, A. A. Bell, and C. H. Beckman, eds.), pp. 487–521. Academic Press, New York.

Bell, A. A. (1980). The time sequence of defense. *In* "Plant Disease" (J. G. Horsfall and E. B. Cowling, eds.), Vol. 5, pp. 53–73. Academic Press, New York.
Campbell, C. L., Huang, J.-S., and Payne, G. A. (1980). Defense at the perimeter: The outer walls and the gates. *In* "Plant Disease" (J. G. Horsfall and E. B. Cowling, eds.), Vol. 5, pp. 103–120. Academic Press, New York.
Cunningham, H. S. (1928). A study of the histologic changes induced in leaves by certain leaf-spotting fungi. *Phytopathology* **18,** 717–751.
Hart, H. (1929). Relation of stomatal behaviour to stem-rust resistance in wheat. *J. Agric. Res. Washington, D.C.)* **39,** 929–948.
Hart, H. (1931). Morphologic and physiologic studies on stem-rust resistance in cereals. *Minn., Agric. Exp. Stn., Tech. Bull.* **266,** 1–76.
Hesler, L. R. (1916). Black rot, leaf spot, and canker of pomaceous fruits. —*N. Y., Agric. Exp. Stn. (Ithaca) Bull.* **379,** 53–148.
Kiraly, Z. (1980). Defenses triggered by the invader: Hypersensitivity. *In* "Plant Disease" (J. G. Horsfall and E. B. Cowling, eds.), Vol. 5, pp. 201–224. Academic Press, New York.
Martin, J. T. (1964). Role of cuticle in the defense against plant disease. *Annu. Rev. Phytopathol.* **2,** 81–100.
Samuel, G. (1927). On the shot-hole disease caused by *Cladosporium carpophilum* and on the "shothole" effect. *Ann. Bot. (London)* **41,** 375–404.
Sherwood, R. T., and Vance, C. P. (1980). Resistance to fungal penetration in Gramineae. *Phytopathology* **70,** 273–279.
Stockwell, V., and Hanchey, P. (1983). The role of the cuticle in resistance of beans to *Rhizoctonia solani. Phytopathology* **73,** 1640–1642.
Weimer, J. L., and Harter, L. L. (1921). Wound-cork formation in the sweet potato. *J. Agric. Res. (Washington, D.C.)* **21,** 637–647.

Metabolic (Biochemical) Defense

Although structural characteristics may provide a plant with various degrees of defense against attacking pathogens, it is becoming increasingly clear that the resistance of a plant against pathogen attacks depends not so much on its structural barriers as on the substances produced in its cells before or after infection. This is apparent from the fact that a particular pathogen will not infect certain plant varieties even though no structural barriers of any kind seem to be present or to form in these varieties. Similarly, in resistant varieties, the rate of disease development soon slows down, and finally, in the absence of structural defenses, the disease is completely checked. Moreover, many pathogens that enter nonhost plants naturally, or that are introduced into nonhost plants artificially, fail to cause infection although no apparent visible host structures inhibit them from doing so. These examples suggest that defense mechanisms of a chemical rather than a structural nature are responsible for the resistance to infection exhibited by plants against certain pathogens.

Preexisting Biochemical Defense

Inhibitors Released by the Plant in Its Environment

Plants exude a variety of substances through the surface of their aboveground parts as well as through the surface of their roots. Some of the compounds released by certain kinds of plants, however, seem to have an inhibitory action against certain pathogens. **Fungitoxic exudates** on the leaves of some plants, for example, tomato and sugar beet, seem to be present in sufficient concentrations to inhibit the germination of spores of the fungi *Botrytis* and

Cercospora, respectively, that may be present in dew or rain droplets on these leaves. Similarly, in the case of onion smudge, caused by the fungus *Colletotrichum circinans,* resistant varieties generally have red scales and contain, in addition to the red pigments, the phenolic compounds protocatechuic acid and catechol. In the presence of water drops or soil moisture containing conidia of the onion smudge fungus on the surface of red onions, these two fungitoxic substances diffuse into the liquid, inhibit the germination of the conidia, and cause them to burst, thus protecting the plant from infection. Both the fungitoxic exudates and the inhibition of infection are missing in white-scaled, susceptible onion varieties.

OH
HO
COOH

Protocatechuic acid

Defense through Lack of Essential Factors

Lack of Recognition between Host and Pathogen Plants of a species or variety may not become infected by a pathogen if their surface cells lack specific **recognition factors** (specific molecules or structures) that can be recognized by the pathogen. If the pathogen does not recognize the plant as one of its host plants, it may not become attached to the plant or it may not produce infection substances, such as enzymes, or structures, such as appressoria, infection pegs, and haustoria, necessary for the establishment of infection. It is not known what types of molecules or structures are involved in the recognition of plants and pathogens, but it is thought that they probably include various types of oligosaccharides and polysaccharides and proteins or glycoproteins (lectins). Also, it is not known to what extent these recognition phenomena are responsible for the success or failure of initiation of infection in any particular host–pathogen combination.

In incompatible host–virus combinations, in which the virus is introduced into wounded cells or into protoplasts upon inoculation, infection may not result because of lack of recognition of the viral nucleic acid by the host ribosomes. The ribosomes, failing to recognize a specific initiation site on the viral nucleic acid, do not produce the enzyme (replicase) necessary to synthesize viral nucleic acid, and therefore, no disease develops.

Lack of Host Receptors and Sensitive Sites for Toxins In host–pathogen combinations in which the pathogen (usually a fungus) produces a host-specific toxin, the toxin, which is responsible for the symptoms, is thought to attach to and react with specific receptors or sensitive sites in the cell. Only plants that have such sensitive receptors or sites become diseased. Plants of other varieties or species that lack such receptors or sites remain resistant to the toxin and develop no symptoms.

Lack of Essential Nutrients for the Pathogen Species or varieties of plants that for some reason do not produce one of the substances essential for the survival of an obligate parasite or for development of infection by any parasite would be resistant to the pathogen that requires it. Thus, for *Rhizoctonia* to infect a plant, the plant must have a substance necessary for formation of a **hyphal cushion** from which the fungus sends into the plant its penetration hyphae. In plants in which this substance is apparently lacking, cushions do not form, infection does not occur, and the plants are resistant. The fungus does not normally form hyphal cushions in pure cultures but forms them when extracts from a susceptible but not a resistant plant are added to the culture. Also, certain mutants of *Venturia inaequalis,* the cause of apple scab, which had lost the ability to synthesize a certain **growth factor,** also lost the ability to cause infection. When, however, the particular growth factor is sprayed on the apple leaves during inoculation with the mutant, the mutant not only survives but it also causes infection. The advance of the infection, though, continues only as long as the growth factor is supplied externally to the mutant. In some host–pathogen combinations, disease develops but the amount of disease may be reduced by the fact that certain host substances are present in lower concentrations. For example, bacterial soft rot of potatoes, caused by *Erwinia carotovora* var. *atroseptica,* is less severe on potatoes with low reducing sugar content than in potatoes high in reducing sugars.

Inhibitors Present in Plant Cells before Infection

It is uncertain whether or not any plant is resistant to a disease only because of an inhibitory compound present in the cell before infection. Several **phenolic compounds** and **tannins,** present in high concentrations in cells of young fruit or leaves, have been proposed as responsible for the resistance of young tissues to pathogenic microorganisms such as *Botrytis.* Many such compounds are potent inhibitors of many hydrolytic enzymes, including the pectolytic, macerating enzymes of plant pathogens. As the young tissues grow older, their inhibitor content and their resistance to infection decrease steadily. Several other types of preformed compounds, for example, the saponins **tomatine** in tomato and **avenacin** in oats, have antifungal activity, but their contribution to the resistance of their hosts to fungal pathogens is not known.

In addition to the simple molecule antifungal compounds listed above, several proteins have been reported to act as inhibitors of pathogen hydrolytic enzymes involved in host cell wall degradation. On the other hand, plant cells also contain hydrolytic enzymes, some of which, such as **glucanases** and **chitinases,** may cause breakdown of pathogen cell wall components and thereby contribute to resistance to infection. For example, in vascular wilt-infected plants, hyphae are sometimes observed to break down. The importance of either of these types of inhibitors to disease resistance, however, is not currently known.

Metabolic Defense Induced by the Attacking Pathogen

Biochemical Inhibitors Produced in Plants in Response to Injury by the Pathogen

Plant cells and tissues respond to injury, whether caused by a pathogen or a mechanical or chemical agent, through a series of biochemical reactions that seem to be aimed at isolating the irritant and at healing the wound. This reaction is often associated with the production of fungitoxic substances around the site of injury as well as the formation of layers of protective tissues such as callus and cork. Some of the compounds thus produced are present in concentrations high enough to inhibit the growth of most fungi and bacteria that cannot infect that host. These compounds include mostly phenolic compounds such as chlorogenic and caffeic acids, oxidation products of phenolic compounds, and also the phytoalexins, most of which are also phenolic compounds.

Defense through the Hypersensitive Reaction

The hypersensitive reaction is one of the most important defense mechanisms in plants. It occurs only in incompatible combinations of host plants and fungi, bacteria, viruses, and nematodes. In such combinations, no difference is observable in the manner of penetration of the epidermis in susceptible and resistant plants. After infection, however, infected cells in resistant varieties rapidly lose turgor, turn brown, and die, while infected cells of susceptible varieties survive considerably longer. In resistant varieties, a number of physiological changes occur in the infected cells and in the cells surrounding them, while in susceptible varieties such changes either do not occur or they occur at a much slower rate. Such changes in hypersensitive reactions include loss of permeability of cell membranes, increased respiration, accumulation and oxidation of phenolic compounds, and production of phytoalexins. The end result of the intense mobilization of internal reactions in these cells is always death and collapse of the infected and a few surrounding cells. Fungal and bacterial pathogens within the area of operation of the hypersensitive reaction are isolated by necrotic tissue and quickly die. In virus diseases, the hypersensitive reaction always results in formation of the so-called local lesions in which the virus may survive for considerable time but is, generally, found in low concentrations, and its spread beyond the lesion is as a rule checked.

Defense through Increased Levels of Phenolic Compounds

Some of the phenolics implicated in disease resistance occur widely in plants and are found in healthy as well as diseased plants, but their synthesis or accumulation seems to be accelerated after infection. Such compounds may be called "common" phenolic compounds. Certain other phenolics, however, are not present in healthy plants but are produced upon stimulation by a pathogen or by a mechanical or chemical injury. Such compounds are known as **phytoalexins.**

"Common" Phenolics It has often been observed that certain "common" phenolic compounds that are toxic to pathogens are produced and accumulate at a faster rate after infection in a resistant variety of plant than in a susceptible variety. Chlorogenic acid, caffeic acid, and scopoletin are examples of such phenolic compounds (Figure 5-8). Although some of the common phenolics may each reach concentrations that could be toxic to the pathogen, it should be noted that several of them appear concurrently in the same diseased tissue, and it is possible that the combined toxic effect of all

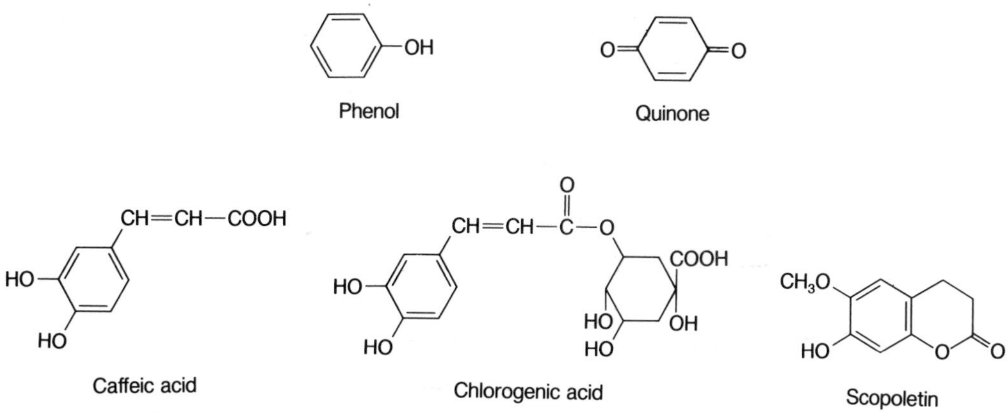

Phenol

Quinone

Caffeic acid

Chlorogenic acid

Scopoletin

FIGURE 5-8 Structures of phenol and guinone and of three common plant phenolics that are toxic to pathogens.

fungitoxic phenolics present, rather than that of each one separately, is responsible for the inhibition of infection in resistant varieties.

Phytoalexins Phytoalexins are toxic substances produced in appreciable amounts in plants only after stimulation by the various types of phytopathogenic microorganisms or by chemical and mechanical injury. Phytoalexins are produced by healthy cells adjacent to localized damaged and necrotic cells in response to materials diffusing from the damaged cells. Phytoalexins are not produced during biotrophic infections. Phytoalexins accumulate around both resistant and susceptible necrotic tissues. Resistance occurs when one or more phytoalexins reach a concentration sufficient to restrict pathogen development. Most known phytoalexins are toxic to and inhibit the growth of fungi pathogenic to plants, but some are also toxic to bacteria, nematodes, and other organisms. Several dozen chemicals with phytoalexinlike properties have been isolated from plants belonging to more than 20 families. Most of these phytoalexins are produced in plants in response to infection by fungi, but a few bacteria, viruses, and nematodes have also been shown to induce production of phytoalexins. Some of the better studied phytoalexins include phaseollin and kievitone in bean, pisatin in pea, glyceollin in soybean, alfalfa, and clover, rishitin in potato, gossypol in cotton, and capsidiol in pepper.

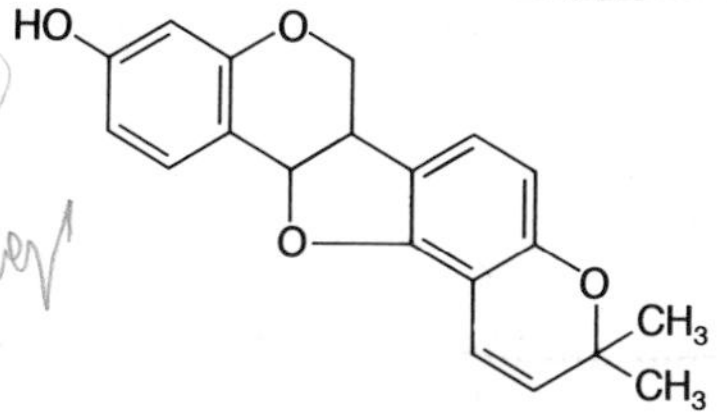

Phaseolin

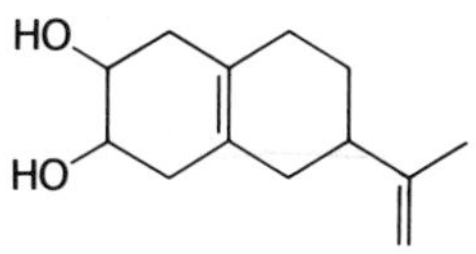

Rishitin

Phytoalexin production is stimulated in a host by the presence of certain pathogen substances called **"elicitors."** Most phytoalexin elicitors are generally high-molecular-weight substances that are constituents of the fungal cell wall, such as glucans, chitosan, glycoproteins, and polysaccharides. Most such elicitors are nonspecific, that is, they are present in both compatible and incompatible races of the pathogen and induce phytoalexin accumulation irrespective of the plant cultivar. A few phytoalexin elicitors, however, are specific, since the accumulation of phytoalexin they cause on certain compatible and incompatible cultivars parallels the phytoalexin accumulation caused by the pathogen races themselves. Although most phytoalexin elicitors are thought to be of pathogen origin, some elicitors are produced by plant cells in response to infection, or are released from plant cell walls after their partial breakdown by cell-wall-degrading enzymes of the pathogen.

Formation of phytoalexins in susceptible (compatible) host–pathogen combinations seems, in some cases, to be prevented by **suppressors** produced by the pathogen. The suppressors seem to also be glucans, glycoproteins, or one of the toxins produced by the pathogen.

The mechanisms by which phytoalexin elicitors, phytoalexin production, phytoalexin suppressors, genes for resistance or susceptibility, and the expression of resistance or susceptibility are connected are still not well understood. Several hypotheses have been proposed to explain the interconnection of these factors, but much more work is needed before a satisfactory explanation can be obtained.

Species or races of fungi pathogenic to a particular plant species seem to stimulate production of generally lower concentrations of phytoalexins than nonpathogens. For example, in the case of pisatin production by pea pods inoculated with the pathogen *Ascochyta pisi,* different varieties of pea produce different concentrations of pisatin, which approximately parallel the resistance of the variety to the pathogen. When the same pea variety is inoculated with different strains of the fungus, the concentration of pisatin produced varies with the fungus strain used for inoculation, and it is approximately inversely proportional to the virulence of each particular strain on the pea variety. Also, in soybean plants infected with the fungus *Phytophthora megasperma* f. sp. *glycinea* combinations of incompatible fungal races and host cultivars resulted in earlier accumulations and higher concentrations of the phytoalexin glyceollin than combinations of compatible races and cultivars. It has been suggested that the higher concentrations of glyceollin in such incompatible host–pathogen combinations is the result of reduced biodegradation rather than increased biosynthesis of the phytoalexin. In some cases, however, for example in the bean—*Colletotrichum lindemuthianum*—and the potato—*Phytophthora infestans*—systems, phytoalexins such as phaseolin and rishitin, respectively, reach equal or higher concentrations in compatible (susceptible) host–pathogen combinations than in incompatible (resistant) ones.

On the other hand, pathogenic races or species of fungi seem to be less sensitive to the toxicity of the phytoalexin(s) produced by their host plant than are nonpathogenic fungi. This has been shown to be true in numerous host–pathogen systems, including the reaction to pisatin in the pea—*Ascochyta pisi*—system. Also, high levels of pisatin and phaseolin accumulated in susceptible combinations of pea and *Fusarium solani* f. sp. *pisi* and of bean and *F. solani* f. sp. *phaseoli,* respectively, without increasing host resistance, apparently because each pathogen could tolerate its host phytoalexin. It has been suggested that these and possibly other pathogens may have an adoptive tolerance mechanism that enables them to tolerate higher concen-

cyanide to the nontoxic formamide ($HCONH_2$). It has been shown that, generally, the pathogenicity of fungal pathogens of cyanogenic plants is proportional to the ability of these pathogens to produce formamide hydro-lyase and to detoxify the cyanide released by the infected tissues. At the same time, the presence of cyanide in cyanogenic plants may play an important role in the defense of these plants against potential pathogens that lack the enzyme(s) needed for detoxification of cyanide.

Defense through Detoxification of Pathogen Toxins

In at least some of the diseases in which the pathogen produces a toxin, resistance to disease is apparently the same as resistance to toxin. However, no satisfactory explanation of the resistance to toxin is yet available.

Detoxification of at least some toxins, for example, fusaric acid and pyricularin, is known to occur in plants, and it may play a role in disease resistance. Some of these toxins appear to be metabolized more rapidly by resistant varieties or are combined with other substances and form less toxic or nontoxic compounds. The amount of the nontoxic compound formed is often proportional to the disease resistance of the variety.

Resistant plants and nonhosts are not affected by the specific toxins produced by *Helminthosporium, Periconia,* and *Alternaria,* but it is not yet known whether the selective action of these toxins depends on the presence of receptor sites in susceptible but not in resistant varieties, the detoxification of the toxins in resistant plants, or some other mechanism.

Defense through Induced Resistance

In plants, **induced resistance** is resistance that develops after preinoculation of plants with various biotic agents or after pretreatment with various chemical or physical agents. Resistance has been induced in a wide variety of plants against fungi, bacteria, viruses, and even insects. Induced resistance is nonspecific since, regardless of the type of agent or pathogen used as an inducer, the level of resistance in the plant is increased against a variety of pathogens. For example, infection of hypersensitive tobacco with tobacco mosaic virus (TMV) induces a systemic resistance to itself, to several other viruses, to fungi *(Phytophthora),* to bacteria *(Pseudomonas tabaci),* and to aphids. On the other hand, infection of tobacco with a root lesion-forming fungus *(Chalara = Thielaviopsis)* or a leaf lesion-causing bacterium *(Pseudomonas syringae)* induces systemic resistance to TMV. Resistance to a pathogen can also be induced by previous inoculation of a plant with an incompatible race of the pathogen, with heat-killed fungal spores or bacteria, and by inoculating the plant with the pathogen while the plant is still young and not yet susceptible to the pathogen.

Induced resistance is at first observed in and restricted to the area surrounding the initial infection sites **(local induced resistance)**, but a few days later it can be detected in uninoculated portions of inoculated leaves and in uninoculated leaves **(systemic induced resistance)**. Induced resistance is expressed either by smaller lesions or by complete resistance to a second infection by the same or a different pathogen. For induced resistance to be expressed, there must be a lag period between the induction treatment (first inoculation) and the challenge inoculation. That time is presumably required for the synthesis and systemic movement of a substance, or substances, from inoculated to uninoculated leaves or other parts of the plant.

In addition to developing as the result of a primary infection, resistance can also be induced by treatment of plants with such natural compounds as viral coat proteins, bacterial or fungal proteins, lipoproteins, polysaccharides, and yeast RNA and by synthetic molecules, mostly polyanions such as polyacrylic acid, salicylic acid (aspirin), and 2-chloroethylphosphonic acid. These substances act as inducers of local resistance in the plant when they are applied by injection or spraying and of systemic resistance when they are absorbed through petioles or roots. In general, the greater the concentration of the inducer, the faster and more effective the induced resistance in the host.

Localized induced resistance in inoculated (treated) leaves usually develops 2–3 days after the primary infection (treatment), while systemic induced resistance, if present, usually develops 7 or more days after the primary infection and may last from 3 to 5 weeks. Induced resistance may be prevented from developing if the plant is also treated with actinomycin D, which prevents transcription of cell DNA to messenger RNA and therefore prevents production of new proteins (enzymes). This suggests that for induced resistance to develop, the host cells must be able to carry out transcription and production of new enzymes, that is, the inducers of resistance activate the part of the plant genome that is responsible for the defense mechanisms of the plant.

Development of induced resistance seems to depend on, or to follow, development of the hypersensitive reaction. Moreover, both the hypersensitive reaction and the induced resistance in plants are accompanied by the production in the resistant tissues of one or more kinds of new soluble host proteins called **b-proteins.** Furthermore, the production of b-proteins, the hypersensitive reaction, and induced resistance are all inhibited by treatment of the plant with actinomycin D and are all suppressed or inhibited at high temperatures (30–35°C). So far, however, neither the role, if any, nor the mechanism by which b-proteins and hypersensitivity are associated with induced resistance is known. An increase in peroxidase, in phenylalanine ammonia lyase activity, in phytoalexins, in lignification, and in proteinase-inhibiting polysaccharides has also been observed in some hosts that develop induced resistance and has been proposed as the mechanism by which induced resistance is brought about. It is thought that plant cell walls may contain bound elicitors of phytoalexins and of inducers of resistance. When the elicitors are released by the affected cells after preinoculation (or pretreatment with chemicals), some of them remain active for a short time and act locally, where they elicit production of effective concentrations of phytoalexins. Other elicitors, however, either are translocated systemically through the plant and induce the appearance of substances like proteinase inhibitors, peroxidase, and b-proteins, which are responsible for induced resistance, or they may initiate a cascading series of reactions that spread throughout the plant and elicit the production of the substances that induce resistance.

SELECTED REFERENCES

Akai, S., and Ouchi, S., eds. (1971). "Morphological and Biochemical Events in Plant-Parasite Interaction." Phytopathol. Soc. Jpn., Tokyo.

Asada, Y., Bushnell, W. R., Ouchi, S., and Vance, C. P., eds. (1982). "Plant Infection: The Physiological and Biochemical Basis." Springer-Verlag, Berlin and New York.

Bailey, J. A., and Deverall, B. J., eds. (1983). "The Dynamics of Host Defense." Academic Press, New York.

Bailey, J. A., and Mansfield, J. W., eds. (1982). "Phytoalexins." Wiley, New York.

Barmore, C. R., and Nguyen, T. K. (1985). Polygalacturonase inhibition in rind of Valencia orange infected with *Diplodia natalensis. Phytopathology* **75,** 446–449.

Bateman, D. F. (1967). Alteration of cell wall components during pathogenesis by *Rhizoctonia solani. In* "The Dynamic Role of Molecular Constituents in Plant-Parasite Interaction" (C. J. Mirocha and I. Uritani, eds.), pp. 58–75. Bruce, St. Paul, Minnesota.

Bell, A. A. (1981). Biochemical mechanisms of disease resistance. *Annu. Rev. Plant Physiol.* **32,** 21–81.

Calow, J. A., ed. (1984). "Biochemical Plant Pathology." Wiley, New Jersey.

Deverall, B. J. (1977). "Defense Mechanisms of Plants." Cambridge Univ. Press, London and New York.

Dickinson, C. H., and Preece, T. F., eds. (1976). "Microbiology of Aerial Plant Surfaces." Academic Press, New York.

Farkas, G. L., and Kiraly, Z. (1962). Role of phenolic compounds in the physiology of plant disease and disease resistance. *Phytopathol. Z.* **44,** 105–150.

Gianinazzi, S. (1984). Genetic and molecular aspects of resistance induced by infections or chemicals. *In* "Plant-Microbe Interactions: Molecular and Genetic Perspectives" (T. Kosuge and E. W. Nester, eds.), Vol. 1, pp. 321–342. Macmillan, New York.

Goodman, R. N., Kiraly, Z., and Wood, K. R. (1987). "The Biochemistry and Physiology of Plant Disease." Univ. of Missouri Press, Columbia.

Heitefuss, R., and Williams, P. H., eds. (1976). "Physiological Plant Pathology," Encycl. Plant Physiol., New Series, Vol. 4. Springer-Verlag, Berlin and New York.

Horsfall, J. G., and Cowling, E. B., eds. (1980). "Plant Disease," Vol. 5. Academic Press, New York.

Kiraly, Z. (1980). Defenses triggered by the invader: Hypersensitivity. *In* "Plant Disease" (J. G. Horsfall and E. B. Cowling, eds.), Vol. 5, pp. 201–224. Academic Press, New York.

Misaghi, I. J. (1982). "Physiology and Biochemistry of Plant-Pathogen Interactions." Plenum, New York.

Muller, K. O. (1959). Hypersensitivity. *In* "Plant Pathology" (J. G. Horsfall and A. E. Dimond, eds.), Vol. 1, pp. 469–519. Academic Press, New York.

Otazu, V., and Secor, G. A. (1981). Soft rot susceptibility of potatoes with high reducing sugar content. *Phytopathology* **71,** 290–295.

Ouchi, S. (1983). Induction of resistance or susceptibility. *Annu. Rev. Phytopathol.* **21,** 289–315.

Patil, S. S. (1980). Defenses triggered by the invader: Detoxifying the toxins. *In* "Plant Disease" (J. G. Horsfall and E. B. Cowling, eds.), Vol. 5, pp. 269–278. Academic Press, New York.

Ryan, C. A. (1984). Systemic responses to wounding. *In* "Plant-Microbe Interactions: Molecular and Genetic Perspectives" (T. Kosuge and E. W. Nester, eds.), Vol. 1, pp. 307–320. Macmillan, New York.

Schlosser, E. W. (1980). Preformed internal chemical defenses. *In* "Plant Disease" (J. G. Horsfall and E. B. Cowling, eds.), Vol. 5, pp. 161–177. Academic Press, New York.

Schoeneweiss, D. F. (1975). Predisposition, stress and plant disease. *Annu. Rev. Phytopathol.* **13,** 193–211.

Sequeira, L. (1980). Defenses triggered by the invader: Recognition and compatibility phenomena. *In* "Plant Disease" (J. G. Horsfall and E. B. Cowling, eds.), Vol. 5, pp. 179–200. Academic Press, New York.

Sequeira, L. (1983). Mechanisms of induced resistance in plants. *Annu. Rev. Microbiol.* **37,** 51–79.

Uritani, I. (1971). Protein changes in diseased plants. *Annu. Rev. Phytopathol.* **9,** 211–234.

Vance, C. P., Kirk, T. K., and Sherwood, R. T. (1980). Lignification as a mechanism of disease resistance. *Annu. Rev. Phytopathol.* **18,** 259–288.

Weinhold A. R., and Hancock, J. G. (1980). Defense at the perimeter: Extruded chemicals. *In* "Plant Disease" (J. G. Horsfall and E. B. Cowling, eds.), Vol. 5, pp. 121–138. Academic Press, New York.

Wheeler, H. (1975). "Plant Pathogenesis." Springer-Verlag, Berlin and New York.

6 GENETICS OF PLANT DISEASE

Introduction

The genetic information of all organisms, that is, the information that determines what an organism can be and can do, is encoded in its DNA (deoxyribose nucleic acid). In RNA viruses, of course, it is encoded in their RNA (ribose nucleic acid). In all organisms, most DNA is present in the chromosome(s). In prokaryotes, that is, organisms that lack an organized, membrane-bound nucleus, for example, bacteria and mycoplasmas, there is only one chromosome and it is present in the cytoplasm, while in the eukaryotes, that is, all other organisms, there are several chromosomes, and they are present in the nucleus. Many if not all prokaryotes, however, and at least some of the lower eukaryotes also carry smaller circular molecules of DNA, called **plasmids,** in the cytoplasm. Plasmid DNA also carries genetic information but multiplies and moves independently of the chromosomal DNA. Furthermore, all cells of eukaryotic organisms carry DNA in their mitochondria. Plant cells, in addition to nuclear and mitochondrial DNA, also carry DNA in their chloroplasts.

Genetic information in DNA is encoded in a linear fashion, in the order of the four bases (A = ademine, C = cytosine, G = guanine, and T = thymine). Each triplet of adjacent bases codes for a particular amino acid. A **gene** is a stretch of a DNA molecule, usually of about one hundred to five hundred or more adjacent triplets, that codes for one protein molecule or, in a few cases, one RNA molecule. Actually, in eukaryotes, the coding region of a gene is often interrupted by noncoding stretches of DNA called **introns.** When a gene is active, that is, it is expressed, one of its DNA strands is copied (transcribed) into RNA. If the gene codes for an RNA and the RNA is a transfer RNA (tRNA) or ribosomal RNA (rRNA), expression of the gene is more or less completed upon transcription. However, if the gene, like the vast majority of them, codes for a protein molecule, then the transcription product is a messenger RNA (mRNA). The mRNA then becomes attached to ribosomes, which with the help of tRNAs, translate the base sequence of the mRNA into a specific sequence of amino acids and eventually construct a particular protein. Each gene codes for a different protein. It is the proteins, which are either part of the structure of cell membranes or act as enzymes, that give cells and organisms their characteristic properties such as shape, size, and color and determine what kinds of chemical substances are produced by the cell.

Of course, not all genes in a cell are expressed at all times, since different kinds of cells and at different times have different functions and needs. Which genes are turned on, when they are turned on or off, and for how long they stay "on" is regulated by additional stretches of DNA called **promoters** or **terminators** that act as signals for genes to be expressed or to stop being expressed, respectively; or act as signals for production of RNAs and proteins that themselves act as inducers, promoters, and enhancers of gene expression or as repressors and terminators of gene expression. In many cases of host–pathogen interaction, genes in the one component are triggered to be expressed by a substance produced by the other component. For example, genes for cell-wall-degrading enzymes in the pathogen are apparently induced by the presence in the host of cell wall macromolecules that are substrates for these enzymes. Also, genes for defense reactions in the host, for example, production of phytoalexins, apparently are triggered to expression by certain inducer molecules (elicitors) produced by the pathogen.

Genes and Disease

When different plants, such as tomato, apple, or wheat, become diseased as a result of infection by a pathogen, the pathogen is generally different for each kind of host plant, and it is often specific for that particular host plant. Thus, the fungus *Fusarium oxysporum* f. sp. *lycopersici* that causes tomato wilt attacks only tomato and has absolutely no effect on apple, wheat, or any other plant. Similarly, the fungus *Venturia inaequalis* that causes apple scab affects only apple, while the fungus *Puccinia graminis* f. sp. *tritici* that causes stem rust of wheat attacks only wheat. What makes possible the development of disease in a host is the presence in the pathogen of one or more genes for specificity and for virulence against the particular host, which in turn is thought to have certain genes for specificity and for susceptibility to the particular pathogen.

The gene or genes for virulence in the pathogen are usually specific for one or a few related kinds of host plants. Also, the genes that make a host plant susceptible to a particular pathogen are present only in that one host and possibly a few related kinds of host plants. It is, therefore, the concurrent occurrence and interaction of specific genes for virulence in the pathogen and of specific genes for susceptibility in the host that determine the initiation and development of disease. A pathogen, then, has a set of genes for virulence that is specifically found in that pathogen and is aimed specifically against its own host or hosts. On the other hand, each host has a set of genes for susceptibility to a particular pathogen, such genes are found only in that particular host plant and apparently accommodate only the specific pathogen. The specificity of genes for virulence and of genes for susceptibility explains why a pathogen that is virulent on one host is not virulent on all other kinds of host plants and why a host plant that is susceptible to one pathogen is not susceptible to all other pathogens of other host plants.

Of course, a few pathogens are able to attack many kinds, sometimes hundreds, of host plants. Such pathogens can attack so many hosts apparently because they either have many diverse genes for virulence or because their genes of virulence somehow have a much wider spectrum of host specificity than those of the commonly more specialized pathogens. Each species of plant, however, seems to be susceptible to a fairly small number of different pathogens, usually less than a hundred for most plants, although some plants seem to be attacked by close to two hundred different pathogens. This means that a single plant species has genes for susceptibility that would allow it to become infected by any of its one to two hundred pathogens.

In spite of the genes for susceptibility that plants have to so many pathogens, countless numbers of individuals of a single plant species, such as corn, wheat, or soybean, survive in huge land expanses year after year. These plants survive either free of disease or with only minor symptoms, even though most of their pathogens are often widespread among the plants. Why aren't all the plants attacked by their pathogens, and why are those that are attacked not usually killed by the pathogens? The answer is complex, but basically it happens because plants, through evolution or through systematic breeding, have acquired in addition to their genes for susceptibility one or more genes for resistance, which protect them from infection or from severe disease. When a new gene for resistance to a pathogen appears or is introduced into a plant, the plant becomes resistant to all the previously existing individuals of the pathogen. Such pathogens contain one and usually more than one gene for virulence, but if they do not contain the additional new gene for virulence that is required to overcome the effect of the new resistance gene in the plant, they cannot infect the plant and the plant remains resistant. Thus, even one new gene for resistance to a pathogen can protect plants that have the gene from the disease caused by many races of the pathogen—at least for several months and possibly for a few to several years.

It has been the experience of researchers with numerous host–pathogen combinations, however, that after a new gene for resistance to a pathogen is introduced into a crop variety and that variety is planted in the fields, a new population of the pathogen appears that contains a new gene for virulence that enables the pathogen to attack the crop plants containing the new gene for resistance. How did this new population of pathogens acquire the new gene for virulence? In most cases the new gene had already been present earlier but only in a few pathogen individuals. Such pathogen individuals may have been but a tiny proportion of the total pathogen population before plants with the new resistance gene were planted widely. After such plants were introduced, however, the new resistance gene excluded all other pathogen individuals except the few containing the new gene for virulence, which could attack these plants. Exclusion of the pathogens that lacked the new gene allowed the few that carried the gene to multiply and take over. Or the new gene may have appeared suddenly *de novo,* that is, through a mutation or by rearrangement of the genetic material of the pathogens through the ever ongoing events of genetic variability in organisms.

Variability in Organisms

One of the most dynamic and significant aspects of biology is that characteristics of individuals within a species are not "fixed" in their morphology and physiology but vary from one individual to another. As a matter of fact, all individuals produced as a result of a sexual process are expected to be different from each other and from their parents in a number of characteristics, although they retain most similarities with them and belong to the same species. This is true of fungi produced from sexual spores such as oospores, ascospores, and basidiospores, of parasitic higher plants produced from seeds, and of nematodes produced from fertilized eggs, as well as of cultivated plants produced from seeds. When individuals are produced asexually, the frequency and degree of variability among the progeny are reduced greatly but even then certain individuals among the progeny will show different characteristics. This is the case in the overwhelmingly asexual reproduction of fungi by means of conidia, zoospores, sclerotia, and uredospores, and in bacteria, mycoplasmas and viruses, as well as in the asexual propagation of plants by means of buds, cuttings, tubers, and so on.

Mechanisms of Variability

In host plants and in pathogens, such as most fungi, parasitic higher plants, and nematodes, which can, and usually do, reproduce by means of a sexual process, variation in the progeny is introduced primarily through segregation and recombination of genes during the meiotic division of the zygote. Bacteria too, however, and even viruses, exhibit variation that seems to be the result of a sexuallike process. In many fungi, heteroploidy and certain parasexual processes lead to variation. On the other hand, all plants and all pathogens, especially bacteria, viruses, and fungi, and probably mycoplasmas, can and do produce variants in the absence of any sexual process by means of mutations and, perhaps, by means of cytoplasmic adaptation.

General Mechanisms of Variability

Two mechanisms of variability—mutation and recombination—occur in both plants and pathogens.

Mutation

A **mutation** is a more or less abrupt change in the genetic material of an organism, which is then transmitted in a hereditary fashion to the progeny. Mutations represent changes in the sequence of bases in the DNA, either through substitution of one base for another, or through addition or deletion of one or many base pairs. Additional changes may be brought about by amplification of particular segments of DNA to multiple copies, by insertion of a movable DNA segment into a coding or regulatory sequence of a gene, and by inversion of a DNA segment.

Mutations occur spontaneously in nature in all living organisms, those that produce only sexually or only asexually and those that reproduce both sexually and asexually. Mutations in single-celled organisms, such as bacteria, in fungi with haploid mycelium, and in viruses are expressed immediately after their occurrence. Most mutant factors, however, are usually recessive; therefore, in diploid or dikaryotic organisms mutations can remain unexpressed until they are brought together in a hybrid.

Mutations for virulence probably occur no more frequently than for any other inherited characteristics, but given the great number of progeny produced by pathogens, it is probable that large numbers of mutants differing in virulence from their parent are produced in nature every year. Besides, considering that only a few genetically homogeneous varieties of each crop plant are planted continuously over enormous land expanses for a number of years, and considering the difficulties involved in shifting from one variety to another on short notice, the threat of new, more virulent, mutants appearing and attacking a previously resistant variety is a real one. Moreover, once a new factor for virulence appears in a mutant, this factor will take part in the sexual or parasexual processes of the pathogen and may produce recombinants possessing virulence quite different in degree or nature from that existing in the parental strains.

Since plants and pathogens contain genetic material (DNA) outside the cell nucleus, mutations in the extranuclear DNA are just as common as in the nuclear DNA and affect whatever characteristics are controlled by the extranuclear DNA. Because the inheritance of characteristics controlled by extranuclear DNA **(cytoplasmic inheritance)** does not follow the Mendelian laws of genetics, however, mutations on that DNA are more difficult to detect and characterize. Through mutations in extrachromosonal DNA, many pathogens acquire the ability to carry out a physiological process that they could not before. Cytoplasmic inheritance presumably occurs in all organisms except viruses and viroids, which lack cytoplasm. Three types of adaptations brought about by changes in the genetic material of the cytoplasm have been shown in pathogens. Pathogens may acquire the ability to tolerate previously toxic substances, to utilize new substances for growth, and to change their virulence toward host plants. Several characteristics of plants are also inherited through the cytoplasm, including the resistance to infection by certain pathogens.

Recombination

Recombination occurs during the sexual reproduction of plants, fungi, and nematodes whenever two haploid (1N) nuclei, containing slightly different genetic material, unite to form a diploid (2N) nucleus, called a **zygote. Recombination** of genetic factors occurs during the meiotic division of the zygote as a result of genetic crossovers in which parts of chromatids (and the genes they carry) of the one chromosome of the pair are exchanged with parts of chromatids of the other chromosome of the pair. In this way a recombination of the genes of the two parental nuclei takes place in the zygote, and the haploid nuclei or gametes resulting after meiosis are different both from gametes that produced the zygote and from each other. In the fungi, the haploid nuclei or gametes often divide mitotically to produce haploid mycelium and spores, which results in genetically different groups of relatively homogeneous individuals that may produce large populations asexually until the next sexual cycle.

Specialized Mechanisms of Variability in Pathogens

Certain mechanisms of variability appear to be operating only in certain kinds of organisms or to be operating in a rather different manner than those described as general mechanisms of variability. These specialized mechanisms of variability are sexuallike or parasexual processes and include heterokaryosis, heteroploidy, and parasexualism in fungi; conjugation, transformation and transduction in bacteria; and genetic recombination in viruses.

Sexuallike Processes in Fungi

Heterokaryosis **Heterokaryosis** is the condition in which, as a result of fertilization or anastomosis, cells of fungus hyphae or parts of hyphae contain two or more nuclei that are genetically different. For example, in the Basidiomycetes, the dikaryotic state may differ drastically from the haploid mycelium and spores of the fungus. Thus in *Puccinia graminis tritici,* the fungus causing stem rust of wheat, the haploid basidiospores can infect barberry but not wheat, and the haploid mycelium can grow only in barberry, while the dikaryotic aeciospores and uredospores can infect wheat but not barberry and the dikaryotic mycelium can grow in both barberry and wheat. Heterokaryosis also occurs in other fungi, but its importance in plant disease development in nature is not known.

Parasexualism **Parasexualism** is the process by which genetic recombinations can occur within fungal heterokaryons. This comes about by the occasional fusion of the two nuclei and the formation of a diploid nucleus. During multiplication, crossing over occurs in a few mitotic divisions and results in the appearance of genetic recombinants by the occasional separation of the diploid nucleus into its haploid components.

Heteroploidy **Heteroploidy** is the existence of cells, tissues, or whole organisms with numbers of chromosomes per nucleus that are different from the normal 1N or 2N for the particular organism. Heteroploids may be haploids, diploids, triploids, or tetraploids, or they may be aneuploids, that is, they have 1, 2, 3, or more extra chromosomes or are missing one or more chromosomes from the normal euploid number. Heteroploidy is often associated with cellular differentiation and represents a normal situation in the development of most eukaryotes. In several studies, spores of the same fungus were found to contain nuclei with chromosome numbers ranging from 2 to 12 per nucleus and also diploids and polyploids. Since it has been shown that the expression of different genes is proportional to, inversely proportional to, or unaffected by dosage, obviously the existence of heteroploid cells or heteroploid whole individuals of some pathogens increases the degree of variability that can be exhibited by these pathogens. Heteroploidy has been repeatedly observed in fungi and has been shown to affect the growth rate, spore size and rate of spore production, hyphal color, enzyme activities, and pathogenicity. It has been shown, for example, that some heteroploids, such as diploids of the normally haploid fungus *Verticillium albo-atrum,* the cause of wilt in cotton, lose the ability to infect cotton plants even when derived from highly virulent haploids. How much of the variability in pathogenicity in nature is due to heteroploidy is still unknown.

Sectoring, that is, the appearance of morphologically distinct sectors in fungus colonies, is a common occurrence when most fungi are cultured on nutrient media. In addition to morphological differences, sectors sometimes also show differences in pathogenicity. Sectoring has yet to be explained

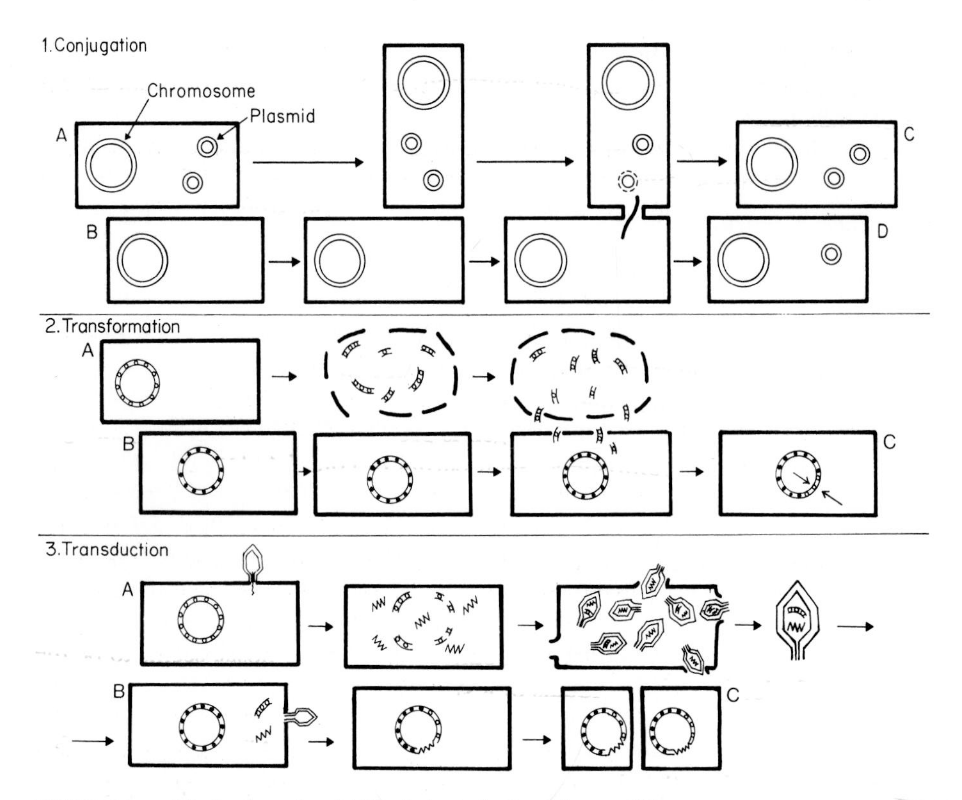

FIGURE 6-1 Mechanism of variability in bacteria through sexuallike processes.

genetically, but it appears that a large proportion of the sectors may be the result of heteroploidy, although mutations, heterokaryosis, and parasexualism are also probably involved.

Sexuallike Processes in Bacteria

New biotypes of bacteria seem to arise with varying frequency by means of at least three sexuallike processes (Figure 6-1). It is probable that similar processes occur in mycoplasmas. (1) **Conjugation,** in which two compatible bacteria come in contact with each other and a small portion of the chromosomal or nonchromosomal (plasmid) genetic material of the one bacterium is transferred to the genetic material of the other. (2) **Transformation,** in which bacterial cells are transformed genetically by absorbing and incorporating in their own cells genetic material secreted by, or released during rupture of, other compatible bacteria. (3) **Transduction,** in which a bacterial virus (phage) transfers genetic material from the bacterium in which the phage was produced to the bacterium it infects next.

Genetic Recombination in Viruses

When two strains of the same virus are inoculated into the same host plant, one or more new virus strains are recovered with properties (virulence, symptomatology, and so on) different from those of either of the original strains introduced into the host. The new strains probably are recombinants, although their appearance through mutation, not hybridization, cannot always be ruled out.

In multipartite viruses consisting of 2, 3, or more nucleic acid components, new virus strains may also arise in host plants or vectors from recombination of the appropriate components of two or more strains of such viruses.

Loss of Pathogen Virulence in Culture

The virulence of pathogenic microorganisms toward one or all of their hosts often decreases when the pathogens are kept in culture for relatively long periods of time or when they are passed one or more times through different hosts. If the culturing of the pathogen is prolonged sufficiently, the pathogen may lose virulence completely. Such partial or complete loss of virulence in pathogens is sometimes called **attenuation,** and it has been shown to occur in bacteria, fungi, and viruses. Pathogens that have experienced partial or complete loss of virulence in culture or in other hosts are often capable of regaining part or all of their virulence if they are returned to their hosts under the proper conditions. Sometimes, however, the loss of virulence may be irreversible. "Loss" of virulence in culture, or in other hosts, seems to be the result of selection of individuals of less virulent or avirulent pathogen strains that happen to be capable of growing and multiplying in culture, or in the other host, much more rapidly than the virulent ones. After several transfers in culture or the other hosts, such attenuated individuals largely, or totally, overtake and replace the virulent ones in the total population so that the pathogen is less virulent or totally avirulent. Upon reinoculation of the proper host, isolates, in which the virulent individuals have been totally replaced by avirulent ones, continue to be avirulent, and therefore loss of pathogenicity is irreversible. On the other hand, upon reinoculation of the proper host with isolates in which at least some virulent individuals survived through the transfers in culture or the other host, the few surviving virulent individuals infect the host and multiply, often in proportion to their virulence. The virulent individuals increase in numbers with each subsequent inoculation, while at the same time nonvirulent individuals are reduced or eliminated with each reinoculation.

Stages of Variation in Pathogens

The entire population of a particular organism on earth, for example, a fungal pathogen, has certain morphological characteristics in common and makes up the **species** of pathogen, such as *Puccinia graminis,* the cause of stem rust of cereals. Some individuals of this species, however, attack only wheat or only barley, or oats, and these individuals make up groups that are called **varieties** or **special forms** *(formae specialis)* such as *P. graminis tritici, P. g. hordei,* and *P. g. avenae.* But even within each special form, some individuals attack some of the varieties of the host plant but not the others, some attack another set of host plant varieties, and so on, each group of such individuals making up a **race.** Thus, there are more than 200 races of *Puccinia graminis* *tritici* (race 1, race 15, race 59, and so on). Occasionally, one of the offspring of a race can suddenly attack a new variety or can cause severe symptoms on a variety that it could barely infect before. This individual is called a **variant.** The identical individuals produced asexually by the variant make up a **biotype.** Each race consists of one or of several biotypes (race 15A, 15B, and so on).

The appearance of new pathogen biotypes may be very dramatic when the change involves the host range of the pathogen. If the variant has lost the ability to infect a plant variety that is widely cultivated, this pathogen simply loses its ability to procure a livelihood for itself and will die without even making its existence known to us. If, on the other hand, the change in the variant pathogen enables it to infect a plant variety cultivated because of its resistance to the parental strain, the variant individual, being the only one that can survive on this plant variety, grows and multiplies on the new variety without any competition and soon produces large populations that spread and destroy the heretofore resistant variety. This is the way the resistance of a plant variety is said to be "broken down," although it was the change in the pathogen, not the host plant, that brought it about.

Types of Plant Resistance to Pathogens

Plants are resistant to certain pathogens either because they belong to taxonomic groups that are immune to these pathogens (nonhost resistance), or because they possess genes for resistance directed against the genes of virulence of the pathogen (true resistance), or because for various reasons, the plants escape or tolerate infection by these pathogens (apparent resistance).

Each kind of plant, for example, potato, corn, orange, is a host to a small and different set of pathogens that make up a small proportion of the total number of known plant pathogens. In other words, each kind of plant is a non-host to the vast majority of known plant pathogens. Non-hosts are immune, that is, completely resistant, to all pathogens of all other plants, even under the most favorable conditions for disease development **(nonhost resistance).** The same plants, however, are susceptible, to a lesser or greater extent, to their own pathogens. Moreover, each plant exhibits specific susceptibility toward each of its own pathogens while it exhibits unspecific immunity (complete or nonhost resistance) to all the other pathogens.

Even within a species of plant that is susceptible to a particular species of pathogen, however, there is considerable variation in both the susceptibility of the various plant cultivars (varieties) towards the pathogen and in the virulence of the various pathogen races towards the plant variety. It is the genetics of such host–pathogen interactions that are of considerable biological interest and of the utmost importance in developing disease control strategies through breeding for resistance.

The variation in susceptibility to the pathogen among plant varieties is due to different kinds and, perhaps, different numbers of genes for resistance that may be present in each variety. The effects of individual resistance genes vary from large to minute, depending on the importance of the functions they control. A variety that is very susceptible to a pathogen isolate obviously has no effective genes for resistance against that isolate. The same variety, however, may (or may not) be susceptible to another pathogen isolate obtained from infected plants of another variety.

	Pathogen isolate	
	1	2
Plant variety X	Susceptible	Resistant

Lack of susceptibility to the second isolate would indicate that the plant variety, which had no genes for resistance against the first pathogen isolate, has one or more genes for resistance against the second isolate. If the same plant variety is inoculated with more pathogen isolates, obtained from still different plant varieties, it is possible that the variety would be susceptible to some of them but not susceptible (it would be resistant) to the other isolates. The latter case would again show that the variety has one or more genes for resistance against each of the isolates to which it is resistant. Although the resistance against some of the isolates might be the result of the same genes for resistance in the variety, it is likely that the variety also contains several genes for resistance, each specific against a particular pathogen isolate.

True Resistance

Disease resistance that is genetically controlled by the presence of one, a few, or many genes for resistance in the plant is known as **true resistance.** In true resistance, the host and the pathogen are more or less incompatible with each other, either because of lack of chemical recognition between the host and the pathogen or because the host plant can defend itself against the pathogen by the various defense mechanisms already present, or activated, in response to infection by the pathogen. There are two kinds of true resistance: horizontal and vertical.

Horizontal Resistance All plants have a certain, but not always the same, level of unspecific resistance that is effective against each of their pathogens. Such resistance is sometimes called nonspecific, general, quantitative, adult-plant, field, or durable resistance, but is most commonly referred to as **horizontal resistance.**

Horizontal resistance is controlled by many (probably dozens, perhaps hundreds of) genes, thereby the name **polygenic** or **multi-gene resistance.** Each of these genes alone may be rather ineffective against the pathogen and may play a minor role in the total horizontal resistance **(minor gene resistance).** The many genes involved in horizontal resistance seem to exert their influence by controlling the numerous steps of the physiological processes in the plant that provide the materials and structures that make up the defense mechanisms of the plant. The horizontal resistance of a plant variety toward all races of a pathogen may be somewhat greater, or smaller, than those of other varieties toward the same pathogen (Figure 6-2), but the differences are usually small and insufficient to routinely distinguish varieties by horizontal resistance **(nondifferential resistance).** In addition, horizontal resistance is affected by, and may vary under, different environmental conditions. Generally, horizontal resistance does not protect plants from becoming infected, but

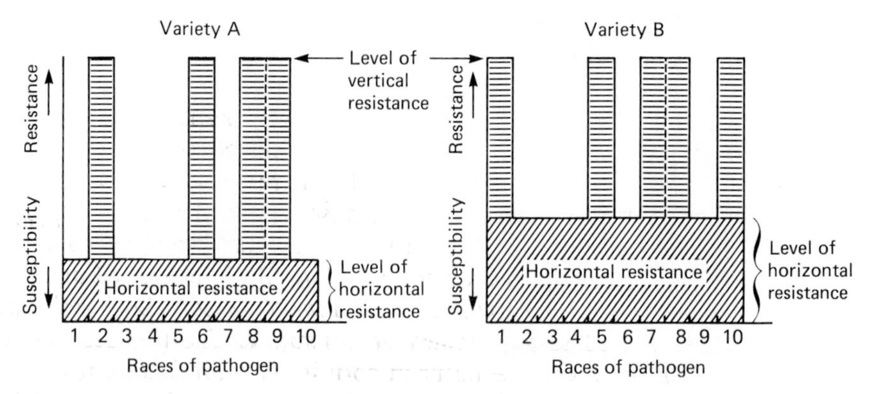

FIGURE 6-2 Levels of horizontal and vertical resistance of two plant varieties toward ten races of a pathogen. (After Vanderplank, 1984).

it slows down the development of individual infection loci on a plant and thereby slows down the spread of the disease and the development of epidemics in the field. Some degree of horizontal resistance is present in plants regardless of whether or not vertical resistance is present.

Vertical Resistance Many plant varieties are quite resistant to some races of a pathogen while they are susceptible to other races of the same pathogen. In other words, depending on the race of the pathogen used to infect a variety, the variety may appear resistant to one pathogen race and susceptible to another race. This type of plant resistance differentiates clearly between races of a pathogen since it is effective against specific races of the pathogen and ineffective against others (Figure 6-2). Such resistance is sometimes called specific, qualitative, or differential resistance, but is most commonly referred to as **vertical resistance.**

Vertical resistance is usually controlled by one or a few genes (thereby the name **monogenic** or **oligogenic resistance**). These genes apparently control a major step in the interaction of the pathogen with the host plant and therefore play a major role in the expression of vertical resistance **(major gene resistance).** In the presence of vertical resistance, the host and pathogen appear incompatible, the host usually responds with a hypersensitive reaction, and so the pathogen fails to become established and multiply in the host. Generally, vertical resistance inhibits the initial establishment of pathogens that arrive at a field from host plants that lack or have different major genes for resistance. Vertical resistance inhibits the development of epidemics by limiting the initial inoculum.

Complete resistance may be provided by a single resistance gene, for example, R1, R2, or R3, but more than one resistance gene may be combined (R1R2, R1R3, R1R2R3) in the same plant, which then is resistant to all the pathogen races to which each of the genes provides resistance. A plant species may have as many as 20–40 resistance genes against a particular pathogen, although each variety may have only one or a few of these genes. For example, wheat has 20–40 genes for resistance against the leaf rust fungus *Puccinia recondita;* barley has a similar number of genes against the powdery mildew fungus *Erysiphe graminis tritici,* and cotton has almost as many against the bacterium *Xanthomonas malvacearum.* Each gene for resistance, such as R2, is in reality unspecific for resistance; that is, it makes the plant resistant to all the races of the pathogen except to the race (such as P2) that contains the corresponding gene for virulence. This pathogen race (P2) and its virulence gene (V2), however, are detected because the pathogen specifically attacks plants that carry this particular gene for resistance (R2).

Whether horizontal or vertical, true resistance is generally controlled by genes located in the plant chromosomes in the cell nucleus. There are, however, several plant diseases in which either type of resistance is controlled by genetic material contained in the cytoplasm of the cell in which genes do not normally follow the Mendelian laws of genetics. Such resistance is sometimes referred to as **cytoplasmic resistance.** The two best known cases of cytoplasmic resistance occur in corn in which resistance to two leaf blights, the southern corn leaf blight caused by *Bipolaris (Helminthosporium) maydis* and the yellow leaf blight caused by *Phyllosticta maydis,* is conferred by characteristics present in normal cytoplasm of various types of corn but absent or suppressed in Texas male-sterile cytoplasm.

Varieties with specific (monogenic or oligogenic) resistance generally show complete resistance to a specific pathogen under most environmental conditions, but a single or a few mutations in the pathogen may produce a new race that may infect the previously resistant variety. On the contrary, varieties with general (polygenic) resistance are less stable and may vary in their reaction to the pathogen under different environmental conditions, but a pathogen will have to undergo many more mutations to completely break down the resistance of the host. As a rule, a combination of major and minor genes for resistance against a pathogen is the most desirable makeup for any plant variety.

Apparent Resistance

In any area and almost every year, limited or widespread plant disease epidemics occur on various crop plants. Under certain conditions or circumstances, however, some very susceptible plants or varieties of these crops may remain free from infection or symptoms and thus appear resistant. Apparent resistance to disease of plants known to be susceptible is generally a result of disease escape or tolerance to disease:

1. *Disease escape* occurs whenever genetically susceptible plants do not become infected because the three factors necessary for disease (susceptible host, virulent pathogen, and favorable environment) do not coincide and interact at the proper time or for sufficient duration.

Plants may escape disease because their seeds germinate faster, or their seedlings harden earlier than others, and before the temperature becomes favorable for the pathogen to attack them. Some plants escape disease because they are susceptible to a pathogen only at a particular growth stage (young leaves, stems, or fruits; at blossoming or fruiting; at maturity and early senescence) and therefore, if the pathogen is absent or inactive at that particular time, such plants avoid becoming infected. For example, young tissues and plants are infected and affected much more severely by *Pythium,* powdery mildews, and most bacteria and viruses than older ones. However, fully grown, mature, and senescent plant parts are much more susceptible to certain pathogens, such as *Alternaria* and *Botrytis,* than are their younger counterparts. Plants may also escape disease because of the distance between fields, the number and position of plants in the field, the spacing of plants in a field, and so on.

In many cases, plants escape disease because they are interspersed with other types of plants that are insusceptible to the pathogen, and the amount of inoculum that reaches them is much less than if they were in monocultural plantations; because their surface hairs and wax repel water and pathogens suspended in it; because their growth habit is too erect or otherwise unfavorable for pathogen attachment and germination; or because their natural openings, such as stomata, are at a higher level than the rest of the leaf surface, or open too late in the day by which time the leaves ared dry and the germ tubes of spores, such as of *Puccinia graminis,* have desiccated. In plant diseases in which pathogens penetrate primarily through wounds caused by heavy winds and rain, dust storms, and insects, lack of such wounds allows disease escape. Also, plants that are unattractive or resistant to the vector of a pathogen escape infection by that pathogen.

Factors that affect the survival, infectivity, multiplication, and dissemination of the pathogen are also likely to allow some plants to escape disease. Such factors include: absence or poor growth of the pathogen at the time the susceptible plant stage is available; destruction or weakening of the pathogen by hyperparasites or by antagonistic microflora at the place of production or at the infection court; misdirection to or trapping of the pathogen by other plants; and lack of pathogen disemination because winds, rain, or vectors are absent.

Several environmental factors play crucial roles in plant disease escape in almost every location. Temperature, for example, determines the geographical distribution of most pathogens, and plants growing outside the range of that temperature escape disease from such pathogens. Most commonly, however, temperature increases plant disease escape in ranges that favor plant growth much more than they do the growth of the pathogen. For example, many plants escape disease from *Pythium* and *Phytophthora* if the temperature is high and the soil moisture low, while some low temperature crops, such as wheat, escape similar diseases from *Fusarium* and *Rhizoctonia* if the temperature is low. Temperatures outside certain ranges inhibit sporulation of fungi as well as spore germination and infection, thereby increasing the chances for disease escape. Low temperatures also reduce the mobility of many insect vectors or pathogens, allowing more plants to escape disease.

Lack of moisture caused by low rainfall or dew or low relative humidity is probably the most common cause of disease escape in plants. Plants in most dry areas or during dry years remain generally free of apple scab, late blight, most downy mildews, and anthracnoses because these diseases require a film of water on the plant or high relative humidity in almost every stage of their life cycle. Similarly, in dry soils such diseases as clubroot of crucifers and damping off induced by *Pythium* and *Phytophthora* are quite rare because such soils inhibit the production and activity of the motile spores of these pathogens. On the other hand, with some diseases, such as common scab of potato caused by *Streptomyces scabies,* plants escape disease in irrigated or wet soils because the plants can defend themselves better in the absence of water stress and because these pathogens are lysed or otherwise inhibited by microorganisms favored by high moisture. Many trees are also in a better position to defend themselves and to escape damage by the canker-causing fungus *Cytospora* sp. in years in which sufficient rainfall or irrigation provides adequate soil moisture in later summer and early fall.

Some other environmental factors also allow plants to escape disease. For example, wind may increase disease escape by blowing from the wrong direction at the right time, thus carrying spores and vectors away from the crop plants, and by drying up plant surfaces quickly, before the pathogen has time to germinate and infect. Also, soil pH increases disease escape in a few diseases, for example, in crucifers from *Plasmodiophora brassicae* at high pH, and in potatoes from *Streptomyces scabies* at low pH, in both cases because the particular pH inhibits survival and growth of the pathogen.

In general, many plant diseases are present some years in some areas and absent on the same kinds of plants in other years or in nearby areas. This suggests that in these areas or years the plants remain free of disease not because they are resistant, but because they escape disease. Earliness is often bred into many wheat and potato varieties to help them escape disease from the rusts and the late blight, respectively. Lateness, rapid growth, resistance to bruising, unattractiveness to vectors, and tolerance to low temperatures are also often bred into crop varieties to help them escape specific diseases. These and many other characteristics, of course, are those that make up "horizontal" resistance. It is true that there is a wide common area between horizontal resistance and disease escape in which the one leads to the other or the two appear identical. Escape from disease depends on environmental conditions as well as on heritable characteristics in the host and the pathogen and is often entirely controlled by the environment. Escape from disease, moreover, is a manageable quality, and farmers, through many of the most common cultural practices actually aim at helping plants escape disease. Such practices include using disease-free, vigorous seed, choosing the proper soil, planting date, depth of sowing, distance between plants and between fields, proper crop rotation, sanitation (roguing, pruning, and so on), interplantings and multilines, insect and vector control, and several others.

2. *Tolerance to disease* is the ability of plants to produce a good crop even when they are infected with a pathogen. Tolerance results from specific, heritable characteristics of the host plant that allow the pathogen to develop and multiply in the host while the host, either by lacking receptor sites for, or by inactivating or compensating for the irritant excretions of the pathogen, still manages to produce a good crop. Tolerant plants are, obviously, susceptible to the pathogen but they are not killed by it and generally show little damage from the pathogen. The genetics of tolerance to disease are not understood; neither is its relationship, if any, to horizontal resistance. Tolerant plants, whether because of exceptional vigor or a hardy structure, probably exist in most host–parasite combinations. Tolerance to disease is most commonly observed in many plant-virus infections in which mild viruses, or mild strains of virulent viruses, infect plants such as potato and apple systemically and yet cause few or no symptoms and have little discernible effect on yield. Generally, however, although tolerant plants produce a good crop even when they are infected, they produce an even better crop when they are not infected.

Genetics of Virulence in Pathogens and of Resistance in Host Plants

Infectious plant diseases are the result of the interaction of at least two organisms, the host plant and the pathogen. The properties of each of these two organisms are governed by their genetic material, the DNA, which is organized in numerous segments making up the genes.

The inheritance of host reaction, that is, the degree of susceptibility or resistance to various pathogens, has been known for a long time and has been used quite effectively in breeding and distributing varieties resistant to pathogens causing particular diseases. The inheritance of pathogen infection type, however, that is, the degree of pathogen virulence or avirulence, has been studied intensively only in the last few decades. It has now become clear that pathogens consist of a multitude of races, each differing from others in its ability to attack certain varieties of a plant species but not other varieties. Thus when a variety is inoculated with two appropriately chosen races of a pathogen, the variety is susceptible to one race but resistant to the other. Conversely, when the same race of a pathogen is inoculated on two appropriately chosen varieties of a host plant, one variety is susceptible while the other is resistant to the same pathogen (Table 6-1). This clearly indicates that, in the first case, one race possesses a genetic characteristic that enables it to attack the plant, while the other race does not, and in the second case, that the one variety possesses a genetic characteristic that enables it to defend itself against the pathogen, so that it remains resistant, while the other variety does not. When several varieties are inoculated separately with one of several races of the pathogen, it is again noted that one pathogen race can infect a certain group of varieties, another race can infect another group of varieties, including some that can and some that cannot be infected by the previous race, and so on (Table 6-1). Thus, varieties possessing certain genes of resistance or susceptibility react differently against the various pathogen races and their genes of virulence or avirulence. The progeny of these varieties react to the same pathogens in exactly the same manner as did the parent plants, indicating that the property of resistance or susceptibility against a pathogen is

Table 6-1. Possible Reactions of Two (Left) and Four (Right) Varieties of a Plant to Two (Left) and Four (Right) Races of a Pathogen. The Plus Sign Indicates Susceptible (Compatible Reaction, Infection); The Minus Sign Indicates Resistant (Incompatible Reaction, Noninfection).

Plant varieties	Pathogen races 1	2
A	−	+
B	+	−

Plant varieties	Pathogen races 1	2	3	4
A	−	+	+	+
B	+	−	−	+
C	+	−	−	+
D	+	−	+	−

genetically controlled (inherited). Similarly, the progeny of each pathogen causes on each variety the same effect that was caused by the parent pathogens, indicating that the property of virulence or avirulence of the pathogen on a particular variety is genetically controlled (inherited).

It appears from the above that, under favorable environmental conditions, the outcome—infection (susceptibility) or noninfection (resistance)—in each host–pathogen combination is predetermined by the genetic material of the host and of the pathogen. The number of genes determining resistance or susceptibility varies from plant to plant, as the number of genes determining virulence or avirulence varies from pathogen to pathogen. In most host–pathogen combinations the number of genes involved and what they control are not yet known. In some diseases, however, particularly those caused by fungi, such as potato late blight, apple scab, powdery mildews, tomato leaf mold, the cereal smuts and rusts, and also in several viral and bacterial diseases of plants, considerable information regarding the genetics of host–pathogen interactions is available.

The Gene-for-Gene Concept

The coexistence of host plants and their pathogens side by side in nature indicates that the two have been evolving together. Changes in the virulence of the pathogens must be continually balanced by changes in the resistance of the host, and vice versa, so that a dynamic equilibrium of resistance and virulence is maintained and both host and pathogen survive. If either the virulence of the pathogen or the resistance of the host increased unopposed, it would have led to the elimination of either the host or the pathogen, respectively, which obviously has not happened. Such a stepwise evolution of resistance and virulence can be explained by the **gene-for-gene concept,** according to which for each gene that confers resistance in the host there is a corresponding gene in the pathogen that confers virulence to the pathogen, and vice versa.

The gene-for-gene concept was first proven in the case of flax and flax rust, but it has since been shown to operate in many other rusts, in the smuts, powdery mildews, apple scab, late blight of potato, and other diseases caused by fungi, as well as some diseases caused by bacteria, viruses, parasitic higher plants, and nematodes. In all these diseases it was shown that whenever a variety is resistant to a pathogen as a result of 1, 2, or 3 resistance genes, the pathogen that can affect it also contains 1, 2, or 3 virulence genes, respectively. Each gene in the host can be detected and identified only by its counterpart gene in the pathogen, and vice versa.

Generally, but not always, in the host, genes for resistance are dominant (R) while genes for susceptibility, that is, lack of resistance, are recessive (r). In the pathogen, on the other hand, it is the genes for avirulence, that is, inability to infect, that are usually dominant (A), while genes for virulence are recessive (a). Thus, when two plant varieties, one carrying the gene for resistance R to a certain pathogen and the other carrying the gene for susceptibility (r) to the same pathogen (lacking the gene R for resistance) are inoculated with two races of the pathogen, one of which carries the gene for avirulence A against the resistance gene R and the other race carries the gene for virulence (v)

Table 6-2 Quadratic Check of Gene Combinations and Disease Reaction Types in a Host–Pathogen System in Which the Gene-for-Gene Concept for One Gene Operates.

	Resistance or susceptibility genes in the plant	
Virulence or avirulence genes in the pathogen	R (resistant) dominant	r (susceptible) recessive
A (avirulent) dominant	AR (−)	Ar (+)
a (virulent) recessive	aR (+)	ar (+)

Minus signs indicate incompatible (resistant) reaction and therefore no infection. Plus signs indicate compatible (susceptible) reaction, and therefore infection develops.

against the resistance gene R, the gene combinations and reaction types shown in Table 6-2 are possible.

Of the four possible gene combinations only *AR* is resistant, that is, when the host has a certain gene for resistance (R) against which the pathogen lacks the specific gene for virulence (A), although it may have other genes for virulence. In the *Ar* combination, infection results because the host lacks genes for resistance (r) and so the pathogen can attack it with its other genes for virulence (after all, it is a pathogen), even though it too lacks a specific virulence gene for the resistance gene the host does not have. In *aR*, infection results because, although the host has a gene for resistance, the pathogen has the gene for virulence that can attack specifically this particular gene for resistance. Finally, in *ar*, infection results because the plant is susceptible (r) and, in addition, the pathogen is virulent (a).

Genes for reistance appear first in hosts through evolution and may later be used in breeding programs. Genes for virulence appear in pathogens after the appearance of particular genes for resistance in the host, and each is specific for attack of a particular gene for resistance. Whenever a new gene for virulence that attacks the existing gene for resistance appears, the resistance of the host breaks down. Plant breeders then introduce in the plant another gene for resistance, which extends the resistance of the host beyond the range of the new gene for virulence in the pathogen. This produces a resistant variety until another gene for virulence appears in the pathogen. When a variety has two or more genes for resistance (R_1, R_2, . . .) against a particular pathogen, the pathogen must also have two or more genes for virulence (a_1, a_2, . . .), each aimed specifically against one of the genes for resistance in the host, for infection to take place. The gene combinations, and disease reaction types, of hosts and pathogens with two genes for resistance or virulence in corresponding loci, respectively, are shown in Table 6-3.

Table 6-3 makes clear several points. First, susceptible (r_1r_2) plants lacking genes for resistance are attacked by all races of the pathogen, even those that do not have specific genes for virulence (A_1A_2). Second, pathogen races or individuals that carry genes for virulence (a_1a_2) for each gene for resistance of the host (R_1R_2) can infect all plants that have any combination of these genes (R_1R_2, R_1r_2, r_1R_2). When a pathogen has one of the two genes for virulence (a_1 or a_2), then it can infect plants that have the corresponding gene for resistance (R_1 or R_2, respectively) but not plants that have a gene for resistance different from the corresponding gene for virulence (for example, patho-

Table 6-3. Complementary Interaction of Two Host Genes For Resistance and The Corresponding Two Pathogen Genes For Virulence and Their Disease Reaction Types.

Virulence (a) or avirulence (A) genes in the pathogen	Resistance (R) or susceptibility (r) genes in the plant			
	R_1R_2	R_1r_2	r_1R_2	r_1r_2
A_1A_2	−	−	−	+
A_1a_2	−	−	+	+
a_1A_2	−	+	−	+
a_1a_2	+	+	+	+

gen with genes A_1a_2 infect plant with r_1R_2 but not with R_1r_2, because a_2 can attack R_2 but A_1 cannot attack R_1).

The gene-for-gene concept has been demonstrated only in plants with vertical (monogenic and oligogenic) types of resistance to a certain disease. Plant breeders apply the gene-for-gene concept every time they incorporate a new resistance gene into a desirable variety that becomes susceptible to a new strain of the pathogen. With the diseases of some crops, new resistance genes must be found and introduced into old varieties at relatively frequent intervals, while in others a single gene confers resistance to the varieties for many years. The gene-for-gene concept presumably applies to horizontal (polygenic or general) resistance as well, although so far proof for this and for polygenic control of virulence in pathogens is lacking.

The Nature of Resistance to Disease

A plant is either immune to a pathogen—that is, it is not attacked by the pathogen even under the most favorable conditions—or it may show various degrees of resistance ranging from almost immunity to complete susceptibility. Resistance may be conditioned by a number of internal and external factors that operate to reduce the chance and degree of infection. The first step in any compatible host–pathogen interaction, in other words, in any infection, is recognition of the host by the pathogen and perhaps the opposite, recognition of the pathogen by the host. Therefore, the absence of recognition factor(s) in the host would make it resistant to the particular pathogen. In addition, any heritable characteristic of the plant that contributes to localization and isolation of the pathogen at the points of entry, to reduction of the harmful effects of toxic substances produced by the pathogen, or to inhibition of the reproduction and, thereby, of the further spread of the pathogen contributes to the resistance of the plant to disease. Furthermore, any heritable characteristic that enables a particular variety to complete its development and maturation under conditions that do not favor the development of the pathogen, also contributes to resistance (disease escape).

The contribution of the genes conditioning resistance in the host seems to consist, primarily, of providing the genetic potential in the plant for development of one or more of the morphological or physiological characters —including those described in the chapters on structural and biochemical defense—that contribute to disease resistance. With the exception of virus and viroid diseases of plants, in which the genes of the host could conceivably come into "face-to-face" confrontation with the "genes" of the viral nucleic acid, the genes of plants infected by other types of pathogens seem never to come in contact with the genes of the pathogen. In general, in all host–pathogen combinations, viruses and viroids included, the interactions between genes of the host and genes of the pathogen are believed to be brought about indirectly through the chemical molecules and physiological processes controlled by the respective genes.

The mechanisms by which genes control the physiological processes that lead to disease resistance or susceptibility are not yet clear, but they are, presumably, no different from the mechanisms controlling any other physiological process in living organisms. Thus, it is known that the genes are each carried by the genetic material (DNA) as successive groups of nucleotide triplets (triplet code), which are first read and transcribed on messenger RNA as the latter is synthesized. The messenger RNA then becomes associated with clusters of ribosomes (polyribosomes) and leads to the production of a specific protein, which is either an enzyme or a structural protein. The produced enzyme participates or initiates biochemical reactions related to one or another of the cellular processes, and may result in the production of a certain morphological characteristic or accumulation of a certain chemical substance.

The genes responsible for determining the kind and properties of a protein are called **structural genes.** The timing of activation of the structural genes, the rate of their activity—protein synthesis—and the timing of their inactivity are controlled by other genes called **regulatory genes.** Furthermore, messenger-RNA formation seems to be initiated only at certain points of the DNA strand, and these points are called **operators.** A single operator may control the transcription into a messenger-RNA of only one structural gene or of a series of structural genes concerned with the different steps of one particular metabolic function, for example the biosynthesis of a fungitoxic phenolic compound or a compound that reacts with and detoxifies a pathogen toxin. The group of genes controlled by such an operator is called an **operon.**

Thus, it is possible that for the production of an inducible enzyme or a fungitoxic substance, a stimulant (inducer) secreted by the pathogen inactivates a repressor molecule, which is the product of a regulatory gene. The function of the repressor was to combine with a specific operator locus and prevent the transcription of that operon, thereby blocking the synthesis and action of the relevant proteins in the absence of infection. After infection, however, and after inactivation of the repressor by the pathogenic stimulant, transcription of the operon can take place, the particular substance is produced and, if this substance is toxic enough to the pathogen, the infection stops and the variety is resistant. On the other hand, if a pathogen mutant appears that does not secrete the particular stimulant (inducer) that inactivates the repressor molecule, the defense reaction does not take place, the pathogen infects the host without opposition and so it causes disease. In that case, the resistance of the host is said to have broken down, but it is actually bypassed by the pathogen rather than broken down. Other possible ways by which a pathogen could "break down" the resistance of a host are through a mutation in the pathogen, which enables it to produce a substance that can react with and neutralize the defensive toxic substance of the host that is directed against the pathogen, and through a mutation in the pathogen that would eliminate or block its receptor site on which the host defensive substance becomes attached. The pathogen then can operate in the presence of that substance and of the defense mechanism that produces it.

Breeding of Resistant Varieties

The value of resistance in controlling plant disease was recognized in the early 1900s. Advances in the science of genetics and the obvious advantages of planting a resistant instead of a susceptible variety made the breeding of resistant varieties possible and desirable. The more recent realization of the dangers of polluting the environment through chemical control of plant diseases gave additional impetus and importance to the breeding of resistant varieties. Thus, breeding resistant varieties, which is but one part of broader plant breeding programs, is more popular and more intensive today than it ever was in the past. Its usefulness and importance are paramount in the production of food and fiber. Yet, some aspects of plant breeding, and of breeding resistant varieties in particular, have shown certain weaknesses and have allowed some plant disease epidemics to occur that could not have developed were it not for the uniformity created in crops through plant breeding.

Natural Variability in Plants

Today's cultivated crop plants are the result of selection, or selection and breeding, of plant lines that evolved naturally in one or many geographic areas over millions of years. The evolution of plants from their ancient ancestors to present day crop plants has occurred slowly and has produced countless genetically diverse forms of these plants. Many such plants still exist as wild types at the point or points of origin or in the areas of natural spread of the plant. Although these plants may appear as useless remnants of evolution that are not likely to play a role in any future advances in agriculture, their diversity and survival in the face of the various pathogens that affect the crop in question indicates that they carry numerous genes for resistance against these pathogens.

Since the beginning of agriculture, some of the wild plants in each locality have been selected and cultivated and thus produced numerous cultivated lines or varieties. The most productive of these varieties were perpetuated in each locality from year to year, and those that survived the local climate and the pathogens continued to be cultivated. Nature and pathogens eliminated the weak and susceptible ones, while the farmers selected the best yielders among the survivors. Surviving varieties had different sets of major and minor genes for resistance. In this fashion, selection of crop plants continued wherever they were grown, each locality independently selecting varieties adapted to its own environment and resistant to its own pathogens. Thus, numerous varieties of each crop plant were cultivated throughout the world and, by their own genetic diversity, contributed to making the crop locally adapted but overall genetically nonuniform and, thereby, safe from any sudden outbreak of a single pathogen over a large area.

Effects of Plant Breeding on Variability in Plants

During the present century, widespread, intensive, and systematic efforts have been made and continue to be made by plant breeders throughout the world to breed plants that combine the most useful genes for higher yields, better quality, uniform size of plants and fruit, uniform ripening, cold hardiness, and disease resistance. In searching for new useful genes, plant breeders cross existing, local, cultivated varieties with each other and with those of other localities, both here and abroad, and with wild species of crop plants from wherever they can be obtained. Furthermore, plant breeders often attempt to generate additional genetic variation by treating their plant material with mutagenic agents. More recently, plant breeders have been generating greater genetic variability, and modifying or accelerating plant evolution in certain directions, by various genetic engineering techniques. Using such techniques, plant breeders can introduce genetic material (DNA) into plant cells directly, via vectors (such as *Agrobacterium tumefaciens*), or via protoplast fusion. Breeders can also obtain plants with different characteristics through culture and regeneration of somatic plant cells, by diploidization of haploid plants, and so on.

The initial steps in plant breeding generally increase the variability of genetic characteristics of plants in a certain locality by combining in such plants genes that were more or less widely separated before. As breeding programs advance, however, and as several of the most useful genes are identified, subsequent steps in breeding tend to eliminate variability by combining the best genes in a few cultivated varieties and leaving behind or discarding plant lines that seem to have no usefulness at the time. In a short time a few "improved" varieties replace most or all others over large expanses of land. The most successful improved varieties are also adopted abroad, and before too long, some of them replace the numerous but commercially inferior local varieties. Occasionally, even the wild types themselves may be replaced by such a variety. Thus, Red Delicious apples, Elberta peaches, certain dwarf wheat and rice varieties, certain genetic lines of corn and potatoes, one or two types of bananas, and sugarcane are grown in huge acreages throughout the world. In almost every crop, relatively few varieties make up the great bulk of the cultivated acreage of the crop throughout a country or throughout the world. The genetic base of these varieties is often narrow, especially since many of them have been derived from crosses of the same or related ancestors. These few varieties are used so widely because they are the best available, they are stable and uniform, and therefore everybody wants to grow them. At the same time, however, because they are so widely cultivated, they carry with them not only the blessings but also the dangers of uniformity. The most serious of these dangers is the vulnerability of large uniform plantings to sudden outbreaks of catastrophic plant disease epidemics.

Plant Breeding for Disease Resistance

Most plant breeding is done for development of varieties that produce greater yields or better quality. While such varieties are being developed, they are tested for resistance against some of the most important pathogens present in the area where the variety is developed and where it is expected to be cultivated. If the variety is resistant to these pathogens, it may be released to growers for immediate production. If, however, it is susceptible to one or more of these pathogens, the variety is usually shelved or discarded; or sometimes it is released for production if the pathogen can be controlled by other means, such as chemical; but more often it is subjected to further breeding in an attempt to incorporate into the variety genes that would make it resistant to the pathogens without changing any of its desirable characteristics.

Sources of Genes for Resistance

The source of genes for resistance is the same gene pool of the crop that provides genes for every other inherited characteristic; that is, other native or foreign commercial varieties, older varieties abandoned earlier or discarded breeders' stock, wild plant relatives, and occasionally, induced mutations.

Often genes of resistance are present in the varieties or species normally grown in the area where the disease is severe and in which the need for resistant varieties is most pressing. With most diseases, a few plants remain virtually unaffected by the pathogen, although most or all other plants in the area may be severely diseased. Such survivor plants are likely to have remained healthy because of resistant characteristics present in them.

If no resistant plants can be found within the local population of the species, plants of the same species from other areas and plants of other species (cultivated or wild) are checked for resistance. If resistant plants are found, they are crossed with the cultivated varieties in an effort to incorporate the resistance genes of the other species into the cultivated varieties. With some diseases, such as late blight of potatoes, it has been necessary to look for resistance genes in species growing in the area where the disease originated. Presumably, plants existing in these areas managed to survive the long, continuous presence of the pathogen because of their resistance to it.

Techniques Used in Classical Breeding for Disease Resistance

The same methods used to breed for any heritable characteristic are also used for breeding for disease resistance and depend upon the mating system of the plant (self- or cross-pollinated). Breeding for disease resistance, however, is considerably more complicated, first because it can be assayed only by making the plants diseased, that is by employing another living and variable organism that must interact with the plants; and second because resistance may not be stable and may break down under certain conditions. For these reasons, several more or less sophisticated systems of screening for resistance have been developed. These screening systems include (1) precise conditions for inoculating the plants with the pathogen, (2) accurate monitoring and control of the environmental conditions in which the inoculated plants are kept, and (3) accurate assessment of disease incidence (percentage of plants, leaves, or fruits infected) and disease severity (proportion of the total area of plant tissue affected by disease). The following techniques are the main ones used for breeding disease resistance.

1. **Mass selection of seed from** the most highly resistant plants surviving in a field where natural infection occurs regularly. It is a simple method but improves plants only slowly, and in cross-pollinated plants there is no control of pollen source.

2. **Pure line or pedigree selection,** in which individual highly resistant plants and their progenies are propagated separately and are inoculated repeatedly to test for resistance. This method is easy and most effective with self-pollinated crops but quite difficult with cross-pollinated ones.

3. **Recurrent selection or back crossing,** in which a desirable but susceptible variety of a crop is crossed with another cultivated or wild relative that carries resistance to a particular pathogen. The progeny is then tested for resistance, and the resistant individuals are back crossed to the desirable variety. This is repeated several times until the resistance is stabilized in the genetic background of the desirable variety. This method is time consuming, and its effectiveness varies considerably with each particular case. It can be applied somewhat more easily in cross-pollinated than in self-pollinated crops.

4. **Other** classical breeding techniques for disease resistance include: the use of F_1 hybrids of two different but homozygous lines carrying different genes for resistance, which allows us to take advantage of the phenomenon of heterosis: the use of natural or artificially induced (UV-light, X-rays) mutants that show increased resistance; and the change of the number of chromosomes in a plant and production of euploids (4n, 6n) or aneuploids (2n ± 1 or 2 chromosomes), by the use of chemicals such as colchicine, and by radiation.

Breeding for Resistance Using Tissue Culture and Genetic Engineering Techniques

Recent advances in plant tissue culture, include meristem tip propagation, callus and single cell culture, haploid plant production, and protoplast isolation, culture, transformation, fusion, and regeneration into whole plants. These advances have opened a whole new array of possibilities and methodologies for plant improvement, including improvement of plant resistance to infection by pathogens. The potential of these techniques is further augmented by combination with recombinant DNA and related technologies (genetic engineering). Genetic engineering techniques allow the detection, isolation, modification, transfer, and expression of single genes, or groups of related genes, from one organism to another. Several tissue culture techniques, for example, regeneration of whole plants from callus, single cells, protoplasts, and microspores or pollen, lead by themselves to plants showing greater variability in many characteristics, including resistance to disease.

Selection of the best among such plants and subsequent application of classical breeding techniques make possible production of improved plants with greater efficiency and at a much greater rate. The application of recombinant DNA technologies in plant improvement depends on the kinds of plant tissue culture with which one is working, but increases their potential tremendously by enabling plant scientists to pinpoint cell genes with specific functions and to transfer them into new cells and organisms.

Tissue Culture of Disease-Resistant Plants Tissue culture of disease-resistant plants is particularly useful with clonally propagated plants such as strawberries, apples, bananas, sugarcane, cassava, and potatoes. Prolific plantlet production from meristem and other tissue cultures facilitate the rapid propagation of plants with exceptional (resistant) genotypes, especially in those crops not easily propagated by seed. An even greater use of tissue culture is for production of pathogen-free stocks of clonally propagated susceptible plants.

Isolation of Disease Resistant Mutants from Plant Cell Cultures Plants regenerated from culture (callus, single cells, or protoplasts) often show considerable variability **(somaclonal variation)**, much of it useless or deleterious. However, plants with useful characteristics may also emerge. For example, when plants were regenerated from leaf protoplasts of a potato variety susceptible to both *Phytophthora infestans* and *Alternaria solani,* some of them (5 of 500) were resistant to *Alternaria solani* and some (20 of 800) were resistant to *Phytophthora infestans.* Similarly, plants exhibiting increased resistance to diseases caused by *Helminthosporium* and *Ustilago* were obtained from tissue cultures of sugarcane.

Production of Resistant Dihaploids from Haploid Plants Immature pollen cells (microspores)—and less often megaspores—of many plants can be induced to develop into haploid (1n) plants in which single copies (alleles) of each gene are present in all sorts of combinations. By vegetative propagation and proper screening for disease resistance, the most highly resistant haploids can be selected. These haploids can be subsequently treated with colchicine, which results in diploitization of the nuclei, that is, doubling of the number of chromosomes and production of dihaploid plants homozygous for all genes, including genes for resistance.

Increasing Disease Resistance by Protoplast Fusion Protoplasts from closely related and even from unrelated plants, under proper conditions, can be made to fuse. The fusion produces **hybrid cells** containing the nuclei (chromosomes) and the cytoplasm of both protoplasts, or perhaps result in **cybrid cells** containing the nucleus of one cell and the cytoplasm of the other cell. Generally, hybrids of unrelated cells sooner or later abort, or may produce callus, but do not regenerate plants. In combinations of more or less related cells, however, while many or most of the chromosomes of one of the cells are eliminated during cell division, one or a few chromosomes of that cell survive and may be incorporated in the genome of the other cell. In this way, plants with more chromosomes and thereby new characteristics can be regenerated from the products of protoplast fusion. Protoplast fusion is particularly useful between protoplasts of different, highly resistant haploid lines of the same variety or species. Protoplast fusion of such lines results in diploid plants that combine the resistance genes of two highly resistant haploid lines.

Genetic Transformation of Plant Cells for Disease Resistance Genetic material (DNA) can be introduced into plant cells or protoplasts by several methods. Such methods include direct DNA uptake, microinjection of DNA, liposome (lipid vesicle)-mediated delivery of DNA, delivery by means of centromere plasmids (minichromosomes), by use of plant viral vectors, and most importantly by use of the natural gene vector system of *Agrobacterium tumefaciens,* the cause of crown gall disease of many plants. In all of these methods, small or large pieces of DNA are introduced into plant cells or protoplasts and the DNA may or may not be integrated in the plant chromosomal DNA. When the introduced DNA carries appropriate regulatory genes recognized by the plant cell, or is integrated near appropriate regulatory genes along plant chromosomes, the DNA is "expressed," that is, it is transcribed into mRNA, which is then translated into protein.

So far, only the *Agrobacterium* system has been used successfully to introduce into plants specific new genes that were then expressed by the plant.This was accomplished by isolating the seed protein gene from beans and splicing it into the appropriate area of the Ti-plasmid (tumor-inducing plasmid) of *Agrobacterium;* the bacteria were then allowed to infect other plants, for example, the sunflower. Upon infection, about one-tenth of the DNA of the plasmid, containing the new gene, is transfered to the plant cell and is incorporated into the plant's genome. There the new gene replicates during plant cell division and is expressed along with the other plant genes.

To date, no gene for disease resistance has been isolated, and therefore no genetic transformation for disease resistance has been accomplished in any host–pathogen combination. It is considered likely, however, that genes for resistance, particularly single genes involved in vertical resistance, will be among the first to be successfully used for genetic transformation of plants, and that breeding for disease resistance will be among the first areas of plant breeding to profit greatly from the application of techniques in genetic engineering.

Advantages and Problems in Breeding for Vertical or Horizontal Resistance

Resistance may be obtained by incorporating into a variety one, a few, or many resistance genes. Some of these genes may control important steps in disease development and may therefore play a major role in disease resistance. Other genes may control peripheral events of lesser importance in disease development and, therefore, play a relatively minor role in disease resistance. Obviously, one or a few major-role genes could be sufficient to make a plant resistant to a pathogen (monogenic, oligogenic, or vertical resistance). However, it would take many minor-effect genes to make a plant resistant (polygenic or horizontal resistance). More importantly, while a plant with vertical resistance may be completely resistant to a pathogen, a plant with horizontal resistance is never completely resistant or completely susceptible. Furthermore, vertical resistance is easy to manipulate in a breeding program, and therefore it is often preferred to horizontal resistance, but both vertical and horizontal have their advantages and limitations.

Vertical resistance is aimed against specific pathogens or pathogen races. Vertical resistance is most effective when (1) it is incorporated in annual crops that are easy to breed, such as small grains; (2) it is directed against pathogens that do not reproduce and spread rapidly, such as *Fusarium;* or pathogens that do not mutate very frequently, such as *Puccinia graminis;* (3) it consists of "strong" genes that confer complete protection to the plant that carries it; and (4) the host population does not consist of a single genetically uniform variety grown over large acreages. If one or more of these, and of several other, conditions are not met, vertical resistance becomes short lived, that is, it breaks down as a result of appearance of new pathogen races that can bypass it or overcome it.

On the other hand, horizontal resistance confers incomplete but durable protection—it does not break down. Horizontal resistance involves host physiological processes that act as mechanisms of defense and that are beyond the limits of the capacity of the pathogen to change, that is, beyond the probable limits of its variability. Horizontal resistance is universally present in wild and domesticated plants but is at its highest in wild plants and at its lowest in greatly "improved" varieties. Horizontal resistance operates against all races of a pathogen including the most pathogenic ones. Actually, the more pathogenic the races present, the greater the selection for general resistance in the host. Horizontal resistance is eroded in the absence of the pathogen because there is no selection pressure for resistance. In many cases, horizontal (polygenic) resistance is an important part of the resistance exhibited by plants possessing vertical (specific) resistance, but it is overshadowed by the latter. Considerable horizontal resistance can be incorporated into cultivated crops by cross-breeding of existing, genetically different varieties. The wider the genetic base of a crop, the greater its horizontal resistance and the smaller the need for the usually temporary vertical resistance.

Varieties with horizontal (polygenic, general, or nonspecific) resistance remain resistant much longer than do varieties with vertical (oligogenic or specific) resistance, but the level of resistance of the second group is much higher while it lasts than that of the first group. Also, varieties with vertical resistance are often attacked suddenly and rapidly by a new virulent race and lead to severe epidemics. These disadvantages can be avoided in some crops by the use of multilines. **Multilines** are either mixtures of individual varieties (lines or cultivars) that are agronomically similar but differ in their resistance genes or varieties that are derived from crossing several to many varieties that contain different resistance genes and then selecting from those that contain the mixtures of genes. Multilines have been developed mostly in small grains against the rust fungi, but their use is likely to increase in these and in other crops as the control of plant diseases with specific resistance and with chemicals becomes more risky or less acceptable.

Incorporating genes for resistance from wild or unsatisfactory plants into susceptible but agronomically desirable varieties is a difficult and painstaking process involving repeated crossings, testings, and backcrossings to the desirable varieties. The feasibility of the method in most cases, however, has been proved repeatedly. Through breeding, varieties of some crops, such as tobacco, have been developed in which genes for resistance against several different diseases have been incorporated.

Vulnerability of Genetically Uniform Crops to Plant Disease Epidemics

Varieties with even complete vertical resistance do not remain resistant forever. The continuous production of mutants and hybrids in pathogens sooner or later leads to the appearance of races that can infect previously resistant varieties. Sometimes, races may exist in an area in small populations and avoid detection until after the introduction of a new variety, or virulent races of the pathogen existing elsewhere may be brought in after introduction of the resistant variety. In all cases, widespread cultivation of a single, previously resistant variety provides an excellent substrate for rapid development and spread of the new race of the pathogen, and usually leads to an epidemic. Thus, genetic uniformity in crops, although very desirable when it concerns horticultural characteristics, is undesirable and often catastrophic when it occurs in the genes of resistance to diseases.

The cultivation of varieties with genetically uniform disease resistance is possible and quite safe if other means of plant disease control, such as chemical, are possible. Thus, a few fruit tree varieties, such as Delicious apples, Bartlett pears, Elberta peaches, and Navel oranges, are cultivated throughout the world in the face of numerous virulent fungal and bacterial pathogens that would destroy them in a short time were it not for the fact that the trees are protected from the pathogens by numerous chemical sprays annually. Even such varieties, however, suffer tremendous losses when affected by pathogens that cannot be controlled with chemicals, as is the case of fire blight of pears and pear decline and of tristeza disease of citrus.

Another case in which varieties with genetically uniform disease resistance are not likely to suffer from severe disease epidemics is when the resistance is aimed against slow-moving soil pathogens such as *Fusarium* and *Verticillium.* Aside from the fact that some pathogens normally produce fewer races than others, even if new races are produced at the same rate, soilborne pathogens lack the dispersal potential of airborne ones. As a result, a new race of a soilborne pathogen would be limited to a relatively small area for a long time, and although it could cause a locally severe disease, it would not spread rapidly and widely to cause an epidemic. The slow spread of such virulent new races of soilborne pathogens allows time for the control of the disease by other means or the replacement of the variety with another one resistant to the new race.

Genetic uniformity in plant varieties becomes a serious disadvantage in the production of major crops because of the potential danger of sudden and widespread disease epidemics caused by airborne or insect-borne pathogens in the vast acreages in which each of these varieties is often grown. Several examples of epidemics that resulted from genetic uniformity are known and some of them have already been mentioned. Southern corn leaf blight was the result of the widespread use of corn hybrids containing the Texas male-sterile cytoplasm; the destruction of the "Ceres" spring wheat by race 56 of *Puccinia graminis* and of the "Hope" and its relative bread wheats by race 15B of *P. graminis* was the result of replacement of numerous genetically diverse varieties by a few uniform ones. The Helminthosporium blight of Victoria oats was

Table 6-4 Acreage of Major U. S. Crops and Extent to Which Small Numbers of Varieties Dominate Crop Acreage (Reproduced with permission of the National Academy of Sciences, 1972)

Crop	Acreage (millions)	Total Varieties	Major Varieties	Acreage (percent)
Bean, dry	1.4	25	2	60
Bean, snap	0.3	70	3	76
Cotton	11.2	50	3	53
Corn[a]	66.3	197	6	71
Millet	2.0		3	100
Peanut	1.4	15	9	95
Peas	0.4	50	2	96
Potato	1.4	82	4	72
Rice	1.8	14	4	65
Sorghum	16.8	?	?	?
Soybean	42.4	62	6	56
Sugar beet	1.4	16	2	42
Sweet potato	0.13	48	1	69
Wheat	44.3	269	9	50

[a] Corn includes only released Agricultural Experimental Station inbreds for seed, forage, and silage.

the result of replacing many varieties with the rust-resistant Victoria oats; coffee rust destroyed all coffee trees in Ceylon because all of them originated from uniform susceptible stock of *Coffea arabica;* tristeza destroyed millions of orange trees in South and North America because they were all propagated on hypersensitive resistant sour orange rootstocks; and pear decline destroyed millions of pear trees in the Pacific coast states because they were propagated on hypersensitive resistant oriental rootstocks. In spite of these and many other well-known examples of plant disease epidemics that occurred because of the concentrated cultivation of genetically uniform crops over large areas, crop production continues to depend on genetic uniformity. A few varieties of each crop used to and for some crops still make up the bulk of the cultivated crop over as vast an area as the United States (Table 6-4). As can be seen in Table 6-4, although a relatively large number of varieties are available for each crop, only a few, often 2 or 3, varieties are grown in more than half the acreage of each crop, and in some they make up more than three-fourths of the crop. It is easy to see that two pea varieties make up almost the entire (96 percent) pea crop of the country, that is, about 400,000 acres, that two varieties account for 42 percent of the sugarbeet crop, that is, about 600,000 acres. But the figures become even more spectacular when one considers the most popular varieties of the truly large acreage crops. Thus, although six corn varieties account for 71 percent, or 47 million acres, one of them alone accounts for 26 percent, or 17 million acres. Furthermore, most of the varieties shared the same male-sterile cytoplasm. Similarly, six varieties of soybean account for about 24 million acres of that crop, and most of these varieties share common ancestors.

It is apparent that several hundreds of thousands or several million acres planted to one variety present a huge opportunity for the development of an epidemic. The variety, of course, is planted so widely because it is resistant to existing pathogens. But this resistance puts extreme survival pressure on the pathogens over that area. It takes one "right" change in one of the zillions of pathogen individuals in the area to produce a new virulent race that can attack the variety. When that happens it is a matter of time—and, usually, of favorable weather—before the race breaks loose, the epidemic develops, and the variety is wiped out! In some cases the appearance of the new race is detected early, and the variety is replaced with another one, resistant to the new race, before a widespread epidemic occurs; this of course requires that varieties of a crop with different genetic base are available at all times. For this reason, most varieties must usually be replaced within about 3–5 years from the time of their widespread distribution.

In addition to the genetic uniformity within one variety, plant breeding often introduces genetic uniformity to several or all cultivated varieties of a crop by introducing one or several genes in all of these varieties or by replacing the cytoplasm of the varieties with a single type of cytoplasm. Induced uniformity through introduced genes includes, for example, the dwarfism gene in the dwarf wheats and rice varieties, the monogerm gene in sugar beet varieties, the determinate gene in tomato varieties, and the stringless gene in bean varieties. Uniformity through replacement of the cytoplasm occurred, of course, in most corn hybrids in the later 1960s when the Texas male-sterile cytoplasm replaced the normal cytoplasm; and cytoplasmic uniformity is commercially employed in several varieties of sorghum, sugar beet, and onions, it is studied in wheat, and is also present in cotton and cantaloupe. Neither the introduced genes nor the replacement cytoplasm, of course, make the plant less resistant to diseases, but if a pathogen appears that is favored by or can take advantage of the characters controlled by that gene or that cytoplasm, then the stage is set for a major epidemic. That this can happen was proved by the southern corn leaf blight epidemic of 1970, the susceptibility of dwarf wheats to new races of *Septoria* and *Puccinia,* of tomatoes with the determinate gene to *Alternaria,* and others.

In recent years, an effort has been made to plant a smaller percent of the total acreage of a crop with a few selected varieties, but for most crops and most areas that acreage is still too great. For example, in 1985 the top 6 soybean varieties made up only 41 percent and the top 9 wheat varieties made up only 34 percent of the total soybean and wheat acreage, respectively. However, during the same year, the 3 most popular cotton varieties made up 54 percent of the total cotton acreage. Furthermore, also in 1985, the 4 most popular potato varieties in each of the 11 leading potato-producing states accounted for 63 to 100 percent of the total potato crop in any one of these states, while the most popular barley varieties in each of the top 6 barley-producing states accounted for 44 to 94 percent of the total crop in each state.

SELECTED REFERENCES

Agrios, G. N. (1980). Escape from disease. *In* "Plant Disease" (J. G. Horsfall and E. B. Cowling, eds.), Vol. 5, pp. 17–37. Academic Press, New York.

Bailey, J. A. (1983). Biological perspectives of host-pathogen interactions. *In* "The Dynamics of Host Defense" (J. A. Bailey and B. J. Deverall, eds.), pp. 1–32. Academic Press, Australia.

Boone, D. M. (1971). Genetics of *Venturia inaequalis. Annu. Rev. Phytopathol.* **9,** 297–318.

Brinkerhoff, L. A., Verhalen, L. M., Johnson, W. M., Essenberg, M., and Richardson, P. E. (1984). Development of immunity to bacterial blight of cotton and its implication for other diseases. *Plant Dis.* **68,** 168–173.

Browder, L. E. (1971). Pathogenic specialization in cereal rust fungi, especially *Puccinia recondita* f. sp. *tritici:* Concepts, methods of study and application. *Kans., Agric. Exp. Stn., Tech. Bull.* **1432,** 1–51.

Browning, J. A., and Frey, K. J. (1969). Multiline cultivars as a means of disease control. *Annu. Rev. Phytopathol.* **7,** 355–382.

Buddenhagen, I. W. (1983). Breeding strategies for stress and disease resistance in developing countries. *Annu. Rev. Phytopathol.* **21,** 385–409.

Bushnell, W. R., and Rowell, J. B. (1981). Suppressors of defense reactions: A model for roles in specificity. *Phytopathology* **71,** 1012–1014.

Daly, J. M. (1983). Current concepts of disease resistance in plants. *In* "Challenging Problems in Plant Health" (T. Kommedahl and P. H. Williams, eds.), pp. 311–323. Am. Phytopathol. Soc., St. Paul, Minnesota.

Day, P. R. (1973). Genetic variability of crops. *Annu. Rev. Phytopathol.* **11,** 293–312.

Day, P. R. (1974). "Genetics of Host-Parasite Interaction." Freeman, San Francisco, California.

Day, P. R., ed. (1977). "The Genetic Basis of Epidemics in Agriculture," Ann. N. Y. Acad. Sci., No. 287. N. Y. Acad. Sci., New York.

Dinoor, A., and Eshed, N. (1984). The role and importance of pathogens in natural plant communities. *Annu. Rev. Phytopathol.* **22,** 443–466.

Downey, K., Voellmy, R. W., Ahmad, F., and Schultz, J., eds. (1984) "Advances in Gene Technology; Molecular Genetics of Plants and Animals." Academic Press, New York.

Dubin, H. J., and Rajoram, S. (1982). The CIMMYT's international approach to breeding disease-resistant wheat. *Plant Dis.* **66,** 967–971.

Dyck, P. L., and Johnson, R. (1983). Temperature sensitive genes for resistance in wheat to *Puccinia recondita. Can. J. Plant Pathol.* **5,** 229–235.

Ellingboe, A. H. (1981). Changing concepts in host-pathogen genetics. *Annu. Rev. Phytopathol.* **19,** 125–143.

Evans, D. A., Sharp, W. R., Ammirato, P. V., and Yamada, Y., eds. (1983). "Handbook of Plant Cell Culture," Vol. 1. Macmillan, New York.

Fincham, J. R. S., and Day, P. R. (1971). "Fungal Genetics," 3rd ed. Davis, Philadelphia, Pensylvania.

Flor, H. H. (1956). The complementary genic systems in flax and flax rust. *Adv. Genet.* **8,** 29–54.

Flor, H. H. (1971). Current status of the gene-for-gene concept. *Annu. Rev. Phytopathol.* **9,** 275–296.

Gallegly, M. E. (1970). Genetics of *Phytophthora. Phytopathology* **60,** 1135–1141.

Gilchrist, D. G., and Yoder, O. C. (1984). Genetics of host-parasite systems: A prospectus for molecular biology. *In* "Plant-Microbe Interactions: Molecular and Genetic Perspectives" (T. Kosuge and E. W. Nester, eds.), Vol. 1, pp. 69–90. Macmillan, New York.

Green, C. J. (1981). Identification of physiologic races of *Puccinia graminis* f. sp. *tritici* in Canada. *Can. J. Plant Pathol.* **3,** 33–39.

Halisky, P. M. (1965). Physiologic specialization and genetics of the smut fungi. III. *Bot. Rev.* **31,** 114–150.

Harris, M. K., and Frederiksen, R. A. (1984). Concepts and methods regarding host plant resistance to arthropods and pathogens. *Annu. Rev. Phytopathol.* **22,** 247–272.

Heath, M. (1981). A generalized concept of host-parasite specificity. *Phytopathology* **71,** 1121–1123.

Holmes, F. O. (1965). Genetics of pathogenicity in viruses and of resistance in host plants. *Adv. Virus Res.* **11,** 139–162.

Holton, C. S., Hoffman, J. A., and Duran, R. (1968). Variation in the smut fungi. *Annu. Rev. Phytopathol.* **6,** 213–242.

Hooker, A. L. (1967). The genetics and expression of resistance in plants to rusts of the genus *Puccinia. Annu. Rev. Phytopathol.* **5,** 163–182.

Hooker, A. L. (1974). Cytoplasmic susceptibility in plant disease. *Annu. Rev. Phytopathol.* **12,** 167–179.

Keen, N. T. (1982). Specific recognition in gene-for-gene host parasite systems. *In* "Advances in Plant Pathology" (D. S. Ingram and P. H. Williams, eds.) Vol. 1, pp. 35–82. Academic Press, London.

Kiyosawa, S. (1982). Genetics and epidemiological modeling of breakdown of plant disease resistance. *Annu. Rev. Phytopathol.* **20,** 93–117.

Kosuge, T., Meredith, C. P., and Hollaender, A., eds. (1983). "Genetic Engineering of Plants." Plenum, New York.

optimum range (18 to 23°C) of temperature for the pathogen. At lower or higher temperatures the minimum wetting period required is higher, for example, 14 hours at 10°C, 28 hours at 6°C. If the length of the wetting period is less than the minimum required for the particular temperature, the pathogen fails to establish itself in the host and to produce disease.

Most fungal pathogens are dependent on the presence of free moisture on the host or high relative humidity in the atmosphere only during germination of their spores and become independent once they can obtain nutrients and water from the host. Some pathogens, however, such as those causing late blight of potato and the downy mildews, must have high relative humidity or free moisture in the environment throughout their development. In these diseases, the growth and sporulation of the pathogen, and also the production of symptoms, come to a halt as soon as dry, hot weather sets in and resume only after a rain or after the return of humid weather.

Though most fungal and bacterial pathogens of aboveground parts of plants require a film of water in order to infect hosts successfully, the spores of the powdery mildew fungi can germinate, penetrate, and cause infection even when there is only high relative humidity in the atmosphere surrounding the plant. In the powdery mildews, spore germination and infection are actually lower in the presence of free moisture on the plant surface then they are in its absence and, in some of them, the most severe infections take place when the relative humidity is rather low (50 to 70 percent). In these diseases, the amount of disease is limited rather than increased by wet weather, as indicated by the fact that the powdery mildews are more common and more severe in the drier areas of the world; their relative importance decreases as rainfall increases.

In many diseases affecting the underground parts of plants such as roots, tubers, and young seedlings—for example in the *Pythium* damping off of seedlings and seed decays—the severity of the disease is proportional to the amount of soil moisture and is greatest near the saturation point. The increased moisture seems to affect primarily the pathogen, which multiplies and moves (zoospores in the case of *Pythium*) best in wet soils, but may also decrease the ability of the host to defend itself through reduced availability of oxygen in water-logged soil and by lowering the temperature of such soils. Many other soil fungi (for example, *Phytophthora, Rhizoctonia, Sclerotinia,* and *Sclerotium*), some bacteria (for example, *Erwinia* and *Pseudomonas*), and most nematodes usually cause their most severe symptoms on plants when the soil is wet but not flooded. Several other fungi, for example, *Fusarium solani,* which is the cause of dry root rot of beans, *Fusarium roseum,* the cause of seedling blights, and *Macrophomina phaseoli,* the cause of charcoal rot of sorghum and of root rot of cotton grow fairly well in rather dry environments, and apparently that enables them to cause more severe diseases in drier soils on plants that are stressed by insufficient water. Similarly, *Streptomyces scabies,* which causes the common scab of potatoes, is most severe in rather dry soils.

Most bacterial diseases, and also many fungal diseases of young tender tissues, are particularly favored by high moisture or high relative humidity. Bacterial pathogens and fungal spores are usually disseminated in water drops splashed by rain, in rainwater moving from the surfaces of infected tissues to those of healthy ones, or in free water in the soil. Bacteria penetrate plants through wounds or natural openings and cause severe disease when present in large numbers. Once inside the plant tissues, the bacteria multiply faster and are more active during wet weather, probably because the plants, through increased water absorption and resulting succulence, can provide the high concentrations of water that favor bacteria. The increased bacterial activity produces greater damage to tissues, and this damage, in turn, helps release greater numbers of bacteria onto, the plant surface, where they are available to start more infections if the wet weather continues.

Effect of Wind

Wind influences infectious plant diseases primarily through its importance in the spread of plant pathogens and, to a smaller extent, through its acceleration of the drying of wet plant surfaces. Most plant diseases that spread rapidly and are likely to assume epidemic proportions are caused by pathogens such as fungi, bacteria, and viruses that are either spread directly by the wind or indirectly by insect vectors that can themselves be carried over long distances by the wind. Some spores, for example, zoosporangia, basidiospores, and some conidia, are quite delicate and do not survive long-distance transport in the wind. Others, for example, uredospores and many kinds of conidia, can be transported by the wind for many miles. Wind is even more important in disease development when it is accompanied by rain. Wind-blown rain helps release spores and bacteria from infected tissue and then carries them through the air and deposits them on wet surfaces which, if susceptible, can be infected immediately. Wind also injures plant surfaces while they are blown about and rub against each other; this facilitates infection by many fungi and bacteria and also by some mechanically transmitted viruses. Wind, however, sometimes helps prevent infection by accelerating the drying of the wet plant surfaces on which fungal spores or bacteria may have landed. If the plant surfaces dry before penetration has taken place, any germinating spores or bacteria present on the plant are likely to desiccate and die and no infection will occur.

Effect of Light

The effect of light on disease development, especially under natural conditions, is far less than that of temperature or moisture, although several diseases are known in which the intensity and/or the duration of light may either increase or decrease the susceptibility of plants to infection and also the severity of the disease. In nature, however, the effect of light is limited to the production of more or less etiolated plants as a result of reduced light intensity. This usually increases the susceptibility of plants to nonobligate parasites, for example, of lettuce and tomato plants to *Botrytis* or of tomato to *Fusarium,* but decreases their susceptibility to obligate parasites, for example, of wheat to the stem rust fungus *Puccinia.*

Reduced light intensity generally increases the susceptibility of plants to virus infections. Holding plants in the dark for one to two days before inoculation increases the number of lesions (that is, infections) appearing after inoculation and this has become a routine procedure in many laboratories. Generally, darkening affects the sensitivity of plants to virus infection if it precedes inoculation with the virus, but seems to have little or no effect on symptom development if it occurs after inoculation. On the other hand, low light intensities following inoculation tend to mask the symptoms of some diseases, which are much more severe when the plants are grown in normal light than when they are shaded.

Effect of Soil pH

The pH of the soil is important in the occurrence and severity of plant diseases caused by certain soil-borne pathogens. For example, the clubroot of crucifers, caused by *Plasmodiophora brassicae,* is most prevalent and severe at about pH 5.7, while its development drops sharply between 5.7 and 6.2, and is completely checked at pH 7.8. On the other hand, the common scab of potato, caused by *Streptomyces scabies,* can be severe at a pH range from 5.2 to 8.0 or above, but its development drops sharply below pH 5.2. It is obvious that such diseases are most serious in areas whose soil pH favors the particular pathogen. In these, and in many other diseases, the effect of soil acidity (pH) seems to be principally on the pathogen, although in some, a weakening of the host through altered nutrition that is induced by the soil acidity may affect the incidence and severity of the disease.

Effect of Host-Plant Nutrition

Nutrition affects the rate of growth and the state of readiness of plants to defend themselves against pathogenic attack.

Nitrogen abundance results in the production of young, succulent growth and may prolong the vegetative period and delay maturity of the plant. These effects make the plant more susceptible to pathogens that normally attack such tissues, and for longer periods. Conversely, plants suffering from a lack of nitrogen would be weaker, slower growing, and faster aging and susceptible to pathogens that are best able to attack weak, slow growing plants. It is known, for example, that fertilization with large amounts of nitrogen increases the susceptibility of the pear to fire blight *(Erwinia amylovora),* and of wheat to rust *(Puccinia)* and powdery mildew *(Erysiphe).* Reduced availability of nitrogen may increase the susceptibility of tomato to *Fusarium* wilt, of many solanaceous plants to *Alternaria solani* early blight and *Pseudomonas solanacearum* wilt, of sugar beets to *Sclerotium rolfsii,* and of most seedlings to *Pythium* damping off.

It is possible, however, that it is the form of nitrogen (ammonium or nitrate) that is available to the host or pathogen that affects disease severity or resistance rather than the amount of nitrogen. Of numerous root rots, wilts, foliar diseases, etc., treated with either form of nitrogen, almost as many decreased or increased in severity when treated with a source of ammonium nitrogen as did when treated with a source of nitrate nitrogen, but each form of nitrogen had exactly the opposite effect on a disease (that is, decrease or increase its severity) than did the other form of nitrogen. For example, *Fusarium* spp., *Plasmodiophora brassicae, Sclerotium rolfsii,* and the diseases they cause (root rots and wilts, clubroot of crucifers, and damping off and stem rots, respectively) increase in severity when an ammonium fertilizer is applied, while *Phymatotrichum omnivorum, Gaeumannomyces graminis, Streptomyces scabies* and the diseases they cause (cotton root rot, take-all of wheat, and scab of potato, respectively) are favored by nitrate nitrogen. It is thought that the effect of each nitrogen form is associated with soil pH influences, since diseases increased by ammonium nitrogen are generally more severe at acid pH while those increased by nitrate nitrogen are generally more severe at neutral to alkaline pH (NH_4^+ ions are absorbed by the roots through exchange with H^+ released by the roots to the surrounding medium, thus reducing soil pH).

Although nitrogen nutrition, because of nitrogen's profound effects on growth, has been studied the most extensively in relation to disease development, studies with other elements such as phosphorus, potassium, and calcium, and also with micronutrients, have revealed similar relationships between levels of the particular nutrients and susceptibility or resistance to certain diseases.

Phosphorus has been shown to reduce the severity of take-all disease of barley (caused by *Gaeumannomyces graminis*) and potato scab (caused by *Streptomyces scabies*), but to increase the severity of cucumber mosaic virus on spinach and glume blotch (caused by *Septoria*) on wheat. Phosphorus seems to increase resistance either by improving the balance of nutrients in the plant or by accelerating the maturity of the crop and allowing it to escape infection by pathogens that prefer younger tissues.

Potassium has also been shown to reduce the severity of numerous diseases, including stem rust of wheat, early blight of tomato, and stalk rot of corn, although high amounts of potassium seems to increase the severity of rice blast (caused by *Pyricularia oryzae*) and root knot (caused by the nematode *Meloidogyne incognita*). Potassium seems to directly affect the various stages of pathogen establishment and development in the host, and to indirectly affect infection by promoting wound healing, by increasing resistance to frost injury and thereby reducing infection that commonly begins in frost-killed tissues, and by delaying maturity and senescence in some crops beyond the periods in which infection by certain facultative parasites can be severely damaging.

Calcium reduces the severity of several diseases caused by root and/or stem pathogens such as the fungi *Rhizoctonia, Sclerotium,* and *Botrytis,* the wilt fungus *Fusarium oxysporum,* and the nematode *Ditylenclus dipsaci,* but it increases the black shank disease of tobacco (caused by *Phytophthora parasitica* var. *nicotianae*) and the common scab of potato (caused by *Streptomyces scabies*). The effect of calcium on disease resistance seems to result from its effect on the composition of cell walls and their resistance to penetration by pathogens.

Reduction in disease levels was also observed when the levels of certain micronutrients were increased. For example, applications of iron to the soil reduced *Verticillium* wilts of mango and of peanuts, and foliar applications of iron compounds reduced the severity of silver leaf of deciduous fruit trees (caused by *Stereum purpureum*). Similarly, applications of manganese reduced potato scab and late blight of potato and stem rot (caused by *Sclerotinia sclerotiorum*) of pumpkin seedlings, while applications of molybdenum reduced late blight of potato and *Ascochyta* blight of beans and peas. The severity of other diseases, however, was raised by the presence of higher levels of these micronutrients, for example, *Fusarium* wilt of tomato by increased iron or manganese and tobacco mosaic of tomatoes by increased manganese.

In general, plants receiving a balanced nutrition, in which all required elements are supplied in appropriate amounts, are more capable of protecting themselves from new infections and of limiting existing infections than when one or more nutrients are supplied in excessive or deficient amounts. However, even a balanced nutrition may affect the development of a disease when the concentration of all the nutrients is increased or decreased beyond a certain range.

Effect of Herbicides

Herbicide use is common and widespread in agriculture. In many cases, herbicides have been shown to increase the severity of certain diseases on crop plants, for example, of *Rhizoctonia solani* on sugar beets and cotton, *Fusarium* wilt of tomatoes and cotton, and *Sclerotium* stem rots of various crops. In other host–pathogen combinations, herbicides appear to decrease disease, for example, *Aphanomyces euteiches* root rot of peas, *Pseudocercosporella herpotrichoides* foot rot of wheat, and *Phytophthora* collar rot of various crops. Herbicides apparently act on plant diseases either directly by affecting (stimulating or retarding) the growth of the pathogen or increasing or decreasing the susceptibility of the host, or indirectly by increasing or decreasing the activity of soil microflora, by eliminating or selecting for certain additional or alternate hosts of the pathogen, or by altering the microclimate (for example, humidity) of the crop plant canopy.

SELECTED REFERENCES

Altman, J., and Campbell, C. L. (1977). Effect of herbicides on plant diseases. *Annu. Rev. Phytopathol.* **15,** 361–385.

Ayres, P. G. (1984). The interaction between environmental stress injury and biotic disease physiology. *Annu. Rev. Phytopathol.* **22,** 53–75.

Baker, R. (1978). Inoculum potential. *In* "Plant Disease" (J. G. Horsfall and E. B. Cowling, eds.), Vol. 2, pp. 137–157. Academic Press, New York.

Chupp, C. (1928). Club root in relation to soil alkalinity. *Phytopathology* **18,** 301–306.

Colhoun, J. (1973). Effects of environmental factors on plant disease. *Annu. Rev. Phytopathol.* **11,** 343–364.

Colhoun, J. (1979). Predisposition by the environment. *In* "Plant Disease" (J. G. Horsfall and E. B. Cowling, eds.), Vol. 4, pp. 75–92. Academic Press, New York.

Dickson, J. G. (1923). Influence of soil temperature and moisture on the development of seedling blight of wheat and corn caused by *Gibberella saubinetii. J. Agric. Res. (Washington, D. C.)* **23,** 837–870.

Galleghly, M. E., Jr., and Walker, J. C. (1949). Plant nutrition in relation to disease development. V. *Am. J. Bot.* **36,** 613–623.

Gunfer, B. M., Touchton, J. T., and Johnson, J. W. (1980). Effects of phosphorus and potassium fertilization on *Septoria* glume blotch of wheat. *Phytopathology* **70,** 1196–1199.

Hepting, G. H. (1963). Climate and forest diseases. *Annu. Rev. Phytopathol.* **1,** 31–50.
Huber, D. M., and Watson, R. D. (1974). Nitrogen form and plant disease. *Annu. Rev. Phytopathol.* **12,** 139–165.
Jacobsen, B. J., and Hopen, H. J. (1981). Influence of herbicides on *Aphanomyces* root rot of peas. *Plant Dis.* **65,** 11–16.
Jones, L. R., Johnson, J., and Dickson, J. G. (1926). Wisconsin studies upon the relation of soil temperature to plant diseases. *Res. Bull.— Wis., Agric. Exp. Stn.* **71.**
Kassanis, B. (1957). Effect of changing temperature on plant virus diseases. *Adv. Virus Res.* **4,** 169–186.
Keitt, G. W., and Jones, K. L. (1926). Studies of the epidemiology and control of apple scab. *Res. Bull.— Wis., Agric. Exp. Stn.* **73.**
MacKenzie, D. R. (1981). Association of potato early blight, nitrogen fertilizer rate and potato yield. *Plant Dis.* **65,** 575–577.
Miller, P. R. (1953). The effect of weather on diseases. *In* "Plant Diseases." Yearb. Agric., pp. 83–93. U. S. Dep. Agric., Washington, D. C.
Palti, J. (1981) "Cultural Practices and Infectious Crop Diseases." Springer-Verlag, Berlin and New York.
Populer, C. (1978). Changes in host susceptibility with time. *In* "Plant Disease" (J. G. Horsfall and E. B. Cowling, eds.), Vol. 2, pp. 239–262. Academic Press, New York.
Schnathorst, W. C. (1965). Environment relationships in the powdery mildews. *Annu. Rev. Phytopathol.* **3,** 343–366.
Schoeneweiss, D. F. (1981). The role of environmental stress in diseases of woody plants. *Plant Dis.* **65,** 308–314.
Shaner, G. (1981). Effects of environment on fungal leaf blights of small grains. *Annu. Rev. Phytopathol.* **19,** 273–296.
Zentmyer, G. A., and Bald, J. G. (1977). Management of the environment. *In* "Plant Disease" (J. G. Horsfall and E. B. Cowling, eds.), Vol. 1, pp. 121–144. Academic Press, New York.

8 PLANT DISEASE EPIDEMIOLOGY

When a pathogen spreads to and affects many individuals within a population over a relatively large area and within a relatively short time, the phenomenon is called an **epidemic.** An epidemic has been defined as any increase of disease in a population. The study of epidemics and of the factors that influence them is called **epidemiology.** Plant disease epidemics, sometimes called **epiphytotics,** occur annually on most crops in many parts of the world. Most epidemics are more or less localized and cause minor to moderate losses because they are kept in check either naturally or by chemical sprays and other control measures. Occasionally, however, some epidemics appear suddenly, go out of control, and become extremely widespread or severe on a particular plant species. Some plant disease epidemics, for example, wheat rusts, southern corn leaf blight, and grape downy mildew, have caused tremendous losses of produce over rather large areas; others, for example, chestnut blight, Dutch elm disease, and coffee rust, have threatened to eliminate certain plant species from entire continents. Still others have caused untold suffering to humans. The Irish potato famine of 1845–1846 was caused by the *Phytophthora* late blight epidemic of potato, and the Bengal famine of 1943 was caused by the *Helminthosporium* brown spot epidemic of rice.

The Elements of an Epidemic

Plant disease epidemics develop as a result of the timely combination of the same elements that result in plant disease: susceptible host plants, a virulent pathogen, and favorable environmental conditions over a fairly long period of time. In addition, however, through their activities humans may unwittingly help to initiate and develop epidemics, or instead may effectively stop the initiation and development of epidemics under situations in which they would almost certainly occur without human intervention.

Thus, the chance of an epidemic increases when the susceptibility of the host and virulence of the pathogen are greater, as the environmental conditions approach the optimum level for pathogen growth, reproduction, and spread, and as the duration of all favorable combinations is prolonged, provided no human intervention occurs to reduce or stop the epidemic.

For the purpose of describing the interaction of the components of plant disease epidemics, the disease triangle that was discussed in Chapter 2 and used to describe the interaction of the components of plant disease can be expanded to include time and humans.

The amount of each of the three components of plant disease, as well as their effect on each other and, therefore, on the development of disease, are

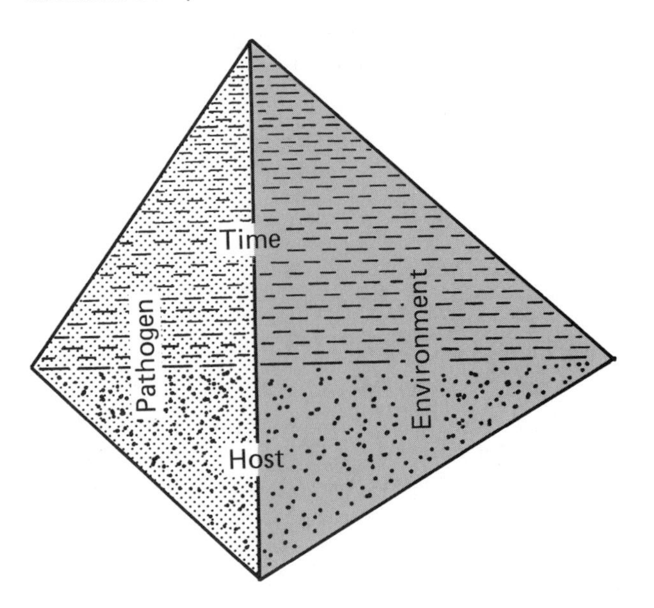

FIGURE 8-1 The disease tetrahedron.

affected by a fourth component: time. Both the specific point in time at which a particular event in disease development occurs and the length of time during which the event takes place affect the amount of disease. The interaction of the four components can be visualized as a tetrahedron, or pyramid, in which each plane represents one of the components. This figure is referred to as the **disease tetrahedron** or **disease pyramid** (Figure 8-1). The effect of time on disease development becomes apparent when one considers the importance of the time of year (that is, the climatic conditions and stage of growth when host and pathogen may coexist), the duration and frequency of favorable temperature and rains, the time of appearance of the vector, the duration of the cycle of a particular disease, the earliness or lateness of maturity of the host, etc. If the four components of the disease tetrahedron could be quantified, the volume of the tetrahedron would be proportional to the amount of disease on a plant or in a plant population.

Disease development in cultivated plants is also greatly influenced by a fifth component: humans. Humans affect the kind of plants grown in a given area, their degree of resistance, the numbers planted, time of planting, and density of the plants. By the resistance of the particular plants they cultivate, humans also determine which pathogens and pathogen races will predominate. By their cultural practices, such as chemical and biological controls, humans affect the amount of primary and secondary inoculum available to attack plants. They also modify the effect of environment on disease development by delaying or speeding up planting or harvesting, by planting in raised beds or in more widely spaced beds, by protecting plant surfaces with chemicals before rains, by regulating the humidity in produce storage areas, etc. The timing of human activities in growing and protecting plants may affect various combinations of these components to a considerable degree, thereby affecting the amount of disease in individual plants and in plant populations. The human component has sometimes been used in place of the component

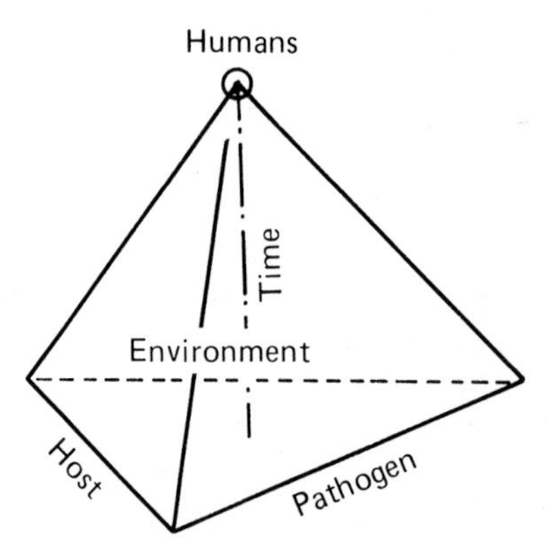

FIGURE 8-2 A schematic diagram of the interrelationships of the factors involved in plant disease epidemics.

"time" in the disease tetrahedron, but it should be considered a distinct fifth component that directly and indirectly influences the development of plant disease.

In the schematic diagram in Figure 8-2, host, pathogen, and environment are each represented by one of the sides of the triangle, time is represented as the perpendicular line arising from the center of the triangle and humans as the peak of the tetrahedron whose base is the triangle and height is the length of time. In this way, humans interact with and influence each of the other four components of an epidemic and thereby increase or decrease the magnitude of the epidemic. Sometimes, of course, humans themselves can be affected to a greater or lesser extent by plant disease epidemics.

Host Factors that Affect Development of Epidemics

Several internal and external factors of particular host plants play an important role in the development of epidemics involving those hosts.

Levels of Genetic Resistance or Susceptibility of the Host

Obviously, host plants carrying high levels of (vertical) resistance do not allow a pathogen to become established in them and thus no epidemic can develop, unless and until a new pathogen race appears that can attack that resistance and the host then becomes susceptible. Host plants carrying lower levels of (horizontal) resistance will probably become infected, but the rate at which the disease and the epidemic will develop depends on the level of resistance and the environmental conditions. Susceptible host plants lacking genes for resistance against the pathogen provide the ideal substrate for establishment and development of new infections and, in the presence of a virulent pathogen and favorable environment, favor the development of disease epidemics.

Degree of Genetic Uniformity of Host Plants

When genetically uniform host plants, particularly with regard to the genes associated with disease resistance, are grown over large areas, a greater likelihood exists that a new pathogen race will appear that can attack their genome and result in an epidemic. This phenomenon has been observed repeatedly, most characteristically in the cases of *Helminthosporium* blight on Victoria oats and of southern corn leaf blight on corn with Texas male-sterile cytoplasm. For similar reasons, the highest rates of development of epidemics generally occur on vegetatively propagated crops and the next highest in self-pollinated crops, while cross-pollinated crops seem to have the lowest rates of epidemics. Thus, this explains why most epidemics in natural populations develop rather slowly.

Type of Crop

In annual crops, such as corn, vegetables, tobacco, and cotton, epidemics generally develop much more rapidly (usually in a few weeks) than they do in perennial woody crops such as fruit and forest trees. Some epidemics of fruit and forest trees, for example, tristeza in citrus, pear decline, Dutch elm disease, and chestnut blight, take years to develop.

Age of Host Plants

Plants change in their susceptibility to disease with age. In some host–pathogen combinations, for example, *Pythium* damping off and root rots, downy mildews, peach leaf curl, systemic smuts, rusts, bacterial blights, and viral infections, the hosts (or their parts) are susceptible only during the growth period and become resistant during the adult period (adult resistance) (Figure 8-3, Ia & b). With several diseases, such as rusts and viral infections,

FIGURE 8-3 Change of susceptibility of plant parts with age. I: Plant susceptible only in the stages of its maximum growth (Ia) or in the earliest stages of its growth (Ib). II: Plant susceptible only after it reaches maturity and susceptibility increases with senescence. III: Plant susceptible while very young and again after it reaches maturity. (After Populer, 1978.)

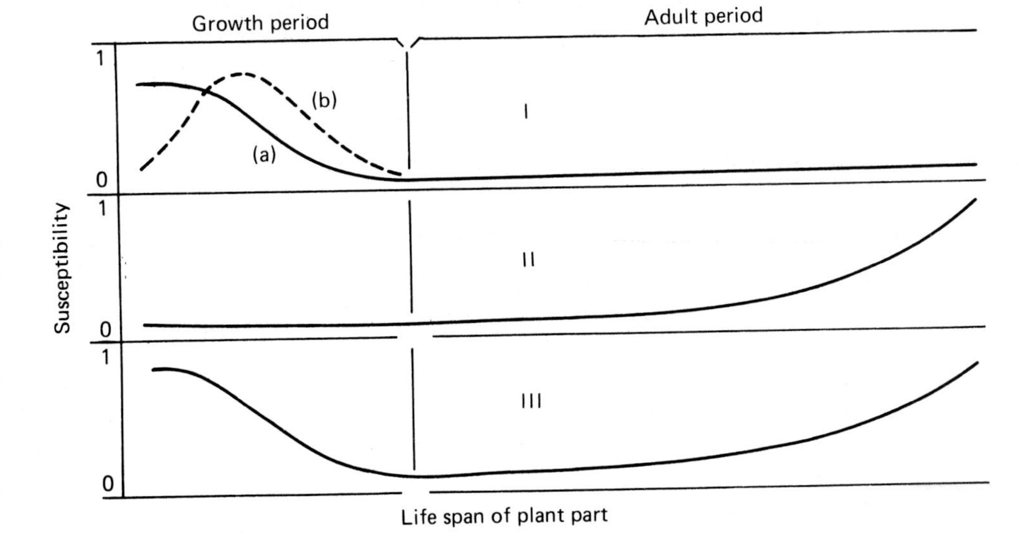

plant parts are actually quite resistant to infection while still very young, become more susceptible later in their growth, and then become resistant again before they are fully expanded (Figure 8-3, Ib). In other diseases, such as infections of blossoms or fruit by *Botrytis, Penicillium, Monilinia,* and *Glomerella,* and in all postharvest infections, plant parts (that is, fruit) are resistant during growth and the early adult period but become susceptible near ripening (Figure 8-3, II). In still other diseases, such as potato late blight (caused by *Phytopthora infestans*) and tomato early blight (caused by *Alternaria solani*), a stage of juvenile susceptibility during the growth period of the plant is followed by a period of relative resistance in the early adult stage and then susceptibility after maturity (Figure 8-3, III).

Apparently then, depending on the particular host–pathogen combination, the age of the host plant at the time of arrival of the pathogen may affect considerably the development of infection and of an epidemic.

Pathogen Factors That Affect Development of Epidemics

Levels of Virulence

Virulent pathogens capable of rapidly infecting the host ensure faster production of larger amounts of inoculum than pathogens of lesser virulence.

Quantity of Inoculum Near Hosts

The greater the number of pathogen propagules (spores, sclerotia, eggs, etc.), within or near fields of host plants, the more inoculum reaches the hosts and at an earlier time, greatly increasing the chances of an epidemic.

Type of Reproduction of the Pathogen

All pathogens produce many offspring but some of them, such as most fungi, bacteria, and viruses, produce incomparably more offspring than others. A few fungi and all nematodes and parasitic plants produce relatively small numbers of offspring. Even more important is the fact that some pathogens (most fungi, bacteria, and viruses) have short reproduction cycles and therefore can produce many reproductive cycles (generations) in a single growing season. These are the polycyclic pathogens that usually cause rusts, mildews, and leaf spots, and are responsible for most of the sudden, catastrophic plant disease epidemics in the world. Some soil fungi, such as *Fusarium* and *Verticillium,* and most nematodes usually have one to a few (2–4), reproductive cycles per growing season. The number of offspring and, especially, the conditions of dispersal of the latter pathogens limit their potential to cause sudden and widespread epidemics in a single season, although they often cause more localized, slower-developing epidemics. Several pathogens, such as the smuts and several short-cycle rusts that lack a repeating stage, require an entire year to complete a life cycle (monocyclic pathogens) and can therefore cause only one series of infections per year. In such diseases, inoculum builds up from one year to the next, and the epidemic develops over several years. Similarly, epidemics caused by pathogens that require more than one year to complete a reproductive cycle are slow to develop. Examples are cedar apple rust (2 years), white pine blister rust (3–6 years), and dwarf mistletoe (5–6 years). Such pathogens produce inoculum and cause a series of infections each year only as a result of overlapping of the polyetic generations.

Ecology of the Pathogen

Some pathogens, such as most fungi and parasitic higher plants, produce their inoculum (spores and seeds, respectively) on the surface of the aerial parts of the host. From there, spores and seeds can be dispersed with ease over a range of distances and can cause widespread epidemics. Other pathogens, such as vascular fungi and bacteria, mycoplasmas, viruses, and protozoa, reproduce inside the plant. In this case, spread of the pathogen is rare or impossible without the help of vectors, and therefore such pathogens can cause epidemics only when vectors are plentiful and active. Still other pathogens, such as soilborne fungi, bacteria, and nematodes, produce their inoculum on infected plant parts in the soil, within which the inoculum disperses slowly, if at all, and presents little danger for sudden or widespread epidemics.

Mode of Spread of the Pathogen

The spores of many plant pathogenic fungi, such as those causing rusts, mildews, and leaf spots, are released into the air and can be dispersed by air breezes or strong winds over variable distances up to several miles. These kinds of fungi are responsible for the most frequent and most widespread epidemics. In terms of their ability to cause sudden and widespread epidemics, the next most important group of pathogens includes those whose inoculum is carried by airborne vectors. Such pathogens are many of the viruses transmitted by aphids and some other insects; mycoplasmas and fastidious bacteria transmitted by leafhoppers, planthoppers, and psyllids; and some fungi (such as the cause of Dutch elm disease), bacteria (such as the cause of bacterial wilt of cucurbits), and even nematodes (such as the cause of pine wilt disease) disseminated primarily by beetles. Pathogens that are transmitted by wind-blown rain (primarily fungi causing diseases like anthracnoses and apple scab, and most bacteria) are almost annually responsible for severe but somewhat localized epidemics within a field, a township, or a valley. Pathogens carried with the seed or other vegetative propagative organs (such as tubers or bulbs) are often placed in the midst of susceptible plants, but their ability to cause epidemics depends on the effectiveness of their subsequent transmission to new plants. Finally, pathogens present in and spreading through the soil, because of the physical restrictions, are generally unable to cause sudden or widespread epidemics but often cause local, slow-spreading diseases of considerable severity.

Environmental Factors That Affect the Development of Epidemics

The majority of plant diseases occur to a lesser or greater extent in most areas where the host is grown and usually do not develop into severe and widespread epidemics. The concurrent presence in the same areas of susceptible plants and virulent pathogens does not always guarantee numerous infections, much less the development of an epidemic, and dramatizes the controlling influence of the environment on the development of epidemics. The environment may affect the availability, growth stage, succulence, and genetic susceptibility of the host. It may also affect the survival, vigor, rate of multiplication, rate of sporulation, ease, direction, and distance of dispersal of the pathogen, and the rate of spore germination and penetration. In addition, the environment may affect the number and activity of the vectors of the pathogen. The most important environmental factors that affect the development of plant disease epidemics are moisture, temperature, and the activities of humans in terms of cultural practices and control measures.

Moisture

As discussed in Chapter 7, abundant, prolonged, or repeated high moisture, whether in the form of rain, dew, or high humidity, is the predominant factor in the development of most epidemics of diseases caused by fungi (blights, downy mildews, leaf spots, rusts, and anthracnoses), bacteria (leaf spots, blights, soft rots), and nematodes. Moisture not only promotes new succulent and susceptible growth in the host, but more importantly, it increases sporulation of fungi and multiplication of bacteria. Moisture facilitates spore release by many fungi and the oozing of bacteria to the host surface, and it enables spores to germinate and zoospores, bacteria, and nematodes to move. The presence of high moisture levels allows all these events to take place constantly and repeatedly and leads to epidemics. In contrast, the absence of moisture for even a few days prevents all of these events from taking place, so that epidemics are interrupted or completely stopped. Some diseases caused by soilborne pathogens, such as *Fusarium* and *Streptomyces,* are more severe in dry, than in wet weather, but such diseases seldom develop into important epidemics. Epidemics caused by viruses and mycoplasmas are only indirectly affected by moisture, primarily by the effect that higher moisture has on the activity of the vector. Such activity may be enhanced, as happens with the fungal and nematode vectors of some viruses, or reduced, as happens with the aphid, leafhopper, and other insect vectors of some viruses and mycoplasmas, whose activity is drastically reduced in rainy weather.

Temperature

Epidemics are sometimes favored by temperatures higher or lower than a certain optimum range for the plant because they reduce the plant's level of (horizontal) resistance and, at certain levels, may even reduce or eliminate the (vertical) resistance conferred by a major gene. Plants growing at such temperatures become "stressed" and predisposed to disease, provided the pathogen remains vigorous or is stressed less than the host.

Temperature also reduces the amount of inoculum of fungi, bacteria, and nematodes that survives cold winters and of viruses and mycoplasmas that survives hot summer temperatures. In addition, low temperatures reduce the number of vectors that survive the winter. Low temperatures occurring during the growing season can reduce the activity of vectors.

The most common effect of temperature on epidemics, however, is its effect on the pathogen during the different stages of pathogenesis, that is,

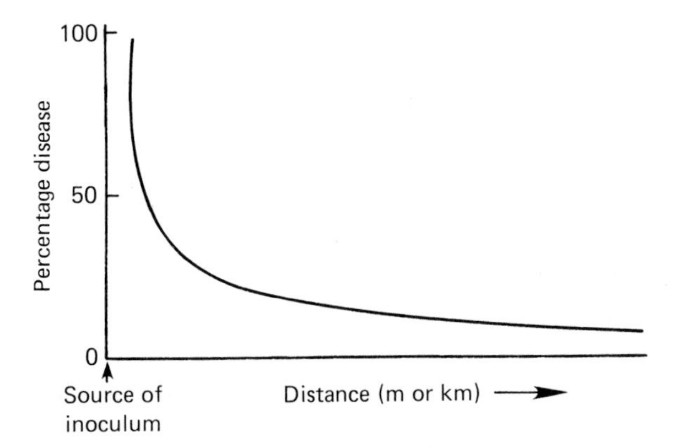

FIGURE 8-5 Schematic diagram of a disease-gradient curve. The percentage of disease and the scale for distance vary with the type of pathogen or its method of dispersal, being small for soilborne pathogens or vectors and larger for airborne pathogens.

time (per day, week, or year) in the plant population under consideration. The patterns of epidemic rates are given by curves called **rate curves** and these curves are different for various groups of diseases (see Figure 8-6). They may be symmetrical (bell shaped)—for example, in late blight of potato (Figure 8-6A); or asymmetrical with the epidemic rate greater early in the season (Figure 8-6B) due to greater susceptibility of young leaves—for example, in apple scab and most downy mildews and powdery mildews; or asymmetrical with the epidemic rate greater late in the season (Figure 8-6C), as observed in the many diseases that start slowly but accelerate markedly as host susceptibility increases late in the season—for example, *Alternaria* leaf blights and *Verticillium* wilts.

Comparison of Epidemics

For better comparison of epidemics of the same disease at different times, different locations, or under different management practices, or to compare

FIGURE 8-6 Schematic diagrams of epidemic rate curves of diseases with a symmetrical epidemic rate (A), with a high epidemic rate early in the season (B), and with a high epidemic rate late in the season (C). Dotted curves indicate possible disease-progress curves that may be produced in each case from the accumulated epidemic rate curves.

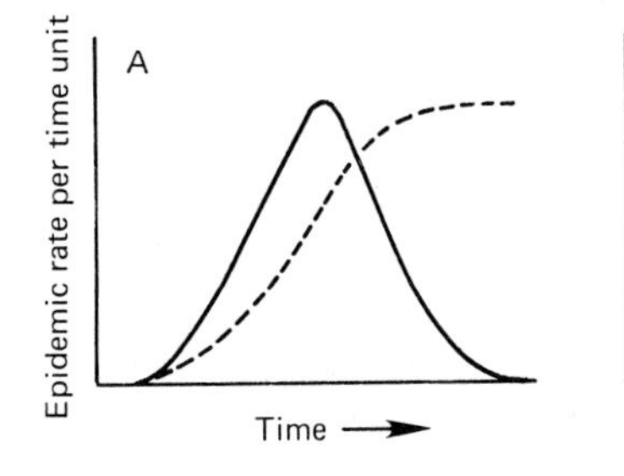

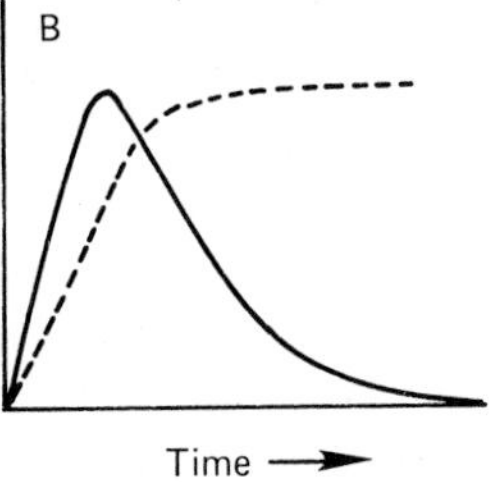

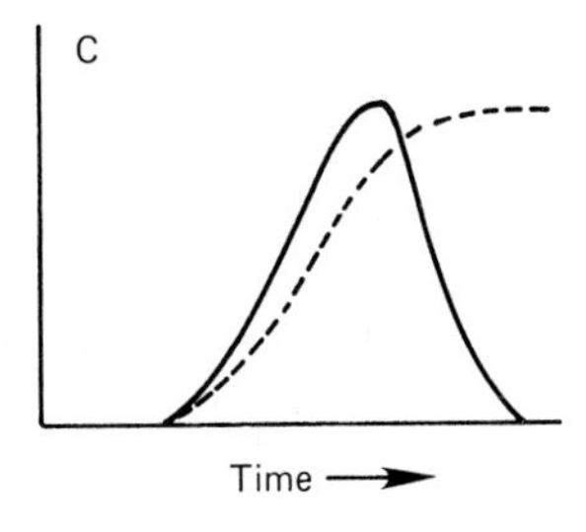

different diseases, the patterns obtained for disease-progress curves and disease-gradient curves are frequently transformed mathematically into straight lines. The slopes of these lines can then be used to calculate epidemic rates.

For monocyclic pathogens, the rate is usually designated r_m and can be calculated from a transformed disease-progress curve if the amount of initial inoculum is known. The final amount of disease is proportional to the amount of initial inoculum and the time during which the pathogen and host are in contact. As already stated, in monocyclic diseases, the amount of inoculum does not increase during the season. In such diseases, therefore, the rate of disease increase is affected only by the inherent ability of the pathogen to induce disease and by the ability of the environmental factors, host resistance, and cultural practices to influence the virulence of the pathogen.

In contrast, the initial inoculum for diseases caused by polycyclic pathogens, although extremely important, has relatively less importance than the number of disease cycles in the final disease outcome. The epidemic rate, usually designated r, can be determined from the slopes of lines of disease-progress curves. Pathogens that have many life cycles also have numerous opportunities to interact with the host. Therefore, the same factors mentioned above—the inherent ability of the pathogen to induce disease, environmental factors, host resistance, and cultural practices—have an opportunity to influence the dispersal, penetration, multiplication, size of lesion, rate of lesion formation, and rate and amount of sporulation several times during the same growth season. The continuous or, sometimes, intermittent increase of the amount of inoculum and disease may result in highly variable infection rates for individual short-time intervals during the growth season and quite variable disease rates (r) for the entire season. In general, the disease rate (r) for polycyclic diseases is much greater than the rate of disease increase (r_m) for monocyclic diseases. For example, the r_m for *Verticillium* wilt of cotton is 0.02 per day and is 1.60 per year for *Phymatotrichum* root rot of cotton, while the r for potato late blight is 0.3–0.5 per day, 0.3–0.6 per day for wheat stem rust, and 0.15 per day for cucumber mosaic virus.

In addition to the monocyclic and polycyclic diseases, there are also those caused by polyetic pathogens. Such pathogens are present for more than one year in the infected plant before they produce effective inoculum—for example, some fungal wilts and viral and mycoplasmal diseases of trees. Because of the perennial nature of their hosts, polyetic diseases behave basically as polycyclic diseases with a lower r. This happens because there are as many diseased trees and almost as much inoculum at the beginning of a year as at the end of the previous one, and both increase exponentially over the years, causing slower but just as severe epidemics. Some well-known polyetic epidemics are chestnut blight ($r = 1.2$ per year), Dutch elm disease ($r = 0.8$ per year), Panama (*Fusarium* wilt) disease of banana ($r = 0.51$ per year), citrus tristeza virus ($r = 0.3-1.2$ per year), and elm phloem necrosis (mycoplasma) (0.6 per year).

Development of Epidemics

For a disease to become significant in a field, and particularly if it is to spread over a large area and develop into a severe epidemic, the right combinations of environmental factors must occur and spread either constantly, or repeatedly and at frequent intervals, over a large area. Even in a single, small field that contains the pathogen, plants almost never become severely diseased from just one set of favorable environmental conditions. It takes repeated disease cycles and considerable time before a pathogen produces enough individuals to cause an economically severe disease in the field. Once large populations of the pathogen are available, however, they can attack, spread to nearby fields, and cause a severe disease in a very short time—just a few days.

A plant disease epidemic can occur in a garden, a greenhouse, or a small field, but "epidemic" generally implies the development and rapid spread of a pathogen on a particular kind of crop plant cultivated over a large area, such as a large field, a valley, a section of a country, the entire country, or even part of a continent. Therefore, the first component of a plant disease epidemic is a large area planted to one more-or-less genetically uniform crop plant, with the plants and the fields being close together. The second component of an epidemic is the presence or appearance of a virulent pathogen at some point among or near the cultivated host plants. Such cohabitations of host plants and pathogens occur, of course, daily in countless locations and cause local diseases of varying severity, but most of these destroy crop plants to a limited extent and do not develop into epidemics. Epidemics develop only when the combinations and progression of the right sets of environmental conditions, that is, moisture, temperature, and wind or insect vector, coincide with the susceptible stage or stages of the plant and with the production, spread, inoculation, penetration, infection, and reproduction of the pathogen.

Thus, for an epidemic to develop, the small original or primary inoculum of the pathogen must be carried by wind or vector to some of the crop plants as soon as they begin to become susceptible to that pathogen. The moisture and temperature must then be right for germination or infection to take place. After infection, the temperature must be favorable for rapid growth and reproduction of the pathogen (short incubation period, short disease cycle) so that numerous new spores will appear as quickly as possible. The moisture (rain, fog, dew) then must be sufficient and should last long enough for abundant release of spores. Winds of the proper humidity and velocity, blowing toward the susceptible crop plants, must then pick up the spores and carry them to the plants while the latter are still susceptible. Most plant disease epidemics are effectively spread from south to north in the northern hemisphere and from north to south in the southern hemisphere. The warmer weather and the growth seasons also move in the same direction, and so the pathogens constantly find plants in their susceptible stage as the season progresses.

In each new location, however, the same set of favorable moisture, temperature, and wind or vector conditions must be repeated so that infection, reproduction, and dispersal of the pathogen can occur as quickly as possible. Furthermore, these conditions must be repeated several times within each location so that the pathogen can multiply, increasing the number of infections it causes on the host plants. These repeated infections usually result in the destruction of almost every plant within the area of an epidemic, although the uniformity of the plants, and the size of the area of cultivation, along with the prevailing weather, determine the final spread of the epidemic.

Fortunately, the most favorable combinations of conditions for disease development do not occur very often over very large areas, and therefore, spectacular plant disease epidemics that destroy crops over large areas are relatively rare. However, small epidemics involving the plants in a field or a valley occur quite frequently. With many diseases—for example, potato late blight, apple scab, and cereal rusts—the environmental conditions seem usually to be favorable, and disease epidemics would occur every year were it not for the control measures (chemical sprays, resistant varieties, and so on) employed annually to avoid such epidemics.

Modeling of Plant Disease Epidemics

An epidemic is a dynamic process. It begins on one or a few plants and then, depending on the kind, magnitude, and duration of environmental factors that influence the host and pathogen, increases in severity and spreads over a larger geographic area until it finally dies down. Epidemics come to a stop when all host plants are either killed by the pathogen, are harvested, or become resistant to the pathogen as they age. In many ways, the appearance, development, and spread of epidemics resemble those of hurricanes. In both cases, humans have been extremely interested in determining the elements and conditions that initiate each, the conditions that influence the rate of increase and the direction of their path, as well as the conditions that bring about their demise. For both phenomena, observations, measurements, mathematical formulas, and computers are used extensively to study the development and to predict the size, path, and time of attack in any given location.

Each plant disease epidemic, for example, of stem rust of wheat, late blight of potato, apple scab, and downy mildew of grape, follows a predictable course in each location each year. The course of the epidemic varies with the host varieties and pathogen races present, with the amount of pathogen inoculum present at the beginning of the epidemic, and with the moisture levels and temperature ranges during the epidemic. The more information we have about each of the components of an epidemic and about each of its subcomponents at any given moment, the better we can understand and describe the epidemic, and the better we can predict its direction and severity at some later point in time or some other place. The ability to predict the direction and severity of an epidemic, of course, has important practical consequences: it allows us to determine whether, when, and often what types of disease management strategies can be employed to slow down, or entirely prevent, the disease in a particular location.

In recent years, in an effort to improve our ability to understand and predict the development of an epidemic, plant pathologists have attempted, more or less successfully, to develop models of potential epidemics of some common and serious diseases. The construction of a model takes into account all of the components and as many of the subcomponents of a specific plant disease for which there is information for quantitative treatment, that is, for treatment by mathematical formulas. The models constructed are generally crude simplifications of real epidemics, roughly analogous, for example, to model toy cars or airplanes as they compare to real cars and airplanes. As with model toys, however, one can get a better picture and understanding of the real thing as the model depicts more and more parts, as the accuracy of the proportions of these parts increases, and as the number of the parts that are interlocked and move increases. The closer the resemblance of the model to the real thing the better we can visualize and understand the functions of the real thing by observing the model. In modeling plant disease epidemics, each component and subcomponent of the epidemic may be considered equivalent to one of the parts of the toy model; and, just as more accurately measured and fitted parts make for a more exact toy model, the more accurately the real subcomponents of an epidemic are measured and fitted together, the more accurately they describe the epidemic. When we have enough information about the values of the subcomponents of an epidemic at different stages and under different conditions, we can then develop a mathematical equation or equations—a mathematical model—that describes the epidemic.

Analysis of mathematical models of epidemics of specific plant diseases provides a great deal of information regarding the amount and efficacy of the initial inoculum, the effects of the environment, the disease resistance of the host, the length of time that host and pathogen may interact, and the effectiveness of various disease management strategies. Attempts to verify models of epidemics with actual observations and experimentation point out areas in which more knowledge is needed, and therefore indicate the directions in which further studies of the particular disease should be pursued.

Computer Simulation of Epidemics

The availability of computers has allowed plant pathologists to write programs that allow simulation of epidemics of several plant diseases. One of the first computer simulation programs, called EPIDEM, was written in 1969 and resulted from modeling each stage of the life cycle of a pathogen as a function of the environment. EPIDEM was designed to simulate early blight epidemics of tomato and potato caused by the fungus *Alternaria solani.* Subsequently, computer simulators were written for *Mycosphaerella* blight of chrysanthemums (MYCOS), for southern corn leaf blight caused by *Helminthosporium maydis* (EPICORN), and for apple scab caused by *Venturia inaequalis* (EPIVEN). A more general and more flexible plant disease simulator, called EPIDEMIC, was written primarily for stripe rust of wheat but could be easily modified for other host–pathogen systems. Computer simulation programs are now available for numerous plant diseases.

In a computer simulation of an epidemic, the computer is given data describing the various subcomponents of the epidemic and control practices at specific points in time (such as at weekly intervals). The computer then provides continuous information regarding not only the spread and severity of the disease over time, but also with regard to the final crop and economic losses likely to be caused by the disease under the conditions of the epidemic as given to the computer.

Computer simulation of epidemics is extremely useful as an educational exercise for students of plant pathology and also for farmers so that they can better understand and appreciate the effect of each epidemic subcomponent on the final size of their crop loss. Computer simulations of epidemics are, however, even more useful in actual disease situations. There, they serve as tools that can evaluate the importance of the size of each epidemic subcomponent, at a particular point in time of the epidemic, by projecting its effect on the final crop loss. By highlighting the subcomponents of an epidemic that are most important at a particular time, the simulation serves to direct attention to mangement measures that are effective against these particular epidemic subcomponents. In subsequent evaluations of the epidemic, the computer evaluates not only the current status of the disease but also the effectiveness of the applied management measures in controlling the epidemic. Computer simulation of epidemics is used fairly extensively today in plant disease forecasting systems. Such systems allow growers to take appropriate control measures against a disease as soon as the conditions that are likely to lead to disease development are present or are likely to be present soon.

Forecasting Plant Disease Epidemics

Being able to forecast plant disease epidemics is intellectually stimulating and also an indication of the success of modeling or computer simulation of particular diseases. It is also extremely useful to farmers in the practical management of crop disease. Disease forecasting allows the prediction of probable outbreaks or increases in intensity of disease and, therefore, allows us to determine whether, when, and where a particular management practice should be applied. In managing the diseases of their crops, growers must always weigh the risks, costs, and benefits of each of numerous decisions. For example, they must decide whether or not to plant a certain crop in a particular field, whether or not to buy more expensive propagating stock free of virus and other pathogens, and whether to plant seed of a more expensive or less-yielding but resistant variety rather than seed of a high-yielding but susceptible variety that needs to be protected by chemical sprays. Most frequently, farmers need forecasts of plant disease development to decide whether to spray a crop right away or to wait for several more days before they spray, since, if they can wait, they can reduce the amounts of chemicals and labor used without increasing the risk of losing their crop.

In developing a plant disease forecast, one must take into account several characteristics of the particular pathogen, host, and, of course, environment. In general, for most monocyclic diseases (such as root rot of peas and Stewart's wilt of corn), and for a few polycyclic diseases that may have a large amount of initial inoculum (such as apple scab), disease development may be predicted by assessing the amount of the initial inoculum. For polycyclic diseases (such as late blight of potato) that have a small amount of initial inoculum but many disease cycles, disease development can best be predicted by assessing the rate of occurrence of the disease cycles. For diseases in which both the amount of initial inoculum and the number of disease cycles are large—for example, beet yellows—both must be assessed for accurate prediction of disease epidemics. Such assessments, however, are often difficult or impossible, and in spite of considerable improvements in equipment and methodology, assessments of initial inoculum or rapidity of disease cycles are seldom accurate. Moreover, monitoring of weather factors is both essential and often difficult in relation to the effects of these factors on plant disease development.

Assessment of Initial Inoculum and of Disease

It is often difficult or impossible, in the absence of the host, to detect small populations of most pathogens. When moderate to large populations of pathogens are present, however, inoculum propagules of soilborne pathogens, such as fungi and nematodes, are estimated after extraction or trapping from soil. Airborne fungal spores and insect vectors are estimated by trapping them in various devices.

Usually it is easier to assess the amount of inoculum present by measuring the number of infections produced on a host within a certain period of time. Even in the presence of a host, however, it is often difficult to find and measure a small amount of disease. Furthermore, in many diseases there is an incubation period during which the host is infected but shows no symptoms. Aerial photography, using films sensitive to near-infrared radiation, has made possible both earlier detection and sharper delineation of diseased areas in crop fields (due to reduced reflectance of diseased foliage tissues that are occupied by water or pathogen cells).

Monitoring Weather Factors that Affect Disease Development

Monitoring weather factors during a plant disease epidemic presents enormous difficulties. The difficulties arise from the need for continuous monitoring of several different factors (temperature, relative humidity, leaf wetness, rain, wind, and cloudiness) at various locations in the crop canopy or on plant surfaces in one or more fields. In the past, measurements were made with mechanical instruments that measured these environmental variables roughly or infrequently and recorded the data inconveniently as ink traces on chart paper. In the last few years, however, several types of electrical sensors have been developed that produce electrical outputs easily recorded by computerized data loggers. Such computerized sensors have greatly improved studies of weather in relation to disease and have facilitated the acceptance and use of predictive disease control systems on the farm.

Several types of traditional and battery-operated electrical instruments are used to measure the various weather factors. Temperature measurements are made with various types of thermometers, hygrothermographs, thermocouples, and especially with thermistors (the latter are semiconductors whose electrical resistance changes considerably with temperature). Relative humidity measurements are made with a hygrothermograph (depends on the contraction and expansion of human hair in relation to relative humidity changes), with a ventilated psychrometer (consisting of a wet- and dry-bulb thermometer or a wet and dry thermistor), or with an electrode-bonded sulfonated polystyrene plate (whose resistance changes logarithmically with relative humidity). Leaf wetness is monitored with string-type sensors that constrict when moistened or slacken when dry and either leave an ink trace in the process or close or break an electrical circuit. Several types of electrical wetness sensors are available that can be either clipped onto leaves or placed among the leaves; they detect and measure the duration of rain or dew because either of the latter helps close the circuit between two pairs of electrodes. Rain, wind, and cloudiness (irradiance) are still measured by the traditional instruments (rain funnels and tipping-bucket gauges for rain, cups and thermal anemometers for wind speed, vanes for wind direction, pyranometers for irradiance). Several of these instruments, however, have become adapted for electronic monitoring.

In modern weather-monitoring systems, the weather sensors are connected to data-logging devices. The data may be read on a digital display, or they are transmitted to a cassette tape recorder or a printer. From the cassette the data may be transferred to a microcomputer. There they may be viewed, processed in several computer languages, organized into separate matrices for each weather variable, plotted, and analyzed. Depending on the particular disease model used, accurate weather information provides the most useful basis for predicting sporulation and infection and therefore provides the best warning for timing disease management practices, such as the application of fungicides.

Examples of Plant Disease Forecast Systems

Generally, it is useful to have the maximum amount of information that is available about a disease before venturing to predict its development. In many cases, however, one or two of the factors that affect disease development predominate so much that knowledge of them is often sufficient for the formulation of a reasonably accurate forecast. Thus, several plant disease forecasting systems use as the criterion the amount of the initial inoculum (for example, Stewart's wilt of corn, blue mold of tobacco, fire blight of apple and pear, pea root rot, and other diseases caused by soilborne pathogens such as *Sclerotium* and cyst nematodes); others use the number of disease cycles or the amount of secondary inoculum (for example, late blight of potato, *Cercospora* and other leaf spots, and downy mildew of grape): still others use as criteria the amount of the initial inoculum as well as the number of disease cycles or the amount of secondary inoculum (apple scab, black rot of grape, cereal rusts, *Botrytis* leaf blight and gray mold, and sugar beet yellows).

Forecasts Based on Amount of Initial Inoculum

In *Stewart's wilt of corn* (caused by the bacterium *Erwinia stewartii*), the pathogen survives the winter in the bodies of its vector, the corn flea beetle. Therefore, the amount of disease that will develop in a growing season can be predicted if we know the number of the pathogen's vectors that survived the previous winter, because that allows us to estimate the amount of inoculum that also survived the previous winter. It has been shown that corn flea beetles are killed by prolonged low winter temperatures. Therefore, when the sum of the mean temperatures for the three winter months December, January, and February at a given location is less than $-1\,°C$, most of the beetle vectors are killed, and so there is little or no bacterial wilt during the following growth season. Warmer winters allow greater survival of beetle vectors, and proportionately more severe wilt outbreaks the following season.

In the *downy mildew (blue mold) of tobacco* (caused by the fungus *Peronospora tabacina*), the disease in most years is primarily a threat to seedbeds in the tobacco-producing states. When January temperatures are above normal, blue mold can be expected to appear early in seedbeds in the following season and to cause severe losses. On the contrary, when January temperatures are below normal, blue mold can be expected to appear late in seedbeds and to cause little damage. If the disease is expected in seedbeds, control measures can be taken to prevent it from becoming established, and subsequent control in the field is made much easier. In recent years, a supplementary blue mold warning system has been operated in North America by the Tobacco Disease Council and the Cooperative Extension Service. The warning system keeps the industry aware of locations and times of appearance and spread of blue mold and helps growers, to some extent, with the timing and intensity of controls.

In *pea root rot* (caused by the fungus *Aphanomyces euteiches*) and in other diseases caused by soilborne fungi and some nematodes, the severity of the disease in a field during a growing season can be predicted by winter tests in the greenhouse. In these tests, susceptible plants are planted in the greenhouse in soil taken from the field in question. If the greenhouse tests show that severe root rot developed in a particular soil, the field from which the soil was obtained is not planted with the susceptible crop. On the contrary, fields whose soil samples allowed development of little or no root rot can be planted and can be expected to produce a crop reasonably free of root rot. With some soilborne pathogens, such as the fungi *Sclerotium* and *Verticillium,* and the cyst nematodes *Heterodera* and *Globodera,* the initial inoculum can be assessed directly by isolating the fungal sclerotia and nematode cysts and then counting them per gram of soil. The greater the number of propagules, the more severe the disease produced.

In *fire blight of apple and pear* (caused by the bacterium *Erwinia amylovora*), the pathogen multiplies much more slowly at temperatures below 15°C than at temperatures above 17°C. In California, a disease outbreak can be expected to occur in the orchard if the daily average temperatures exceeded a "disease prediction line" obtained by drawing a line from 16.7°C on March 1 to 14.4°C on May 1. Therefore, when such conditions occur, application of a bactericide during bloom is indicated to prevent an epidemic.

Forecasts Based on Weather Conditions Favoring Development of Secondary Inoculum

In late blight of potato and tomato (caused by the fungus *Phytophtora infestans*), the initial inoculum is usually low and generally too small to detect and measure directly. Even with low initial inoculum, the initiation and development of a late blight epidemic can be predicted with reasonable accuracy if the moisture and temperature conditions in the field remain within certain ranges favorable to the fungus. When constant cool temperatures between 10 and 24°C prevail, and the relative humidity remains over 75 percent for at least 48 hours, infection will take place and a late blight outbreak can be expected from 2 to 3 weeks later. If within that period and afterward, several hours of rainfall, dew, or relative humidity close to saturation point occur, they will serve to increase the disease and will foretell the likelihood of a major late blight epidemic. In recent years, computerized predictive systems for late blight epidemics have been developed. In one such system, called BLITECAST, moisture and temperature are monitored continuously. From this information disease severity values are calculated and predicted, and recommendations are issued to growers as to when to begin spraying. More recent refinements in late blight forecasting include, in addition to data on moisture and temperature, information on the level of resistance of the potato variety to late blight and the effectiveness of the fungicide used. Information on all these parameters is, of course, very useful in the formulation of recommendations for fungicide applications.

Several *leaf spots* [such as those caused by the fungi *Cercospora* on peanuts and celery and *Exserohilum (Helminthosporium) turcicum* on corn] can be predicted by taking into account the number of spores trapped daily, the temperature, and the duration of periods with relative humidity near 100 percent. An infection period is predicted if high (95–100 percent) relative humidity lasts for more than 10 hours, and growers are then urged to apply chemical sprays immediately.

Forecasts Based on Amounts of Initial and Secondary Inoculum

In *apple scab* (caused by the fungus *Venturia inaequalis*), the amount of initial inoculum (ascospores) is usually large and is released over a period of 1–2 months following bud break. Infections from primary inoculum must be prevented with well-timed fungicide applications during blossoming, early leafing, and fruit development, otherwise the entire crop is likely to be lost. After primary infections, however, secondary inoculum (conidia) is produced, which multiplies itself manyfold with each succeeding generation. The pathogen can infect wet leaf or fruit surfaces at a range of temperatures from 6 to 28°C. The length of time that leaves and fruit need to be wet, however, is much shorter at the optimum temperatures than at either extreme (9 hours at 18–24°C versus 28 hours at 6 or 28°C). By combining temperature and leaf wetness duration data, the apple scab forecast system can predict not only whether an infection period will occur, but also whether the infection periods will result in light, moderate, or severe disease. Such information, collected and analyzed by individuals or by weather-sensing microcomputers, is used to make recommendations to growers. The latter are advised of the need and timing of fungicide application and about the kind of fungicide (protective or eradicant) that should be used to control the disease.

In the *wheat leaf and stem rusts* (caused by the fungi *Puccinia recondita* and *Puccinia graminis*), short (1 to 2 week) forecasts of subsequent disease intensity can be obtained by taking into account disease incidence, stage of plant growth, and spore concentration in the air.

In many *insect-transmitted virus diseases* of plants (for example, barley yellow dwarf, cucumber mosaic virus, and sugar beet yellows), the likelihood, and sometimes the severity, of epidemics can be predicted. This is accomplished by determining the number of aphids, especially viruliferous ones, coming into the field at certain stages of the host growth. A number of the aphids caught in traps placed in the field are tested for virus by allowing them to feed on healthy plants, or by testing them for virus serologically through the ELISA technique. The more numerous the viruliferous aphids and the earlier they are detected, the more rapid and more severe will be the virus infection. Such predictions can be improved by taking into account late winter and early spring temperatures, which influence the population size of the overwintering aphid vectors.

Farmer-Warning Systems

In many states and countries, different types of warning systems are in place for one or more important plant diseases. The purpose of these systems is to warn farmers of the impending onset of an infection period or that an infection period has already occurred, so that they can take immediate appropriate control measures to stop recent infections from developing or prevent further infections from occurring.

In most cases, the warning system begins with a grower, an extension agent, or a private consultant surveying certain fields on a regular basis, or when the weather conditions are likely to favor maturation of the primary inoculum or appearance of the particular disease. When mature inoculum (such as ascospores in apple scab) or traces of disease (for example, in potato late blight) are found, the county extension office is notified. The county extension office in turn notifies the state extension plant pathologist, who collates all reports about the disease from around the state and by telephone or in writing notifies all concerned county agents. They, in turn, by telephone, radio, or letter, notify all farmers in their county. For diseases of potential regional or national epidemic consequences, the state extension plant pathologist notifies the federal plant disease survey office of the U.S. Department of Agriculture, which in turn notifies all extension plant pathologists in adjacent and other states that may be affected by that plant disease.

Recently, computerized warning systems have been tried and are in use for certain diseases in some states. Some of them (such as BLITECAST) use centrally located computers that process weather data collected on the farm by individual growers and are phoned in when certain weather conditions prevail or at certain intervals. The computer then processes the data, determines whether an infection period is imminent, likely to occur, or cannot occur, and a recommendation is made to the grower as to whether or not to spray and what materials to apply.

More recently, small special-purpose computers have been developed that have field sensors and can be mounted on a post in a farmer's field. Such units (such as the Apple Scab Predictor) monitor and collect data in the field on temperature, relative humidity, duration of leaf wetness, and rainfall amounts, analyze the data automatically, make predictions of disease occurrence and intensity and, on the spot, make recommendations for disease control measures. The same unit can be used for any disease for which a prediction program is available and can either be reprogrammed into the unit or the program circuit boards can be interchanged. Predictions from such units are obtained by using a simplified keyboard and display right in the field, or the unit can be linked to a personal computer if additional processing of the data is desired.

REFERENCES

Adams, P. B. (1981). Forecasting onion white rot disease. *Phytopathology* **71,** 1178–1181.

Anonymous (1983). Symposium on estimating yield reduction of major food crops of the world. (Several papers.) *Phytopathology* **73,** 1575–1600.

Berger, R. D. (1977). Application of epidemiological principles to achieve plant disease control. *Annu. Rev. Phytopathol.* **15,** 165–183.

Berger, R. D., and Bartz, J. A. (1982). Analysis of monocyclic pathosystems with *Erwinia-Lycopersicon* as the model. *Phytopathology* **72,** 365–369.

Castor, L. L, Ayers, J. E., MacNab, A. A., and Krause, R. A. (1975). Computerized forecasting system for Stewart's bacterial disease on corn. *Plant Dis. Rep.* **59,** 533–536.

Day, P. R., ed. (1977). "The Genetic Basis of Epidemics in Agriculture," Ann. N. Y. Acad. Sci., Vol. 287. N. Y. Acad. Sci., New York.

Easton, G. D. (1982). Late blight of potatoes and prediction of epidemics in arid central Washington state. *Plant Dis.* **66,** 452–455.

Fleming, R. A. (1983). Development of a simple mechanistic model of cereal rust progress. *Phytopathology* **73,** 308–312.

Fry, W. E. (1982). "Principles of Plant Disease Management." Academic Press, New York.

Gibbs, J. N. (1978). Intercontinental epidemiology of Dutch elm disease. *Annu. Rev. Phytopathol.* **16,** 287–307.

Gilligan, C. A. (1983). Modeling of soilborne pathogens. *Annu. Rev. Phytopathol.* **21,** 45–64.

Gutierrez, A. P., DeVay, J. E., Pullman, G. S., and Freibertshauser, (1983). A model of *Verticillium* wilt in relation to cotton growth and development. *Phytopathology* **73,** 89–95.

Hatfield, J. L., and Thomason, I. J., eds. (1982). "Biometerology in Integrated Pest Management." Academic Press, New York.

Horsfall, J. G., and Cowling, E. B., eds. (1978). "Plant Disease, Vol. 2. Academic Press, New York.

Huismann, O. C. (1982). Interactions of root growth dynamics to epidemiology of root-invading fungi. *Annu. Rev. Phytopathol.* **20,** 303–327.

James, W. C. (1974). Assessment of plant disease. *Annu. Rev. Phytopathol.* **12,** 27–48.

Jones, A. L., Lillevik, S. L., Fisher, P. D., and Stebbins, T. C. (1980). A microcomputer-based instrument to predict primary apple scab infection periods. *Plant Dis.* **64,** 69–72.

Jones, A. L., Fisher, P. D., Seem, R. C., Kroon, J. C., and Van De Motter, J. (1984). Development and commercialization of an in-field microcomputer delivery system for weather-driven predictive models. *Plant Dis.* **68,** 458–463.

Kampmeijer, P., and Zadoks, J. C. (1977). "EPIMUL. A Simulator of Foci and Epidemics in Mixtures of Resistant and Susceptible Plants, Mosaics and Multilines." Pudoc, Wageningen, The Netherlands.

Kranz, J., and Hau, B. (1980). Systems analysis in epidemiology. *Annu. Rev. Phytopathol.* **18,** 67–83.

Krause, R. A., Massie, L. B., and Hyre, R. A. (1975). BLITECAST, a computerized forecast of potato late blight. *Plant Dis. Rep.* **59,** 95–98.

MacHardy, W., and Sondei, J. (1981). Weather-monitoring instrumentation for plant disease management programs and epidemilogical studies. *N. H., Agric. Exp. Stn., Bull.* **519,** 1–40.

MacKenzie, D. R. (1981). Scheduling fungicide applications for potato late blight with BLITECAST. *Plant Dis.* **65,** 394–399.

Madden, L. V. (1983). Measuring and modeling crop losses at the field level. *Phytopathology* **73,** 1591–1593.

Main, C. E. (1977). Crop destruction—The *raison d' etre* of plant pathology. *In* "Plant Disease" (J. G. Horsfall and E. B. Cowling, eds.), Vol.1 1, pp. 55–78. Academic Press, New York.

Manzer, F. E., and Cooper, G. R. (1967). Aerial photographic methods of potato disease detection. *Maine Agric. Exp. Stn., Bull.* **646.**

Miller, P. R., and O'Brien, M. J. (1957). Prediction of plant disease epidemics. *Annu. Rev. Microbiol.* **11,** 77–110.

Mills, W. D. (1944). Efficient use of sulfur dusts and sprays during rain to control apple scab. *Cornell Ext. Bull.* **630.**

Minogue, K. P., and Fry, W. E. (1983). Models for the spread of disease: Model description. *Phytopathology* **73,** 1168–1173.

Minogue, K. P., and Fry, W. E. (1983). Models for the spread of a plant disease: Some experimental results. *Phytopathology* **73,** 1173–1176.

Nesmith, W. C. (1984). The North American blue mold warning system. *Plant Dis.* **68,** 933–936.

Ostergaard, H. (1983). Predicting development of epidemics on cultivar mixtures. *Phytopathology* **73,** 166–172.

Pedgley, D. E. (1982). "Windborne Pests and Diseases—Meteorology of Airborne Organisms." Halsted Press, Wiley, New York.

Plumb, R. T., and Thresh, J. M., eds. (1983). "Plant Virus Epidemiology: The Spread and Control of Insect-Borne Viruses." Blackwell, Oxford.

Populer, C. (1978). Changes in host susceptibility with time. *In* "Plant Disease" (J. G. Horsfall and E. B. Cowling, eds.), Vol. 2, pp. 239–262. Academic Press, New York.

Rapilly, F. (1979). Yellow rust epidemiology. *Annu. Rev. Phytopathol.* **17,** 59–73.

Sall, M. A. (1980). Epidemiology of grape powdery mildew: A model. *Phytopathology* **70,** 338–342.

Scott, P. R., and Bainbridge, A., eds. (1978). "Plant Disease Epidemiology." Blackwell, Oxford.

Seem, R. C. (1984). Disease incidence and severity relationships. *Annu. Rev. Phytopathol.* **22,** 133–150.

Sutton, J. C., Gillespie, T. J., and Hildebrand, P. D. (1984). Monitoring weather factors in relation to plant disease. *Plant Dis.* **68,** 78–84.

Teng, P. S., and Zadoks, J. C. (1980). Computer simulation of plant disease epidemics. *In* "Yearbook of Science and Technology." pp. 23–31. McGraw-Hill, New York.

Thomson, S. V., Schroth, M. N., Moller, W. J., and Reil, W. O. (1982). A forecasting model for fire blight of pear. *Plant Dis.* **66,** 576–579.

Toler, R. W., Smith, B. D., and Harlan, J. C. (1981). Use of aerial color infrared photography to evaluate crop disease. *Plant Dis.* **65,** 24–31.

Vanderplank, J. E. (1963). "Plant Diseases: Epidemics and Control." Academic Press, New York.

Vanderplank, J. E. (1975). "Principles of Plant Infection." Academic Press, New York.

Young, H. C., Jr., Prescott, J. M., and Saari, E. E. (1978). Role of disease monitoring in preventing epidemics. *Annu. Rev. Phytopathol.* **16,** 263–285.

Zadoks, J. C. (1984). A quarter century of disease warning. *Plant Dis.* **68,** 352–355.

Zadoks, J. C., and Schein, R. D. (1979). "Epidemiology and Plant Disease Management." Oxford Univ. Press, London and New York.

9 CONTROL OF PLANT DISEASES

Apart from being intellectually interesting and scientifically justified, the study of the symptoms, causes, and mechanisms of plant diseases allows the development of methods to combat plant diseases and, thus, increases the quantity and improves the quality of plant products.

Methods of control vary considerably from one disease to another, depending on the kind of pathogen, the host, and the interaction of the two. In controlling diseases, plants are generally treated as populations rather than as individuals, although certain hosts (especially trees, ornamentals, and sometimes, virus-infected plants) may be treated individually. With the exception of trees, however, the damage or loss of one or a few plants is usually considered insignificant, and control measures are generally aimed at saving the populations rather than a few individual plants.

Considering the regularity with which most serious diseases of crop plants appear in an area year after year, the rapid spread of most plant diseases, and the difficulty of curing a disease after it has begun to develop, it is easy to understand why almost all control methods are aimed at protecting plants from becoming diseased rather than at curing them after they have become diseased. As a matter of fact, few infectious plant diseases can be satisfactorily controlled in the field by therapeutic means, although certain diseases can be cured under experimental conditions.

The various control methods can be classified as regulatory, cultural, biological, physical, and chemical, depending on the nature of the agents employed. **Regulatory control measures** aim at excluding a pathogen from a host or from a certain geographic area. Most **cultural control methods** aim at helping plants avoid contact with a pathogen and at eradicating or reducing the amount of a pathogen in a plant, a field, or an area. **Biological** and some cultural **control methods** aim at improving the resistance of the host or at favoring microorganisms antagonistic to the pathogen. Finally, **physical** and **chemical methods** aim at protecting the plants from pathogen inoculum that is likely to arrive and at curing an infection that is already in progress.

Epidemiological studies can help determine how effective various controls might be for a particular disease. In general, reducing or excluding the initial inoculum is most effective for the management of monocyclic pathogens. Controls such as crop rotation, removal of alternate hosts, and soil fumigation reduce the initial inoculum. Because polycyclic pathogens increase many times the amount of initial inoculum during the growing season, reducing the initial inoculum must usually be accompanied by a control (such as chemical protection or horizontal resistance) that also reduces the infection rate. Many controls, for example totally excluding a pathogen from an area, are useful for both monocyclic and polycyclic pathogens.

Control Methods That Exclude the Pathogen from the Host

As long as plants and pathogens can be kept away from each other, no disease will develop. Many plants are grown in areas of the world where certain pathogens are still absent. They are, therefore, free of the diseases caused by such pathogens.

To prevent the import and spread of plant pathogens into areas where they are absent, national and state laws regulate the conditions under which certain crops may be grown and distributed between states and countries. Such regulatory control is applied by means of quarantines, inspections of plants in the field or warehouse, and occasionally by voluntary or compulsory eradication of certain host plants. Furthermore, plants are sometimes intentionally grown in areas from which a pathogen is largely or entirely excluded by unfavorable climatic conditions such as low rainfall and low relative humidity or by lack of vectors. This type of exclusion is called **evasion.**

Quarantines and Inspections

When plant pathogens are introduced into an area in which they did not previously exist, they are likely to cause much more catastrophic epidemics than do the existing pathogens. Plants that develop in the absence of a pathogen have no opportunity to select resistance factors specific against the pathogen and are, therefore, extremely vulnerable to attack. Some of the worst plant disease epidemics that have occurred throughout the world, for example the downy mildew of grapes in Europe and the bacterial canker of citrus, the chestnut blight, the Dutch elm disease, and the soybean cyst nematode in the United States, are all diseases caused by pathogens that were introduced from abroad. It has been estimated, for example, that if soybean rust were introduced into the United States in the mid-1980s, it would result in total losses to consumers and other sectors of the U.S. economy of more than seven billion dollars per year.

In order to keep out foreign plant pathogens and to protect the nation's farms, gardens, and forests, the Plant Quarantine Act of 1912 was passed by Congress. This act prohibits or restricts entry into or passage through the United States from foreign countries of plants, plant products, soil, and other materials carrying or likely to carry plant pathogens not known to be widely established in this country. Similar quarantine regulations exist in most other countries.

Experienced inspectors stationed at all points of entry into the country enforce quarantines of produce likely to introduce new pathogens. Plant quarantines are already credited for the interception of numerous foreign plant pathogens and, thereby, with saving the country's plant world from potentially catastrophic diseases. Plant quarantines are considerably less than foolproof, however, because pathogens may be introduced in the form of spores or eggs on unsuspected carriers, and latent infections of seeds and other plant propagative organs may exist even after treatment. Various steps taken by plant quarantine stations, such as growing plants under observation for certain periods of time before they are released to the importer and repeated inspection of imported nursery stock in the grower's premises, tend to reduce the chances of introduction of harmful pathogens. With the annual imports of flower bulbs from Holland, U.S. quarantine inspectors may visit the flower fields in Holland and inspect them for diseases. If they find the fields to be disease free, they issue inspection certificates allowing the import of such bulbs into the United States without further tests.

Similar quarantine regulations govern the interstate, and even the intrastate, sale of nursery stock, tubers, bulbs, seeds, and other propagative organs, especially of certain crops, such as potatoes and fruit trees. The movement and sale of such materials within and between states is controlled by the regulatory agencies of each state.

Several voluntary inspection systems are also in effect in various states in which appreciable amounts of nursery stock and potato seed tubers are produced. Growers interested in producing and selling disease-free plants submit to a voluntary inspection or indexing of their crop in the field and in storage by the state regulatory agency, by experiment station personnel, or others. If, after certain procedures recommended by the inspecting agency are carried out, the plant material is found to be free of certain, usually virus, diseases, the inspecting agency issues a certificate indicating that the plants are free from these specific diseases, and the grower may then advertise the produce as disease free.

Evasion of Pathogen

In several plant diseases, control depends largely on attempts to evade pathogens. For example, bean anthracnose, caused by the fungus *Colletotrichum lindemuthianum,* and the bacterial blights of bean, caused by the bacteria *Xanthomonas phaseoli* and *Pseudomonas phaseolicola,* are transmitted through the seed. In most areas where beans are grown, at least a portion of the plants and the seeds become infected with these pathogens. In the dry, irrigated regions of the western United States, however, the conditions of low humidity are unsuitable for these pathogens, and therefore the plants and the seeds are free of them. Using western-grown seeds free of these pathogens is the main recommendation for control of these diseases. Similarly, in order to produce potato seed tubers free of viruses, potatoes are grown in remote locations in the cooler, northern states (Maine, Wisconsin, and Idaho) and at higher elevations, where aphids, the vectors of these viruses, are absent or their populations are small and can be controlled. In many cases, a susceptible crop is planted at a great enough distance from other fields containing possibly diseased plants that the pathogen would not likely infect the crop. This type of **crop isolation** is practiced mostly with perennial plants, such as peach orchards isolated from x-disease mycoplasma-infected chokecherry shrubs or trees. Also, during much of this century, banana production in Central America depended on evading the fungus *Fusarium oxysporum* f. *cubense,* the cause of Fusarium wilt (Panama disease) of banana, by moving on to new, previously uncultivated fields as soon as older banana fields became infested with *Fusarium* and yields became unprofitable.

Growers carry out numerous activities aimed at helping the host evade the pathogen. Such activities include using vigorous seed, selecting proper (early or late) planting dates and proper sites, maintaining proper distances between fields and between rows and plants, and using proper insect and weed control. All these practices increase the chances that the host will remain free of the pathogen or go through its susceptible stage before the pathogen reaches the host.

Use of Pathogen-Free Propagating Material

When a pathogen is excluded from the propagating material (seed, tubers, bulbs, nursery stock) of a host, it is often possible to grow the host free of that pathogen for the rest of its life. Examples are woody plants affected by nonvectored viruses. In most crops, if the host can be grown free of the pathogen for a considerable period of its early life, during which the plant can attain normal growth, it can then produce a fairly good yield in spite of a potential later infection. Examples are crops affected by vectored viruses and mycoplasmas, fungal, bacterial, and nematode pathogens.

There is absolutely no question that every host plant and every crop grows better and produces a greater yield if the starting propagating material is free of pathogens, or at least free of the most important pathogens. For this reason, every effort should be made to obtain and use pathogen-free seed or nursery stock, even if the cost is considerably greater than for propagating material of unknown pathogen content.

All types of pathogens can be carried in or on propagating material. True seed, however, is invaded by relatively few pathogens, although several may contaminate its surface. Seed may carry internally one of a few fungi (such as those causing anthracnoses and smuts), certain bacteria (bacterial wilts, spots, and blights), and one of several viruses (tobacco ringspot in soybean, bean common mosaic, lettuce mosaic, barley stripe mosaic, squash mosaic, and prunus necrotic ringspot). On the other hand, vegetatively propagating material such as buds, grafts, rootstocks, tubers, bulbs, corms, cuttings, and rhizomes are expected to carry internally almost every virus, viroid, mycoplasma, protozoan, and vascular fungus or bacterium present systemically in the mother plant, in addition to any fungi, bacteria, and nematodes that may be carried on these organs externally. Some nematodes may also be carried internally in some below-ground propagating organs (tubers, bulbs, corms, and rhizomes) and in or on the roots of nursery stock.

Pathogen-Free Seed

Seed that is free of fungal, bacterial, and some viral pathogens is usually obtained by growing the crop and producing the seed in (1) an area free of or isolated from the pathogen, (2) an area not suitable for the pathogen (such as the arid western regions where bean seed is produced free of anthracnose and bacterial blights), or (3) an area not suitable for the vector of the pathogen (such as the northern or high altitude fields where aphids, the vectors of many viruses, are absent or rare).

It is very important, and with seed-transmitted and aphid-borne viruses it is indispensable, that seed be essentially free of the pathogen, especially virus. Because the pathogen will be present in the field at the beginning of the growth season, even a small proportion of infected seeds is sufficient to provide enough inoculum to spread and infect many plants early, thus causing severe losses. It has been shown, for example, that to control lettuce mosaic virus, only seed lots that contain less than 1 infected seed per 30,000 lettuce seeds must be used. For this purpose seed companies have their lettuce seed tested for lettuce mosaic virus every year. In past years, seeds were tested (indexed) by growing out hundreds of thousands of lettuce seedlings in insect-proof greenhouses, observing them over several weeks for lettuce mosaic symptoms, and attempting to transmit the virus from suspect plants to healthy plants. Later, indexing was done by inoculating a local lesion indicator plant (in this case *Chenopodium quinoa*) with sap from ground samples of groups of seeds and observing it for virus symptoms. At present, testing for lettuce mosaic virus in seed is done with serological techniques, particularly via ELISA, which is faster, more sensitive, and less expensive than the other methods.

Testing seed for fungal and bacterial pathogens is done by symptomatology, microscopically, and by culturing the pathogen on general or selective nutrient media. To detect and identify bacteria, serological tests are now being used with increasing frequency and accuracy. If seed free of fungal and bacterial pathogens cannot be obtained by other means, hot water (50°C) treatment of the seed can free it from the pathogen. Examples are cabbage seed freed from *Xanthomonas campestris,* the cause of black rot of cabbage, and from *Phoma lingam,* the cause of black leg of cabbage, and seed of wheat and other cereals freed from *Ustilago* sp., the cause of loose smuts of cereals.

Pathogen-Free Vegetative Propagating Materials

To obtain vegetative propagating material free of pathogens that are systemically distributed throughout the plant (viruses, viroids, mycoplasmas, fastidious bacteria, and some wilt-inducing fungi and bacteria) it must be obtained from mother plants that had tested free of the particular pathogen or pathogens. To ensure continuous production of pathogen-free buds, grafts, cuttings, rootstocks, and runners of trees, vines, and other perennials, the mother plant is indexed for the pathogen at regular (1- to 2-year) intervals. Indexing is usually done by taking grafts or sap from the plant and inoculating susceptible indicator plants to observe possible symptom development. Futhermore, the new plants must be planted in pathogen- and vector-free soil. Moreover, the new plants must be protected from airborne vectors of the pathogen if they are to remain free of the pathogen for a considerable time. Indexing of mother plants for viruses (and some mycoplasmas) is now done in several states for most pome, stone, and citrus fruits, as well as for grapes, strawberries, raspberries, and several ornamentals, such as roses and chrysanthemums. Some viruses are now indexed by serological (ELISA) or biochemical tests rather than via bioassay. It is anticipated that before too long most perennial plants will be produced from pathogen-free propagating material.

For certain crops, such as potato, complex programs of inspection, indexing, and certification have evolved to produce pathogen-free seed potatoes. In every state where seed potatoes are produced, they must meet a

Table 9-1. Maximum Tolerances for Diseases in Certified Seed Potatoes Allowed in Various States

Disease	Tolerance levels allowed (%)
Leaf roll virus	0.5–1
Mosaic viruses	1–2
Spindle tuber viroid	0.1–2
Total virus content	0.5–3
Fusarium and/or *Verticillium* wilt	1–5
Ring rot *(Corynebacterium sepedonicum)*	0
Root knot (Meloidogyne sp.)	0–0.1
Late blight *(Phytophthora infestans)*	0

slightly varying maximum allowable tolerance for various diseases (Table 9-1). The initial mother plants that test free of these diseases are propagated for a few years by state agencies in isolated farms, usually at high altitudes, where aphids are absent or rare. The plants and the tubers are inspected and tested repeatedly each season to ensure continued freedom from each pathogen. When enough pathogen-free seed potatoes are produced, they are turned over to commercial seed potato producers, who further multiply them and finally sell them to farmers. While in the fields of the commercial seed producers, the potato plants are inspected repeatedly, infected plants are rogued, and insect vectors are controlled. For the seed to be "certified," the plants in the field must show disease levels no higher than those allowed by the particular state (Table 9-1). In several certification programs, samples of the harvested tubers are sent to a southern state, where they are grown during the winter and further checked for symptoms. In some states, serological tests (ELISA) are now replacing some of the bioassays.

With some crops, such as carnation and chrysanthemum, greenhouse growers need cuttings free of the vascular wilt–causing fungi *Fusarium* and *Verticillium* each time, but it is almost impossible to keep these two fungi from the production beds. It was noted early, however, that short cuttings taken from the tips of rapidly growing shoots were usually free of either of these fungi, and this became a common practice to control these diseases.

Sometimes, it is impossible to find even a single plant of a variety that is free of a particular pathogen, especially viruses. In that case, one or a few healthy plants are initially obtained by tissue culture of the upper millimeter or so of the growing meristematic tip of the plant, which most viruses do not invade. In some cases, healthy plants can be obtained from virus-infected plants by eliminating the virus through heat treatment.

Dormant plant material, such as budwood, dormant trees, and tubers, is usually treated with hot water at temperatures ranging from 35 to 54°C, with treatment times lasting from a few minutes to several hours. Actively growing plants are sometimes treated with hot air, which allows better survival of the plant and better elimination of the pathogen than does hot water. Temperatures of 35 to 40°C seem to be optimal for air treatment of growing plants. For hot-air treatment, the infected plants are usually grown in growth chambers for varying periods, generally lasting 2–4 weeks. Some viruses, however, require treatment for 2–8 months, while others may be eliminated in just one week. All mycoplasmas, all fastidious bacteria, and many viruses can be eliminated from their hosts by heat treatment, but for some viruses such treatment has not always been dependable.

In recent years, with some crops such as strawberries and orchids, once one or a few pathogen-free plants have been obtained by any of the above methods, they are subsequently used as foundation material from which thousands, hundreds of thousands, and even millions of pathogen-free plants are produced through tissue culture techniques in the laboratory. These plants are later set out in the greenhouse or the field before they are sold to growers or retailers as pathogen-free plants at a premium price.

Control Methods That Eradicate or Reduce the Pathogen Inoculum

Many different types of control methods aim at eradicating or reducing the amount of pathogen present in an area, a plant, or plant parts (such as seeds). Many such methods are cultural, that is, they depend primarily on certain actions of the grower, such as host eradication, crop rotation, sanitation, improving plant growing conditions, creating conditions unfavorable to pathogens, polyethylene mulching, trickle irrigation, ecofallow, and reduced tillage farming. Some methods are physical, that is, they depend on a physical factor, such as heat, or cold. Examples are soil sterilization, heat treatment of plant organs, refrigeration, and radiations. Several methods are chemical, that is, they depend on the use and action of a chemical substance to reduce the pathogen. Examples are soil treatment, soil fumigation, and seed treatment. Some methods are biological, that is, they use living organisms to reduce the pathogen inoculum. Examples are use of trap crops and antagonistic plants against nematodes and use of amendments that favor microflora antagonistic to the pathogen.

Cultural Methods That Eradicate or Reduce the Inoculum

Host Eradication

When a pathogen has been introduced into a new area in spite of quarantine, a plant disease epidemic frequently follows. To prevent such an epidemic, all the host plants infected by or suspected of harboring the pathogen may have to be removed and burned. This eliminates the pathogen and prevents greater losses from the spread of the pathogen to additional plants. Beginning in 1915, this type of host eradication controlled the bacterial canker of citrus in Florida and other southern states, where more than three million trees had to be destroyed. Another outbreak of citrus canker occurred in Florida in 1984, and again the disease was apparently brought under control through the painful destruction of millions of nursery and orchard trees in that state. In the last quarter century, a campaign to contain and eradicate witchweed *(Striga asiatica)* in the eastern Carolinas in the United States has been successful. On the other hand, attempts by several European countries to eradicate fire blight of apple and pear (caused by the bacterium *Erwinia amylovora*) and plum pox virus of stone fruits and attempts by several South American countries to eradicate coffee rust (caused by the fungus *Hemileia vastatrix*) have not been successful, and the pathogens continue to spread. Host eradication (roguing) is also carried out routinely in many nurseries, greenhouses, and fields to prevent the spread of numerous diseases by eliminating infected plants that provide a ready source of inoculum within the crop.

Certain pathogens of annual crops, for example cucumber mosaic virus, overwinter only or mainly in other perennials, usually wild plants. Eradication of the host in which the pathogen overwinters is sometimes enough to eliminate completely or to reduce drastically the amount of inoculum that can cause infections the following season. In some crops, such as potatoes, pathogens of all types may overwinter in infected tubers that are left in the field. Many such tubers produce plants in the spring that allow the pathogen to come above ground from where it can be spread further by insects, rain, and wind. Eradication of such volunteer plants helps greatly to reduce the inoculum of these pathogens.

Some pathogens require two alternate hosts to complete their life cycles. For example, *Puccinia graminis tritici* requires wheat and barberry, *Cronartium ribicola* requires pine and currant *(Ribes),* and *Gymnosporangium juniperi-virginianae* requires cedar and apple. In these cases, eradication of the wild or economically less important alternate host interrupts the life cycle of the pathogen and leads to control of the disease. This has been carried out somewhat successfully with stem rust of wheat and white pine blister rust through eradication of barberry and currant, respectively, although, owing to other factors, both diseases are still widespread and catastrophic. In cases like the cedar-apple rust, however, in which both hosts may be important, control through eradication of the alternate host is impractical.

Crop Rotation

Soil-borne pathogens which attack plants of one or a few species or even families of plants, can sometimes be eliminated from the soil by planting, for 3 or 4 years, crops belonging to species or families not attacked by the particular pathogen. Complete control through crop rotation is possible with pathogens that are soil invaders, that is, survive only on living plants or only as long as the host residue persists as a substrate for their saprophytic existence. When the pathogen is a soil inhabitor, however,—that is, it produces long-lived spores or can live as saprophytes for more than 5 or 6 years—crop rotation becomes ineffective or impractical. In the latter cases, crop rotation can still reduce although not eliminate the pathogen (for example, *Verticillium*) populations in the soil. In such cases appreciable yields from the susceptible crop can be obtained every third or fourth year of the rotation.

In some cropping systems the field is tilled and left fallow for a year or part of the year. During fallow, debris and inoculum are destroyed by microorganisms with little or no replacement. In areas with hot summers, fallow allows greater heating and drying of the soil and marked reduction of nematodes and some other pathogens. Other cropping systems utilize herbicides and reduced tillage or reduced tillage and fallow (ecofallow). In some such systems, certain diseases (for example, stalk rot of grain sorghum, caused by

Fusarium moniliforme) have been dramatically reduced, but other diseases (for example, Septoria leaf blotch of wheat) have been increased.

Sanitation

Sanitation consists of all activities aimed at eliminating or reducing the amount of inoculum present in a plant, a field, or a warehouse and at preventing the spread of the pathogen to other healthy plants and plant products. Thus, plowing under or removing and properly disposing of infected leaves, pruning infected or dead branches, and removing or destroying any other plant debris that may harbor the pathogen reduce the amount of disease that will develop later. In the Pacific Northwest and in parts of California, infected crop debris of grass seed and rice crops was destroyed by burning, which reduced or eliminated the surface inoculum of several pathogens.

By washing their hands before handling certain kinds of plants, such as tomato, workers who smoke may reduce the spread of tobacco mosaic virus. Washing the soil off farm equipment before moving it from one field to another may also help to prevent the spread of any pathogens present in the soil. Similarly, by washing the produce, its containers, and the walls of storage houses, the amount of inoculum and subsequent infections may be reduced considerably.

Creating Conditions Unfavorable to the Pathogen

Stored products should be properly aerated to hasten the drying of their surfaces and inhibit germination and infection by any fungal or bacterial pathogens present on them. Similarly, spacing plants properly in the field or greenhouse prevents the creation of high humidity conditions on plant surfaces and inhibits infection by certain pathogens, such as *Botrytis*. Good soil drainage also reduces the number and activity of certain fungal pathogens (for example, *Pythium*) and nematodes and may result in significant disease control. Appropriate choice of fertilizers or soil amendments may also lead to changes in the soil pH, which may unfavorably influence the development of the pathogen. Flooding fields for long periods of time or dry fallowing may also reduce the number of certain pathogens in the soil (for example, *Fusarium* and nematodes) by starvation, lack of oxygen, or desiccation.

In the production of many crops, particularly containerized nursery stock, using composted tree bark in the planting medium has resulted in the successful control of diseases caused by several soil-borne pathogens, for example, *Phytophthora, Pythium* and *Thielaviopsis* root rots, and *Rhizoctonia* damping-off and crown rot, *Fusarium* wilt, and some nematode diseases of several crops, especially of Easter lily, poinsettia, and rhododendron. The suppressive effect is apparently a result of the release from the bark of substances with fungicidal activity.

Polyethylene Traps and Mulches

Many plant viruses, such as cucumber mosaic, are brought into crops, such as peppers, by airborne aphid vectors. When vertical, sticky, yellow polyethylene sheets are erected along the edges of susceptible crops, a considerable number of aphids are attracted to and stick to the plastic, thereby reducing the amount of virus inoculum reaching the crop. On the other hand, if reflectant aluminum or whitish-gray polyethylene sheets are used as mulches between the plants or rows in the field, incoming aphid vectors are repelled and misled away from the field. Reflectant mulches, however, cease to function as soon as the crop canopy covers them.

When clear polyethylene is placed over moist soil, during sunny summer days the temperature at the top 5 cm of soil may reach as high as 52°C compared to a maximum of 37°C in unmulched soil. If sunny weather continues for several days or weeks, the increased soil temperature from solar heat, known as **solarization,** inactivates many soil-borne pathogens, for example, *Verticillium,* and thereby reduces the inoculum and the potential for disease.

Biological Methods That Eradicate or Reduce the Inoculum

Biological control of pathogens, that is, the total or partial destruction of pathogen populations by other organisms, occurs routinely in nature. There are, for example, several diseases in which the pathogen cannot develop in certain areas either because the soil, called **suppressive soils** contains microorganisms antagonistic to the pathogen, or because the plant that is attacked by a pathogen has also been naturally inoculated, before or after the pathogen attack, with antagonistic microorganisms. Sometimes, the antagonistic microorganisms may consist of avirulent strains of the same pathogen that destroy or inhibit the development of the pathogen, as happens in **hypovirulence** and **cross protection**. In some cases, even higher plants reduce the amount of inoculum either by trapping available pathogens (trap plants) or by releasing into the soil substances toxic to the pathogen. In recent years, humans have been trying to take advantage of such natural biological antagonisms and have been developing strategies by which biological control can now be used effectively against several plant diseases. Biological antagonisms, although subject to numerous ecological limitations, can be expected to become an important part of the control measures employed against many more diseases.

Suppressive Soils

Several soil-borne pathogens, such as *Fusarium oxysporum* (the cause of vascular wilts), *Gaeumannomyces graminis* (the cause of take-all of wheat), *Phytophthora cinnamomi* (the cause of root rots of many fruit and forest trees), *Pythium* sp. (the cause of damping-off), and *Heterodera avenae* (the oat cyst nematode), develop well and cause severe diseases in some soils, known as **conducive soils,** while they develop much less and cause much milder diseases in other soils, known as **suppressive soils.** The mechanisms by which soils are suppressive to different pathogens are not always clear but may involve biotic and/or abiotic factors and may vary with the pathogen. In most cases, however, it appears that they operate primarily by the presence in such soils of one or several microorganisms antagonistic to the pathogen. Such antagonists, through the antibiotics they produce, through competition for food, or through direct parasitizing of the pathogen, do not allow the pathogen to reach high enough populations to cause severe disease.

Numerous kinds of antagonistic microorganisms have been found to increase in suppressive soils; most commonly, however, pathogen and disease suppression has been shown to be caused by fungi, such as *Trichoderma, Penicillium,* and *Sporodesmium,* or by bacteria of the genera *Pseudomonas, Bacillus,* and so on. Supressive soil added to conducive soil can reduce the amount of disease by introducing microorganisms antagonistic to the pathogen. Suppressive, virgin soil has been used, for example, to control *Phytophthora* root rot of papaya by planting papaya seedlings in suppressive soil placed in holes in the orchard soil, which was infested with the root rot fungus *Phytophthora palmivora.* On the other hand, in several diseases, continuous cultivation (monoculture) of the same crop in a conducive soil, after some years of severe disease, eventually leads to reduction in disease through increased populations of microorganisms antagonistic to the pathogen. For example, continuous cultivation of wheat or cucumber leads to reduction of take-all of wheat and of *Rhizoctonia* damping-off of cucumber, respectively.) Such soils are suppressive to future disease development. That suppressiveness is due to antagonistic microflora can be shown by pasteurization of the soil at 60°C for 30 minutes, which completely eliminates the suppressiveness.

A sort of "soil suppressiveness" develops after crop rotation of appropriate crops for sufficient duration. Crop rotations not only result in lack of positive selection for the pathogen but they also provide the time required for biological destruction of pathogen inoculum by resident antagonists in the soil. Moreover, the tillage associated with crop rotation further contributes to biological destruction of inoculum by fragmenting and accelerating the breakdown process of crop residues infested with pathogens.

Some Examples of Antagonistic Microorganisms Reducing the Amount of Pathogen Inoculum

The mycelium and resting spores (oospores, sclerotia) of several phytopathogenic soil fungi such as *Pythium, Phytophthora, Rhizoctonia, Sclerotinia,* and *Sclerotium,* are invaded and parasitized **(mycoparasitism)** or are lysed **(mycolysis)** by several fungi, which as a rule are not pathogenic to plants. Several fungi, including some oomycetes, chytridiomycetes, and hyphomycetes and some pseudomonad and actinomycetous bacteria, infect the resting spores of several fungi. Among the most common mycoparasitic fungi are *Trichoderma* sp., mainly *T. harzianum,* which has been shown to parasitize mycelium of *Rhizoctonia* and *Sclerotium,* to inhibit the growth of many others, for example, *Pythium, Fusarium,* and *Fomes,* and to reduce the diseases caused by most of these pathogens. Other common mycoparasitic fungi are *Laetisaria arvalis* (*Corticium* sp.), a mycoparasite and antagonist of *Rhizoctonia* and *Pythium;* also, *Sporidesmium sclerotivorum, Gliocladium virens,* and *Coniothyrium minitants,* all destructive parasites and antagonists of *Sclerotinia sclerotiorum* and all effectively controlling several of the *Sclerotinia* diseases.

Many other fungi have been shown to antagonize and inhibit numerous fungal pathogens of aerial plant parts. For example, *Chaetomium* sp. suppressed *Venturia inaequalis* ascospore and conidia production in fallen and growing leaves, respectively. *Tuberculina maxima* parasitizes the white pine blister rust fungus *Cronartium ribicola; Darluca filum* and *Verticillium lecanii* parasitize several rusts; *Ampelomyces quisqualis* parasitizes several powdery mildews; *Tilletiopsis* sp. parasitizes the cucumber powdery mildew fungus *Spaerotheca fuligena;* and *Nectria inventa* and *Gonatobotrys simplex* parasitize two pathogenic species of *Alternaria.*

In addition to fungi, bacteria of the genus *Streptomyces* and *Pseudomonas* have been shown to parasitize and/or inhibit the pathogenic fungi *Pythium* sp. and *Gaeumannomyces tritici,* the mycophagous nematode *Aphelenchus avenae* parasitizes *Rhizoctonia* and *Fusarium,* and the amoeba *Vampyrella* parasitizes the pathogenic fungi *Cochliobolus sativus* and *Gaeumannomyces graminis.*

Plant pathogenic nematodes are also parasitized by other microorganisms. For example, the cyst nematodes *Heterodera* and *Globodera* are parasitized by the nematophagous fungi *Catenaria auxiliaris, Nematophthora gynophila,* and *Verticillium chlamydosporium,* while the root knot nematode *Meloidogyne* sp. is parasitized by the fungus *Dactylella oviparasitica. Meloidogyne javanica* is also parasitized by the bacterium *Bacillus (Pasteuria) penetrans.*

The mechanisms by which antagonistic microorganisms affect pathogen populations are not always clear, but they are generally attributed to one of four effects: (1) direct parasitism and death of the pathogen, (2) competition with the pathogen for food, (3) direct toxic effects on the pathogen by antibiotic substances released by the antagonist, and (4) indirect toxic effects on the pathogen by volatile substances, such as ethylene, released by the metabolic activities of the antagonist.

Many of the above antagonistic microorganisms are naturally present in crop soils and exert a certain degree of biological control over one or many plant pathogens, regardless of human activities. Humans, however, have been attempting to increase the effectiveness of antagonists either by introducing new and larger populations of antagonists, for example, *Trichoderma harzianum* and *Bacillus penetrans,* in fields where they are lacking, and/or by adding soil amendments that serve as nutrients for, or otherwise stimulate growth of, the antgonistic microorganisms and increase their inhibitory activity against the pathogen. Unfortunately, while both approaches are effective in the laboratory and in the greenhouse, neither has been successful in the field. New microorganisms added to the soil of a field cannot compete with the existing microflora and cannot maintain themselves for very long. Also, soil amendments, so far, have not been selective enough to support and build up only the populations of the introduced or existing antagonists; so the eventual disease control is quite limited. There are several cases of successful biological control of plant pathogens when the antagonistic microorganism is used for direct protection of the plant from infection by the pathogen. These are discussed on pages 201–205.

Control through Trap Plants

If a few rows of rye, corn, or other tall plants are planted around a field of beans, peppers, or squash, many of the incoming aphids carrying viruses that attack the beans, peppers, and squash will first stop and feed on the peripheral taller rows of rye or corn. Because most of the aphid-borne viruses are nonpersistent in the aphid, many of the aphids lose the bean-, pepper-, or squash-infecting viruses by the time they move onto these crops. In this way trap crops considerably reduce the amount of inoculum that reaches a crop.

Trap plants are also used against nematodes, although in a different way. Some plants that are not actually susceptible to certain sedentary plant-parasitic nematodes produce exudates that stimulate eggs of these nematodes to hatch. The larvae enter these plants but are unable to develop into adults, and eventually they die. Such plants are also called **trap crops.** By using trap crops in a crop rotation program, growers can reduce considerably the nematode population in the soil. For example, *Crotalaria* plants trap the larvae of the root-knot nematode *Meloidogyne* sp., and black nightshade plants *(Solanum nigrum)* reduce the populations of the golden nematode, *Heterodera rostochiensis.* Similar results can be obtained by planting highly susceptible plants, which after infection by the nematodes are destroyed (plowed under) before the nematodes reach maturity and begin to reproduce.

Unfortunately, trap plants do not give a sufficient degree of disease control to offset the expense involved and, therefore, they have been little used in practical control of nematode diseases of plants.

Control through Antagonistic Plants

A few kinds of plants, for example, asparagus and marigolds, are antagonistic to nematodes because they release substances in the soil that are toxic to several plant-parasitic nematodes. When interplanted with nematode-susceptible crops, antagonistic plants decrease the number of nematodes in the soil and in the roots of the susceptible crops. Antagonistic plants are not used on a large scale for practical control of nematode diseases of plants for the same reasons that trap plants are not used.

Physical Methods That Eradicate or Reduce the Inoculum

The physical agents most commonly used in controlling plant diseases are temperature (high or low), dry air, unfavorable light wavelengths, and various types of radiation.

Control by Heat Treatment

Soil Sterilization by Heat Soil is usually sterilized in greenhouses, and sometimes in seed beds and cold frames, by the heat carried in live or aerated steam or hot water. The soil is steam sterilized either in special containers (soil sterilizers), into which steam is supplied under pressure, or on the greenhouse benches, in which case steam is piped into and is allowed to diffuse through the soil. At about 50°C, water molds, nematodes, and some oomycetous fungi are killed, while most plant pathogenic fungi and bacteria, along with some worms, slugs, and centipedes are usually killed at temperatures between 60 and 72°C. At about 82°C, most weeds, the rest of plant pathogenic bacteria, most plant viruses in plant debris, and most insects are killed. Heat-tolerant weed seeds and some plant viruses such as tobacco mosaic virus (TMV) are killed at or near the boiling point, that is, between 95 and 100°C. Generally, soil sterilization is completed when the temperature in the coldest part of the soil has remained for at least 30 minutes at 82°C or above, at which temperature almost all plant pathogens in the soil are killed. Heat sterilization of soil is frequently achieved by heat produced electrically rather than supplied by steam or hot water.

It is important to note, however, that excessively high or prolonged high temperatures should be avoided during soil sterilization because not only do they destroy all normal saprophytic microflora in the soil, but they also result in release of toxic levels of some (such as manganese) salts and in accumulation of toxic levels of ammonia (by killing the nitrifying bacteria before they kill the more heat-resistant ammonifying bacteria).

Hot-Water Treatment of Propagative Organs Hot-water treatment of certain seeds, bulbs, and nursery stock is commonly used to kill any pathogens with which they are infected or which may be present inside seed coats, bulb scales, and so on. In some diseases, seed treatment with hot water was for many years the only means of control, as in the loose smut of cereals, in which the fungus overwinters as mycelium inside the seed where it cannot be reached by chemicals. Similarly, treatment of bulbs and nursery stock with hot water frees them from nematodes that may be present within them, such as *Ditylenchus dipsaci* in bulbs of various ornamentals and *Radopholus similis* in citrus rootstocks.

The effectiveness of the method is based on the fact that dormant plant organs can withstand higher temperatures than those their respective pathogens can survive for a given period of time. The temperature of the hot water used and the duration of the treatment varies with the different host–pathogen combinations. Thus, in the loose smut of wheat the seed is kept in hot water at 52°C for 11 minutes, whereas bulbs treated for *Ditylenchus dipsaci* are kept at 43°C for 3 hours.

Hot-Air Treatment of Storage Organs Treatment of certain storage organs with hot air removes the excess moisture from their surfaces and hastens the healing of wounds, thus preventing their infection by certain weak pathogens. For example, keeping sweet potatoes at 28–32°C for 2 weeks helps the wounds to heal and prevents infection by *Rhizopus* and by soft-rotting bacteria. Also, hot-air "curing" of harvested tobacco leaves removes most moisture from them and protects them from attack by fungal and bacterial saprophytes.

Control by Eliminating Certain Light Wavelengths

Alternaria, Botrytis, and *Stemphylium* are examples of plant pathogenic fungi that sporulate only when they receive light in the ultraviolet range (below 360 nm). It has been possible to control diseases on greenhouse vegetables caused by several species of these fungi by covering or constructing the greenhouse with a special UV-absorbing vinyl film that blocks transmission of light wavelengths below 390 nm.

Drying Stored Grains and Fruit

All grains, legumes, and nuts carry with them a sufficient variety and number of fungi and bacteria that can cause decay of these organs in the presence of sufficient moisture. Such decay, however, can be avoided if seeds and nuts are harvested when properly mature, are allowed to dry in the air sufficiently (to about 12 percent moisture) before storage, and are stored under conditions of ventilation that do not allow buildup of moisture to levels (above 12 percent) that would allow storage fungi to become activated. All fleshy fruit, such as peaches and strawberries, is generally harvested later in the day, after the dew is gone, to ensure that it does not carry surface moisture with it during storage and transit, which could result in the decay of the fruit by fungi and bacteria.

Many fruits also can be stored dry for a long time and can be kept free of disease if they are dried sufficiently before storage and if moisture is kept below a certain level during storage. For example, grapes, plums, dates, and figs can be dried in the sun or through warm air treatment to produce raisins, prunes, and dried dates and figs, respectively, that are generally unaffected by fungi and bacteria as long as they are kept dry. Even slices of fleshy fruit such as apples, peaches, and apricots can be protected from infection and decay by fungi and bacteria if they are sufficiently dried by exposure to the sun or to warm air currents.

Disease Control by Refrigeration

Refrigeration is probably the most widely used method of controlling postharvest diseases of fleshy plant products. While low temperatures at or slightly above the freezing point do not kill any of the pathogens that may be on or in the plant tissues, they do inhibit or greatly retard the growth and activities of all such pathogens and thereby prevent the spread of existing infections and the initiation of new ones. Most perishable fruits and vegetables are refrigerated immediately after harvest, transported in refrigerated vehicles, and kept refrigerated until they are used by the consumer. Regular refrigeration of especially succulent fruits and vegetables is sometimes preceded by a quick hydrocooling or air cooling of these products, aimed at removing the excess heat, carried in them from the field, as quickly as possible to prevent any new infections. The magnitude of disease control through refrigeration and its value to the growers and the consumers can hardly be exaggerated.

Disease Control by Radiations

Various types of electromagnetic radiations, such as ultraviolet (UV) light, X-rays, and γ-rays, as well as particulate radiation, such as α-particles and β-particles have been studied for their ability to control postharvest diseases of fruits and vegetables by killing the pathogens present on them. Some satisfactory results were obtained in experimental studies using γ-rays to control postharvest infections of peaches, strawberries, and tomatoes by some of their fungal pathogens. Unfortunately, with many of these diseases the dosage of radiation required to kill the pathogen also injures the plant tissues on which the pathogens exist. So far, no plant diseases are commercially controlled by radiations.

Chemical Methods That Eradicate or Reduce the Inoculum

Chemical pesticides are generally used for direct protection of plant surfaces from infection or for eradicating a pathogen that has already infected a plant. A few chemical treatments, however, are aimed at reducing the inoculum before it comes in contact with the plant. They include soil treatments (such as fumigation), disinfection of warehouses, and control of insect vectors of pathogens.

Soil Treatment with Chemicals

Soil to be planted with vegetables, ornamentals, trees, or other high-value crops such as tobacco and strawberries is frequently treated with chemicals for control primarily of nematodes but occasionally also of soil-borne fungi, such as *Verticillium,* weeds, and bacteria. Certain fungicides are applied to the soil as dusts, drenches, or granules to control damping off, seedling blights, crown and root rots, and other diseases. In fields where irrigation is possible, the fungicide is sometimes applied with the irrigation water, particularly sprinkler irrigation. Fungicides used for soil treatments include metalaxyl, metam-sodium, captan, diazoben, PCNB, and chloroneb. Most soil treatments, however, are aimed at controlling nematodes, and the materials used are volatile (fumigants) so that their fumes penetrate (fumigate) the soil throughout. Some nematicides, however, are not volatile but, instead, dissolve in soil water and are then distributed through the soil.

Fumigation The most promising method of controlling nematodes and certain other soil-borne pathogens and pests in the field has been through the use of chemicals called nematicides. Some of them, including chloropicrin, methyl bromide, Mylone, Vapam, and Vorlex, give off gases after application to the soil and are general-purpose preplant fumigants; they are effective against a wide range of soil microorganisms including, in addition to all nematodes, many fungi, insects, and weeds. Other nematicides, such as Dasanit, Furadan, Mocap, and Temik, are of low volatility, are effective against nematodes and insects, and can be applied before and after planting of many, particularly nonfood, crops, which are tolerant to these chemicals.

Nematicides used as soil fumigants are available as liquids, emulsifiable concentrates, and granules. Nematicides are applied to the soil either by spreading the chemical evenly over the entire field (broadcast) or by applying it only to the rows to be planted with the crop (row treatment). In both cases the fumigant is applied through delivery tubes attached at the back of tractor-mounted chisel-tooth injection shanks or disks spaced at variable widths and usually reaching 6 inches below the soil surface. The nematicide is covered instantly by a smoothing and firming drag or can be mixed into the soil with disk harrows or rototillers. Highly volatile nematicides should be immediately covered with polyethylene sheeting (Fig. 9-1), and this should be left in place for at least 48 hours. When small areas are to be fumigated, the most convenient method is through injection of the chemical with a hand applicator or by placement of small amounts of granules in holes 6 inches deep, 6 to 12 inches apart, and immediately covering the holes with soil. In all cases of preplant soil fumigation with phytotoxic nematicides, at least 2 weeks must

FIGURE 9-1 Soil fumigation for the control of nematodes. Plastic sheet covers soil to keep volatile nematicides from escaping too soon. (Photo courtesy U.S.D.A.)

elapse from the time of treatment to seeding or planting in the field to avoid plant injury.

In the above types of nematicide application, only a small portion of the soil and its microorganisms immediately come in contact with the chemical. The effectiveness of the fumigants, however, is based on the diffusion of the nematicides in a gaseous state through the pores of the soil throughout the area in which nematode control is desired. The distance the vapors move is influenced by the size and continuity of soil pores, by the soil temperature (the best range is between 10 and 20°C), by soil moisture (best at about 80 percent of field capacity), and by the type of soil (more material is required for soils rich in colloidal or organic matter). Nematicides with low volatility, such as Furadan, do not diffuse through the soil to any great extent and must be mixed with the soil mechanically or by irrigation water or rainfall. Except for the highly volatile ones, most nematicides can be applied in irrigation water as soaks or drenches, but only low-volatility nematicides can be applied through overhead sprinkler systems.

In practice, nematode control in the field is generally obtained by preplant soil fumigation with one of the nematicides applied only before planting. These chemicals are nonspecific, that is, they control all types of nematodes, although some nematodes are harder to control than others no matter what the nematicide. Chloropicrin, methyl bromide, Mylone, and Vapam are expensive, broad spectrum nematicides that must be covered upon application and are therefore used for treatment of seedbeds and small areas. On the other hand, DD is cheaper, need not be covered upon application, and is used for treatment of large fields. DD, however, controls only nematodes and, therefore, mixtures of DD with Vapam,Chloropicrin, or methyl bromide are often used to increase its fungicidal action. The contact nematicides applied before or after planting can be used as preplant soil treatments for all types of crops, but their use after planting is limited to nonfood crops such as turf, ornamentals, nurseries, and young nonbearing orchard trees, and to a few, specific food crops for which each chemical has received clearance by the Food and Drug Administration. It should be noted that most of the contact nematicides were developed as insecticides, and nematicidal dosages are much higher than insecticidal ones. All nematicides are extremely toxic to humans and animals and should be handled with great caution.

Disinfestation of Warehouses

To prevent stored products from being infected by pathogens left over in the warehouse from previous years, the storage rooms are first cleaned thoroughly and the debris is removed and burned. The walls and floors are washed with a copper sulfate solution (1 pound in 5 gallons of water) or sprayed with a 1 : 240 solution of formaldehyde. Warehouses that can be closed airtight and in which the relative humidity can be kept at nearly 100 percent while the temperature is between 25 and 30°C can be effectively fumigated with chloropicrin (tear gas) used at 1 pound per each 1000 cubic feet. Fumigation of warehouses can also be carried out by burning sulfur in the warehouse at the rate of 1 pound per 1000 cubic feet of space, or with formaldehyde gas generated by adding 23 ounces of potassium permanganate to 3 pints of formaldehyde per 1000 cubic feet of space. In all cases the fumigants should be allowed to act for at least 24 hours before the warehouse doors are opened for aeration.

Control of Insect Vectors

When the pathogen is introduced or disseminated by an insect vector, control of the vector is as important as, and sometimes easier than, the control of the pathogen itself. Application of insecticides for the control of insect carriers of fungus spores and bacteria has been fairly successful and is a recommended procedure in the control of several such insect-carried pathogens.

In the case of viruses, mycoplasmas, and fastidious bacteria, however, of which insects are the most important disseminating agents, insect control has been helpful in controlling the spread of their diseases only when it has been carried out at the area and on the plants on which the insects overwinter or feed before they enter the crop. Controlling such diseases by killing the insect vectors with insecticides after they have arrived at the crop has seldom proved adequate. Even with good insect control, enough insects survive for sufficiently long periods to spread the pathogen. Nevertheless, appreciable reduction in losses from certain such diseases has been obtained by controlling their insect vectors, and the practice of good insect control is always desirable.

In recent years, success in virus transmission by insects has been achieved by interfering with the ability of the aphid vector to acquire and to transmit the virus rather than by killing the insects. The interference is provided by spraying the plants several times each season with fine-grade mineral oils that seem to have little effect on the probing and feeding behavior of the aphids and are not particularly toxic to the aphids but interfere with the transmission of nonpersistent, semipersistent, and even some persistent aphid-borne viruses. Control of aphid-borne viruses by oil sprays has been more successful with some viruses (for example, cucumber mosaic virus on cucumber and pepper, and potato virus Y on pepper) and in some areas (such as Florida) than in others.

Disease Control by Immunizing, or Improving the Resistance of, the Host

Unlike humans and animals, plants lack an antibody-producing system and cannot be "immunized" by vaccination the way humans can. Treatment of plants with certain pathogens, however, often leads to temporary or permanent "immunization" of the plants, that is, induced resistance to a pathogen to which the plants are normally susceptible. Some of these treatments involve only viruses and are known as cross protection; others may involve different kinds of pathogens and are known as induced resistance. In some cases, the resistance of a host plant to a particular pathogen can be improved by simply improving the growing conditions (fertilizer, irrigation, drainage, and so on) of the host. By far the most common improvement of host resistance to almost any pathogen, however, is brought about by improving the genetic resistance of the host, that is, by breeding and using resistant varieties.

Cross Protection

The term **cross protection** specifically applies to the protection of a plant by a mild strain of a virus from infection by a strain of the same virus that causes much more severe symptoms. This appears to be a general phenomenon among virus strains. Its application in controlling virus diseases has met with some success in cross protecting tomatoes with mild strains of tobacco mosaic virus and citrus with mild strains of citrus tristeza virus. Cross protection has not gained widespread use, however, because appropriate mild strains of viruses are not available, the method for field crops is laborious, and there is a danger of mutations toward new, more virulent strains, double infections, and the spread to other crops in which virulence might be greater.

Induced Resistance

There are many examples in which plants infected with one pathogen become more resistant to subsequent infection by another pathogen and, also, of plants becoming resistant to a pathogen if they have been inoculated with the same pathogen at an earlier growth stage at which they are resistant to the pathogen. (There are, however, also many examples of plants being more susceptible to a pathogen if they are already infected with another pathogen).

For example, bean and sugarbeet inoculated with virus exhibit a greater resistance to infection by certain obligate fungal pathogens causing rust and powdery mildews than do virus-free plants. Also, in tobacco, tobacco mosaic virus induces a systemic resistance not only to itself but also to unrelated viruses, to fungi like *Phytophthora parasitica* var, *nicotianae,* to bacteria like *Pseudomonas tabacci,* and even to certain aphids! Inversely, inoculation of tobacco with a root lesion-inducing fungus such as *Chalara elegans (Thielaviopsis basicola)* or a leaf lesion-inducing bacterium such as *Pseudomonas syringae,* induces systemic resistance to TMV. Additional examples of induced resistance include fire blight of pear, in which induction was by inoculation with a nonpathogenic bacterium, and cucurbit anthracnose, in which resistance to the fungus *Colletotrichum lagenarium* was induced by inoculating the plants while young with the same fungus.

It later became apparent that resistance to pathogens could be induced in their hosts by treating (rubbing, infiltrating, or injecting) the latter with naturally occurring compounds obtained from the pathogen, such as the coat protein of tobacco mosaic virus (TMV), a proteinaceous component or a glycoprotein fraction from the bacterium *(Pseudomonas solanacearum),* a lipid component from the fungus *(Phytophthora infestans),* and so on, or resistance to pathogens could be induced by treating plants with unrelated natural compounds such as a water-soluble fraction of the nonpathogenic bacterium *Nocardia,* or a polysaccharide from a nonpathogenic fungus, or a proteinaceous compound isolated from an unrelated plant, all inducing complete resistance to TMV infection.

Even more significantly, resistance in plants, primarily to TMV, but also to several other viruses and to fungi, such as *Peronospora tabacina,* and bacteria, such as *Pseudomonas syringae,* can be induced with several types of synthetic compounds, particularly polyacrylic acid, which is applied by injecting into the plant, spraying onto leaves, or absorption through the petiole or through the roots. Other synthetic chemicals reported as effective inducers of resistance, primarily against TMV, include salicylic acid (aspirin) and 2-chloroethylphosphonic acid.

The scientific developments in the area of induced resistance are quite exciting and certainly promising; however, no disease is currently controlled by any of the mechanisms of induced resistance listed above. Still, a great deal of work is going on in these areas, including work by private industry interested in developing and marketing compounds that would induce resistance to plant disease. It is expected that some truly effective compounds will be commercially available before too long.

Improving the Growing Conditions of Plants

Cultural practices aiming at improving the vigor of the plant often help increase its resistance to pathogen attack. Thus, proper fertilization, field drainage, irrigation, proper spacing of plants, and weed control improve the growth of plants and may have a direct or indirect effect on the control of a particular disease. The most important measures for controlling *Valsa* canker of fruit and other trees, for example, are adequate irrigation and proper fertilization of the trees.

Use of Resistant Varieties

The use of resistant varieties is the cheapest, easiest, safest, and most effective means of controlling plant diseases in crops for which such varieties are available. Cultivation of resistant varieties not only eliminates losses from disease but also eliminates expenses for sprays and other methods of disease control, and avoids the contamination of the environment with toxic chemicals that would otherwise be used to control plant diseases. Moreover, for many diseases, such as those caused by vascular pathogens and viruses, which cannot be adequately controlled by any available means, and for others, such as cereal rusts and root rots, which are economically impractical to control in other ways, the use of resistant varieties provides the only means of producing acceptable yields.

Varieties of crops resistant to some of the most important or most difficult to control diseases are made available to growers by federal and state experiment stations and by commercial seed companies. More than 75 percent of the total agricultural acreage in the United States is planted with varieties that are resistant to one or more diseases. With some crops, such as small grains and alfalfa, varieties planted because they are resistant to certain disease(s) make up 95 to 98 percent of the crop. Growers and consumers alike have gained the most from the use of varieties resistant to the fungi causing rusts, smuts, powdery mildews, and vascular wilts, but several other kinds of fungal diseases, and many diseases caused by viruses, bacteria, and nematodes are controlled through resistant varieties.

Resistant varieties have been used in only a few cases for disease control in fruit and forest trees. Examples are blister rust and fusiform rust of pine. It is difficult to replace susceptible varieties with resistant ones and to keep the resistant ones from being attacked by new races of the pathogen that are likely to develop over the long life span of trees.

Although it is always preferable to use varieties that have both vertical (initial inoculum-limiting) and horizontal (rate-limiting) resistance, most resistant varieties carry only one or a few (2 or 3) major genes of (vertical) resistance either because it is difficult to combine high levels of the two types of resistance or because the mechanism of inheritance of horizontal resistance genes is usually unknown. Such varieties are, of course, resistant only to some of the races of the pathogen and, if the pathogen is airborne and new races can be brought in easily, as happens with the cereal rusts, the powdery mildews, and *Phytophthora infestans,* new races virulent to the resistant variety appear quickly and become widespread. Appearance of new races leads to "breakdown of resistance" of the old variety, and as the new race takes over, the variety must be replaced with another variety that has different genes for resistance. As a result, varieties with vertical resistance need to be replaced every few (3, 5, or 10) years, depending on the genetic plasticity of the pathogen, the particular gene or combination of genes involved, the degree and manner of deployment of the gene(s), and the favorableness of the weather conditions toward disease development (disease pressure).

Several techniques are employed to increase the useful "life span" of a resistant variety. To begin with, varieties are tested for resistance against as many pathogens and as many races of the pathogens as are available. Second, before they are released, varieties are tested for resistance in as many locations as possible, often, for some important crops like the cereals, in many countries and continents. Local breeding programs may carry out additional adaptation breeding to incorporate resistance to local pathogens. In this way, only resistant varieties to almost all races of all pathogens in almost all places are released.

Even after a resistant variety is released, however, measures can be taken to prolong its resistance. Any management strategy such as sanitation, seed treatment, or fungicide application that reduces inoculum pressure on the variety is likely to increase its useful life span. For slowly dispersing patogens (such as soil-borne pathogens), rotation of varieties with different resistances keeps down pathogen populations compatible with each variety, so each variety can last longer. For large area crops, such as wheat, varieties could last longer against airborne pathogens (for example, stem rust) if they were each deployed in one of three or four regional zones of the epidemiological region. In this way, even if a new race that could attack a variety in one region did appear, it could not spread to the other varieties in the other regions because they have a different set of genes for resistance than the one whose resistance "broke down." A still different approach involves the use of varietal mixtures or multilines. Multilines are composed of ten or so near isogenic lines, each possessing a different gene for vertical resistance and, therefore, being resistant to a large proportion of the pathogen population. This results in overall reduction of the pathogen reproduction rate, which reduces the rate of disease increase and the inoculum pressure on each of the other varieties. If one of the isogenic lines is attacked severely by a race of a pathogen, the following year the isoline is substituted in the multiline seed with a different isogenic line that can resist the new race.

Direct Protection of Plants from Pathogens

Some pathogens are endemic in an area, for example, the apple scab fungus *Venturia inaequalis,* the crown gall bacterium, *Agrobacterium tumefaciens,* and the cucumber mosaic virus; others are likely to arrive annually from southern areas, for example, the wheat stem rust fungus *Puccinia graminis.* If experience has shown that none of the other methods of control is likely to prevent a major epidemic, then the plants must be protected directly from infection by such a pathogen that is likely to arrive on the plant surfaces in rather large numbers. The most common means of direct protection of plants from pathogens include some biological controls (fungal and bacterial antagonists) but primarily the chemical control measures, that is, the use of chemicals for foliar sprays and dusts, seed treatments, treatment of tree wounds, and control of postharvest diseases of produce.

Direct Protection by Biological Controls

Biological control practices for direct protection of plants from pathogens involve the deployment of antagonistic microorganisms at the infection court before or after infection takes place.

Fungal Antagonists

Biocontrol of Heterobasidion (Fomes) annosum by Peniophora gigantea. *Heterobasidion annosum,* the cause of root and butt rot of conifers, infects freshly cut pine stumps and then spreads into the roots. Through the root contacts, it then spreads into the roots of standing trees, which it kills. If we inoculate the stump surface with oidia of the fungus *Peniophora gigantea* immediately after the tree is felled, *Peniophora* occupies the cut surface and spreads through the stump into the lateral roots. There, it successfully competes with and excludes or replaces the pathogenic *Heterobasidion* in the stump, thereby protecting nearby trees. The oidia are applied to the cut surface either as a water suspension or as a powder, or they are added to the lubricating oil placed on the chain saw and are thus deposited on the surface as it is cut.

Biocontrol of Chestnut Blight with Hypovirulent Strains of the Pathogen Chestnut blight, caused by the fungus *Endothia parasitica,* is controlled naturally in Italy, and artificially in France through inoculation of cankers, caused by the normal pathogenic strains of the fungus, with hypovirulent strains of the same fungus. The hypovirulent strains carry viruslike double-stranded (ds) RNAs that apparently limit their pathogenicity. The ds RNAs apparently pass through mycelial anastomoses from the hypovirulent to the virulent strains, the latter are rendered hypovirulent, and the development of the canker slows down or stops. Hypovirulent strains are being tested in the United States also, but so far control of chestnut blight has been limited to experimental trees.

Biological Control of Fusarium Wilt with a Nonpathogenic Strain Fusarium wilt of sweet potato, caused by *Fusarium oxysporum* f. sp. *batatas,* was recently reported to be successfully controlled by inoculating sweet potato cuttings with a nonpathogenic strain of the same fungus. This strain had been isolated from the vascular tissue of a sweet potato plant that remained healthy among other plants that had been killed by the wilt-inducing strains of the fungus.

Biological Control of Diseases of Aerial Plant Parts with Fungi Many filamentous fungi and yeasts have been shown to be effective antagonists of fungi infecting the aerial parts of plants. For example, inoculation of postbloom, dead, tomato flowers with conidia of *Cladosporium herbarum* or *Penicillium* sp. almost completely suppressed subsequent infection of developing fruits by *Botrytis cinerea.* Similarly, spraying spores of common bark saprophytes, such as *Cladosporium* sp. and *Epicoccum* sp., and of the soil fungus *Trichoderma* on pruning cuts of fruit trees has prevented infection by canker-causing pathogens such as *Nectria galligena* and *Cytospora* sp. Sprays with *Trichoderma* in the field also reduced *Botrytis* rot of strawberries and of grapes at the time of harvest and in storage. Several foliar diseases have also been reduced significantly (by more than 50 percent) when the leaves were sprayed with spores of common phylloplane fungi, for example, *Alternaria, Cochliobolus, Septoria, Colletotrichum,* and *Phoma,* or with spores of hyperparasites, for example, the cucumber powdery mildew fungus *Sphaerotheca fuliginea* with spores of *Ampelomyces quisqualis,* of the wheat leaf rust fungus *Puccinia recondita* with spores of *Darluca filum,* and of the carnation rust fungus with *Verticillium lecanii.* None of the above, and of numerous other known cases of fungal antagonism by fungi, is used as yet for practical control of any disease of aerial plant parts.

Biological Control of Postharvest Diseases Postharvest rots of several fruits could be reduced considerably by spraying with spores of antagonistic fungi at different stages of fruit development or by dipping the harvested fruit in the inoculum. For example, significant reduction of citrus green mold (caused by *Penicillium digitatum*) was obtained by treating the fruit with the antagonist *Trichodernia viride,* while preharvest and postharvest Botrytis rot of strawberries was reduced by several sprays of *Trichoderma* spores on strawberry blossoms and young fruit. Similarly, Penicillium rot of pineapple was reduced considerably by spraying the fruit with nonpathogenic strains of the pathogen.

Biocontrol of Root Pathogens by Mycorrhizae Roots of most plants form a symbiotic relationship with certain kinds of zygomycete, ascomycete, and basidiomycete fungi (mycorrhizae). Mycorrhizae colonize roots intercellularly (ectomycorrhizae) or intracellularly (endomycorrhizae). Although mycorrhizae obtain organic nutrients from the plant, they benefit the plant by promoting nutrient uptake and enhancing water transport by the plant, thus increasing growth and yield, and sometimes by providing the plant with considerable protection against several soil-borne pathogens. Mycorrhizae have been shown to provide considerable protection to pine seedlings from *Phytophthora cinnamomi,* to tomato and douglas fir seedlings from *Fusarium oxysporum,* to cotton from *Verticillium* wilt and the root knot nematode, and to soybean from *Phytophthora megasperma* and *Fusarium solani.* Commercial preparations of certain mycorrhizal fungi are available against some of these pathogens, but problems of production, specificity, and application of mycorrhizae to plants remain.

Bacterial Antagonists

Biocontrol of Crown Gall with Agrobacterium Radiobacter Strain K84 Crown gall of pome, stone, and several small fruits (grapes, raspberries) and ornamentals (rose and euvonymous) is caused by the bacterium *Agrobacterium tumefaciens.* For several years now, crown gall has been controlled commercially by treating the seeds, seedlings, and cuttings with a suspension of strain K84 of the related but nonpathogenic bacterium *Agrobacterium radiobacter.* Control is based on production by strain 84 of a bacteriocin, that is, an antibiotic specific against related bacteria and called agrocin 84. The bacteriocin selectively inhibits most pathogenic agrobacteria that arrive at surfaces occupied by strain 84. Strains of *A. tumefaciens* insensitive to agrocin 84 have been reported in several countries.

Seed Bacterization Treatment of seeds such as cereals, sweet corn, and carrots with water suspensions, slurries, or powders containing the bacteria *Bacillus subtilis* strain A13 or *Streptomyces* sp. has protected the plants against root pathogens and has resulted in better growth and yield of these crops.

Pseudomonas rhizobacteria, primarily of the *P. fluorescens* and *P. putida* groups, applied to seeds, seed pieces, and roots of plants have resulted in less soft rot and in consistent increases in growth and yield in several crops. For example, in some experiments, treated potato seed tubers produced from 5 to 33 percent greater yield; treated sugarbeet seeds produced 4–6 more tons of sugarbeets per hectare, corresponding to an increase of from 955 to 1227 kilograms of sugar per acre; treated radish seeds produced from 60 to 144 percent more root weight than untreated ones, and treated wheat seed planted in *Gaeumannomyces graminis tritici* (take-all of wheat)-infected soil, produced 27 percent more yield than untreated seed. The mechanism or mechanisms by which these plant growth-promoting rhizobacteria increase yield is not clear. It appears, however, that inhibition of harmful, toxic microorganisms and of soil-borne pathogens by antibiotics or by competition for iron, are at least some of the determinants of their effectiveness.

Biocontrol of Postharvest Diseases When several kinds of stone fruits, that is, peaches, nectarines, apricots, and plums, were treated after harvest with suspensions of the antagonistic bacterium *Bacillus subtilis,* they remained free of brown rot, caused by the fungus *Monilinia fructicola,* for at least nine days.

Biocontrol with Bacteria of Bacteria-mediated Frost Injury Frost-sensitive plants are injured when temperatures drop below 0 °C because ice forms within their tissues. Small volumes of pure water can be supercooled to −10 °C or below without ice formation, provided no catalyst centers or nuclei are present to influence ice formation. It has been shown recently, however, that certain strains of at least three species of epiphytic bacteria (*Pseudomonas syringae, P. fluorescens,* and *Erwinia herbicola*), which are present on many plants, serve as ice-nucleaction-active catalysts for ice formation at as high temperatures as −1 °C. Such bacteria usually make up a small proportion (0.1–10 percent) of the bacteria found on leaf surfaces. By isolating, culturing, mass producing, and applying on the plant surfaces non-ice-nucleation active bacteria antagonistic to the ice-nucleation-active bacteria, it has been possible to reduce and replace numbers of ice-nucleation-active bacteria on treated plant surfaces with non-ice-nucleation active bacteria. This treatment protects frost-sensitive plants from injury at temperatures at which untreated plants may be severely injured.

Biological Control of Diseases of Aerial Plant Parts with Bacteria Numerous bacteria, most of them saprophytic Gram-negative bacteria of the genera *Erwinia, Pseudomonas,* and *Xanthomonas,* and fewer of the Gram-positive genera *Bacillus, Lactobacillus,* and *Corynebacterium,* are found on aerial plant surfaces, particularly early in the growing season. Some pathogenic bacteria, such as *Pseudomonas syringae* pv. *syringae, P.s.* pv. *morsprunorum, P.s.* pv *glycinea, Erwinia amylovora,* and *E. carotovora* also live epiphytically (on the surface) on leaves, buds, and so on before they infect and cause disease. In several cases, spraying leaf surfaces with preparations of saprophytic bacteria, or with avirulent strains of pathogenic bacteria, has reduced considerably the number of infections caused by bacterial and fungal pathogens. For example, fire blight of apple blossoms, caused by *Erwinia amylovora,* was partially controlled with sprays of *Erwinia herbicola;* bacteria leaf streak of rice, caused by *Xanthomonas translucens* ssp. *oryzicola,* was reduced with sprays of isolates of *Erwinia* and of *Pseudomonas.* Similarly, sprays with non-ice-nucleating strains of *Pseudomonas syringae* and *Erwinia herbicola* reduced the numbers and protected from frost plants carrying the normal complement of ice-nucleating strains of the same species of bacteria.

Several cases of added epiphytic bacteria inhibiting plant infections by fungi are also known. For example, spraying grass plants with *Pseudomonas fluorescens* reduced infection by *Drechslera (Helminthosporium) dictyoides,* spraying with *Bacillus subtilis* reduced infection of apple leaf scars by *Nectria galligena,* and spraying of peanut or tobacco plants with *Pseudomonas cepacia* or *Bacillus* sp. reduced *Cercospora* and *Alternaria* leaf spot on these hosts, respectively. None of these biological controls is used in practice to control any disease of aerial plant parts as well.

Viral Parasites of Plant Pathogens

All pathogens—fungi, bacteria, mycoplasmas, and nematodes—are attacked by viruses. So far, however, only for bacterial pathogens have viruses been tested as possible biological controls. **Bacteriophages** or **phage** (bacteria-destroying viruses) are known to exist in nature for most plant pathogenic bacteria. Successful experimental control of several bacterial diseases was obtained when the bacteriophages were mixed with the inoculated bacteria, when the plants were first treated with bacteriophages and then inoculated with bacteria, and when the seed was treated with the phage. In practice, however, not one bacterial disease is controlled effectively by bacteriophage. Also, no plant disease caused by a bacterium has been cured yet by treatment with phage after the disease has developed.

Direct Protection by Chemical Controls

The most commonly known means of controlling plant diseases in the field and in the greenhouse and, sometimes, in storage, is through the use of chemical compounds that are toxic to the pathogens. Such chemicals either inhibit germination, growth, and multiplication of the pathogen or are outright lethal to the pathogen. Depending on the kind of pathogens they affect, the chemicals are called fungicides, bactericides, nematicides, viricides, or for the parasitic higher plants, herbicides. Some chemicals are toxic to all or most kinds of pathogens, others affect only one kind of pathogen, and certain compounds are toxic to only a few or a single specific pathogen. About 60 percent of all the chemicals (mostly fungicides) used to control plant diseases is applied to fruit and about 25 percent to vegetables.

Most of the chemicals are used to control diseases of the foliage and of other aboveground parts of plants. Others are used to disinfest and protect from infection seeds, tubers, and bulbs. Some are used to disinfest the soil, others to disinfest warehouses, to treat wounds, or to protect stored fruit and vegetables from infection. Still others (insecticides) are used to control the insect vectors of some pathogens.

The great majority of older chemicals applied on plants or plant organs could only protect them from subsequent infection and could not stop or cure a disease after it had started. The great majority of these chemicals are effective only in the plant area to which they have been applied (local action) and are not absorbed or translocated by the plants. Many new chemicals, however, do have a therapeutic (eradicant) action, and several are absorbed and systemically translocated by the plant (systemic fungicides and antibiotics).

Methods of Application of Chemicals for Plant Disease Control

Foliage Sprays and Dusts Chemicals applied as sprays or dusts on the foliage of plants are usually aimed at control of fungus diseases and to a lesser extent control of bacterial diseases. Most fungicides and bactericides are **protectants** and must be present on the surface of the plant in advance of the pathogen in order to prevent infection. Their presence usually does not allow fungus spores to germinate or they may kill spores upon germination. Contact of bacteria with bactericides may inhibit their multiplication or cause their death.

Some newer fungicides also have a direct effect on pathogens that have already invaded the leaves, fruit, and stem, and in this case they act as **eradicants,** that is they kill the fungus inside the host, or they may suppress the sporulation of the fungus without killing it. Some fungicides, such as dodine, have a partial systemic action because they can be absorbed by a part of the leaf tissues and translocated internally into the whole leaf area. Several fungicides—benomyl, thiabendazole, carboxin, and metalaxyl—are clearly systemics and can be translocated internally throughout the host plant. Some bactericides, for example, streptomycin, are also systemics, as are most antibiotics.

Some newer systemic fungicides, such as the **sterol inhibitors** triadimefon and fenarimol, and metalaxyl, are so effective in postinfection applications that they can be used as **rescue treatments** of crops; in other words, they can be applied effectively after an epidemic is well under way.

Fungicides and bactericides applied as sprays appear to be more efficient than when applied as dusts. Although it may appear that dusts are preferable to sprays if application is to be made during a rain because they adhere better to wet plant tissues, in reality, neither sticks well when applied during a rain. Sometimes other compounds, for example, **lime,** may be added to the active chemical in order to reduce its phytotoxicity and make it safer for the plant. Compounds with a low surface tension, such as **detergents,** are often added to fungicides so that they spread better, thereby increasing the contact area between fungicide and the sprayed surface. Finally, some compounds, for example, **starch and oils,** are added to increase the adherence of the fungicide to the plant surface.

In fields with sprinkler irrigation available, satisfactory control of many foliar diseases on a variety of crops has recently been obtained by applying protectant or systemic fungicides to the foliage, and somewhat to the roots, through the irrigation system **(fungigation).**

Because many fungicides and bactericides are protectant in their action, it is important that they be at the plant surface before the pathogen arrives or at least before it has had time to germinate, enter, and establish itself in the host. Because spores require a film of water on the leaf surface or at least atmospheric humidity near saturation before they can germinate, sprays or dusts seem to be most effective when they are applied before or immediately after every rain. Considering that many fungicides and bactericides are effective only upon contact with the pathogen, it is important that the whole surface of the plant be covered with the chemical in order to be protected. For this reason, young, expanding leaves, twigs, and fruits must be sprayed more often than mature tissues, since small, growing leaves may outgrow protection 3 to 5 days after spraying. The interval between sprays of mature tissue may vary from 7 to 14 days or longer, depending on the particular disease, the frequency and duration of rains, the persistence or residual life of the fungicide, and the season of the year. The same factors also determine the optimal number of sprays per season, which may vary from 2 or 3 to 15 or more. Figure 9-2 shows some types of equipment used for spraying and dusting plants and for injecting chemicals into plants or into the soil.

In the last several years, several systemic fungicides have become available, and their number, ease of application, duration of effectiveness, and even the number of diseases they control are increasing steadily. Systemic chemicals are gradually replacing many of the contact, preventive fungicides because of their effectiveness, long-lasting activity, and the limited number of applications required to protect a crop from one or many diseases.

The number and variety of chemicals used for foliar sprays and dusts is quite large. Some of these compounds are specific against certain diseases, others are effective against a wide spectrum of pathogens. Sprays with these materials usually contain 0.5–2 lb of the compound per hundred gallons of water, although some, for example, sulfur, are applied at 4–6 lb per 100 gallons of water. Some of the fungicides used for foliar sprays or dusts are also used for seed treatments.

Seed Treatment Seeds, tubers, bulbs, and roots are usually treated with chemicals to prevent their decay after planting or damping off of young seedlings by controlling pathogens carried on them or existing in the soil where they will be planted. More recently, seeds have been treated with systemic fungicides in order to inactivate pathogens in infected seeds (for example, carboxin for control of loose smut), or in order to provide the foliage of the developing plant with systemic protection against the pathogen (for example, metalaxyl for control of downy mildews of oats and sorghum and triadimenol for control of leaf rust and *Septoria* leaf blotch of wheat and of *Pyrenophora* net blotch of barley). Chemicals can be applied on the seed as dusts or as thick water suspensions mixed with the seed. The seed can also be soaked in a water or solvent solution of the chemical and then allowed to dry. Tubers, bulbs, corms, and roots can be treated in similar ways.

In treating seeds or any other propagative organs with chemicals, precautions must be taken so that their viability is not lowered or destroyed. At the same time, enough chemical must stick to the seed to protect it from attacks of pathogens and, when the seed is planted, to diffuse into, and disinfest a sphere of soil around the seed in which the new plant will grow without being attacked at this particularly vulnerable period of growth.

Chemicals used to treat seeds, bulbs, corms, tubers, and roots include some inorganic copper and zinc compounds, but mostly organic protectant compounds such as captan, chloroneb, chloranil, dichlone, hexachlorobenzene, maneb, zineb, mancozeb, thiram, pentachloronitrobenzene (PCNB), and the systemic compounds carboxin, benomyl, thiabendazole, metalaxyl,

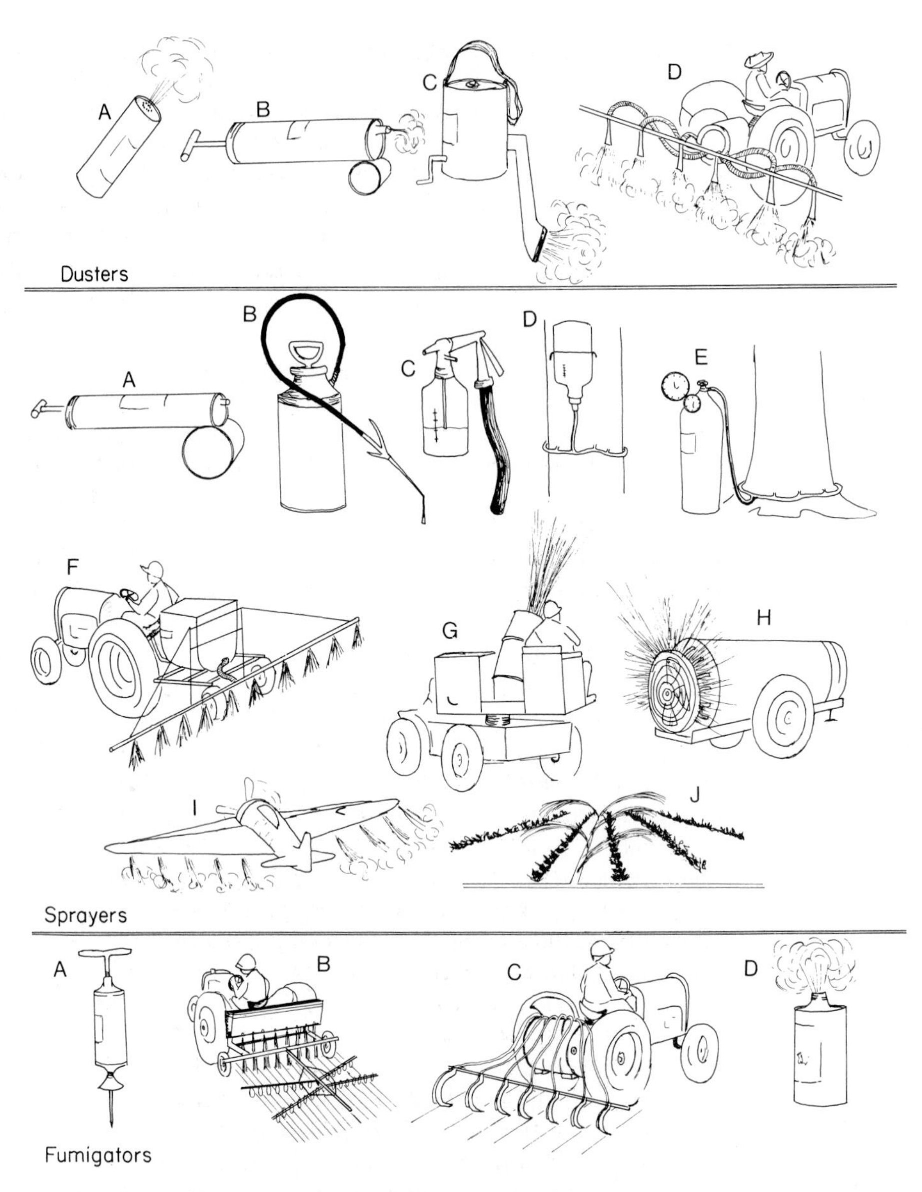

FIGURE 9-2 Various types of equipment used for dusting, spraying, injection, or fumigation for control of plant diseases. Dusters: (A–C) Portable dusters, (D) Tractor mounted. Sprayers: (A–C) Portable sprayers, (D) Tree injection gravity flow apparatus, (E) Tree injection under pressure, (F–H) Tractor-mounted sprayers for annuals (F) and for trees (G,H), (I) Airplane spraying (or dusting), (J) Spraying through the irrigation system. Fumigators: (A) Handgun fumigator, (B,C) Tractor mounted gravity-flow or pump-driven injectors, (D) Fumigation can for greenhouse or warehouse.

triadimenol, and streptomycin. Some chemicals may control specific diseases of some plants while others are more general in their action and may control many diseases of a number of plants.

Soil Treatment In addition to treating (fumigating) soil to be planted with vegetables, ornamentals, or trees with volatile chemicals (fumigants) for control of nematodes, fungi, and bacteria by reducing the inoculum, certain fungicides are applied to the soil as dusts, drenches, or granules to control damping off, seedling blights, crown and root rots, and other diseases. Such fungicides include captan, diazoben, PCNB, and chloroneb, metalaxyl, fosetyl-Al, triadimefon, ethazol, and propamocarb. Many of the systemic fungicides, for example, metalaxyl, fosetyl-Al, are so effective even in small amounts that a single preplant broadcast soil spray followed by disking can provide a full season's protection to the crop from several of the pathogens it normally controls. In many cases, foliar diseases (such as downy mildews and rusts) can be controlled by impregnating the fungicide (for example, metalaxyl, triadimenol) into the fertilizer and applying them together before planting.

In recent years, protective and systemic fungicides have been applied to the soil (and to the foliage) through the irrigation water (fungigation) for control of soil-borne diseases.

Treatment of Tree Wounds Large pruning cuts and wounds made on the bark of branches and trunks accidentally or in the process of removing infections of fungi and bacteria need to be protected from drying and from becoming ports of entry for new pathogens. Drying of the margins of large tree wounds is usually prevented by painting them with shellac or any commercial wound dressing. The exposed wood is then sterilized by swabbing it with a solution of either 0.5–1.0 percent sodium hypochlorite (10–20 percent Clorox), or with 70 percent ethyl alcohol. Finally, the entire wound is painted with a permanent, tree wound dressing, such as a 10:2:2 mixture of lanolin, rosin, and gum, or Cerano, or Bordeaux paint, or an asphalt-varnish tree paint. Some wound dressings, such as Cerano and Bordeaux paint, are themselves disinfectants, while most others require the addition of a disinfectant, such as 0.25 percent phenyl mercuric nitrate or 6 percent phenol. It must be kept in mind, however, that many commercial wound dressings, especially those that are asphalt-based, are phytotoxic enough to prevent, rather than promote, wound healing.

Control of Postharvest Diseases The use of chemicals for the control of postharvest diseases of fruits and vegetables is complicated enormously by the fact that most compounds effective against storage diseases leave on the produce high concentrations of residues that are toxic to consumers. Many chemicals also cause injury to the products under storage conditions and give off undesirable odors.

A number of fungitoxic chemicals, however, most of them used specifically for control of postharvest diseases, have been developed. Most of them are used as dilute solutions into which the fruits or vegetables are dipped before storage, or as solutions used for washing or hydrocooling of fruits and vegetables immediately after harvest. Some chemicals, for example, elemental sulfur, are used as dusts or crystals that undergo sublimation in storage, and others, such as SO_2, as gases. Finally, some chemicals are impregnated in the boxes or wrappers containing the fruit. Among the compounds used for commercial control of postharvest diseases of, primarily, citrus fruits but also of other fruits are borax, biphenyl, sodium *o*-phenylphenate, benomyl, thiabendazole, and imazalil. Certain other chemicals, such as elemental sulfur,

sulfur dioxide, dichloran, captan, and benzoic acid, have been used mostly for the control of storage rots of stone and pome fruits, bananas, grapes, strawberries, melons, and potatoes.

Types of Chemicals Used for Plant Disease Control

Many hundreds of chemicals have been advanced to date for crop protection as fumigants, soil treatments, sprays, dusts, paints, pastes, and systemics. The most important of these chemicals and some of their properties and uses are described below.

Copper Compounds *Bordeaux mixture,* the product of reaction of copper sulfate and calcium hydroxide (lime), is the most widely used copper fungicide throughout the world. It controls many fungus and bacterial leaf spots, blights, anthracnoses, downy mildews, and cankers but causes burning of leaves or russeting of fruit such as apples when applied in cool, wet weather. The phytotoxicity of Bordeaux is reduced by increasing the ratio of lime to copper sulfate. Copper is the only ingredient in the Bordeaux mixture that is toxic to pathogens and, sometimes, to plants, while lime's role is primarily that of a "safener." For dormant sprays, concentrated Bordeaux is made by mixing 10 pounds of copper sulfate, 10 pounds of lime, and 100 gallons of water; it has the formula 10:10:100. The most commonly used formula for Bordeaux is 8:8:100. For spraying young, actively growing plants, the amounts of copper sulfate and lime are reduced, and the formulas used may be 2:2:100, 2:6:100, and so on. For plants known to be sensitive to Bordeaux, a much greater concentration of lime may be used, as in the formula 8:24:100.

In the "fixed" or "insoluble" copper compounds the copper ion is only slightly soluble, and these compounds are, therefore, less phytotoxic than Bordeaux, but also less effective as fungicides. The fixed coppers are used for control of the same diseases as Bordeaux and they can also be used as dusts. The fixed coppers contain either basic copper sulfate or basic copper chlorides, or copper oxides, or miscellaneous other formulations. Most of them are recommended as sprays at the rate of 4 lb per 100 gallons of water or as 7 percent copper dusts.

Kocide is a copper formulation fungicide and bactericide that contains copper hydroxide. The latter is formulated so that it exhibits a large surface area, which probably accounts for the increased availability of the soluble cupric ion at the plant surface. It controls the same diseases as the other copper compounds, but because it is more soluble and it is used in smaller amounts, it does not clog spray nozzles as often.

Sulfur Compounds Several inorganic sulfur formulations and numerous organic sulfur compounds have proved to be excellent fungicides and are used to control a variety of diseases.

• Inorganic Sulfur Compounds

The element **sulfur** as a dust, wettable powder, paste, or liquid is used primarily to control powdery mildews on many plants, but it is also effective against certain rusts, leaf blights, and fruit rots. Sulfur, in its different forms, is available under a variety of trade names. Most sulfur formulations are applied at the rate of 1 to 6 lb per 100 gallons of water and may cause injury in hot (temperatures above 30°C), dry weather, especially to sulfur-sensitive plants such as tomato, melons, and grape.

By boiling lime and sulfur together lime-sulfur, self-boiled lime-sulfur, and dry lime-sulfur are produced, which are used as sprays for dormant fruit trees to control blight or anthracnose, powdery mildew, apple scab, brown rot of stone fruits, and peach leaf curl and are sometimes used for summer control of the same diseases. The various lime-sulfurs are applied at the rate of 2–10 gallons per 100 gallons of water.

- **Organic Sulfur Compounds: Dithiocarbamates**

The organic sulfur compounds are unquestionably the most important, most versatile, and most widely used group of modern fungicides. They include thiram, ferbam, nabam, maneb, and zineb and are all derivatives of dithiocarbamic acid. It is believed that the dithiocarbamates are toxic to fungi because they are metabolized to the isothiocyanate radical (—N=C=S), which inactivates the sulfhydryl groups (—SH) in amino acids and enzymes within pathogen cells and thereby inhibits the production and function of these compounds.

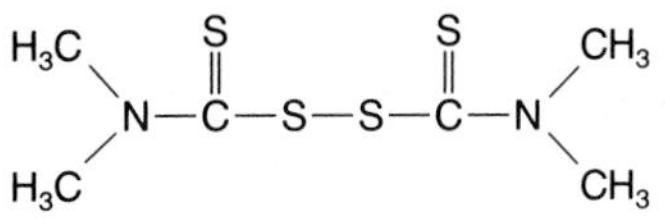

$(H_3C)_2N-C(=S)-S-S-C(=S)-N(CH_3)_2$

Thiram

Thiram consists of two molecules of dithiocarbamic acid joined together. It is used mostly for seed and bulb treatment for vegetables, flowers, and grasses, but also for the control of certain foliage diseases, such as rusts of lawns, fruits, and vegetables. Thiram is also good as a soil drench for control of damping off and seedling blights. Thiram, in various formulations, is sold under many trade names. Thiram and Tersan are two examples.

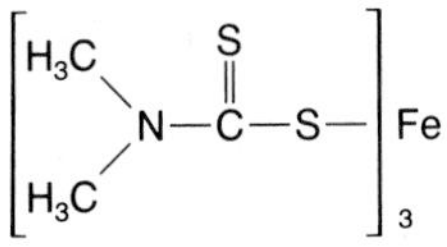

$[(H_3C)_2N-C(=S)-S-]_3Fe$

Ferbam

Ferbam consists of three molecules of dithiocarbamic acid reacted to one atom of iron. Ferbam is used to control many foliage diseases of fruit trees and ornamentals.

Another group of dithiocarbamic acid derivates with different molecular configurations contains the fungicides nabam (Na), zineb (Zn), and maneb (Mn). Nabam and even ferbam have been largely replaced by newer fungicides.

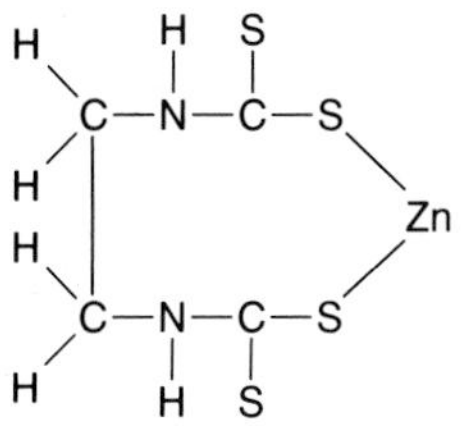

$H_2C(-NH-C(=S)-S-)-CH_2(-NH-C(=S)-S-)\ Zn$

Zineb

Zineb is sold as Dithane Z-78; it is an excellent, safe, multipurpose foliar and soil fungicide for the control of leaf spots, blights, and fruit rots of vegetables, flowers, fruit trees, and shrubs.

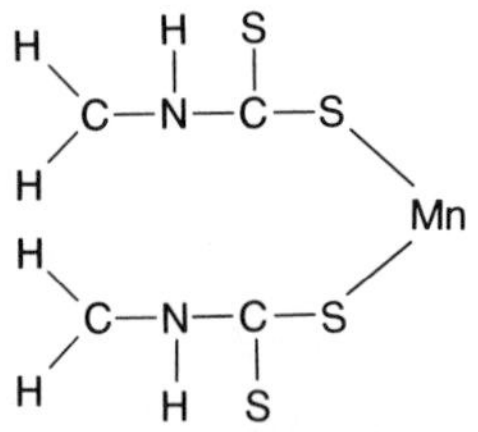

Maneb

Maneb contains manganese; it is sold as Manzate, Dithane M-22, and Tersan LSR and is an excellent, broad-spectrum fungicide for the control of foliage and fruit diseases of many vegetables, especially tomato, potato, and vine crops, and of flowers, trees, turf, and some fruits. Maneb is one of the most frequently used fungicides for control of vegetable diseases. Maneb is often mixed with zinc or with zinc ion and results in the formulations known as maneb-zinc (sold as Manzate D) and as zinc ion maneb called mancozeb (sold as Manzate 200, Dithane M-45). The addition of zinc reduces the phytotoxicity of maneb and improves its fungicidal properties.

Quinones Quinones, which occur naturally in many plants and are also produced upon oxidation of plant phenolic compounds, often show antimicrobial activity and are often considered to be associated with the innate resistance of plants to disease. Only two quinone compounds, chloranil and dichlone, however, have been developed, and only dichlone is still used commercially as a fungicide.

Dichlone, sold as Phygon, is used mainly as a seed treatment for certain vegetables and grasses. Dichlone is also used as a protectant or eradicant spray for certain blights, fruit rots, and cankers of vegetables and fruit.

Aromatic Compounds Many rather unrelated compounds that have an aromatic ring are toxic to microorganisms, and several have been developed into fungicides and are used commercially. Most seem to inhibit production of compounds that have $—NH_2$ and —SH groups, that is, amino acids and enzymes.

Hexachlorobenzene, or HCB, is used as seed treatment for the control of seed- and soil-borne bunt in wheat and other grains.

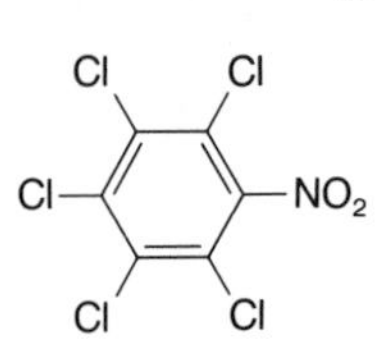

Pentachloro-
nitrobenzene

Pentachloronitrobenzene, sold as PCNB and Terrachlor, is a longlasting soil fungicide. It controls various soil-borne diseases of vegetables, turf, and ornamentals and is applied as a dip or in the furrow at planting time. It is used primarily against *Rhizoctonia, Sclerotinia,* and *Plasmodiophora* but has no effect on *Pythium.*

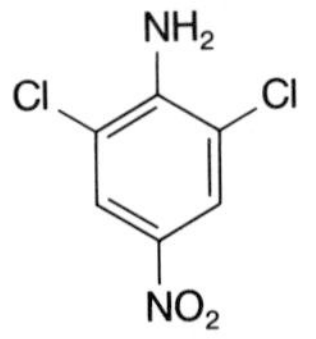

Dichloran

Dichloran, sold as Botran and DCNA, is used as a foliar, fruit, and soil fungicide for diseases of vegetables and flowers caused mostly by sclerotia-producing fungi, and as postharvest dip or spray for fruits, vegetables, and flowers affected by the same fungi or by *Rhizopus* and *Penicillium.*

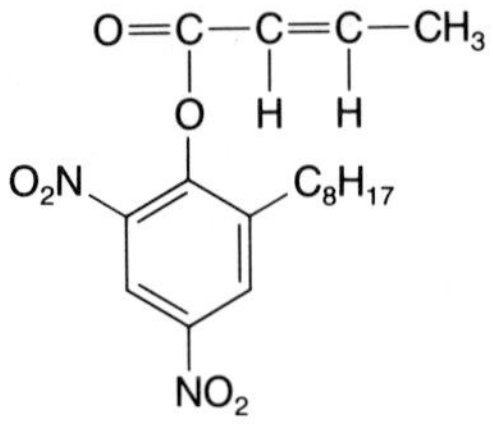

Dinocap

Dinocap, sold as Karathane and Mildex is specific against powdery mildews. It also suppresses mites.

Diazoben, sold as Dexon, is used as a seed and soil fungicide against damping-off and root rots of many ornamentals, vegetables, and fruits caused by *Pythium, Aphanomyces,* and *Phytophthora.*

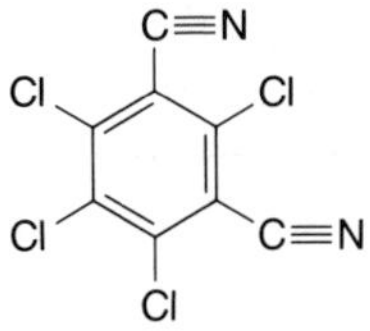

Chlorothalonil

Chlorothalonil, available as Bravo, is an excellent broad-spectrum fungicide against many leaf spots, blights, downy mildews, rusts, anthracnoses, scabs, fruit rots of many vegetables, field crops and ornamentals, and even trees. Another formulation of chlorothalonil is sold as Daconil 2787 and is used primarily against foliage diseases of turf grasses and of some ornamentals. A tablet formulation called Termil is thermally dispersed in greenhouses for control of *Botrytis* on many ornamentals and for several leaf molds and blights of tomato.

Biphenyl, sold as biphenyl, is used widely for control of postharvest diseases of citrus caused by *Penicillium, Diplodia, Botrytis,* and *Phomopsis.* Biphenyl is volatile and is applied by impregnating shipping materials with it which then volatilizes in storage and protects the stored fruit.

Heterocyclic Compounds The heterocyclic compounds are a rather heterogeneous group but include some of the best fungicides, for example, the related captan, folpet, and captafol and the related iprodione and vinclozolin. Most of them also inhibit production of essential compounds containing $—NH_2$ and —SH groups (amino compounds and enzymes).

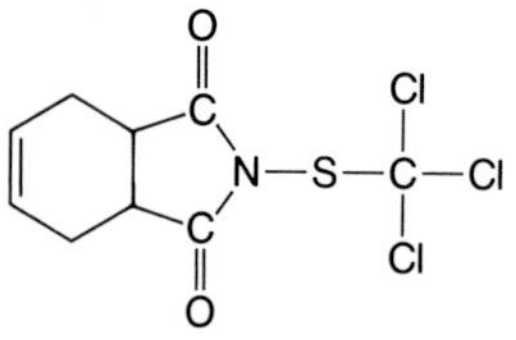

Captan

Captan is an excellent fungicide for control of leaf spots, blights, and fruit rots on fruits, vegetables, ornamentals, and turf. It is also used as a seed protectant for vegetables, flowers, and grasses and as a postharvest dip for certain fruits and vegetables.

Folpet is similar to captan in spectrum and effectiveness. In addition, it controls many powdery mildews.

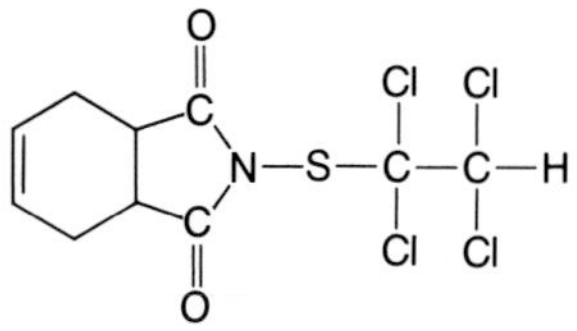

Captafol

Captafol is sold as Difolatan, and has properties similar to those of captan and folpet. Moreover, Difolatan exhibits unusual resistance to weathering, which provides extended redistribution and residual activity. These properties, combined with its low phytotoxicity, also allow the use of up to three times the regular amount of Difolatan as a single application treatment (SAT) on apples against apple scab, cherry leaf spot, citrus melanose, and scab, and against several foliage diseases of tomato. Such concentrated sprays may provide protection for longer periods and reduce the number of sprays needed.

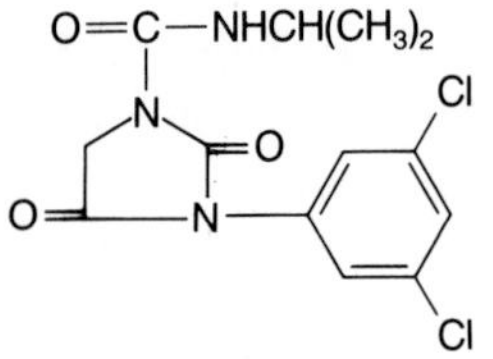

Iprodione

Iprodione, sold as Rovral or Chipco-26019, is a broad-spectrum, foliage-contact fungicide. It inhibits spore germination and mycelial growth but shows mostly preventative and only early curative activity. It is effective against *Botrytis, Monilinia,* and *Sclerotinia* and also against *Alternaria, Helminthosporium,* and *Rhizoctonia.* It is applied most often as a foliar spray and also as a postharvest dip and as a seed treatment. Iprodione is used primarily on turf, stone fruits, grapes, and lettuce.

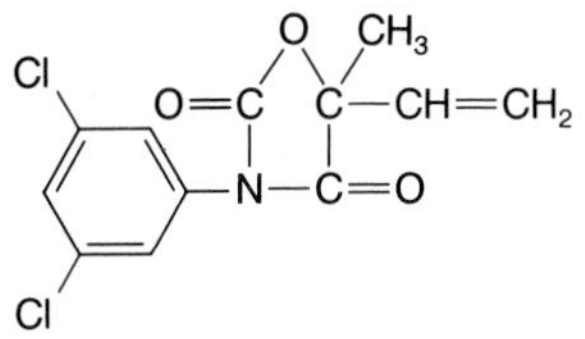

Vinclozolin

Vinclozolin, sold as Ornalin, Ronilan, or Vorlan, is a contact, protective fungicide, effective against sclerotia-producing ascomycetes *(Botrytis, Monilinia, Sclerotinia),* and other fungi. It is used mostly as a spray on strawberries, lettuce, turf, and ornamentals and is tested on many other fruits and vegetables.

Glyodin is a liquid fungicide with excellent wetting and sticking properties. It is effective against apple scab and certain other foliar diseases of fruit trees and ornamentals. It is often combined with dodine (Glyodex).

Dyrene (anilazine) is used for spraying ornamentals, turf, and vegetables.

Systemic Fungicides Systemic fungicides are absorbed through the foliage or roots and are translocated upward internally by the plant through the xylem. Systemic fungicides generally move upward in the transpiration stream and may accumulate at the leaf margins, while downward translocation in the phloem is rare or does not occur at all. These fungicides are not reexported to new growth. Some of them become systemically translocated when sprayed on herbaceous plants, but most are only locally systemic on the sprayed leaves. Many systemics are most effective when applied as seed treatments, root-dip, in-furrow treatment or soil drench, and in trees when injected into the trunks.

Several systemic fungicides are currently available, and many more are in the experimental stage. Systemic fungicides belong to many different groups of compounds.

Almost all systemic fungicides act by inhibiting one or a few specific steps in the metabolism of the fungi they control. As a result, new strains of these fungi, resistant to each systemic fungicide, appear within a few years of the introduction of each such compound. For this reason, a systemic fungicide must either be abandoned after appearance of a pathogen strain resistant to it, or it must be used in combination with another broad-sprectrum contact fungicide under various schemes of application.

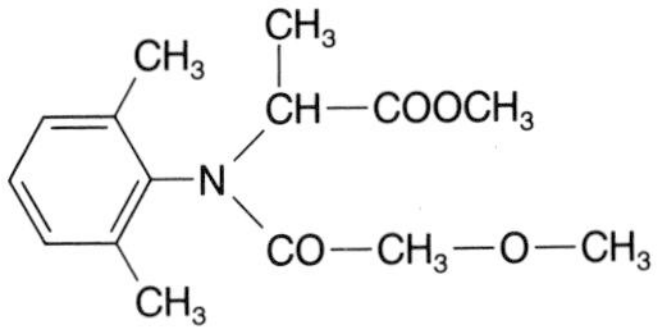

Metalaxyl

- **Acylalanines**

The most important acylalanine fungicide is metalaxyl. It is effective against *Pythium, Phytophthora,* and several downy mildews. It is sold as Ridomil for use in the soil or foliage, as Apron for use as a seed dressing, and as Subdue for use on ornamentals and turf. Metalaxyl is the first and one of the best systemic fungicides against oomycetes. It is long-lasting and widely used as a soil or seed treatment for control of *Pythium* and *Phytophthora* seed rot and damping off and as soil treatment for control of *Phytophthora* stem rots and cankers in annuals and perennials and of certain downy mildews (for example, of tobacco). It is also effective as a curative treatment when applied after infection has begun. Metalaxyl is quite water soluble and is readily translocated from roots to the aerial parts of most plants, but its lateral translocation is slight. Because the use of metalaxyl has already resulted in the appearance of strains resistant to metalaxyl in some pathogens, it is recommended that it be used in combination with other, broad-spectrum fungicides.

- **Benzimidazoles**

Benzimidazoles include some important systemic fungicides such as benomyl, carbendazim, thiabendazole, and thiophanate. They are effective against numerous types of diseases caused by a wide variety of fungi. Most benzimidazoles are converted at the plant surface to methyl benzimidazole carbamate (MBC, carbendazim), and this compound interferes with nuclear division of sensitive fungi.

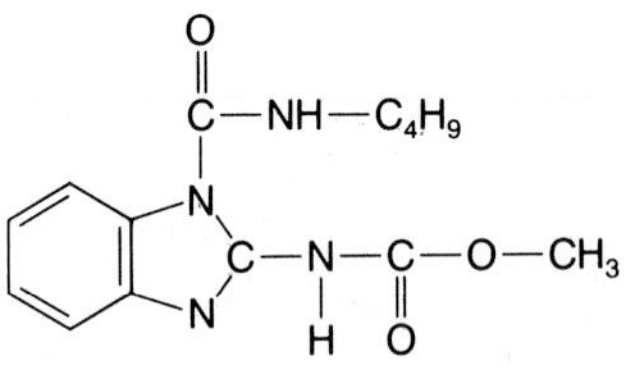

Benomyl

Benomyl is sold as Benlate, Tersan 1991, and others. It is a safe, broad-spectrum fungicide, effective against a large number of important fungus pathogens, and it also suppresses mites. It controls a wide range of leaf spots and blotches, blights, rots, scabs, and seed- and soil-borne diseases. Benomyl is particularly effective for powdery mildew of all crops; scab of apples, peaches, and pecans; brown rot of stone fruits; fruit rots in general; *Cercospora* leaf spots; cherry leaf spot; black spot of roses; blast of rice; various *Sclerotinia* and *Botrytis* diseases; and loose and covered smuts of wheat. It is highly active against and suppresses infections by *Rhizoctonia, Thielaviopsis, Ceratocystis, Fusarium,* and *Verticillium.* It has no effect on oomycetes, on some dark-spored Imperfects such as *Helminthosporium* and *Alternaria,* on some Basidiomycetes, and on bacteria. Benomyl may be applied as a seed treatment, foliar spray, trunk injection, root dip or row treatment, and as a fruit dip. Benomyl seems to be mutagenic and to hasten the appearance of pathogen races resistant to it.

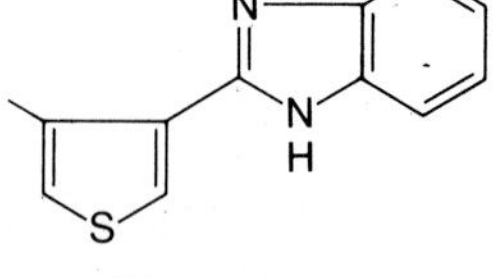

Thiabendazole

Thiabendazole is sold as Mertect. It is also a broad-spectrum fungicide and effective against many Imperfect fungi causing leaf spot diseases of turf and ornamentals and diseases of bulbs and corms. It is commonly used as a postharvest treatment for the control of storage rots of citrus, apples, pears, bananas, potatoes, and squash.

Thiophanate, under the trade name Topsin, is effective against several root and foliage fungi affecting turf grasses.

Thiophanate methyl, under the trade names Fungo and Topsin M, is a broad-spectrum preventive and curative fungicide for use on turf and as a foliar spray to control powdery and downy mildews, *Botrytis* diseases, numerous leaf and fruit spots, scabs, and rots. It is also used as a soil drench or dry soil mix to control soil-borne fungi attacking bedding plants, foliage plants, and container-grown plants.

- **Oxathiins**

Oxathiins were the first systemic fungicides to be discovered (1966). They include primarily carboxin and oxycarboxin and are effective against some smut and rust fungi and against *Rhizoctonia.* Oxathiins are selectively concentrated in cells of these fungi and inhibit succinic hydrogenase, an enzyme important in mitochondrial respiration.

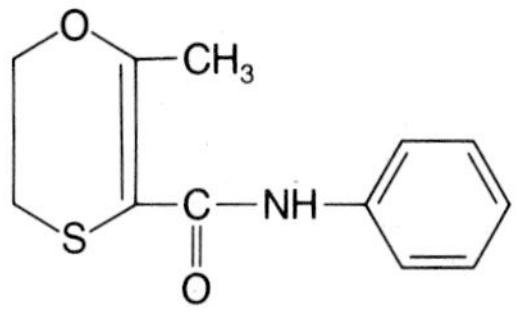

Carboxin

Carboxin is sold as Vitavax. It is used as a seed treatment and is effective against damping-off diseases caused by *Rhizoctonia* and against the various smuts of grain crops.

Oxycarboxin is sold as Plantvax. It is sometimes used as a seed or foliar treatment and is effective in controling a wide variety of rust diseases.

- **Morpholines**

Morpholines include the fungicides dodemorph, sold as Meltatox, and tridemorph, sold as Calixin. They are preventative and eradicant foliar fungicides effective against powdery mildews and leaf spots on cereals, ornamentals, and tropical plants.

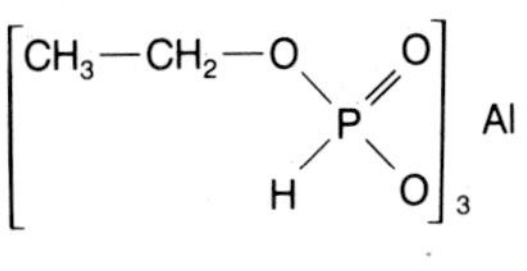

Fosetyl-Al

- **Organic Phosphates**

The organic phosphates include fosetyl-Al, sold as Aliette, which is very effective against foliar, root, and stem diseases caused by oomycetes such as

Phytophthora, Pythium, and the downy mildews in a wide variety of crops. It is applied as a foliar spray, soil drench, root- or postharvest dip, and as a soil incorporation. Treatments may be effective for from 2 to 6 months, depending on the crop. Fosetyl-Al has been reported to stimulate defense reactions and the synthesis of phytoalexins against oomycetes. Three other compounds are also included: Kitazin (IBP) and edifenphos (Hinosan), both effective against rice blast and several other diseases, and pyrazophos (Afugan), which is effective against powdery mildews and *Helminthosporium* diseases on various crops.

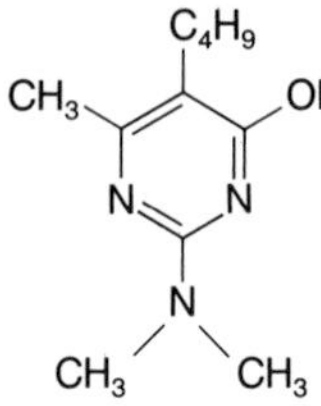

Dimethirimol

- **Pyrimidines**

The pyrimidines include diamethirimol (Milcurb), ethirimol (Milstem), and bupirimate (Nimrod), all effective against powdery mildews of various crop plants, and fenarimol (Rubigan) and nuarimol (Trimidal), effective against powdery mildews and also several other leaf spot, rust, and smut fungi.

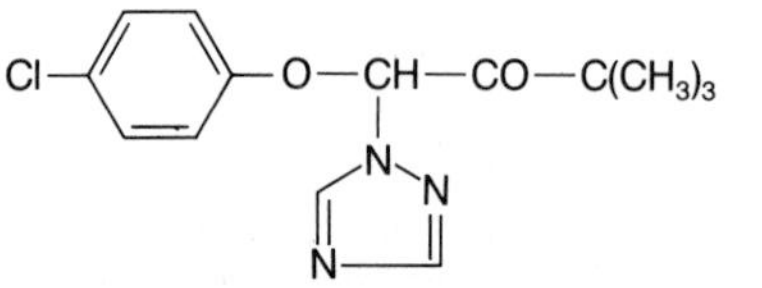

Triadimefon

- **Triazoles**

The triazoles include several excellent systemic fungicides such as triadimefon (Bayleton), triadimenol (Baytan), bitertanol (Baycor), boutrizol (Indar), propiconazole (Tilt), and etaconazole (Vangard). They show long protective and curative activity against a broad spectrum of foliar, root, and seedling diseases such as leaf spots, blights, powdery mildews, rusts, smuts, and others, caused by many ascomycetes, imperfects, and basidiomycetes. They are applied as foliar sprays and as seed and soil treatments.

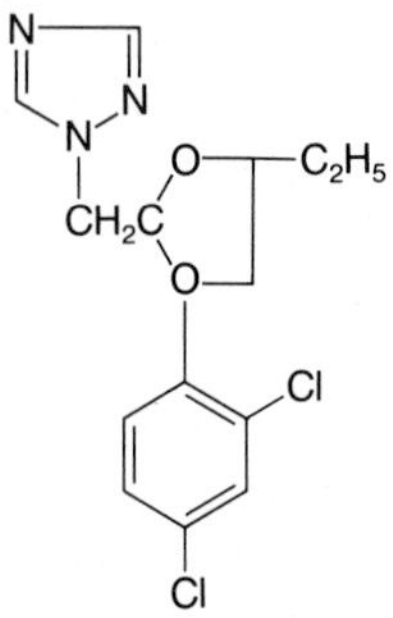

Etaconazole

- **Miscellaneous Systemics**

Several excellent systemic fungicides of different chemical composition and affiliation are included in the miscellaneous category.

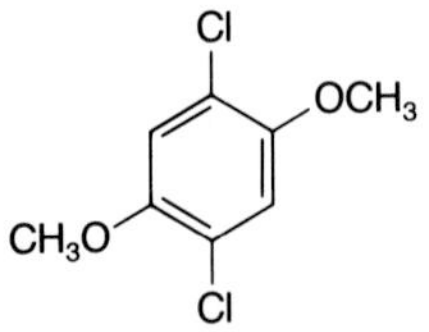

Chloroneb

Chloroneb, sold as Demosan, is used as a seed and soil fungicide for turf and ornamentals. It is sometimes applied as a seed overcoat to seed treated with standard fungicides. It does not leach from the soil, and it is effective against seedling blights in cotton, beans, beets, and so on.

Cyprofuram, sold as Vinicur, is effective against the oomycetes and is applied as a foliar spray, a soil surface spray, and as a seed treatment. It has many excellent properties but is still experimental.

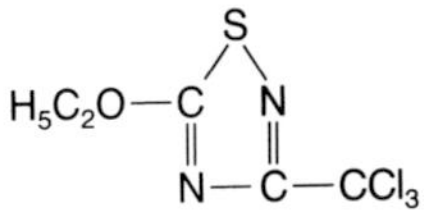

Ethazol

Ethazol, sold as Truban, Terrazole, or Koban, is a seed, soil, and turf fungicide effective against damping-off and root and stem rots caused by *Pythium* and *Phytophthora.* It is often sold combined with PCNB or with Thiophanate methyl (Banrot) for broader spectrum application, particularly against *Fusarium* and *Rhizoctonia.*

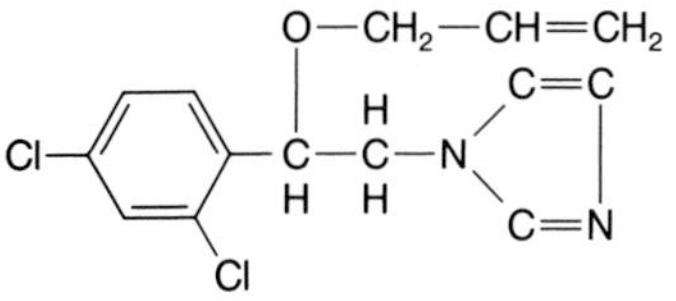

Imazalil

Imazalil, sold as Fungaflor, is effective against many ascomycetes and imperfect fungi causing powdery mildews, leaf spots, fruit rots, and vascular wilts. It is applied as a foliar spray, a seed treatment, and as a postharvest treatment. It has excellent curative and preventive properties. Many of its uses are still being developed.

Prochloraz, sold as Prochloraz, is effective against ascomycetes and imperfects causing powdery mildews, leaf spots, and blights and fruit rots. Still experimental, it is used as a spray or a seed treatment.

Propamocarb, sold as Banol and Previcur, is effective against *Pythium, Phytophthora,* the downy mildews, some rusts, and others. It is applied as a seedling dip, soil drench, seed treatment, soil surface spray, and as a foliar spray. Most of its uses still are under development.

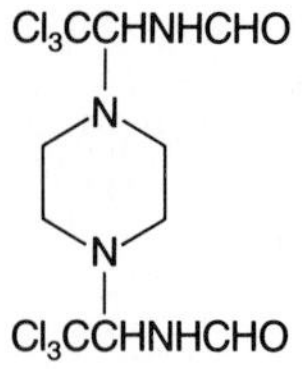

$$\begin{array}{c} Cl_3CCHNHCHO \\ | \\ \text{piperazine ring (N, N)} \\ | \\ Cl_3CCHNHCHO \end{array}$$

Triforine

Triforine, sold as Cela, Funginex, or Saprol, is effective against many ascomycetes and imperfects causing powdery mildews, foliar and fruit spots, fruit rots, anthracnose, and some basidiomycetes causing rusts. It is used as a foliar spray.

Miscellaneous Organic Fungicides A number of other, chemically diverse compounds are excellent protectant fungicides for certain diseases or groups of diseases.

$$C_{12}H_{25}NH{-}\underset{\underset{NH}{\|}}{C}{-}NH_2 \cdot CH_3COOH$$

Dodine

Dodine is sold as Cyprex. It is an excellent fungicide against apple scab, and it also controls certain foliage diseases of cherry, strawberry, pecan, and roses. It gives long-lasting protection and is also a good eradicant. It appears to have local systemic action in leaves. Strains of the apple scab fungus resistant to dodine have appeared and predominate in some areas.

Fentin hydroxide, sold as Du-Ter, is a broad-spectrum fungicide with activity against many leaf spots, blights, and scabs. It also has suppressant or antifeeding properties on many insects.

Polyram is a foliar and seed protectant fungicide. It controls rusts, downy mildews, and leaf spots and blights of vegetables, ornamentals, and certain trees.

Oxyquinoline sulfate (also benzoate and citrate) is used as a soil drench to control damping off and other soil-borne diseases. An oxyquinoline–copper complex has also been used as a seed treatment, as a foliar spray against certain diseases of fruits and vegetables, and as a wood preservative for packing boxes, baskets, and crates.

Two cadmium-containing fungicides, Caddy (cadmium chloride) and Cadminate (cadmium succinate), are used for control of turf diseases.

Zinc is sometimes used as zinc naphthenate for disinfestation and preservation of wood.

Antibiotics Antibiotics are substances produced by one microorganism and are toxic to another microorganism. Most antibiotics known to date are products of Actinomycetes and some fungi, for example *Penicillium,* and are toxic mostly to bacteria, including fastidious bacteria, mycoplasmas, and also to some fungi. The chemical formulas of most antibiotics are complex and are not, as a rule, related to each other. Antibiotics used for plant disease control are generally absorbed and translocated systemically by the plant. Antibiotics may control plant diseases by acting on the pathogen or on the host directly, or after undergoing transformation within the host. In some cases, the application of antibiotics to control bacterial plant diseases has led to the development of bacterial strains resistant to the antibiotic.

Among the most important antibiotics in plant disease control are streptomycin, tetracyclines, and cycloheximide.

Streptomycin is produced by the actinomycete *Streptomyces griseus.* It binds to bacterial ribosomes and prevents protein synthesis, the initiation of peptide chains, and the recognition of normal triplets. Streptomycin or streptomycin sulfate is sold as Agrimycin and Phytomycin and as a spray shows activity against a broad range of bacterial plant pathogens causing spots, blights, wilts, and rots. Streptomycin is also used as a soil drench, for example, in the control of geranium foot rot caused by *Xanthomonas* sp., as a dip for potato seed pieces against various bacterial rots of tubers, and as a seed disinfectant against bacterial pathogens of beans, cotton, crucifers, and cereals. Moreover, streptomycin is effective against several oomycetous fungi, especially *Pseudoperonospora humuli,* the cause of downy mildew of hops.

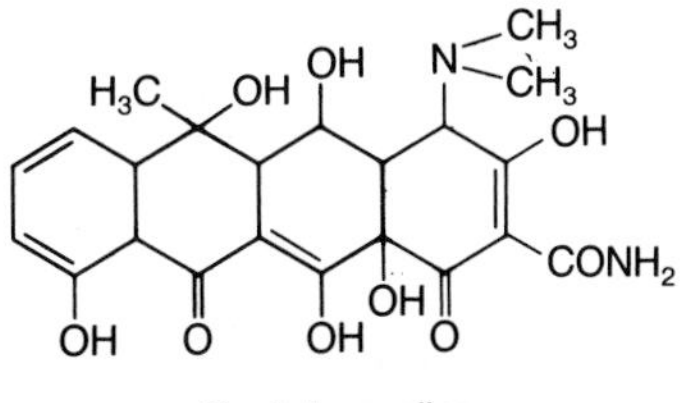

Oxytetracycline

Tetracyclines are antibiotics produced by various species of *Streptomyces* and are active against many bacteria, against all mycoplasmas, and against some fastidious bacteria. Tetracyclines also bind to bacterial ribosomes and block the binding of aminoacyl tRNAs to amino acids, which results in inhibition of protein synthesis. Of the tetracyclines, Terramycin (oxytetracycline), Aureomycin (chlortetracycline), and Achromycin (tetracycline) have been used to some extent for plant disease control. Oxytetracycline is often used with streptomycin in the control of fire blight of pome fruits. Tetracyclines, when injected into trees infected with mycoplasmas or fastidious bacteria, stop the development of the disease and induce remission of symptoms, that is, the symptoms disappear and the trees resume growth as long as some tetracycline is present in the trees. Usually one injection at the end of the growing season is sufficient for normal growth of the tree during the following season.

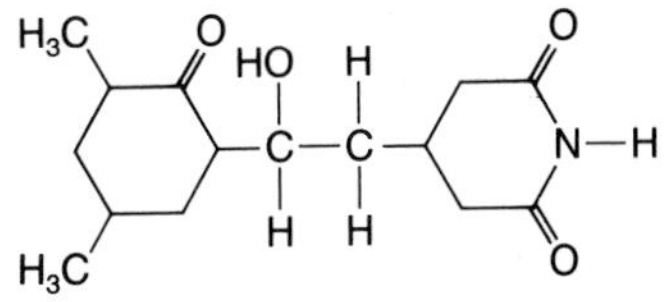

Cycloheximide

Cycloheximide is produced by *Streptomyces griseus* and is obtained as a by-product in the production of streptomycin. Sold as Actidione, it is effective against many phytopathogenic fungi. Cycloheximide is used for the control of many turf diseases and of cherry leafspot, caused by *Coccomyces hiemalis.* It is also effective against powdery mildews of many crop and ornamental plants, but its high phytotoxicity appreciably limits its usefulness.

Several more antibacterial and antifungal antibiotics are used in Japan and some other countries in the Orient. Of these the most common are Blasticidin, used against the rice blast fungus *Pyricularia oryzae,* kasugamycin, polyoxin, and piomycin, used against rice blast and also many other leaf, stem, and fruit spots of fruits and vegetables, and mildiamycin, used against powdery mildews.

Growth Regulators Certain plant hormones have been shown to reduce infection of plants by certain pathogens, for example, tomato by *Fusarium,* potato by *Phytophthora,* through the increase by these substances of the disease resistance of the host. In tobacco plants treated with maleic hydrazide, a growth retardant, the rootknot nematode, *Meloidogyne,* is unable to induce giant cell formation and is thereby prevented from completing its life cycle and from causing disease. Kinetin treatment of leaves, before or shortly after inoculation with virus, also reduces virus multiplication and the number and size of lesions on local-lesion hosts and postpones the onset of systemic symptoms and death of the plant. Stunting and axillary bud suppression associated with certain virus and mycoplasma diseases of plants can be overcome with sprays of gibberellic acid. Although treatments with various growth regulators have given encouraging control of some plant diseases in experimental trials, only gibberellic acid sprays are used somewhat for the field control of sour cherry yellows virus on cherries.

Nematicides Most of the nematicides are volatile soil fumigants and are active against not only nematodes but also insects, fungi, bacteria, weed seeds, and almost anything else living in the soil. Several newer chemicals are nonfumigant granular or liquid substances active mostly against nematodes and insects. The four main groups of nematicides are halogenated hydrocarbons, organophosphates, isothiocyanates, and carbamates.

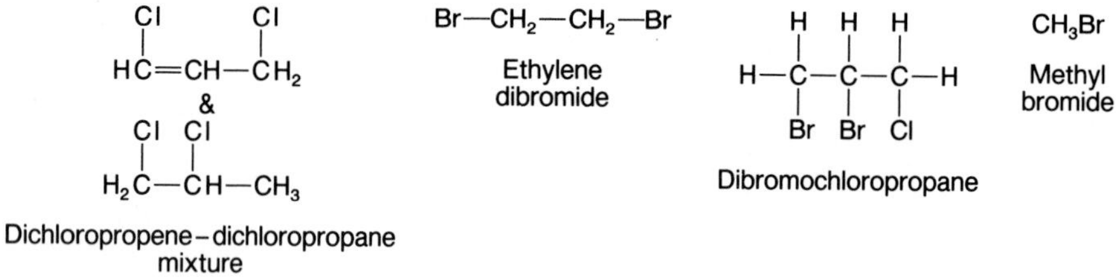

$$\underset{|}{\overset{Cl}{}}HC{=}CH{-}CH_2(Cl) \quad \& \quad H_2C(Cl){-}CH(Cl){-}CH_3$$

Dichloropropene-dichloropropane mixture

$$Br{-}CH_2{-}CH_2{-}Br$$

Ethylene dibromide

$$H{-}CH(Br){-}CH(Br){-}CH(Cl){-}H$$

Dibromochloropropane

$$CH_3Br$$

Methyl bromide

• Halogenated Hydrocarbons

Halogenated hydrocarbons include dichloropropene-dichloropropane (D-D), ethylene dibromide (EDB), dibromochloropropane (DBCP) and methyl bromide (MB). Discovered in the 1940's, they are excellent nematicides and were used extensively into the 1980's. Recently, however, because of their high toxicity, contamination of produce and ground water, and because one of them (DBCP) has caused sterility in male workers in the manufacturing plant, the use of most of them as nematicides has been banned in the United States. A small amount (1–2 percent) of chloropicrin is added to these chemicals to serve as a warning agent. They are applied to the soil by injection at least two weeks before planting. They all kill nematodes and insects, and at the higher dosages most kill soil-borne pathogens and weed seeds. Methyl bromide is a broad spectrum fumigant against soil-borne pathogens and is also used for above-ground control of dry-wood termites and for fumigation of agricultural produce.

Halogenated hydrocarbons affect organisms because they are soluble in lipids and disrupt the function of membranes and nervous systems.

- **Organophosphates**

Organophosphates include prorate (Thimet), disulfoton (Disyston), ethoprop (Mocap), fensulfothion (Dasanit), fenamiphos (Nemacur), fosthietan (Nem-A-Tak), and a few others. Many of the organophosphates were developed initially as insecticides but are taken up and are distributed systemically through the plant and are effective nematicides. They are available as water-soluble liquids or granules, have low volatility, can be applied before or after planting, and are effective only against nematodes and soil fungi through contact or by digestion. Like the organophosphate insecticides, these nematicides inhibit the nerve-transmitter enzyme cholinesterase and result in paralysis and ultimately death of affected nematodes.

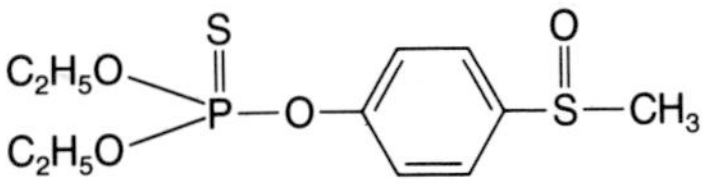

Fensulfothion

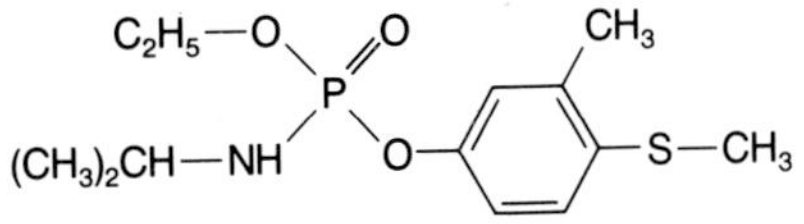

Fenamiphos

- **Isothiocyanates**

The isothiocyanates include metam-sodium (Vapam), vorlex (Vorlex), and dazomet (Mylone). They are active against nematodes, soil insects, weeds, and most soil fungi. Applied by injection into the soil at least two weeks before planting, they act by releasing either isothyocyanate (metam-sodium) or methylisothiocyanate (Vorlex and dazomet), which inactivates the —SH group in enzymes.

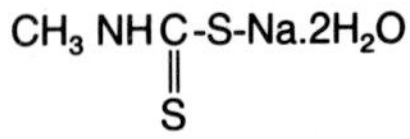

$CH_3NHC(=S)\text{-S-Na}.2H_2O$

Metam-sodium

$CH_3{-}N{=}C{=}S$

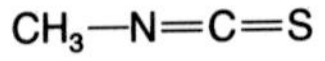

Vorlex ingredient

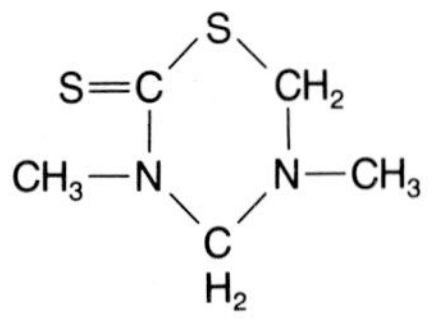

Dazomet

- **Carbamates**

The carbamates include aldicarb (Temik), carbofuran (Furadan), oxamyl (Vydate), and carbosulfan (Advantage). They are active against nematodes and soil insects, as well as some foliage insects. Available as granules or liquids of low volatility, they are easily soluble in water and can be taken up and translocated systemically by the plant. They are either drilled into the soil or are broadcast and disked into the soil before or after planting. They act by inhibiting the enzyme cholinesterase and causing paralysis and death of affected nematodes and insects.

$CH_3SC(CH_3)_2CH{=}NOC(=O)N(H)CH_3$

Aldicarb

$H_3C{-}N(H){-}C(=O){-}O{-}$(2,2-dimethyl-2,3-dihydrobenzofuran-7-yl)

Carbofuran

$(CH_3)_2NC(=O){-}C(SCH_3){=}NOC(=O)NHCH_3$

Oxamyl

- **Miscellaneous Nematicides**

Chloropicrin (Cl_3CNO_2), the common tear gas, is highly volatile and effective against nematodes, insects, fungi, and weed seeds. It can be used alone or mixed with other nematicides.

Avermectins are a new class of natural compounds that are obtained as fermentation products of *Streptomyces avermitilis* and exhibit nematicidal properties. Avermectins are still used only for experiments.

Mechanisms of Action of Chemicals Used to Control Plant Diseases

The complete mechanisms by which the various chemicals applied to plants control plant diseases are as yet unknown for most of the chemicals. Some of the chemicals, for example, fosetyl-Al, seem to reduce infection by increasing the resistance of the host to the pathogen. They may do this by altering the constitution of the cell walls of the host, by limiting the availability of essential coenzymes in the host, or by altering the rate or the direction of metabolism in the host, which may thus be in a better position to defend itself against the pathogen.

The great majority of chemicals are used for their toxicity directly to the pathogen and are effective only as protectants at the pathogens' points of entry. Such chemicals act by inhibiting the pathogen's ability to synthesize certain of its cell wall substances, by acting as solvents of, or otherwise damaging, the cell membranes of the pathogen, by forming complexes with, and thus inactivating, certain essential coenzymes of the pathogen, or by inactivating enzymes and causing general precipitation of the proteins of the pathogen. For example, sulfur interferes with electron transport along the cytochromes of fungi and is then reduced to hydrogen sulfide (H_2S), which is toxic to most cellular proteins. Copper ion (Cu^{++}) is toxic to all cells because it reacts with sulfhydryl groups (—SH) of certain amino acids and causes denaturation of proteins and enzymes. Many organic fungicides also are toxic because they inactivate proteins and enzymes through reaction with their —SH groups. For example, the dithiocarbamates and ethazol, when taken up by fungal cells, release thiocarbonyl (—N=C—S), which inactivates —SH groups. Similarly, the chlorinated aromatic and heterocyclic compounds such as PCNB, chlorothalonil, chloroneb, captan, and vinclozolin react with $-NH_2$ and —SH groups and inactivate enzymes that have such groups. Furthermore, some nematicides, such as the halogenated hydrocarbons, disrupt the functions of membranes and nervous systems, while others, such as the organophosphates, inhibit the nerve-transmitter enzyme cholinesterase and cause paralysis and death of the pathogen.

The systemic fungicides and antibiotics are absorbed by the host, are translocated internally through the plant, and are effective against the pathogen at the infection locus both before and after infection has become established. Chemicals that can cure plants from infections that have already become established are called **chemotherapeutants,** and control of plant diseases with such chemicals is called **chemotherapy.** Once in contact with the pathogen, chemotherapeutants seem to affect pathogens in ways similar to those mentioned above for the nonsystemic chemicals, but systemic fungicides are much more specific in that they apparently affect only one function in the pathogen rather than a variety of them. For example, oxathiins inhibit the enzyme succinic dehydrogenase, which is essential in mitochondrial respiration, while benzimidazoles interfere with nuclear division by binding to protein subunits of the spindle microtubules. Moreover, the polyoxin antifungal antibiotics and the organophosphate fungicides kitazin and edifenphos act primarily by inhibiting chitin synthesis in the pathogen. As a result, new pathogen races resistant to one or another of the systemic fungicides have already appeared.

Several systemic fungicides have been shown to inhibit ergosterol biosynthesis and are commonly referred to as sterol inhibitors or sterol-inhibiting fungicides. The sterol inhibitors include bitertanol, fenapanil, imazalil, prochloraz, triadimefon, triarimol, triforine, and etaconazole. Although chemically these compounds have several structural similarities, they do not form a homogeneous group. Ergosterol is a cellular compound that plays a crucial role in the structure and function of the membranes of many fungi, and chemicals that inhibit ergosterol biosynthesis have effective fungicidal action. The sterol-inhibiting fungicides penetrate the leaf cuticle and therefore are highly effective in curative postinfection applications.

Resistance of Pathogens to Chemicals

Just as human pathogens resistant to antibiotics and insects and mites resistant to certain insecticides and miticides appeared rather soon after continuous and widespread use of these chemicals, several plant pathogens have also developed strains that are resistant to certain fungicides. For many years, when only protectant fungicides such as thiram, maneb, or captan were used, no such resistant strains were observed, presumably because these fungicides affect several vital processes of the pathogen and too many gene changes would be necessary to produce a resistant strain. Resistance to some fungicides, all of which contained a benzene ring, began to appear in the 1960s when *Penicillium* strains resistant to diphenyl, *Tilletia* strains resistant to hexachlorobenzene, and *Rhizoctonia* strains resistant to PCNB were found to occur naturally. In some areas these strains became major practical problems. Later, a strain of *Venturia inaequalis* (the cause of apple scab) appeared that was resistant to dodine and that excellent chemical became ineffective against the fungus over a large area.

Strains of *Erwinia amylovora,* the fire blight bacterium, that were resistant to the systemic antibiotic streptomycin, had been known for several years. It was the introduction and widespread use of the systemic fungicides, especially benomyl, however, that really triggered the appearance of strains of numerous fungi resistant to one or more of these fungicides. In some cases, strains resistant to the fungicide appeared and became widespread after only two years of use of the chemical. To date, several of the important fungal pathogens, for example, *Botrytis, Cercospora, Colletotrichum, Fusarium, Verticillium, Sphaerotheca, Aspergillus, Penicillium,* and *Ustilago,* are known to have produced strains resistant to one or more of the systemic fungicides, and it appears that resistant strains of other fungi can be expected to develop in the future. This is apparently because systemic fungicides are specific in their action. That is, they affect only one or perhaps two steps in a genetically controlled event in the metabolism of the fungus and, as a result, a resistant population can arise quickly either by a single mutation or by selection of resistant individuals in a population.

The most common mechanisms by which pathogens develop resistance to various fungicides, bactericides, and so on, is by (1) decreased permeability of pathogen cell membranes to the chemical, (2) detoxification of the chemical through modification of its structure or through binding it to a cell constituent, (3) decreased conversion into the real toxic compound, (4) decreased affinity at the reactive site in the cell (for example, of benomyl to spindle protein subunits), (5) bypassing a blocked reaction through a shift in metabolism, and (6) compensation for the effect of inhibition by producing more of the inhibited product (for example, an enzyme).

Good systemic or nonsystemic fungicides that become ineffective because of the appearance of new resistant strains can still be saved, and the resistant strains can still be controlled to a practical level through changes in the methods of deployment of the fungicide. This can be achieved either by using mixtures of specific systemic and wide-spectrum protectant fungicides, such as benomyl and either captan or dichloran or iprodione for control of *Botrytism* or *Sclerotinia,* or metalaxyl and maneb or zineb for control of downy mildews; by alternating sprays with systemic and protectant fungicides, or by spraying during half the season with systemic and the other half with protectant fungicides. In each of these schedules, the systemic or specific action chemical carries most of the weight in controlling the disease, while the protectant or nonspecific chemical eliminates any strains of the pathogen that may develop resistance to the systemic or specific action chemical.

Restrictions on Chemical Control of Plant Diseases

Although most chemicals used to control plant diseases are much less toxic than most insecticides, they are, nevertheless, poisonous substances, and some of them, especially the nematicides, are extremely poisonous. Also, many have adverse genetic effects in lower and higher organisms, including humans. For this reason, a number of restrictions are imposed in the licensing, registration, and use of each chemical.

In the United States, both the Food and Drug Administration (FDA) and the Environmental Protection Agency (EPA) keep a close watch on the registration, production, and use of pesticides. It is estimated that only 1 out of 10,000 new compounds synthesized by the pesticide industry turns out to be a successful pesticide, and it takes 7 to 9 years and $15 to 25 million from initial laboratory synthesis to government registration.

In the meantime, exhaustive biological tests, field testing, crop residue analyses, toxicological tests, and environmental impact studies are carried out. If the compound meets all requirements it is then approved for use on specific food or nonfood crops for which data have been obtained. Clearance must be obtained separately for each crop and each use (seed treatment, spray, soil drench) for which the chemical is recommended.

Once a chemical is approved for a certain crop, then two important restrictions on the use of the chemical must be observed: (1) the number of days before harvest that use of a particular chemical on the crop must stop; and (2) the amount of the chemical that can be used per acre must not exceed a certain amount. If either of these restrictions is not observed it is likely that, at harvest, the crop, especially vegetables and fruits, carries on it a greater amount than is allowed for the particular chemical, and the crop can be seized. All recommendations contained in bulletins published by the Extension Service are within the tolerances established by FDA and EPA and should be followed carefully.

Integrated Control of Plant Diseases

Control of plant diseases is most successful and economical when all available pertinent information regarding the crop, its pathogens, the environmental conditions expected to prevail, locality, availability of materials, and costs are taken into account in developing the control program. Usually, an integrated control program is aimed against all diseases affecting a crop. Sometimes an integrated control program is aimed against a particularly destructive and common disease, for example, apple scab, and potato late blight.

Control in a Perennial Crop

In an integrated control program of an orchard crop, such as apple, peach, or citrus, one must first consider the nursery stock to be used and the location where it will be planted. If the fruit tree is susceptible to certain viruses, mycoplasmas, crown gall bacteria, or nematodes, the nursery stock (both the rootstock and the scion) must be free of these pathogens. Stock free of certain viruses and other diseases can usually be bought from selected nurseries whose crops are inspected and certified. If the possibility of nematodes on the roots exists, the stock must be fumigated. The location where the trees will be planted must not be infested with fungi such as *Phytophthora, Armillaria,* or serious and numerous nematodes; if it is, it should be treated with fumigants before planting, and varieties grafted on rootstocks resistant to these pathogens should be preferred. The drainage of the location should be checked and improved, if necesary. Finally, the young trees should not be planted between or next to old trees that are heavily infected with canker fungi and bacteria, insect-transmitted viruses and mycoplasmas, pollen-transmitted viruses or with other pathogens.

Once the trees are in place and until they begin to bear fruit, they should be fertilized, irrigated, pruned, and sprayed for the most common insects and diseases so that they will grow vigorously and free of infections. Later on, when the trees bear fruit, the care should increase, as should the vigilance to detect and control diseases that affect any part of the tree. Any trees that develop symptoms of a disease caused by a systemic pathogen such as a virus or mycoplasma should be removed as soon as possible.

Disease control in an orchard may begin in the winter, when dead twigs, branches or fruit are removed during pruning operations and are buried or burned to reduce the amount of fungal or bacterial primary inoculum that will start infections in the spring. Pruning shears and saws should be disinfected before moving to new trees to avoid spreading any pathogens from tree to tree. Because many fungi and bacteria (as well as insects and mites) are activated in the spring by the same weather conditions that make buds open, a "dormant" spray, containing a fungicide–bactericide (such as Bordeaux mixture), or a plain fungicide plus a miticide–insecticide (such as Superior oil), is applied before bud break. After that, as the buds open, the blossoms and leaves that are revealed are usually very susceptible to either fungal or bacterial pathogens, or both, depending on what is present in the particular area. Therefore, these organs (blossoms and leaves) must be protected with sprays containing a fungicide and/or a bactericide and, possibly, an insecticide and/or miticide that does not harm bees. It is usually possible to find effective materials compatible with each other so all of them can be mixed in the same tank and sprayed at once. If one compound, however, must be used to control an existing disease but is incompatible with the other compounds, then a separate spray will be needed. Because flowers appear over a period of several days and the leaves enlarge rapidly at that stage, and because many fungi release their spores and bacteria ooze out most abundantly during and soon after bloom, the blossoms and leaves may have to be sprayed with a systemic fungicide, or, if only protectant fungicides are available, the trees must be sprayed frequently (every 3–5 days) so that they will be protected by the fungicide or bactericide (or both), especially if it rains often and stays wet for many hours. Insecticides and miticides may still have to be used with the fungicide, but these insecticides must not be toxic to bees, which must be allowed to pollinate the flowers. The frequent sprays usually continue as long as there are spores being released by fungi, or bacteria oozing out, as long as the weather stays wet, and as long as there are growing plant tissues. Combining the use of weather forecasts with disease control is most helpful.

Once blossoming is over, young fruit appear, which may or may not be affected by the same pathogens and insects as the flowers and leaves. If they are, the same spray schedule with the same materials continues as long as there is inoculum around. If a systemic fungicide had been used early in the season, later sprays should be made with a broad spectrum protectant fungicide to forestall the appearance of fungicide-resistant strains of the pathogen. But often new pathogens and insects may attack the fruit, and the schedule must be adjusted and materials must be included that control the new pathogens.

Usually, fruit becomes susceptible to several fruit-rotting fungi that attack fruit from the stage of early maturity through harvest and storage. Therefore, fruit must be sprayed every 10 to 14 days with materials that will control these fungi until harvest. Most fruit rots start at wounds made by insects, and therefore insect control must continue. Also, wounding of fruit during harvesting and handling must be avoided to prevent fungus infections. Fruit-picking baskets and crates must be clean, free of rotten debris, which may harbor fruit-rotting fungi, and the packinghouse and warehouse must also be clean, free of debris, and preferably fumigated with formaldehyde, sulfur dioxide, or some other fumigant. Harvested fruit is often washed in a water solution containing a fungicide to further protect the fruit during storage and transportation. During packing, infected fruit is removed and discarded. The fruit should be refrigerated during storage and transportation, of course, so that any existing infections will develop slowly and no new infections will get started.

Control in an Annual Crop

In an integrated control program of an annual crop, such as potatoes, one must again start with healthy stock and must plant it in a suitable field. Potato tuber seed may carry several viruses, the late blight fungus, ring rot bacteria, as well as several other fungi, bacteria, and nematodes. Therefore, starting with clean, disease-free seed is of paramount importance. Certified potato seed is usually free of most such important pathogens and is produced under strict quarantine and inspection rules that guarantee seed free of these pathogens. Healthy seed must then be planted in a field free of old potato tubers that may harbor some of the above pathogens, free of *Verticillium, Fusarium,* and the root-knot nematode. It is best not to follow a potato crop with another, and rotation with legumes, corn, or other unrelated crops will usually reduce the populations of potato pathogens. Any potato cull piles should be destroyed or sprayed to ensure that no *Phytophthora* sporangia will be blown from there to the potato plants in the field later on. Tubers are cut with disinfected knives to reduce spread of ring rot among seed pieces and the seed pieces are usually treated with a fungicide, a bactericide, and an insecticide to protect them from pathogens on their surface or in the soil. The soil may have to be treated with a fumigant if it is known to be infested with the root knot or other nematodes, *Fusarium,* or *Verticillium.* The seed pieces are planted at a date when their sprouts are expected to grow quickly since slow-growing sprouts in cool weather are particularly susceptible to *Rhizoctonia* attack. The field must, of course, have good drainage to avoid damping off, seed-piece rot, and root rots.

A few weeks after the young plants have emerged they become susceptible to attack by early blight *(Alternaria)* and late blight *(Phytophthora infestans).* If the diseases occur regularly year after year, in addition to using resistant varieties, the grower should start spraying with the appropriate fungicides as soon as the disease appears, or even before, and should continue the sprays, especially for late blight, throughout the season whenever the weather is cool and damp. Insecticide sprays control insects and may reduce the spread of viruses. Using weather data to forecast disease appearance and development can help in spraying at the right time and in not wasting any sprays. Before harvest, the infected vines must be killed with chemicals to destroy late blight inoculum that could come in contact with the tubers when they are dug up. Tubers must be harvested carefully to avoid wounding that would allow storage-rot fungi such as *Fusarium* and *Pythium* to gain entrance into the tuber. The tubers must then be sorted, and the damaged ones discarded. The healthy tubers are stored at about 15°C for the wounds to heal and then at about 2°C to prevent development of fungus rots in storage. Storage rooms must of course be cleaned and disinfested before the tubers are brought in. Potato cull piles should not be kept near the field but should either be burned or buried as soon as possible.

Thus, in an integrated control program, several control methods are employed including regulatory inspections for healthy seed or nursery crop production, cultural practices (crop rotation, sanitation, pruning), biological control (resistant varieties), physical control (storage temperature), and chemical controls (soil fumigation, seed or nursery stock treatment, sprays, disinfestation of cutting tools, crates, warehouses, washing solution). Each one of these measures must be taken for best results, and the routine use of each of them makes all of them that much more effective.

SELECTED REFERENCES

Anagnostakis, S. L. (1982). Biological control of chestnut blight. *Science* **215,** 466–471.

Anonymous (1984). "Fungicide and Nematicide Tests," Vol. 40. Am. Phytopathol. Soc., St. Paul, Minnesota.

Backman, P. A. (1978). Fungicide formulation: Relationship to biological activity. *Annu. Rev. Phytopathol.* **16,** 211–237.

Baker, K. F., and Cook, R. J. (1974). "Biological Control of Plant Pathogens." Freeman, San Francisco, California.

Barksdale, T. H., and Stoner, A. K. (1981). Levels of tomato anthracnose resistance measured by reduction of fungicide use. *Plant Dis.* **65,** 71–72.

Bird, L. S. (1982). The MAR (multi-adversity resistanc) system for genetic improvement of cotton. *Plant Dis.* **66,** 172–176.

Blakeman, J. P., and Fokema, N. J. (1982). Potential for biological control of plant diseases on the phylloplane. *Annu. Rev. Phytopathol.* **20,** 167–192.

Brown, M. E. (1974). Seed and root bacterization. *Annu. Rev. Phytopathol.* **12,** 181–197.

Browning, J. A., and Frey, K. J. (1969). Multiline cultivars as a means of disease control. *Annu. Rev. Phytopathol.* **7,** 355–382.

Browning, J. A., Simons, M. D., and Torres, E. (1977). Managing host genes: Epidemiologic and genetic concepts. *In* "Plant Disease" (J. G. Horsfall and E. B. Cowling, eds.). Vol. 1, pp. 191–212. Academic Press, New York.

Bruehl, G. W., ed. (1975). "Biology and Control of Soil-Borne Plant Pathogens." Am. Phytopathol. Soc., St. Paul, Minnesota.

Cook, R. J. (1977). Management of the associated microbiota. *In* "Plant Disease" (J. G. Horsfall and E. B. Cowling, eds.), Vol. 1 pp. 145–166. Academic Press, New York.

Cook R. J., and Baker, K. F. (1983). "The Nature and Practice of Biological Control of Plant Pathogens." Am. Phytopathol. Soc., St. Paul, Minnesota.

Costa, A. S., and Muller, G. W. (1980). Tristeza control by cross protection: A U. S.-Brazil cooperative success. *Plant Dis.* **64,** 538–541.

Davidse, L. C., and de Waard, M. A. (1984). Systemic fungicides. *Plant Pathol.* **2,** 191–257.

Dekker, J. (1976). Acquired resistance to fungicides. *Annu. Rev. Phytopathol.* **14,** 405–428.

Delp, C. J. (1980). Coping with resistance to plant disease control agents. *Plant Dis.* **64,** 652–657.

Dennis, C., ed. (1983). "Post-Harvest Pathology of Fruits and Vegetables." Academic Press, London.

Doupnick, B. J., and Boosalis, M. G. (1980). Ecofallow—a reduced tillage system—and plant diseases. *Plant Dis.* **64,** 31–35.

Dowling, C. F., Jr., Graham, A. E., and Alfieri, S. A., Jr. (1982). Plant inspection and certification. *Plant Dis.* **66,** 345–351.

Dubin, H. J., and Rajaram, S. (1982). The CIMMYT's international approach to breeding disease resistant wheat. *Plant Dis.* **66,** 967–971.

Eckert, J. W., and Sommer, N. F. (1967). Control of diseases of fruits and vegetables by postharvest treatment. *Annu. Rev. Phytopathol.* **5,** 391–432.

Edgington, L. V. (1981). Structural requirements of systemic fungicides. *Annu. Rev. Phytopathol.* **19,** 107–124.

Edgington, L. V., Martin, R. A., Bruin, G. C., and Parsons, I. M. (1980). Systemic fungicides: A perspective after 10 years. *Plant Dis.* **64,** 19–23.

Erwin, D. C. (1973). Systemic fungicides: Disease control, translocation, and mode of action. *Annu. Rev. Phytopathol.* **11,** 389–422.

Fry, W. E. (1982). "Principles of Plant Disease Management." Academic Press, New York.

Georgopoulos, S. G., and Zaracovitis, C. (1967). Tolerance of fungi to organic fungicides. *Annu. Rev. Phytopathol.* **5,** 109–130.

Gianinazzi, S. (1984). Genetic and molecular aspects of resistance induced by infections or chemicals. *In* "Plant-Microbe Interactions: Molecular and Genetic Perspectives" (T. Kosuge and E. W. Nester, eds.) Vol. 1, pp. 321–342. Macmillan, New York.

Grogan, R. G. (1980). Control of lettuce mosaic with virus-free seed. *Plant Dis.* **64,** 446–449.

Hardison, J. R. (1976). Fire and flame for plant disease control. *Annu. Rev. Phytopathol.* **14,** 355–379.

Harvey, J. M. (1978). Reduction of losses in fresh market fruits and vegetables. *Annu. Rev. Phytopathol.* **16,** 321–341.

Hoitink, H. A. J. (1980). Composted bark, a lightweight growth medium with fungicidal properties. *Plant Dis.* **64,** 142–147.

Hollings, M. (1965). Disease control through virus-free stock. *Annu. Rev. Phytopathol.* **3, 367–396.**

Hornby, D. (1983). Suppressive soils. *Annu. Rev. Phytopathol.* **21,** 65–85.

Horsfall, J. G., and Cowling, E. B., eds. (1977). "Plant Disease," Vol 1. Academic Press, New York.

James, J. R. (1984). Concepts for developing new plant protection compounds. *Plant Dis.* **68,** 651–652.

Kahn, R. P. (1977). Plant quarantine: Principles, methodology and suggested approaches. *In* "Plant Health and Quarantine in International Transfer of Genetic Resources" (W. B. Hewitt and L. Chiarappa, eds.), pp. 289–307. CRC Press, Cleveland, Ohio.

Katan, J. (1981). Solar heating (solarization) of soil for control of soilborne pests. *Annu. Rev. Phytopathol.* **19,** 211–236.

Kerr, A. (1980). Biological control of crown gall through production of Agrocin 84. *Plant Dis.* **64,** 24–30.

Kerry, B. (1981). Fungal parasites: A weapon against cyst nematodes. *Plant Dis.* **65,** 390–394.

Kingsolver, C. H., Melching, J. S., and Bromfield, K. R. (1983). The threat of exotic plant pathogens to agriculture in the United States. *Plant Dis.* **67,** 595–600.

Ko, W.-H. (1982). Biological control of Phytophthora root of papaya with virgin soil. *Plant Dis.* **66,** 446–448.

Kuchler, F., Duffy, M., Shrum, R. D., Dowler, W. M. (1984). Potential economic consequences of the entry of an exotic fungal pest: The case of soybean rust. *Phytopathology* **74,** 916–920.

Lewis, F. H., and Hickey, K. D. (1972). Fungicide usage on decidous fruit trees. *Annu. Rev. Phytopathol.* **10,** 399–428.

Linderman, R. G., Moore, L. W., Baker, K. F., and Cooksey, D. A. (1983). Strategies for detecting and characterizing systems for biological control of soilborne plant pathogens. *Plant Dis.* **67,** 1058–1064.

Lindow, S. E. (1983). Methods of preventing frost injury caused by epiphytic ice-nucleation-active bacteria. *Plant Dis.* **67,** 327–333.

McCoy, R. E. (1982). Use of tetracycline antibiotics to control yellows diseases. *Plant Dis.* **66,** 539–542.

Mankau, R. (1980). Biological control of nematode pests by natural enemies. *Annu. Rev. Phytopathol.* **18,** 415–440.

Marx, D. H. (1972). Ectomycorrhizae as biological deterrents to pathogenic root infections. *Annu. Rev. Phytopathol.* **10,** 429–454.

Mathre, D. E., Metz, S. G., and Johnson, R. H. (1982). Small grain cereal seed treatment in the postmercury era. *Plant Dis.* **66,** 526–531.

Moller, W. J., Schroth, M. N., and Thomson, S. V. (1981). The scenario of fire blight and streptomycin resistance. *Plant Dis.* **65,** 563–568.

Munnecke, D. E., and VanGundy, S. D. (1979). Movement of fumigants in soil, dosage responses, and differential effects. *Annu. Rev. Phytopathol.* **17,** 405–429.

National Academy of Sciences (1968a). "Plant Disease Development and Control." Natl. Acad. Sci., Washington, D. C.

National Academy of Sciences (1968b). "Control of Plant Parasitic Nematodes." Natl. Acad. Sci., Washington, D. C.

National Academy of Sciences (1968c). "Effects of Pesticides on Fruit and Vegetable Physiology." Natl. Acad. Sci., Washington, D. C.

National Academy of Sciences (1972a). "Genetic Vulnerability of Major Crops." Natl. Acad. Sci., Washington, D. C.

National Academy of Sciences (1972b). "Pest Control Strategies for the Future." Natl. Acad. Sci., Washington, D. C.

Nesmith, W. C. (1984). Changes in fungicide use patterns. *Plant Dis.* **68,** 834–835.

Okabe, N., and Goto, M. (1963). Bacteriophages of plant pathogens. *Annu. Rev. Phytopathol.* **1,** 397–418.

Palti, J. (1981). "Cultural Practices and Infectious Crop Diseases." Springer-Verlag, Berlin and New York.

Papavizas, G. C., and Lumsden, R. D. (1980). Biological control of soilborne fungal propagules. *Annu. Rev. Phytopathol.* **18,** 389–413.

Raju, B. C., and Olson, C. J. (1985). Indexing systems for producing clean stock for disease control in commercial floriculture. *Plant Dis.* **69,** 189–192.

Rast, A. T. B. (1972). M11-16, an artificial symptomless mutant of tobacco mosaic virus for seedling inoculation of tomato crops. *Neth. J. Plant Pathol.* **78,** 110–112.

Rathmell, W.-G. (1984). The discovery of new methods of chemical disease control: Current developments, future prospects and the role of biochemical and physiological research. *Adv. Plant Pathol.* **2,** 259–288.

Roberts, D. A. (1978). "Fundamentals of Plant-Pest Control." Freeman, San Francisco, California.

Rodriguez-Kabana, R., and Curl, E. A. (1980). Nontarget effects of pesticides on soilborne pathogens and disease. *Annu. Rev. Phytopathol.* **18,** 311–332.

Sasaki, T., Honda, Y., Umekawa, M., and Nemoto, M. (1985). Control of certain diseases of greenhouse vegetables with ultraviolet-absorbing vinyl film. *Plant Dis.* **69,** 530–533.

Sasser, J. N., Kirkpatrick, T. L., and Dybas, R. A. (1982). Efficacy of avermectins for root-knot control in tobacco. *Plant Dis.* **66,** 691–692.

Schenck, N. C. (1981). Can mycorrhizae control root disease? *Plant Dis.* **65,** 230–234.

Schneider, R. W., ed. (1982). "Suppressive Soils and Plant Disease." Am. Phytopathol. Soc., St. Paul, Minnesota.

Schroth, M. N., and Hancock, J. G. (1982). Disease-suppressive soil and root-colonizing bacteria. *Science* **216,** 1376–1381.

Sharvelle, E. G. (1969). "Chemical Control of Plant Diseases." University Publishing, College Station, Texas.

Shepard, J. F., and Claflin, L. E. (1975). Critical analyses of the principles of seed potato certification. *Annu. Rev. Phytopathol.* **13,** 271–293.

Shurtleff, M. C., and Taylor, D. P. (1964). Soil disinfestation. Methods and materials. *Circ.—Univ. Ill., Coop. Ext. Serv.* **893,** 1–23.

Sijpesteijn, A. K., and van der Kerk, G. J. M. (1965). Fate of fungicides in plants. *Annu. Rev. Phytopathol.* **3, 127–152.**

Slabaugh, W. R., and Grove, M. D. (1982). Postharvest diseases of bananas and their control. *Plant Dis.* **66,** 746–750.

Sommer, N. F. (1982). Postharvest handling practices and postharvest diseases of fruit. *Plant Dis.* **66,** 351–364.

Staub, T., and Sozzi, D. (1984). Fungicide resistance: A continuing challenge. *Plant Dis.* **68,** 1026–1031.

Sumner, D. R., Doupnik, B., Jr., and Boosalis, M. G. (1981). Effect of reduced tillage and multiple cropping on plant diseases. *Annu. Rev. Phytopathol.* **19,** 167–187.

Szkolnik, M. (1978). Techniques involved in greenhouse evaluation of deciduous tree fruit fungicides. *Annu. Rev. Phytopathol.* **16,** 103–129.

Thomson, W. T. (1984). "Agricultural Chemicals. Book IV. Fungicides." Thomson Publications, Fresno, California.

Tuite, J., and Foster, G. H. (1979). Control of storage diseases of grains. *Annu. Rev. Phytopathol.* **17,** 343–366.

Vidaver, A. K. (1983). Bacteriocins: The lure and the reality. *Plant Dis.* **67,** 471–474.

Ware, G. W. (1982). "Fundamentals of Pesticides: A Self-Instruction Guide." Thomson Publications, Fresno, California.

Waterworth, H. E., and White, G. A. (1982). Plant introductions and quarantine: The need for both. *Plant Dis.* **66,** 87–90.

Weller, D. M., and Cook, R. J. (1983). Suppression of take-all of wheat by seed treatments with fluorescent Pseudomonads. *Phytopathology* **73,** 463–469.

Wellman, R. H. (1977). Problems in development, registration and use of fungicides. *Annu. Rev. Phytopathol.* **15,** 153–163.

Wilhelm, S., and Paulus, A. O. (1980). How soil fumigation benefits the California strawberry industry. *Plant Dis.* **64,** 264–270.
Wilson, C. L., and Lawrence, P. P. (1985). Potential for biological control of postharvest plant diseases. *Plant Dis.* **69,** 375–379.
Windels, C. E., and Lindow, S. E. (1985). "Biological Control in the Phylloplane." Am. Phytopathol. Soc., St. Paul, Minnesota.
Zitter, T. A., and Simons, J. N. (1980). Management of viruses by alteration of vector efficiency and by cultural practices. *Annu. Rev. Phytopathol.* **18,** 289–310.

II

Specific Plant Diseases

10 ENVIRONMENTAL FACTORS THAT CAUSE PLANT DISEASES

Introduction

Plants grow best within certain ranges of the various factors that make up their environment. Such factors include temperature, soil moisture, soil nutrients, light, air and soil pollutants, air humidity, soil structure, and pH. Although these factors affect all plants growing in nature, their importance is considerably greater for the cultivated plants, which are often grown in areas barely meeting the requirements for normal growth. Moreover, cultivated plants are frequently grown or kept in completely artificial environments (greenhouses, homes, warehouses) or are subjected to a number of cultural practices (fertilization, irrigation, spraying with pesticides) that may affect their growth considerably.

General Characteristics

The common characteristic of noninfectious diseases of plants is that they are caused by the lack or excess of something that supports life. Noninfectious diseases occur in the absence of pathogens, and cannot, therefore, be transmitted from diseased to healthy plants. Noninfectious diseases may affect plants in all stages of their lives, such as seed, seedling, mature plant, or fruit, and they may cause damage in the field, in storage, or at the market. The symptoms caused by noninfectious diseases vary in kind and severity with the particular environmental factor involved and with the degree of deviation of this factor from its normal. Symptoms may range from slight to severe, and affected plants may even die.

Diagnosis

The diagnosis of noninfectious diseases is sometimes made easy by the presence on the plant of characteristic symptoms known to be caused by the lack or excess of a particular factor (Figure 10-1). At other times diagnosis can be arrived at by carefully examining and analyzing the weather conditions prevailing before and during the appearance of the disease; recent changes in the atmospheric and soil contaminants at or near the area where the plants are growing; and the cultural practices, or possible accidents in the course of these

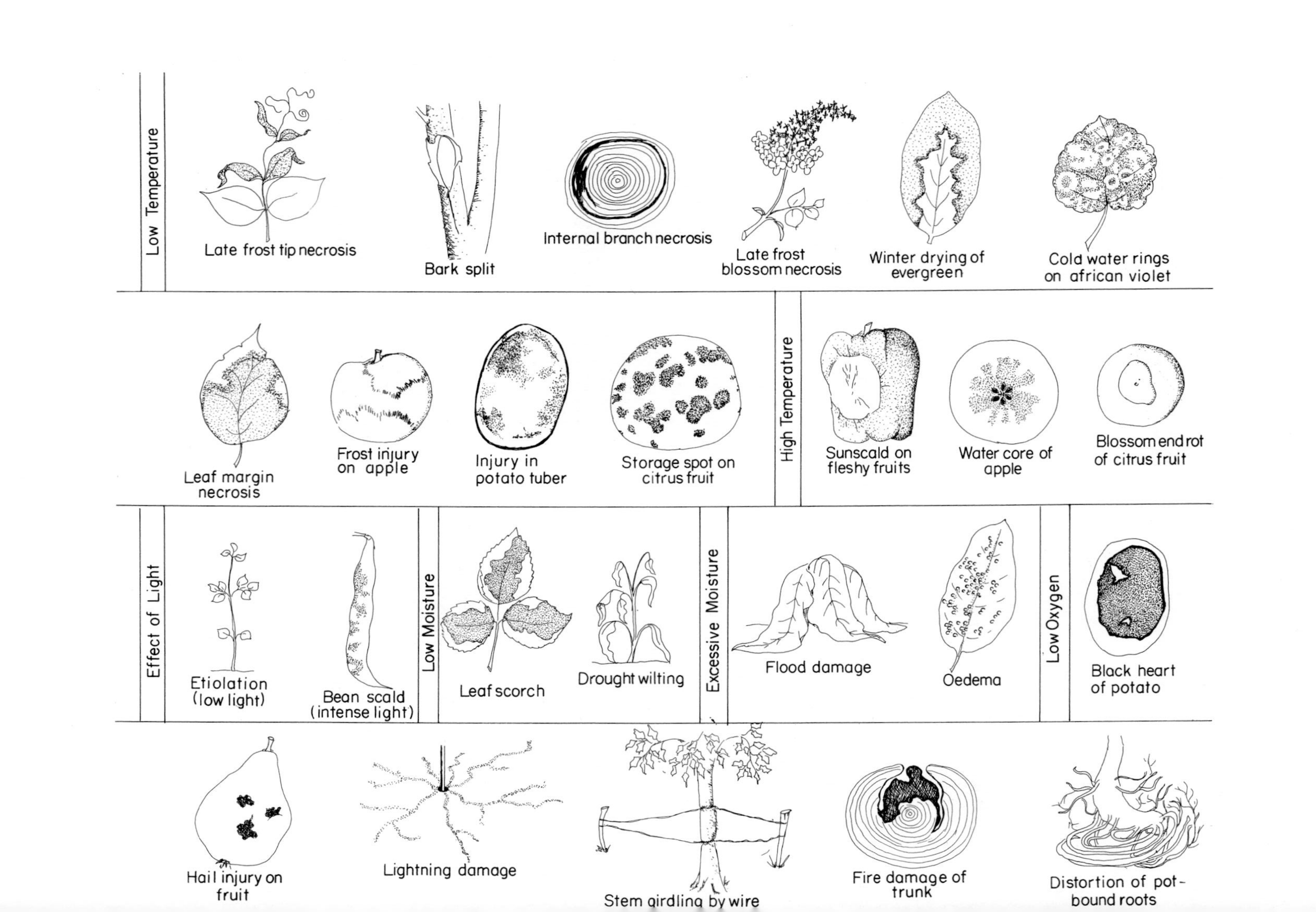
Low Temperature
Late frost tip necrosis
Bark split
Internal branch necrosis
Late frost blossom necrosis
Winter drying of evergreen
Cold water rings on african violet
Leaf margin necrosis
Frost injury on apple
Injury in potato tuber
Storage spot on citrus fruit
High Temperature
Sunscald on fleshy fruits
Water core of apple
Blossom end rot of citrus fruit
Effect of Light
Etiolation (low light)
Bean scald (intense light)
Low Moisture
Leaf scorch
Drought wilting
Excessive Moisture
Flood damage
Oedema
Low Oxygen
Black heart of potato
Hail injury on fruit
Lightning damage
Stem girdling by wire
Fire damage of trunk
Distortion of pot-bound roots

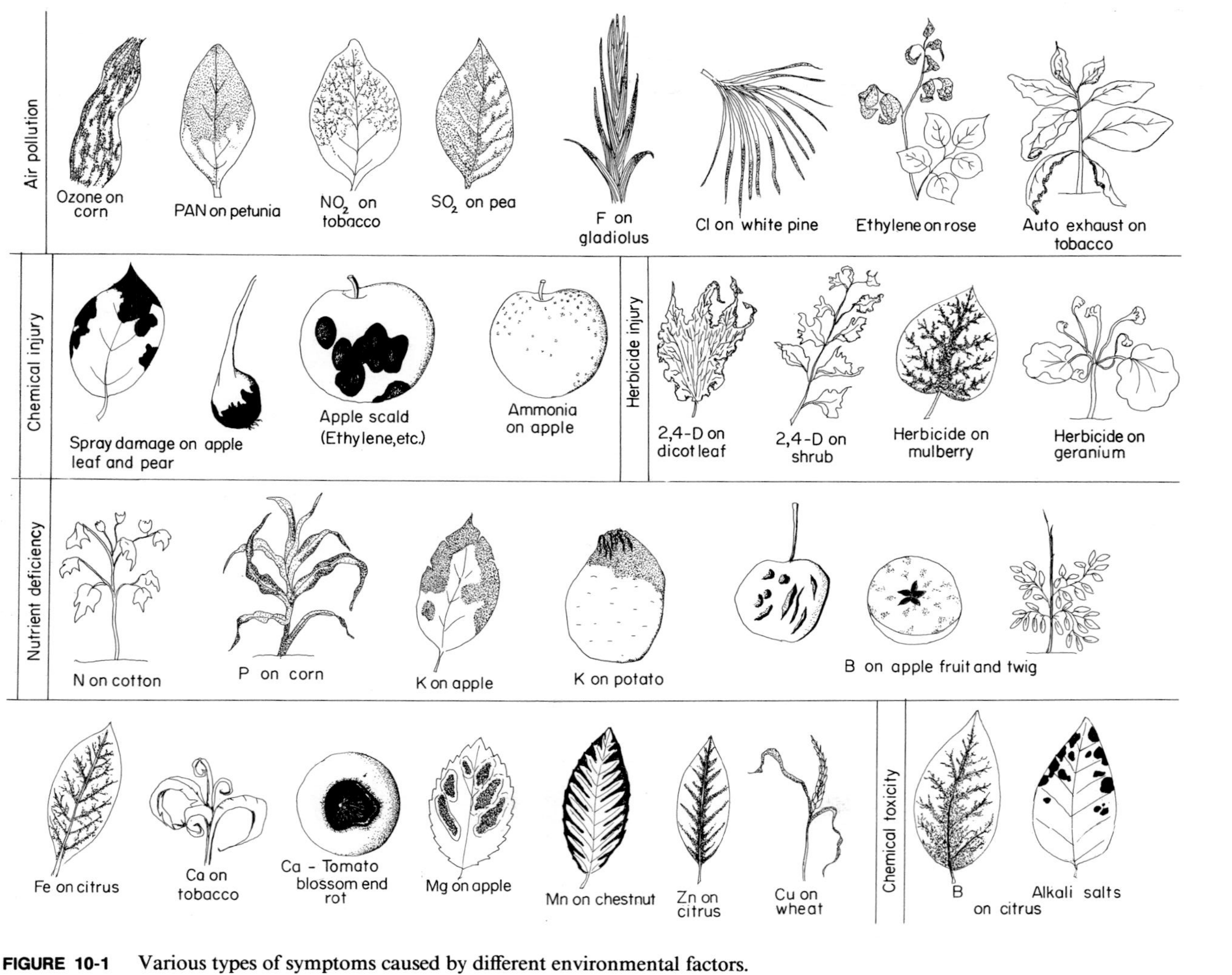

FIGURE 10-1 Various types of symptoms caused by different environmental factors.

practices, preceding the appearance of the disease. Often, however, the symptoms of several noninfectious diseases are too indistinct and closely resemble those caused by several viruses, mycoplasmas, and many root pathogens. The diagnosis of such noninfectious diseases then becomes a great deal more complicated. One must obtain proof of absence from the plant of any of the pathogens that could cause the disease, and one must reproduce the disease on healthy plants after subjecting them to conditions similar to those thought of as the cause of the disease. To distinguish further among environmental factors causing similar symptoms, the investigator must cure the diseased plants, if possible, by growing them under conditions in which the degree or the amount of the suspected environmental factor involved has been adjusted to normal.

Control

Noninfectious plant diseases can be controlled by ensuring that plants are not exposed to the extreme environmental conditions responsible for such diseases or by supplying the plants with protection or substances that would bring these conditions to levels favorable for plant growth.

Temperature Effects

Plants normally grow at a temperature range from 1 to 40°C, most kinds of plants growing best between 15 and 30°C. Perennial plants and dormant organs, such as seeds and corms, of annual plants may survive temperatures considerably below or above the normal temperature range of 1 to 40°C. The young, growing tissues of most plants, however, and the entire growth of many annual plants are usually quite sensitive to temperatures near or beyond the extremes of this range.

The minimum and maximum temperatures at which plants can still produce normal growth vary greatly with the plant species and with the stage of growth the plant is in during the low or high temperatures. Thus, plants such as tomato, citrus, and other tropical plants grow best at high temperatures and are injured severely when the temperature drops to near or below freezing. On the other hand, plants such as cabbage, winter wheat, alfalfa, and most perennials of the temperate zone can withstand temperatures considerably below freezing without any apparent ill effects. Even the latter plants, however, are injured and finally killed if the temperature drops too low.

A plant may also differ in its ability to withstand extremes in temperature at different stages of its growth. Thus, older, hardened plants are more resistant to low temperatures than are young seedlings. Also, different tissues or organs on the same plant may vary greatly in their sensitivity to the same low temperature. Buds are more sensitive than twigs, flowers and newly formed fruit are more sensitive than leaves, and so on.

High-Temperature Effects

Plants are generally injured faster and to a greater extent when temperatures become higher than the maximum for growth than when they are lower than the minimum. However, too high a temperature rarely occurs in nature. High temperature seems to cause its effects on the plant in conjunction with the effects of other environmental factors, particularly excessive light, drought, lack of oxygen, or high winds accompanied by low relative humidity. High temperatures are usually responsible for **sunscald** injuries (Figure 10-2A) appearing on the sun-exposed sides of fleshy fruits and vegetables, such as peppers, apples, tomatoes, onion bulbs, and potato tubers. On hot, sunny days the temperature of the fruit tissues beneath the surface facing the sun may be much higher than that of those on the shaded side and of the surrounding air. This results in discoloration, a water-soaked appearance, blistering, and a desiccation of the tissues beneath the skin, which leads to sunken areas on the fruit surface. Succulent leaves of plants may also develop sunscald symptoms, especially when hot sunny days follow periods of cloudy, rainy weather. Irregular areas on the leaves become pale green at first but soon collapse and form brown, dry spots. This is a rather common symptom of fleshy leaved house plants kept next to windows with a southern exposure in early spring and summer when the sun's rays heat the fleshy leaves excessively. Too high a soil temperature at the soil line sometimes kills young seedlings (Figure 10-2B) or causes cankers at the crown on the stems of older plants. High temperatures also seem to be involved in the **water core** disorder of apples (Figure 10-2C) and, in combination with reduced oxygen, in the **blackheart** of potatoes.

Low-Temperature Effects

Far greater damage to crops is caused by low than by high temperatures. Low temperatures, even if above freezing, may damage warm-weather plants such as corn and beans. They may also cause excessive sweetening and, upon frying, undesirable caramelization of potatoes due to hydrolysis of starch to sugars at the low temperatures.

Temperatures below freezing cause a variety of injuries to plants. Such injuries include the damage caused by late frosts to young meristematic tips (Figure 10-3A and C) or entire herbaceous plants, the frost-killing of buds of peach, cherry, and other trees, and the killing of flowers, young fruit, and sometimes, succulent twigs of most trees. Frost bands, consisting of discolored, corky tissue in a band or large area of the fruit surface, are often produced on apples, pears, and so on, after a late frost (Figure 10-3D). Low winter temperatures may kill the young roots of trees, such as apple, and may also cause bark-splitting and canker development (Figures 10-3B and 10-4) on trunks and large branches, especially on the sun-exposed side, of several kinds of fruit trees. Cross sections of limbs may show a black ring or a "blackheart" condition in the wood. Fleshy tissues, such as potato tubers, may be injured at subfreezing temperatures. The injury varies depending on the degree of temperature drop and the duration of the low temperature. Early injury affects only the main vascular tissues and appears as a ringlike necrosis; injury of the finer vascular elements that are interspersed in the tuber gives the appearance of netlike necrosis. With more general injury, large chunks of the tuber are damaged, creating the so-called "blotch-type" necrosis (Figure 10-3E).

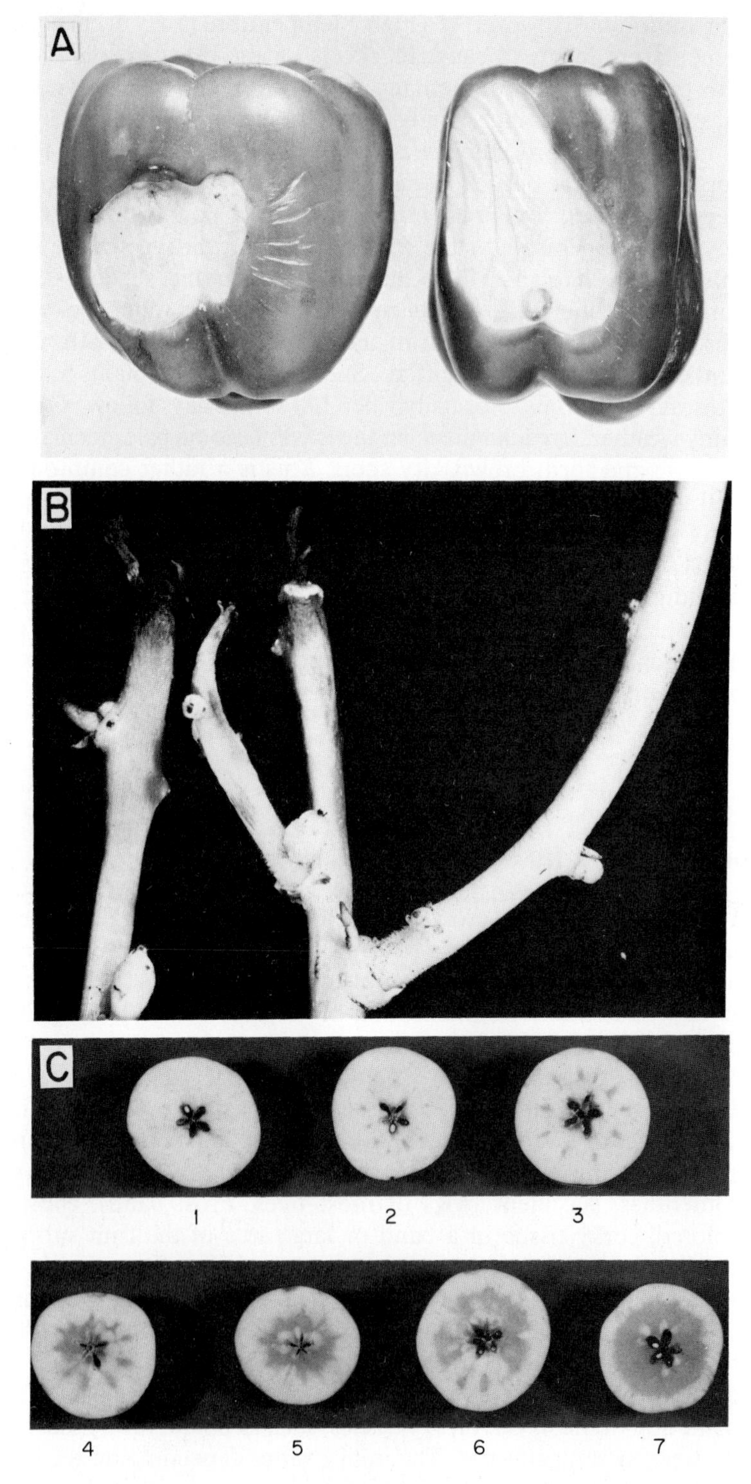

FIGURE 10-2 (A) Sunscald injury on pepper fruits. (B) Potato sprouts killed at the soil line by excessively high temperatures. (C) Stages of watercore development in Delicious apples. 1 = healthy. (Photos: A—courtesy USDA, B—courtesy Dept. Plant Pathol., Cornell Univ., C—courtesy W. J. Lord.)

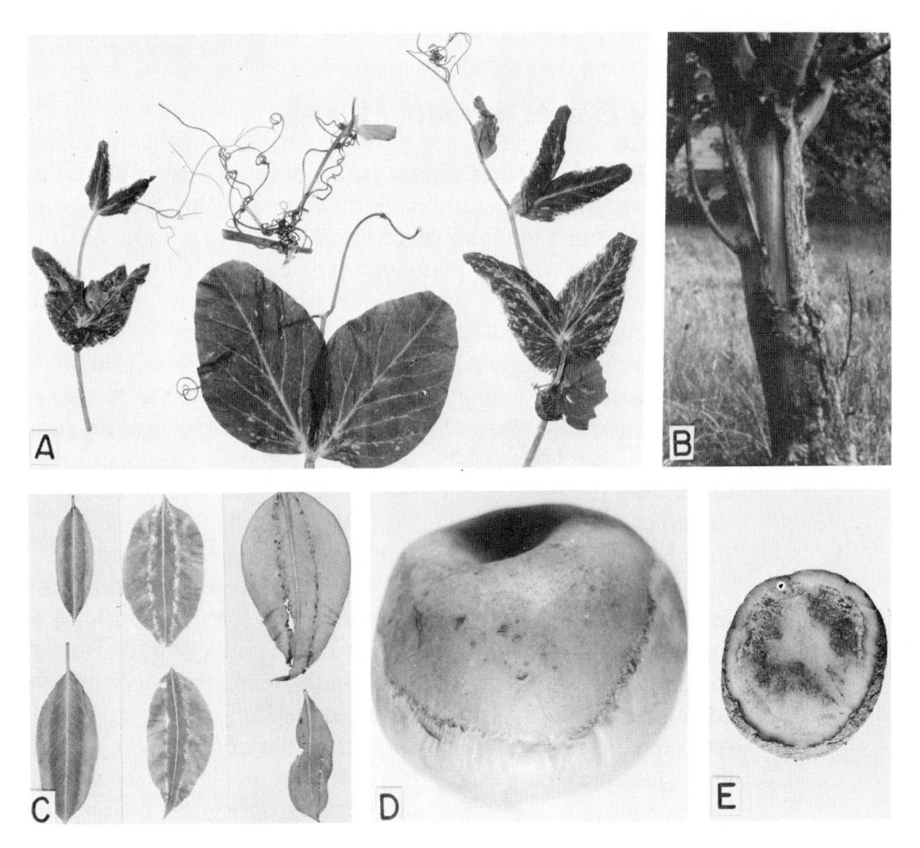

FIGURE 10-3 (A) Chilling injury on leaves and tips of young pea plants due to late frost. (B) Bark split on apple tree trunk due to low winter temperature. (C) Late frost injury on emerging pear leaves. (Left) Discoloration of upper side, (middle) discoloration and necrotic line as seen on upper side of leaf, (right) necrotic line on lower side. (D) Frost injury on apple fruit. (E) Low temperature injury on potato tuber in storage. (Photos A and E courtesy Dept. Plant Pathol., Cornell Univ.)

FIGURE 10-4 (A) Frost damage on young growth of rhododendron. (B) Cracking of rhododendron stem caused by frost. (Photos courtesy Dept. Plant Pathol., Cornell Univ.)

Low-Temperature Effects on Indoor Plants

Indoor plants, whether grown in a home or a greenhouse, are particularly sensitive to low temperatures, both where they are growing and during transportation from a greenhouse or florist's shop to a home or from one home to another. Often, indoor plants are tropical plants grown far away from their normal climate. Exposure of such plants to low, not necessarily freezing, temperatures may cause stunting, yellowing, leaf or bud drop, and so on. Similarly, when grown indoors, even local plants remain in a succulent vegetative state and are completely unprepared for the stresses of low, particularly subfreezing, temperatures. Plants near windows or doors during cold winter days and, especially, nights are subject to temperatures that are much lower than those away from the window. Also, cracks or breaks in windows or the holes of electrical outlets on outside walls let in cold air that may injure the plants. A drop of night temperatures below 12°C may cause leaves and particularly flower buds of many plants to turn yellow and drop. Exposure of indoor plants to subfreezing temperatures for a few minutes or a few hours, for example, while they are carried in the trunk of a car from the greenhouse to the house, may result in the death of many shoots and flowers, or in a sudden shock to the plants from which they may take weeks or months to recover completely. Such a shock is often observed on plants that had been kept indoors and are then transplanted in the field in the spring when temperatures outdoors, although not freezing, are nevertheless much lower than those in the greenhouse. Even without the shock effect, plants growing at temperatures that are generally near the lower—or near the upper—limit of their normal range grow poorly and produce fewer and smaller blossoms and fruits.

Mechanisms of Low and High Temperature Injury to Plants

The mechanisms by which high and low temperatures injure plants are quite different. High temperatures apparently inactivate certain enzyme systems and accelerate others, thus leading to abnormal biochemical reactions and cell death. High temperature may also cause coagulation and denaturation of proteins, disruption of cytoplasmic membranes, suffocation, and possibly release of toxic products into the cell.

Low temperatures, on the other hand, injure plants primarily by inducing ice formation between or within the cells. The rather pure water of the intercellular spaces freezes first and normally at about 0°C, while the water within the cell contains dissolved substances that, depending on their nature and concentration, depress the freezing point of water several degrees. Furthermore, when the intercellular water becomes ice, more vapor (water) moves out of the cells and into the intercellular spaces, where it also becomes ice. The reduced water content of the cells depresses further the freezing point of the intracellular water. This could continue, up to a point, without damaging the cell, but below a certain point, ice crystals form within the cell, disrupt the plasma membrane, and cause injury and death to the cell.

Ice formation in supercooled water within leaves is greatly influenced by the kinds and numbers of epiphytic bacteria that may be present on the surface of the leaves. Certain strains of some pathogenic (for example, *Pseudomonas syringae*) bacteria and of some saprophytic bacteria, when present on or in the substomatal cavities of leaves, act as catalysts for ice nucleation. By their presence alone, such ice-nucleation-active bacteria induce the supercooled water around them and in the leaf cells to form crystals, thereby causing frost injury to the leaves, blossoms, and so on at temperatures considerably higher ($-1°C$) than would have happened in the absence of such bacteria (≈ -5 to $-10°C$). The freezing point of water in cells varies with the tissues and species of the plant; in some tissues of the winter-hardy species of the north, ice probably never forms within the cells regardless of how low the temperatures become. Even when ice forms only in the intercellular spaces, cells and tissues may be damaged either by the inward pressure exerted by the ice crystals, or by loss of water from their protoplasm to the intercellular spaces. This loss causes plasmolysis and dehydration of the protoplasm, which may cause coagulation. The rapidity of the temperature drop in a tissue is also important, since this affects the amount of water remaining in a cell and, therefore, the freezing point of the cell contents. Thus, a rapid drop in temperature may result in intracellular ice formation where a slow drop to the same low temperature would not. The rate of thawing may have similarly variable effects, since rapid thawing may flood the area between cell wall and protoplast and may cause tearing and disruption of the protoplast if the latter is incapable of absorbing the water as fast as it becomes available from the melting of ice in the intercellular spaces.

Moisture Effects

Low Soil Moisture Effects

Moisture disturbances in the soil are probably responsible for more plants growing poorly and being unproductive annually, over large areas, than any other single environmental factor. Small or large territories may suffer from drought over periods of time. The subnormal amounts of water available to plants in these areas may result in reduced growth, diseased appearance, or even death of the plants. Lack of moisture may also be localized in certain types of soil, slopes, or thin soil layers underlaid by rock or sand and may result in patches of diseased-looking plants, while the immediately surrounding areas appear to contain sufficient amounts of moisture and the plants in them grow normally. Plants suffering from lack of sufficient soil moisture usually remain stunted, are pale green to light yellow, have few, small and drooping leaves, flower and fruit sparingly and, if the drought continues, wilt and die (Figure 10-5). Although annual plants are considerably more susceptible to short periods of insufficient moisture, even perennial plants and trees are damaged by prolonged periods of drought and produce less growth, small, scorched leaves and short twigs, dieback, defoliation, and finally wilting and death. Plants weakened by drought are also more susceptible to certain pathogens and insects.

FIGURE 10-5 (A) Healthy fuchsia plant (left), stunted plant due to insufficient water (middle) and plant wilting due to lack of water. (B) Leaf scorch due to insufficient water reaching the leaf. (C) Stunted, wilted and dead corn plants in low part of a field flooded for several days because of heavy rains. (D) Brown, sunken, dry area on orange caused by reduced oxygen during storage.

Low Relative Humidity Effects

Lack of moisture in the atmosphere, that is, low relative humidity, is usually temporary and seldom causes damage. When combined with high wind velocity and high temperature, however, it may lead to excessive loss of water from the foliage and may result in leaf scorching or burning, shrivelled fruit, and temporary or permanent wilting of plants.

Conditions of low relative humidity are particularly common and injurious to house plants during the winter. In modern homes and apartments, heating provides comfortable temperatures for plant growth, but it often dries the air to relative humidities of 15 to 25 percent, which are equivalent to that of desert environments. The air is particularly dry over or near the sources of dry heat, such as radiators. Potted plants kept under these conditions not only use up the water much faster, grow poorly, and may begin to wilt sooner, but the leaves, especially the lower ones, of many kinds of plants become spotted or scorched and fall prematurely, while their flowers suddenly wither and drop off. These effects are particularly noticeable when plants are brought into such a hot, dry house directly from a cool, moist greenhouse or florist's shop. Generally, all house plants prefer high humidity, and certain ones require high humidity if they are to grow properly and produce flowers. Therefore, house plants should never be placed over radiators, and humidity should be increased with a commercial humidifier, by occasionally dampening the leaves with water by placing the pot on a brick or pebbles in a large pan of water, in a plastic case, or some other container.

High Soil Moisture Effects

Excessive soil moisture occurs much less often than drought where plants are grown, but poor drainage or flooding of planted fields, gardens, or potted plants may result in more serious and quicker damage, or death, to plants (Figure 10-5C) than that from lack of moisture. Poor drainage results in plants that lack vigor, wilt frequently, and have leaves that are pale green or yellowish green. Flooding during the growth season may cause permanent wilting and death of succulent annuals within 2 to 3 days. Trees, too, are killed by waterlogging, but the damage usually appears more slowly and after their roots have been continually flooded for several weeks.

As a result of excessive soil moisture caused by flooding or by poor drainage, the fibrous roots of plants decay, probably because of the reduced supply of oxygen to the roots. Oxygen deprivation causes stress, asphyxiation, and collapse of many root cells. Wet, anaerobic conditions favor the growth of anaerobic microorganisms that, during their life processes, form substances, such as nitrites, that are toxic to plants. Besides, the root cells damaged directly by the lack of oxygen lose their selective permeability and may allow toxic metals or other poisons to be taken up by the plant. Also, once parts of roots are killed, more damage is done by faculative parasites that may be greatly favored by the new environment. Thus, the wilting of the plants, which soon follows flooding, is probably the result of lack of water in the aboveground parts of plants caused by the death of the roots, although it appears that translocated toxic substances may also be involved.

In addition, many plants, particularly potted house plants, show several symptoms that are the result of incorrect watering: Either the soil is allowed to dry out too much before it is then repeatedly flooded with water or the plant is almost constantly overwatered. In either case, overwatered plants may suddenly drop their lower leaves, or their leaves may turn yellow. Sometimes they develop brown or black wet patches on the leaves or stems, or the roots and lower stem may turn black and rot as a result of infection by pathogenic microorganisms encouraged by the excessive watering. Such symptoms can be avoided or corrected by watering only when the topsoil feels dry and then applying enough water to saturate thoroughly the whole mass of soil. Plants should never be watered when the soil is still wet, especially during the winter. When watering, any excess water should be drained through the drainage hole, which should always be present in the bottom of the pot. A period of dryness should not be followed with repeated heavy watering but by a gradual return to normal watering. Generally, the supply of water should be maintained as uniform as possible.

Another common symptom of house plants, and sometimes of outdoor plants, that is caused by excessive moisture is the so-called edema (swelling). **Edema** appears as numerous small bumps on the lower side of leaves or on stems. The "bumps" are small masses of cells that divide, expand, and break out of the normal leaf surface and at first form greenish-white swellings or galls. Later, the exposed surface of the swelling becomes rusty colored and has a corky texture. Edema is caused by overwatering, especially during cloudy, humid weather, and can be avoided by reduced watering and providing better lighting and air circulation to the plant. Many other disorders are caused by excessive or irregular watering. It is known, for example, that tomatoes grown under rather low moisture conditions at the time they are ripening often crack if they are suddenly supplied with abundant moisture by overwatering or by a heavy rainfall. Also, bitter pit of apples, consisting of small, sunken, black spots on the fruit, is the result of irregular supply of moisture, although excessive nitrogen and low calcium fertilzation seem also to be involved in bitter pit development.

Inadequate Oxygen

Low oxygen conditions in nature are generally associated with high soil moisture or high temperatures. Lack of oxygen may cause desiccation of roots of different kinds of plants in waterlogged soils, as was mentioned under moisture effects. A combination of high soil moisture and high soil or air temperature causes root collapse in plants. The first condition, apparently, reduces the amount of oxygen available to the roots, while the other increases the amount of oxygen required by the plants. The two effects together result in an extreme lack of oxygen in the roots and cause their collapse and death.

Low oxygen levels may also occur in the centers of fleshy fruit or vegetables in the field, especially during periods of rapid respiration at high temperatures, or in storage of these products in fairly bulky piles (Figure 10-5d). The best known such case is the development of the so-called blackheart of potato, in which fairly high temperatures stimulate respiration and abnormal enzymatic reactions in the potato tuber. The oxygen supply of the cells in the interior of the tuber is insufficient to sustain the increased respiration, and the cells die of suboxidation. Enzymatic reactions activated by the high temperature and suboxidation go on before, during, and after the death of the cells. These reactions abnormally oxidize normal plant constituents into dark melanin pigments. The pigments spread into the surrounding tuber tissues and finally make them appear black.

Light

Lack of sufficient light retards chlorophyll formation and promotes slender growth with long internodes, thus leading to pale green leaves, spindly growth, and premature drop of leaves and flowers. This condition is known as **etiolation.** Etiolated plants are found outdoors only when plants are spaced too close together or when they are growing under trees or other objects. Etiolation of various degrees, however, is rather common in house plants, and also in greenhouses, seedbeds, and cold frames, where plants often receive inadequate light. Etiolated plants are usually thin and tall and are susceptible to lodging.

Excess light is rather rare in nature and seldom injures plants. Many injuries attributed to light are probably the result of high temperatures accompanying high light intensities. Excessive light, however, seems to cause sunscald of pods of beans grown at high altitudes where, due to the absence of dust, more light of short wavelengths reaches the earth. The pods develop small water-soaked spots, which quickly become brown or reddish brown and shrink.

The amount of light is considerably more important in relation to house plants. Some of them prefer shade or semishade during the growth season but full sunlight during the winter. Others prefer shade throughout the year, while still others must have sunlight all year long. As a rule, house plants with deep green leaves prefer or tolerate shade much better than do plants with colored leaves, the latter generally doing better when they receive considerable sunlight. Most flowering house plants grow and flower best with full exposure to sunlight at all seasons. Lack of sufficient light for any of these kinds of plants has the same effects as on the outdoor plants, that is, pale green leaves, spindly growth, leaf drop, few or no flowers, and flower drop. On the other hand, excessive sunlight on plants that prefer less light often results in the appearance of yellowish-brown or silvery spots on their leaves. Plants suddenly moved to an area with strikingly different light intensity than the previous one often respond with general defoliation.

Air Pollution

The air at the earth's surface consists primarily of nitrogen and oxygen (78 and 21 percent, respectively). Much of the remaining 1 percent is water vapor and carbon dioxide. The activities of humans in generating energy, manufacturing goods, and disposing of wastes result in the release into the atmosphere of a number of pollutants that may alter plant metabolism and induce disease. Air pollution damage to plants, especially around certain types of factories, has been recognized for about a century. Its extent and importance, however, increased with the industrial revolution and will, apparently, continue to increase with the world's increasing population, industrialization, and urbanization.

Air Pollutants and Kinds of Injury to Plants

Almost all air pollutants causing plant injury are gases, but some particulate matter or dusts may also affect vegetation. Some gas contaminants, such as ethylene, ammonia, chlorine, and sometimes mercury vapors, exert their injurious effects over limited areas. Most frequently they affect plants or plant products stored in poorly ventilated warehouses in which the pollutants are produced by the plants themselves (ethylene), or from leaks in the cooling system (ammonia).

More serious and widespread damage is caused to plants in the field by chemicals such as ozone (Figure 10-6), sulfur dioxide, hydrogen fluoride, nitrogen dioxide, peroxyacyl nitrates, and particulates. In many localities, for example, the Los Angeles basin, air pollutants spread into the area surrounding the source(s) of pollution, become trapped, and cause serious plant damage. More frequently, most air pollutants are transported downwind from the urban or industrial centers in which they are produced and may be carried by

FIGURE 10-6 Flecking on the upper surface of tobacco leaf caused by naturally occurring high concentrations of ozone in the atmosphere.

the wind to areas that are several miles, often hundreds of miles and sometimes thousands of miles, from the source. High concentraions of or long exposure to these chemicals cause visible and sometimes characteristic symptoms (such as necrosis) on the affected plants. More important economically, however, is the fact that even when plants are exposed to dosages less than those that cause acute damage, their growth and productivity may still be suppressed because of interference by the pollutants with the metabolism of the plant. Moreover, prolonged exposure to air pollutants seems to weaken plants and to predispose them to attack by insects and by some pathogens. The main pollutants, their sources, and their effects on plants are given in Table 10-1.

Main Sources of Air Pollutants

Some air pollutants, such as sulfur dioxide and hydrogen flouride, are produced as such directly from a source, such as refineries, combustion of fuel, and ore and fertilizer processing. Others, such as ozone and peroxyacyl nitrates, are produced in the atmosphere as secondary products of photochemical reactions involving NO_2, O_2, hydrocarbons, and sunlight.

Automobile exhaust in the streets and highways and exhausts of other internal combustion engines in factories are probably the most important sources of ozone and other phytotoxic pollutants. Thousands of tons of incompletely burned hydrocarbons and NO_2 are released into the atmosphere daily by automobile exhausts. In the presence of ultraviolent light from the sun, this nitrogen dioxide reacts with air oxygen and forms ozone and nitric oxide. The ozone may react with nitric oxide to form the original compounds:

$$NO_2 + O_2 \underset{}{\overset{\text{sunlight}}{\rightleftharpoons}} O_3 + NO$$

In the presence of unburned hydrocarbon radicals, however, the nitric oxide reacts with these instead of ozone, and therefore the ozone concentration builds up.

$$O_3 + [NO + \text{Unburned hydrocarbons from automobiles, etc.}] \rightarrow O_3 + \text{Peroxyacyl nitrates}$$

Ozone, too, can react with vapors of certain unsaturated hydrocarbons, but the products of such reactions (various organic peroxides) are also toxic to plants. Normally, the noxious fumes produced by automobiles and other engines are swept up by the warm air currents from the earth's surface rising into the cooler air above, where the fumes are dissipated. During periods of calm, stagnant weather, however, an inversion layer of warm air is formed above the cooler air, and this prevents the upward dispersion of atmospheric pollutants. The pollutants then are trapped near the ground, where after sufficient buildup, they may seriously damage living organisms.

Peroxyacyl nitrate (PAN) injury has been observed primarily around metropolitan areas where large amounts of hydrocarbons are released into the air from automobiles. The problem is especially serious in areas like Los Angeles and New Jersey, where the atmospheric conditions are conducive to the formation of inversion layers. Many different kinds of plants are affected by PAN over large geographical areas surrounding the locus of PAN formation, due to diffusion or to dispersal of the pollutant by light air currents.

How Air Pollutants Affect Plants

The concentration at which each pollutant causes injury to a plant varies with the plant and even with the age of the plant or the plant part. As the duration that the plant is exposed to the pollutant is increased, damage can be caused by increasingly smaller concentrations of the pollutant until a minimum dose–injury threshold is reached. Plant injury by air pollutants generally increases with increased light intensity, increased soil moisture and air relative humidity, increased temperature, and with the presence of other air pollutants.

Ozone injures the leaves of plants exposed for even a few hours at concentrations of 0.1 to 0.5 parts per million (ppm). Ozone is taken into leaves through stomata and injures primarily pallisade but also other cells by disrupting the cell membrane. Affected cells near stomata collapse and die, and white (bleached) necrotic flecks appear, first on the upper side and later on either leaf surface. Many crop plants, such as alfalfa, bean, citrus, grape, potato, soybean, tobacco, and wheat, and many ornamentals and trees, such as ash, lilac, several pines, and poplar, are quite sensitive to ozone, while some other crops, such as cabbage, peas, peanuts, and pepper, are of intermediate sensitivity, and some, such as beets, cotton, lettuce, strawberry, and apricot, are tolerant.

Sulfur dioxide may injure plants in concentrations as low as 1 to 5 ppm. Since sulfur dioxide is absorbed through the leaf stomata, conditions that favor or inhibit the opening of stomata similarly affect the amount of sulfur dioxide absorbed. After absorption by the leaf, sulfur dioxide reacts with water and forms phytotoxic sulfite ions. The latter, however, are slowly oxidized in the cell to produce harmless sulfate ions. Thus, if the rate of sulfur dioxide absorption is slow enough, the plant may be able to protect itself from the buildup of phytotoxic sulfites.

Table 10-1. Air Pollution Injury to Plants

Pollutant	Source	Susceptible Plants	Symptoms	Remarks
Ozone (O_3)	Automobile exhausts. Other internal combustion engines. (Released NO_2 combines with O_2 in sunlight $\rightarrow O_3$) From stratosphere. From lightning, from forests.	Expanding leaves of all plants, especially tobacco, bean, cereals, alfalfa, petunia, pine, citrus, corn	Stippling, mottling, and chlorosis of leaves, primarily on upper leaf surface. Spots are small to large, bleached white to tan, brown, or black (Fig. 10-6). Premature defoliation and stunting occurs in plants such as citrus, grapes, and vines.	Enters through stomata. It is the most destructive air pollutan to plants. A major component of smog.
Peroxyacyl nitrates (PAN)	Automobile exhausts or other internal combustion engines. (Gasoline vapors and incompletely burned gasoline $\pm O_3$ or $NO_3 \rightarrow$ PAN).	Many kinds of plants, including spinach, petunia, tomato, lettuce, dahlia.	Causes "silver leaf" on plants, i.e., bleached white to bronze spots on lower surface of leaves that may later spread throughout leaf thickness and resemble ozone injury.	Particularly severe near metropolitan areas with smog and inversion layers.
Sulfur dioxide (SO_2)	Stacks of factories. Automobile exhausts, and other internal combustion engines.	Many kinds of plants, including alfalfa, violet, conifers, pea, cotton, bean. Toxic at 0.3 to 0.5 ppm.	Low concentrations cause general chlorosis. Higher concentrations cause bleaching of interveinal tissues of leaves.	Also combines with moisture and forms toxic acid droplets (acid rain).
Nitrogen dioxide (NO_2)	From oxygen and nitrogen in the air by hot combustion sources, e.g., furnaces, internal combustion engines.	Many kinds of plants including beans, tomatoes. Toxic at 2 to 3 ppm.	Causes bleaching and bronzing of plants similar to that caused by SO_2. At low concentration it also suppresses growth of plants.	

Hydrogen fluoride (HF)	Stacks of factories processing ore or oil.	Many kinds of plants, including corn, peach, tulip. Actively growing, especially wet leaves, are most sensitive. Toxic at 0.1 to 0.2 ppb.	Leaf margins of dicots and leaf tips of monocots turn tan to dark brown, die and may fall from the leaf. Some plants tolerate HF up to 200 ppm.	HF may evaporate or be washed out of plant and plant recovers slowly.
Chlorine (Cl_2) and Hydrogen chloride (HCl)	Refineries, glass factories, incineration of plastics.	Many kinds of plants, usually near the source. Toxic at 0.1 ppm.	Leaves show bleached, necrotic areas between veins. Leaf margins often appear scorched. Leaves may drop prematurely. Damage resembles that caused by SO_2.	
Ethylene (CH_2CH_2)	Automobile exhausts. Burning of gas, fuel oil, and coal. From ripening fruit in storage.	Many kinds of plants. Toxic at 0.05 ppm.	Plants remain stunted, their leaves develop abnormally and senesce prematurely. Plants produce fewer blossoms and fruit. Fruit, e.g., apples, develop depressed, necrotic, dark areas (scald).	Ethylene is a plant hormone with numerous functions.
Particulate matter (dusts)	Dust from roads, cement factories. Burning of coal, etc.	All plants.	Forms dust or crusty layers on plant surfaces. Plants become chlorotic, grow poorly, and may die. Some dusts are toxic and burn leaf tissues directly or after dissolving in dew or rainwater.	

PAN is also taken into leaves through stomata and causes injury at concentrations as low as 0.01 to 0.02 ppm. In large urban areas, concentrations of 0.02 to 0.03 are not uncommon, and in the downtown areas of some cities PAN concentrations of 0.05 to 0.21 have been measured. Once inside leaves, PAN attacks preferentially the spongy parenchyma cells, which collapse and are replaced by air pockets that give the leaf a glazed or silvery appearance. The symptoms on broad-leaved plants appear on the lower leaf surface, while monocot leaves show symptoms on both sides. Young leaves and tissues are more sensitive to PAN, and periodic exposures of leaves to PAN often cause "banding" and in some plants even margin "pinching" of leaves because of discoloration and death of the most sensitive affected cells, respectively.

Acid Rain

Normal, unpolluted rain would contain almost pure water (H_2O) in which there would be dissolved some carbon dioxide (CO_2), some ammonia (NH_3) originating from organic matter and existing in water as NH_4^+, and varying but small amounts of cations (Ca^{++}, Mg^{++}, K^+, and Na^+) and anions (Cl^-, SO_4^{--}). Although the pH of pure water is a neutral 7.0, the pH of normal, "unpolluted" rain is usually 5.6; in other words, it is already acidic. Such rain, however, is considered "normal," and only when the pH of rain or snow is below 5.6 is it considered acidic ("acid rain").

Acid rain is the result of human activities, primarily the combustion of fossil fuels (oil, coal, natural gas) and the smelting of sulfide ores. These activities release in the atmosphere large quantities of sulfur and nitrogen oxides, which when in contact with atmospheric moisture are converted into two of the strongest acids known (sulfuric and nitric) and fall to the ground in rain and snow. The pH of rain and snow over large regions of the world ranges from 4.0 to 4.5, which is from 5 to 30 times more acid than the lowest pH (5.6) expected for unpolluted areas. The lowest rain pH's reported so far (2.4 in Scotland, 1.5 in West Virginia and 1.7 in Los Angeles) are more acidic than vinegar (pH 3.0) and of lemon juice (pH 2.2). It is estimated that about 70 percent of the acid in acid rain is sulfuric acid, with nitric acid contributing about 30 percent. In addition to sulfur contained in the acids carried in the rain, it is believed that an approximately equal amount of sulfur reaches all surfaces through dry deposition of particulate sulfur. In humid or wet weather, this sulfur too is oxidized to sulfuric acid.

Acid rain exerts a variety of influences by greatly increasing the solubility of all kinds of molecules and by directly (through the low pH and the toxicity of the $—SO_4^{--}$ and $—NO_3^-$ ions) or indirectly (through the dissolved molecules) affecting many forms of life. The adverse effects of acid rain on the microorganisms, plants, and fishes of rivers and lakes have been well documented. The effects of acid rain on crop plants have been more difficult to document. Experiments in which acidic rain (pH 3.0) was applied to plants showed that, under some conditions, treated leaves developed pits, spots, and curling and that treated plants, with or without symptoms, sometimes showed reductions in dry weight. Also, more seeds of some plant species germinated when the soil in which they were planted received acid rain than when it did not, while the opposite was observed for other species. Experiments conducted to determine the effect of acid rain on the initiation and development of plant diseases have shown that in some diseases, such as *Cronartium fusiforme* rust of oak, only 14 percent as many telia formed under acidic (pH 3.0) rain treatment than under a pH 6.0 rain treatment and that beans treated with acidic rain (pH 3.2) had only 34 percent as many nematode egg masses than they did under a pH 6.0 rain treatment. On the other hand, a bacterial disease (halo blight) and the rust disease of bean were sometimes more severe and others milder with the acidic rain than with the pH 6.0 rain. In general, although some evidence exists that acid rain causes variable amounts of damage to at least some plants, consistent quantitative data are still insufficient to determine the extent of such damage on various crops in the areas where they occur.

Nutritional Deficiencies in Plants

Plants require several mineral elements for normal growth. Some elements, such as nitrogen, phosphorus, potassium, calcium, magnesium, and sulfur, needed in relatively large amounts, are called "major" elements, while others, like iron, boron, manganese, zinc, copper, molybdenum, and chlorine, needed in very small amounts, are called "trace" or "minor" elements or "micronutrients." Both major and trace elements are essential to the plant. When they are present in the plant in amounts smaller than the minimum levels required for normal plant growth, the plant becomes diseased and exhibits various external and internal symptoms. The symptoms may appear on any or all organs of the plant, including leaves, stems, roots, flowers, fruits, and seeds.

The kinds of symptoms produced by deficinecy of a certain nutrient depend primarily on the functions of that particular element in the plant. These functions presumably are inhibited or interfered with when the element is limiting. Certain symptoms are the same in deficiency of any of several elements, but other diagnostic features usually accompany a deficiency of a particular element. Numerous plant diseases occur annually in most agricultural crops in many locations as a result of reduced amounts or reduced availability of one or more of the essential elements in the soils where the plants are grown. The presence of lower-than-normal amounts of most essential elements usually results in merely a reduction in growth and yield. When the deficiency is greater than a certain critical level, however, the plants develop acute or chronic symptoms and may even die. Some of the general deficiency symptoms caused by each essential element, the possible functions affected, and some examples of common deficiency disorders are given in Table 10-2 and shown in Figures 10-7 and 10-8.

Soil Minerals Toxic to Plants

Soils often contain excessive amounts of certain essential or nonessential elements, both of which at high concentration may be injurious to the plants.

Table 10-2. Nutrient Deficiencies in Plants

Deficient Nutrient	Functions of Element	Symptoms
Nitrogen N	Present in most substances of cells.	Plants grow poorly and are light green in color. The lower leaves turn yellow or light brown and the stems are short and slender (Fig. 10-7A).
Phosphorus P	Present in DNA, RNA, phospholipids (membranes), ADP, ATP, etc.	Plants grow poorly and the leaves are bluish-green with purple tints. Lower leaves sometimes turn light bronze with purple or brown spots. Shoots are short and thin, upright, and spindly.
Potassium K	Acts as a catalyst of many reactions.	Plants have thin shoots which in severe cases show dieback. Older leaves show chlorosis with browning of the tips, scorching of the margins, and many brown spots usually near the margins. Fleshy tissues show end necrosis (Fig. 10-7C and E).
Magnesium Mg	Present in chlorophyll and is part of many enzymes.	First the older, then the younger leaves become mottled or chlorotic, then reddish. Sometimes necrotic spots appear. The tips and margins of leaves may turn upward and the leaves appear cupped. Leaves may drop off (Fig. 10-7D).
Calcium Ca	Regulates the permeability of membranes. Forms salts with pectins. Affects activity of many enzymes.	Young leaves become distorted, with their tips hooked back and the margins curled. Leaves may be irregular in shape and ragged with brown scorching or spotting. Terminal buds finally die. The plants have poor, bare root systems. Causes blossom end rot of many fruits (Fig. 10-7F). Increases fruit (e.g., apple) decay in storage. May be responsible for tipburn in mature detached lettuce heads at high (24–35°C) temperatures.
Boron B	Not really known. Affects translocation of sugars and utilization of calcium in cell wall formation.	The bases of young leaves of terminal buds become light green and finally break down. Stems and leaves become distorted. Plants are stunted (Fig. 10-8). Fruit, fleshy roots or stems, etc., may crack on the surface and/or rot in the center. Causes many plant diseases, e.g., heart rot of sugar beets, brown heart of turnips, browning or hollow stem of cauliflower, cracked stem of celery, corky spot, dieback and rosette of apples, hard fruit of citrus, top sickness of tobacco.
Sulfur S	Present in some amino acids and coenzymes.	Young leaves are pale green or light yellow without any spots. The symptoms resemble those of nitrogen deficiency.
Iron Fe	Is a catalyst of chlorophyll synthesis. Part of many enzymes.	Young leaves become severely chlorotic, but their main veins remain characteristically green. Sometimes brown spots develop. Part of or entire leaves may dry. Leaves may be shed (Fig. 10-7B).
Zinc Zn	Is part of enzymes involved in auxin synthesis and in oxidation of sugars.	Leaves show interveinal chlorosis. Later they become necrotic and show purple pigmentation. Leaves are few and small, internodes are short and shoots form rosettes, and fruit production is low. Leaves are shed progressively from base to tip. It causes "little leaf" of apple, stone

Table 10-2. Nutrient Deficiencies in Plants *(continued)*

Deficient Nutrient	Functions of Element	Symptoms
		fruits and grape, "sickle leaf" of cacao, "white tip" of corn, etc.
Copper Cu	Is part of many oxidative enzymes.	Tips of young leaves of cereals wither and their margins become chlorotic. Leaves may fail to unroll and tend to appear wilted. Heading is reduced and the heads are dwarfed and distorted. Citrus, pome, and stone fruits show dieback of twigs in the summer, burning of leaf margins, chlorosis, rosetting, etc. Vegetable crops fail to grow.
Manganese Mn	Is part of many enzymes of respiration, photosynthesis, and nitrogen utilization.	Leaves become chlorotic but their smallest veins remain green and produce a checked effect. Necrotic spots may appear scattered on the leaf. Severely affected leaves turn brown and wither.
Molybdenum Mo	Is an essential component of the nitrate reductase enzyme	Melons, and probably other plants, exhibit severe yellowing and stunting and they fail to set fruit.

Of the essential elements, those required by plants in large amounts, such as nitrogen and potassium, are usually much less toxic when present in excess than are the elements required only in trace amounts, such as manganese, zinc, and boron. Even among the latter, however, some trace elements such as manganese and magnesium have a much wider range of safety than do others, such as boron or zinc. Besides, not only do the elements differ in their ranges of toxicity, but various kinds of plants also differ in their susceptibility to the toxicity to a certain level of a particular element. Concentrations at which nonessential elements are toxic also vary among elements, and plants in turn vary in their sensitivity to them. For example, some plants are injured by very small amounts of nickel but can tolerate considerable concentrations of aluminum.

The injury occurring from the excess of an element may be slight or severe and is usually the result of direct injury by the element to the cell. On the other hand, the element may interfere with the absorption or function of another element and thereby lead to the symptoms of a deficiency of the element being interfered with. Thus, excessive sodium induces a deficiency of calcium in the plant, while the toxicity of copper, manganese, or zinc is both direct on the plant and induces a deficiency of iron in the plant.

Excessive amounts of sodium salts, especially sodium chloride, sodium sulfate, and sodium carbonate, raise the pH of the soil and cause what is known as alkali injury. This injury varies in the different plants and may range from chlorosis to stunting, leaf burn, wilting, to outright killing of seedlings and young plants. Some plants, such as wheat and apple, are very sensitive to alkali injury, while others, such as sugar beets, alfalfa, and several grasses, are quite tolerant. On the other hand, when the soil is too acidic, the growth of some kinds of plants is impaired and various symptoms may appear. Plants usually grow well in a soil pH range from 4 to 8, but some

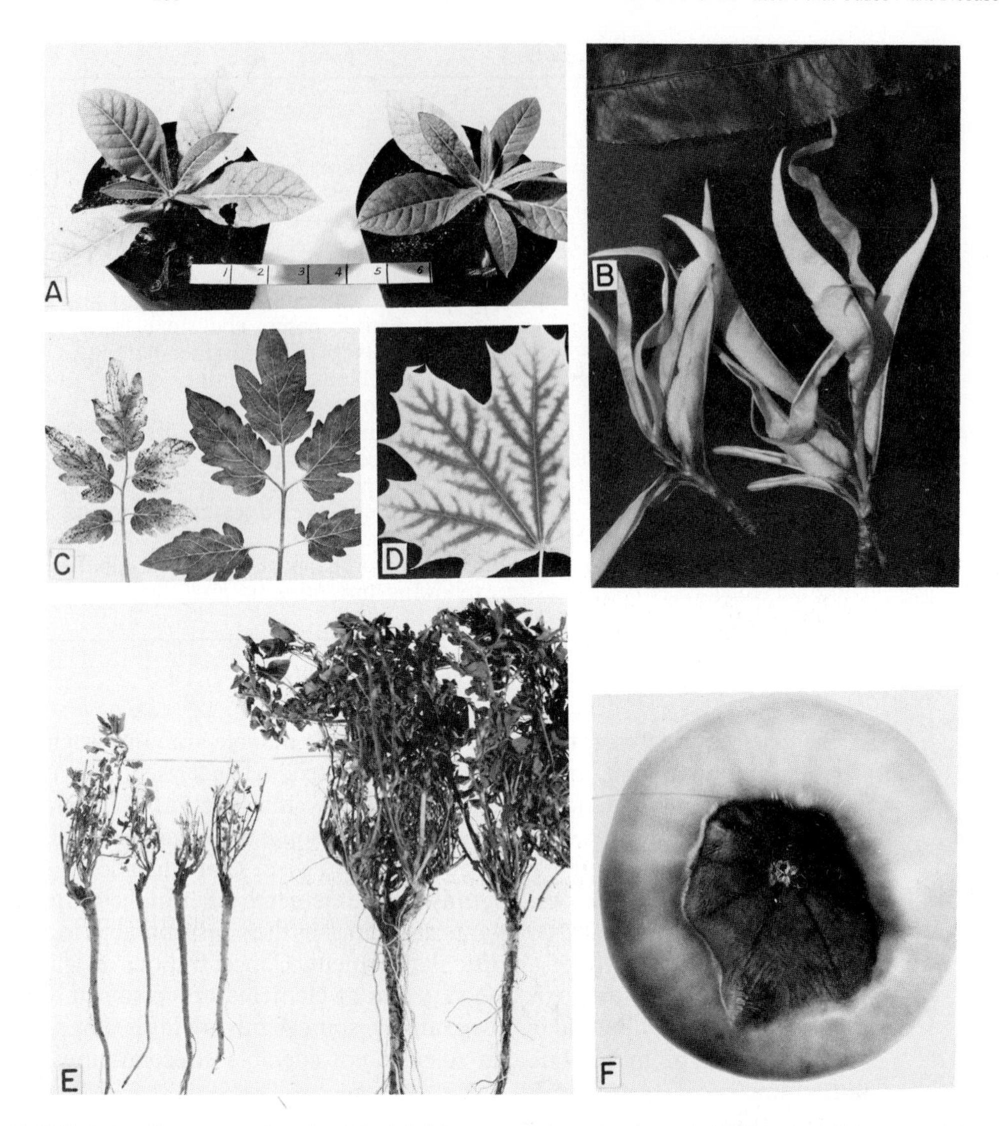

FIGURE 10-7 Some examples of nutrient deficiency symptoms in plants. (A) Nitrogen deficiency on tobacco (left) and one week after fertilization. (B) Iron deficiency on peach. Note uniform yellowing of affected leaves compared to normal leaf at top of photo. (C) Potassium deficiency on tomato (left). Healthy leaf at right. (D) Magnesium deficiency symptoms on maple. (E) Healthy (right) and potassium-deficient alfalfa plants. (F) Blossom end rot of tomato caused in part by calcium deficinecy.

plants grow better in the lower pH than others, and vice versa. Thus, blueberries grow well in acid soils, while alfalfa grows best in alkaline soils. The injury caused by low pH is, in most cases, brought about by the greater solubility of mineral salts in acid solutions. These salts then become available in concentrations that, as was pointed out above, either are toxic to the plants or interfere with the absorption of other necessary elements and so cause symptoms of mineral deficiency.

Boron, manganese, and copper have been most frequently implicated in mineral toxicity diseases, although other minerals, such as aluminum and iron, also damage plants in acid soils. Excess boron is toxic to many vegeta-

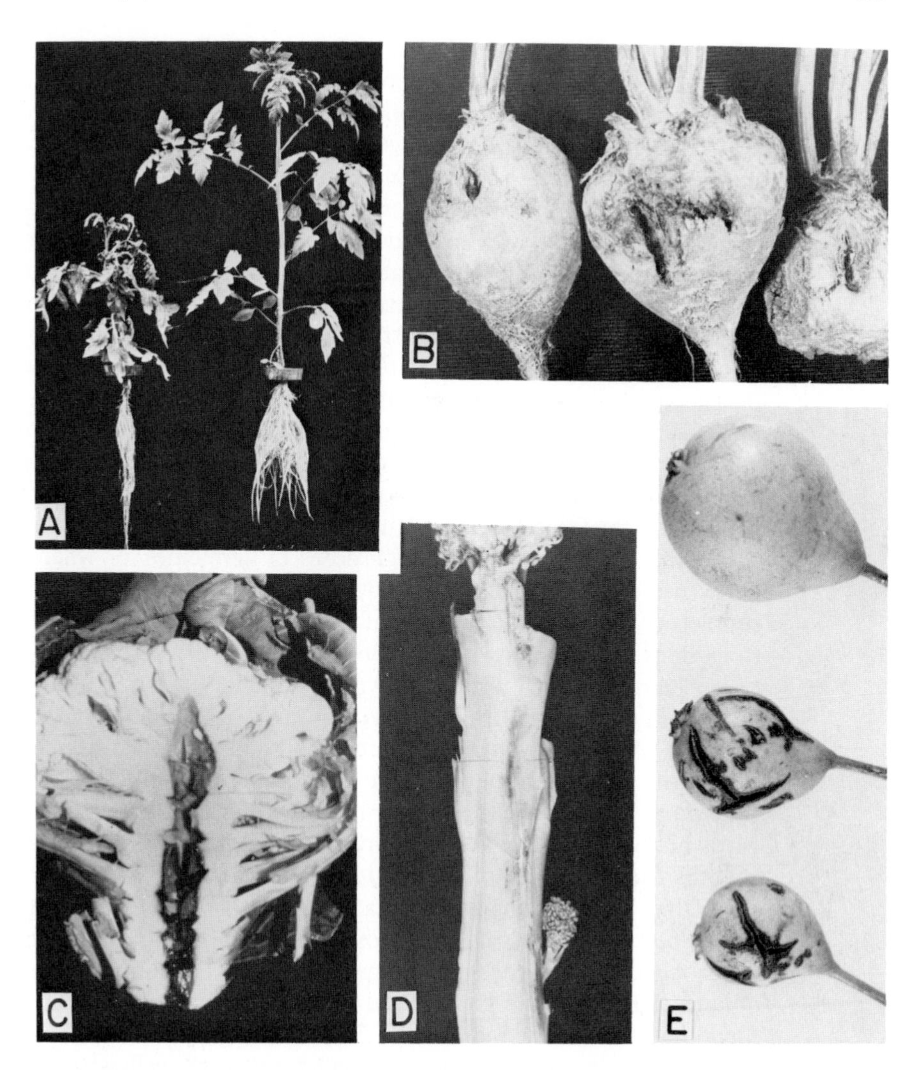

FIGURE 10-8 Boron deficiency symptoms on plants. (A) Healthy (right) and stunted tomato plant. (B) Cracking and breakdown of beets. (C) Internal breakdown of cauliflower stem. (D) Corky neck surface and internal breakdown of broccoli stem. (E) Healthy (top) and cracked pears due to boron deficiency aggravated by prolonged drought.

bles and trees. Excess manganese is known to cause a crinkle-leaf disease in cotton and has been implicated in the internal bark-necrosis of Red Delicious apple and in many other diseases of several crop plants. Sodium and chlorine ions also have been shown to cause symptoms of poor growth and decline, like those shown by some of the trees along roads in northern areas where heavy salting is carried out in the winter to remove ice from roads.

Herbicide Injury

Some of the most frequent plant disorders seem to be the result of the extensive use of herbicides (weed killers). The constantly increasing number of herbicides in use by more and more people for general or specific weed control is creating numerous problems among those who use them, their neighbors, and those who use soil that has been treated with herbicides.

Herbicides are either specific against broad-leaved weeds [atrazine, simazine, 2,4-D, dicamba (Banvel-D)] and are applied in corn and other small grain fields and on lawns, or they are specific against grasses and some broad-leaved weeds [DCPA (Dacthal), trifluralin (Treflan)] and are applied in pastures, orchards, and in vegetable and truck crop fields. In addition, some herbicides are general weed or shrub killers [glyphosate (Roundup], paraquat, terbacil (Sinbar), picloram]. Most herbicides are safe as long as they are used to control weeds among the right crop plants, at the right time, at the correct dosage, and when the correct environmental conditions prevail. When any one of the above conditions is not met, abnormalities develop on the cultivated plants with which the herbicides come in contact. Affected plants show various degrees of distortion or yellowing of leaves (Figure 10-9), browning, drying and shedding of leaves, stunting and even death of the plant. Much of this damage is caused by too high doses of herbicides, or when applied too early in the season or on too cold or too hot a day, or when dust or spray droplets of an herbicide are carried by the wind to nearby sensitive plants or to gardens or fields on which plants sensitive to the herbicide are grown. Of course, direct application of the wrong pesticide in a field with a particular crop plant will kill the crop just as if it were a weed.

Use of preplant or preemergence herbicides through application to the soil before or at planting time often affects seed germination and growth of the young seedlings if too much or the wrong herbicide has been applied. Most herbicides are used up or are inactivated within a few days to a few months from the time of application; some, however, persist in the soil for more than a year. Sensitive plants planted in fields previously treated with such a persistent herbicide may grow poorly and may produce various symptoms. Also, home owners, home gardeners, and greenhouse operators often obtain what looks like good, weed-free soil from fields that, unbeknown to them, had been treated with herbicides. Such soil when used to grow potted, bench, or garden plants results in smaller, distorted, yellowish plants (Figure 10-9D), which sometimes shed some or all of their leaves and either die or finally recover.

Other Improper Agricultural Practices

As with herbicides, a variety of other agricultural practices improperly carried out may cause considerable damage to plants and significant financial losses. Almost every agricultural practice can cause damage when applied the wrong way, at the wrong time, or with the wrong materials. Most commonly, however, losses result from the application of chemicals, such as fungicides, insecticides, nematicides, and fertilizer, at too high concentrations or on plants sensitive to them. Spray injury resulting in leaf burn or spotting or russeting of fruit is common on many crop plants (Figure 10-10).

Excessive or too deep cultivation between rows of growing plants may be more harmful than useful because it cuts or pulls many of the plants' roots.

FIGURE 10-9 (A, B, and C) Injury on trees from drift of herbicide applied to lawn or orchard. (A) The leaves are smaller with greatly narrowed interveinal areas. (B) Leaves are rolled and petioles are distorted. Normal leaf at bottom. (C) Yellowing of veins or entire leaves caused by herbicide injury. (D) Leaf distortion of geranium cuttings after transplanting in soil contaminated with herbicide. Note two normal leaves (bottom) developed before transplanting. (E) Frenching of tobacco caused by accumulation of toxic substances in the soil produced by bacteria *(Bacillus cereus)* and fungi *(Aspergillus)* and interfering with amino acid metabolism in plants. (Photo C courtesy W. J. Lord).

Road or other construction often cuts a large portion of the roots of nearby trees and results in their dieback and decline. Inadequate or excessive watering may cause wilting or any of the symptoms described earlier. In the case of African violets, droplets of cold water on the leaves cause the appearance of rings and ringlike patterns reminiscent of virus ringspot diseases. Potatoes stored next to hot water pipes under the kitchen sink often develop black heart. Trees frequently grow poorly and their leaves are chlorotic, curled, or reddened because their trunk is girdled by the fence wire.The roots of plants

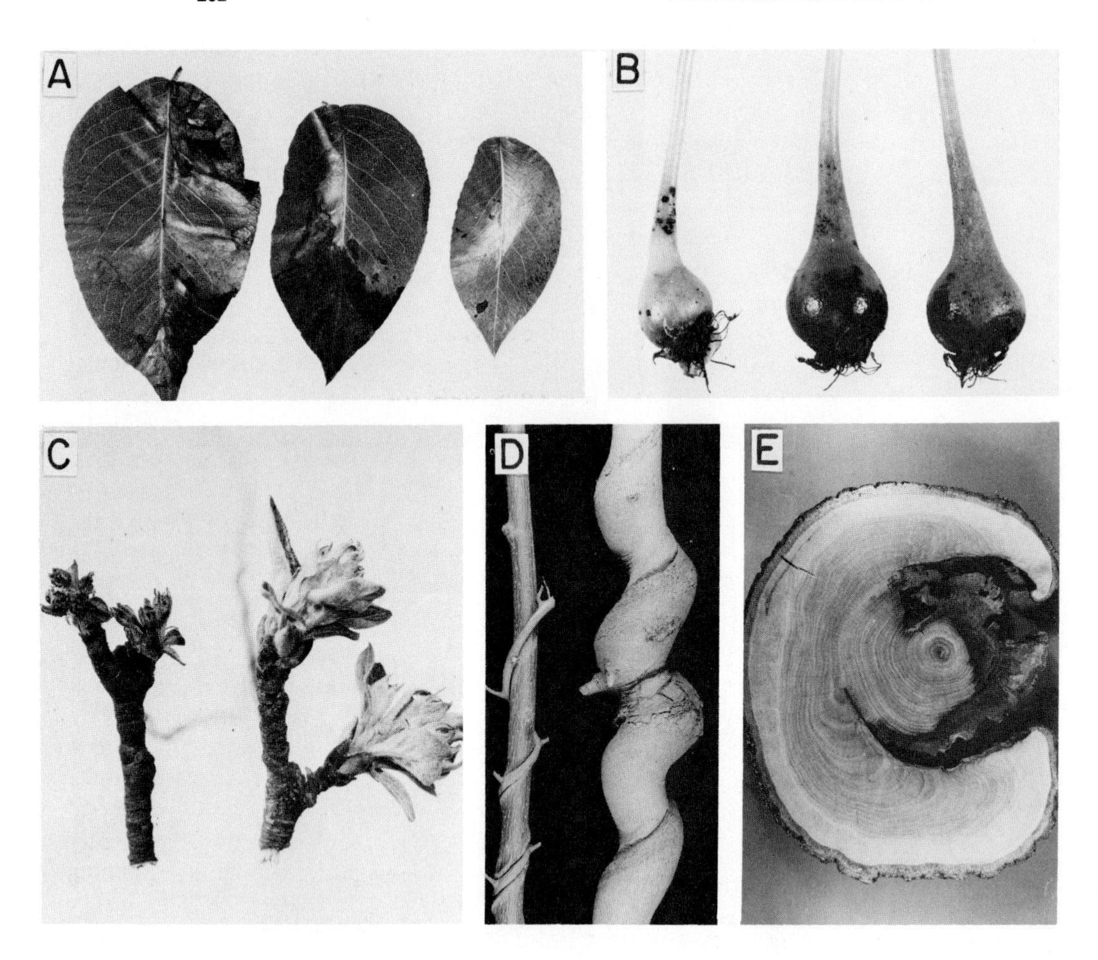

FIGURE 10-10 Spray injury on pear leaf (A) and fruit (B), and on apple blossom (C, left). (D) Distortion of maple stem and twig by the climbing vine of bittersweet, *Celastrus scandens*. (E) Fire damage on oak trunk.

potted in pots that are too small for their size are often badly distorted and twisted and the whole plant grows poorly (Figure 10-1).

The Often Confused Etiology of Stress Diseases

Diagnosis of an abiotic disease is often every bit as difficult as the diagnosis of a biotic disease. When combinations of single or multiple abiotic and biotic diseases occur on the same plant or in an entire area, however, the diagnosis of the diseases and the determination of the relative importance of each become extremely difficult and often impossible.

When plants are adversely affected by an environmental factor, such as low moisture, nutrient deficiency, air pollution, or freezing, they are generally and concurrently weakened and predisposed to infection by one or more weakly parasitic pathogens. For example, all three conditions mentioned above predispose annual plants to infection by the fungus *Alternaria* and many perennial plants to infection by canker-causing fungi such as *Cytospora* and *Botryosphaeria*. A late blossom frost is often followed by infection with *Botrytis, Alternaria,* or *Pseudomonas*. Herbicide injury is likely to be followed by root rots caused by *Fusarium* and *Rhizoctonia*. Flooding injury is often followed by *Pythium* root infections.

Obviously, many of the stresses discussed in this chapter are often complicated by biotic diseases that follow. As a matter of fact, many epidemic disease problems, such as stalk rot of corn, tree declines, and stand depletions in forage legumes, although thought of as being caused by one or more biotic agents, they are in reality set off by one or another of the environmental factors discussed in this chapter. Thus, stalk rot of corn, although caused by one of several common fungi *(Fusarium, Diplodia, Gibberella),* actually occurs or becomes important only under conditions of low potassium and low moisture stress in early season. Similarly, the additional stress caused by some herbicides on soybean, sugar beet, and cotton seedlings increases the susceptibility of these crops to the *Thielaviopsis basicola* and *Rhizoctonia* root rots and damping off.

A striking example of the often confused etiology of stress diseases has been developing in recent years in Europe, where many different forest tree species, shurbs, and herbs have been exhibiting various degrees of yellowing, reduced growth, defoliation, abnormal growth, decline, and eventually death. This widespread general decline of forests (called "waldsterben") has been occurring and spreading in large areas of Central Europe since about 1980. Such decline seems to be triggered by the stress caused by atmospheric depositions of toxic or growth-altering air pollutants but is subsequently aggravated by additional abiotic and biotic predisposing or stress-inducing factors. The air pollutants themselves, such as ozone, cause some direct injury and reduction in photosynthesis, but the mixture of deposited acidic pollutants may also cause acidification of soils. This may result in leaching out and therefore deficiency in certain elements such as magnesium, or in increases in solubility of certain toxic elements, such as aluminum, thereby causing aluminum toxicity in plants. The latter then causes necrosis of fine roots, which leads to increased moisture or nutrient stress and eventually drying out and death of trees, particularly during dry periods. In addition to the effects caused by these abiotic factors, affected trees show increased susceptibility to insects and to foliage and root pathogens such as *Lophodermium, Phytophthora,* and *Armillaria,* which further increase the moisture and water stress and reduce photosynthesis in the plant.

REFERENCES

Berg, A., Clulo (Berg), G., and Orton, C. R. (1958). Internal bark necrosis of apple resulting from manganese toxicity. *Bull.—W. Va., Agric. Exp. Stn.* **414,** 22.

Carne, W. M. (1948). The non-parasitic disorders of apple fruits in Australia. *Bull.—C.S.I.R.O. (Aust.)* **238,** 1–83.

Colhoun, J. (1979). Predisposition by the environment. *In* "Plant Disease" (J. G. Horsfall and E. B. Cowling, eds.), Vol. 4, pp. 75–96. Academic Press, New York.

Daines, R. H., Leone, I. A., and Brennan, E. (1960). Air pollution as it affects agriculture in New Jersey. *Bull.—N. J. Agric. Expt. Stn.* **794,** 1–14.

Darley, E. F., and Middleton, J. T. (1966). Problems of air pollution in plant pathology. *Annu. Rev. Phytopathol.* **4,** 103–118.

Dodd, J. L. (1980). The role of plant stresses in development of corn stalk rots. *Plant Dis.* **64,** 533–537.

Evans, L. S. (1984). Acidic precipitation effects on terrestrial vegetation. *Annu. Rev. Phytopathol.* **22,** 397–420.

Jacobson, J. S., and Hill, A. C., eds. (1970). "Recognition of Air Pollution Injury to Vegetation: A Pictorial Atlas." Air Pollut. Control Assoc., Pittsburgh, Pennsylvania.

Krupa, S. V., Pratt, G. C., and Teng, P. S. (1982). Air pollution: An important issue in plant health. *Plant Dis.* **66,** 429–434.

Lacasse, N. L., and Treshow, M., eds. (1976). "Diagnosing Vegetation Injury Caused by Air Pollution." Applied Science Associates, Inc., Washington, D. C.

Laurence, J. A., and Weinstein, L. H. (1981). Effects of air pollutants on plant productivity. *Annu. Rev. Phytopathol.* **19,** 257–271.

Levitt, J. (1972). "Responses of Plants to Environmental Stresses." Academic Press, New York.

McMurtrey, J. E., Jr. (1953). "Environmental, Nonparasitic Injuries," Yearb. Agric., U. S. Dept. Agric., Washington, D. C. pp. 94–100.

Manning, W. J., and Feder, W. A. (1980). "Biomonitoring Air Pollutants with Plants." Applied Science Publishers, London.

Pell, E. J. (1979). How air pollutants induce disease. *In* "Plant Disease" (J. G. Horsfall and E. B. Cowling, eds.), Vol. 4, pp. 273–292. Academic Press, New York.

Schoenweiss. D. F. (1981). The role of environmental stress in diseases of woody plants. *Plant Dis.* **65,** 308–314.

Schutt, P., and Cowling, E. B. (1985). Waldsterben, a general decline of forests in central Europe: Symptoms, development and possible causes. *Plant Dis.* **69,** 548–558.

Wallace, T. (1961). "The Diagnosis of Mineral Deficiencies in Plants by Visual Symptoms." Stationery Office, London.

11 PLANT DISEASES CAUSED BY FUNGI

INTRODUCTION

Fungi are small, generally microscopic, eukaryotic, usually filamentous, branched, spore-bearing organisms that lack chlorophyll and have cell walls that contain chitin, cellulose, or both. Most of the 100,000 fungus species known are strictly saprophytic, living on dead organic matter, which they help decompose. Some, about 50, species cause diseases in humans, and about as many cause diseases in animals, most of them superficial diseases of the skin or its appendages. More than 8000 species of fungi, however, can cause diseases in plants. All plants are attacked by some kinds of fungi, and each of the parasitic fungi can attack one or many kinds of plants. Some of the fungi can grow and multiply only by remaining in association with their host plants during their entire life; these fungi are known as **obligate parasites** or **biotrophs.** Others require a host plant for part of their life cycles but can complete their cycles on dead organic matter or can grow and multiply on dead organic matter as well as on living plants; these fungi are called **nonobligate parasites.**

Characteristics of Plant Pathogenic Fungi

Morphology

Most fungi have a plantlike vegetative body consisting of more or less elongated, continuous, branched, microscopic filaments, which have definite cell walls. The body of the fungus is called **mycelium,** and the individual branches or filaments of the mycelium are called **hyphae** (Figure 11-1). Each hypha or mycelium is generally uniform in thickness. Hyphae of some fungi are only 0.5 μm in diameter, while in others they may be more than 100 μm thick. The length of the mycelium may be only a few micrometers in some fungi, but in others it may produce mycelial strands several meters long.

In some fungi the mycelium consists of many cells containing one or two nuclei per cell. In others the mycelium is coenocytic, that is, it contains many nuclei and either the entire mycelium is one continuous, tubular, branched or unbranched multinucleate cell or it is partitioned by several cross walls (septa), each segment being a multinucleate hypha. Growth of the mycelium occurs at the tips of the hyphae.

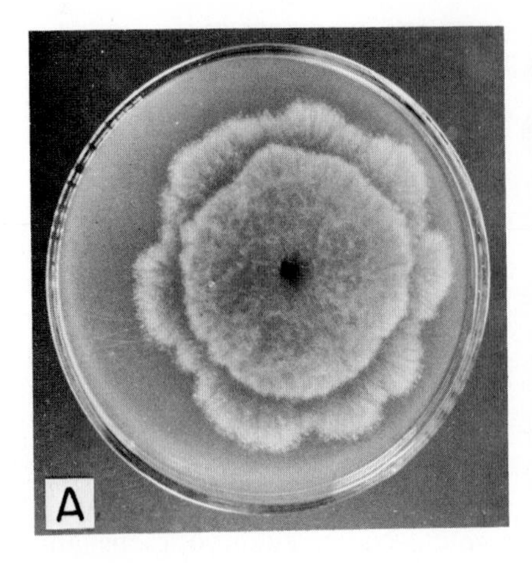

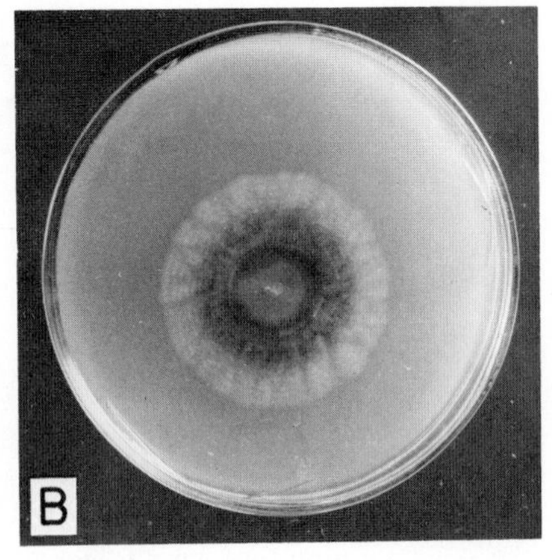

FIGURE 11-1 Appearance of the vegetative body (mycelium) of two fungi in culture. (A) *Physalospora.* (B) *Phoma.*

Some lower fungi lack true mycelium and produce instead a naked, amoeboid, multinucleate **plasmodium** (for example, Myxomycetes) or a system of strands of grossly dissimilar and continuously varying diameter called a **rhizomycelium** (for example, Chytridiomycetes).

Reproduction

Fungi reproduce chiefly by means of spores (Figure 11-2). **Spores** are specialized propagative or reproductive bodies consisting of one or a few cells. Spores may be formed asexually (through the production by the mycelium of separate, specialized cells, the spores, without involvement of karyogamy and meiosis) or as the result of a sexual process.

In the lower fungi, asexual spores are produced inside a sac called a **sporangium** and are released through an opening of the sporangium or upon its rupture. Some of these spores are motile by means of flagella and are, therefore, called **zoospores.** Other fungi produce asexual spores called **conidia** by the cutting off of terminal or lateral cells from special hyphae called **conidiophores.** In some fungi terminal or intercalary cells of a hypha enlarge, round up, form a thick wall, and separate to form **chlamydospores.** In still other fungi, asexual spores (conidia) are produced inside thick-walled structures called **pycnidia.**

Sexual reproduction, or processes resembling it, occur in most groups of fungi. In some, two cells (gametes) of similar size and appearance unite and produce a zygote, called a **zygospore.** In others, the gametes are of unequal size and the zygote they form is called an **oospore.** In some fungi no definite gametes are produced, but instead one mycelium may unite with another compatible mycelium. In one group of fungi (Ascomycetes) the sexual spores, usually eight in number, are produced within the zygote cell, the **ascus,** and the spores are called **ascospores.** In another group of fungi (Basidiomycetes), sexual spores are produced on the outside of the zygote cell called the **basidium,** and the spores are called **basidiospores.**

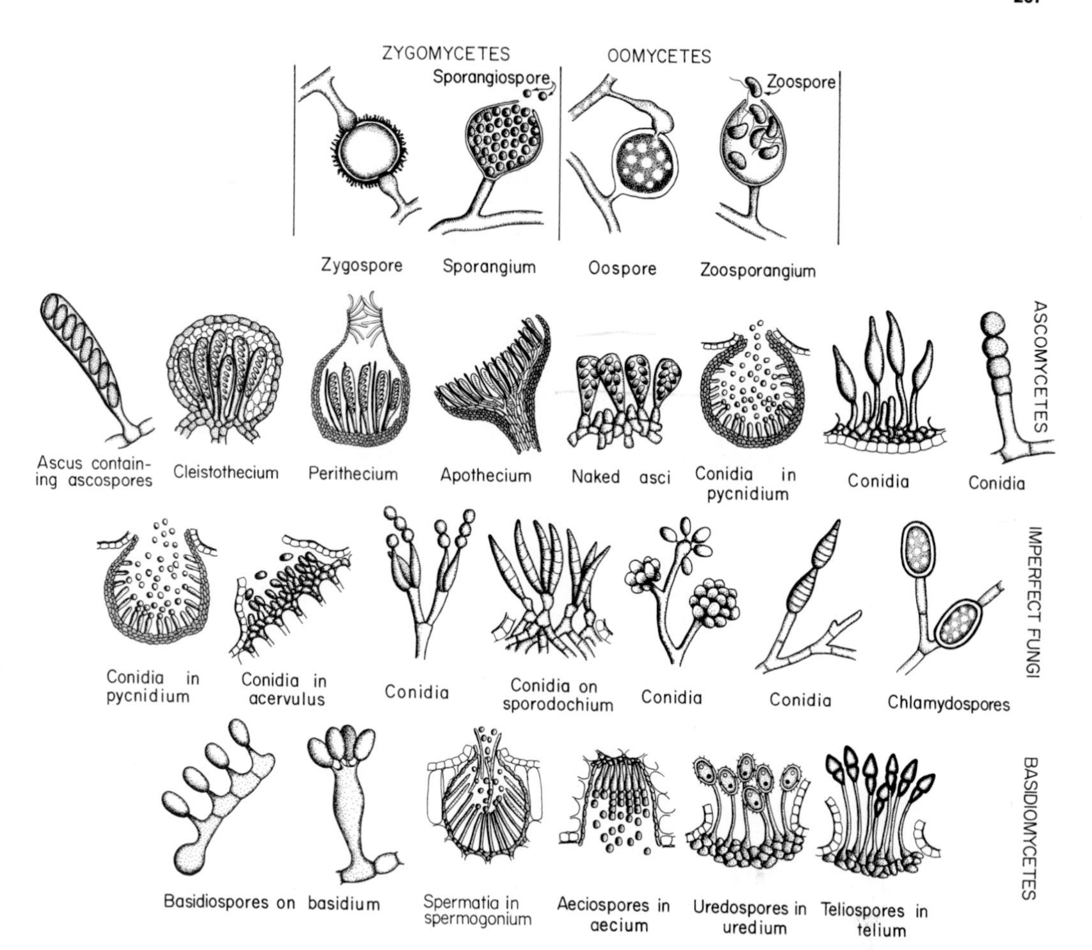

FIGURE 11-2 Representative spores and fruiting bodies of the main groups of fungi.

For a large group of fungi (Fungi Imperfecti or Deuteromycetes) no sexual reproduction is known either because they do not have one or because it has not yet been discovered. Apparently these fungi reproduce only asexually.

The union of the sexual nuclei in the zygote produces a diploid ($2N$) nucleus. Usually the first divisions of this nucleus are meiotic so that throughout its life the fungus contains haploid ($1N$) nuclei. In some groups of fungi, however, especially in the Basidiomycetes, the cells of the entire mycelium contain two haploid nuclei that remain separate inside the cell. Such mycelium is called **dikaryotic** but behaves very much as though it were a diploid mycelium (in which the two nuclei are united).

In most fungi both male and female gametes are produced on the same mycelium (**hermaphroditic** fungi). When the male gametes can fertilize the female ones of the same mycelium, the fungus is called **homothallic.** In many cases, however, the male gametes can fertilize only the female gametes of another, sexually compatible mycelium, and the fungus then is called **heterothallic.**

Ecology and Spread

Almost all plant pathogenic fungi spend part of their lives on their host plants and part in the soil or on plant debris on the soil. Some fungi pass all of their lives on the host, and only the spores may land on the soil, where they remain inactive until they are again carried to a host on which they grow and multiply. Other fungi (such as *Venturia*) must pass part of their lives on the host as parasites and part on dead tissues on the ground as saprophytes in order to complete their life cycle in nature. The latter group of fungi, however, remain continually associated with host tissues, whether living or dead, and in nature do not grow on any other kind of organic matter. A third group of fungi grow parasitically on their hosts, but they continue to live, grow, and multiply on the dead tissues of the host after its death and may further move out of the host debris into the soil or other decaying plant material on which they grow and multiply as strict saprophytes. The dead plant material that they colonize need not be related at all to the host they can parasitize. These fungi are usually soil pathogens, have a wide host range, and can survive in the soil for many years in the absence of their hosts. They, too, however, may need to infect a host from time to time in order to increase their populations, since protracted and continuous growth of these fungi as saprophytes in the soil results in more or less rapid reduction in their numbers.

During their parasitic phase fungi assume various positions in relation to the plant cells and tissues. Some fungi (such as powdery mildews) grow outside the plant surface but send their feeding organs (haustoria) into the epidermal cells of the plant. Some (such as *Venturia*) grow only between the cuticle and the epidermal cells. Others grow between the cells in the intercellular spaces and may or may not send haustoria into the cells. Still others grow between and through the cells indiscriminately. Obligate parasites (biotrophs) can grow only in association with living cells, being unable to feed on dead cells. On the other hand, the mycelium of some nonobligate parasites never comes in contact with living plant cells, because their enzymes macerate and kill the plant cells ahead of the mycelium. In most cases, however, regardless of the position of the mycelium in the host, the reproductive bodies (spores) of the fungus are produced at or very near the surface of the host tissues to ensure their prompt and efficient dissemination.

The survival and performance of most plant pathogenic fungi depend greatly on the prevailing conditions of temperature and moisture or the presence of water in their environment. Free mycelium survives only within a certain range of temperatures (−5 to +45°C) and in contact with moist surfaces, inside or outside the host. Most kinds of spores, however, can withstand broader ranges of both temperature and moisture and carry the fungus through the low winter temperatures and the dry summer periods. Spores, however, also require favorable temperatures and moisture in order to germinate. Furthermore, lower fungi producing zoospores require free water for the production, movement, and germination of the zoospores.

Zoospores are the only fungus structures that can move by themselves. Zoospores, however, can move for only very short distances (a few millimeters or centimeters, perhaps). Besides, only myxomycetes, chytridiomycetes, and oomycetes produce zoospores. The great majority of plant pathogenic fungi depend for their spread from plant to plant and to different parts of the same plant on chance distribution by such agents as wind, water, birds, insects, other animals, and humans. Fungi are disseminated primarily in the form of spores. Fragments of hyphae and hard masses of mycelium known as **sclerotia** may also be desseminated by the same agents although to a much lesser extent.

Spore dissemination in almost all fungi is passive, although their initial discharge in some fungi is forcible. The distance to which spores may be disseminated varies with the agent of dissemination. Wind is probably the most important disseminating agent of spores of most fungi and may carry spores over great distances. For specific fungi, other agents such as water or insects may play a much more important role than wind in the dissemination of their spores.

Classification of Plant Pathogenic Fungi

The fungi that cause diseases on plants are a diverse group. Because of their large numbers and diversity, only a sketchy classification of some of the most important phytopathogenic genera will be presented here. Fungi belong to the kingdom **Mycetae** (fungi).

The Lower Fungi

DIVISION I: MYXOMYCOTA—Produce a plasmodium or plasmodiumlike structure.

CLASS I: MYXOMYCETES (The slime molds)—Lack mycelium. Their body is a naked, amorphous plasmodium. They produce zoospores.

Order: Physarales—Saprophytic plasmodium that gives rise to crusty fructifications containing spores. They produce zoospores that have two flagella.

Genus: *Fuligo, Mucilago,* and *Physarum* cause slime molds on low-lying plants.

CLASS 2: PLASMODIOPHOROMYCETES

Order: Plasmodiophorales—Plasmodia produced within cells of roots and stems of plants. They produce zoospores that have two flagella.

Genus: *Plasmodiophora, P. brassicae* causing clubroot of crucifers.

Polymyxa, P. graminis parasitic in wheat and other cereals.

Spongospora, S. subterranea causing powdery scab of potato tubers.

DIVISION II: EUMYCOTA (eumycetes)—Produced mycelium, not plasmodium.

Subdivision 1: MASTIGOMYCOTINA—Produce zoospores.

CLASS I: CHYTRIDIOMYCETES—Have round or elongated mycelium that lacks cross walls.

Order: Chytridiales—Have cell wall but lack true mycelium, at most a rhizomycelium. Zoospores have a single flagellum.

Genus: *Olpidium, O. brassicae* being parasitic in roots of cabbage and other plants.

Physoderma, P. maydis causing brown spot of corn.

Synchytrium, S. endobioticum causing potato wart.

Urophlyctis, U. alfalfae causing crown wart of alfalfa.

CLASS 2: OOMYCETES (The water molds, white rusts, and downy mildews) — Have elongated mycelium. Produce zoospores in zoosporangia. Zoospores have two flagella. Sexual resting spores (oospores) produced by the union of morphologically different gametes.

Order: Saprolegniales—Have well-developed mycelium. Zoospores produced in long, cylindrical zoosporangia attached to mycelium. Oospores.

Genus: *Aphanomyces,* causing root rot of many vegetables.

Order: Peronosporales—Sporangia, usually zoosporangia produced along somatic hyphae or at tips of hyphae and set free. Oospores.

Family: Pythiaceae. Sporangia on somatic hyphae or on sporangiophores of indeterminate growth. Facultative parasites.

Genus: *Pythium,* causing damping off of seedlings, seed decay, root rots, and cottony blight of turf grasses.

Phytophthora, P. infestans causing late blight of potato, others causing mostly root rots.

Family: Albuginaceae (The white rusts). Sporangia borne in chains. Obligate parasites.

Genus: *Albugo, A. candida* causing white rust of crucifers.

Family: Peronosporaceae (The downy mildews)—Sporangia borne on sporangiophores of determinate growth. Sporangia wind-borne. Obligate parasites.

Genus: *Plasmopara, P. viticola* causing downy mildew of grape.

Peronospora, P. nicotianae causing downy mildew (blue mold) of tobacco.

Bremia, B. lactucae causing downy mildew of lettuce.

Sclerospora, S. graminicola causing downy mildew of grasses.

Pseudoperonospora, P. cubensis causing downy mildew of cucurbits.

Subdivision 2: ZYGOMYCOTINA

CLASS: ZYGOMYCETES (The bread molds)—Terrestrial fungi. Produce nonmotile asexual spores in sporangia. No zoospores. Their resting spore is a zygospore, produced by the fusion of two morphologically similar gametes.

Order: Mucorales—Nonmotile asexual spores formed in terminal sporangia.

Genus: *Rhizopus,* causing soft rot of fruits and vegetables.

Choanephora, C. cucurbitarum causing soft rot of squash.

Order: Endogonales—Mycorrhizal fungi. Produce spores singly in soil, or in sporocarps containing zygospores, chlamydospores, or sporangia.

Genus: *Endogone,* and others.

The Higher Fungi

Subdivision 3: ASCOMYCOTINA (ascomycetes, the sac fungi)—Produce sexual spores, called ascospores, generally in groups of eight within an ascus.

CLASS I: HEMIASCOMYCETES—Asci naked, no ascocarps produced.

Order: Endomycetales (the yeasts).

Genus: *Saccharomyces, S. cerevisiae,* the bread yeast.

Order: Taphrinales—Asci arising from binucleate ascogenous cells.

Genus: *Taphrina*—causing peach leaf curl, plum pocket, oak leaf blister, etc.

CLASS 2: PYRENOMYCETES (The perithecial fungi)—Asci in fruiting bodies completely closed (cleistothecia) or in fruiting bodies with an opening (perithecia).

Order: Erysiphales (The powdery mildews)—Mycelium and cleistothecia on surface of host plant.

Genus: *Erysiphe,* causing powdery mildew of grasses, cucurbits, etc.

Microsphaera, one species causing powdery mildew of lilac.

Podosphaera, P. leucotricha causing powdery mildew of apple.

Spaerotheca, S. pannosa causing powdery mildew of roses and peach.

Uncinula, U. necator causing powdery mildew of grape.

Order: Sphaeriales—Perthecia with dark-colored, usually firm walls.

Genus: *Botryosphaeria (Physalospora) B. obtusa,* causing black rot and frogeye leafspot of apple.

Ceratocystis, causing vascular wilts and sapwood stains of trees.

Diaporthe, causing bean pod blight, soybean stem canker, citrus melanose, and fruit rot of eggplant.

Endothia, E. parasitica causing chestnut blight.

Eutypa armeniacae, causing dieback in grape and apricot.

Glomerella, G. cingulata causing many anthracnose diseases, and bitter rot of apple.

Gnomonia, causing anthracnose or leaf spot diseases.

Hypoxylon, H. mammatum causing a severe canker of poplars.

Rosellinia, causing root diseases of fruit trees and vines.

Valsa, causing canker diseases of peach and other trees.

Xylaria, causing tree cankers and wood decay.

Order: Hypocreales—Perithecia light-colored, or red or blue.

Genus: *Claviceps, C. purpurea* causing ergot of rye.

Gibberella, causing foot or stalk rot of corn and small grains.

Nectria, causing twig and stem cankers of trees.

CLASS 3: LOCULOASCOMYCETES (The ascostromatic fungi)—Produce pseudothecia, i.e., peritheciumlike stromata with asci in separate or single large cavities.

Order: Myriangiales—Cavities arranged at various levels and containing single asci.

Genus: *Elsinoe,* causing anthracnose of grape and raspberry, and scab of citrus.

Order: Dothideales—Cavities arranged in a basal layer and containing many asci. Pseudothecia lack pseudoparaphyses.

Genus: *Capnodium,* one of many epiphytic genera causing the sooty molds.

Didymella (Mycosphaerella), causing leaf spots of many plants.

Guignardia, G. bidwellii causing black rot of grapes.

Microcyclus (Dothidella), M. ulei causing the leaf spot of rubber trees.

Plowrightia (Dibotryon), P. morbosum causing black knot of cherries and plums.

Order: Pleosporales—Cavities arranged in a basal layer and containing many asci. Pseudothecia have pseudoparaphyses.

Genus: *Cochliobolus, C. sativus* causing leaf spots and root rots on grain crops.

Gaeumannomyces (Ophiobolus), G. graminis causing the take-all disease of wheat.
Pyrenophora causing leaf spots of cereals and grasses.
Venturia, V. inaequalis causing apple scab.

CLASS 4: DISCOMYCETES (The cup fungi)—Asci produced at the surface of fleshy cup- or saucer-shaped apothecia.

Order: Phacidiales—Apothecia develop within a stroma.
Genus: *Rhytisma, R. acerinum* causing tar spot of maple leaves.

Order: Helotiales—Apothecia not in a stroma but sometimes developing from a stroma or sclerotium. Asci release spores through an apical, circular perforation.
Genus: *Diplocarpon, D. rosae* causing black spot of roses.
Higginsia (Coccomyces), H. hiemalis causing cherry leaf spot.
Lophodermium causing pine needle blight.
Monilinia, M. fructicola causing brown rot of stone fruits.
Pseudopeziza, P. trifolii (medicaginis) causing alfalfa leaf spot.
Sclerotinia, S. sclerotiorum causing watery soft rot of vegetables.

Subdivision 4: DEUTEROMYCOTINA—The imperfect fungi. Asexual fungi—Sexual reproduction and structures lacking or unknown.

CLASS 1: COELOMYCETES—Conidia borne in pycnidia or acervuli.

Order: Sphaeropsidales—Asexual spores produced in pycnidia.
Genus: *Ascochyta, A. pisi* causing pea blight.
Coniothyrium causing cane blight on raspberry.
Cytospora causing canker diseases on peach and other trees (its sexual stage is *Valsa*).
Diplodia, D. maydis causing stalk and ear rot of corn.
Phoma, P. lingam causing black leg of crucifers.
Phomopsis, causing blights and stem cankers of trees.
Phyllosticta, causing leaf spots of many plants.
Septoria, S. apii causing late blight of celery.

Order: Melanconiales—Asexual spores produced in acervulus.
Genus: *Colletotrichum,* causing anthracnose on many field crops.
Coryneum (Stigmina), C. beijerincki (S. carpophilum) causing blight on stone fruits.
Cylindrosporium, causing leaf spots on many kinds of plants.
Gloeosporium, similar if not identical to *Colletotrichum,* causing anthracnose on many plants.
Marssonina, causing leaf and twig blight of poplar, strawberry leaf scorch, and anthracnose of walnuts.
Melanconium, M. fuligenum causing bitter rot of grape.
Sphaceloma, causing anthracnose of grape, raspberry, and scab of citrus and avocado.

CLASS 2: HYPHOMYCETES.

Order: Hyphales (Moniliales)—Asexual spores produced on or within hyphae freely exposed to the air.
Genus: *Alternaria,* causing leaf spots and blights on many plants.
Asperigillus, causing rots of stored seeds.
Bipolaris, Drechslerea, Exserohilum (Helminthosporium), causing leaf spots, root rots, and blights of cereals and turf grasses.

Botrytis, B. cinerea causing gray mold and blights on many plants.
Cercospora, one species causing early blight of celery.
Fulvia (Cladosporium), F. fulva causing leaf-mold of tomato.
Fusarium, causing wilt and root rot diseases of many annual plants and cankers of forest trees.
Geotrichum, G. candidum causing sour rot of fruits and vegetables.
Graphium, G. ulmi causing Dutch elm disease (its sexual stage is *Ceratocystis*).
Penicillium, causing blue mold rot of fruits and other fleshy organs.
Phymatotrichum, P. omnivorum causing root rot of cotton and other plants.
Pyricularia, causing rice blast and gray leaf-spot of turf grasses.
Spilocaea (Fusicladium), causing apple scab (its sexual stage is *Venturia*).
Strumella, causing cankers on oak.
Thielaviopsis, T. basicola causing black root rot of tobacco.
Trichoderma, antagonsitic to many plant pathogenic fungi.
Verticillium, causing wilt of many annuals and perennials.

CLASS 3: AGONOMYCETES (Mycelia Sterilia).

Order: Agonomycetales (Myceliales)—No sexual or asexual spore forms common or known.

Genus: *Rhizoctonia,* causing root rots and crown rots of annuals and brown-patch of turf grasses (sexual stage *Thanatephorus*).
Sclerotium causing root and stem rots of many plants (sexual stage *Aethalia*).

Subdivision 5: BASIDIOMYCOTINA (Basidiomycetes, the club fungi) — Sexual spores, called basidiospores or sporidia, are produced externally on a one- or four-celled spore-producing structure called a basidium.

CLASS 1: HEMIBASIDIOMYCETES (TELIOMYCETIDAE) (The rust and smut fungi)—Basidium with cross walls or being the promycelium of a teliospore. Teliospores single or united into crusts or columns, remaining in host tissue or bursting through the epidermis.

Order: Ustilaginales—Fertilization by means of union of compatible spores, hyphae, etc. Only teliospores and basidiospores (sporidia) are produced.

Genus: *Sphacelotheca,* several species causing loose smut of sorghum.
Tilletia, several species causing bunt, or stinking smut, of wheat.
Urocystis, U. cepulae causing smut of onion.
Ustilago, causing smut of corn, wheat, barley, etc.

Order: Uredinales—Sperm cells called spermatia fertilize special receptive hyphae in spermogonia. Produce aeciospores, uredospores (repeating spores), teliospores, and basidiospores.

Genus: *Cronartium,* several species causing stem rusts of pines.
Gymnosporangium, G. juniperi-virginianae causing cedar apple rust.

Melampsora, M. lini causing rust of flax.
Phragmidium, one species causing rust of roses.
Puccinia, several species causing rust of cereals and of other plants.
Uromyces, U. appendiculatus causing rust of beans.

CLASS 2: HYMENOMYCETES (The wood decay and root rot fungi)—Basidium without cross walls. Basidiocarp lacking or present. Include the mushrooms, shelf fungi, puff balls. Basidia produced in definite layers (hymenia) and becoming exposed to the air before the spores are shot off from the sterigmata.

Order: Exobasidiales—Basidiocarp lacking: basidia produced on surface of parasitized tissue.
Genus: *Exobasidium,* causing leaf, flower, and stem galls on ornamentals.

Order: Aphyllochorales (Polyporales)—Hymenium lining the surfaces of small pores or tubes.
Genus: *Aethalia (Sclerotium),* causing root and stem rots of many plants.
Corticium, one species causing the red thread disease of turf grasses.
Heterobasidium (Fomes), causing heart rot of many trees.
Lenzites, causing brown rot of conifers and decay of wood products.
Peniophora, causing decay of conifer logs and pulpwood.
Polyporus, causing root and stem rot of many trees.
Poria, causing wood and root rots of forest trees.
Schizophyllum, causing white rot in deciduous forest trees.
Stereum, causing wood decay and silver leaf disease of trees.

Order: Tulasnellales—Saprobic, mycorrhizal, or parasitic. Basidiocarps weblike, inconspicuous, often waxy in appearance.
Genus: *Thanatephorus (Rhizoctonia),* causing root and stem rots of many annual plants and brown patch of turf grasses.
Typhula, one species causing snowmold or blight of turf grasses.

Order: Agaricales—Hymenium on radiating gills or lamellae. Many are mycorrhizal fungi.
Genus: *Armillaria, A. mellea* causing root rots of forest and fruit trees.
Marasmius, causing the fairy ring disease of turf grasses.
Pholiota, causing brown wood rot in deciduous forest trees.
Pleurotus, causing white rot on many deciduous forest trees.

Identification

Because each fungal disease of plants is usually caused by only one fungus, and because there are more than 100,000 different species of fungi, the identification of the fungus species on a diseased plant specimen or culture of a fungus means that all but one of all the known fungus species must be eliminated as being the fungus in question.

The most significant fungus characteristics used for identification are spores and fructifications, or spore-bearing structures, and to some extent the characteristics of the fungus body (plasmodium or mycelium). These items are examined under the compound microscope directly after removal from the specimen. The specimen is often kept moist for a few days to promote fructification development, or the fungus may be isolated and grown on artificial media, and identification is made on the basis of the fructifications produced on the media. For some fungi, special nutrient media have been developed that allow selective growth only of the particular fungus, allowing quick identification of the fungus.

The shape, size, color, and manner of arrangement of spores on the sporophores or the fruiting bodies, as well as the shape and color of the sporophores or fruiting bodies, are sufficient characteristics to suggest, to one somewhat experienced in the taxonomy of fungi, the class, order, family, and genus to which the particular fungus belongs. In any case, these characteristics can be utilized to trace the fungus through published analytical keys of the fungi to the genus and, finally, to the species to which it belongs. Once the genus of the fungus has been determined, specific descriptions of the species are found in monographs of genera or in specific publications in research journals.

Since there are usually lists of the pathogens affecting a particular host plant, one may use such host indexes as short cuts in quickly finding names of fungus species that might apply to the fungus at hand. Host indexes, however, merely offer suggestions in determining identities, which must ultimately be determined by reference to monographs and other more specific publications.

Symptoms Caused by Fungi on Plants

Fungi cause local or general symptoms on their hosts, and such symptoms may occur separately on different hosts, concurrently on the same host, or follow one another on the same host. In general fungi cause local or general necrosis or killing of plant tissues, hypotrophy and hypoplasia (stunting) of plant organs or entire plants, and hyperplasia (excessive growth) of plant parts or whole plants.

The most common necrotic symptoms are:

Leaf spots—Localized lesions on host leaves consisting of dead and collapsed cells.

Blight—General and extremely rapid browning of leaves, branches, twigs, and floral organs resulting in their death.

Canker—A localized wound or necrotic lesion, often sunken beneath the surface of the stem of a woody plant.

Dieback—Extensive necrosis of twigs beginning at their tips and advancing toward their bases.

Root rot—Disintegration or decay of part or all of the root system of a plant.

Damping off—The rapid death and collapse of very young seedlings in the seed bed or field.

Basal stem rot—Disintegration of the lower part of the stem.

Soft rots and dry rots—Maceration and disintegration of fruits, roots, bulbs, tubers, and fleshy leaves.

Anthracnose—A necrotic and sunken ulcerlike lesion on the stem, leaf, fruit, or flower of the host plant.

Scab—Localized lesions on host fruit, leaves, tubers, etc., usually slightly raised or sunken and cracked, giving a scabby appearance.

Decline—Plants growing poorly; leaves small, brittle, yellowish, or red; some defoliation and dieback present.

Almost all of the above symptoms may also cause pronounced stunting of the infected plants. In addition, certain other symptoms such as leaf rust, mildews, wilts, and even certain diseases causing hyperplasia of some plant organs, such as clubroot, may cause stunting of the plant as a whole.

Symptoms associated with hypertrophy or hyperplasia and distortion of plant parts include:

Clubroot—Enlarged roots appearing like spindles or clubs.
Galls—Enlarged portions of plants usually filled with fungus mycelium.
Warts—Wartlike protuberances on tubers and stems.
Witches'-brooms—Profuse, upward branching of twigs.
Leaf curls—Distortion, thickening, and curling of leaves.

In addition to the above, three goups of symptoms may be added:

Wilt—Usually a generalizd secondary symptom in which leaves or shoots lose their turgidity and droop because of a disturbance in the vascular system of the root or of the stem.
Rust—Many, small lesions on leaves or stems, usually of a rusty color.
Mildew—Chlorotic or necrotic areas on leaves, stems, and fruit, usually covered with mycelium and the fructifications of the fungus.

In many diseases, the pathogen grows or produces various structures on the surface of the host. These structures, which include mycelium, sclerotia, sporophores, fruiting bodies, and spores, are called **signs** and are distinct from **symptoms,** which refer only to the appearance of infected plants or plant tissues. Thus, in the mildews, for example, one sees mostly the signs consisting of a whitish, downy growth of fungus mycelium and spores on the plant leaves, fruit, or stem, while the symptoms consist of chlorotic or necrotic lesions on leaves, fruit, and stem, reduced growth of the plant, and so on.

Isolation of Fungi (and Bacteria)

Most plant diseases can be diagnosed by observation with the naked eye or with the microscope, and for these, isolation of the pathogen is not necessary. There are many fungal and bacterial diseases, though, in which the pathogen cannot be identified because it is mixed with one or more contaminants, because it has not yet produced its characteristic fruiting structures and spores, because the same disease could be caused by more than one similar-looking pathogen and perhaps by some environmental factor, or because the disease is caused by a new, previously unknown pathogen that must be isolated and studied. Just as often, pathogens of even known diseases must be isolated from diseased plant tissues whenever a study of the characteristics and habits of these pathogens is to be undertaken.

Preparing for Isolation

Even before attempting to isolate the fungus or bacterium pathogen from a diseased plant tissue, one must perform several preliminary operations, including the following.

1. Sterilize glassware, such as Petri dishes, test tubes, and pipettes, by dry heat (150–160°C for 1 hour or more) or autoclaving or by dipping for 1 minute or

more in a potassium dichromate–sulfuric acid solution, or in 5 percent formalin or 95 percent ethyl alcohol. All chemically treated glassware should be rinsed through at least three changes of sterile (boiled or autoclaved) water.

2. Prepare solutions for treating the surface of the infected or infested tissue to eliminate or markedly reduce surface contaminants that could interfere with the isolation of the pathogen. These solutions can be used either as a surface wipe or as a dip. The most commonly used surface sterilants are: 0.525 percent sodium hypochlorite (1 part Clorox to 9 parts water) solution, used both for wiping infected tissues or dipping sections of such tissues in it and for wiping down table or bench surfaces before making isolations; and 95 percent ethyl alcohol, which is mild and is used for leaf dips for 3 seconds or more. The tissues must be blotted dry with a sterile paper towel.
3. Prepare culture media on which the isolated fungal or bacterial pathogens will grow. An almost infinite number of culture media can be used to grow plant pathogenic fungi and bacteria. Some of them are entirely synthetic—made up of known amounts of certain chemical compounds—and are usually quite specific for certain pathogens. Some are liquid or semiliquid and are used primarily for growth of bacteria but also of fungi in certain cases. Most media contain an extract of a natural source of carbohydrates and other nutrients, such as potato, corn meal, lima bean, or malt extract, to which variable amounts of agar are added to solidify the medium and form a gel on or in which the pathogen can grow and be observed. The most commonly used media are potato dextrose agar (PDA), which is good for most, but not all fungi, water agar or glucose agar (1–3 percent glucose in water agar) for separating some fungi (*Pythium* and *Fusarium*) from bacteria, and nutrient agar, which contains beef extract and peptone and is good for isolating bacterial plant pathogens. Fungi can also be separated in culture from bacteria by adding 1 or 2 drops of a 25 percent solution of lactic acid, which inhibits growth of bacteria, to 10 ml of the medium before pouring it in the plate. Solutions of culture media are prepared in flasks, which are then plugged and placed in an autoclave at 120°C and 15 pounds pressure for 20 minutes (Figure 11-3). The sterilized media are then allowed to cool somewhat and are subsequently poured from the flask into sterilized Petri dishes, test tubes, or other appropriate containers. If agar was added to the medium, the latter will soon solidify and is then ready to be used for growth of the fungus or bacterium. The pouring of the culture medium into Petri dishes, tubes, and so on is carried out as aseptically as possible either in a separate culture room or in a clean room free from drafts and dust. In either case, the work table should be wiped with a 10 percent Clorox solution, hands should be clean, and tools such as scalpels, forceps, and needles should be dipped in alcohol and flamed to prevent introduction of contaminating microorganisms.

Although most fungi and most bacteria can be cultured on nutrient media with ease, some of them have specific and exacting requirements and will not grow on most commonly used nutrient media. Some groups of fungi, namely the Erysiphales, the causes of the powdery mildew diseases, and the Peronosporales, which cause the downy mildews, are considered strictly obligate parasites and cannot be grown on culture media. Another group of fungi, the Uredinales, which cause the rust diseases of plants, were, until recently, also thought to be strictly obligate parasites. In the last several years, however, it has become possible to grow in culture some stages of some rust fungi by adding certain components to the media, and so the rust fungi are no longer considered to be obligate parasites, although they are, of course, obligate parasites in nature. The fastidious phloem- and xylem-limited bacteria also are either so far impossible to grow in culture or are grown on special complex

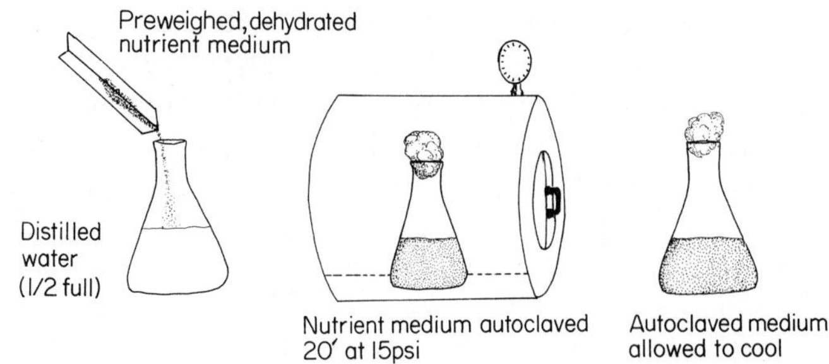

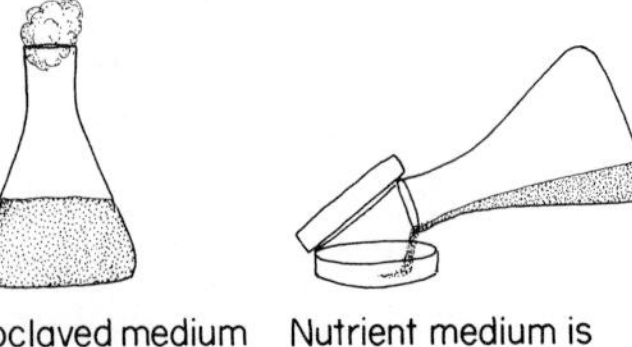

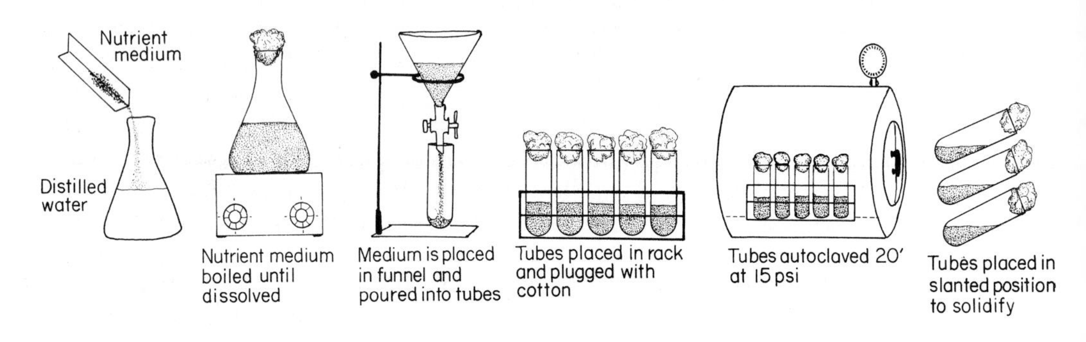

FIGURE 11-3 Preparation of solid nutrient media in plates (petri dishes) and in test tube slants.

nutrient media. Of the other pathogens, only some spiroplasmas have been grown in culture. None of the other 100 or so mycoplasmalike organisms and none of the viruses, nematodes, or protozoa has been grown on nutrient culture media, so far, although it is expected that media for culturing mycoplasmalike organisms will soon be discovered.

Isolating the Pathogen

From Leaves

If the infection of the leaf is still in progress in the form of a fungal leaf spot or blight and if there are spores present on the surface, a few spores may be shaken loose over a petri plate containing culture medium or picked up at the point of a sterile needle or scalpel and placed on the surface of the culture medium. If the fungus does grow in culture, isolated colonies of mycelium will appear in a few days as a result of germination of the spores. These colonies can be subcultured on separate plates and thus secure some plates that will contain the pathogen free of any contaminants.

Sometimes, isolation of the pathogen from fungal or bacterial leaf spots and blights is made by surface sterilizing the area to be cut with Clorox solution, removing a small part of the infected tissue with a sterile scalpel, and placing it in a plate containing a nutrient medium.

The most common method, however, for isolating pathogens from infected leaves as well as other plant parts is the one in which several small sections 5- to 10-mm square are cut from the margin of the infected lesion to contain both diseased and healthy-looking tissue (Figure 11-4). These are placed in one of the surface sterilant solutions, making sure that the surfaces do get wet, and after about 15 to 30 seconds the sections are taken out aseptically one by one and at regular, for example, 10- to 15-second intervals, so that each of them has been surface-sterilized for different times. The sections are then blotted dry on clean, sterile paper towels or are washed in three changes of sterile water and are finally placed on the nutrient medium, usually three to five per dish. Those sections surface-sterilized the shortest time usually contain contaminants along with the pathogen, while those surface-sterilized the longest produce no growth at all because all organisms have been killed by the surface sterilant. Some of the sections left in the surface sterilant for intermediate periods of time, however, will allow only the pathogen to grow in culture in pure colonies, since the sterilant was allowed to act long enough to kill all surface contaminants but not long enough to kill the pathogen that was advancing alone from the diseased to the healthy tissue. These colonies of the pathogen are then subcultured aseptically for further study.

If fruiting structures (pycnidia, perithecia) are present on the leaf, it is sometimes possible to pick them out, drop them in the surface sterilant for a few seconds, and then plate them on the nutrient medium. This procedure, however, requires that most of the work be done under the stereoscopic microscope (binoculars) since the fruiting structures are generally too small to see and to handle with the naked eye. Fruiting structures, after surface

FIGURE 11-4. Isolation of fungal pathogens from infected plant tissue.

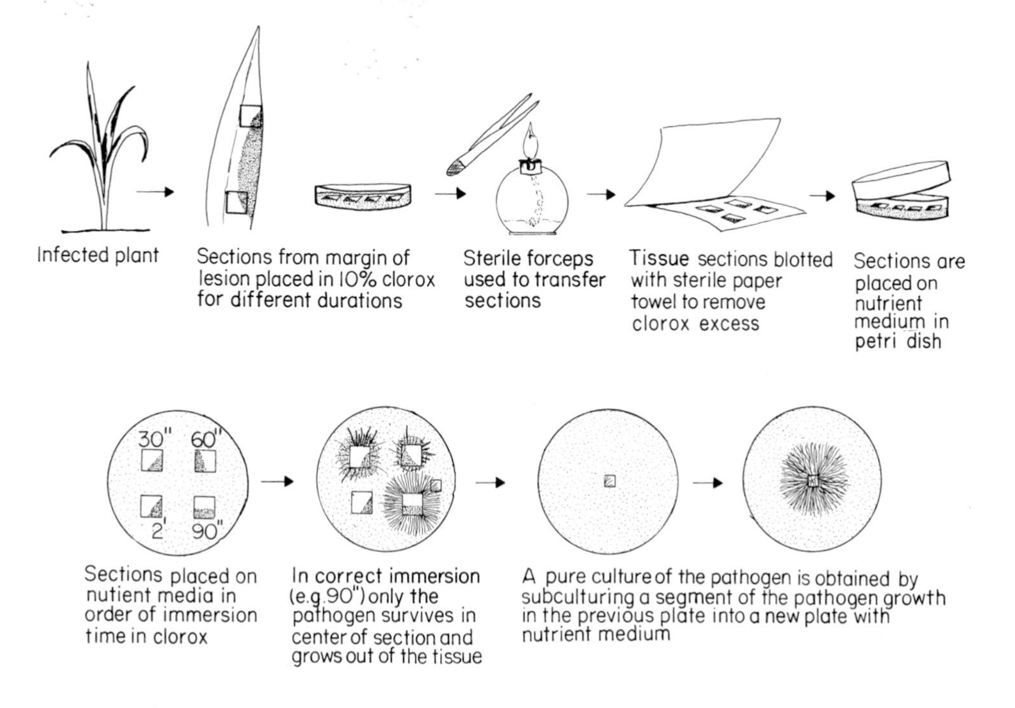

sterilization, may also be crushed in a small drop of sterile water, and then the spores in the water may be diluted serially in small tubes or dishes containing sterile water. Finally, a few drops from each tube of the serial dilution are placed on a nutrient medium, and single colonies free of contaminants develop from germinating spores obtained from some of the serial dilution tubes.

The serial dilution method is often used to isolate pathogenic bacteria from diseased tissues contaminated with other bacteria. After surface sterilization of sections of diseased tissues from the margin of the infection, the sections are ground aseptically but quite thoroughly in a small volume of sterile water, and then part of this homogenate is diluted serially in equal volumes or ten times the volume of the initial water. Finally, plates containing nutrient agar are streaked with a needle or loop dipped in each of the different serial dilutions, and single colonies of the pathogenic bacterium are obtained from the higher dilutions that still contain bacteria.

From Stems, Fruits, and Other Aerial Plant Parts

Almost all the methods described for isolating fungal and bacterial pathogens from leaves can also be used to isolate these pathogens from superficial infections of the stems, fruits, and other aerial plant parts. In addition to these methods, however, pathogens can often be isolated easily from infected stems and fruits in which the pathogen has penetrated fairly deeply by splitting the stem or breaking the fruit from the healthy side first and then tearing it apart toward and past the infected margin, thus exposing tissues not previously exposed to contaminants and not touched by hand or knife and therefore not contaminated. Small sections of tissue can be cut from the freshly exposed area of the advancing margin of the infection with a flamed scalpel and can be plated directly on the culture medium.

From Roots, Tubers, Fleshy Roots, and Vegetable Fruits in Contact With Soil

Isolating pathogens from any diseased plant tissue in contact with soil presents the additional problems of numerous saprophytic organisms invading the plant tissue after it has been killed by the pathogen. For this reason, repeated, thorough washing of such diseased tissues to remove all soil and most of the loose, decayed plant tissue, in which most of the saprophytes are present, is the first step in isolating the pathogen. If the infected root is small, once it is washed thoroughly pathogens can be isolated from it by following one of the methods described for isolating pathogens from leaves. If isolation is attempted from fleshy roots or other fleshy tissues and penetration of the pathogen is slight resulting only in surface lesions, the tissue is washed free from adhering soil, and several bits of tissue from the margin of the lesions are placed in Clorox solution. The tissue sections are picked from the solution one by one, blotted or washed in sterile water, and placed on agar in petri plates. If the pathogen has penetrated deeply into the fleshy tissue, the method described above for stems and fruit, that is, by breaking the specimens from the healthy side first, then tearing toward the infected area and plating bits taken from the previously unexposed margin of the rot, can be used most effectively.

Life Cycles of Fungi

Although the life cycles of the fungi of the different groups vary greatly, the great majority of them go through a series of steps that are quite similar (Figure 11-5). Thus almost all fungi have a spore stage with a simple, haploid (possessing one set of chromosomes or $1N$) nucleus. The spore germinates into a hypha, which also contains haploid nuclei. The hypha may either produce simple, haploid spores again (as is always the case in the Imperfect Fungi) or it may fuse with another hypha to produce a fertilized hypha in which the nuclei unite to form one diploid nucleus, called a zygote (containing two sets of chromosomes, or $2N$). In the Oomycetes the zygote will divide to produce simple, haploid spores, which close the cycle. In a brief phase of most Ascomycetes, and generally in the Basidiomycetes, the two nuclei of the fertilized hypha do not unite but remain separate within the cell in pairs (dikaryotic or $N + N$) and divide simultaneously to produce more hyphal cells with pairs of nuclei. In the Ascomycetes, the dikaryotic hyphae are found only inside the fruiting body, in which they become the ascogenous hyphae, since the two nuclei of one cell of each hypha unite into a

FIGURE 11-5 Schematic presentation of the generalized life cycles of the main groups of phytopathogenic fungi.

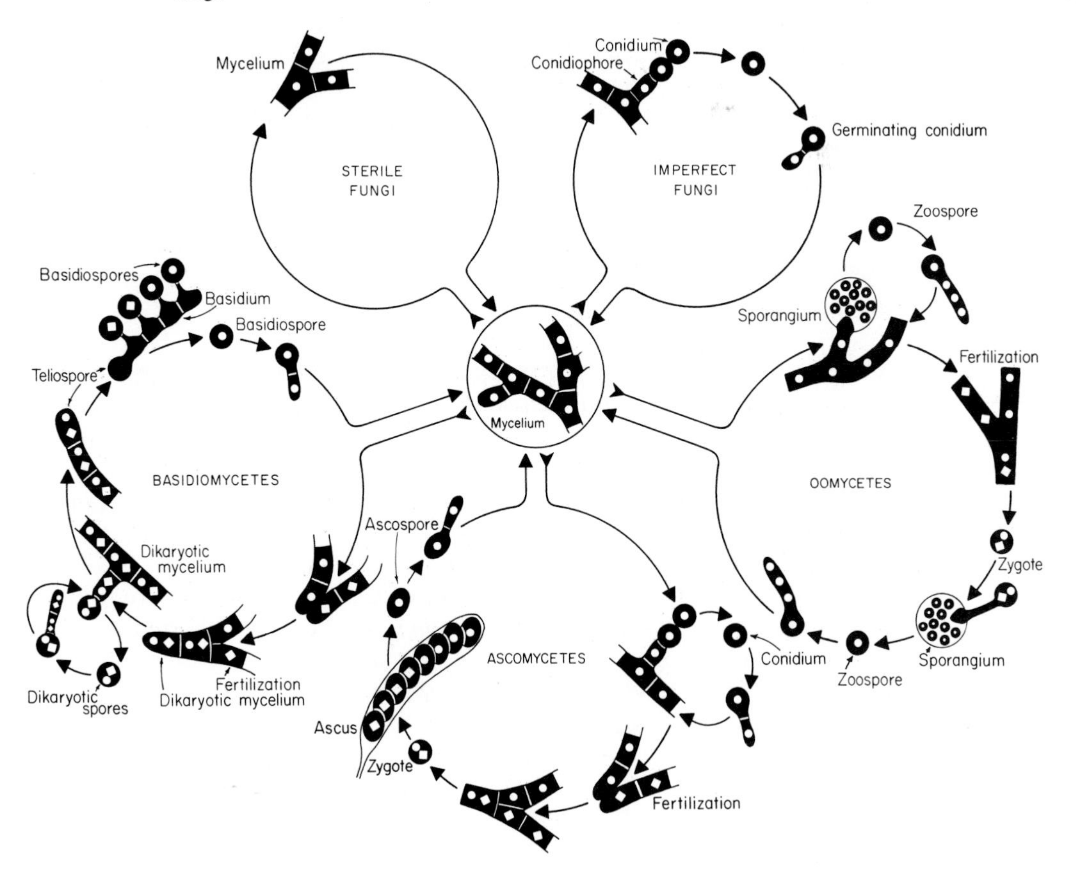

zygote ($2N$), which divides meiotically to produce ascospores that contain haploid nuclei.

In the Basidiomycetes haploid spores produce only short haploid hyphae. Upon fertilization, dikaryotic ($N + N$) mycelium is produced, and this develops into the main body of the fungus. Such dikaryotic hyphae may produce, asexually, dikaryotic spores that will grow again into a dikaryotic mycelium. Finally, however, the paired nuclei of the cells unite and form zygotes. The zygotes divide meiotically and produce basidiospores that contain haploid nuclei.

In the Imperfect Fungi, of course, only the asexual cycle (haploid spore → haploid mycelium → haploid spore) is found. Even in the other fungi, however, a similar asexual cycle is the most common one by far, since it can be replaced many times during each growth season. The sexual cycle usually occurs only once a year.

Control of Fungal Diseases of Plants

The endless variety and the complexity of the many fungus diseases of plants have led to the development of a correspondingly large number of approaches for their control. The particular characteristics in the life cycle of each fungus, its habitat preferences, and its performance under certain environmental conditions are some of the most important points to be considered in attempting to control a plant disease caused by a fungus. Although some diseases can be controlled completely by just one type of control measure, a combination of measures is usually necessary for satisfactory control of most diseases.

The most common control measures include: The use of pathogen-free seed or propagating stock; destruction of plant parts or refuse harboring the pathogen; destruction of volunteer plants or alternate hosts of the pathogen; use of clean tools and containers; proper drainage of fields and aeration of plants; crop rotation; and the use of resistant plant varieties.

The most effective method, however, and sometimes the only one available for controlling most of the fungus diseases of plants, is the application of chemical sprays or dusts on the plants, their seeds, or into the soil where the plants are to be grown. Soil-inhabiting fungi may be controlled by steam or electric heat and by volatile liquids.

In some diseases the fungus is carried in the seed, and control can be obtained only through treatment of the seed with systemic fungicides or hot water. In others, control of the insect vectors may be the only available possibility. In general, great advances have been made toward controlling fungus diseases of plants, especially through resistant varieties and through chemicals, and as a result, these diseases are probably much easier to control than any other group of plant diseases, although the losses caused by fungus diseases of plants are still very great.

SELECTED REFERENCES

Ainsworth, G. C., Sparrow, F. K., and Sussman, A. S., eds. (1965–1973). "The Fungi: An Advanced Treatise," Vols. 1–4. Academic Press, New York.

Alexopoulos, C. J., and Mims, C. W., (1979). "Introductory Mycology." 3rd ed. Wiley, New York.
Barnett, H. L., and Binder, F. L., (1973). The fungal host-parasite relationship. *Annu. Rev. Phytopathol.* **11**, 273–292.
Barnett H. L., and Hunter, B. B., (1972). "Illustrated Genera of Imperfect Fungi." Burgess, Minneapolis, Minnesota.
Buczacki, S. T., ed. (1983). "Zoosporic Plant Pathogens: A Modern Perspective." Academic Press, London.
Clements, F. E., and Shear, C. L., (1957). "The Genera of Fungi." Hafner, New York.
Cole, G. T., and Kendrick, B. (1981). "Biology of Conidial Fungi," 2 vols. Academic Press, New York.
Cummins, G. B., (1959). "Illustrated Genera of Rust Fungi." Burgess, Minneapolis, Minnesota.
Fergus, C. L. (1960). "Illustrated Genera of Wood Destroying Fungi." Burgess, Minneapolis, Minnesota.
Garraway, M. O., and Evans, R. C. (1984). "Fungal Nutrition and Physiology." Wiley, New York.
Griffin, D. M. (1969). Soil water in the ecology of fungi. *Annu. Rev. Phytopathol.* **7**, 289–310.
Kendrick, B. ed. (1971). "Taxonomy of Fungi Imperfecti." Toronto Univ. Press, Toronto.
Meredith, D. S. (1973). Significance of spore release and dispersal mechanisms in plant disease epidemiology. *Annu. Rev. Phytophathol.* **11**, 313–342.
Moore-Landecker, E. (1982). "Fundamentals of the Fungi," 2nd ed. Prentice-Hall, Englewood Cliffs, New Jersey.
Scott, K. J., and Chakravorty, A. K., eds. (1982). "The Rust Fungi." Academic Press, London.
Shoemaker, R. A. (1981). Changes in taxonomy and nomenclature of important genera of plant pathogens. *Annu. Rev. Phytopathol.* **19**, 297–307.
Stevens, F. L. (1913). "The Fungi Which Cause Plant Disease." Macmillan, New York.
Subramanian, C. V. (1983). "Hyphomycetes; Taxonomy and Biology." Academic Press, New York.
Tsao, P. H. (1970). Selective media for isolation of pathogenic fungi. *Annu. Rev. Phytopathol.* **8**, 157–186.

DISEASES CAUSED BY THE LOWER FUNGI

Diseases Caused by Myxomycetes

Myxomycetes are fungi whose vegetative body is a plasmodium, that is, an amoeboid mass of protoplasm that has many nuclei and no definite cell wall. In the true Myxomycetes, also called slime molds, the plasmodium does not invade plant cells, and at a certain point of its life cycle, it is used up to form superficial fructifications that contain resting spores (Figure 11-6). The slime molds produce zoospores that have two flagella (Figure 11-7).

Only one group of Myxomycetes causes diseases in plants by simply growing externally on the surface of leaves, stems, and fruits without parasitizing the plant. The order Physarales includes the true slime mold genera *Fuligo, Mucilago,* and *Physarum,* which cause slime molds on plants growing low on the ground, such as turf grasses, strawberries, vegetables, and small ornamentals (Figure 11-6). They are most common in warm weather after heavy rains or watering. All aboveground parts of plants in some areas, and even the soil between plants, may be covered by a creamy white or colored slimy growth, which later changes to distinct, crusty, ash-gray, or colored fruiting structures. The latter give the affected plants a dull gray appearance.

Slime molds are saprophytic members of the true slime molds (Myxomycetes). Their plasmodium creeps like an amoeba and feeds on decaying

FIGURE 11-6 (A) Begonia leaf covered with fructifications (sporangia) of a slime mold. (B) Slime mold fructifications on the blades of a turfgrass.

organic matter and microorganisms such as bacteria, which it simply engulfs and digests. There are many species of slime mold fungi, the most common of which are *Physarum, Fuligo,* and *Mucilago.*

The plasmodium grows mostly in the upper layer of the soil and in the thatch, but during or after warm wet weather it comes to the soil surface and creeps over low-lying vegetation. On the plant surface, which these fungi use merely as a means of support, the plasmodium produces the crusty fruiting structures, which vary in size, shape, and color depending on the species of slime mold (Figures 11-6, 11-8, and 11-9). The fruiting structures are sporangia, that is, containers filled with dark masses of powdery spores, and are easily rubbed off the plant. The spores are spread by wind, water, mowers, or other equipment and can survive unfavorable weather. In cool, humid weather, the spores absorb water, their cell wall cracks open, and a single, naked, motile swarm spore emerges from each. The swarm spores feed like the plasmodium while they undergo several divisions and various changes. Finally they unite in pairs to form amoeboid zygotes, which enlarge, become multinucleate, and become the plasmodium.

No control is usually considered necessary against slime molds. When they become too numerous and unsightly, breaking up of the spore masses by raking, brushing, or hosing down with water in dry weather, and removal of affected leaves or mowing of grass corrects the problem. Slime mold fungi are generally quite sensitive to many fungicides, so if the problem becomes serious, spraying with any fungicide, such as captan or thiram, used to control other diseases of the particular plant should also control the slime molds.

SELECTED REFERENCES

Alexopoulos, C. J., and Mims, C. W. (1979). "Introductory Mycology," 3rd ed. Wiley, New York.

Couch, H. B. (1973). "Diseases of Turfgrasses," 2nd ed. Krieger Pub. Co., New York.

Diseases Caused by Plasmodiophoromycetes

Three common and often severe diseases of plants are caused by plasmodiophoromycetes. The following fungi are involved:

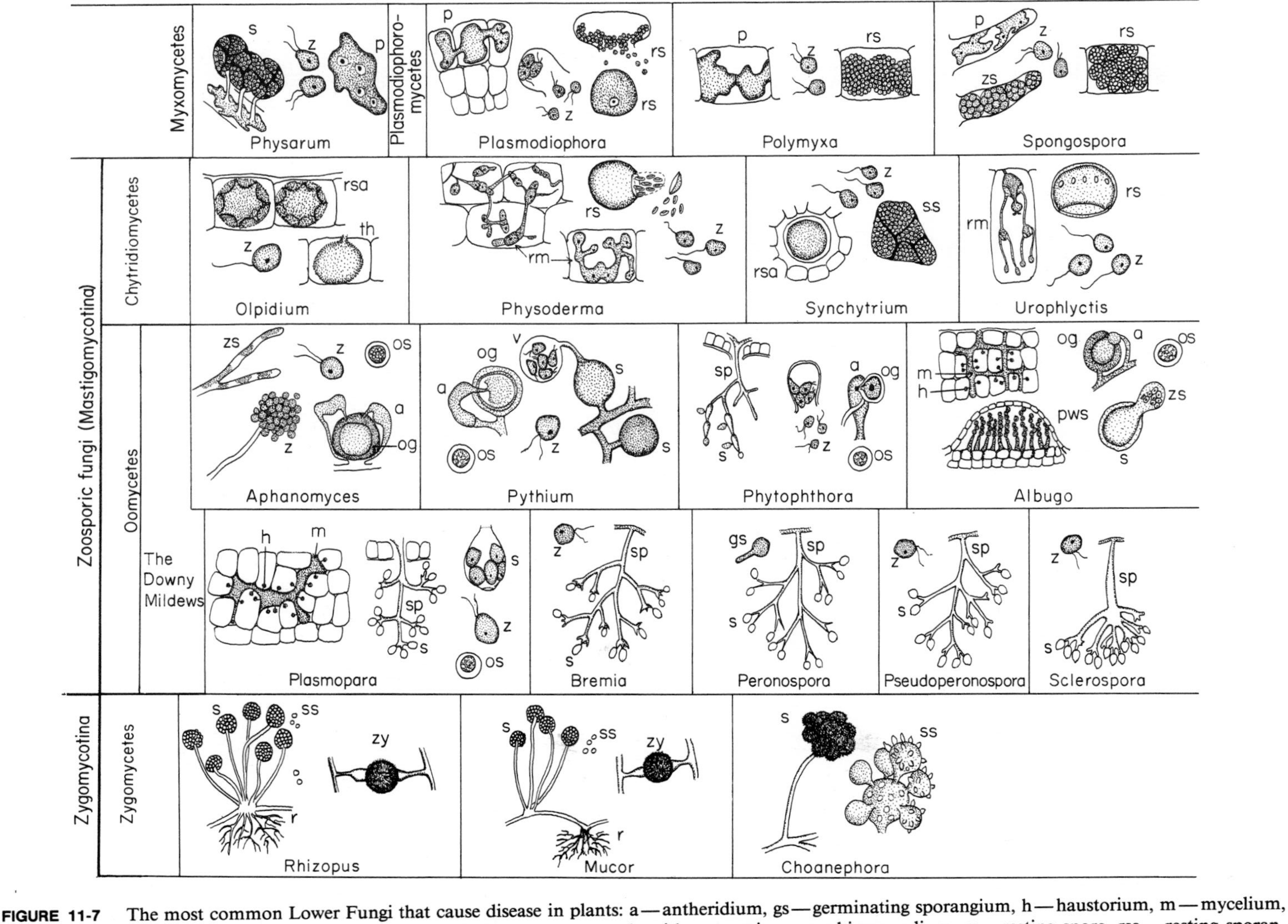

FIGURE 11-7 The most common Lower Fungi that cause disease in plants: a—antheridium, gs—germinating sporangium, h—haustorium, m—mycelium, og—oogonium, os—oospore, p—plasmodium, pws—pustule with sporangia, rm—rhizomycelium, rs—resting spore, rsa—resting sporangium, s—sporangium, sp—sporangiophore, ss—sporangiospore, th—thallus, z—zoospore, zs—zoosporangium, zy—zygospore.

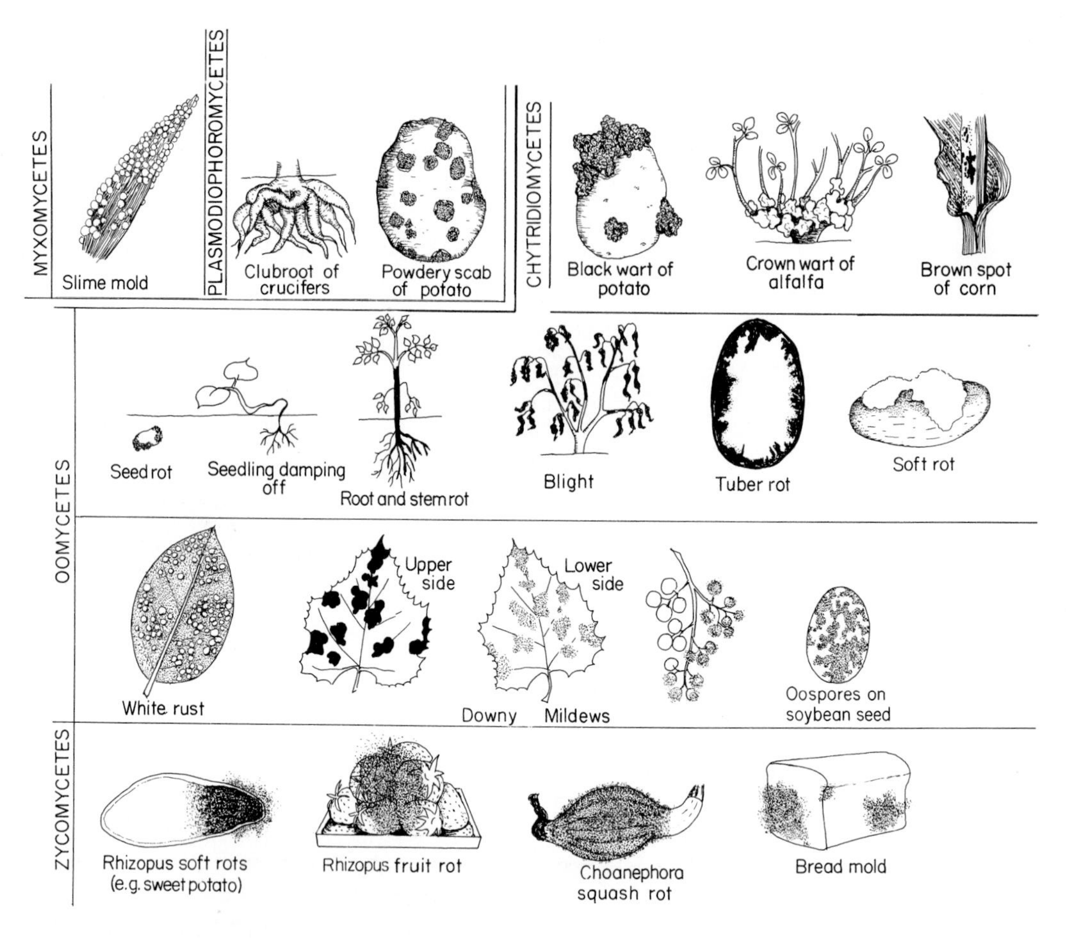

FIGURE 11-8 The most common symptoms caused by the Lower Fungi.

Plasmodiophora, causing clubroot of crucifers
Polymyxa, causing a root disease of cereals and grasses
Spongospora, causing the powdery scab of potato (Figure 11-10)

These fungi are widespread in soils in which they overwinter as resting spores. When the temperature is favorable and moisture is plentiful, the resting spore produces one zoospore, which infects a root hair and produces a plasmodium. The latter is transformed into zoosporangia, which produce numerous secondary zoospores that, probably after pairing, enter root or tuber tissues, produce plasmodium, and cause the typical disease. The plasmodium spreads into the host tissues and is finally transformed into overwintering resting spores.

The pathogens are obligate parasites, and although they can survive in the soil as resting spores for many years, they can only grow and multiply in a rather limited number of hosts. The plasmodium lives off the host cells it invades and it does not kill these cells. On the contrary, in some diseases many invaded and adjacent cells are stimulated by the pathogen to enlarge and divide, thus making available more nutrients for the pathogen. These pathogens spread from plant to plant by means of zoospores and by anything that moves soil or water containing spores, by infected transplants, and so on.

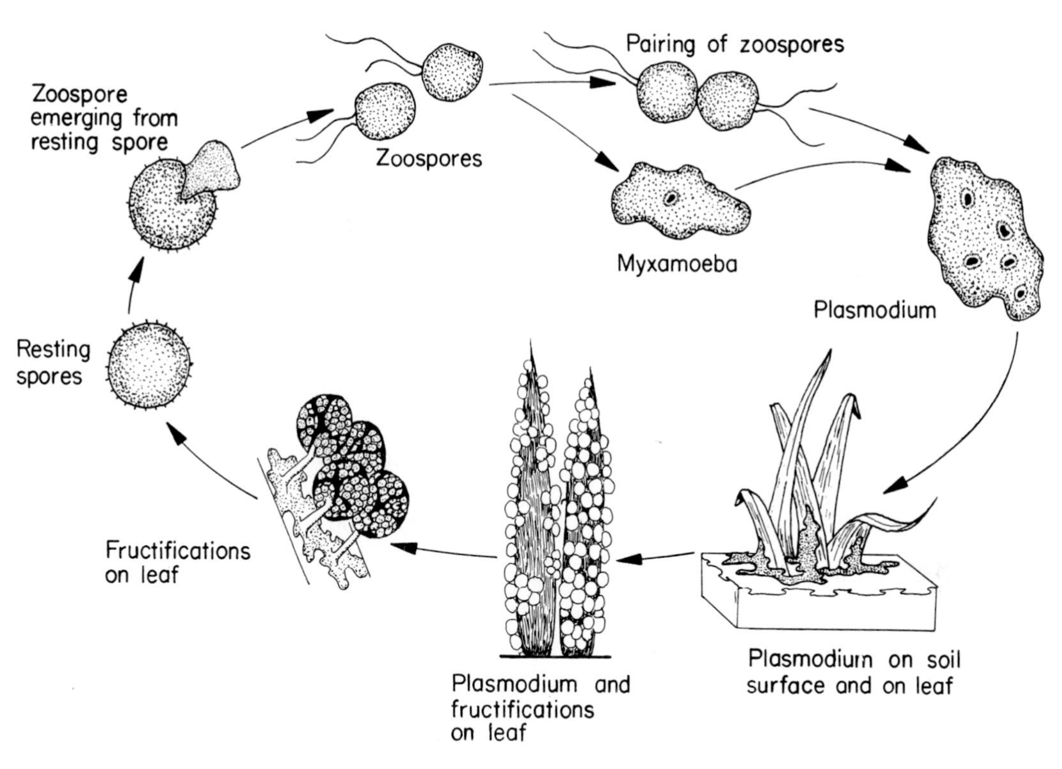

FIGURE 11-9 Life cycle of slime mold fungi.

Control of these pathogens is difficult and depends mostly on avoiding contamination of pathogen-free soils, use of healthy transplants, tubers, and crop rotation with nonhost plants, and adjustment of soil pH.

Polymyxa and *Spongospora,* in addition to the diseases they cause, can also transmit destructive plant viruses. *Polymyxa graminis* is a vector of several viruses, including soilborne wheat mosaic virus, barley yellow mosaic,

FIGURE 11-10 Mature stage of powdery scab of potato caused by *Spongospora subterranea.*

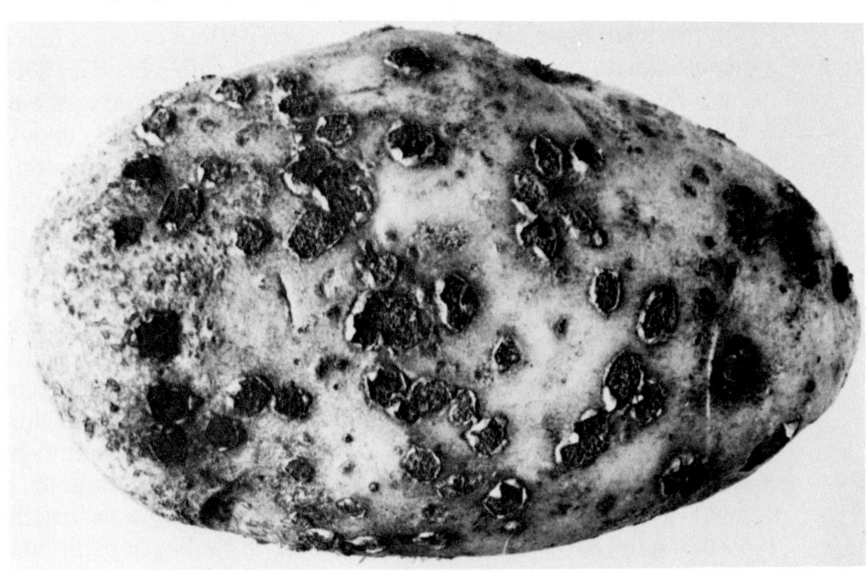

oat mosaic, wheat spindle streak, and peanut clump viruses, while *P. betae* is a vector of beet necrotic yellow vein virus. *Spongospora* is a vector of the potato mop-top virus.

Clubroot of Crucifers

The clubroot disease of cruciferous plants, such as cabbage and cauliflower, is widely distributed all over the world, found wherever plants of the mustard family grow. It has been observed most frequently in Europe and North America.

Clubroot causes serious losses when susceptible varieties of any cruciferous species are grown in infested fields. Fields once infested with the clubroot pathogen remain so indefinitely and become unfit for cultivation of crucifers.

- *Symptoms.* Infected plants may have pale green to yellowish leaves, which may show flagging and wilting in the middle of hot, sunny days but may recover during the night (Figure 11-11A). Affected plants show almost normal vigor at first, but then gradually become stunted. Young plants may be killed by the disease within a short time after infection, while older plants may remain alive but fail to produce marketable heads.

  The most characteristic symptoms of the disease appear on the roots and sometimes the underground part of the stem (Figure 11-11B). The symptoms consist of small or large spindlelike, spherical, knobby, or club-shaped swellings on the roots and rootlets. These malformations may be isolated and cover only part of some roots or they may coalesce and cover the entire root system of the plant. The older and usually the larger clubbed roots disintegrate before the end of the season due to invasion by bacteria and other weakly parasitic soil microorganisms.

- *The Pathogen: Plasmodiophora brassicae. Plasmodiophora brassicae's* body is a plasmodium. The plasmodium gives rise to zoosporangia or to resting spores. Upon germination, they produce zoospores. The single zoospore from resting spores penetrates host root hairs and there develops into a plasmodium. After a few days, the plasmodium cleaves into multinucleate portions surrounded by separate membranes; each portion develops into a zoosporangium. Each zoosporangium contains four to eight secondary zoospores which are discharged outside the host through pores dissolved in the host cell wall. Some of these zoospores fuse in pairs to produce probably dikaryotic zygotes, which can cause new infections and produce new plasmodium containing many, paired "dikaryotic" nuclei. These nuclei undergo fusion (karyogamy) followed by meiosis. The plasmodium finally turns into resting spores (Figure 11-12), which are released into the soil upon disintegration of the host cell walls by secondary microorganisms.

- *Development of Disease.* The plasmodium resulting from the germination of the secondary zoospores penetrates young root tissues directly; it can also penetrate secondarily thickened roots and underground stems through wounds. From these points of primary infection the plasmodium spreads to cortical cells and reaches the cambium through direct penetration of host cells (Figure 11-13). From the point of infection of the cambium, the plasmodium spreads in all directions in the cambium, outward into the cortex and inward toward the xylem and into the medullary rays. Single-point infections result

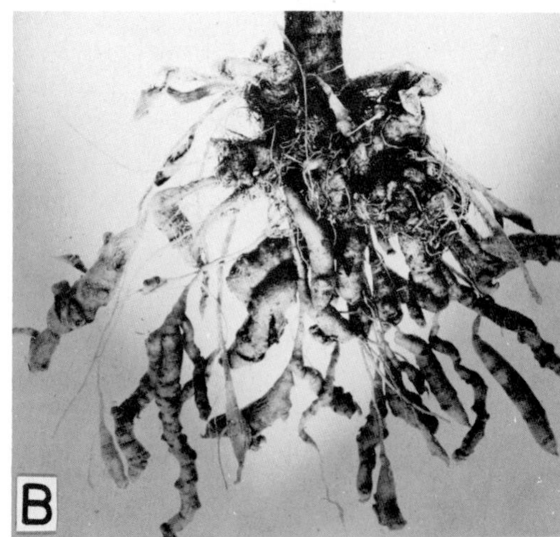

FIGURE 11-11 (A) Midday wilting of half-grown cabbage plants that have severely clubbed roots. (B) Malformed, spindlelike or clubbed roots infected with *Plasmodiophora brassicae.*

FIGURE 11-12 Scanning electron micrograph of resting spores of *Plasmodiophora brassicae* within cells of club roots (Photo courtesy M. F. Brown and H. G. Brotzman) × 1000.

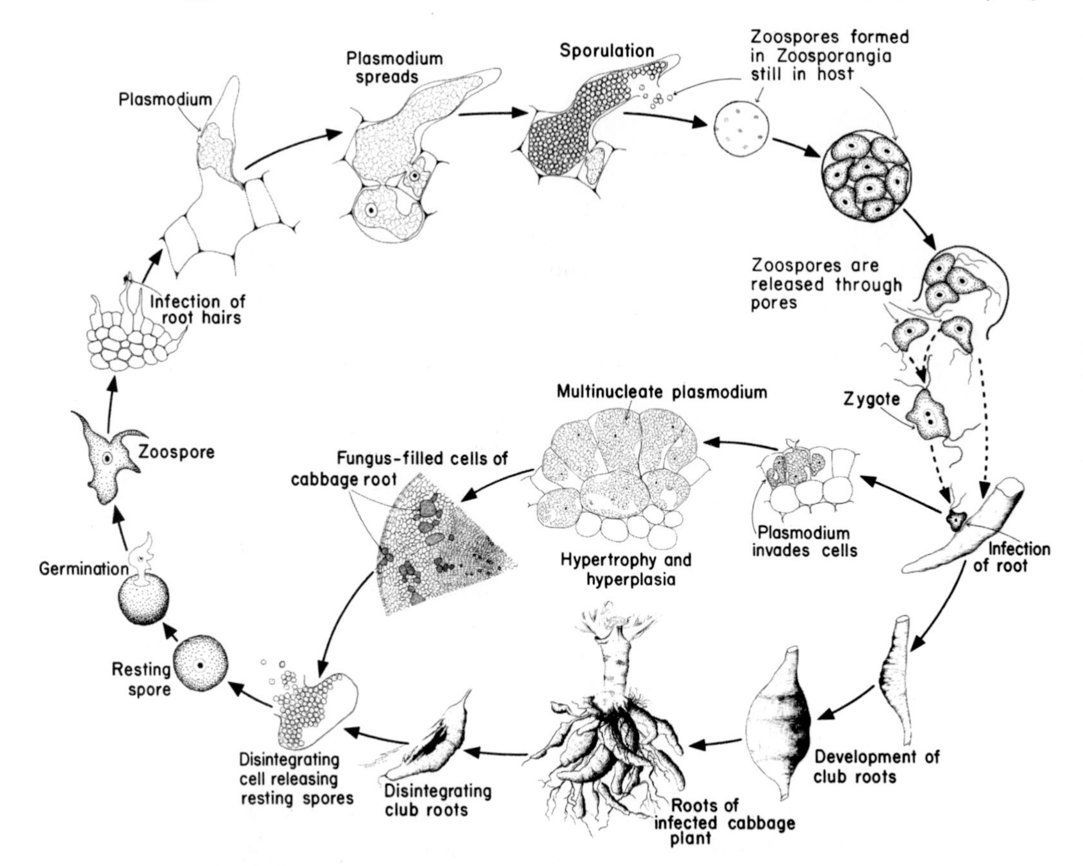

FIGURE 11-13 Disease cycle of clubroot of crucifers caused by *Plasmodiophora brassicae.*

in spindle-shaped clubs, being widest at the point of invasion and tapering off away from it.

As the plasmodia pass through cells, they become established in some of them and stimulate these cells to abnormal enlargement (hypertrophy) and abnormal division (hyperplasia). Infected cells may be five or more times as large as adjacent uninfected ones. The infected cells of a club are distributed in small groups throughout the diseased tissue, and the groups are usually separated by uninfected cells. The stimulus that is responsible for the abnormal growth of the cells appears to diffuse in advance of the pathogen and acts on the noninvaded cells of diseased tissues as well as on the infected ones. Actively growing and dividing cells, that is, cambial cells, are more easily invaded by the pathogen and are more responsive to the stimulus than other cells.

In most cases many cells of infected clubs remain free from plasmodia, but in rare instances almost all the cells of a club may be infected. When few cells are infected the plasmodia become large, whereas when many cells are infected, they remain relatively small. Thus, there seems to be a fairly constant ratio between the volume occupied by the plasmodium and that of the diseased tissue, the former being approximately 30 percent of the latter.

The plasmodium-infected clubs not only utilize much of the food required for the normal growth of the plant, but they also interfere with the absorption and translocation of mineral nutrients and water through the root system, resulting in gradual stunting and wilting of the aboveground parts of the plant. Furthermore, the rapid growing and greatly enlarged cells of the club tissues are unable to form a cork layer at the surface and are easily ruptured and invaded by secondary, weakly parasitic microorganisms. The invasion of clubs by bacteria and the subsequent disintegration by them lead to formation of substances toxic to the plant that are partly responsible for the wilting of the tops.

◦ *Control.* Growing cruciferous crops in fields known to be infested with the clubroot pathogen should be avoided. If that is not possible, cabbage and the other susceptible cruciferous crops should be planted in well-drained fields with a pH slightly above neutral (usually about 7.2) or in fields in which the pH is adjusted to 7.2 by the addition of the proper amount of hydrated lime. The use of soil liming in the control of clubroot is based on the fact that spores of the clubroot organism germinate poorly or not at all in alkaline media.

Although soil fumigants capable of disinfecting fields from the clubroot organism are available, the cost of materials and application is as yet prohibitive. However, seedbed areas can be kept free of clubroot by treating the soil with chloropicrin, methyl bromide, dazomet, sodium-metam, or Vorlex approximtely 2 weeks before planting. The clean, clubroot-free seedlings should, upon transplanting, be watered with a solution of PCNB (pentachloronitrobenzene).

The search for varieties of cruciferous hosts resistant to clubroot has been only partially successful. Such resistance has been most highly developed in varieties of rutabaga and turnip, but the extensive use of resistant varieties in infested soil has resulted in the appearance of highly virulent new races of the pathogen, sometimes within three years of the time of the release of the resistant varieties. Although some varieties of the most popular cruciferous hosts are resistant to certain races of the clubroot organism and can be grown in areas infested with these races, no varieties of cabbage, cauliflower, Brussels sprouts, or broccoli resistant to all the races of *P. brassicae* are currently available.

SELECTED REFERENCES

Buczacki, S. T. (1983) *Plasmodiophora:* An inter-relationship between biological and practical problems. *In* "Zoosporic Plant Pathogens" (S. T. Buczacki, ed.), pp. 161–191. Academic Press, London.

Colhoun, J. (1958). Clubroot disease of crucifers caused by *Plasmodiophora brassicae. Commonw. Mycol. Inst. Phytopathol. Pap.* **3**, 1–108.

Dobson, R. L. and Gabrielson, R. L. (1983). Role of primary and secondary zoospores of *Plasmodiophora brassicae* in the development of clubroot in Chinese cabbage. *Phytopathology* **73**, 559–561.

Reyes, A. A., Davidson, T. R., and Marks, C. F. (1974). Races, pathogenicity and chemical control of *Plasmodiophora brassicae* in Ontario. *Phytopathology* **64**, 173–177.

Williams, P. H., Keen, N. T., Strandberg, J. O., and McNabola, S. S. (1968). Metabolite synthesis and degradation during clubroot development in cabbage hypocotyls. *Phytopathology* **58**, 921–928.

Woronin, M. (1878). *Plasmodiophora brassicae.* Urheber der Kohlplfanzen-Hernie. *Jahrb. Wiss. Bot.* **11**, 548–574; Eng. transl. by C. Chupp in *Phytopathol. Classics* **4** (1934).

they fail to germinate, become soft and mushy, then turn brown, shrink, and finally disintegrate. Seed infections taking place in the soil cannot be observed, and the only manifestations of the disease are poor stands. Poor stands, however, are also the result of infections of the seedlings by the damping-off fungus after the seed has germinated but before the seedling has emerged above the soil line. Tissues of such young seedlings can be attacked at any point. The initial infection appears as a slightly darkened, water-soaked spot. The infected area enlarges rapidly, the invaded cells collapse, and the seedling is overrun by the fungus and dies shortly after the beginning of infection. In both cases infection takes place before the seedlings emerge, and this phase of the disease is called preemergence damping off.

Seedlings that have already emerged are usually attacked at the roots and sometimes at or below the soil line (Figure 11-14A-E). The succulent tissues of the seedling are easily penetrated by the fungus, which invades and kills the cells very rapidly. The invaded areas become water soaked and discolored, and the cells soon collapse. At this stage of infection the basal part of the seedling stem is much thinner and softer than the above, yet uninvaded, parts; owing to loss of firmness and supporting power, the invaded portion of the stem cannot support the part of the seedling above it, whereupon the seedling falls over on the soil. The fungus continues to invade the seedling after it has fallen to the ground, and the seedling quickly withers and dies. This phase of the disease is called postemergence damping off.

When older plants are attacked by the damping-off fungus they usually show only small lesions on the stem; these, however, if sufficiently large or numerous, can girdle the plant and cause stunting or death. More commonly, infections on older plants are limited to rootlets, which are damaged and frequently killed by the fungus; this results in stunting, wilting, and death of the aboveground part of the plant.

Soft fleshy organs of some vegetables, such as cucurbit fruits, green beans, potatoes, and cabbage heads, are sometimes infected by the damping-off fungi during extended wet periods in the field, in storage, and in transit. Such infections result in a cottony fungus growth on the surface of the fleshy organ, while the interior turns into a soft, watery, rotten mass, called "leak" (Figure 11-4F).

The Pathogen: Pythium Sp. *Pythium* is the most important cause of the pre- and postemergence phases of damping off. Several species of *Pythium* are involved, namely *P. aphanidermatum, P. debaryanum,. P. irregulare,* and *P. ultimum.* The effect of each one of them on its hosts, however, is usually similar to that of the others. It should be noted, however, that several different fungi, such as *Phytophthora, Rhizoctonia,* and *Fusarium,* can and often do cause symptoms quite similar to one or the other phase of those described above. Several other fungi, such as *Cercospora, Septoria, Mycosphaerella, Glomerella, Colletotrichum, Helminthosporium, Alternaria,* and *Botrytis,* and even some bacteria, such as *Pseudomonas* and *Xanthomonas,* when they are carried in or on the seed, also cause damping off and kill seedlings.

Pythium produces a white, slender, profusely branching, and rapidly growing mycelium. The mycelium gives rise to terminal, or intercalary sporangia, which may be spherical, filamentous, or variously shaped. Sporangia germinate directly by producing one to several germ tubes, or by producing a short hypha at the end of which a vesicle is formed (Figure 11-15). The

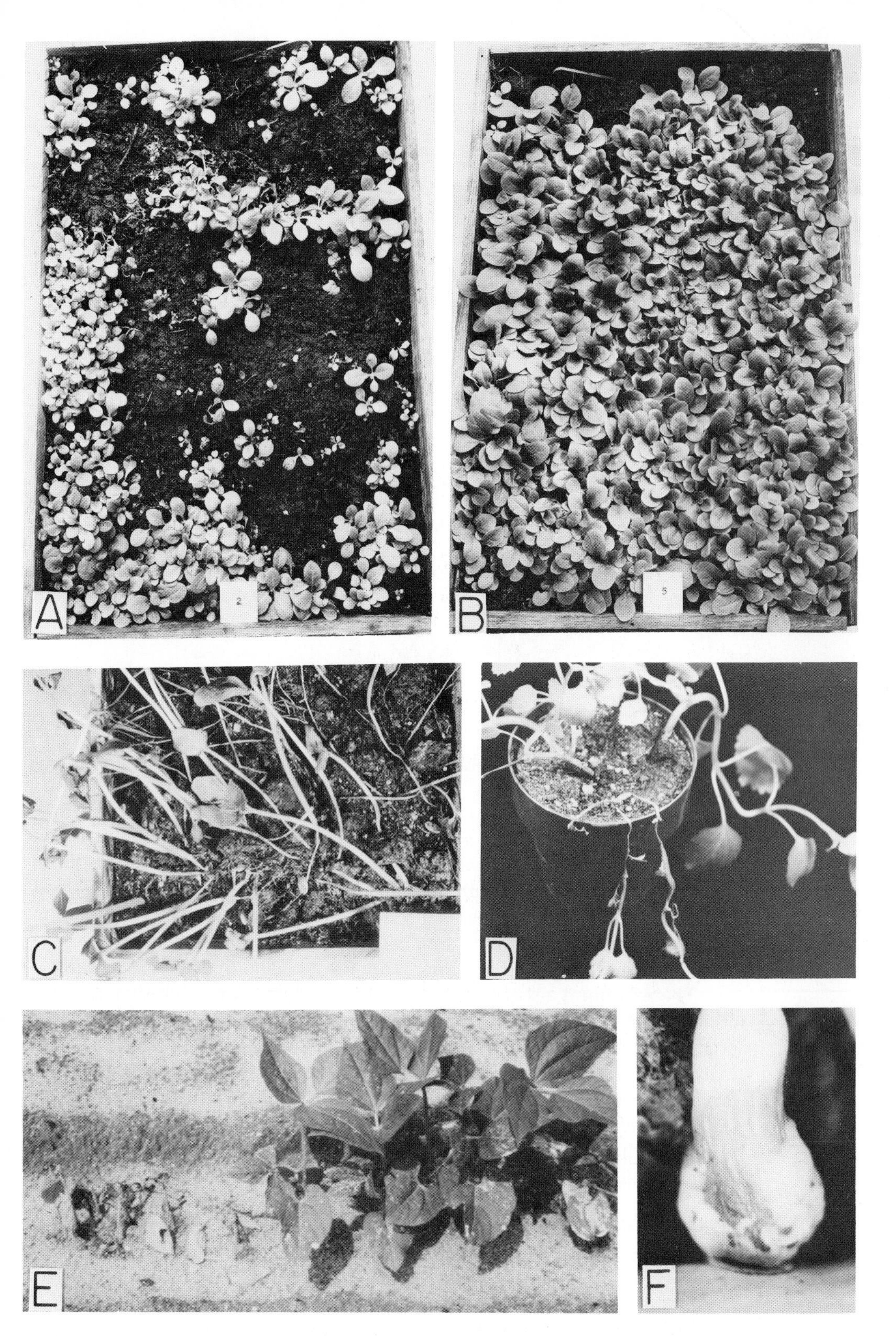

FIGURE 11-14 Damping-off symptoms (A) on tobacco seedlings compared to control (B). (C) Damping-off symptoms on cucumber seedlings. (D) Root and stem rot of Swedish ivy caused by *Pythium*. (E) Damping off on bean. (F) Soft rot of young butternut squash caused by *Pythium*. (Photo E courtesy G. C. Papavizas.)

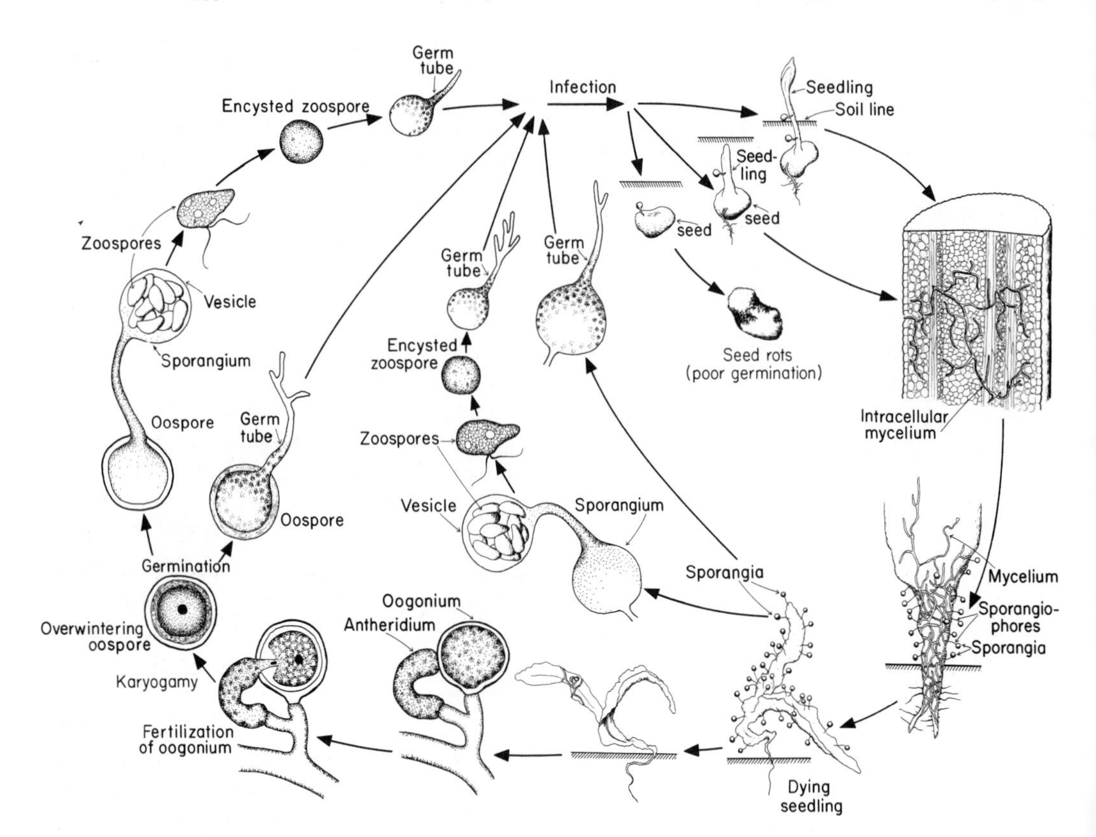

FIGURE 11-15 Disease cycle of damping off and seed decay caused by *Pythium* sp.

protoplasm passes from the sporangium into the vesicle, and there it forms more than 100 zoospores. When the zoospores are released they swarm about in the water for a few minutes, come to rest, encyst by rounding off, and germinate by producing a germ tube. The germ tube usually penetrates the host tissue and starts new infection, but sometimes it produces another vesicle in which several secondary zoospores are formed, and this may be repeated.

The mycelium also gives rise to spherical oogonia and club-shaped antheridia at the ends of short hyphae. The hypha bearing the antheridium may originate from the hypha bearing the oogonium or from another hypha of the mycelium. Upon contact with the oogonium the antheridium produces a fertilization tube, which enters the oogonium. Through this tube the male nuclei of the antheridium move toward the female nuclei of the oogonium, unite with them, and form the zygote. A thickening of the wall of the fertilized oogonium takes place, and the thick-walled structure containing the zygote is called an oospore. Oospores are resistant to high or low temperatures and other adverse factors and serve as the overwintering stage of the fungus. Since oospores require a resting period before they germinate, they are also called resting spores. Oospores, too, germinate either by producing germ tubes that develop into mycelium or by producing vesicles in which zoospores are formed in a way similar to that described for sporangia. The type of germination of both sporangia and oospores is determined primarily by the temperature of the medium, temperatures above 18°C favoring germination by germ tubes, while temperatures between 10 and 18°C induce germination by means of zoospores.

Pythium species are widely distributed in waters and soils throughout the world. They live on dead plant and animal substrata as saprophytes or as low-grade parasites attacking the fibrous roots of plants. When a wet soil is heavily infested with *Pythium,* any seeds, or young plants emerging from seeds, in such a soil may be attacked by the fungus.

Development of Disease. Spore germ tubes or saprophytic mycelium of *Pythium* come in contact with seeds or seedling tissues of host plants either by chance or because the exudates of these plants serve as nutrients and chemotropic stimulants to its zoospores and mycelium that move or grow toward the plants. The fungus enters the seeds by direct penetration of the moistened, swollen seed coats or through cracks and further penetrates the embryo or emerging seedling tissues through mechanical pressure and dissolution by means of enzymes. Pectinolytic enzymes secreted by the fungus dissolve the middle lamella, which holds the cells together, resulting in maceration of the tissues. Further invasion and breakdown of tissues occurs as a result of growth of the fungus between and through the cells. At the points where the hyphae pass through the cell walls they are constricted to approximately half their normal diameter. Proteolytic enzymes break down the protoplasts of invaded cells, while physical forces and, in some cases, fungal cellulolytic enzymes bring about complete collapse and disintegration of the cell walls. The fungus consumes many of the plant cell substances and the products of their breakdown and uses them as building blocks for its own body or as an energy source for its own metabolic activities. Thus, infected seeds are killed and turn into a rotten mass consisting primarily of fungus and substances such as suberin and lignin, which this fungus cannot break down.

The infection of roots and stems of young, tender seedlings progresses in essentially the manner described above. The initial infection usually occurs at or slightly below the surface of the soil, depending on moisture level and depth of planting. The mycelium penetrates the epidermal and cortical cells of the stem directly, consumes part or all of their contents, and breaks down their cell walls, bringing about the collapse of cells and tissues. In this area vascular tissues may also be invaded, in which case they become discolored even beyond the extent of the cortical lesion. Seedlings so invaded die quickly. When the invasion of the fungus is limited to the cortex of the belowground stem of the seedling, the latter may continue to live and grow for a short time until the lesion extends above the soil line. Then, the invaded, collapsed tissues cannot support the seedling, which falls over and dies (Figures 11-14 and 11-15).

If the initial infection occurs when the seedling is already well developed and has well-thickened and lignified cell walls and active cambium, the advance of the fungus is checked at or near the point of infection, and only relatively small lesions develop. Well-developed, mature tissues present considerable resistance to the mechanical pressure and to the enzymatic activity of the fungus by means of their increased thickness and modified chemical and physical structure of their cell walls.

Rootlets of most plants can be attacked by *Pythium* at almost any stage of plant growth. The fungus enters root tips and proliferates in the young cells, causing a rapid collapse and death of the rootlet. Further advance of the fungus into older roots is usually limited to the cortex of the root. Fleshy or relatively young roots are invaded to a much greater extent, the lesions extending to several centimeters in length and through the thickness of the cortex in depth.

Infections by *Pythium* of fleshy vegetable fruits and other organs can take place in the field, in storage, in transit, and in the market. Field infections begin at the point of contact of the fruit with wet soil heavily infested with the fungus. Enzymes secreted by the fungus macerate the tissue, which becomes soft and watery. An entire cucumber fruit may be invaded by the fungus within 3 days of inoculation. As the mycelium advances through the fleshy tissue it also bursts through the confining epidermis and forms, at first, small patches of aerial mycelium; these soon enlarge, coalesce, and form a continuous, luxuriant, cottony mycelial weft encasing most or all of the fruit. In storage, the aerial mycelium of an infected fruit grows out between adjacent fruits, invests them partially or completely, and soon penetrates, infects, and destroys these fruits.

During the development of infection, only mycelium can be found in the infected tissues. As the infection progresses, however, sporangia begin to appear, followed by the production of oospores. Sporangia and oospores can be produced inside or outside the host tissues or both, depending on the particular fungus species involved.

The severity of the diseases and amount of losses caused by *Pythium* infections are greater when the soil is kept wet (when its water content is at least 50 percent of its water-holding capacity) for prolonged periods; when the temperature is unfavorable for the host plant (too low for plants requiring high temperatures for optimum growth or too high for plants requiring relatively low temperatures for best growth); when there is an excess of nitrogen in the soil; and when the same crop is planted in the same field for several consecutive years.

○ *Control.* Pythium diseases in the greenhouse can be controlled through the use of soil sterilized by steam or dry heat or with volatile chemicals, such as methyl bromide with or without chloropicrin, and through the use of chemically treated seed. Greenhouse benches and containers must also be sterilized, however, or they must be treated with a 1 percent copper sulfate solution. Even so, recontamination often and easily occurs from infested soil on the greenhouse floor, tools, water hoses, and other items.

Few crops are known in which varieties showing any resistance to *Pythium* have been found, and so far no commercial varieties resistant to this pathogen are available. In recent years, however, control of *Pythium* seed rot and damping off has been obtained by treating the seeds with conidia of the antagonistic fungus *Trichoderma,* sp., *Penicillium oxalicum,* or *Gliocladium virens,* or by incorporating conidia of *Trichoderma* or *Streptomyces* sp. into commercial soilless mixes used in greenhouses and by nursery owners, or by incorporating into naturally infected soils organic matter carrying the antagonistic fungus *Laetisaria arvalis (Corticium* sp.).

Certain cultural practices are sometimes helpful in reducing the amount of infection. Good soil drainage is the most important of all. Improvement of heavy soils and improvement of air circulation among plants are advisable; planting should be done when temperatures are favorable for fast plant growth; application of excessive amounts of nitrate forms of nitrogen fertilizers should be avoided. No one crop should be planted in the same field for more than two consecutive years since that would increase the fungus population in the soil unless the soil is sterilized periodically. For container-grown nursery crops and ornamentals, including composted tree bark as a replacement for most of the peat markedly reduced the root rots caused by *Pythium* and several other root pathogens.

In the field, where soil sterilization is difficult and expensive, seed treatment with one or more of a number of chemicals is the most important disease preventive measure. The most commonly used materials for seed or bulb treatment are chloranil, thiram, captan, dichlone, ferbam, and diazoben. The systemic fungicide metalaxyl gives excellent control of damping off, seedling blights, and root rots caused by *Pythium* and *Phytophthora.* It can be applied as soil or seed treatment. Although more than one of these chemicals may be used on any crop with good results, there are usually some chemical-crop combinations that are much more effective than the rest, while some other combinations may be phytotoxic or ineffective under certain conditions.

Seed treatment is sometimes followed by spraying of seedlings with metalaxyl, ziram, chloranil, captan, or soluble coppers. This is especially importrant when the soil is heavily infested with *Pythium* or when the soil stays wet for prolonged periods during the early stages of plant growth.

SELECTED REFERENCES

Anonymous (1974). Symposium on the genus *Pythium.* Several papers. *Proc. Am. Phytopathol. Soc.* **1**, 200–223.

Buchholtz, W. F. (1938). Factors influencing the pathogenicity of *Pythium debaryanum* on sugar beet seedlings. *Phytopathology* **28**, 448–475.

Buczacki, S. T., Ed. (1983). "Zoosporic Plant Pathogens: A Modern Perspective." Academic Press, London.

Conway K. (1985). Selective medium for isolation of *Pythium* spp. from soil. *Plant Dis.* **69**, 393–395.

Drechler, C. (1925). The cottony leak of cucumbers caused by *Pythium aphanidermatum. J Agric. Res. (Washington, D. C.).* **30**, 1035–1042.

Hendrix, F. F., Jr., and Campbell, W. A. (1973). Pythiums as plant pathogens. *Annu. Rev. Phytopathol.* **11**, 77–98.

Middleton, J. T. (1943). The taxonomy, host range and geographic distribution of the genus *Pythium, Torrey Bot. Club. Mem.* **20**, 1–171.

Mulder, J. (1969). The pathogenicity of several *Pythium* species to rootlets of apple seedlings. *Neth. J. Plant Pathol.* **75**, 178–181.

Waterhouse, G. M. (1968). The genus *Pythium. Mycol. Pap.* **110,** 1–171.

Phytophthora Diseases

Species of *Phytophthora* cause a variety of diseases on many different types of plants ranging from seedlings of annual vegetables or ornamentals to fully developed fruit and forest trees. Most species cause root rots, damping off of seedlings, and rots of lower stems, tubers, and corms similar to those caused by *Pythium* sp. Others cause rots of buds or fruits, and some cause foliar blights that attack the foliage, young twigs, and fruit. Some species are host specific, that is, they attack only one or two species of host plants, but others have a wide host range and may cause similar or different symptoms on many different kinds of host plants. The best-known species is *Phytophthora infestans,* the cause of late blight of potatoes and tomatoes, but several other species cause extremely destructive diseases on their hosts. Some of the other important species and their better-known diseases are listed here.

P. cactorum, causes collar or trunk rot of apple, foot rot of lily and stock, blight of peony, dieback of azalea, stem rot and wilt of snapdragon, root rot of sweet clover, and blossom blight of tulip.
P. capsici, causes root rot of pepper, carrot, and pumpkin, and fruit rot of pepper, cucurbits, eggplant, tomato, etc.
P. cambivora, causes root and crown rot of trees.
P. cinnamomi, causes root rot of avocado, azalea, chestnut, cinnamon, oak, pine, pineapple, and many other trees and shrubs, the "jarrah dieback" of natural forests in Australia.
P. citrophthora, causes root rot and fruit rot of citrus.
P. cryptogea, causes root rot of tomato and of calla lily, and leaf and stem rot of gloxinia.
P. erythroseptica, causes pink rot of potato, soft rot of sugarcane cuttings, and leaf blight of calla lily.
P. fragariae, causes red stele root rot of strawberry.
P. megasperma, causes root rot of crucifers, carrot, potato, spinach, beet, hollyhock, etc.
P. megasperma var. *sojae,* causes Phytophthora stem rot of soybeans.
P. palmivora, causes bud rot of coconut and stem rot of dieffenbachia and peperomia.
P. parasitica, causes damping off, leaf blight, stem canker, and buckeye rot of fruit in tomato, crown rot of rhubarb, soft rot of cucurbits, foot rot of citrus.
P. parasitica var. *nicotianae,* causes black shank of tobacco.
P. phaseoli, causes downy mildew of lima bean.
P. syringae, causes lilac blight, pruning wound cankers of almond, and fruit rot of apples.

Phytophthora Root Rots

Several species of *Phytophthora,* particularly *P. cinnamomi, P. cryptogea, P. fragariae,* and *P. megasperma,* causes root rots on numerous species of plants that include fruit, forest, and ornamental trees and shrubs, annual vegetables and ornamentals, and strawberries. The most common *Phytophthora* root rots are littleleaf disease of pine, root rot of avocado and pineaple, root rots of alfalfa, azalea, calla lily, crucifers, tomato, carrot, red stele of strawberries, and many others (Figure 11-16). The losses caused by *Phytophthora* root rots are great, especially on trees and shrubs, but the pathogen often goes undetected or unidentified. Plants suffering from such root rots often begin by showing symptoms of drought and starvation, and become weakened and susceptible to attack by other pathogens or various other causes that are then mistakenly taken as the causes of the death of the plants.

Phytophthora root rots cause damage in their hosts in nearly every part of the world where the soil becomes too wet for good growth of susceptible plants and the temperature remains fairly low, that is, between 15 and 23°C.

Young seedlings of trees and annual plants may be killed by the disease within a few days, weeks, or months, while in older plants the killing of roots may be slow or rapid, depending on the amount of fungus present in the soil and the prevailing environmental conditions. As a result, older trees show

FIGURE 11-16 (A) Strawberry plant (right) affected with the red stele disease caused by *Phytophthora fragariae*. Almost all feeder roots have been destroyed. Normal plant at left. (B) Red stele-affected strawberry roots being killed from the tip upwards (left), and roots split lengthwise to show the reddened central portion (middle). Healthy roots split lengthwise at right. (C) Root and stem rot of soybean seedlings caused by *P. megasperma*. Healthy plant at left. (D) Phytophthora root and stem rot of coleus. (E) Tobacco black shank caused by *P. parasitica* var. *nicotiana*. (Photos A, B, and E courtesy U.S.D.A. Photo C courtesy G. C. Papavizas).

sparse foliage, shorter, cupped, and yellow leaves, and dieback of twigs and branches. Such trees increase very little in height and diameter and usually die within 3 to 10 years after infection. Fewer and smaller fruit and seeds are produced each succeeding year.

On all hosts affected by *Phytophthora* root rot, many of the small roots are dead, and necrotic brown lesions are often present on the larger roots. On young plants, or on older succulent plants, the whole root system may decay, followed by a more or less rapid death of the plant. In strawberries, as in the other plants, most of the small rootlets rot away, while the larger ones show progressive stages of browning beginning at the tips. In addition, in late spring and before or through harvest, affected larger strawberry roots show a red-colored core or stele, a symptom diagnostic of the strawberry red stele root rot caused by *Phytophthora fragariae* (Figure 11-16, A, B).

The behavior of the various *Phytophthora* species that cause root rots is generally similar. The fungus overwinters as oospores, chlamydospores, or mycelium in infected roots, and these spores may also overwinter in the soil. In the spring, the oospores and chlamydospores germinate by means of zoospores, while the mycelium grows further and produces zoosporangia that release zoospores. The zoospores swim around in the soil water and infect roots of susceptible hosts with which they come in contact. More mycelium and zoospores are produced during wet, cool weather and spread the disease to more plants. In dry, hot, or too cold weather the fungus survives as oospores, chlamydospores, or mycelium that can start infections again when the soil is wet and the temperature favorable.

Control of *Pytophthora* root rots depends on planting susceptible crops in soils free of the pathogen or in soils that are light and drain well and quickly. All planting stock should be free of infection and, when available, only resistant varieties should be planted. For plants in pots, greenhouses, or seedbeds, the soil and containers should be sterilized with steam before planting.

In recent years, excellent control of *Phytophthora* root and lower stem rots has been obtained through the use of several systemic fungicides such as metalaxyl, fosetyl Al, ethazol, and propamocarb, which are applied as seed treatments, soil treatments, transplant dips, and as sprays, or with overhead irrigation water. For some crops, such as strawberries, *Phytophthora* root rot has also been effectively controlled through soil fumigation with a mixture of methyl bromide and chloropicrin. In some cases, *Phytophthora* root rots have been controlled by planting seedlings in suppressive soil that contains either microorganisms antagonistic to *Phytophthora* or inorganic substances toxic to the fungus. Several fungi and bacteria have been shown to parasitize *Phytophthora* oospores or to be antagonistic to *Phytophthora,* but none of them has been used to effectively control *Phytophthora* so far. In the last several years, however, composted tree bark mixed with soil or soilless mixes used in production of container-grown plants or in greenhouse beds has significantly reduced plant infections by *Phytophthora.*

• Phytophthora Foot, Crown, Collar, Stem, or Trunk Rots

Most species of *Phytophthora* listed in the general part cause diseases primarily of the lower stem that are described by one or another of the terms listed

FIGURE 11-17 (A) Declining citrus tree as a result of girdling by the foot rot fungus *Phytophthora parasitica* or *P. citrophthora.* (B) Close-up of the rotting and of the trunk base caused by the same pathogens. (Photos courtesy Agric. Res. and Educ. Center, Lake Alfred, Fla.)

here. The most common and most serious of such diseases include the trunk canker or collar rot of apple trees, foot rot of citrus trees (Figure 11-17), root and crown rot of cherry trees (Figure 11-18) black shank of tobacco (Figure 11-16E), foot rot of lily and stock, stem rot and wilt of snapdragon and soybeans, pink rot of potato, soft rot of sugarcane cuttings, stem rot of dieffenbachia and peperomia, and others. In many of these diseases the fungus also attacks the roots, it may attack and kill seedlings before or after emergence above the soil line, thus causing damping-off symptoms, and in some cases the fungus also attacks and causes partial or complete rot of the fruit, as for example, in tomato, pepper, cucurbits, citrus, and cacao (Figure 11-19).

The general characteristics of *Phytophthora* rots of the lower stem of plants, considering the wide variety of host plants and fungus species involved, appear quite similar. Like the other *Phytophthora* diseases, these are also favored by rather low temperatures and by high soil moisture and atmospheric humidity, and are therefore more common and most severe in low-lying, poorly drained areas. In some host-fungus combinations, the fungus attacks the stem below the soil line or it may first attack the main root and thus cause droughtlike symptoms and general decline of the aboveground parts of the plant before any direct lesions or cankers appear above the soil line. In most cases, however, the fungus attacks the plant at or near the soil line where it causes a water soaking of the bark that appears as a dark area on the trunk. The dark area enlarges in all directions and, if the plant is small and succulent, the darkening may soon encircle the entire stem, after which the lower leaves drop and eventually the whole plant wilts. On larger plants and on trees, the darkening may be on one side of the stem and soon becomes a depressed canker below the level of healthy bark. In early stages, the diseased

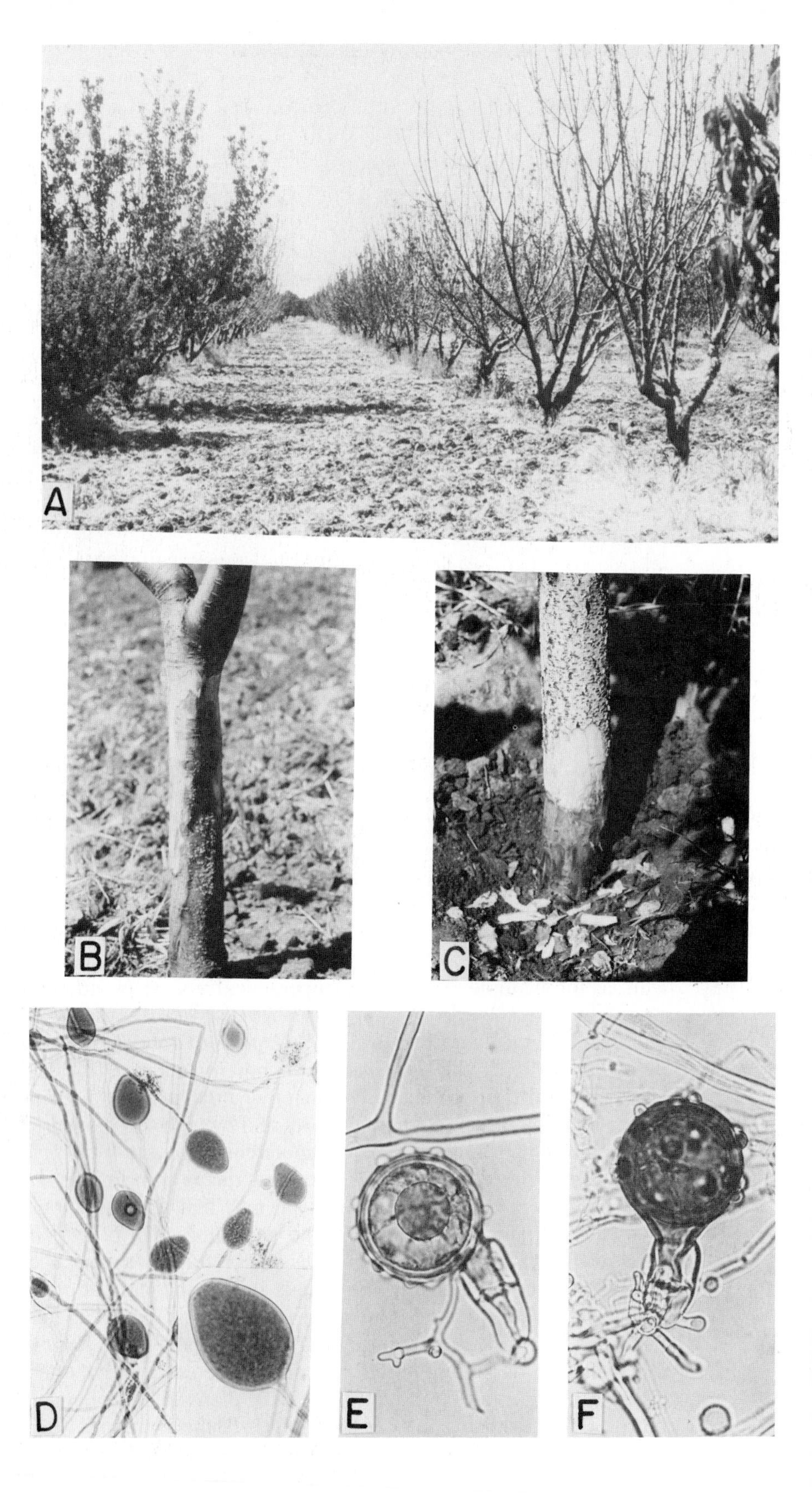
A
B
C
D
E
F

FIGURE 11-19 Black pod disease of cacao caused by *Phytophthora palmivora.* (Photos courtesy M. O. K. Adegbola and A. Adebayo, Cacao Res. Inst., Nigeria.)

bark is firm and intact while the inner bark is slimy and may produce a moist, gummy exudate. Later, the affected area becomes shrunken and cracked. The collar rot canker may spread through the tissue up into the trunk (Figure 11-18) and sometimes the branches or down into the root system. Invasion of the root usually begins at the crown area or at ground level. As the cankers spread and enlarge, they may girdle the trunk, limbs, or roots, at first causing the plant or tree to grow poorly, produce fewer and smaller fruit, show sparse foliage and dieback of twigs, and finally killing the plant parts beyond the infected area.

The *Phytophthora* species that cause these diseases live and reproduce primarily in the soil and usually attack susceptible plants at or below the soil line. Sometimes, however, spores of the fungus may be splashed into injured aboveground bark or low-lying fruit and may cause infections at these points. Fleshy fruits, such as cucurbits, lying on wet soil infested with the fungus are also attacked directly. The fungus overwinters in infected tissues as mycelium, oospores, or chlamydospores.

Control of *Phytophthora* diseases of the lower stem requires all precautions described for *Phytophthora* root rots. In addition, they can be controlled by applying metalaxyl, fosetyl Al, propamocarb, or ethazol to the soil, seeds,

---

FIGURE 11-18 (A) Sweet cherry trees dying from Phytophthora root and crown rot. (B) Extensive trunk canker of cherry on Mahaleb rootstock caused by *P. cambivora.* (C) Typical crown rot symptoms of cherry on Mazzard rootstock caused by *P. cambivora* and *P. megasperma.* Sporangia (D), oospore with antheridium (E), and oospore (F) of *P. cambivora.* (Photos courtesy S. M. Mircetich, from S. M. Mircetich and Matherton, *Phytopathology* **66**, 549–558.)

or transplants, and by injections of fosetyl Al into the trunks of trees. Also, on dormant, susceptible plants, primarily trees, and in the soil around them, application of a solution of copper oxychloride or Bordeaux mixture seems to greatly inhibit the growth and activity of the fungus. Resistant varieties should always be preferred, especially for heavy, poorly drained soils. With fruit trees, resistant rootstocks and sometimes interstocks offer the most effective means of controlling foot rot or collar rot.

• Late Blight of Potatoes

The late blight disease of potatoes is found in nearly all areas of the world where potatoes are grown. It is most destructive, however, in the eastern half of North America and in northwestern Europe, where potatoes are grown in large acreage and where cool, moist weather favors both potato production and the late blight disease. Late blight is also very destructive to tomatoes and to several other species in the family Solanaceae.

Late blight may kill the foliage and stems of potato and tomato plants at any time during the growing season. It also attacks potato tubers and tomato fruits in the field, which rot either in the field or while in storage, in transit, and at market.

Late blight may cause total destruction of all plants in a field within a week or two when weather conditions are favorable and when no control measures are applied. Losses, however, vary from one area to another and

FIGURE 11-20 Late blight symptoms on potato (A) and tomato (B) leaves. The whitish zone surrounding the necrotic area of (A) consists of conidiophores and conidia of *Phytophthora infestans.* (C) Exterior and cross-section view of late blight symptoms on potato tubers. (D) Late blight on the stem of a young potato plant originating from mycelium that overwintered in the infected potato seed piece. (E, F) Exterior and cross-section views of late blight symptoms on tomato fruit. (Photos A-D courtesy Dept. Plant Path., Cornell Univ.)

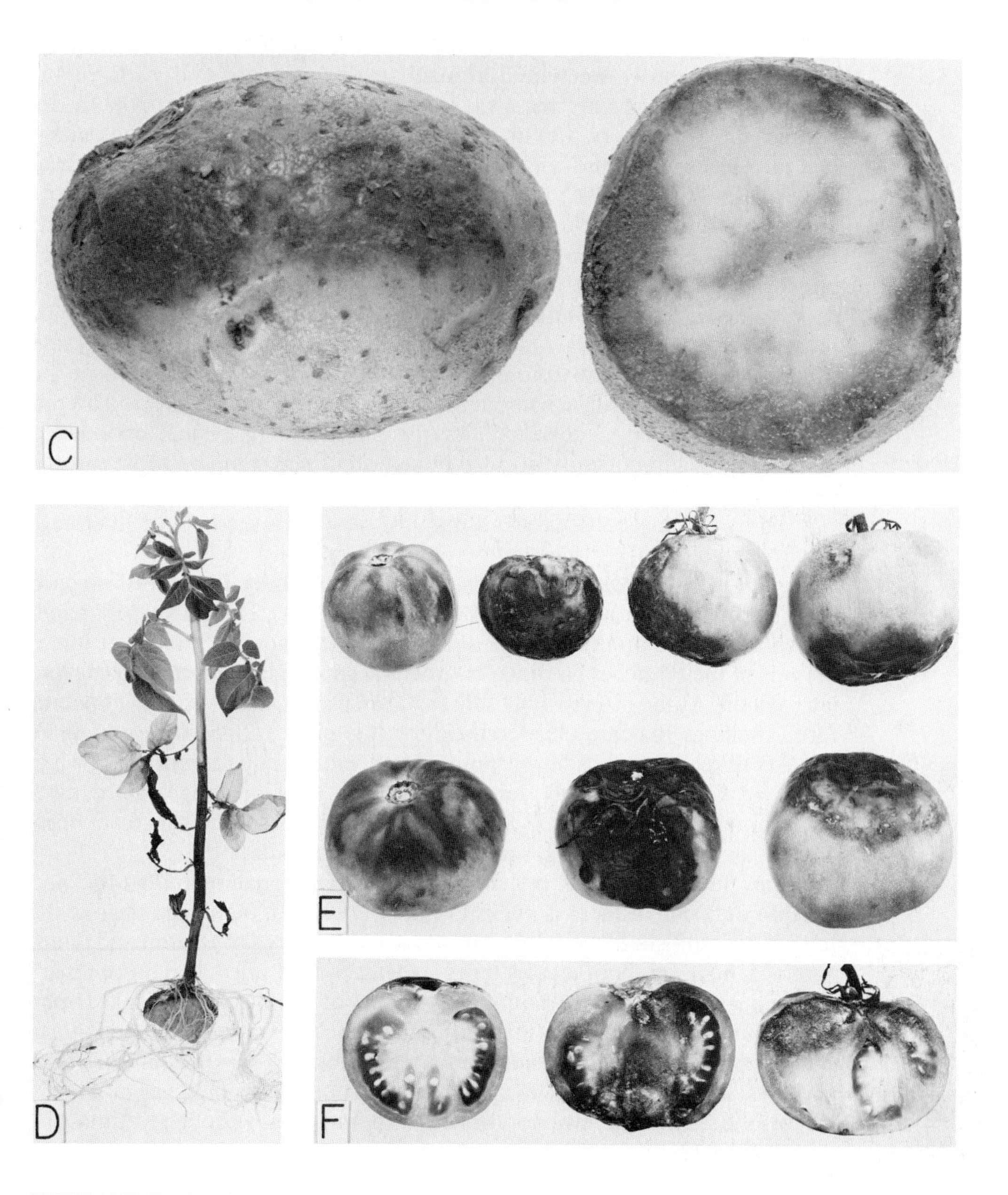

FIGURE 11-20 *(Continued)*

from year to year, depending on the prevailing temperature and moisture at certain periods of the growing season and on the control measures practiced. Even when losses in the field are small, potatoes may become infected during harvest and may rot in storage.

○ *Symptoms.* Symptoms appear at first as circular or irregular water-soaked spots, usually at the tips or edges of the lower leaves. In moist weather the spots enlarge rapidly and form brown, blighted areas with indefinite borders. A zone of white, downy fungus growth 3–5 mm wide appears at the border of the lesions on the undersides of the leaves (Figure 11-20 A, B). Soon the entire leaflet and then all the leaflets on a leaf are infected, die, and become limp. Under continuously wet conditions all tender, aboveground parts of the plants blight and quickly rot away, giving off a characteristic odor. In dry weather the activities of the fungus are checked. Existing lesions stop enlarging, turn black, curl, and wither, and no fungus appears on the underside of the leaves. When the weather becomes moist again the fungus resumes its activities and the disease once again develops rapidly.

Affected tubers at first show more or less irregular, purplish-black or brownish blotches. When cut open, the affected tissue appears water-soaked, dark, somewhat reddish brown, and extends 5–15 mm into the flesh of the tuber (Figure 11-20 C, D). Later the affected areas become firm and dry and somewhat sunken. Such lesions may be small or may involve almost the entire surface of the tuber without spreading deeper into the tuber. The rot, however, continues to develop after the tubers are harvested, or infected tubers may be subsequently invaded by secondary fungi and bacteria, causing soft rots and giving the rotting potatoes a putrid, offensive odor.

Tomato fruit is attacked and may rot rapidly in the field or in storage (Figure 11-20 E, F).

◦ *The Pathogen: Phytophthora Infestans.* The mycelium produces branched sporangiophores of unrestricted growth (Figure 11-21). Lemon-shaped, papillate sporangia are produced at the tips of the sporangiophore branches, but as the tips of the branches continue to grow the sporangia are pushed aside and later fall off. At the places where sporangia are produced, the sporangiophores form swellings that are characteristic for this fungus. Sporangia germinate almost entirely by means of zoospores at temperatures up to 12 or 15°C, while above 15°C sporangia may germinate directly by producing a germ tube. Each sporangium produces 3–8, sometimes more, zoospores, which are liberated by the bursting of the sporangial wall at the papilla.

This fungus requires two mating types for sexual reproduction, and because only one of them is present in most countries, the sexual stage of the fungus has rarely been found. In Mexico and other areas of Central and South America, however, both mating types are widely distributed, and oospores are common. When the two mating types grow adjacently, the female hypha grows through the young antheridium and develops into globose oogonium above the antheridium. The antheridium then fertilizes the oogonium, which develops into a thick-walled and hardy oospore. Oospores germinate by means of a germ tube, which produces a sporangium, although at times the germ tube grows directly into mycelium.

◦ *Development of Disease.* The pathogen overwinters as mycelium in infected potato tubers. The mycelium spreads in the tissues of the potato tubers and it finally reaches a few of the shoots produced from infected tubers used as seed, volunteer plants that develop from diseased tubers left in the field, or sprouts produced by infected potatoes in cull piles or dumps. The mycelium spreads up the stem most rapidly in the cortical region causing discoloration and collapse of the cells (Figure 11-20 D). Later, the mycelium grows between the pith cells of the stem, but it is seldom found in the vascular system. The mycelium grows through the stem and travels up to the surface of the soil. When the mycelium reaches the aerial parts of plants, it produces sporangiophores, which emerge through the stomata of the stems and leaves and project into the air (Figure 11-21). The sporangia produced on the sporangiophores

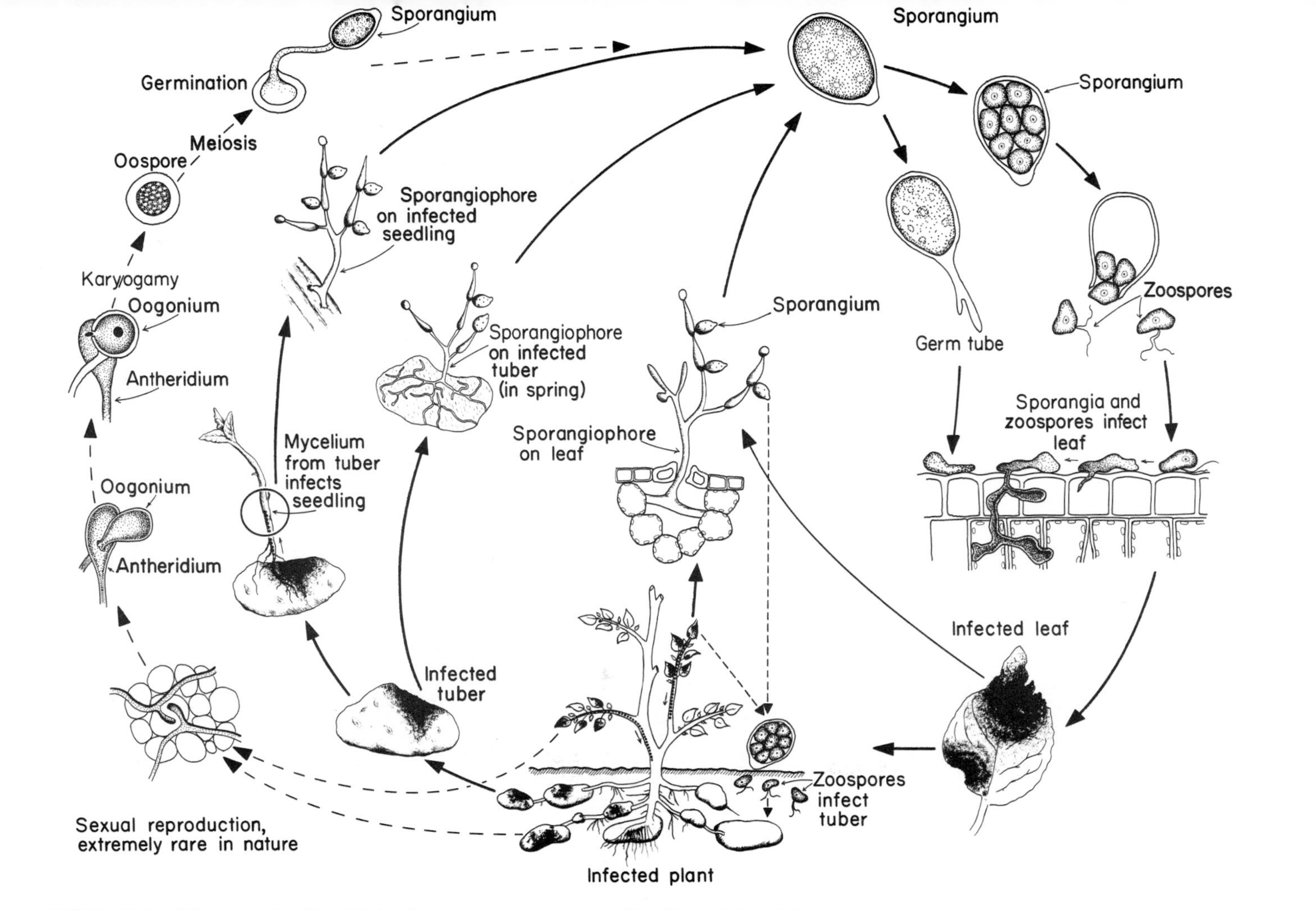

FIGURE 11-21 Disease cycle of late blight of potato and tomato caused by *Phytophthora infestans*.

become detached and drift off when ripe, or are dispersed by rain. When the sporangia land on wet potato leaves or stems, they germinate and cause new infections. The germ tube penetrates the leaf cuticle or enters through a stoma and produces mycelium, which grows profusely between the cells and sends long, curled haustoria into the cells. The cells on which the mycelium feeds are killed, and as they begin to decay, the mycelium spreads peripherally into fresh tissue. A few days after infection, new sporangiophores emerge from the stomata of the leaves and produce numerous sporangia, which are spread by the wind and infect new plants. In favorable weather the period from infection to sporangia formation may be as short as 4 days and, therefore, a large number of asexual generations and new infections may be produced in one growing season. With the advance of the disease the established lesions increase in area and new ones develop resulting in premature killing of the foliage and proportional reduction in potato tuber yields.

The second phase of the disease, the infection of tubers, begins in the field when, during wet weather, sporangia are washed down from the leaves and are carried into the soil. The tubers near the surface of the soil are attacked by the emerging zoospores, which germinate and penetrate the tubers through lenticels or through wounds. In the tuber the mycelium grows mostly between the cells and sends its long, sicklelike haustoria into the cells. Tubers are rarely infected by mycelium growing down the stem of a diseased mother plant. If, however, at harvest, the tubers are contaminated with living sporangia still present on the soil or if the tubers are exposed while the fungus is still sporulating on partially diseased foliage, further infection will occur, which may not be apparent at the time, but it will develop in storage. Most of the blighted tubers rot in the ground or during storage.

The development of late blight epidemics depends greatly on the effect of humidity and temperature on the different stages of the life cycle of the fungus. The fungus sporulates most abundantly at relative humidity near 100 percent and at temperatures between 16 and 22°C. Sporangia lose their viability in 3–6 hours at relative humidities below 80 percent. Germination of sporangia takes place only when free water or dew is present on the leaves and, at 10–15°C, it may be completed within 0.5–2 hours. After germination a period of 2–2.5 hours at 15–25°C is required for penetration of the germ tubes into the host tissue. After penetration, the mycelium develops most rapidly at 17–21°C, which is also optimal for sporulation. Temperatures above 30°C check the growth of the fungus in the field but do not kill it, and the fungus can start to sporulate again when the temperature becomes favorable, provided of course that the relative humidity is sufficiently high.

◦ *Control.* Late blight of potatoes can be successfully controlled by a combination of sanitary measures, resistant varieties, and well-timed chemical sprays. Only disease-free potatoes should be used for seed. Potato dumps or cull piles should be burned before planting time in the spring or sprayed with strong herbicides to kill all sprouts or green growth. All volunteer potato plants in the area, whether in the potato or other fields should be destroyed, since any volunteer potato plant can be a source of late blight infection.

Only the most resistant potato varieties available should be planted. Unfortunately, most popular commercial potato varieties are more or less susceptible to late blight. The blight fungus comprises a number of races, which differ from each other in the potato varieties that they can attack.

Several potato varieties resist one or more races of the late blight fungus. Some of them are resistant to vine infection but not to tuber infection. New varieties, derived from crosses with *Solanum demissum,* have one or more of the R genes for resistance to late blight and have withstood attack by all known races of the fungus for a short while but were attacked by other races not previously distinguished or perhaps not previously existent. Many varieties possess so-called "field resistance," which is only a partial horizontal resistance of varying degrees but which is effective against all races of the blight fungus. However, it is not sufficient to rely on varietal resistance to control late blight, since in favorable weather, blight can severely infect these varieties unless they are sprayed with a good proctective fungicide. Even resistant varieties should be regularly sprayed with fungicides to eliminate, as much as possible, the possibility of becoming suddenly attacked by races of the fungus to which they are not resistant or by entirely new races. On the other hand, it is always advisable to use resistant varieties, even when sprays with fungicides are considered the main control strategy, since resistant varieties delay the onset of the disease or reduce its rate of development, so that fewer sprays on a resistant variety may be needed to obtain a satisfactory level of control of the disease.

Chemical sprays with fungicides, if applied properly, generally will keep late blight under control. Spraying should start when potato plants are 15–30 cm high or at least 10 days before the date late blight usually appears in the area. Sprays should be applied once every 5–10 days when the weather is damp, misty, or rainy and when the nights are cool and should continue until the foliage dies naturally or is killed artificially by "vine-killers." Proper timing and thorough coverage of old and new foliage are essential if plants are to be protected from the disease. Once late blight becomes established, it is extremely difficult to control unless the weather turns hot (35°C and above) and dry.

During the last several years, several computerized (Blitecast is an example) and noncomputerized programs have been developed for forecasting periods of occurrence of late blight and for scheduling timely applications of appropriate fungicides. All these programs depend on the fact that late blight is favored by moderate temperatures and abundant moisture. An abundant supply of fungal inoculum is assumed. By frequent (weekly or daily) field scouting for the initial appearance of disease symptoms in each location, a relationship between temperature, moisture, and infection is established. By recording daily occurrence and duration of temperature, relative humidity, and rainfall, and by analyzing all the data through the computer, it is possible to predict the first and subsequent infection periods and, therefore, to schedule timely application of appropriate fungicides to prevent or reduce infection. Some forecasting schemes are even programmed to take into account the degree of resistance of the variety to the fungus and the effectiveness (protectant, eradicant, systemic) of the fungicide in formulating the recommendation for fungicide application.

Materials used for late blight control include mancozeb, metalaxyl, a combination of metalaxyl and mancozeb, captafol, chlorothalonil, polyram, fentin hydroxide, and several copper materials, such as Kocide, copper oxychloride, and Bordeaux mixture. Protective spraying of foliage usually effects a considerable reduction in tuber infection. When, however, partially blighted leaves and stems are surviving at harvest time, it is necessary to remove the aboveground parts of potato plants or destroy them by chemical sprays or mechanical means. Herbicides used for this purpose include dinoseb, diquat, paraquat, endothall, ametryn, several inorganic compounds such as copper sulfate, sodium and potassium arsenites, sulfuric acid, and certain dinitro compounds.

SELECTED REFERENCES

Bain, H. F., and Demaree, J. B. (1945). Red stele root disease of the strawberry caused by *Phytophthora fragariae. J. Agric. Res. (Washington, D. C.)* **70**, 11–30.

Baines, R. C. (1939). *Phytophthora* trunk canker or collar rot of apple trees. *J. Agric. Res. (Washington, D. C.)* **59**, 159–184.

Benson, D. M. (1985). Fungicides for control of *Phytophthora* root rot of azalea in landscape beds. *Plant Dis.* **69**, 697–699.

Berg, A. (1926). Tomato late blight and its relation to late blight of potato. *W. Va., Agric. Exp. Stn., Tech. Bull.* **205**, 1–31.

Bonde, R., and Schultz, E. S. (1943). Potato cull piles as a source of late-blight infection. *Am. Potato J.* **20**, 112–118.

Coffey, M. D., *et al.* (1984). Chemical control of *Phytophthora cinnamomi* on avocado rootstocks. *Plant Dis.* **68**, 956–958.

Cox, A. E., and Large, E. C. (1960). Potato blight epidemics throughout the world. *U. S., Dep. Agric., Agric. Handb.* **174**, 1–230.

Darvas, J. M., Toerien, J. C., and Milne, D. L. (1984). Control of avocado root rot by trunk injection with phosethyl-Al. *Plant Dis.* **68**, 691–693.

Debruyn, H. L. G. (1951). Pathogenic differentiation in *Phytophthora infestans. Phytopathol. Z.* **18**, 339–359.

Duniway, J. M. (1979). Water relations of water molds. *Annu. Rev. Phytopathol.* **17**, 431–460.

Erwin, D. C., Bartnicki-Garcia, S., and Tsao, P. H., eds. (1983) *"Phytophthora:* Its Biology, Taxonomy, Ecology, and Pathology." Am. Phytophathol. Soc. St. Paul, Minnesota.

Frosheiser, F. I. (1980). Conquering *Phytophthora* root rot with resistant alfalfa cultivars. *Plant Dis.* **64**, 909–912.

Fry, W. E., Apple, A. E., and Bruhn, J. (1983). Evaluation of potato late blight forecasts modified to incorporate host resistance and fungicide weathering. *Phytopathology* **73**, 1054–1059.

Gallegly, M. E. (1968). Genetics of pathogenicity of *Phytophthora infestans. Annu. Rev. Phytopathol.* **6**, 375–396.

Hickman, C. J. (1970). Biology of *Phytophthora zoospores. Phytopathology* **60**, 1128–1135.

Klotz, L. J., and Calavan, E. C. (1969). Gum diseases of citrus in California. *Calif., Agric. Exp. Stn., Ext. Serv. Circ.* **396**, 1–26.

Ko, W. H. (1982). Biological control of Phytophthora root rot of papaya with virgin soil. *Plant Dis.* **66**, 446–448.

Ko, W. H. (1985). Nature of suppression of *Phytophthora capsici* in a Hawaiian soil. *Phytopathology* **75**, 683–685.

Krause, R. A., Massie, L. B., and Hyre, R. A. (1975). Blitecast: A computerized forecast of potato blight. *Plant Dis. Rep.* **59**, 95–98.

MacHardy, W. E. (1979). A simplified, noncomputerized program for forecasting potato late blight. *Plant Dis. Rep.* **63**, 21–25.

MacKenzie, D. R. (1980). Scheduling fungicide application for potato late blight with BLITECAST. *Plant Dis.* **64**, 394–399.

Newhook, F. J., and Podger, F. D. (1972).The role of *Phytophthora ciinnamomi* in Australian and New Zealand forests. *Annu. Rev. Phytopathol.* **10**, 299–326.

Platt, H. W. (1983). Effects of metaloxyl, mancozeb, and chlorothalonil on blight, yield, and tuber rot of potato. *Can. J. Plant Pathol.* **5**, 38–42.

Reddick, D., and Mills, W. (1938). Building up virulence in *Phytophthora infestans. Am. Potato J.* **15**, 29–34.

Schmitthenner, A. F. (1985). Problems and progress in control of Phytophthora root rot of soybean. *Plant Dis.* **69**, 362–368.

Stevenson. W. R. (1983). An integrated program for managing potato late blight. *Plant Dis.* **67**, 1047–1048.

Sutherland, E. D., and Lockwood, J. L. (1984). Hyperparasitism of oospores of some Peronosporales by *Actinoplanes missuriensis* and *Humicola fiscoatra* and other Actinomycetes and fungi. *Can. J. Plant Pathol.* **6**, 139–145.
Waterhouse, G. M. (1956). The genus *Phytophthora. Commonw. Mycol. Inst., Misc. Publ.* **12.**
Waterhouse, G. M. (1970). Taxonomy in *Phytophthora. Phytopathology.* **60**, 1141–1143.

The Downy Mildews

The downy mildews are primarily foliage blights that attack and spread rapidly in young, tender green tissues including leaves, twigs, and fruit. Their development and severity, in areas where susceptible hosts and the respective downy mildew fungus are present, depend greatly on the presence of a film of water on the plant tissues and on high relative humidity in the air during cool or warm, but not hot, periods. The reproduction and spread of the downy mildew fungi are rapid and their diseases can cause heavy losses in short periods of time.

Although even the late blight of potato and tomato looks like and is often called a downy mildew, the true downy mildews are caused by a group of oomycetous fungi that belong to the family Peronosporaceae. All species of this family are obligate parasites of higher plants and cause downy mildew diseases on a large number of plants including most of the cultivated grain crops and vegetables, and many field crops, ornamentals, shrubs, and vines.

The downy mildews of several crops have caused spectacular and catastrophic epidemics in the past, and some of them continue to the present to cause epidemics and severe losses. The best known of the downy mildew epidemics is probably the one caused by the downy mildew of grapes, which soon after it was imported to Europe from the United States, almost completely destroyed the grape and wine industry in France and most of the rest of Europe and resulted in the discovery of the first fungicide, Bordeaux mixture, in 1885. In recent years, the downy mildew of sorghum has appeared and spread in the United States and has raised fears of future introduction of other, even more serious, downy mildews of gramineous crops now present in Asia and Africa. In 1979, a devastating epidemic of downy mildew (blue mold) of tobacco spread rapidly from Florida up the eastern states to New England and into Canada and destroyed much of the tobacco in its path, causing losses to growers of more than 250 million dollars. Similar epidemics also began in the following three years but were kept in check with heavy uses of fungicides, although not before they spread over several states.

The downy mildew fungi produce sporangia on sporangiophores that are distinct from the mycelium in their way of branching. The sporangia are located at the tips of the branches. Each genus of downy mildew fungi has its own distinctive type of branching of its sporangiophores, and this is used for their identification. The sporangiophores are usually long, white at first, emerging in groups from the plant tissues through the stomata. Later on they appear grayish or light brown and form a visible mat of fungus growth on the lower or both sides of leaves and on other affected tissues. Each sporangiophore grows until it reaches maturity and then produces its crop of sporangia, all at about the same time.

In most downy mildews, the sporangia germinate generally by producing zoospores or, at higher temperatures, by producing germ tubes. In the genus

Bremia, however, the sporangia germinate most commonly by means of a germ tube, and in the genera *Peronospora* and *Peronosclerospora* the sporangia germinate only by means of a germ tube. Whenever sporangia germinate by producing a germ tube, they are considered spores in themselves rather than sporangia, and in that case they are often called conidia, which always germinate by germ tubes.

The oospores of the downy mildews usually germinate by germ tubes, but in a few cases they may produce a sporangium that releases zoospores.

In most downy mildews, for example, of alfalfa, corn, lettuce, onion, sorghum, spinach, sugarcane, sunflower, and tobacco, the pathogen routinely causes systemic shoot infection of its host when it is carried in the seed or bulb or when infection takes place at the seedling or young plant stage. When organs of older plants are attacked they may develop localized, although not necessarily small, infected areas, or they may allow the fungus to spread into young tissues and become locally systemic. In some downy mildews, for example, of grape, soybean, and some grasses, infections usually result in small or large lesions on the leaves and stem, but even in these the fungus often becomes systemic when it infects young shoots and fruit stalks.

The downy mildews can and often do cause rapid and severe losses of crop plants while still in the seedbed or later on in the field. They often destroy from 40 to 90 percent of the young plants or young shoots in the field, causing heavy or total losses of crop yields. The amount of losses depends somewhat on the initial amount of inoculum but, most importantly, on the prolonged presence of wet, cool weather during which the downy mildew fungi sporulate profusely, cause numerous new infections, and spread into and rapidly kill young succulent tissues. The spread and destructiveness of downy mildews in cool, wet weather used to be uncontrollable. For some field crops, and in many less developed countries, downy mildews are still uncontrollable, checked only when the weather turns hot and dry. Since the early 1980s, however, several systemic fungicides, such as metalaxyl, propamocarb, and fosetyl-Al, have become available and have improved considerably our ability to control these diseases, although downy mildews are still very difficult to control.

Some of the most common or most serious downy mildew fungi and the diseases they cause are listed below. The structure of their sporangiophores is given in Figure 11-7.

Bremia lactucae, causing downy mildew of lettuce.

Peronospora, causing downy mildew of snapdragon *(P. antirrhini),* of onion *(P. destructor)* (Figure 11-22, C-D), of spinach *(P. effusa),* of soybeans *(P. manchurica),* mildew (blue mold) of tobacco *(P. tabacina)* and of alfalfa and clover *(P. trifoliorum).*

Peronosclerospora, causing downy mildew of sorghum and corn *(P. Sorghi),* of corn *(P. maydis* and *(P. philippinensis),* of corn and sugarcane *(P. sacchari).*

Plasmopara, causing downy mildew of grape *(P. viticola).* (Figure 11-23, A-C), and of sunflower *(P. halstedii).*

Pseudoperonospora, causing downy mildew of cucurbits *(P. cubensis)* (Figure 11-23, D-F), of hops *(P. humuli).*

Sclerophthora macrospora, causing downy mildew of cereals (corn, rice, wheat) and grasses.

Sclerospora, causing downy mildew of grasses and millets *(S. graminicola).*

• Downy Mildew of Grape

Downy mildew of grape occurs in most parts of the world where grapes are grown under humid conditions. The fungus that causes downy mildew of grape is native to North America, and although it attacks the native American grape vines it does not affect them very seriously. When the fungus, however, was introduced inadvertently into Europe at about 1875, the European grape, *Vitis vinifera,* which had evolved in the absence of the downy mildew fungus, was extremely susceptible to it, and the fungus began to spread among vineyards throughout France and throughout most of Europe, destroying the crop and the vineyards in its path. Even today, although other grapes may be attacked by downy mildew, it is the popular *Vinifera* grapes that are affected by far the most severely. Downy mildew is still most destructive in Europe and in the eastern half of the United States, where it may cause severe epiphytotics year after year, but it is also known to have caused serious losses in some years in northern Africa, in South Africa, in parts of Asia, Australia, and South America. Dry areas are usually free of the disease.

Downy mildew affects the leaves, fruit, and vines of grape plants and causes losses through killing of leaf tissues and defoliation, through production of low quality, unsightly, or entirely destroyed grapes, and through weakening, dwarfing, and killing of young shoots. When weather is favorable and no protection against the disease is provided, downy mildew can easily destroy 50–75 percent of the crop in one season.

◦ *Symptoms.* Downy mildew is usually first observed as small, pale yellow spots with indefinite borders on the upper surface of the leaves, while on the under surface of the leaves, and directly under the spots, a downy growth of the sporophores of the fungus appears (Figure 11-23 A-C). Later, the infected leaf areas are killed and turn brown, while the sporophores of the fungus on the under surface of the leaves become dark gray. The necrotic lesions are irregular in outline, and as they enlarge they may coalesce to form large dead areas on the leaf, frequently resulting in defoliation.

During blossom or early fruiting stages, entire clusters or parts of them may be attacked, are quickly covered with the downy growth, and die. If infection takes place after the berries are half-grown, the fungus grows mostly internally, the berries become leathery and somewhat wrinkled and develop a reddish-marbling to brown coloration.

Infection of green young shoots, tendrils, leaf stems, and fruit stalks results in stunting, distortion, and thickening (hypertrophy) of the tissues. Entire shoots may be covered with the downy growth of the fungus. Later the fungus growth breaks down and disappears, and the infected tissues turn brown and die. In late or localized infections the shoots may not be killed, but they show various degrees of distortion.

◦ *The Pathogen: Plasmopara viticola.* The mycelium diameter varies from 1 to 60 μm because the hyphae take up the shape of the intercellular spaces of the infected tissues. The mycelium grows between the cells but sends numerous, globose haustoria into the cells (Figure 11-24). In humid weather the mycelium produces sporangiophores, which emerge on the underside of the leaves and on the stems through the stomata or rarely by pushing directly through the epidermis. In the young fruit, sporangiophores emerge through lenticels. Usually 4–6 sporangiophores arise through a single stoma, but sometimes there may be as many as 20. Each produces 4–6 branches at nearly right

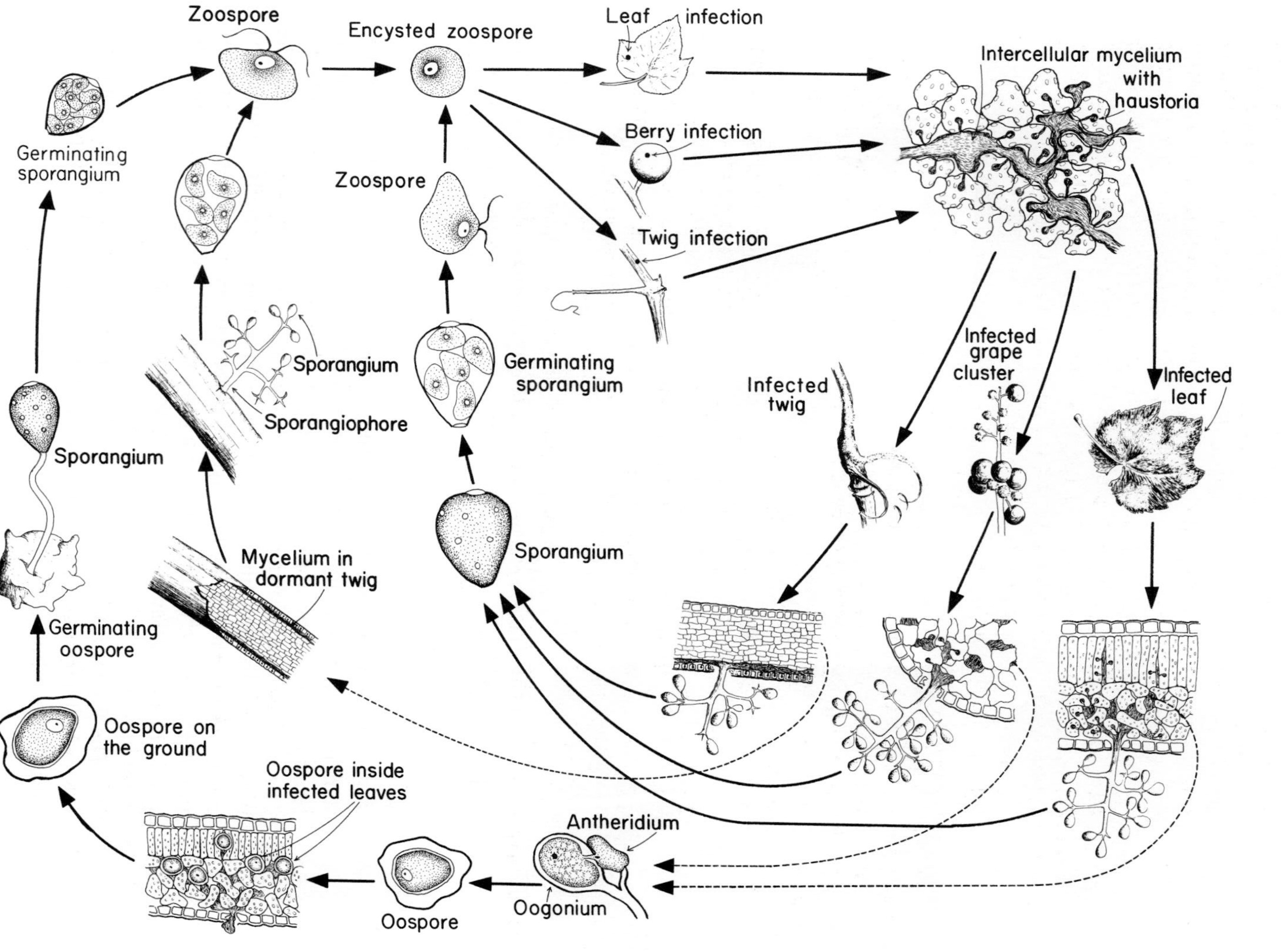

FIGURE 11-24 Disease cycle of downy mildew of grapes caused by *Plasmopara viticola*.

◦ *Development of Disease.* The pathogen overwinters as oospores embedded in the dead leaves, and occasionally, in dead berries and shoots (Figure 11-24). In certain areas, the fungus may also overwinter as mycelium in infected, but not killed, twigs. The dead leaves containing the oospores disintegrate during the winter and liberate the oospores. During rainy periods in the spring the oospores germinate either on the ground or on parts of grape plants to which they are carried by wind or splashing rain drops. The produced sporangium or its zoospores are transported by wind or water to the wet leaves near the ground, which they infect. Penetration takes place through stomata of the lower surface of the leaves. The mycelium then spreads into the intercellular spaces of the leaf, obtaining food through globose haustoria, which it sends into the cells. The mycelium continues to spread into the tissues, and when it reaches the substomatal cavity it forms a cushion of mycelium from which the sporangiophores arise and emerge through the stoma. On these primary lesions, great numbers of sporangia are produced, which may be carried by wind or rain to nearby healthy plants, germinate quickly, and produce many zoospores. The zoospores then cause secondary infections through stomata or lenticels and thus rapidly spread the disease. The period from infection to new sporangia formation varies from 5 to 18 days, depending on temperature, humidity, and varietal susceptibility.

In the stems the fungus invades the cortex, ray parenchyma, and pith. The distortion and hypertrophy of infected stems is caused by the enlargement of the affected cells and the large volume of mycelium present in the intercellular spaces. Finally, the affected stem cells are killed and collapse, producing brown sunken areas in the stem. In the young berries, infection is also intercellular, the chlorophyll breaks down and disappears, the cells collapse and turn brown.

At the end of the growing season the fungus forms oospores in the infected old leaves, and sometimes in the shoots and berries.

◦ *Control.* Several American grape varieties show considerable resistance to downy mildew, but most European *(vinifera)* varieties are quite susceptible; even the relatively resistant varieties, however, require protection through chemicals.

The most effective fungicides for control of downy mildew have been Bordeaux mixture, ferbam, zineb, mancozeb, folpet, and captan. The applications begin before bloom and are continued at 7- to 10-day intervals, although the time and number of applications vary with the local conditions, particularly the frequency and duration of rainfall during the growing season. Disease prediction systems, based on duration of leaf wetness, relative humidity, and temperature are used in Europe and elsewhere to identify infection periods and time fungicide applications.

In recent years, soil applications or sprays of the systemic fungicides metalaxyl, fosetyl Al, cyprofuram, and others, alone or in combination with copper oxychloride or mancozeb, have given excellent control of grape downy mildew.

SELECTED REFERENCES

Anderson, P. J. (1937). Downy mildew of tobacco, *Bull—Conn. Agric. Exp. Stn., New Haven* **405**.

Arens, K. (1929). Physiologische Untersuchungen an *Plasmopara viticola,* unter besonderer Berucksichtingung der Infektionsbedingungen. *Jahrb. Wiss. Bot.* **70**, 93–157.

Blaeser, M., and Weltzien, H. C. (1978). Die Bedutung von Spangienbildung, -ausbreitung and -keimung fur die Epidemiebildung von *Plasmopara viticola. Z. Pflanzenkr. Pflanzenschutz* **85,** 155–161.

Cook, H. T. (1932). Studies of the downy mildew of onions and the causal organism, *Peronospora destructor. Mem.—N. Y., Agric. Exp. Stn. (Ithaca)* **143.**

Davis, J. M., Main, C. E., and Bruck, R. I. (1981). Analysis of weather and the 1980 blue mold epidemic in the United States and Canada. *Plant Dis.* **65**, 508–512.

DeCastella, F., and Brittlebank, C. C. (1924). Downy mildew of the vine *(Plasmopara viticola* B. and de T.). *Dep. Agric., Victoria, Aust. Bull.* **49**, 1–45.

Doran, W. L. (1932). Downy mildew of cucumbers. *Mass. Agric. Exp. Stn., Bull.* **283,** 1–22.

Frederiksen, R. A. (1980). Sorghum downy mildew in the United States: Overview and outlook. *Plant Dis.* **64**, 903–908.

Frederiksen, R. A., *et al.* (1973). Sorghum downy mildew, a disease of maize and sorghum. *Tex. Agric. Exp. Stn.* Res. Monogr. **2**, 1–32.

Gäumann, E. (1923). Beitrage zu einer Monographie der gattung *Peronospora. Beitr. Kryptogamonoflora Schweiz.* **5**, 1–360.

Gregory, C. T. (1915). Studies on *Plasmopara viticola. Int. Congr. Viticult. Rep.*, pp. 126–150.

Klinkowski, M. (1962). Die europaische Pandemic von *Peronospora tabacina,* dem Erreger des Blauschimmels des Tabaks. *Biol. Zentralbl.* **81**, 75–89.

Lee, T. C., and Wicks, T. (1982). Dual culture of *Plasmopara viticola* and grapevine and its application to systemic fungicide evaluation. *Plant Dis.* **66**, 308–310.

Lucas, G. B. (1980). The war against blue mold. *Science* **210,** 147–153.

Millardet, P. M. A. (1885). (1) Traitement du mildiou et du rot. (2) Traitement du mildiou par le melange de sulphate de cuivre et de chaux. (3) Sur l'histoire du traitment du mildiou par le sulphate de cuivre. *J. Agric. Prat.* **2**, 513–516, 707–719, 801–805; Engl. transl. by F. L. Schneiderhan in *Phytopathol. Classics* **3** (1933).

Royle, D. J., and Thomas, G. G. (1973). Factors affecting zoospore responses towards stomata in hop downy midlew *(Pseudoperonospora humuli)* including some comparisons with grapevine downy mildew *(Plasmopara viticola). Physiol. Plant Pathol.* **3**, 405–417.

Smith, R. W., Lorbeer, J. W., and Abd-Elrazik, A. A. (1985). Reappearance and control of onion downy mildew epidemics in New York. *Plant Dis.* **69**, 703–706.

Spencer, D. M., ed. (1981). "The Downy Mildews." Academic Press, New York.

Thomas, C. E. (1977). Disease development and sporulation of *Pseudoperonospora cubensis* on susceptible and resistant cantaloups. *Plant Dis. Rep.* **61**, 375–377.

Waterhouse, G. M. (1964). The genus *Sclerospora.* Diagnosis (or descriptions) from the original papers and a key. *Commonw. Mycol. Inst., Misc. Publ.* **17**.

Wicks, T., and Lee, T. C. (1982). Evaluation of fungicides applied after infection for control of *Plasmopara viticola* on grapevine. *Plant Dis.* **66**, 839–841.

Williams, R. J. (1984). Downy mildews of tropical cereals. *Adv. Plant Pathol.* **2**, 2–103.

Zachos, D. G. (1959). Recherches sur la biologie et l'épidemiologie du mildiou de la vigne en Grèce. Bases de prévisions ou d'avertissements. *Ann. Inst. Phytopathol. Benaki* **2**, 196–355.

Diseases Caused by Zygomycetes

The Zygomycetes have well-developed mycelium, also without cross wall, produce nonmotile spores in sporangia, and their resting spore is a thick-walled zygospore produced by the union of two morphologically similar gametes. The Zygomycetes are strictly terrestrial fungi, their spores often floating around in the air, and are either saprophytes or weak parasites of plants and plant products on which they cause soft rots or molds. Some, for example, *Rhizopus,* are opportunistic pathogens of humans.

Two genera of Zygomycetes are known to cause disease in living plants or living plant tissue (Figures 11-7 and 11-8). These are: (1) *Choanephora,* which attacks the withering floral parts of many plants after fertilization and from them invades the fruit and causes a soft rot of squash, pumpkin, pepper, and okra, but most severely of summer squash (Figure 11-25C); and (2) *Rhizopus,* the common bread mold fungus, which in addition causes soft rot of many fleshy fruits, vegetables, flowers, bulbs, corms, and seeds (Figures 11-25,A, B). Another genus, *Mucor,* also causes molding of bread and other processed plant products, and rarely, a rot of sweet potatoes stored at low temperatures. Other genera, for example, *Glomus* and *Gigaspora,* are fungi that become associated with roots of plants and form endomycorrhizae that are beneficial to the plant.

The plant-pathogenic Zygomycetes are weak parasites. They grow mostly as saprophytes on dead or processed plant products, and even when they infect living plant tissues, they first attack injured or dead plant tissues. In the latter, the fungi build up large masses of mycelium. This secretes enzymes that diffuse into the living tissue and disrupt and kill the cells. The mycelium then grows into the "living" tissue.

The Zygomycetes have typical elongated mycelium without cross walls. Their asexual spores are produced within sporangia, are nonmotile, and are called sporangiospores, or conidia. They are generally spread by air currents. Their sexual spores are produced through the fusion of two more or less similar-looking sex cells, they are called zygospores. The zygospores are thick walled and can overwinter or withstand other adverse conditions. Under favorable moisture and temperature conditions the zygospores germinate by producing a sporangium containing sporangiospores.

• Rhizopus Soft Rot of Fruits and Vegetables

Rhizopus soft rot of fruits and vegetables occurs throughout the world on harvested fleshy organs of vegetable, fruit, and flower crops and is important only during storage, transit, and marketing of these products. Among the crops affected most by this disease are sweet potatoes, strawberries, all cucurbits, peaches, cherries, peanuts, and several other fruits and vegetables. Corn and some other cereals are affected under fairly high conditions of moisture. Bulbs, corms, and rhizomes of flower crops, for example, gladiolus and tulips, are also susceptible to this disease.

When conditions are favorable, the disease spreads rapidly throughout the containers, and losses can be great in a short period of time. (See Figure 11-81, page 438).

◦ *Symptoms.* Infected areas of fleshy organs appear water soaked at first, and they are very soft. If the skin of the infected tissues remains intact, the softened fleshy organ gradually loses moisture until it shrivels into a mummy. More frequently, however, the softened skin ruptures during handling or under pressure, for example, from surrounding fruits, and a whitish-yellow liquid drops out. Soon fungus hyphae grow outward through the wounds and cover the affected portions by producing tufts of whiskerlike gray sporangiospores bearing black sporangia at their tips (Figure 11-25). The bushy growth of the fungus extends to the surface of the healthy portions of affected fruit when they become wet with the exuding liquid and even to the surface of the containers. Affected tissues at first give off a mildly pleasant smell, but soon yeasts and bacteria move in and a sour odor develops. When loss of moisture

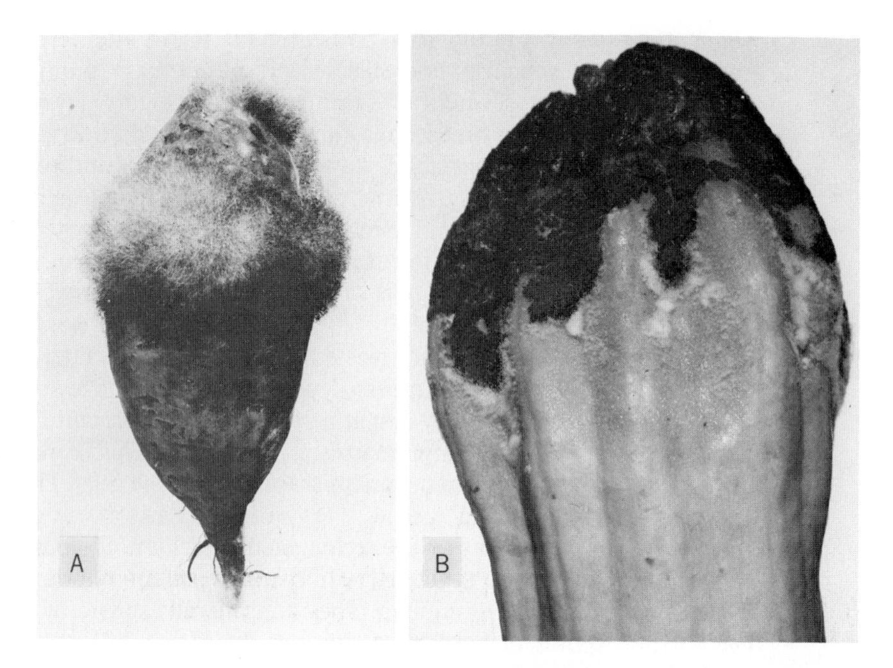

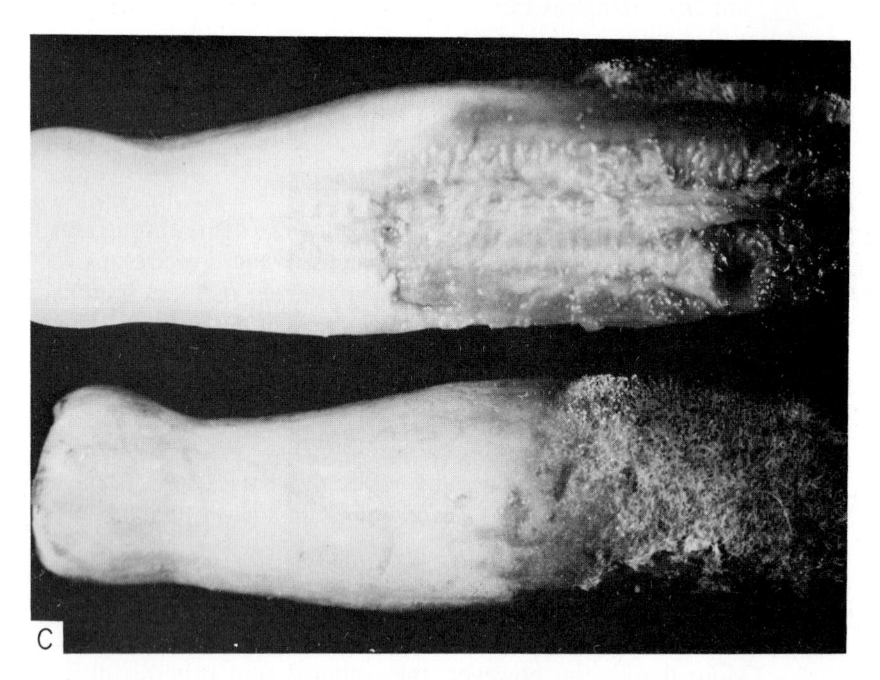

FIGURE 11-25 *Rhizopus* soft rot on sweet potato (A) and on squash (B). *Choanephora* soft rot of young summer squash in the field (C). Sporangiophores and sporangia growing on the surface of sweet potato and squash in the presence of high relative humidity. The maceration and softening of infected tissue can be seen in the longitudinal section of squash (C, upper). (Photo A courtesy of U.S.D.A. Photo B courtesy Dept. Plant Path., Cornell Univ.)

is rapid, infected organs finally dry up and mummify; otherwise they break down and disintegrate in a watery rot.

◦ *The Pathogen: Rhizopus sp.* Rhizopus is found almost everywhere in nature. It lives usually as a saprophyte and sometimes as a weak parasite on stored organs of plants. The mycelium of the fungus has no cross walls and produces long, aerial sporangiophores at the tips of which black spherical sporangia develop (Figures 11-25 and 11-26). These consist of a thin membrane containing thousands of small spherical sporangiospores. When the membrane of the sporangium is ruptured, the liberated sporangiospores are released and float about in the air or drop to the surface. If they fall on a moist surface or wound of a susceptible plant organ, the sporangiospores germinate and produce mycelium again. When the mycelium grows on a surface, it produces stolons, that is, hyphae that arch over the surface and at the next point of contact with the surface produce both rootlike hyphae, called rhizoids, which grow toward the surface, and more aerial sporangiophores bearing sporangia. From each point of contact more stolons are produced in all directions. Adjacent hyphae produce short branches called progametangia, which grow toward each other. When they come in contact, the tip of each hyphae is separated from the progametangium by a cross wall. The terminal cells are the gametangia. These fuse, their protoplasts mix, and their nuclei pair. The cell formed by the fusion enlarges and develops a thick, black, and warty cell wall. This sexually produced spore is called a zygospore and is the overwintering or resting stage of the fungus. When it germinates it produces a sporangiophore bearing a sporangium full of sporangiospores.

◦ *Development of Disease.* Throughout the year, sporangiospores float about and if they land on wounds of fleshy fruits, roots, corms, or bulbs they

FIGURE 11-26 Disease cycle of soft rot of fruits and vegetables caused by *Rhizopus* sp.

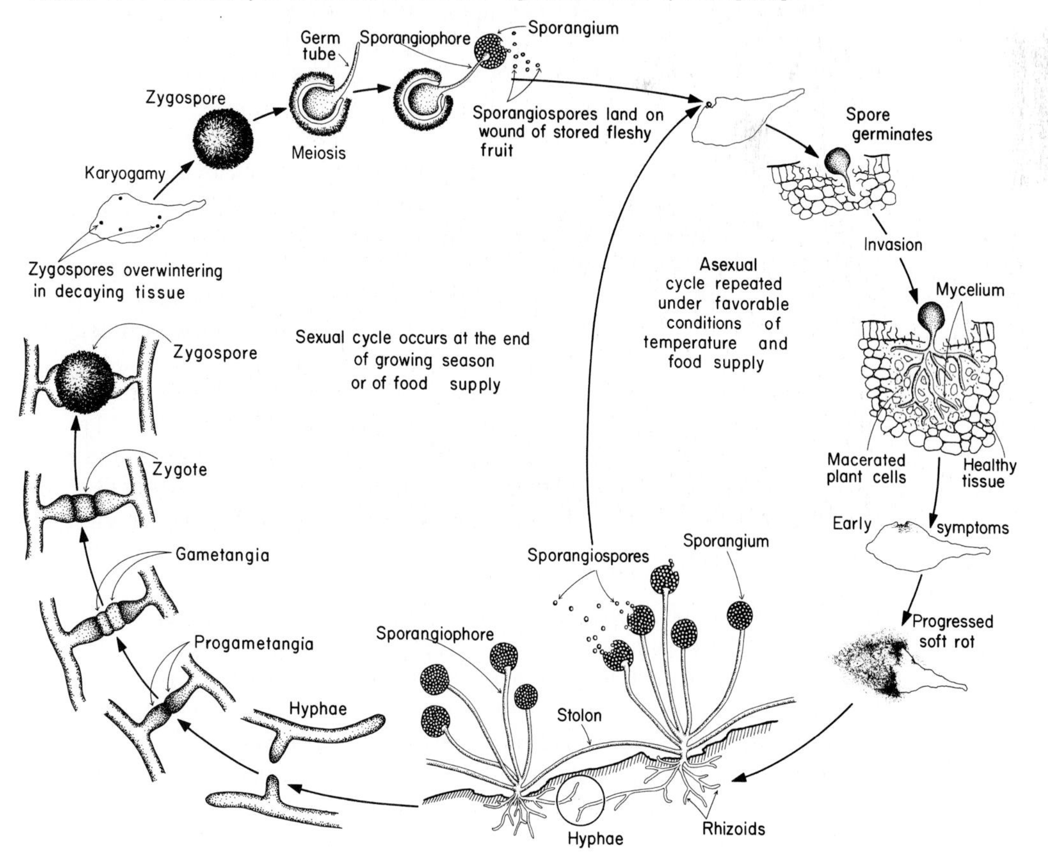

germinate (Figure 11-26). The hyphae thus produced secrete pectinolytic enzymes, which break down and dissolve the pectic substances of the middle lamella, that is, the substances that hold the cells in place in the tissues. This results in loss of cohesion among the cells, since they are now surrounded by liquefied substance, and upon pressure they move freely among each other, resulting in "soft rot."

The pectinolytic enzymes secreted by the fungus advance ahead of the mycelium and separate the cells before the mycelium gets there. Cells in the macerated tissues are then attacked by the cellulolytic enzymes of the fungus, which break down the cellulose of the cell wall, and the cells disintegrate. The mycelium grows intercellularly and does not seem to invade cells until after they are killed and have begun to disintegrate. Thus the mycelium seems never to be in contact with living host cells but is instead surrounded by dead cells and nonliving organic substances, living more like a saprophyte than a parasite.

The epidermis of infected organs is not usually broken down by the fungus, which continues to grow inside the tissues. The epidermis, however, is softened greatly and breaks easily with the slightest pressure during handling of the product or by the weight of the fruits above it. The fungus then emerges through wounds already existing on the fruit or through subsequent ruptures of the epidermis and produces aerial sporangiophores, sporangia, stolons, and rhizoids, the latter capable of piercing the softened epidermis. In extremely fleshy fruits, such as peaches and strawberries, the mycelium arising from an infected fruit or from a sporangiospore can penetrate healthy fruit in contact with the infected fruit in the absence of visible wounds. In some cases the epidermis of such fruits has already been softened by the fungal enzymes present in the liquid exuding from the infected fruit, but this is not always necessary.

Sporangiospores are produced on infected tissue within a few days and can cause new infections immediately after they are released. Zygospores, however, are produced later, when the food supply in the infected tissues begins to diminish and, with heterothallic species, only when a compatible strain is also present. Since zygospores germinate after a rest period, most infections in storage are caused by asexual sporangiospores, and within a package of fruit, most infections occur from the hyphae growing from the surface of the previously rotted fruit rather than from spores.

The initiation of infection and the invasion of the tissues by the fungus are influenced greatly by temperature and humidity and by the stage of ripeness of the tissue. Unfavorable combinations of temperature and humidity or insufficient maturity of the fruit slow down the growth and activity of the fungus, and they may allow the host to form layers of cork cells and other histological barriers that retard or completely inhibit further infections by the fungus.

◦ *Control.* Since the spores of the fungus are present everywhere, wounds of any size can be points of infection. Wounding of fleshy fruits, roots, tubers, and bulbs should, therefore, be avoided as much as possible during harvest, handling, and transportation of these products. Wounded organs should be discarded or packed and stored separately from healthy ones.

Storage containers and warehouses should be cleaned before use and should be disinfected with a copper sulfate solution, formaldehyde, sulfur fumes or chloropicrin.

Temperature control of storage rooms and shipping cars is important. With very succulent fruits, such as strawberries, picking fruit in the morning when it is cool and keeping it at temperatures below 10°C reduces losses considerably. Sweet potatoes and some other not so succulent organs may be protected from the disease by keeping them at 25–30°C and 90 percent humidity for 10–14 days, during which the cut surfaces cork over and do not allow subsequent penetration by the fungus. After this period is over, temperature is lowered again to about 12°C.

Quite effective control of the disease can be obtained by wrapping susceptible fruit in paper impregnated with various fungicidal substances such as dichloran.

SELECTED REFERENCES

Adams, J. F., and Russell, A. M. (1920). *Rhizopus* infection of corn on the germinator. *Phytopathology* **10**, 535–543.

Anderson, H. W. (1925). *Rhizopus* rot of peaches. *Phytopathology* **15**, 122–124.

Beraha, L., Ramsey, G. B., Smith, M. A., Wright, W. R. (1959). Effects of gamma radiation on brown rot and *Rhizopus* rot of peaches and the causal organisms. *Phytopathology* **49**, 354–356.

Luepschen, N. S. (1964). Effectiveness of 2,6-dichloro-4-nitroaniline-impregnated peach wraps in reducing *Rhizopus* decay losses. *Phytopathology* **54**, 1219–1222.

Mirocha, C. J., and Wilson, E. E. (1961). Hull rot disease of almonds. *Phytopathology* **51**, 843–847.

Spaulding, D. H. (1963). Production of pectinolytic and cellulolytic enzymes by *Rhizopus stolonifer*. *Phytopathology* **53**, 929–931.

Srivastava, D. N., and Walker, J. C. (1959). Mechanisms of infection of sweet potato roots by *Rhizopus stolonifer*. *Phytopathology* **49**, 400–406.

Stevens, N. E., and Wilcox, R. B. (1917). *Rhizopus* rot of strawberries in transit. *U. S., Dept. Agric., Bull.* **531**, 1–22.

Yang, S. M., Morris, J. B., Unger, P. W., and Thompson, T. E. (1979). Rhizopus head rot of cultivated sunflower in Texas. *Plant Dis. Rep.* **63**, 833–835.

DISEASES CAUSED BY THE HIGHER FUNGI

Diseases Caused by Ascomycetes and Imperfect Fungi

Ascomycetes and Imperfect Fungi (Deuteromycetes) are two groups of fungi that closely resemble each other in the structure of the mycelium, the production of asexual spores, and in the kinds of diseases they cause in plants, as well as in the way they induce these diseases. Thus, both produce a haploid mycelium that has cross walls, both produce conidia in identical types of conidiophores or fruiting bodies, and both cause plant diseases that may appear as leaf spots, blights, cankers, fruit spots, fruit rots, anthracnoses, stem rots, root rots, vascular wilts, or soft rots.

The only difference between Ascomycetes and Imperfect Fungi is that the former also produce, regularly or rarely, sexual spores, known as ascospores. In many Ascomycetes, however, ascospores are seldom found in nature and seem to play little or no role in the survival of the fungus and in its ability to cause disease in plants. Thus, such Ascomycetes reproduce, spread, cause disease, and overwinter as mycelium or conidia, or both, so that they

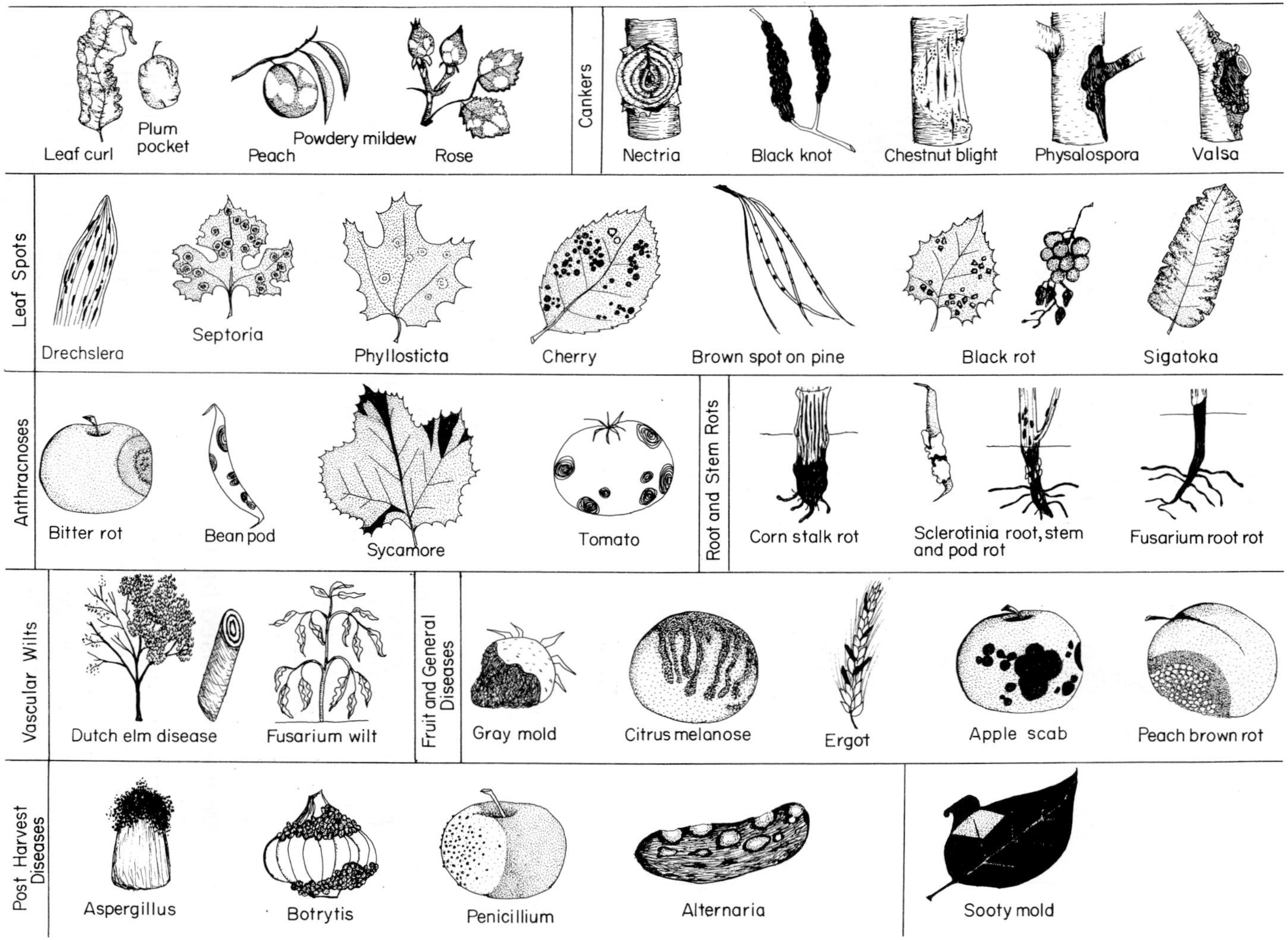

FIGURE 11-33 Common symptoms caused by some important Ascomycetes and Imperfect Fungi.

FIGURE 11-34 Sooty mold on orange leaves. The fungus forms a leathery mat that can be peeled off or washed off. (Photo courtesy Agric. Res. Educ. Center, Lake Alfred, Fla.)

The *Taphrina* diseases are best known in Europe and North America but probably occur all over the world. *Taphrina* causes defoliation of peach trees, which may lead to small fruit or fruit drop. In plum, 50 percent or more of the fruit may be affected and lost in years when the disease is severe. In both peach and plum, buds and twigs may also be affected, thus devitalizing the tree.

◦ *Symptoms.* In peach and nectarine parts of or entire infected leaves are thickened, swollen, distorted, and curled downward and inward (Figure 11-35). Affected leaves at first appear reddish or purplish but later turn reddish yellow or yellowish gray. At this stage the fungus produces its spores on the swollen areas, which appear powdery gray. The leaves later turn yellow to brown and drop. Blossoms, young fruit, and the current year's twigs may also be attacked. Infected blossoms and fruit generally fall early in the season. The infected twigs are always swollen and stunted and die during the summer.

In plum, the disease first appears as small white blisters on the fruit. The blisters enlarge rapidly as the fruit develops and soon involve the entire fruit. The fruit increases abnormally in size and is distorted, the flesh becoming spongy. The seed ceases to develop, turns brown, and withers, leaving a hollow cavity. The fruit appears reddish at first, but later becomes gray and covered with a grayish powder. Leaves and twigs may also be affected, as in peach.

FIGURE 11-35 (A) Peach leaf curl caused by *Taphrina deformans*. (B) Oak leaf blister caused by *T. coerulescens*.

- *The Pathogen: Taphrina sp.* The mycelial cells of *Taphrina* contain two nuclei. These cells may develop into an ascus, usually containing eight uninucleate ascospores. The ascospores multiply by budding inside or outside the ascus producing blastospores (conidia). The latter may bud again to produce more thin- or thick-walled conidia or may germinate to produce mycelium. Upon germination, the conidial nucleus divides and the two nuclei move into the germ tube. As the mycelium grows, both nuclei divide concurrently, producing the binucleate cells of the mycelium. Mycelial cells near the plant surface separate from each other and produce the asci.
- *Development of Disease.* The fungus apparently overwinters as ascospores or thick-walled conidia on the tree, perhaps on the bud scales. In the spring, these spores are washed, splashed, or blown onto young tissues, germinate, and penetrate the developing leaves and other organs directly through the cuticle or through stomata (Figure 11-36). The binucleate mycelium then grows between cells and invades the tissues extensively, inducing excessive cell enlargement and cell division, which result in the enlargement and distortion of the plant organs. Later, numerous hyphae grow outward in the area between the cuticle and epidermis. There they break into their component cells, which produce the asci. The asci enlarge, exert pressure on the host cuticle from below, and eventually break through to form a compact, feltlike layer of naked asci. The ascospores are released into the air, carried to new tissues, and bud to form conidia. Infection occurs mainly during a short period after the buds open. All organs become resistant to infection as they grow older. Infection is favored by low temperature and a high humidity from the time of bud swell until young shoots and leaves develop, that is, the period during which the new tissues are susceptible.
- *Control.* *Taphrina* diseases are easily controlled by a single fungicide spray, preferably in late fall after the leaves have fallen or in early spring before leaf buds swell. The fungicides most commonly used are ferbam, elgetol, and Bordeaux mixture (8:8:100). Difolatan may also be used as one application only, before leaf drop is complete, while chlorothalonil controls the disease if applied twice, in late fall and in early spring.

SELECTED REFERENCES

Fitzpatrick, R. E. (1934). The life history and parasitism of *Taphrina deformans*. *Sci. Agric.* **14**, 305–306.

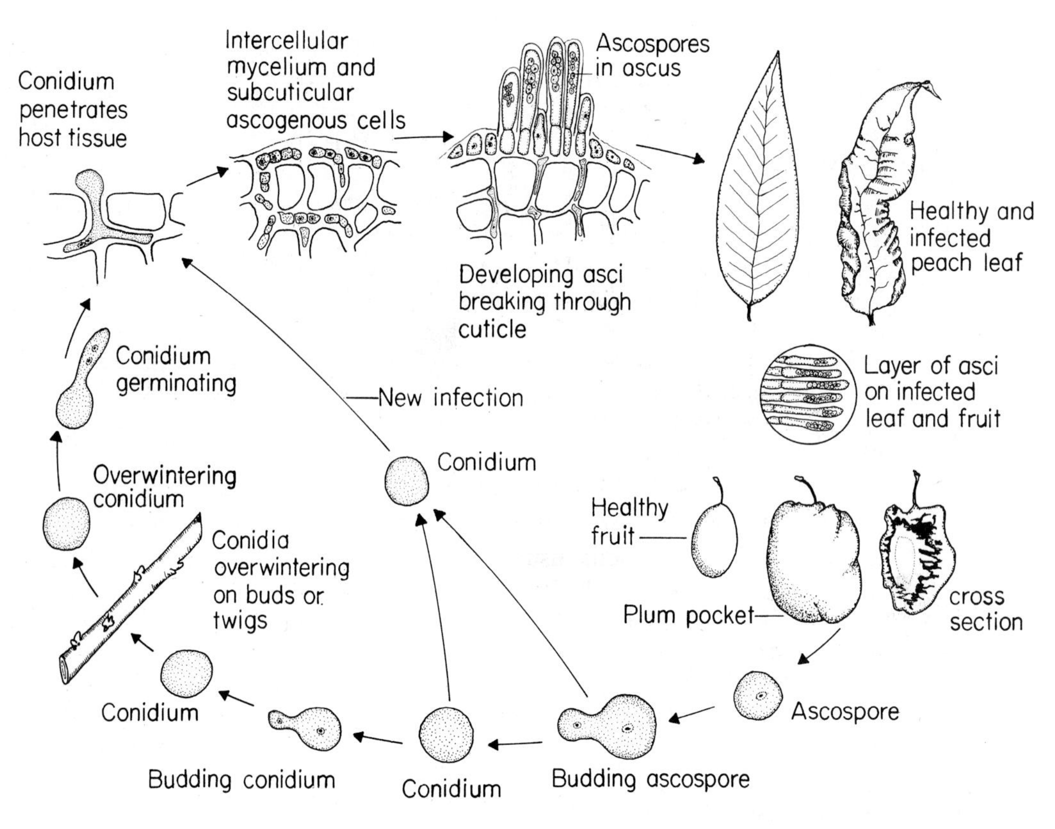

FIGURE 11-36 Disease cycle of peach leaf curl and plum pocket caused by *Taphrina* sp.

Martin, E. M. (1940). The morphology and cytology of *Taphrina deformans*. *Am. J. Bot.* **27**, 743–751.
Mix, A. J. (1935).The life history of *Taphrina deformans*. *Phytopathology* **25**, 41–46.
Mix, A. J. (1949). A monograph of the genus *Taphrina*. *Univ. Kans. Sci. Bull.* **33**, 1–167.
Ritchie, D. F., and Werner, D. J. (1981). Susceptibility and inheritance of susceptibility to peach leaf curl in peach and nectarine cultivars. *Plant Dis.* **65**, 731–734.
Weber, G. F. (1941). Leaf blister of oaks. *Bull.—Fla., Agric. Exp. Stn.* **558,** 1–2.

The Powdery Mildews

Powdery mildews are probably the most common, conspicuous, wide-spread, and easily recognizable plant diseases. They affect all kinds of plants: cereals and grasses, vegetables, ornamentals, weeds, shrubs, fruit trees, and broad-leaved shade and forest trees.

Powdery mildews are characterized by the appearance of spots or patches of a white to grayish, powdery, mildewy growth on young plant tissues, or of entire leaves and other organs being completely covered by the white powdery mildew. Tiny, pinhead-sized, spherical fruiting bodies, at first white, later yellow-brown, and finally black cleistothecia may be present singly or in groups on the white to grayish mildew in the older areas of infection. Powdery mildew is most commonly observed on the upper side of the leaves, but it also affects the under side of leaves, young shoots and stems, buds, flowers, and young fruit.

The fungi causing powdery mildews are obligate parasites—they cannot be cultured on artificial nutrient media. They produce mycelium that grows only on the surface of plant tissues, never invading the tissues themselves. They obtain nutrients from the plant by sending haustoria (feeding organs) into the epidermal cells of the plant organs. The mycelium produces short conidiophores on the plant surface. Each canidiophore produces chains of rectangular, ovoid, or round conidia that are carried by air currents. When environmental conditions or nutrition become unfavorable, the fungus may produce one or a few asci inside a closed ascocarp, the cleistothecium. The powdery mildew fungi, although they are common and cause serious diseases in cool or warm, humid areas, are even more common and severe in warm, dry climates. This happens because their spores can be released, germinate, and cause infection even when the relative humidity in the air is fairly high but there is no film of water on the plant surface. Once infection has begun, the mycelium continues to spread on the plant surface regardless of the moisture conditions in the atmosphere.

Powdery mildews are so common, widespread, and ever present among crop plants and ornamentals that the total losses, in plant growth and crop yield they cause each year on all crops, probably surpass the losses caused by any other single type of plant disease. Powdery mildews seldom kill their hosts but utilize their nutrients, reduce photosynthesis, increase respiration and transpiration, impair their growth, and reduce yields, sometimes by as much as 20–40 percent. Among the plants most severely affected by powdery mildew are the various cereals, such as wheat and barley, primarily because in these crops chemical control of plant diseases is difficult or impractical. Other crops that suffer common and severe losses from powdery mildew are the cucurbits, especially cantaloupe, squash and cucumber, sugar beets, strawberries, and clovers, many ornamentals such as rose, begonia, delphinium, azalea and lilac, grape, and many trees, particularly apple, catalpa, and oak. The control of powdery mildews in cereals and several other annual crops has been primarily through the use of resistant varieties but, more recently, also with systemic fungicides such as ethirimol, triadimenol, and triforine used as seed treatments, or prochloraz, triadimefon, tridemorph, and triforine, and others, used as foliar sprays. The same chemicals are used for control of powdery mildews in the other crops and in ornamentals, although elemental sulfur and dinocap have been and still are used extensively and effectively. Powdery mildew on trees, such as apple, is effectively controlled with sprays of any of several sterol-inhibiting systemic fungicides such as triadimefon, etaconazole, bitertanol, and triforine.

The powdery mildew diseases of the various crop or other plants are caused by many species of fungi of the family Erysiphaceae grouped into six main genera. These genera are distinguished from each other by the number (one versus several) of asci per cleistothecium and by the morphology of hyphal appendages growing out of the wall of the cleistothecium. The main genera are illustrated in Figure 11-28, and the most important diseases they cause are listed below.

Erysiphe cichoracearum, causes powdery mildew of begonia, chrysanthemum, cosmos, cucurbits, dahlia, flax, lettuce, phlox, zinnia.

E. graminis, causes powdery mildew of cereals and grasses (Figure 11-37G).
E. polygoni, causes powdery mildew of beans, soybeans, clovers and other legumes, beets, cabbage and other crucifers, cucumber and cantaloupe, delphinium, hydrangea.
Microsphaera alni, causes powdery mildew of blueberries, catalpa, elm, lilac, linden, oak, rhododendron, and sweet pea.
Phyllactinia sp., causes powdery mildew of catalpa, elm, maple, oak.
Podosphaera leucotricha, causes powdery mildew of apple (Figure 11-37F), pear and quince.
P. oxyacanthae, causes powdery mildew of apricot, cherry, peach and plum.
Spaerotheca macularis, causes powdery mildew of strawberry.
S. mors-uvae, causes powdery mildew of gooseberry and currant.
S. pannosa, causes powdery mildew of peach and rose (Figure 11-37, A-C).
Uncinula necator, causes powdery mildew of grape, horsechestnut, linden (Figure 11-37, D, E).

• Powdery Mildew of Rose

Powdery mildew of roses occurs everywhere in the world where roses are grown. Powdery mildew is one of the most important diseases of roses, both in the garden and in the greenhouse. The disease appears on roses year after year and causes reduced flower production and weakening of the plants by attacking the buds, young leaves, and growing tips of the plants.

◦ *Symptoms.* On the young leaves the disease appears at first as slightly raised blisterlike areas that soon become covered with a grayish white, powdery fungus growth, and as the leaves expand they become curled and distorted (Figure 11-37). On the older leaves, large white patches of fungus growth appear, but there is usually little distortion. Lesions on leaves may appear more or less discolored and may eventually become necrotic.

White patches of fungus growth, similar to those on the leaves, usually appear on young, green shoots, and they may coalesce and cover the entire terminal portions of the growing shoots, which may become arched or curved at their tip. Sometimes buds may be attacked and become covered with white mildew before they open. They then either fail to open or open improperly, the infection spreading to the flower parts, which become discolored, dwarfed, and eventually die.

◦ *The Pathogen: Sphaerotheca pannosa* f. sp. *rosae.* The pathogen that causes powdery mildew on roses seems to be a distinct special form of *S. pannosa,* since it has been shown in some cases that *S. pannosa* fungus from roses does not attack peaches, and vice versa.

The mycelium is white and grows on the surface of the plant tissues, sending globose haustoria into the epidermal cells of the plant (Figure 11-38). The mycelium forms a weft of hyphae on the surface, some of which develop into short, erect, conidiophores. At the tip of each conidiophore egg-shaped conidia are produced that cling together in chains.

With the coming of cool weather late in the season, conidia production ceases and cleistothecia may be formed. The young cleistothecia are globose and at first white, then brown, and finally black when mature. The mature cleistothecia also have several mycelioid appendages, which are flaccid, indefinite hyphae arising from cells of the cleistothecium. The cleistothecia are more or less buried in the mycelial wefts on the plant tissues. The ascospores

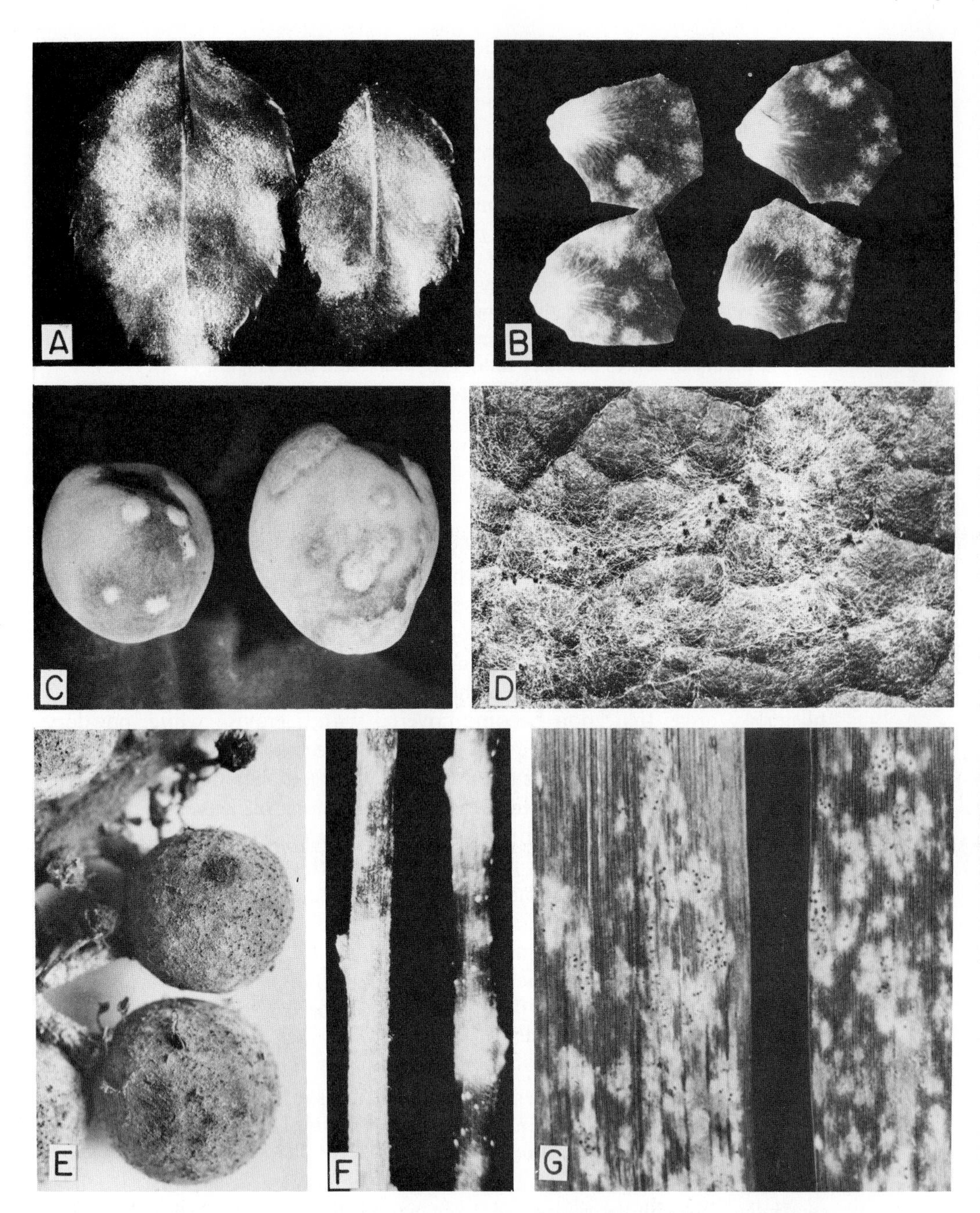

FIGURE 11-37 Powdery mildew on rose leaves (A) and petals (B), caused by *Sphaerotheca pannosa* f. sp. *rosae* and on peach fruit (C) caused by *Sphaerotheca pannosa.* (D, E) Mycelium, conidia and dark cleistothecia on grape leaves (D) and fruit cluster (E) caused by the powdery mildew fungus *Uncinula necator.* (F) Powdery mildew on apple twigs caused by *Podosphaera leucotricha.* (G) Powdery mildew on wheat leaves caused by *Erysiphe graminis.* The dark dots are cleistothecia. (Photos D, E courtesy Shade Tree Lab, Univ. of Mass. Photo G courtesy U.S.D.A.)

continue to develop during the fall, and in the spring they are mature and ready for dissemination. In the spring the cleistothecia absorb water and crack open. The single ascus in each cleistothecium protrudes its tip, bursts open, and discharges its eight mature ascospores, which the wind carries away. The

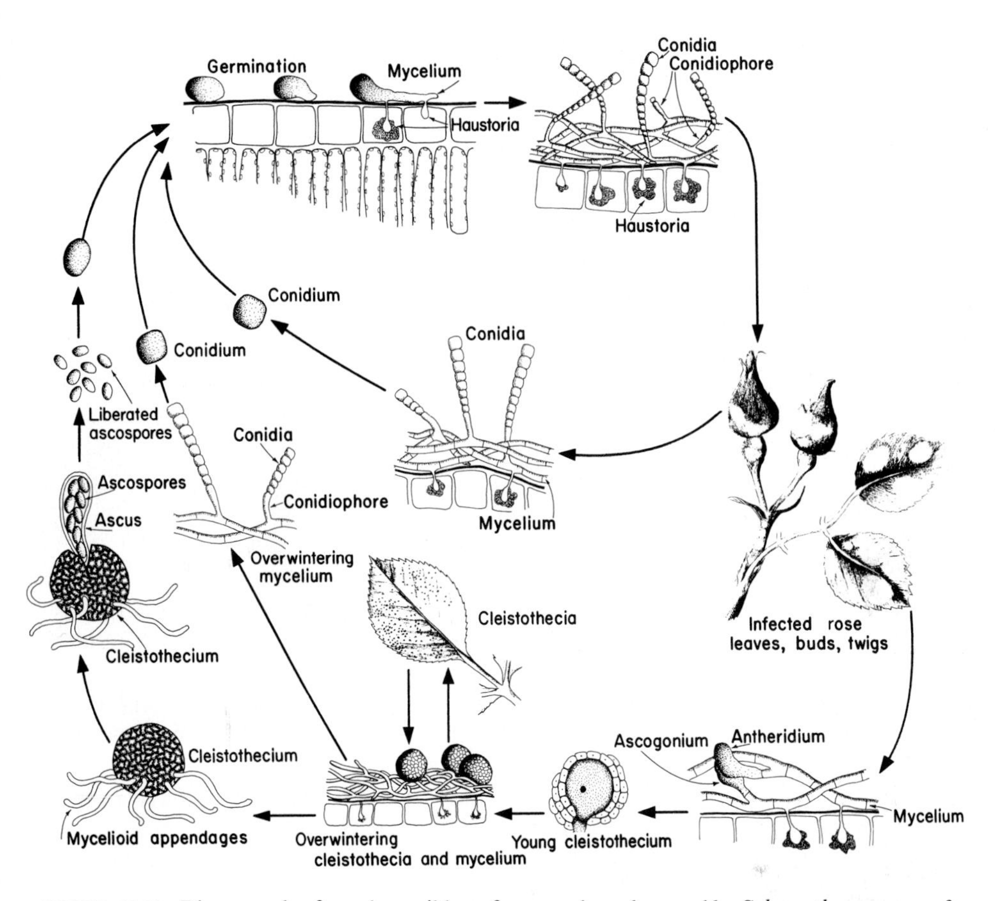

FIGURE 11-38 Disease cycle of powdery mildew of roses and peach caused by *Sphaerotheca pannosa* f. sp. *rosae.*

ascospores are about the size of conidia and behave exactly like conidia with respect to germination, infection, and formation of subsequent structures.

◦ *Development of Disease.* On outdoor roses the powdery mildew of rose fungus apparently overwinters mostly as mycelium in the buds, although cleistothecia form occasionally on leaves, petals, and stems, especially around the thorns, toward the end of the season. On greenhouse roses the pathogen perpetuates itself almost exclusively as mycelium and conidia.

When the fungus overwinters as mycelium in dormant buds, shoots arising from such buds become infected and provide inoculum for subsequent secondary mycelial or spore infection and disease development on foliage and flowers. When the fungus overwinters as cleistothecia the discharged mature ascospores also serve as primary inoculum (Figure 11-38). Ascospores or conidia are carried by wind to young green tissues, and if temperature and relative humidity are sufficiently high the spores germinate by putting out a germ tube. The germ tube quickly produces a short, fine hypha growing directly through the cuticle and the epidermal cell wall into the epidermal cells. The penetrating hypha enlarges immediately upon entrance into the cell lumen and forms a globose haustorium by which the fungus obtains its nutrients. The germ tube continues to grow and branch on the surface of the plant tissue producing a network of superficial mycelium. As the mycelium spreads on the plant, it continues to send haustoria into the epidermal cells.

The absorption of nutrients from the cells depletes their food supply, weakens them, and may sometimes lead to their death. Photosynthesis in the affected areas is greatly reduced, and the other functions of the cells are also impaired. Infection of young leaves also causes irritation and uneven growth of the affected and the surrounding cells, resulting in slight raised areas on the leaf and distortion of the leaf. The aerial mycelium produces short, erect conidiophores, each bearing a chain of 5–10 conidia. The conidia are disseminated by air currents and cause new infections on the expanding leaves and shoots. Greenhouse roses are susceptible throughout the year. In the field, however, expanding tissues seem to be the most susceptible ones, and only under quite favorable humidity and temperature do fully developed tissues become infected. Growth of severely infected shoots is inhibited. Infected buds often do not open. If they do open, the flowers become infected and do not develop.

◦ *Control.* Many new rose varieties show a moderately high level of resistance, but the resistance is not stable since some of them are resistant in some geographical areas but susceptible in others, or even in the same location, they are resistant some years and susceptible others. The variable resistance is presumably due to the presence or prevalence of different races of the pathogen in different geographical areas or during different seasons. Most popular rose varieties are highly susceptible to powdery mildew, and therefore additional protection by fungicides is necessary.

Powdery mildew has been controlled in the past by application of sulfur, dinocap, benomyl, and certain other fungicides. Sulfur may be used as a spray, as a dust, and in the greenhouse, as a vapor. Dinocap, benomyl, and cycloheximide are used as sprays. Under most conditions, weekly applications give adequate protection, but during rapid development of new growth, temperature fluctuations, and frequent rains, more frequent applications may be necessary. In the last few years, newer, more effective systemic fungicides such as triforine, fenarimol, triadimefon, dodemorph, and etaconazole have replaced many of the older fungicides in the control of powdery mildew. Some of these, for example, fenarimol, triadimefon, and etaconazole, are excellent for controlling powdery mildew in greenhouses by volatilization at normal greenhouse temperatures or after heating.

Several fungi such as *Ampelomyces quisqualis, Cladosporium oxysporum, Tilletiopsis* sp., and *Verticillium lecanii,* and the insect *Thrips tabaci* have been reported to parasitize or antagonize the rose powdery mildew and the powdery mildew fungi of several other crops. Although this control approach appears promising, so far it has not been developed sufficiently to be used for practical control of powdery mildews.

SELECTED REFERENCES

Cherewick, W. J. (1944). Studies on the biology of *Erysiphe graminis. Can. J. Res.* **22**, 52–86.

Coyier, D. L. (1983). Control of rose powdery mildew in the greenhouse and field. *Plant Dis.* **67**, 919–923.

Hills, F. J., Hall, D. H., and Kontaxis, D. G. (1975). Effect of powdery mildew on sugarbeet production. *Plant Dis. Rep.* **59**, 513–515.

Horst, R. K. (1983). "Compendium of Rose Diseases." Am. Phytopathol. Soc., St. Paul, Minnesota.

Massey, L. M. (1948). Understanding powdery mildew. *Am. Rose Annu.* **33**, 136–145.

Moseman, J. G. (1966). Genetics of powdery mildews. *Annu. Rev. Phytopathol.* **4**, 269–290.

Schnathorst, W. C. (1965). Environmental relationships in the powdery mildews. *Annu. Rev. Phytopathol.* **3**, 343–366.

Spencer, D. M., ed. (1978). "The Powdery Mildews." Academic Press, New York.
Szkolnik, M. (1983). Unique vapor activity by CGA-64251 (Vangard) in the control of powdery mildews roomwide in the greenhouse. *Plant Dis.* **67**, 360–366.
Yarwood, C. E. (1957). Powdery mildews. *Bot. Rev.* **23**, 235–300.
Yoder, K. S., and Hickey, K. D. (1983). Control of apple powdery mildew in the mid-Atlantic region. *Plant Dis.* **67**, 245–248.

Foliar Diseases Caused by Ascomycetes and Imperfect Fungi

Many species of Ascomycetes and of Imperfect Fungi cause primarily foliage diseases, but some may also affect blossoms, young stems, fruit, and even roots.

Most of the foliar Ascomycetes reproduce by means of conidia formed on free hyphae or in pycnidia, but a few produce conidia in sporodochia or in acervuli. In many, the conidia overwinter; others reproduce by means of conidia during the growing season and by their perfect ascigerous stage at the end of the season and over winter. Some produce their ascocarps and ascospores, along with conidia, throughout the growing season. The primary inoculum of these fungi, therefore, may be either ascospores or conidia and usually originates from infected fallen or hanging leaves of the previous year.

Some of the most common Ascomycetes causing primarily foliar diseases are the following:

Higginsia (Coccomyces) sp., causing leaf spot or shot-hole of cherries and plums.
Microcyclus (Dothidella ulei), causing South American leaf blight of rubber.
Elytroderma deformans, causing a leaf spot and witches'-broom of pines.
Lophodermium pinastri, causing needle blight of pines.
Guignardia, causing leaf spot and black rot of grape, *G. bidwellii,* leaf blotch of horsechestnut *(G. aesculi),* and leaf spots on Boston ivy and Virginia creeper.
Didymella (Mycosphaerella), causing the extremely destructive Sigatoka disease of banana *(D. musicola),* leaf spots of strawberry *(D. fragariae),* of pear *(D. sentina),* leaf spot and black rot of cucurbits *(D. melonis),* citrus greasy spot *(D. citri),* and other diseases.
Pseudopeziza, causing the common leaf spot of alfalfa (Figure 11-39) and clovers and the yellow leaf blotch of alfalfa.
Rhabdocline pseudotsugae, causing needle cast of Douglas fir.
Rhytisma, causing tar spot of maple and willow.
Scirrhia, causing brown spot needle blight of pine *(S. aciola,* the conidial stage of which is *Lecanosticta* or *Septoria),* and Dothistroma needle blight of pine *(S. pini,* the conidial stage of which is *Dothistroma pini).*

Several other ascomycetes causing primarily foliar diseases, such as *Cochliobolus, Diplocarpon, Gnomonia, Pyrenophora,* and *Venturia,* could be listed here, but they either cause important additional symptoms (*Venturia* causes scab on apple fruit) or are discussed with another more cohesive group (*Gnomonia* and *Diplocarpon* are included in the anthracnose diseases) or the pathogens are discussed under their imperfect stage (*Cochliobolus* and *Pyrenophora* are discussed under *Helminthosporium* diseases).

Some of the most common Imperfect Fungi causing primarily foliar but also other symptoms on a large variety of host plants are: *Alternaria, Ascochyta, Cercospora* (Figure 11-39B). *Cladosporium* (Figure 11-39C). *Helminthosporium, Phylosticta, Pyricularia* (Figure 11-40), and *Septoria.* Many other

FIGURE 11-39 (A) Alfalfa leaf spot caused by *Pseudopeziza trifolii.* (B) Sugar beet leaf spot caused by *Cercospora beticola.* (C) Tomato leaf mold caused by *Fulvia (Cladosporium) fulvum.* (Photo C courtesy U.S.D.A.).

less common imperfect fungi causing leaf spots would be included here, for example, *Curvularia, Kabatiella,* and *Rhynchosporium,* causing leaf spots on grasses, cereals, and corn. On the other hand, several other fungi, such as *Botrytis* and *Colletotrichum,* could be listed here, but they so frequently affect other plant parts that they are discussed elsewhere.

The foliar spots and blights caused by Imperfect Fungi affect numerous hosts and appear in many variations. The disease cycles and controls of these diseases are quite similar, however, although considerable variability may exist among diseases caused by specific fungi on different hosts, especially

FIGURE 11-40 Foliar symptoms on rice leaves caused by the rice blast fungus *Pyricularia oryzae.* (Photo courtesy U.S.D.A.)

when the diseases develop under different environmental conditions. Thus, these fungi attack primarily the foliage of annual or perennial plants, or both, by means of conidia produced on free, single, or grouped hyphae *(Alternaria, Cercospora, Cladosporium, Helminthosporium,* and *Pyricularia)* or in pycnidia *(Ascochyta, Phyllosticta,* and *Septoria).* On the infected areas numerous conidia are produced that spread by wind, wind-blown rain, water, and insects to other plants and cause more infections. In most cases these fungi overwinter primarily as conidia or mycelium in fallen leaves or other plant debris. Some, however, can overwinter as conidia or mycelium in or on seed of infected plants, or as conidia in the soil. When perennial plants are infected, they may overwinter as mycelium in infected tissues of the plant. When these fungi are carried with the seed of annual plants, damping off of seedlings may develop. Control of such diseases is accomplished by using resistant varieties and fungicidal sprays or seed treatments, but disease-free seed or removal and destruction of contaminated debris or both may be most important in some diseases.

• Alternaria Diseases

The Alternaria diseases are among the most common diseases of many kinds of plants throughout the world. They affect primarily the leaves, stems, flowers, and fruits of annual plants, especially vegetables and ornamentals, but also may affect parts of trees such as citrus and apple. Alternaria diseases

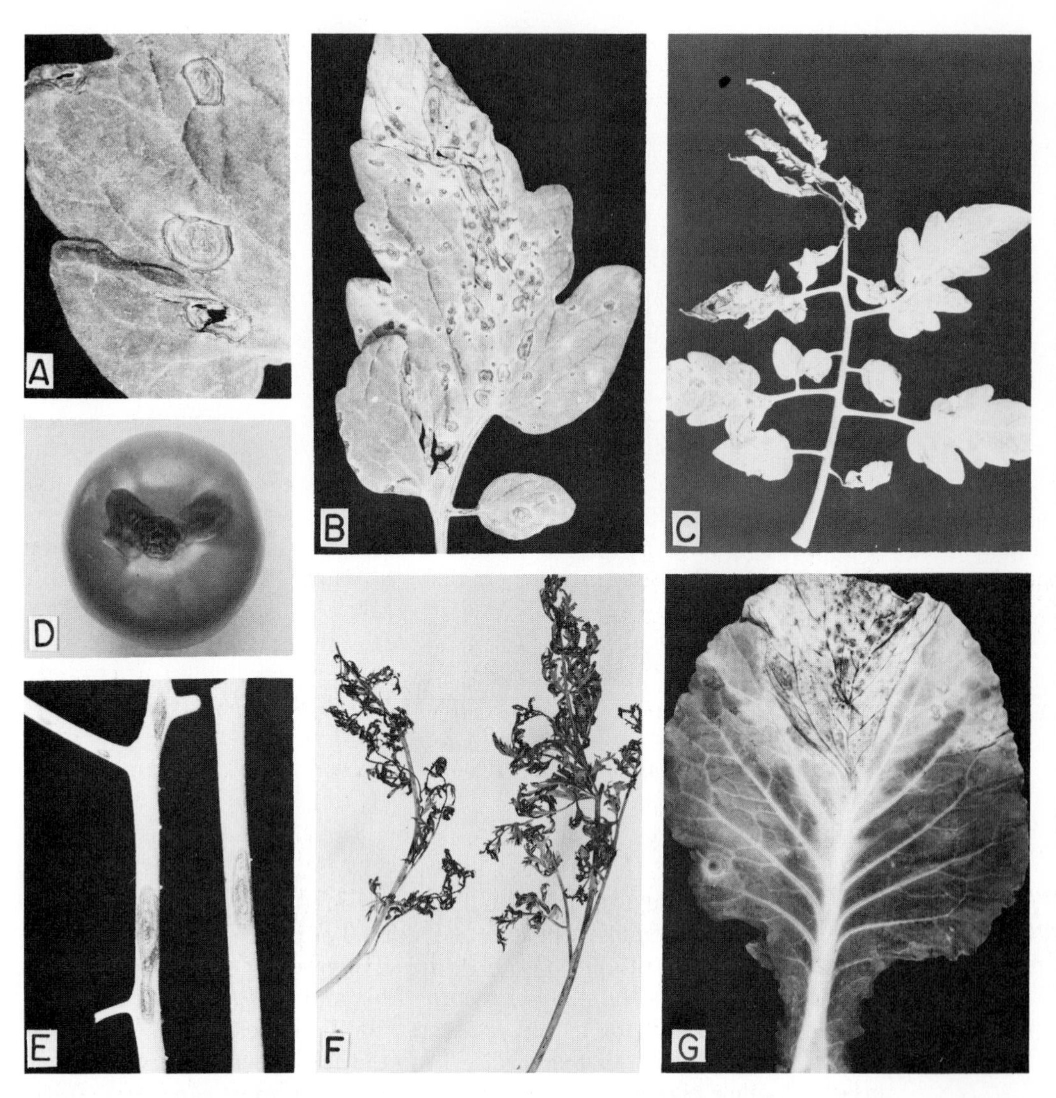

FIGURE 11-41 Symptoms caused by species of *Alternaria* (A–E) Early blight of tomato caused by *A. solani.* (A) Typical target-shaped lesions on tomato leaf. (B, C) Leaf spots (B) and blight (C) on tomato. (D) Typical *Alternaria* lesion at the stem end of tomato fruit. (E) Lesions on tomato stems. (F) Carrot blight caused by *A. dauci.* (G) Cabbage leaf spot caused by *A. brassicae.*

appear usually as leaf spots and blights, but they may also cause damping off of seedlings, collar rots, and tuber and fruit rots. Some of the most common diseases caused by *Alternaria* include early blight of potato and tomato (Figure 11-41, A-E), leaf spot of bean, tobacco, geranium, and stock, blight of carrot (Figure 11-41F), carnation, chrysanthemum, petunia and zinnia, leaf spot and blight of crucifers (Figure 11-41G), onion, purple blotch, leaf spot and fruit spot on squash and on apple, core rot of apple, and rot of lemons and oranges.

The leaf spots are generally dark brown to black, often numerous and enlarging and usually developing in concentric rings, which give the spots a targetlike appearance (Figure 11-41, A-C). Lower, senescent leaves are usually attacked first, but the disease progresses upward and makes affected leaves go into a yellowish, senescent condition and either dry up and droop or fall off. Dark, often targetlike, sunken spots develop on branches and stems of plants such as tomato (Figure 11-41E). Stem lesions developing on seedlings may form cankers, which may enlarge, girdle the stem, and kill the plant, or if present near the soil line they may develop into a collar rot. In belowground parts, such as potato tubers, dark, slightly sunken, circular, or irregular lesions develop that may be up to 2 cm in diameter and 5–6 mm in depth. Fruits affected by *Alternaria* are usually attacked when they approach maturity, in some hosts at the stem end while in others at the blossom end or at other points through wounds, growth cracks, and so on (Figure 11-41D). The spots are brown to black and may be small, sunken, with well-defined margins, or they may enlarge to cover most of the fruit, have a leathery consistency and a black, velvety surface layer of fungus growth and spores. In some fruits, such as citrus and tomato, a small lesion at the surface may indicate an extensive spread of the infection in the central core and the segments of the fruit.

The pathogen, *Alternaria* sp., has dark-colored mycelium and in older diseased tissue it produces short, simple, erect conidiophores that bear single or branched chains of conidia (Figure 11-42). The conidia are large, dark, long, or pear shaped and multicellular, with both tranverse and longitudinal cross walls. The conidia are detached easily and are carried by air currents. *Alternaria* occurs in many species throughout the world. Their spores are present in the air and dust everywhere and are one of the most common fungal causes of hay fever allergies. *Alternaria* spores also land and grow as contaminants in laboratory cultures of other microorganisms and on dead plant tissue killed by other pathogens or other causes. Actually, many species of *Alternaria* are mostly saprophytic, that is, they cannot infect living plant tissues but they grow only on dead or decaying plant tissues and, at most, on senescent or old tissues such as old petals, old leaves, and ripe fruit. Therefore, it is often difficult to decide whether an *Alternaria* fungus found on a diseased tissue is the cause of the disease or a secondary contaminant.

Many species of *Alternaria* produce toxins, some of which, such as tentoxin, produced by *A. tenuis* are host nonspecific, while others, such as AK-toxin and AM-toxin, produced by *A. kikuchiana* and *A. mali,* respectively, are host specific.

Plant pathogenic species of *Alternaria* overwinter as mycelium in infected plant debris and as mycelium or spores in or on seeds. If the fungus is carried with the seed, it may attack the seedling, usually after emergence, and cause damping off or stem lesions and collar rot. More frequently, however, spores that are produced abundantly, especially during heavy dews and frequent rains, are blown in from mycelium growing on debris, infected cultivated plants, or weeds. The germinating spores penetrate susceptible tissue directly or through wounds and soon produce new conidia that are further spread by wind, splashing rain, or tools. With few exceptions, *Alternaria* diseases are more prevalent on older, senescing tissues and particularly on plants of low vigor or poor nutrition or those under some other kind of stress caused by unfavorable environmental conditions, insects, or other diseases.

Alternaria diseases are controlled primarily through the use of resistant varieties, of disease-free or treated seed, and through chemical sprays with fungicides such as chlorothalonil, maneb, captafol, mancozeb, and fentin

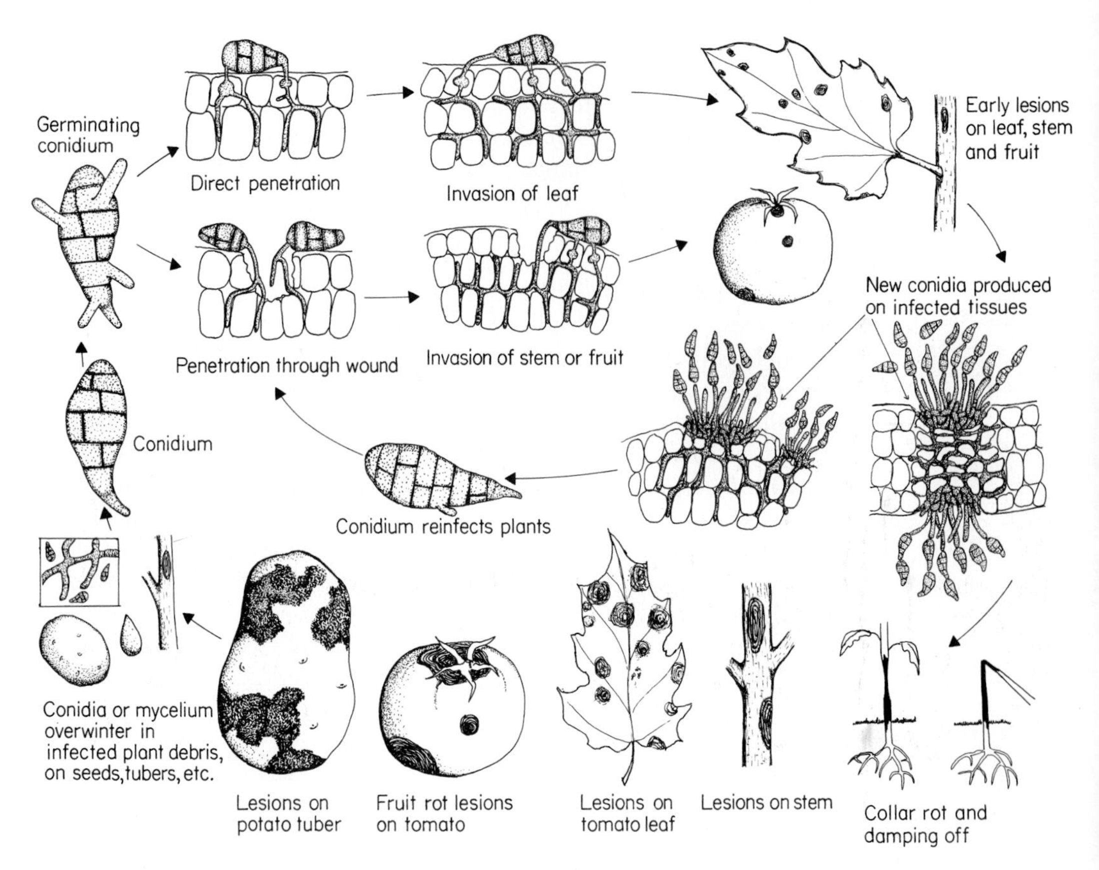

FIGURE 11-42 Development and symptoms of diseases caused by *Alternaria.*

hydroxide. The sprays should begin soon after the seedlings emerge or are transplanted and should be repeated at 1- to 2-week intervals, depending on disease prevalence and severity and frequency of rains. Applications of increased rates of nitrogen fertilizer generally reduce the rate of infection by *Alternaria* on potato and also the final amount of disease. Crop rotation, removal and burning of plant debris, if infected, and eradication of weed hosts help reduce the inoculum for subsequent plantings of susceptible crops. Several mycoparasitic fungi are known to parasitize various species of *Alternaria,* but so far none has been developed into an effective biological control of the pathogen. In the greenhouse, infections by at least some *Alternaria* species can be drastically reduced by covering the greenhouse with special UV-light-absorbing film, since filtering out the UV light inhibits spore formation by these fungi.

• Coccomyces Leaf Spot or Shot-hole of Cherries and Plums

Widespread and serious throughout the world, especially in areas with a humid climate, Coccomyces leaf spot is most severe on sour cherries and less so on sweet cherries and plums. In cherries, the symptoms appear as numerous minute, red to purplish-black spots on the upper sides of leaves and, in severe infections, on the petioles, fruit, and pedicels (Figure 11-43A). On the

FIGURE 11-43 (A) Cherry leaf spot caused by *Higginsia hiemalis.* (B) Horsechestnut leaf blotch caused by *Guignardia aesculi.* (C–G) Black rot of grape caused by *Guignardia bidwellii.* (C, D, E) Foliar symptoms of black rot on grape showing damage to leaves (C, D) and the characteristic circular arrangement of the *Phyllosticta*-type pycnidia (E). (F, G) Black rot symptoms on young (F) and completely rotten berries (G).

under sides of the spots on the leaves, slightly raised, waxy pustules appear that after heavy dews or rains produce a white, mildewlike growth. Infected leaves usually turn yellow and fall, or the spots may drop out and the leaves have a shot-hole appearance. The shot-hole effect is even more common and more pronounced on plum and prune leaves, which may appear completely skeletonized. Severe leaf drop and shot-holing early in the season are common and weaken the trees, which may die back during the same or subsequent years. The disease is especially common and serious in nursery plantings.

The fungus, *Coccomyces* sp. (or *Higginsia* sp.), produces ascospores in apothecia formed in fallen infected leaves and *Cylindrosporium*-type conidia formed in acervuli on the under side of infected leaves. The conidia are colorless, threadlike, straight, or curved and consist of one or several cells. The ascospores are produced in the spring over a period of 6–7 weeks and are forcibly ejected when the leaves are thoroughly soaked. The ascospores are then carried by air currents and cause the primary infections on the leaves, which then produce large numbers of conidia that are spread by rain from leaf to leaf and cause all subsequent infections.

Control of the disease is achieved by 4–5 sprays starting at petal-fall, and one postharvest spray with benomyl, captan, thiophanate methyl, or certain other fungicides such as dodine, sulfur, and ferbam.

• Black Rot of Grape

Black rot of grape is present in Europe and in the United States and Canada east of the Rocky Mountains. It is probably the most serious disease of grapes where it commonly occurs, particularly in warm, humid regions. In the absence of control measures and in favorable weather, the crop is usually destroyed completely, either through direct rotting of the berries or through blasting of the blossom clusters.

The disease causes numerous scattered, circular, red necrotic spots on leaves in late spring (Figure 11-43, C-G). The spots form usually between the veins and are most apparent on the upper side of the leaves. Later, when the spots are about 2–6 mm or more in diameter, the main area of the spots appears brown to grayish-tan, while the margins appear as a black line. Black dotlike, *Phyllosticta*-type pycnidia are formed on the upper side of the spots in a ring near the outer edge of the brown area of the spot. On the shoots, tendrils, the leaf and flower stalks, and on leaf veins, the spots are purple to black, somewhat depressed and elongated, and bear scattered pycnidia. Spots begin to appear on berries when the berries are about half grown. These spots are at first whitish but are soon surrounded by a rapidly widening brown ring with a black margin. The central area of the spot remains flat or becomes depressed, and dark pycnidia appear near the center. The whole berry soon becomes rotten and shrinks, and it becomes coal black as the surface becomes studded with numerous black pycnidia.

The fungus, *Guignardia bidwellii,* in addition to conidia-bearing pycnidia, also produces ascospores in globose perithecia forming in rotten, mummified fruit. The perithecia supposedly develop from transformed pycnidia. The fungus overwinters mostly as ascospores in perithecia, but conidia can also survive the winter in most locations where grapes grow, so both ascospores and conidia can cause primary infections in the spring (Figure 11-44). The release of ascospores and conidia takes place only when the perithecia and pycnidia become thoroughly wet, but while the ascospores are shot out forcibly and may then be carried by air currents, the conidia are exuded in a viscid mass from which they can be washed down or splashed away by rain.

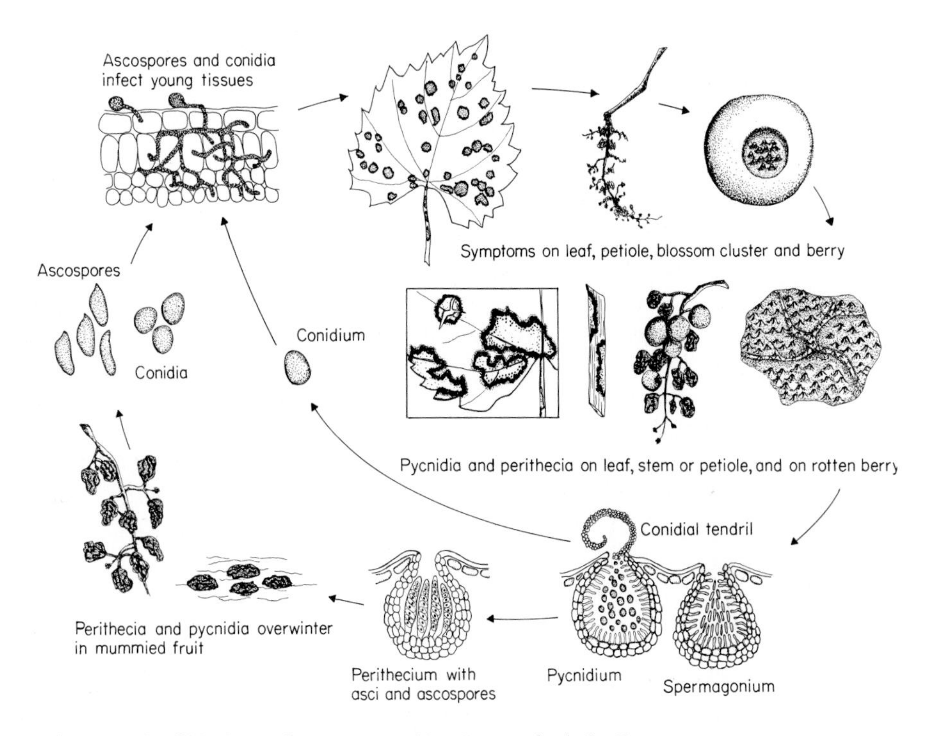

FIGURE 11-44 Disease cycle of black rot of grapes caused by *Guignardia bidwellii.*

Ascospores may be discharged continually through the spring and summer, although most of them are discharged in the spring.

Primary infections, whether from ascospores or conidia, take place on young, rapidly growing leaves and on fruit pedicels. In the ensuing spots, pycnidia are produced rapidly, and these pycnidia provide the conidia for the secondary infections of berries, stems, and so on.

Control of the grape leaf spot and black rot depends primarily on timely sprays of grapevines with fungicides. Sprays with ferbam or benomyl just before bloom, immediately after bloom, and 10–14 days later give good control of the disease. Captan, folpet, triadimefon, and copper fungicides are also effective against black rot and are used when downy mildew or powdery mildew are also to be controlled. Where black rot has been severe, another ferbam or captan application should be made during the first part of June.

• Needle Casts and Blights of Conifers

Several ascomycetous fungi such as *Elytroderma, Hypoderma, Lophodermium,* and *Scirrhia* cause leaf diseases on pine, while *Rhabdocline* infects leaves of Douglas fir. Many other fungi also affect the various conifer hosts and may be more or less damaging depending on the importance of the tree species in the particular area and on the prevailing environmental conditions.

All needle cast and blight diseases have certain common characteristics, although each differs from all others in some respects. Thus, the needlelike leaves of the conifers are infected by the conidia and occasionally by the ascospores of these fungi at some time during the growing season. The type

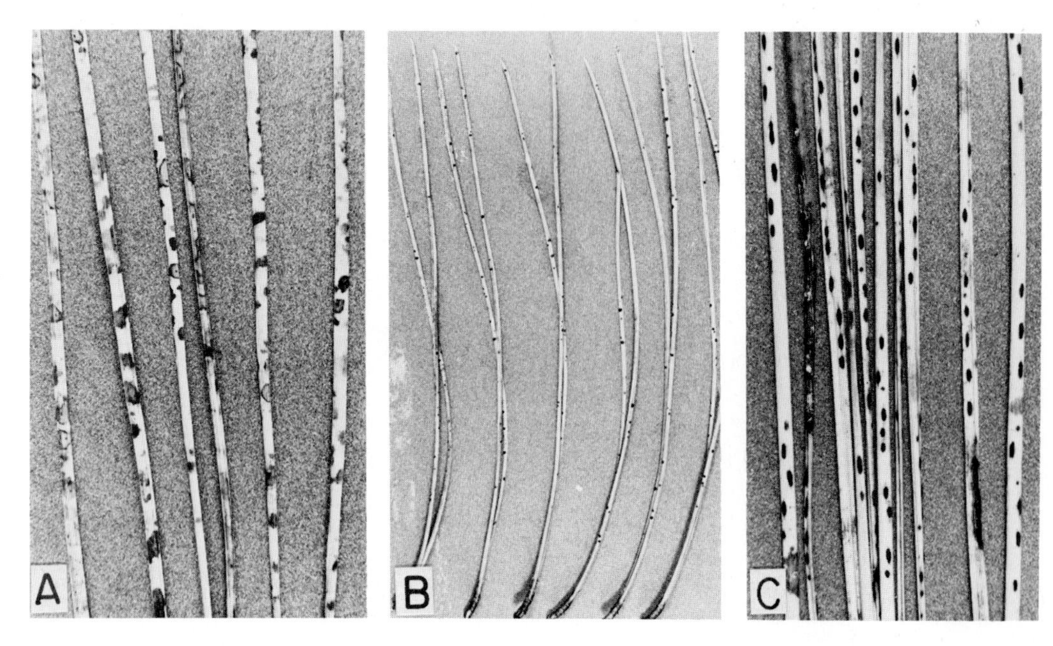

FIGURE 11-45 Needle spots and blights of conifers. (A) *Scirrhia* brown spot on long-leaf pine. (B) *Lophodermium* needle cast on red pine. (C) *Hypoderma* needle cast on long-leaf pine.

and time of infection may vary with the location in which the particular species grows. The fungus enters the needle and usually causes a light green to yellow spot that sooner or later turns brown or red, encircles the needle and kills the part of the needle beyond the spot (Figure 11-45). The fungus may spread into the needle, or separate new infections may develop on the needle. As a result, the entire needle is often killed and either clings to the tree for a while, giving the tree a reddish-brown, burned appearance, or it is shed, resulting in partial or total defoliation of the tree. On the infected needles, whether on the tree or on the ground, the fungus produces its conidia and, occasionally, its ascospores in perithecia, which are either released into the air or are exuded during wet weather and are washed down or splashed by the rain into other needles and trees. In some needle blights, the fungus may overwinter as mycelium in infected but still living needles, while in most cases the fungi overwinter as ascospores or conidia in dead needles on the tree or on the ground.

Needle casts and blights can be destructive on mature trees, especially in plantations of single species, which may be killed following repeated defoliations. Every year, thousands of trees are cut when dead or dying from foliage diseases. These same diseases, however, can be devastating in young or nursery trees, which they can kill by the millions in a relatively short time if the weather is favorable and no adequate control measures are practiced.

Most, but not all, needle casts and blights can be controlled with fungicidal sprays, especially in the nursery and in young plantation trees. Larger trees are either cut before they die (salvage cutting), or they too may be protected, when possible, with fungicides applied from airplanes. The fungicides most commonly used against leaf cast and blight diseases are Bordeaux mixture (usually a 4:4:100 mixture), other copper fungicides such as tribasic copper sulfate, benomyl, maneb, and chlorothalonil. In some needle diseases, two sprays either early or late in the season, whenever most of the infections with the particular fungus takes place, are sufficient to keep the disease in check, especially in large trees. In most cases, however, nurseries must be sprayed at least every 2 weeks from May through October if the seedlings are to survive the needle attacks by fungi and to grow.

Mycosphaerella Diseases

• Banana Leaf Spot, or Sigatoka Disease

Banana leaf spot, or Sigatoka disease occurs throughout the world and is one of the most destructive diseases of banana, although its economic importance varies with the region. If causes losses by reducing the functional leaf surface of the plant, which results in small, unevenly ripened bananas that may fall and fail to ripen.

The disease first appears as small, indistinct, longitudinal, light yellow spots parallel to the side veins of leaves that unfurled about a month earlier. A few days later, the spots become 1–2 cm long, turn brown with light grey centers, and become readily visible. Such spots cause little damage, but they soon enlarge further, the tissue around them turns yellow and dies, and adjacent spots coalesce forming large dead areas on the leaf (Figure 11-46, A, B). In severe infections, entire leaves die within a few weeks. Since it takes at least 12 healthy leaves on mature banana plants to carry fruit to maturity, destruction of most mature leaves by the leaf spot disease may leave only a few functioning leaves that are insufficient to bring the fruit to maturity. As a result, immature fruit bunches on such plants fail to fill out and ripen and may fall. If the fruit is nearing maturity at the time of heavy infection, the flesh ripens unevenly, individual bananas appear undersized and angular in shape, their flesh develops a buff pinkish color, and they store poorly.

The causal fungus is *Mycosphaerella musicola.* It produces spermatia in spermogonia, ascospores in perithecia, and conidia of the *Cercospora* type in sporodochia. Sporodochia appear while the spots are still light yellow, but successive crops of abundant conidia are produced by the same sporodochia only during the brown spot stage of the disease (Figure 11-47). Conidia are produced on both sides of the leaf but are usually more abundant on the upper side and are spread by wind and dripping or splashing water. Although conidia are produced throughout the year, their release and germination depends on water or high humidity. Perithecia, formed as a result of fertilization of sexual hyphae by compatible spermatia, are produced during warm humid weather, and their ascospores are shot out violently in response to wetting of the perithecia. Ascospores are spread by air currents and are responsible for the long-distance spread of the disease, while conidia are generally the most important means of the local spread of the disease. Infection by either ascospores or conidia produces the same type of spot and subsequent development of the disease.

In addition to the common Sigatoka disease, caused by *Mycosphaerella musicola,* there is also the black leaf streak disease caused by *M. fijiensis,* and the black Sigatoka disease caused by *M. fijiensis* var. *difformis.* The black Sigatoka pathogen was discovered in Honduras in 1972. It causes spotting 8–10 days faster than *M. musicola,* and after severe outbreaks of black Sigatoka in 1973 and 1974, *M. f.* var. *difformis* replaced *M. musicola* within 2

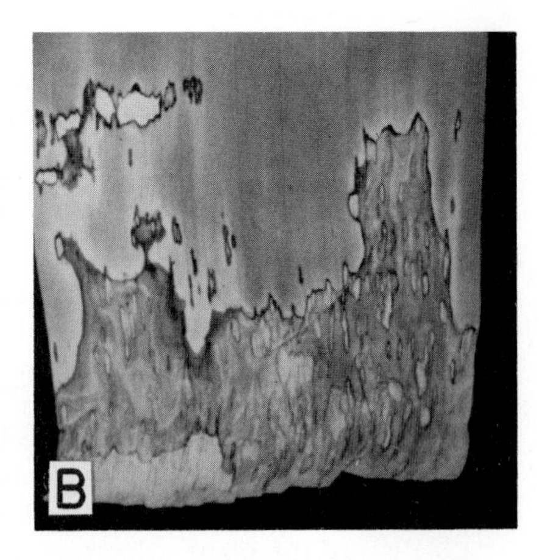

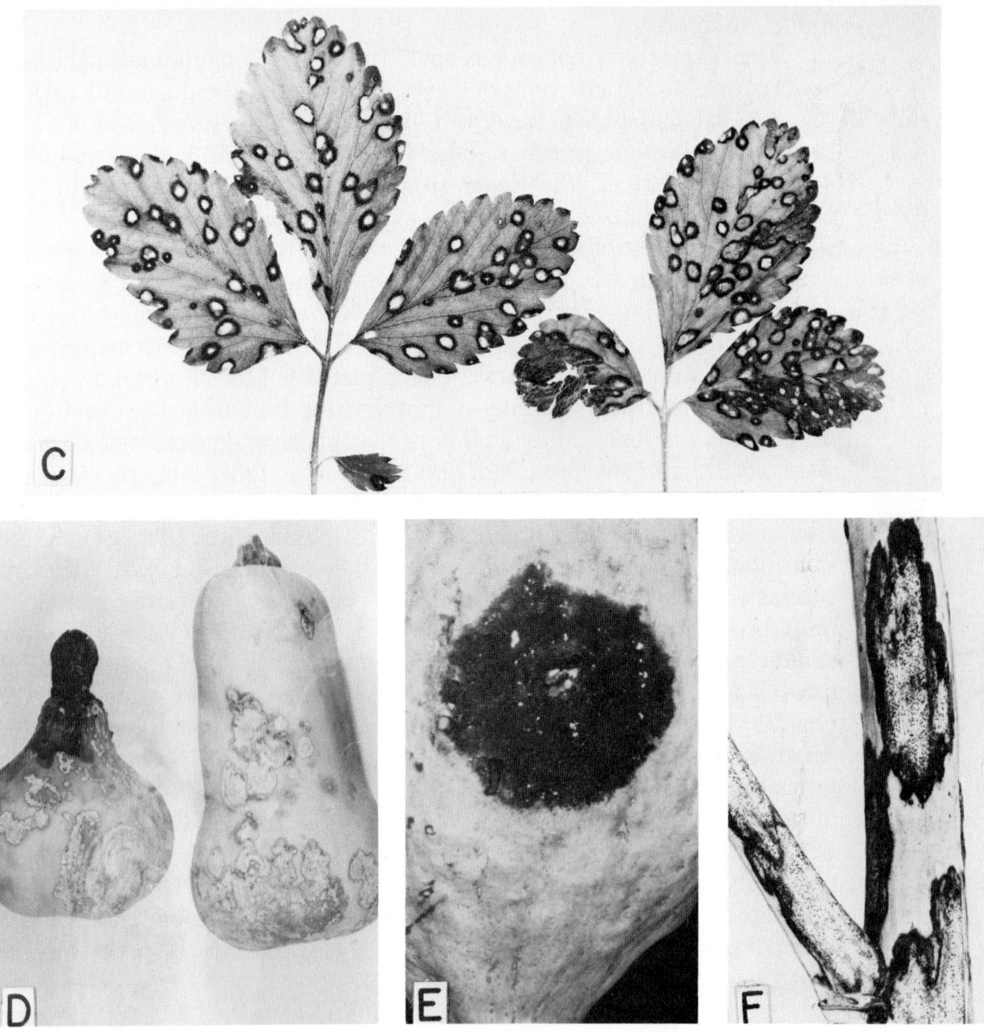

FIGURE 11-46 (A, B) Banana leaf spot (Sigatoka disease) caused by *Mycosphaerella musicola*. (C) Strawberry leafspot caused by *M. fragariae*. (D) Superficial and galloping type of black rot on butternut squash caused by *M. melonis*. (E) Black rot lesion on winter squash. (F) Lesions and pycnidia of *Mycosphaerella brassicola* on stem of young cabbage plant. (Photos A and B, courtesy R. H. Stover. Photo C, courtesy U.S.D.A.)

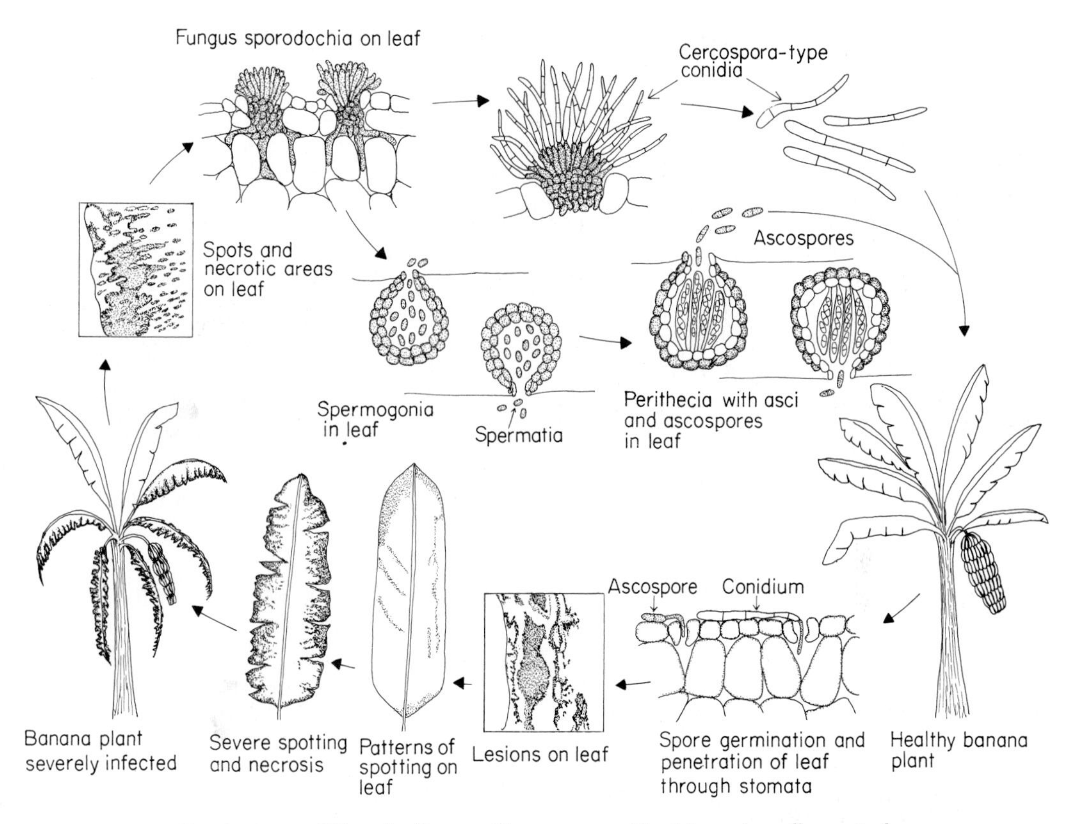

FIGURE 11-47 Development of Sigatoka disease of banana caused by *Mycosphaerella musicola.*

years. The three fungi have similar life cycles and morphology except that *M. f.* var. *difformis* produces sporodochia in young spots and its hyphae move from one stoma to another and cause lesions over entire leaves much more commonly than *M. musicola.*

The Sigatoka diseases are controlled by a combination of measures including quarantine, sanitation by removal and destruction of badly infected leaves, and mainly by frequent, year-round application of fungicidal sprays. For many years Bordeaux mixture or copper oxychloride with or without zineb were the fungicides used. Later it was shown that zineb or copper oxychloride suspended in mineral oil or mineral oil alone (consisting largely of saturated hydrocarbons) gave better and cheaper control than Bordeaux. The oil sprays, however, are phytotoxic under certain conditions, may reduce fruit yields by as much as 10 percent, may cause patchy, retarded ripening when applied directly to fruit, and have resulted in an increase of certain other leaf spot and fruit spot and rot diseases previously controlled by Bordeaux mixture. To date, chlorothalonil, mancozeb, or other fungicides are routinely included in oil–water–fungicide emulsions for best all-around results. In some areas it is necessary to apply ground or airplane sprays every 10–12 days throughout the year, especially for control of black leaf streak and black Sigatoka diseases, while in other areas one application every 3, 4 or even 6 weeks suffices to maintain control.

• Cucurbit Leaf Spot, Gummy Stem Blight, and Black Rot

Cucurbit leaf spot, gummy stem blight, and black rot is probably worldwide in distribution, and in favorable weather the pathogen infects all parts of all cucurbits. Usually, however, it attacks the leaves and stems of watermelons, cucumber, and cantaloupe and the fruit of squash, pumpkin, and gourd. When the fungus is carried in the seed it also causes damping off, killing the seedlings. On the leaves, petioles, and stems, pale brown or gray spots develop. On the stems, the spots usually start at the joints, become elongated and cracked, and exude an amber-colored gummy material. The spots on the leaves enlarge along with the spots on stems and petioles and make the leaves turn yellow and die. Even the whole plant may wilt and die. On the fruit, the spots appear at first as yellowish, irregularly circular areas that later turn gray to brown and may have a droplet of gummy exudate in the center. The spot finally turns black. In some kinds of squash the black rot lesions are superficial and spread over the surface or appear as a dry, brownish, circular mass, the enlargement of which is arrested by a defensive barrier of squash tissue (Figure 11-46 D,E). Very often, however, especially in storage, the fungus penetrates through the rind and spreads widely throughout the squash and into the seed cavity. On all the spots, whether on the leaf, stem, or fruit, the fungus produces closely spaced groups of pale-colored pycnidia and dark, globular perithecia that are sometimes arranged in rings and are visible with the naked eye.

The fungus, *Didymella bryoniae (Mycosphaerella melonis),* produces ascospores in perithecia and conidia in pycnidia. Both conidia and ascospores are short-lived after they are released, and the fungus generally overwinters in diseased plant refuse as chlamydospores and in or on the seed. Thus, either type of spore or infected seed can result in primary infections, with the subsequently profusely produced conidia causing the secondary infections. Cucurbit plants seem to be predisposed to infection by *Mycosphaerella melonis* by previous infestation with beetles or infection with powdery mildew. The striped cucumber beetle, in addition, appears to serve as a vector of *M. melonis* among cucurbit plants in the field.

Control of black rot of cucurbits is difficult, requiring the use of clean or treated seed, long crop rotations, and frequent applications of such fungicides as mancozeb, chlorothalonil, and captan. Good control of the leaf and stem infections reduces fruit infections both in the field and in storage. However, further care is needed to avoid infection in storage. Wounding of stored fruit must be avoided. Curing of squash at 23–29°C for two weeks to heal the wounds and subsequent storage at 10–12°C are very helpful. If the inoculum present in the field is heavy, dipping the squash fruit in formaldehyde or Clorox before curing and storage is also helpful.

• Cercospora Diseases

Cercospora diseases are almost always leaf spots that either stay relatively small and separate or they may enlarge and coalesce, resulting in leaf blights. The diseases are generally widespread, and among the most commonly affected hosts are sugar beet, carrot, celery, eggplant, pea, peanut, tomato, rice, corn, sugarcane, most other cereals and grasses, azalea, Boston ivy, dahlia, geranium, tobacco, and many trees and other crops in temperate zones and in the tropics.

The leaf spots in some plants, such as beet and tobacco, are brown, small, about 3–5 mm in diameter and roughly circular with reddish-purple borders (Figure 11-39B). Later, their centers become ashen-gray, thin, papery, and brittle and may drop out leaving a ragged hole, or the spots, if sufficiently numerous, may coalesce, causing large necrotic areas. On most other hosts, such as celery, carrot, and geranium, the leaf spots are small, reddish, or yellowish at first, but they enlarge rapidly, the affected tissue changing to an ashen-gray color and a dry, papery texture. The spots are irregularly circular to angular, with or without a distinct border, and often coalesce to form large blighted areas. In monocotyledonous plants the spots are narrow and long, usually 0.5 by 5.0 cm, and may coalesce and kill leaves. In humid weather the affected leaf surface on all hosts is covered with an ashen-gray mold barely visible to the naked eye. In severe attacks, all the foliage is destroyed and may fall off. On fleshy plants, similar lesions are produced on stems and leaf petioles.

Several species of *Cercospora* are responsible for the diseases on the various hosts. The fungus produces long, slender, colorless to dark, straight to slightly curved, multicellular conidia on long, dark conidiophores. The conidiophores arise from the plant surface in clusters through the stomata and form conidia successfully on new growing tips. The conidia are easily detached and often blown long distances by the wind. The fungus is favored by high temperatures and therefore is most destructive in the summer months and in warmer climates. Most *Cercospora* species produce the nonspecific toxin cercosporin, which acts as a photosensitizing agent in the plant cells, that is, it kills cells only in the light. The toxin results in the production of atomic oxygen in the cells, which causes disruption of cell membranes and loss of electrolytes from cells. Although *Cercospora* spores need water to germinate and penetrate, heavy dews seem to be sufficient for abundant infection. It overwinters in or on the seed and as minute black stromata in old affected leaves.

Cercospora diseases are controlled by using disease-free seed, or seed at least 3 years old, by which time the fungus in the seed has died; crop rotations with hosts not affected by *Cercospora* or by the same *Cercospora* species; and by spraying the plants, both in the seedbed and in the field, with fungicides such as benomyl, dyrene, chlorothalonil, Bordeaux mixture, maneb, and dodine.

• Septoria Diseases

Septoria diseases occur throughout the world and affect numerous crops on which they cause mostly leaf spots and blights. The most common and serious diseases they cause are leaf blotch and glum blotch of wheat and other cereals and grasses and leaf spots of celery, beet, carrot, cucurbits, lettuce, tomato, soybean, brambles, aster, azalea, carnation (on which they also cause a corm rot), chrysanthemum, marigolds, and many others.

On cereals and grasses, the leaf spots appear as light green to yellow or brown spots, first between the veins but soon becoming darker and spreading rapidly to form irregular blotches (Figure 11-48C). These spots may be restricted or may coalesce and cover the entire blade and sheath, depending on the variety and humidity. The blotches often appear speckled due to the more or less abundant, small, submerged brown pycnidia forming on them. In favorable weather, plants bcome defoliated and the fungus invades the culm

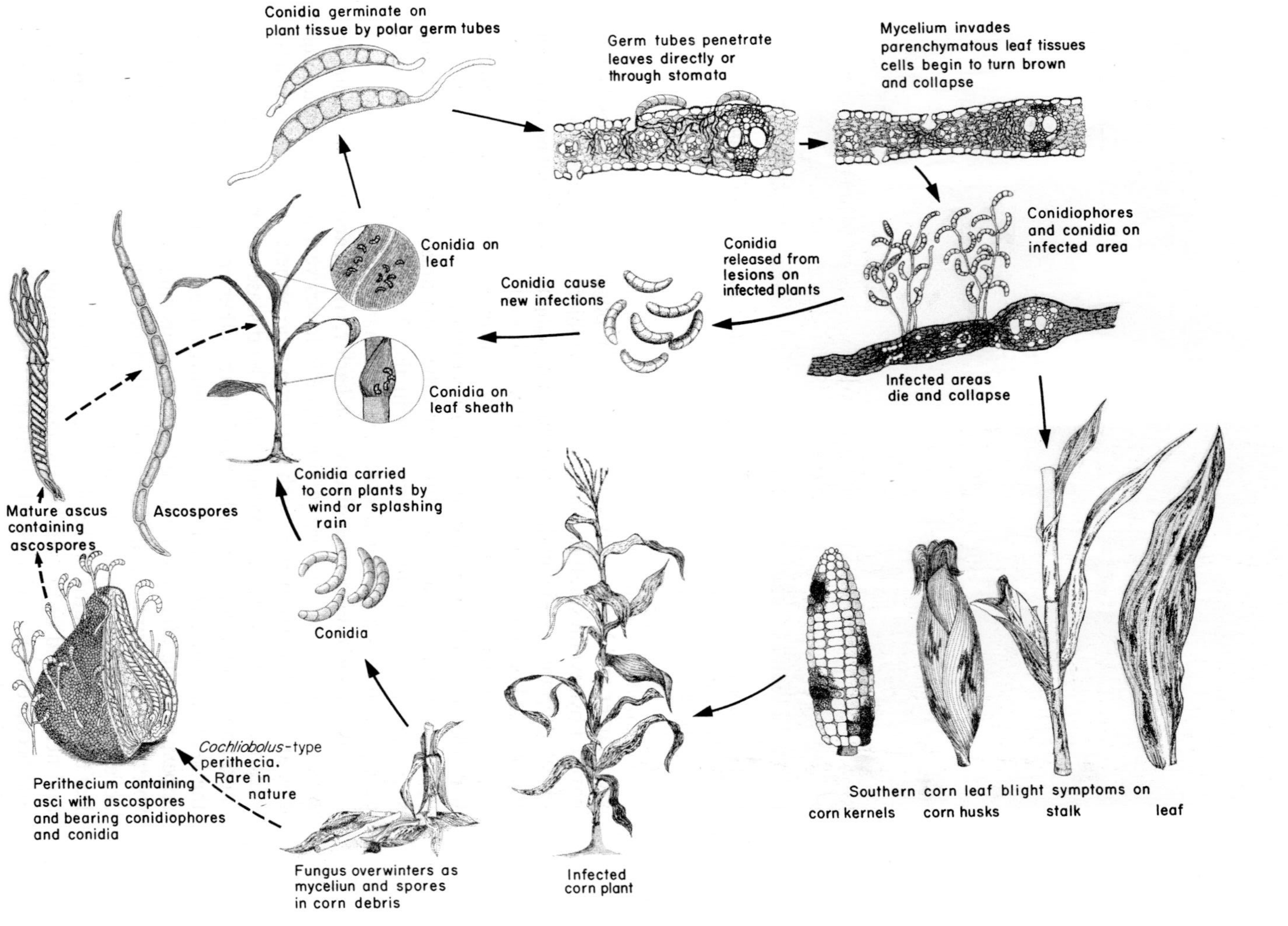

FIGURE 11-51 Disease cycle of southern corn leaf blight caused by *Bipolaris (Helminthosporium) maydis* race T.

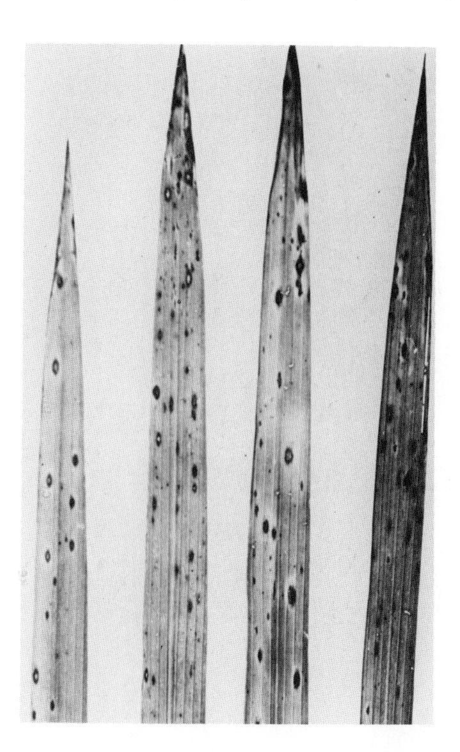

FIGURE 11-52 Rice leaves with spots caused by the brown spot fungus *Helminthosporium oryzae.* (Photo courtesy U.S.D.A.)

and may spread into the leaves. The seedlings may be blighted and killed before or after emergence, or they may survive but their growth is retarded. Crowns are infected at or just below the soil line and show a reddish-brown decay that destroys the tiller buds and advances into the root system, which it kills. Winter survival of root rot-infected wheat and barley plants is reduced considerably, in some cases by 10-30 percent in wheat and 20–60 percent in barley. The leaf spots are brown and elongate with definite margins and frequently enlarge and coalesce to form brown irregular stripes or blotches that cover large areas of the leaf blade. Older lesions are covered with a layer of olive-colored conidiophores and conidia of the fungus. Floral parts and kernels also develop lesions, or their entire surface may appear dark brown. The embryo end of the kernel is often black, and this symptom is characteristic of this disease. Due to crown and root rot and to leaf blotch, surviving plants are shorter; the spikes may be only partially emerged and may be sterile or have poorly filled kernels. Blight of floral parts and kernels also causes sterility or death of individual kernels.

In the **Victoria blight of oats,** which affects only types of oats bred from the cultivar "Victoria," the fungus causes necrotic lesions at the base of the shoot of seedlings, which then develop reddish stripes on the young leaves. Severely affected seedlings die. Those that survive later develop a rot at the base of the stem and in the root and brown lesions on the leaves. The stem is weakened in the infected region and is easily toppled over, while a velvety spore mass develops on its lower blackened nodes. The fungus causing

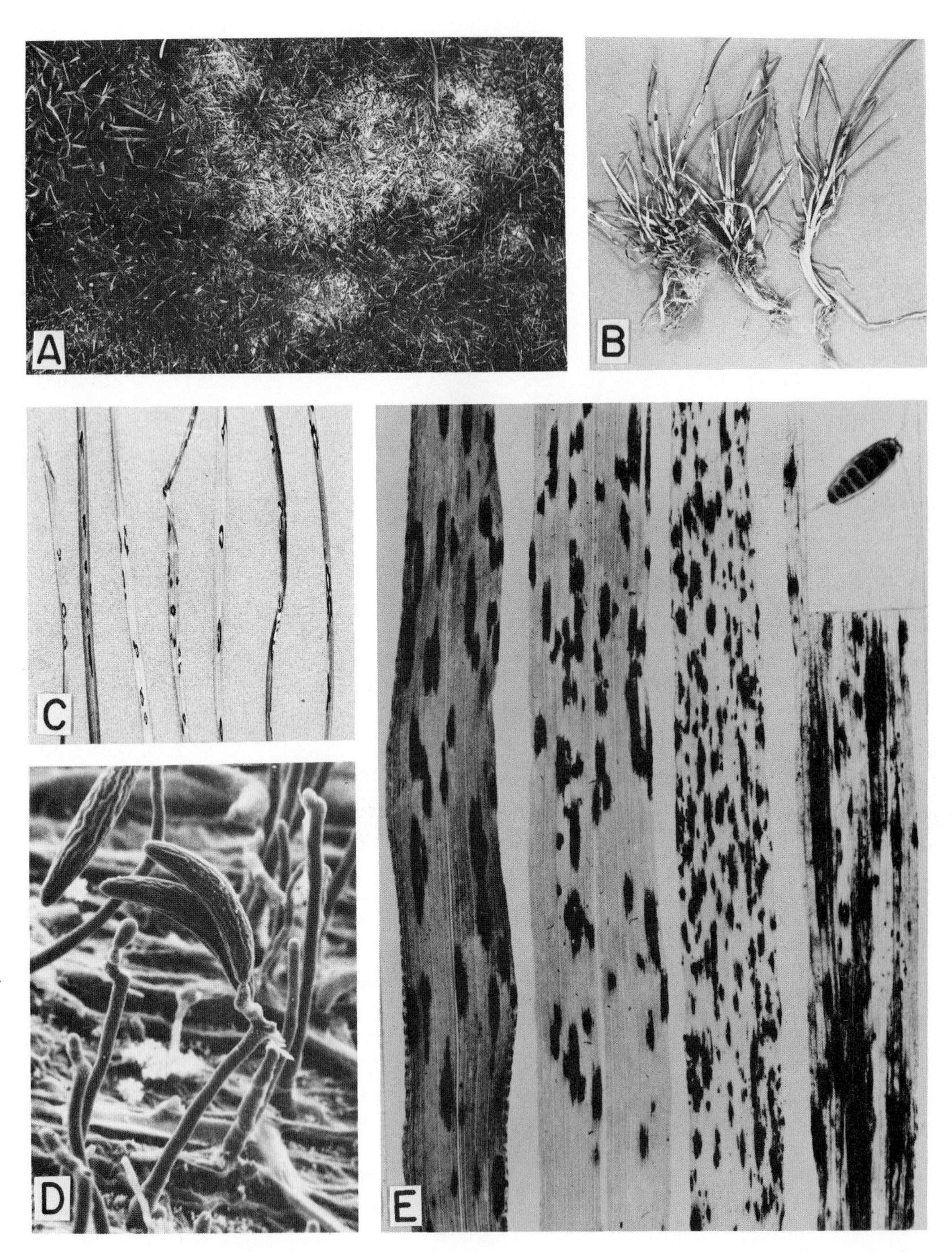

FIGURE 11-53 *Helminthosporium* diseases of turfgrasses and cereals. (A) A small patch of dead turf showing early stages of melting out. (B, C) Plants and leaves from area in (A) showing spotting of leaves and death of sheaths and roots. (D) Scanning electron micrograph of *H. maydis* conidiophores and conidia on corn leaf surface. (E) Spot blotch of cereals and grasses caused by various species of *Helminthosporium* and a spore of one of the species. (Photo D courtesy M. F. Brown and H. G. Brotzman. Photo E courtesy U.S.D.A.)

Victoria blight of oats produces the specific toxin victorin that affects and produces the typical stunting and chlorosis symptoms of the disease only on "Victoria" oats.

Lawn and fine golf course grasses are frequently attacked by various *Helminthosporium* species, which cause the most common and serious group of diseases known as leaf spots, blights, crown (foot) rots and root rots—**melting out**—of these grasses (Figure 11-53A-C). These diseases resemble in most respects those described above for the small grain crops, although on turf grasses one or more leaf spots often merge and girdle the leaf blade, which then turns yellow or reddish-brown and dies back from the tip. In severe infections leaves become completely blighted, wither, die, and drop off. Furthermore, the perennial nature of the turf grasses and the fact that they are mowed and irrigated several times each year create additional opportunities for the diseases to be spread and to become established. The diseases are usually most destructive during wet or humid weather or where the turf is sprinkle-irrigated frequently, especially late in the day. In advanced cases of infection turf areas of various sizes and shapes turn yellow, then brown to straw colored, the plants in these areas are killed and thinned out, and finally all plants die and decay (melting out). The disease, if unchecked and if the weather is favorable, may spread and kill the entire turf in the area.

Quite distinctly from the above diseases, another species of *Helminthosporium* causes a pitted or scaly bark condition and small, black, sunken spots on apple and pear and a blister canker on pear.

The pathogen in the above diseases is one or several of the many species of *"Helminthosporium"* affecting plants. The fungus produces large, cylindrical, dark, 3- to many-celled (usually 5–10) conidia that have thick walls and sometimes are slightly curved (Figure 11-53). The conidia are produced successively on new growing tips of quite dark. septate, irregular conidiophores.

In the last several years, most plant pathogenic species of *Helminthosporium* have been reclassified under different genera, depending on the shape and manner of germination of the conidium and on the kind of perfect (ascomycetous) stage they produce. Several species have been shown to produce, with more or less regularity, black perithecia containing cylindrical asci, within which are formed colorless, threadlike, 4- to 9-celled, coiled ascospores. Thus, former *Helminthosporium* species that produce a *Cochliobolus* perfect stage are now known as *Bipolaris* species, those that produce a *Pyrenophora* perfect stage are known as *Drechslera* species, and those that produce a *Setosphaeria* perfect stage are known as *Exserohilum.* Only species that do not produce a perfect stage are still called *Helminthosporium,* but only one such species *(H. solani)* is a plant pathogen. However, because this classification has not yet been universally accepted, because all the older and much of the recent literature can be found under *Helminthosporium,* and more importantly, because of the biological and plant pathological similarities of all these species and genera, the diseases can be effectively discussed under the name *"Helminthosporium"* diseases.

The various *"Helminthosporium"* species survive the winter as mycelium or spores in or on infected or contaminated seed, in plant debris, and in infected crowns or roots of susceptible plants. Some species of the fungus are weak parasites, but several are potent pathogens. When in the soil, however, all of them are weak as saprophytes, probably because of antagonism by soil microorganisms, especially at high nitrogen content. Many

Helminthosporium species, for example, *H. victoriae* and *H. maydis* race T, produce potent host-specific toxins such as victorin and HM T-toxin that play important roles in the development of the respective diseases. Most *"Helminthosporium"* species are favored by moderate to warm (18–32°C) temperatures and particularly by humid, damp weather. Most *Helminthosporium* diseases, especially leaf spots, are retarded by dry weather, while crown- and root-affecting fungi may continue their invasion of diseased plants, killing out the plants (melting out) in irregular areas. Spread of the fungus is through the seed and infected debris blown or carried away. Over short distances during the growing season the fungus spreads through its numerous conidia, which may be carried by air currents, splashing rain, or by clinging to cultivating equipment, feet, animals, and so on.

Control of *Helminthosporium* diseases depends on the use of resistant varieties, pathgen-free seed, seed treatment with fungicides, proper crop rotation and fertilization, plowing under of infected plant debris, and by fungicides. In turf grasses, control of *Helminthosporium* diseases is facilitated by mowing at the recommended maximum height, reducing or removing the accumulated dense thatch, supplying sufficient fertilizer, and by irrigating quickly and sufficiently but in widely spaced (7- to 10-day) intervals. If fungicides are necessary, several of them, including cycloheximide, cycloheximide-thiram, chlorothalonil, Dyrene, and maneb, can be applied beginning in early spring and continuing at 1- to 2-week intervals for as long as necessary to get the disease under control. Several systemic fungicides, such as imazalil, nuarimol, etaconazole, and fenapanil, when applied as seed treatments or, in turf grasses, with the irrigation water, give good control of the root rot and several of the other symptoms and are being tested for grower application.

SELECTED REFERENCES

Anonymous (1970). Southern corn leaf blight. *Plant Dis. Rep.* **54**, *(Spec. Issue)*, 1099–1136.

Berger, R. D. (1973). Early blight of celery: Analysis of disease spread in Florida. *Phytopathology* **63**, 1161–1165.

Bergström, G. C., Knavel, D. E., and Kuć, J. (1982). Role of insect injury and powdery mildew in the epidemiology of the gummy stem blight disease of cucurbits. *Plant Dis.* **66**, 683–686.

Calvert, O. H., and Zuber, M. S. (1973). Ear-rotting potential of *Helminthosporium maydis* race T in corn. *Phytopathology* **63**, 769–772.

Chieu, W., and Walker, J. C. (1949). Physiology and pathogenicity of the cucurbit black-rot fungus. *J. Agric. Res. (Washington, D. C.)* **78**, 589–615.

Childs, T. W. (1968). Elytroderma disease of ponderosa pine in the Pacific Northwest. *U. S., For. Serv., Res. Pap. PNW* **69**, 1–45.

Chupp, C. (1953). "A Monograph of the Fungus Genus *Cercospora*." Published by the Author, Ithaca, New York.

Chupp C., and Sherf, A. F. (1960). "Vegetable Diseases and Their Control." Ronald Press, New York.

Cochran, L. C. (1932). A study of two *Septoria* leaf spots of celery. *Phytopathology* **22**, 791–812.

Couch, H. B. (1973). "Diseases of Turfgrasses." Krieger Publ. Co., New York.

Daub, M. E. (1982). Cercosporin, a photosensitizing toxin from *Cercospora* species. *Phytopathology* **72**, 370–374.

Eisensmith, S. P., and Jones, A. L. (1981). Infection model for timing fungicide applications to control cherry leaf spot. *Plant Dis.* **65**, 955–958.

Eyal, Z. (1981). Integrated control of *Septoria* diseases of wheat. *Plant Dis.* **65**, 763–768.

Eyal, Z., and Ziv, O. (1974). The relationship between epidemics of Septoria leaf blotch and yield losses in spring wheat. *Phytopathology* **64**, 1385–1389.

Forsberg, J. L. (1975). "Diseases of Ornamental Plants," Spec. Publ. No. 3 Rev. University of Illinois College of Agriculture, Urbana-Champaign.

Frank, J. A. (1985). Influence of root rot on winter survival and yield of winter barley and winter wheat. *Phytopathology* **75**, 1039–1041.
Gibson, I. A. S. (1972). Dothistroma blight of *Pinus radiata*. *Annu. Rev. Phytopathol.* **10**, 51–72.
Jewell, F. F., Sr. (1983). Histopathology of the brown spot fungus on longleaf pine needles. *Phytopathology* **73**, 854–858.
Keitt, G. W., Blodgett, E. G., Wilson, E. E., and Magie, R. O. (1937). The epidemiology and control of cherry leaf spot. *Res. Bull.— Wis., Agric. Exp. Stn.* **132**, 1–117.
Klotz, L. J. (1923). A study of the early blight fungus, *Cercospora apii. Mich., Agric. Exp. Stn., Tech. Bull.* **63**, 1–43.
Mackenzie, D. R. (1981). Association of potato early blight, nitrogen fertilizer rate, and potato yield. *Plant Dis.* **65**, 575–577.
Meehan, F. L., and Murphy, H. C. (1946). A new *Helminthosporium* blight of oats. *Science* **104**, 413–414.
Meredith, D. S. (1970). Banana leaf spot disease (Sigatoka) caused by *Mycosphaerella musicola. Commonw. Mycol. Inst. Phytopathol. Pap.* **11**, 1–147.
Nagel, C. M. (1945). Epiphytology and control of sugar beet leaf spot caused by *Cercospora beticola. Iowa Agric. Exp. Stn., Res. Bull.* **338**, 680–705.
Nichols, T. H., and Brown, H. D. (1972). How to identify *Lophodermium* and brown spot diseases on pines. *U. S., For. Serv., North Cent. For. Exp. Stn., Leafl.*, pp. 1–5.
Ou, S. H. (1980). A look at worldwide rice blast disease control. *Plant Dis.* **64**, 439–445.
Padnamadhan, S. Y. (1973). The great Bengal famine. *Annu. Rev. Phytopathol.* **11**, 11–26.
Peterson, G. W. (1967). Dothistroma needle blight of Austrian and ponderosa pines: Epidemiology and control. *Phytopathology* **57**, 437–441.
Reddick, D. (1911). The black rot of grapes. *Bull.—N.Y., Agric. Exp. Stn. (Ithaca)* **293**, 289
Rozier, A. (1931). Le black-rot. *Rev. Vitic.* **74**, 5–10, 21–25, 37–40, 53–59, 69–71.
Scharen, A. L., ed. (1985). "Septoria of Cereals," Proc. Workshop, USDA, ARS-12. Montana State University, Bozeman.
Schenck, N. C., and Stelter, T. J. (1974). Southern corn leaf blight development relative to temperature, moisture and fungicide applications. *Phytopathology* **74**, 619–624.
Shipton, W. A., Boyd, W. R. J., Rosielle, A. A. and Shearer, B. L. (1971). The common Septoria diseases of wheat. *Bot. Rev.* **37**, 231–262.
Shurtleff, M. C., and Randell, R. (1974). "How to Control Lawn Diseases and Pests." Intertec Publ. Corp., Kansas City, Missouri.
Shurtleff, M. C., *et al.* (1973). "A Compendium of Corn Diseases." Am. Phytopathol. Soc. St. Paul, Minnesota.
Singh, T., and Sinclair, J. B. (1985). Histopathology of *Cercospora sojina* in soybean seeds. *Phytopathology* **75**, 185–189.
Skolnik, M. (1974). Unusual post-infection activity of a piperazine derivative fungicide for the control of cherry leaf spot. *Plant Dis. Rep.* **58**, 326–329.
Smith, D. H., and Littrell, R. H. (1980). Management of peanut foliar diseases with fungicides. *Plant Dis.* **64**, 356–361.
Sprague, R. (1944). Septoria diseases of gramineae in western United States. *Oreg. State Monogr. Stud. Bot.* **6**, 1–151.
Stover, R. H. (1980). Sigatoka leaf spots of bananas and plantains. *Plant Dis.* **64**, 750–756.
Taber, R. A. (1979). Recent trends in the indentification of plant pathogenic fungi. *Proc.: Opening Sess. Plenary Int. Sess. Symp., Congr. Plant Prot., 9th, 1979,* Vol. 1, pp. 314–317.
Ullstrup, A. J. (1972). The impacts of the southern corn leaf blight epidemics of 1970–1971. *Annu. Rev. Phytopathol.* **10**, 37–50.
Western, J. H., ed. (1971). "Diseases of Crop Plants." Macmillan, New York.

Stem and Twig Cankers Caused by Ascomycetes and Imperfect Fungi

Cankers are localized wounds or dead areas of the bark that are often sunken beneath the surface of the stem or twigs of woody plants. In some cankers, the healthy tissues immediately next to the canker may increase in thickness and appear higher than the normal surface of the stem.

Innumerable kinds of pathogens cause cankers on trees, and cankers can also be caused by factors other than pathogens. Most cankers, however, have many similarities. The most common causes of tree cankers are Ascomycetous fungi, although some other fungi, particularly among the Imperfects, some bacteria, and some viruses also cause cankers.

The basic characteristics of cankers are that they are visible dead areas more or less localized that develop in the bark and, sometimes, in the wood of the tree. Cankers generally begin at a wound or at a dead stub. From that point, they expand in all directions but much faster along the main axis of the stem, branch, or twig. Depending on the host-pathogen combination and the prevailing environmental conditions, the host may survive the disease by producing callus tissue around the dead areas and thus limiting the canker. In infections of large limbs of perennial hosts, concentric layers of raised callus tissue may form. If, however, the fungus grows faster than the host can produce its defensive tissues, either no callus layers form and the canker appears diffuse and spreads rapidly, or the fungus invades each new callus layer and the canker grows larger each year. Young twigs are often girdled by the canker and killed soon after infection, but on larger limbs and stems cankers may extend to one or several meters in length, although their width extends to only part of the perimeter of the limb. Eventually, however, the limb or entire tree may be killed through girdling either by the original canker or by additional cankers that develop from new infections caused by the spores of the original canker.

Cankers are generally much more serious on fruit trees such as apple and peach, which they debilitate and kill. On forest trees, with the exception of chestnut blight, *Hypoxylon* canker, and *Dothichiza* canker, cankers deform, but do not kill their hosts. They do, however, reduce tree growth and the quality of lumber, result in greater wind breakage, and weaken the trees so that other more destructive wood- or root-rotting fungi can attack the trees.

Although most canker-causing fungi are Ascomycetes, only some of them, for example, *Plowrightia* and *Nectria,* produce their sexual ascigerous stage regularly. The other canker fungi produce primarily asexual conidia, usually in pycnidia partially or mostly embedded in the bark, and only occasionally do they produce perithecia. For this reason many of these fungi are known by the name that was given them while they were classified as Imperfect fungi, before their sexual stage had been found.

Some of the canker-causing fungi and their most important host plants are listed below.

Botryosphaeria dothidea, causes canker on apple, peach, pecan, hickory, sweetgum, redbud, many other trees, and on currant and gooseberry.
Ceratocystis fimbriata, causes canker diseases on cacao, coffee, stone fruits, rubber, poplar, London plane, and sycamore.
Cryptodiaporthe populea (Imperfect stage *Dothichiza populea*) causes the *Dothichiza* canker of poplar.
Plowrightia (Dibotryon) morbosum, causes black knot of plum and cherry.
Endothia parasitica, causes chestnut blight.
Eutypa armeniacae, causes dieback in grape and apricot.
Eutypella parasitica, causes *Eutypella* trunk canker of sugar maple (Figure 11-54), red maple and boxelder.
Gremmeniella abietina (Scleroderris lagerbergii) causes the Scleroderris canker of conifers.

FIGURE 11-54 *Eutypella* trunk canker of maple. Note size of canker. (Photo courtesy U.S.D.A.)

Hypoxylon mammatum, causes hypoxylon canker of aspen.
Nectria galligena, causes canker of apple and pear, aspen, beech, birch, basswood, black walnut, elm, maple, oak, and other trees.
Urnula craterium (Imperfect stage *Strumella coryneoidea*), causes *Strumella* canker primarily of red and black oak, but also of hickory, beech, and maple.
Valsa sp. (Imperfect stage *Cytospora* sp.), causes *Cytospora* canker of peach and many other fruit trees, poplar, willow, and more than 70 species of hardwood trees and shrubs, as well as spruce and some other conifers.

A brief description of the main characteristics of the canker diseases caused by some of the above fungi, as well as control measures, when possible, are given below.

• Dothichiza Canker of Poplar

One of the most important diseases of poplar, especially black, Lombardy, and Simon poplar, and of cottonwood, in Europe and North America, Dothichiza canker of poplar causes cankers on twigs, branches, and stems (Figure

11-55, A–C). All the young trees in nurseries and plantations may be killed rapidly by cankers that girdle the stem. On older trees, branch and stem cankers result in a typical dieback. The fungus enters the tree through lenticels, buds, and bark cracks. The developing canker is discolored and sunken. The host checks the advance of the fungus during favorable conditions, but during low temperatures or drought when the host vigor is reduced, the fungus resumes its activity, the canker enlarges, and may girdle and kill the branch or tree. The fungus produces conidia in pycnidia on the diseased parts throughout the growing season. The conidia are spread by rain, water, and wind and can probably cause new infections throughout the season, but most infections seem to occur in late spring when conidia are also most abundant. Perithecia and ascospores are seldom produced.

• Black Knot of Plum and Cherry

Black knot of plum and cherry occurs on cultivated and wild plums and cherries, primarily in the eastern half of the United States and New Zealand. It causes conspicuous, 2–25 cm or more long, coal-black knotty swellings on one side of, or encircling, twigs and branches (Figure 11-55, D, E). They may be several times the diameter of the limbs and make heavily infected trees appear quite grotesque. Infected plants become worthless after a few years as a result of limb death and stunting of the trees. The fungus, *Plowrightia (Dibotryon) morbosum,* produces conidia on free hyphae and ascospores in perithecia formed on the black knots. Both conidia and ascospores are spread by wind and rain, and in early spring they can penetrate healthy and injured woody tissue of the current season's growth. Large limbs are also attacked, especially at points of developing small twigs. The fungus grows into the cambium and xylem parenchyma and along the axis of the twig. After 5 or 6 months, excessive parenchyma cells are produced and pushed outward forming the swelling. The following spring conidia are produced on the knot surface, giving it a temporary olive-green velvety appearance. The knots enlarge rapidly during the second summer, and in their surface layer, perithecia are formed that develop during the winter and release ascospores the following spring. The knots continue to expand in following years. The disease can be controlled by pruning and burning of all black knots and destruction of black knots or of all affected wild plums and cherries near the orchard. Spraying the orchard trees before and during bloom with sulfur or captan, each alone or combined with thiophanate methyl, or with fixed copper fungicides to which hydrated lime is added, protects trees from infection.

• Chestnut Blight

After it was introduced in New York City in 1904, chestnut blight spread and by 1940 destroyed practically all American chestnut trees throughout their natural range in the eastern third of the United States from the Canadian border south nearly to the Gulf of Mexico. American chestnuts killed by the blight composed 50 percent of the overall value of eastern hardwood timber stands. The fungus, *Endothia parasitica,* which some call *Cryphonectria parasitica,* also attacks oak, red maple, hickory, and, sporadically, other trees, but not nearly as severely as it attacks the American chestnut. It is now

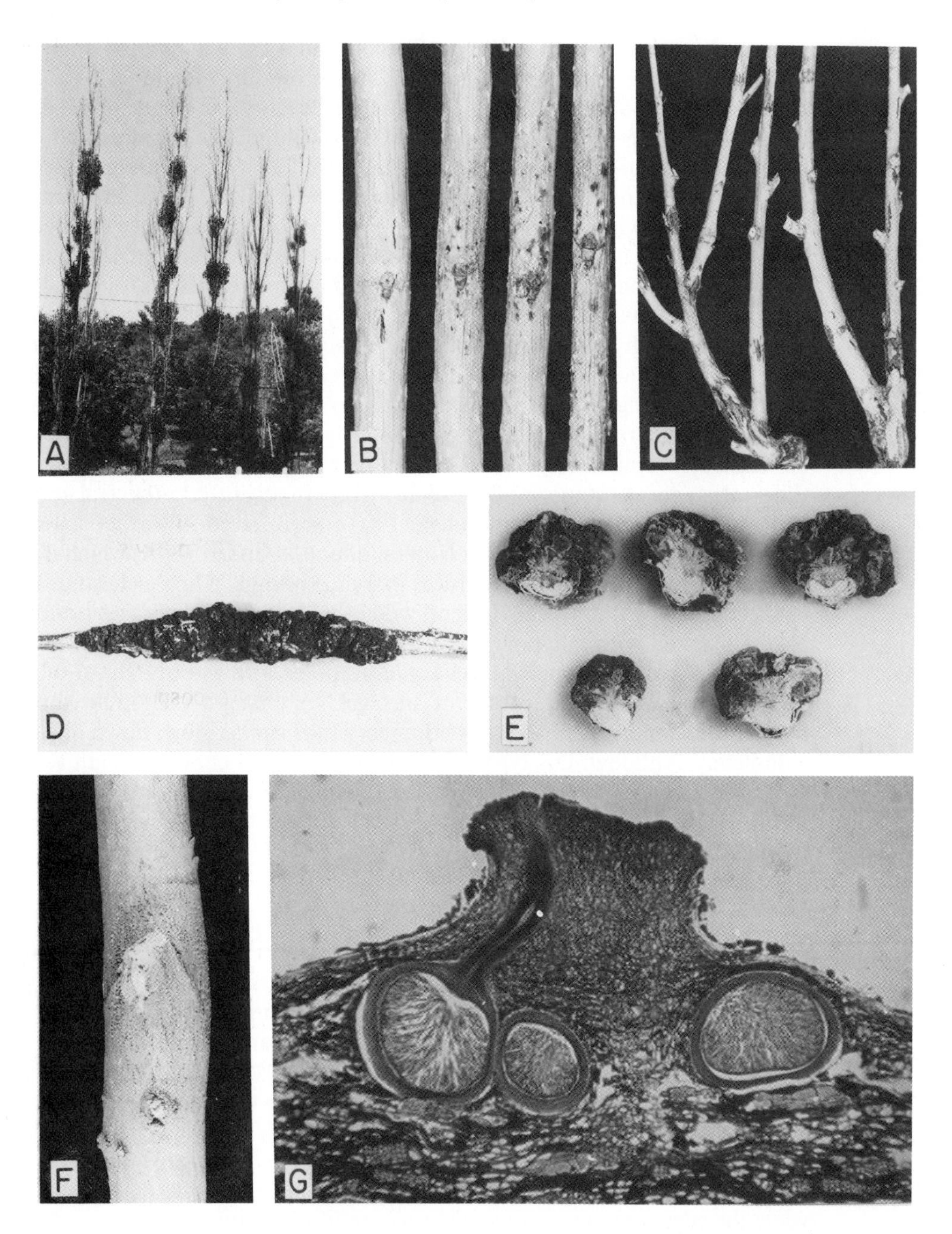

FIGURE 11-55 (A–C) *Dothichiza* canker on Lombardy poplar. (A) A row of trees whose twigs have been killed by cankers like those shown in B and C. Bunches of new twigs have sprouted in some parts of the trees. (B) Stages in the development of a canker along the twig. Infection is generally centered around a leaf scar. (C) Advanced cankers, mostly at the base of twigs. (D, E) External and cross-section views of black knot canker on cherry caused by *Plowrightia morbosum.* (F, G) Chestnut blight canker caused by *Endothia parasitica.* (F) Canker on young chestnut stem, apparently started at broken branch. (G) Perithecia of *Endothia parasitica* embedded in chestnut bark.

present throughout North America, Europe, and Asia. The fungus penetrates the bark of stems through wounds and then grows into the inner bark and cambium. Soon a swollen or sunken canker develops, the bark of which is reddish-orange to yellow-green and covered by pimplelike pycnidia and perithecia (Figure 11-55, F, G). Cankers often have long cracks on their surface, may be several inches to many feet long, and eventually girdle the stem or branch causing wilting and death of the parts beyond the canker. The pycnidia produce tiny conidia that ooze out as long orange curllike masses during moist weather and are spread by birds, crawling or flying insects, or splashing rain. The perithecia produce ascospores that are shot forcibly into the air and may be carried by wind over long distances. The fungus survives and continues to invade and produce its spores in trees or parts of trees already killed by the blight. Blighted trees almost always produce sprouts below the basal cankers, but the resulting saplings become blighted in turn by new infections. No control is available against chestnut blight, although some new systemic fungicides appear promising for isolated trees. In recent years several strains with reduced virulence (hypovirulence) have been found in Europe and the United States. All of these strains contain double stranded RNA, which is the kind of RNA present in many viruses that infect fungi. When a chestnut tree canker caused by the typical virulent, wild, dsRNA-free *Endothia parasitica* is also inoculated with a hypovirulent dsRNA-containing strain of the fungus, the dsRNA passes through mycelial anastomoses into the mycelium of the virulent strain. The acquisition of the dsRNA changes this strain into a hypovirulent one, and further development of the canker slows down or stops completely. Although this type of biological control of chestnut blight works well on isolated trees, it has not yet been possible to use it on a large scale in the forest. So far, no completely resistant American chestnuts have been found.

- **Nectria Canker**

Nectria canker is one of the most important diseases of apples and pears and of many species of hardwood forest trees in most parts of the world. Losses are greater in young trees because in these the fungus girdles the trunk or scaffold branches, while in older trees only small branches are usually killed directly (Figure 11-56, A–D). Cankers on the main stem of older trees, however, reduce the vigor and value or productivity of the tree, and such trees are subject to wind breakage. *Nectria* cankers usually develop around bud scars, wounds, twig stubs, or in the crotches of limbs. Young cankers are small, circular, brown areas. Later, the central area becomes sunken and black, while the edges are raised above the surrounding healthy bark. In many hosts and under favorable conditions for the host, the fungus grows slowly, the host produces callus tissue around the canker, and the margin of the canker cracks. The tissues under the black bark in the canker are dead, dry and spongy, flake off, and fall out, revealing the dead wood and the callus ridge around the cavity. In subsequent years the fungus invades more healthy tissue and new, closely packed, roughly concentric ridges of callus tissue are produced every year, resulting in the typical open, target-shaped *Nectria* canker. In some hosts, however, and under conditions that favor the fungus, invasion of the host is more rapid, the bark in the cankered area is roughened and cracked but does not fall off, and the successive callus ridges are some distance apart.

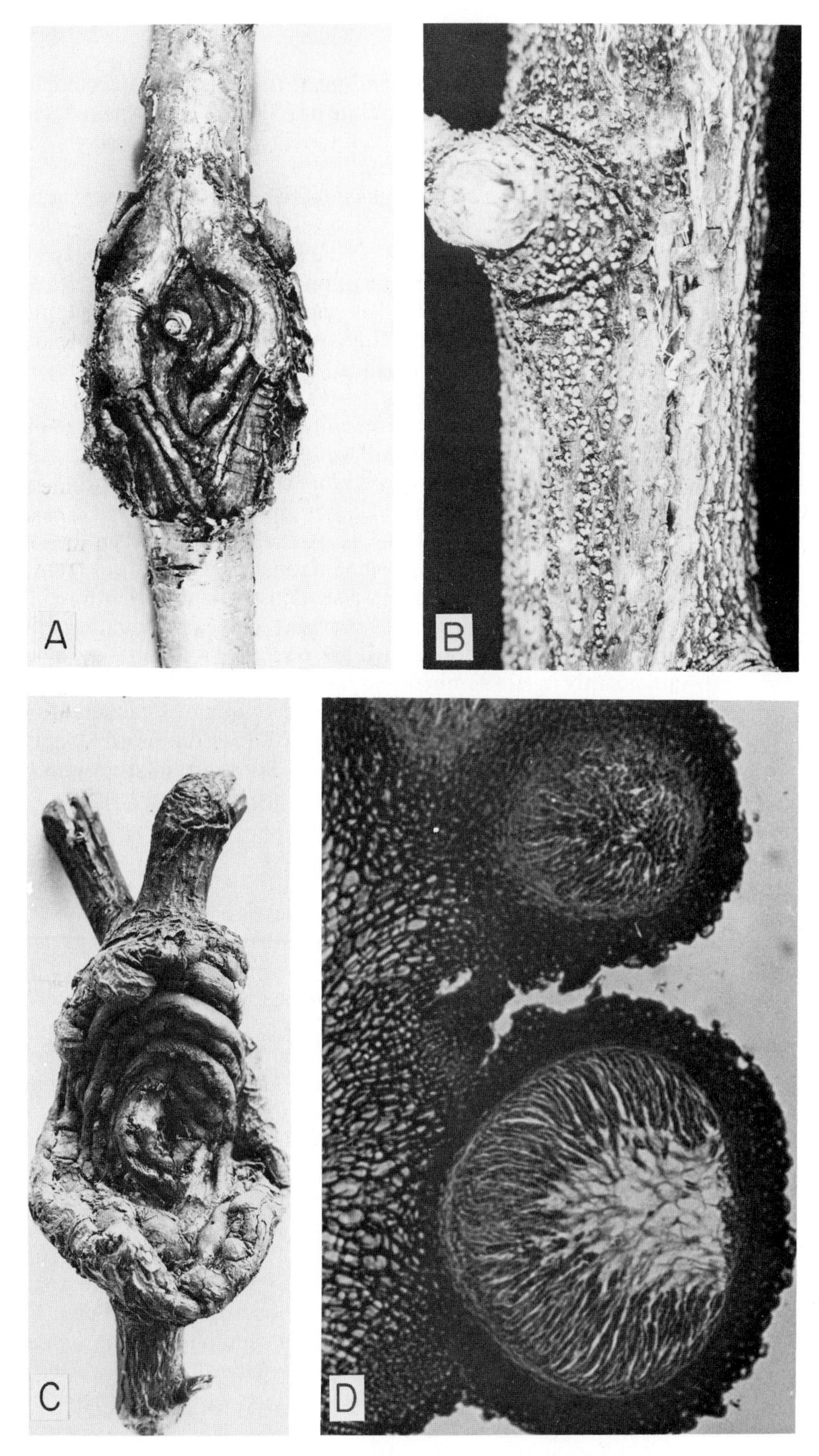

FIGURE 11-56 *Nectria* cankers and fruiting bodies on maple (A) and apple (B-D). (B) Conidial stage of *Nectaria galligena* on apple branch. (D) *Nectria* perithecia. (Photo A courtesy U.S.D.A.)

In hosts such as apple and pear, fruits are also infected and develop a circular, sunken, brown rot. White or yellowish pustules producing numerous conidia form on rotted areas. Internally the rotted tissue is soft and has a striated appearance.

The fungi, *Nectria galligena* and some related species, attack the many different tree hosts. All *Nectria* species produce similar, two-celled ascospores in brightly colored perithecia on the surface of a cushion-shaped stroma, but different *Nectria* species produce conidia, which in themselves do not appear to be related and are classified as various species of Imperfect fungi. Thus, the conidia of *N. galligena* are either single-celled microconidia or more commonly 2- to 4-celled, cylindrical to crescent-shaped macroconidia of the *Cylindrocarpon* type (Figure 11-57). The conidia are produced soon after infection on small, white, or creamy-yellow or bright orange-pink sporodochia, which appear on the surface of the bark over the infected area or on fruit. The conidia are produced more commonly early in the season but also in the summer and early fall. They are spread by wind and by washing during rainy periods and perhaps by insects. Perithecia appear in the cankers in late summer and fall and in the same stroma that earlier produced the conidia, which they eventually replace. The ascospores are either forcibly discharged and carried by wind or, in moist weather, they ooze from the perithecium and are washed by rain or carried by insects. Ascospores are dispersed more abundantly in late summer and fall but are also released at other times of the year. Sanitation, that is, removal and burning of cankered limbs or trees where possible, is often the only control measure possible. Spraying with a fungicide such as captafol or 8:8:100 Bordeaux mixture immediately after leaf fall helps reduce *Nectria* infections in fruit trees.

• Valsa, or Cytospora, Canker and Dieback

Valsa, or Cytospora, canker and dieback is a worldwide canker disease that probably affects more species of trees than any of the canker diseases previously described. It is estimated that more than 70 species of fruit trees, harwood forest and shade trees, shrubs, and conifers are attacked by one of several species of the pathogen. The fungus, *Valsa* sp., which some call *Leucostoma* sp., is most commonly found in its Imperfect stage, *Cytospora* sp., and therefore the disease is usually known as *Cytospora* canker.

Cytospora canker is most serious on peach and the other stone fruits, on apple and pear, and on poplar and willow, but it is often serious on many other shade or forest trees (Figure 11-58). Actually, few orchards are free from its damaging effects. Many trees are seriously injured by cankers on the trunk, in the main crotch, on the limbs, and on the branches. Infected branches of fruit trees often break from the weight of the crop or during storms. *Cytospora* canker is most severe on fruit or shade trees growing under stress, such as those growing on an unfavorable site, or injured by drought, frost, fire, or severe pruning. The fungus is mostly a saprophyte living on dead bark but becomes parasitic when the tree is weakened. However, the presence of *Cytospora* on a dead twig or branch does not mean that this fungus has killed it.

Infection of small twigs and branches results in dieback without definite cankers. The cankers occur and are most pronounced on trunks and large

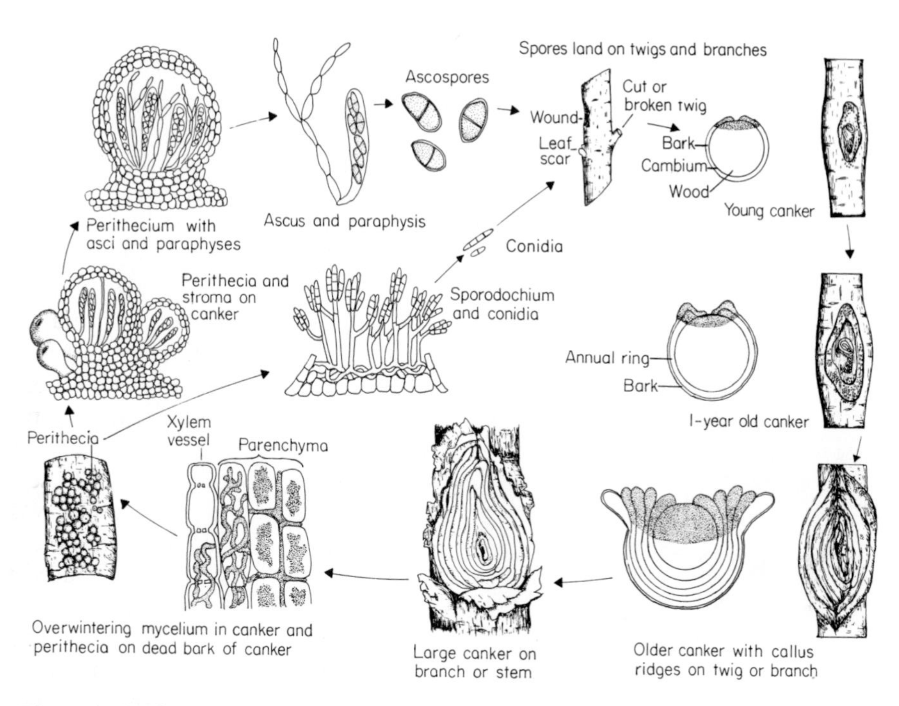

FIGURE 11-57 Disease cycle of *Nectria* canker caused by *Nectria galligena*.

branches. A canker appears first as a gradual circular killing of the bark of a limb or stem. The infected area soon becomes brownish and sunken, and the host often produces raised callus tissue around it. In some hosts, particularly the stone fruit trees, the inner diseased bark becomes dark and odorous, and a copious flow of gum exudes from the dead tissues. The cankers may be long and slightly sunken or short, deeply sunken, with conspicuously raised callus borders. Later the bark shrivels and separates from the underlying wood and from the surrounding healthy bark. Small, pimplelike pycnidia of the fungus appear on the dead bark, and later the shriveled bark may slough off exposing dead wood beneath. The cankers increase in size each year and become unsightly, rough swellings. Many twigs and branches die back as a result of cankers that girdle them completely (Figures 11-58 and 11-59).

Cytospora cankers result from infections by conidia of the fungus *Cytospora.* The fungus produces *Valsa*-type ascospores in perithecia, but the perithecial stage is not common. The conidia are produced inside pycnidia consisting of many connecting cavities and one pore through which the conidia are exuded. The spores are small, hyaline, one-celled, and slightly curved. They are produced in a gelatinous matrix, which during wet weather, absorbs water, swells, and oozes out of the pycnidium, carrying with it the masses of spores. The spores may be washed away or splashed by rain or may be spread by insects and humans. If it is moist but not rainy, the exuded conidia may form coiled threads of spores that dry out and harden and remain on the canker for several days or weeks. Most infections take place in the dormant season, particularly in late fall or early winter and in late winter or early spring. Weakened injured trees, however, may be infected throughout

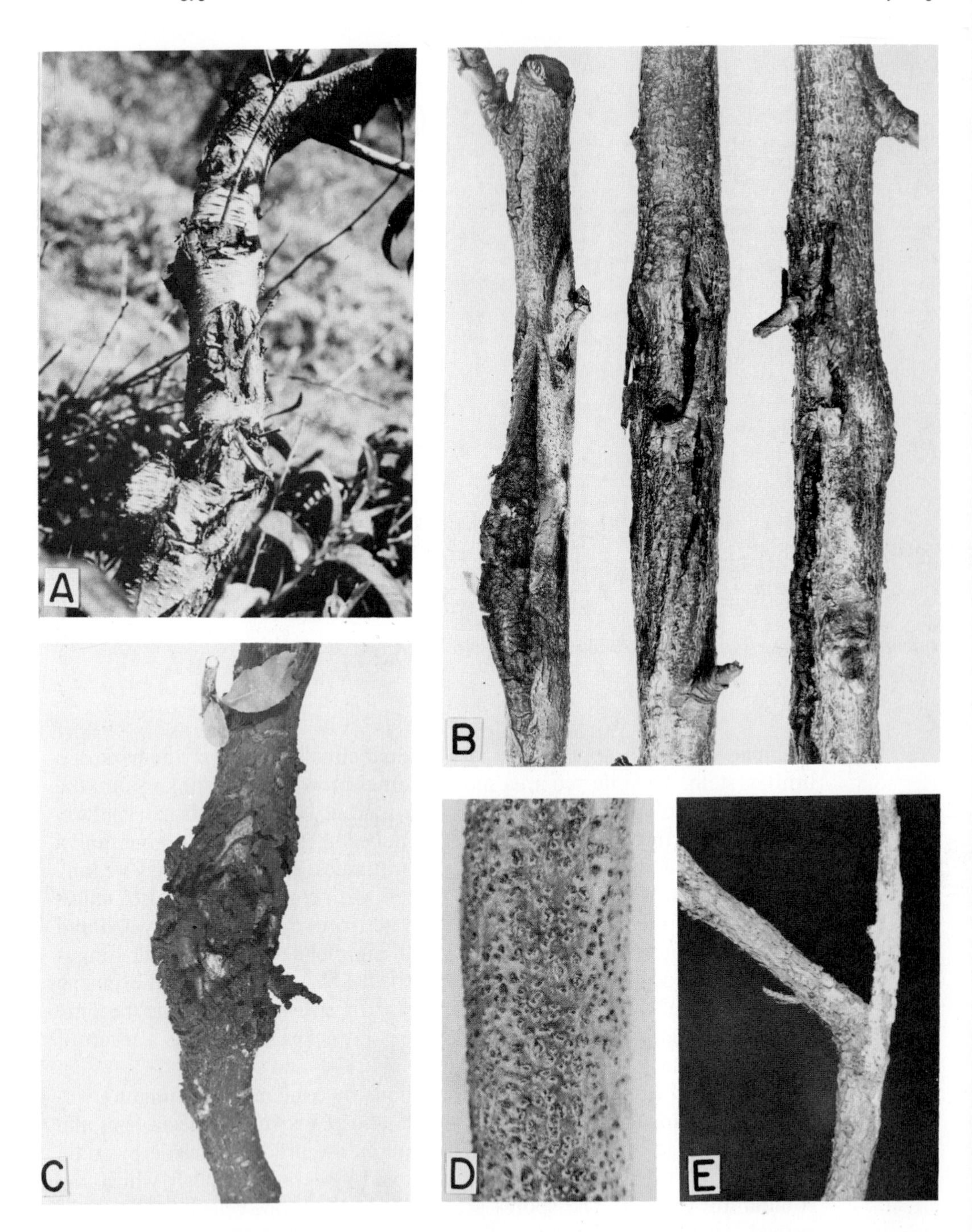

FIGURE 11-58 *Cytospora* canker on peach trunk (A) and twigs (B), and on plum twig (C). (D) *Cytospora* pycnidia embedded in the bark of apple twig. (E) *Cytospora* canker of spruce. Note white pitch flowing out of canker area.

the growing season. Both the mycelium and the conidia of the fungus can live over winter on the infected parts.

Infection of small twigs can occur through injuries or leaf scars. In larger branches, wounds of any kind, broken branches and pruning stubs, winter injuries, and sunscalds also make ideal points of entry for the fungus. The

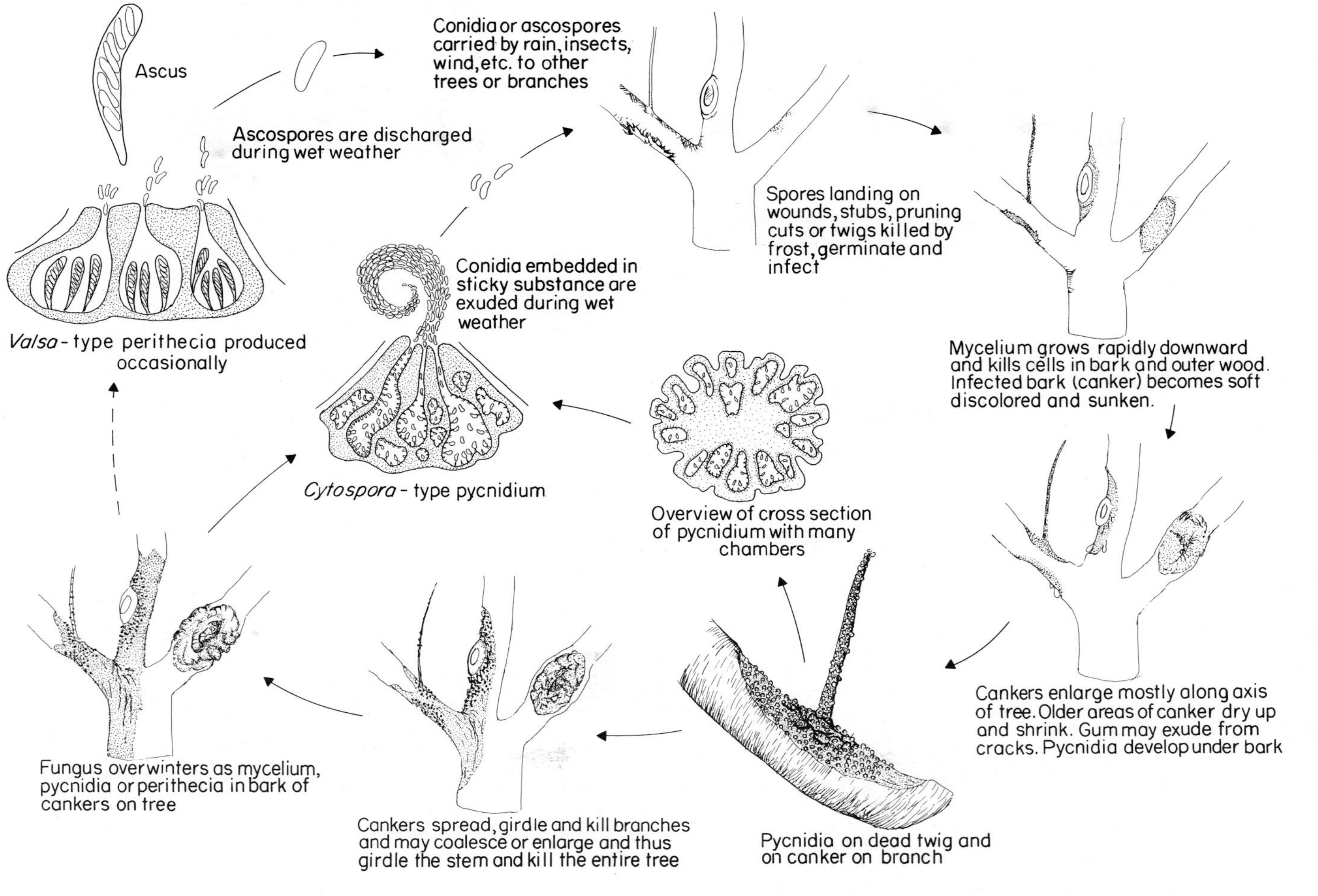

FIGURE 11-59 Disease cycle of the *Cytospora (= Valsa)* canker of peach and most other trees.

fungus becomes established in dead bark and wood and invades the surrounding tissues to form a canker. The fungus grows through the cells in the bark and the outer few rings of the wood.

Control measures for the *Cytospora* canker depend on: good cultural practices such as watering and fertilization to keep the trees in good vigor; avoiding wounding and severe pruning of trees; removing cankers from trunks and large branches during dry weather and treating the wound and all pruning cuts with a disinfectant and a wound dressing; removing and burning cankered and dead branches and twigs; pruning as late in the spring as possible and spraying with Phygon XL immediately after pruning and before it rains. The above practices help but do not completely prevent *Cytospora* canker.

SELECTED REFERENCES

Anagnostakis, S. L. (1982). Biological control of chestnut blight. *Science* **215,** 466–471.

Bolay, A., and Moller, W. J. (1977). *Eutypa armeniacae,* agent d'un grave déperissement de vignes en production. *Rev. Suisse Vitic., Arboric., Hortic.* **9,** 241–251.

Brandt, R. W. (1964). *Nectria* canker of hardwoods. *U.S., For. Serv., For. Pest Leafl.* **84,** 1–7.

Davidson, A. G., and Prentice R. M., eds. (1967). "Important Forest Insects and Diseases of Mutual Concern to Canada, the United States and Mexico." Dept. For. Urban Dev., Canada.

Dhanvantari, B. N. (1982). Relative importance of *Leucostoma cincta* and *L. persoonii* canker of peach in southwestern Ontario. *Can. J. Plant Pathol.* **4,** 221–225.

Diller, J. D. (1965). Chestnut blight. *U.S. For. Serv., For. Pest Leafl.* **94,** 1–7.

Dubin, H. J., and English, H. (1974). Factors affecting control of European apple canker by Difolatan and basic copper sulfate. *Phytopathology* **64,** 300–306.

Dubin, H. J., and English, H. (1975). Epidemiology of European apple canker in California. *Phytopathology* **65,** 542–550.

Fulbright, D. W. (1984). Effect of eliminating dsRNA in hypovirulent *Endothia parasitica. Phytopathology* **74,** 722–724.

Griffin, G. J., Hebard, F. V., Wendt, R. W., and Elkins, J. R. (1983). Survival of American chestnut trees: Evaluation of light resistance and virulence in *Endothia parasitica. Phytopathology* **73,** 1084–1092.

Gross, H. L. (1967). *Cytospora* canker of black cherry. *Plant Dis. Rep.* **51,** 261–263.

Hampson, M. C. and Sinclair, W. A. (1973). Xylem dysfunction in peach caused by *Cytospora leucostoma. Phytopathology* **63,** 676–680.

Hickey, K. D. (1966). Black knot disease of plum and its control. *N. Y. State Coll. Agric. (Cornell), Ext. Bull.* **1173,** 1–7.

Hubbes, M. (1959). Untersuchungen uber *Dothichiza populea,* den Erreger des Rindenbrandes der Pappel. *Phytopathol. Z.* **35,** 58–96.

Lortie, M. (1964). Pathogenesis in cankers caused by *Nectria galligena. Phytopathology* **54,** 261–263.

McCubbin, W. A. (1918). Peach canker. *Can., Dept. Agric., Bull.* **37,** 1–20.

McPartland, J. M., and Schoeneweiss, D. F. (1984). Hyphal morphology of *Botryosphaeria dothidea* in vessels of unstressed and drought-stressed stems of *Betula alba. Phytopathology* **74,** 358–362.

Moller, W. J., and Kassimatis, A. N. (1978). Dieback of grapevines caused by *Eutypa armeniacae. Plant Dis. Rep.* **62,** 254–258.

Roane, M. K., Griffin, G. J., and Elkins, J. R. (1986). "Chestnut Blight, Other Endothia Diseases, and the Genus Endothia." *APS Monograph Series.* APS Press, St. Paul, Minnesota.

Rosenberger, D. A., and Gerling, W. D. (1984). Effect of black rot incidence on yield of Stanley prune trees and economic benefits of fungicide protection. *Plant Dis.* **68,** 1060–1065.

Schoenenweiss, D. F. (1981). The role of environmental stress in diseases of woody plants. *Plant Dis.* **65,** 308–314.

Skilling, D. D. (1981). Scleroderis canker—development of strains and potential damage in North America. *Can. J. Plant Pathol.* **3,** 263–265.

Swinburne, T. R. (1975). European canker of apple *(Nectria galligena). Rev. Plant Pathol.* **54,** 787–799.

Van Alfen, N. K. (1982). Biology and potential for disease control of hypovirulence of *Endothia parasitica*. *Annu. Rev. Phytopathol.* **20,** 349–362.
Wainwright, S. H., and Lewis, F. H. (1970). Developmental morphology of the black knot pathogen on plum. *Phytopathology* **60,** 1238–1244.
Weaver, D. J. (1974). A gummosis disease of peach trees caused by *Botryosphaeria dothidea*. *Phytopathology* **64,** 1429–1432.

Anthracnose Diseases Caused by Ascomycetes and Imperfect Fungi

Anthracnoses are diseases of the foliage, stems, or fruits that typically appear as small or large, dark-colored spots or slightly sunken lesions with a slightly raised rim. In addition to leaf spots, anthracnoses often have a prolonged latent stage in fruit infection and may cause twig or branch dieback. In some anthracnoses, particularly serious on some fruit crops, the symptoms appear as small spots with raised, corky surfaces. Anthracnose diseases of fruit often result in fruit drop and fruit rot. Anthracnoses are caused by fungi that produce their asexual spores, conidia, within small, black acervuli arranged in the lesion concentrically or in a scattered manner.

Four ascomycetous fungi, *Diplocarpon, Elsinoe, Glomerella,* and *Gnomonia,* belong to this group, although a few of their species produce conidia in pycnidia rather than acervuli. Actually, these fungi are found in nature mostly in their conidial stage and can overwinter as mycelium or conidia, yet, some more frequently than others also produce ascospores in perithecia, which may overwinter. The main anthracnose and related diseases caused by these fungi are listed below.

Diplocarpon, causing black spot of rose *(D. rosae),* leaf scorch of strawberry *(D. earliana),* and leaf spot of hawthorn *(D. maculata).*

Elsinoe (conidial stage: *Sphaceloma*), causing anthracnose of grape *(E. ampelina),* of dogwood *(E. corni),* of rasberry *(E. veneta),* and scab of citrus *(E. australis* and *E. fawcetti*), of avocado *(E. perseae),* of pansy violet, and of poinsettia.

Glomerella (conidial stage: *Colletotrichum* or *Gloeosporium*), causing anthracnose of azalea, cyclamen, and sweet pea, bitter rot of apple and cranberry, dieback and canker of camellia and privet, and ripe rot of grape and other fruits *(G. cingulata),* anthracnose of soybean *(G. glycines),* of cotton *(G. gossypii),* and of bean *(G. lindemuthiana).*

Gnomonia, causing anthracnose of walnut *(G. leptostyla),* of oak *(G. veneta)* (Figure 11-60A), of sycamore *(G. platani)* (Figure 11-60B), of linden *(G. tiliae),* and leaf spot of elm *(G. ulmea)* (Figure 11-60C) and of hickory *(G. caryae).*

The acervulus-producing Imperfect Fungi make up the order Melanconiales. The most important plant pathogenic fungi that produce acervuli are *Colletotrichum (Gloeosporium), Coryneum, Cylindrosporium, Marssonina, Melanconium,* and *Sphaceloma.* Some of these are the conidial stages of some of the Ascomycetes that cause anthracnose diseases. Thus, several *Colletotrichum* or *Gloeosporium* species produce a *Glomerella*-type perfect stage, and the diseases caused by these species are discussed as Glomerella diseases; some *Cylindrosporium* species have a *Higginsia*-type perfect stage and some a *Mycosphaerella*-type perfect stage; some *Marssonina* species have a *Diplocaron-* or *Gnomonia*-type perfect stage; and most *Sphaceloma* species have an *Elsinoe*-type perfect stage. For some species of these same fungi, however, and for all species of *Coryneum* and *Melanconium,* no perfect stage is yet known.

FIGURE 11-60 (A) Young oak leaves and twigs killed by the anthracnose fungus *Gnomonia veneta.* (B) Defoliated and dead twigs of sycamore tree showing anthracnose caused by *Gnomonia veneta (G. platani).* (C) Elm leaf spot caused by *Gnomonia ulmea.* (D) Black spot of rose caused by *Diplocarpon rosae.* (E) Scanning electron micrograph of *Marssonina juglandis* acervulus with conidia. (F) Anthracnose of poplar caused by *Marssonina* sp. Note lesions on veins and petiole. (G) Anthracnose symptoms on maple leaf caused by *Gloeosporium apocryptum.* (Photo E courtesy M. F. Brown and H. G. Brotzman.)

In addition to the diseases described or listed as caused by acervulus producing fungi that have perfect stages, some of the other important plant diseases caused strictly by the imperfect stages of the fungi are:

Colletotrichum (Gloeosporium), causing anthracnose of cereals and grasses (*C. graminicola,* rarely producing a *Glomerella* perfect stage), of cucurbits *(C. lagenarium),* of lima bean *(C. truncatum),* anthracnose of peony, anthracnose or fruit rot of eggplant and of tomato *(C. phomoides),* red rot of sugarcane *(C. falcatum),* anthracnose of pepper, spinach, turnip, and cauliflower, onion smudge *(C. circinans),* and black scale of lily bulbs, anthracnose of citrus, fig, olive, avocado, and of many other plants.
Coryneum (Stigmina), causing "Coryneum blight," "shot hole" or "fruit spot" of stone fruits, especially peach and apricot.
Melanconium, causing bitter rot of grapes.

Anthracnose diseases, particularly those caused by *Colletotrichum (Gloeosporium)* or *Glomerella* group, are very common and destructive on numerous crop and ornamental plants, and their geographic distribution is in most cases worldwide. Although severe everywhere, anthracnose diseases cause their most severe losses in the tropics and subtropics.

Black Spot of Rose

The black spot of rose appears as small to large circular, black lesions on the leaves (Figure 11-60), and on susceptible varieties, as raised, purple-red, later blackened and blistered irregular blotches on immature wood of first-year canes. The leaf spots are consistently black, have fringed margins and may coalesce to produce large, irregular, black lesions. The leaf tissue around the lesions turns yellow, and often, when severely infected, entire leaves become yellow and fall off prematurely, leaving the canes almost completely defoliated.

The fungus, *Diplocarpon rosae,* produces ascospores in tiny apothecia formed in old lesions, and *Marssonina*-type conidia in acervuli forming between the outer wall and cuticle of the epidermis (Figure 11-60E). The conidiophores arise from a thin black stroma, are short, and give rise to successive crops of conidia. The abundant conidia production pushes up and ruptures the cuticle. The fungus overwinters as mycelium or as ascospores and conidia in infected leaves and canes. Both kinds of spores can cause primary infections of leaves in the spring by direct penetration. The mycelium grows in the mesophyll but within two weeks forms acervuli and conidia at the upper surface. Conidia are produced throughout the growing season and cause repeated infections during warm, wet weather.

Control of *Diplocarpon* diseases is through sanitation, such as removing and burning infected leaves, cutting back the canes of diseased rose plants, and spraying with benomyl, chlorothalonil, zineb, or mancozeb or applying sulfur-copper dust. Applications should begin as soon as new leaves appear in the spring or at the first appearance of black spot on the foliage and then repeated at 7- to 10-day intervals or within 24 hours after each rain.

Glomerella Diseases

At least four species of *Glomerella* cause serious anthracnose diseases of numerous important annual crop and ornamental plants, while one, *G. cingulata,* also causes cankers and dieback of woody plants such as camellia and privet, bitter rot of apples, and ripe rot of grape, pears, peaches, and other fruits.

- *Glomerella Anthracnose Diseases of Annual Plants.* The most important Glomerella anthracnose diseases are those on bean, cotton, soybean, and sweet pea. Although different species of the fungus are responsible for the disease on each of these hosts, the life cycles of the fungi, the symptoms, the disease development, and control are approximately the same in all cases.

The diseases are present wherever their hosts are grown, although they are more severe in warm to cool, humid areas, and generally are not a problem under dry conditions. Plants in all stages of growth are subject to anthracnose. The fungus is often present in or on the seed produced in infected pods or bolls. Infected seed may show yellowish to brown sunken lesions of various sizes. When infected seeds are planted, many of the germinating seedlings are killed before emergence. Dark-brown, sunken lesions with pink mass of spores in the center are often present on the cotyledons of young seedlings. The fungus may destroy one or both of the cotyledons, while its spores spread onto the hypocotyl and the mycelium moves into the stem. On the stem the fungus produces numerous small, shallow, reddish-brown specks that subsequently enlarge, become elongated, and finally sunken. The lesions are covered with myriads of pink- to rust-colored spores. If conditions are humid, the lesions may be so numerous that they girdle and weaken the stem to the point where it cannot support the top of the plant. The fungus also attacks the petioles and the veins of the under side of the leaves, on which it causes long, dark, brick-red to purplish colored lesions that later turn dark brown to almost black. Few lesions are produced between the veins in bean, but they are rather common on cotton, and in sweet pea they may involve the entire leaf (Figure 11-61, B, C).

The anthracnose fungi also attack and cause their most characteristic symptoms on the bean pods and on the cotton bolls. On the latter, the fungus seems to produce numerous lesions that spread and coalesce and cover most or all the surface of young cotton bolls, with almost continuous masses of spores covering the infected area. On pods, small flesh- to rust-colored elongated lesions appear, which later become sunken, circular, and about 5–8 mm in diameter. Lesions developing on young pods may extend through the pod and even to the seed, while in older pods the lesions do not extend beyond the pod. As the pods mature, the margin of the lesions is generally slightly raised, while the pink spore masses of the lesions dry down to gray, brown, or black granulations or to small pimplelike protrusions.

- *Glomerella Fruit Rots.* The most important of the Glomerella fruit rots are bitter rot of apple and ripe rot of grape.

Bitter rot is worldwide in distribution and in warm, humid weather may cause enormous losses by destroying an entire crop of apples just a few weeks before harvest. Bitter rot symptoms may appear when the fruit is half grown but more frequently when it approaches its full size. The rot starts as small, light brown areas that enlarge rapidly and become circular and somewhat sunken in the center. The surface of the spots is smooth at first and may be dark brown or black until the spots are 1–2 cm in diameter. Then numerous, slightly raised cushions appear mostly near the center of the spots and some extending outward toward the edge of the spots. In humid weather, the cushions produce creamy masses of pink-colored spores, sometimes arranged

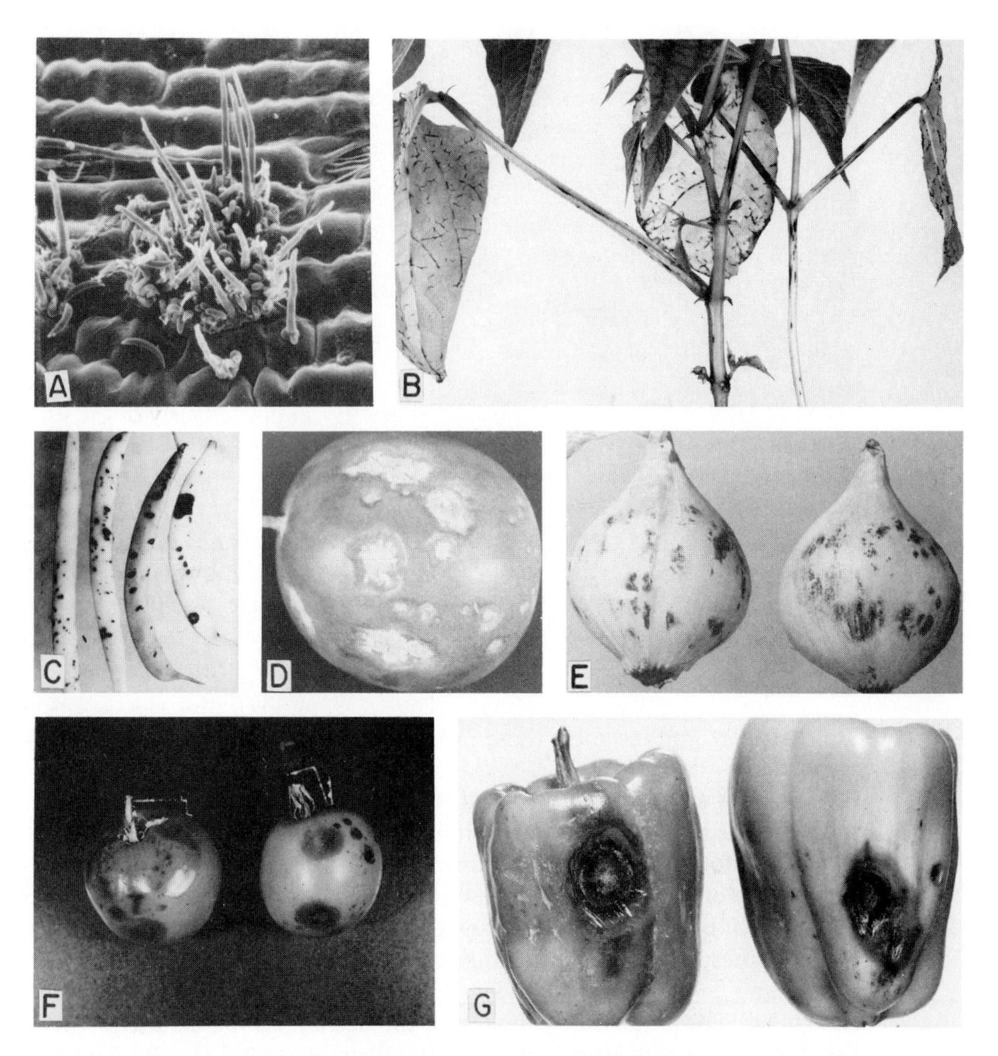

FIGURE 11-61 (A) Scanning electron micrograph of acervulus of *Colletotrichum graminicola* showing setae (the long structures) and conidia. (B) Bean anthracnose lesions on stems, petioles and leaves caused by *Glomerella (Colletotrichum) lindemuthiana.* (C) Bean anthracnose on pods. (D) Watermelon anthracnose caused by *Colletotrichum circinans.* (F) Tomato anthracnose caused by *Colletotrichum phomoides.* (G) Anthracnose on peppers caused by *Glomerella cingulata.* (Photo A courtesy M. F. Brown and H. G. Brotzman. Photo B courtesy G. C. Papavizas. Photo E courtesy U.S.D.A.)

in concentric circles, the rotted area expands rapidly, and more rings of spore masses appear (Figure 11-62). In older rotted areas the pink masses disappear and the tissue becomes dark brown to black, wrinkled, and sunken. The rot also spreads inward toward the apple core forming a cone of somewhat watery, rotted tissue that may or may not be bitter. When, as it is common, several spots develop on a fruit, they usually enlarge, fuse, and rot the entire fruit, which may drop and mummify or mummify and cling on the twig. Bitter rot infections may occur in the fall but fail to develop appreciably during cold storage. When, however, the fruit is marketed and kept at room temperature, bitter rot may develop very rapidly. Bitter rot cankers on the

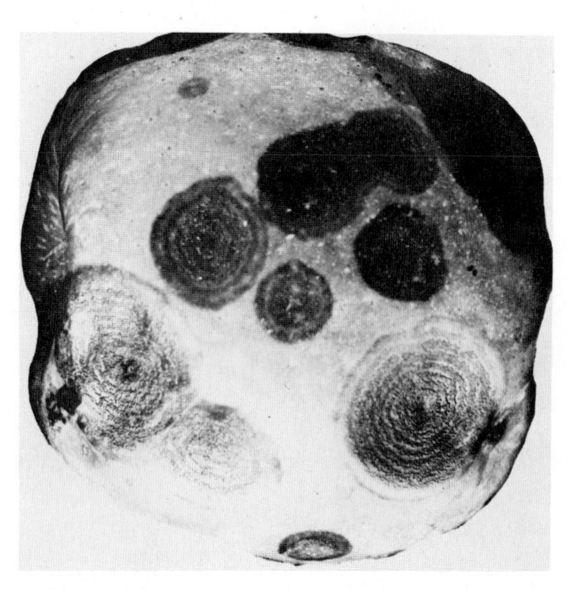

FIGURE 11-62 Bitter rot of apple caused by *Glomerella cingulata.* Note the concentric arrangement of *Gloeosporium*-type acervuli. (Photo courtesy U.S.D.A.)

limbs are rare and resemble those produced on branches and stems described above.

Ripe rot of grape and other fruits is also worldwide in distribution but is most serious in areas with warm, humid weather during the ripening of the fruit. Ripe rot appears when the fruit is nearly mature and may continue its destruction of fruit after it has been picked and during shipment and marketing. On white or light-colored grapes, a small reddish-brown spot appears at first. It soon spreads to over half the berry and becomes darker with a purplish center and a light-brown border. In dark-colored grapes no appreciable change of color is caused by the rot on the berries. As the disease progresses, the whole berry rots, usually in a continuous manner but sometimes marked by a series of concentric zones. By this time, the rotted berry becomes more or less densely covered with numerous, small, slightly raised pustules from which, in humid weather, pinkish slimy masses of spores ooze out. Later, the spore masses become darker, almost reddish-brown. The rotted berries then develop a sunken area at the point of infection and gradually become more or less shriveled and mummified, while the pustules continue to produce spores. Infected berries often "shell" or drop off before the rot causes them to dry up.

The fungus, *Glomerella* sp., occasionally produces ascospores in perithecia on overwintering fruit and cankers of some hosts, but in other hosts ascospores and perithecia almost never develop. In all hosts, however, the fungus reproduces profusely by forming pink masses of *Colletotrichum-* or *Gloeosporium*-type conidia in acervuli (Figure 11-63). The acervuli consist of a cushion stroma of mycelium several to many cells thick and develop just beneath the cuticle, which is ruptured by the upward pressure of the developing mass of conidiophores and conidia. The conidia are held together in a sticky mass, which is firm and horny in dry weather, but under humid, moist weather the conidia are released in a pinkish mass and may be washed or spattered by raindrops or carried by wind-blown mist and insects.

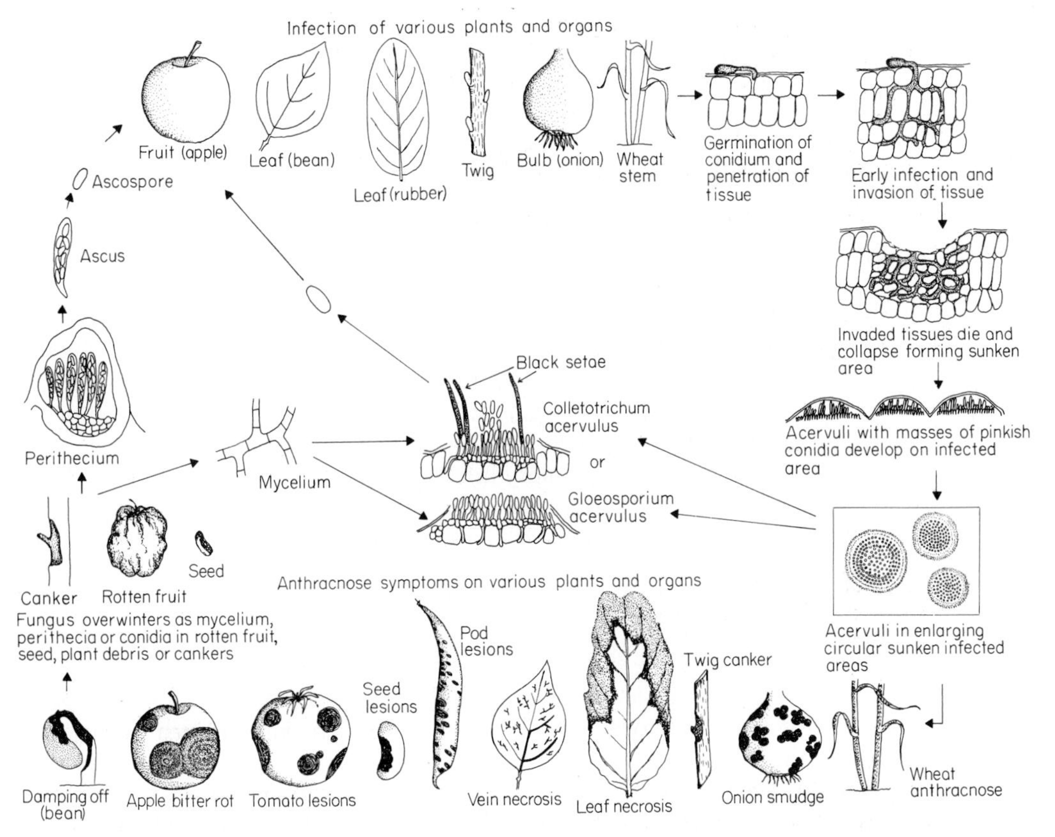

FIGURE 11-63 Disease cycle of the anthracnose diseases caused by *Glomerella cingulata* and *Colletotrichum* or *Gloeosporium* sp.

The fungus usually overwinters in diseased stems, leaves, and fruit as mycelium or spores, in the seed of most affected annual hosts, and in cankers of perennial hosts. In the cases in which ascospores are produced in mummified fruit, cankers, and so on, they may cause primary infections in the spring, but generally these are not necessary. The surviving mycelium quickly produces conidia in the spring that can themselves cause primary infections and subsequent continuous secondary infections during the entire season as long as temperature and humidity are favorable. Infection is by direct penetration of uninjured tissue, and the mycelium grows intercellularly and may remain latent for some time before the cells begin to collapse and rot. The mycelium then produces acervuli and conidia just below the cuticle, which upon rupture again release conidia that cause more infections. The fungus generally requires high temperatures and humidity for best activity, and although it can attack young cotyledons, stems, leaves, and fruit, attacks of young fruit generally remain latent until the fruit are past a certain stage of development and maturity, at which point the infections develop fully.

Control of *Glomerella* diseases depends on the use of disease-free seed grown in the arid west or use of treated seed and also crop rotation of hosts like bean, soybean, sweet pea, and cotton, the use of resistant varieties when available, removal and burning of dead twigs, branches, and fruit infected with the fungus in woody plants, and finally, on spraying with fungicides. The fungicides most commonly used include Bordeaux mixture, captan, ferbam,

Dikar, mancozeb, captafol, benomyl, and thiophanate-methyl. The kind of fungicide used and the timing of its application vary considerably with the host treated.

• Gnomonia Anthracnose and Leaf Spot Diseases

In Europe, North America, and other parts of the world, various species of *Gnomonia* attack mostly forest and shade trees on which they cause symptoms primarily on leaves, for example, elm (Figure 11-60C) and hickory, or more general anthracnose symptoms on the leaves and young shoots or twigs, for example, linden, oak (Figure 11-60A), sycamore, and walnut (Figure 11-60E). Gnomonia diseases are favored by wet, humid weather. The various species of the fungus overwinter primarily in fallen leaves or on infected twigs as immature perithecia, which mature and produce ascospores in the spring. Ascospores produce most primary infections on young leaves and twigs; then conidia are produced in acervuli, generally of the *Gloeosporium* type, and cause all subsequent infections. Both ascospores and conidia are disseminated only during rainy weather.

Sycamore anthracnose is the most important disease of sycamore (*Platanus* sp.). The disease may appear as twig blight before leaf emergence at which time it kills the tips of small, one-year-old twigs. During bud expansion, the disease may appear as bud blight, killing the buds before they open. The most frequently observed symptom is shoot blight in which the disease causes the sudden death of expanding shoots and also young leaves. The killing of leaves while they are still very small has often been confused with frost damage. Later the fungus may cause a leaf blight in which irregular brown areas develop adjacent to the midrib, veins, and leaf tips. In moist weather, small, cream-colored acervuli form on the under side of the leaf on dead tissue along the veins. In some years, sycamores may be completely defoliated in the spring by this disease and produce new leaves in the summer. From the buds or leaves, the fungus spreads into the twigs on which it produces cankers, or the fungus may spread through and kill small twigs and form cankers on the branches around the bases of the dead twigs (Figure 11-60B). In trees severely affected with anthracnose for several successive years, many branches may die. The fungus overwinters as mycelium and as immature perithecia in twig cankers and in fallen leaves. As in the other anthracnose diseases, this too is favored by rainy weather and rather cool temperatures.

Control of *Gnomonia* diseases in trees under forest conditions is not economically feasible. In shade and ornamental trees, however, burning of leaves, pruning of infected twigs, fertilization and watering help reduce the disease. Valuable trees should be sprayed two to four times at 10- to 14-day intervals starting as soon as the buds begin to swell (sycamore) or soon after the buds open. The fungicides most commonly recommended are zineb, Bordeaux mixture, and dodine.

• Colletotrichum (Gloeosporium) Diseases

Colletotrichum diseases are the most common anthracnoses and are very similar, if not identical, to the diseases caused by *Glomerella*. The latter is probably the sexual stage of most or all species of *Colletotrichum (Gloeosporium).*

On **cereals and grasses,** the fungus, which lives saprophytically on crop residue, may attack young seedlings, but usually it attacks the tissues at the crown and the bases of stems of more developed plants. Infected areas first appear bleached but later become brown. Brownish blotches often develop at or near diseased joints. Toward maturity of the plant, numerous small black acervuli appear on the stems, lower leaf sheaths, and sometimes on the leaves and on the chaff and spikes of diseased heads (Figure 11-61A). Depending on how early the plant is attacked and on the severity of the disease, the plant may show a general reduction in vigor, premature ripening or dying of the head, and shriveled grain. The fungus occasionally causes superficial infection on seeds, and it can also overwinter as mycelium on the seed. When seedborne, the fungus may result in root rot and crown rot of the developing plant. Anthracnose of corn and other cereals has become a major problem in areas where reduced tillage, practiced to minimize loss of soil or water, allows greater survival of inoculum of the pathogen on the crop residue.

Anthracnose of cucurbits is probably the most destructive disease of these crops everywhere, being most severe on watermelon, cantaloupe, and cucumber. All aboveground parts of the plants are affected. Symptoms appear first on the leaves as small, water-soaked, yellowish areas that enlarge from several millimeters to 1–2 cm and become black in watermelon, while in all other cucurbits they turn brown. Infected tissues dry up and break. Lesions also develop on the petioles, which may result in defoliation of the vine; on the fruit pedicel, which causes the fruit to turn dark, shrivel, and die; and on the stem, which weaken or kill whole vines. The fruit becomes susceptible to infection at about the time of ripening. Circular, watery, dark, sunken lesions appear on the surface of the fruit that may be from 5 mm to 10 cm in diameter and up to 8 mm in depth (Figure 11-61D). The lesions expand rapidly in the field and during transit or storage and may coalesce to form larger ones. The sunken lesions have dark centers, which in moist weather are filled with pink spore masses exuding from acervuli that break through the cuticle. Severely affected fruits are often tasteless or even bitter and are often invaded by soft-rotting bacteria and fungi that enter through the broken rind. The fungus overwinters in infected debris in the soil and on or in the seed.

Anthracnose or **ripe rot of tomato,** and other vegetables and fruits results in serious losses of fruit and occasionally in damage to stems and foliage. Canning tomatoes are particularly susceptible to anthracnose before and after harvest, but other tomatoes, as well as eggplant, pepper, apple, pear, and banana, may be attacked in a similar manner from the time ripening begins through harvest and in storage. In early stages of tomato infection, the symptoms appear as small, circular, sunken, water-soaked spots resembling indentations caused by burnt circular objects. As the fruit softens, the spots enlarge up to 2–3 cm in diameter, and their central portion becomes dark and slightly roughened as a result of black acervuli developing just beneath the skin (Figure 11-61, F, G). The spots are often numerous and coalesce, leading to the eventual watery softening of the fruit and finally rotting of the fruit, sometimes accelerated by other invading microorganisms. Enormous numbers of conidia are present in acervuli below the skin even in the smallest spots, and under some conditions, pink or salmon-colored masses of spores are also produced on the surface of the spots. The fungus overwinters in infected plant debris and in or on the seed. The fungus may cause light infections of foliage and young stems that may go unnoticed, but they enable the fungus to survive and multiply somewhat until the fruit begins to ripen and become susceptible to infection. High temperatures and high relative humidity or wet weather at the time of ripening favor infection and spread of the fungus and often lead to destructive epidemics.

In **onion smudge,** dark-green or black smudges appear on the outer scales or neck of the bulbs primarily of white onions (Figure 11-61E), although most colored varieties are also partially susceptible, especially in the colorless region of the bulb neck. The smudgy spots first appear as tiny black stromata or visible mycelium beneath the cuticle of the scale and may be scattered over the surface of the bulb; more commonly, they congregate in uniformly black, smudgy, circular areas or are arranged in concentric circles, the outer one being up to two or more centimeters in diameter. In moist weather the stromata produce acervuli filled with cream-colored spore masses and containing numerous black, stiff, bristlelike hairs (setae) visible with a hand lens. The fungus attacks inner, living scales only under conditions of favorable high moisture and temperature and, usually, underneath spots on outer scales or where the outer scales have cracked and exposed the inner scales to the soil. The fungus overwinters as mycelium or stromata on infected onions, on sets, and in the soil as a saprophyte.

In **citrus anthracnose,** especially of orange, grapefruit, and lemon, the disease affects all mature, weakened, or injured aboveground plant parts, including leaves, twigs, and fruit. Anthracnose may occur on trees of any size, in the nursery or in the orchard, but it rarely develops on vigorously growing trees. Anthracnose is common, however, on trees that are weakened or injured from inadequate fertilization, drought, low temperatures, spray injury, insects, other diseases, and so on. Anthracnose leaf spots bring about death and drying of infected tissues with minute black acervuli produced in concentric circles in the dead areas. During severe infections, defoliation may occur. Weakened twigs are invaded by the anthracnose fungus and die back slowly or rapidly, resulting in what is called withertip. Leaves in affected twigs turn yellow, wither and drop, or they die quickly and dry before they can fall off. Affected branches also lose fruit. Numerous, small, black, pimplelike acervuli develop on the dead portions of twigs. On dying or dead areas of the surface of citrus fruits, anthracnose fruit spots develop that are circular and sunken and vary from tiny specks to dark-brown or black areas 5–10 mm in diameter. The spots become dry and hard and sometimes are dotted with small black acervuli that in humid weather exude masses of pinkish spores. Although anthracnose fruit spots generally affect only the rind, the disease often extends into the pulp and gives it a disagreeable or bitter taste. Some strains of the anthracnose fungus cause a soft decay that results in fruit drop. Also, often other fungi invade the fruit through anthracnose spots and speed up the rate of fruit decay. Overripe fruit is particularly susceptible to anthracnose infection. When spores of the anthracnose fungus from dead twigs are washed over fruits and then germinate on the fruit, they cause anthracnose russeting of the fruit. The russeting may appear as a large blotch or as a tear stain.

The fungus, *Colletotrichum,* or *Gloeosporium* sp., produces colorless, one-celled, ovoid, cylindrical, and sometimes curved or dumbbell-shaped conidia in acervuli. Masses of conidia appear pink or salmon colored. The acervuli are subepidermal and break out through the surface of the plant tissue; they are disk- or cushion-shaped and waxy, with simple short, erect conidiophores. *Colletotrichum* has been distinguished from *Gloeosporium* by the fact that *Colletotrichum* acervuli have dark, long spines, or sterile hairlike hyphae while *Gloeosporium* acervuli do not. Because, however, this is not always so, the distinction of the two genera is often impossible, and they are often considered as the same fungus. As mentioned earlier, many *Colletotrichum* species also produce an ascigerous perfect stage, usually *Glomerella,* occasionally *Physalospora* or others, while many *Gloeosporium* species have as perfect stage the ascomycetes *Glomerella* or *Gnomonia,* and occasionally others.

The fungus is favored by high temperatures and humid or moist weather. The conidia are released and spread only when the acervuli are wet and are generally spread by splashing and blowing rain or by coming in contact with insects, other animals, tools, and so on. The conidia germinate only in the presence of water. Upon germination they produce an appressorium and penetration peg and penetrate the host tissues directly (Figure 11-63). In the beginning the hyphae may grow rapidly, both intercellularly and intracellularly, but cause little or no visible discoloration or other symptoms of disturbance. Then more or less suddenly, especially when fruits begin to ripen, the fungus becomes more aggressive, and symptoms appear. In many hosts the fungus reaches the seed and is either carried on the seeds or, in some, it may even invade a small number of seeds without causing any apparent injury to them. There is considerable variability in the kinds of host plants each species of *Colletotrichum* or *Cloeosporium* can attack, and there may be several races with varying pathogenicity within each species of the fungus.

Control of *Colletotrichum* or *Gloeosporium* diseases depends on the use of disease-free seed or seed treated with chemicals and hot water, on a two- to three-year crop rotation when possible, on the use of resistant varieties that are available for several annual crops, and the use of fungicides such as benomyl, maneb, zineb, mancozeb, chlorotholonil, captafol, and folpet.

SELECTED REFERENCES

Baker, K. F. (1948). The history, distribution, and nomenclature of the rose blackspot fungus. *Plant Dis. Rep.* **32,** 260–274, 397.

Barksdale, T. H., and Stoner, A. K. (1981). Levels of tomato anthracnose resistance measured by reduction of fungicide use. *Plant Dis.* **65,** 71–72.

Brook, P. J. (1977). *Glomerella cingulata* and bitter rot of apple. *N. Z. J. Agric. Res.* **20,** 547–555.

Daykin, M. E., and Milholland, R. D. (1984). Ripe rot of muscadine grape caused by *Colletotrichum gloeosporioides* and its control. *Phytopathology* **74,** 710–714.

Daykin, M. E., and Milholland, R. D. (1984). Histopathology of ripe rot caused by *Colletotrichum gloeosporioides* on muscadine grape. *Phytopathology* **74,** 1339–1341.

Drake, C. R. (1968). Grape diseases and their control in Virginia. *Va. Polytech. Inst. Publ.* **32,** 1–10.

Forsberg, J. L. (1975). "Diseases of Ornamental Plants," Spec. Publ. No. 3 Rev. University of Illinois, College of Agriculture, Urbana-Champaign.

Kendrick, J. B., Jr., and Walker, J. C. (1948). Anthracnose of tomato. *Phytopathology* **32,** 247–260.

Layton, D. V. (1937). The parasitism of *Colletotrichum lagenarium. Iowa, Agric. Exp. Stn., Res. Bull.* **223,** 37–67.

Politis, D. J., and Wheeler, H. (1973). Ultrastructural study of penetration of maize leaves by *Colletotrichum graminicola*. *Physiol. Plant Pathol.* **3,** 465–471.
Roberts, R. G., and Snow, J. P. (1984). Histopathology of cotton boll rot caused by *Colletotrichum capsici*. *Phytopathology* **74,** 390–397.
Schneider, R. W., *et al.* (1974). *Colletotrichum truncatum* borne within the seedcoat of soybean. *Phytopathology* **64,** 154–155.
Schuldt, P. H. (1955). Comparison of anthracnose fungi on oak, sycamore, and other trees. *Contrib. Boyce Thompson Inst.* **18,** 85–107.
Sutton, B. C. (1966). Development of fructifications on *Colletotrichum graminicola* and related species. *Can. J. Bot.* **44,** 887–897.
Sutton, T. B., and Shane W. W. (1983). Epidemiology of the perfect stage of *Glomerella cingulata*. *Phytopathology* **73,** 1179–1183.
Taylor, J. (1971). Epidemiology and symptomatology of apple bitter rot. *Phytopathology* **61,** 1028–1029.
Thompson, D. C., and Jenkins, S. F. (1985). Effects of temperature, moisture, and cucumber cultivar resistance on lesion size increase and conidial production by *Colletotrichum lagenarium*. *Phytopathology* **75,** 828–832.
Thompson, D. C., and Jenkins, S. F. (1985). Pictorial assessment key to determine fungicide concentrations that control anthracnose development on cucumber cultivars with varying resistance levels. *Plant Dis.* **69,** 833–836.
Walker, J. C. (1921). Onion smudge. *J. Agric. Res. (Washington, D.C.)* **20,** 685–722.
Weber, G. F. (1973). "Bacterial and Fungal Diseases of Plants in the Tropics." Univ. of Florida Press, Gainesville.
Wellman, F. L. (1972). "Tropical American Plant Disease," pp. 236–273. Scarecrow Press, Metuchen, New Jersey.
Wheeler, H., Politis, D. J., and Poneleit, C. G. (1974). Pathogenicity, host range and distribution of *Colletotrichum graminicola* in corn. *Phytopathology* **64,** 293–295.
Zaumeyer, W. J., and Thomas, H. R. (1957). A monographic study of bean diseases and methods for their control. *U. S., Dep. Agric., Tech. Bull.* **868,** 1–255.

Fruit and General Diseases Caused by Ascomycetes and Imperfect Fungi

These groups of diseases are caused by Ascomycetes and Imperfect Fungi other than those already described and are most commonly found on the fruit or cause most of their damage by their effect on the fruit although they may affect other parts of the plant as well. Most of these fungi differ considerably from each other in fruiting structures, life cycles, and in the diseases they cause. Most of these Ascomycetes produce ascospores in perithecia and both, Ascomycetes and Imperfects, produce conidia on free hyphae. The most common ascomycetous fungi and the most important diseases they cause are the following.

Claviceps purpurea, causing ergot of cereals and grasses.
Diaporthe citri, causing pod blight of lima beans, stem canker of soybeans and melanose of citrus fruits.
Physalospora (Botryosphaeria), causing black rot, frogeye leaf spot, and canker of apple *(P. obtusa),* oak canker and dieback *(P. glandicola),* blight and black canker of willow *(P. miyabeana),* canker and dieback of many tropical trees such as citrus, cacao, coconut, rubber, and tropical forest trees *(P. rhodina).*
Venturia inaequalis, causing apple scab.
Monilinia fructicola, causing brown rot of stone fruits.

The most common Imperfect Fungi causing fruit and general diseases on plants are the following:

Botrytis, causing blossom blights and fruit rots, but also damping off, stem cankers or rots, leaf spots, and tuber, corm, bulb, and root rots of many vegetables, flowers, small fruits, and other fruit trees.

Phomopsis, causing blights of eggplant, carrot, azalea, juniper, stem canker of gardenia, cane and leaf spot and berry rot of grape, and stem-end rot of citrus fruits.

• Ergot of Cereals and Grasses

Ergot occurs throughout the world, most commonly on rye but also on wheat and less frequently on barley and oats. It is also very common on certain wild and cultivated grasses. An ergot disease affecting corn occurs in Mexico. The disease causes loss of some of the grains in infected heads, occasionally resulting in losses of up to 5 percent in rye and 10 percent in wheat, but its main importance is due to the fact that the characteristic fungal sclerotia that replace the grains are poisonous to humans and animals that eat feed or bread made from grain contaminated with sclerotia.

The symptoms of ergot appear at first as creamy to golden-colored droplets of a sticky liquid exuding from young florets of infected heads. These droplets may go unnoticed, but they are each soon replaced by a hard, horn-shaped, purplish-black fungal mass that is usually a few millimeters in diameter but may be from 0.2 to 5.0 cm long. These are the sclerotia of the fungus, sometimes called ergots, that grow in place of the kernel and consist of a hard compact mass of fungal tissue, which in cross section appears white or faintly purple (Figure 11-64A,B).

Ergot is caused by *Claviceps purpurea* and other *Claviceps* species. The fungus overwinters as sclerotia on or in the ground, or mixed with the seed. In the spring, about the time cereals are in bloom, sclerotia on or near the surface of the soil germinate by forming from one to sixty flesh-colored stalks 0.5–2.5 cm tall (Figure 11-64C). The tip of each stalk consists of a spherical head at the periphery of which are embedded numerous perithecia, each containing many asci. Each ascus contains eight long, multicellular ascospores. The ascospores are either extruded in a viscous fluid in humid weather and are spread to flowers by insects or are discharged forcibly into the air, are picked up by air currents, and are carried by wind to young open flowers. There they germinate and infect the ovaries directly or by way of the stigma. Within about a week the fungus in the ovary tissues produces spreading, stromalike sporodochia that produce numerous conidia of the *Sphacelia* type. The conidia are floating in a sticky liquid and exude from the young florets as creamy or golden droplets, which are known as the "honey dew" stage. The liquid of the "honey dew" stage is sugary and attractive to insects, which visit such infected flowers, become smeared with the conidia of the fungus present in the "honey dew," and carry them to other healthy flowers, which the conidia promptly infect. Conidia are probably spread to other flowers by splashing rain, resulting in more infections. Gradually, the secretion of the "honey dew" ceases and each infected ovary, instead of producing normal seed, becomes replaced by a hard mass of fungal mycelium, which eventually forms the characteristic ergot sclerotium. The sclerotia mature about the same time as the healthy seeds and either fall to the ground where they overwinter, or they are harvested with the grain and may be returned to the land with the seed.

FIGURE 11-64 (A, B) (A) Ergot of rye caused by *Claviceps purpurea.* The long, hornlike structures on the rye head are the sclerotia of the pathogen. (B) Ergot sclerotia of various shapes and sizes. *(Figure continued)*

Susceptibility of the various hosts and varieties seems to be related to the duration the flowers remain open. This characteristic has gained significance in the last few years because of the efforts of breeders to produce hybrid varieties in several cereal crops, since the flowers of male-sterile lines of self-pollinated cereals remain open, and therefore susceptible to ergot infection, for a long time. For example, in some experimental plots, 76 percent of the heads and 36 percent of the flowers of male-sterile barley lines became infected with ergot.

Although ergot does not cause ergotism in humans nearly as often or as severely as it used to, it is probable that it is involved in many otherwise unexplainable poisonings of humans, and it certainly continues to be of economic importance as an animal disease. Grain containing more than 0.3 percent by weight of the ergot sclerotia is classed as ergoty, that is, can cause ergotism and therefore it may not legally be sold and milled for flour and human consumption. Although most of the sclerotia can be removed from ergoty grain with modern cleaning machinery, it is costly and quite often difficult to remove enough sclerotia to meet the legal standards, particularly in poorer countries, and the remaining traces have often proven toxic to

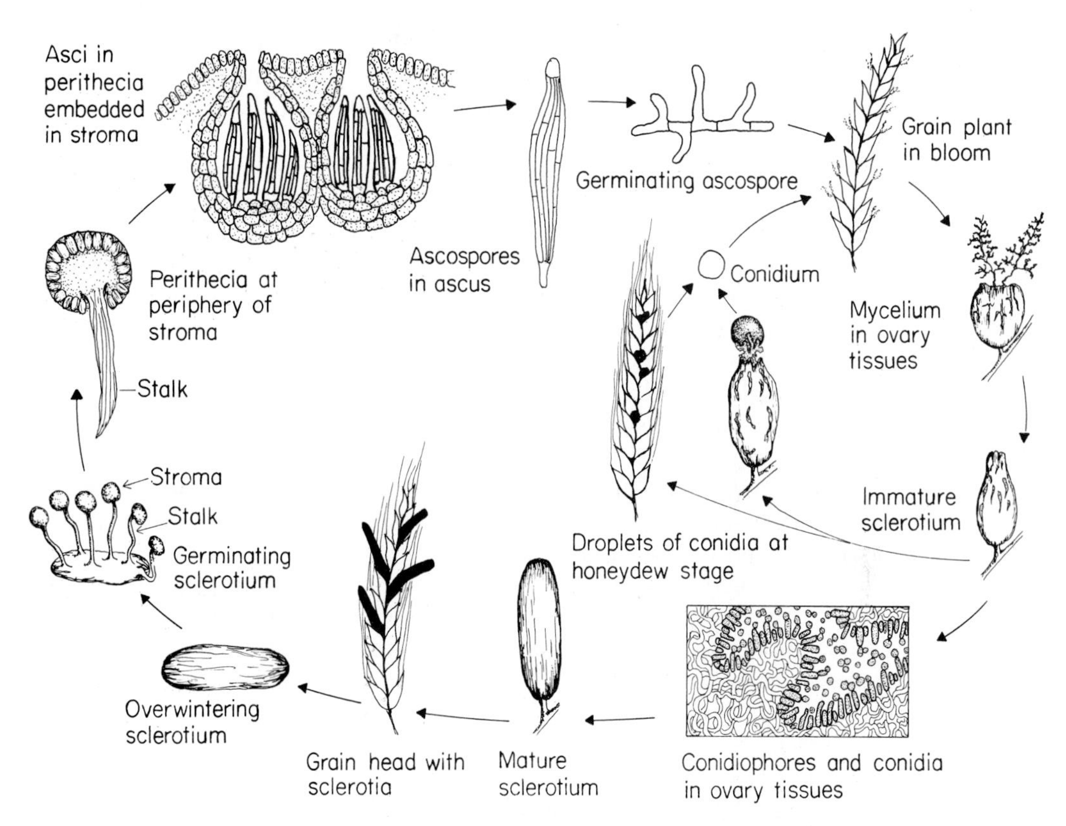

FIGURE 11-64 (C) Disease cycle of ergot of grains caused by *Claviceps purpurea.*

livestock. Feeding livestock with cleanings from contaminated grain or grazing pastures that have infected grass heads can also lead to reproductive failure or gangrene of the peripheral parts of the animals.

Control of ergot depends entirely on cultural and sanitary procedures. Only clean seed or seed that has been freed from ergot should be sown. Sclerotia may be removed from seed by machinery or by soaking contaminated seed for 3 hours in water and then floating off the sclerotia in a solution of about 18 kg salt in 100 liters of water. Ergot sclerotia do not survive for more than a year and do not germinate if buried deep in the ground. Therefore, deep ploughing or crop rotation with a noncereal for at least a year helps eliminate the pathogen from a particular field. Wild grasses should be mowed or grazed before flowering to prevent production of ergot sclerotia on them and avoid poisoning of livestock through them and also to prevent the spread of the fungus to cultivated cereals and grasses. On male-sterile barley, reasonable control of ergot was obtained with benomyl sprays before and during flowering. Some fungi, for example, *Fusarium roseum,* are hyperparasites of the ergot fungus and appear promising as biological control agents but have yet to be developed.

• Apple Scab

Apple scab exists in every country where apples are grown. It is most severe in areas with cool, moist springs and summers, however, and may be completely absent in areas with very dry or warm climates. In the United States it is most serious in the north central and northeastern states. The disease affects all apples in the genus *Malus*, but similar scab diseases affect pears and hawthorns.

Scab is the most important disease of apples. Its primary effect is the reduction of the quality of infected fruit, but it also reduces fruit size and the length of time infected fruit can be kept in storage. Infection of the stem of young fruit results in premature fruit drop. Severe leaf infections result in reduction of the functioning leaf surface, defoliation, and poor fruit bud development for the next year's crop. Losses from apple scab may be 70 percent or more of the total fruit value, and in many apple-producing areas no marketable fruit can be harvested if scab control measures are not taken.

○ *Symptoms.* The first symptoms of apple scab appear on the under surface of sepals or young leaves of the flower buds as light, somewhat olive-colored, irregular, spots. Soon after, the lesions become olive green with a velvety grayish-dark surface and more circular in outline. Later, the velvety surface disappears and the lesions appear metallic black in color and may be slightly raised. Lesions on leaves that have already unfolded are generally on the upper surface of the leaves (Figure 11-65A). The number of lesions per leaf varies with the severity of infection. Lesions may remain distinct or they may coalesce. After severe early leaf infection, leaves may become dwarfed and curled and may later fall off.

Infections of the fruit appear as distinct, almost circular scab lesions, which at first are velvety and olive green but later become darker, scabby, and sometimes cracked (Figure 11-65B). The cuticle of the fruit is ruptured at the margin of the lesions. Severe early fruit infections result in misshapen, cracked fruits, which frequently drop prematurely. Infections late in the season when the fruit approaches maturity result in small lesions that may even be too small to be visible at harvest but develop into dark scab spots during storage.

Twig and blossom infections appear as small scab spots, but they are uncommon and of little importance.

○ *The pathogen: Venturia inaequalis.* The mycelium is at first light in color, but later turns brownish in the host tissues. In young leaf lesions the mycelium develops radially in branched ribbons of hyphae, but in older leaves and on the fruit the mycelial strands are compact and thick and in several superimposed layers. In living tissues the mycelium is located only between the cuticle and the epidermal cells and produces short, erect, brownish conidiophores, which successively give rise to several, one- or two-celled, reddish brown *Spilocaea (Fusicladium)*-type conidia of variable, but rather characteristic, shape (Figure 11-66). In dead leaves the mycelium grows through the leaf tissues. Fertilization takes place by means of ascogonia and antheridia and pseudothecia form. The latter, when mature, are dark brown to black with a slight beak and a distinct opening (ostiole). Inside the pseudothecium 50–100 asci are formed, each containing eight ascospores. Each ascospore consists of two cells of unequal size, which are hyaline at first but turn brown when mature.

○ *Development of disease.* The pathogen overwinters in dead leaves on the ground as immature pseudothecia. Pseudothecia initially develop in the fall and early winter and probably continue to grow during warm periods in the

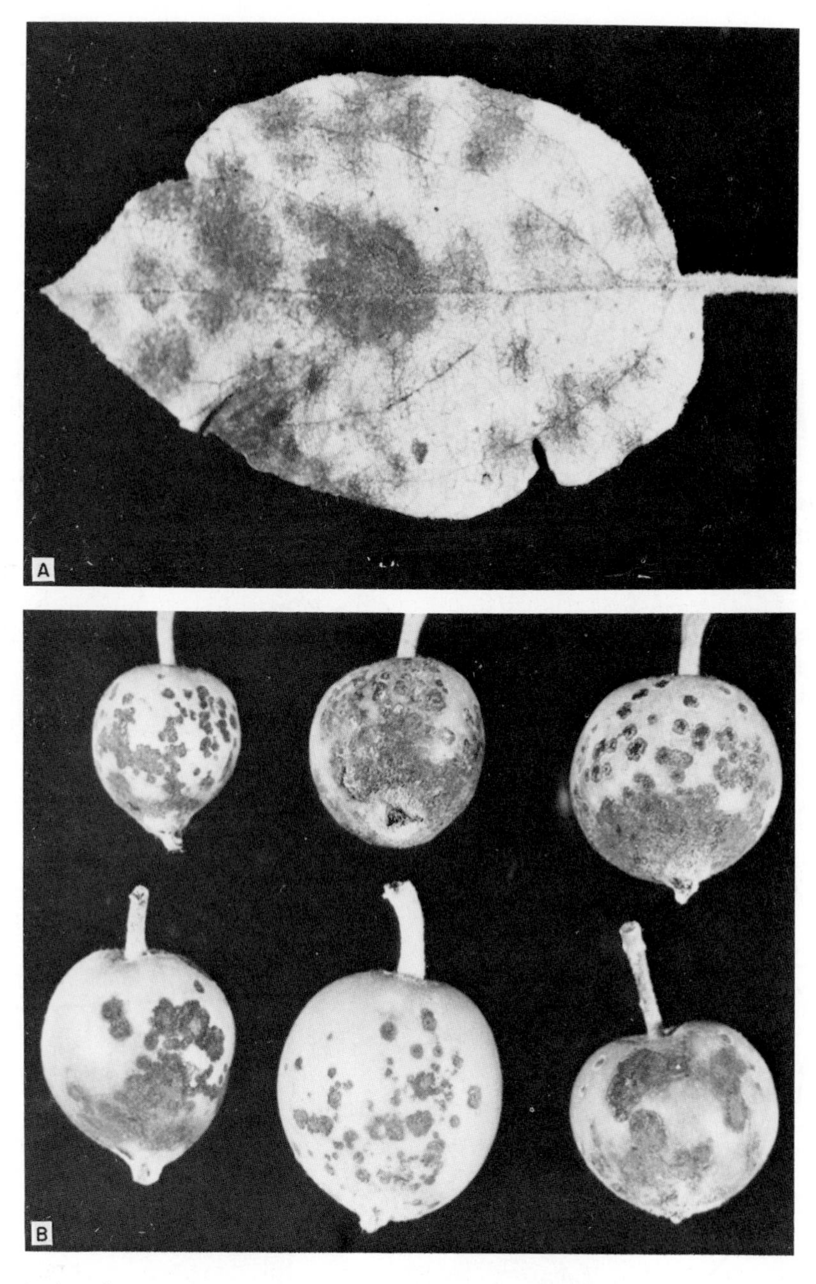

FIGURE 11-65 Apple scab symptoms on leaves (A) and young fruit (B) caused by *Venturia inaequalis*.

winter and early spring, but rapid growth and ascospore maturation occur with the resumption of favorable weather for growth and development of the host (Figure 11-66). Neither all the pseudothecia nor all the asci of a pseudothecium mature simultaneously, and some may have ascospores before the apple buds start to open in the spring. Most of the ascospores in the pseudothecia, however, mature in the period during which the fruit buds open.

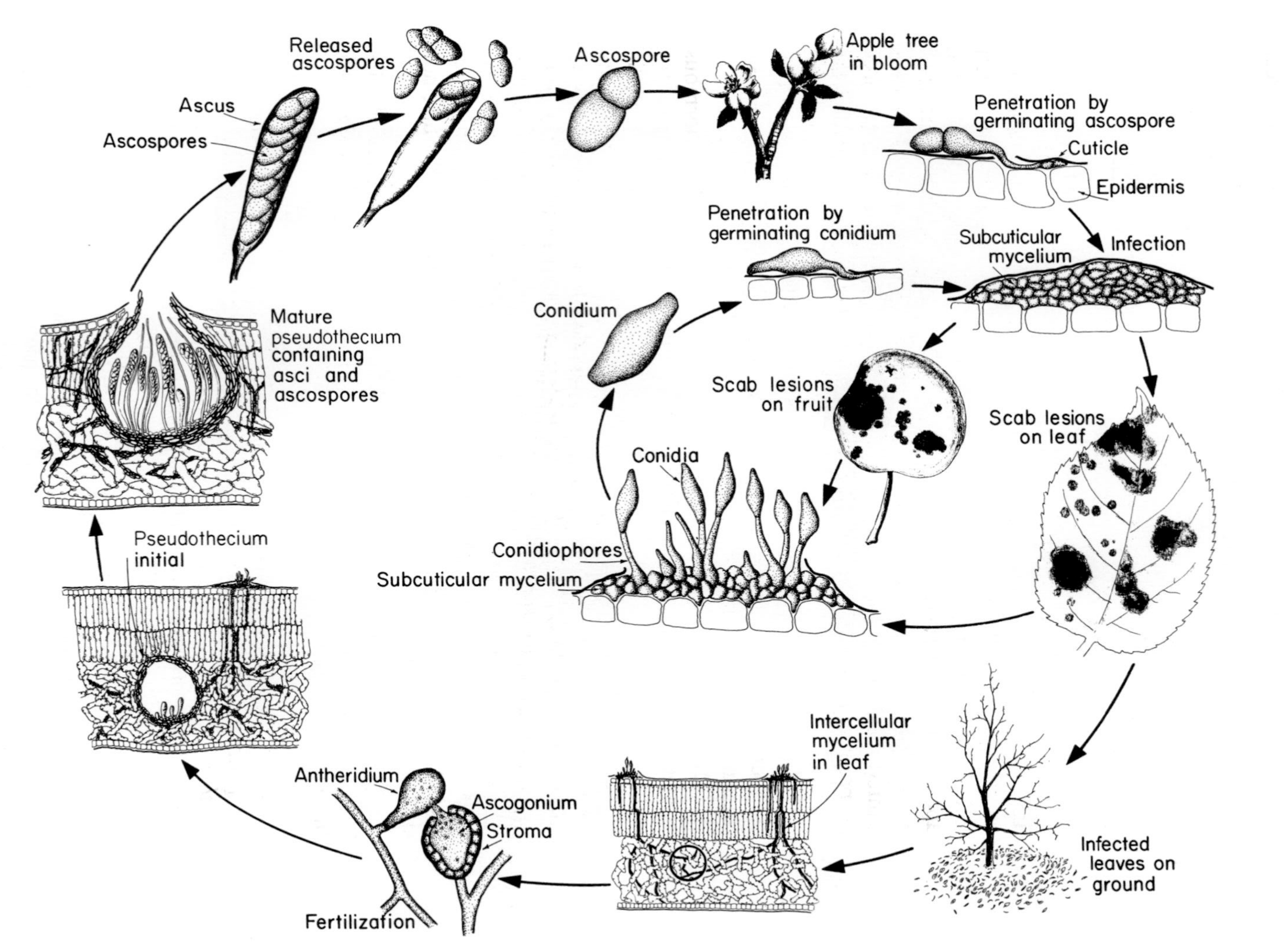

FIGURE 11-66 Disease cycle of apple scab caused by *Venturia inaequalis*.

When dead leaves containing pseudothecia become thoroughly soaked in the spring, the asci elongate, push through the ostiole, and forcibly discharge the ascospores into the air; air currents may carry them to susceptible green apple tissues. Ascospore discharge may continue for 3–5 weeks after petal fall.

The ascospores can germinate and cause infection only when they are kept wet over a certain minimum period of time at temperatures ranging roughly from 6 to 26°C. Thus, for infection to occur, the spores must be continuously wet for 28 hours at 6°C, for 14 hours at 10°C, for 9 hours at 18–24°C, or 12 hours at 26°C.

Upon germination on an apple leaf or fruit, the ascospore produces a disklike appressorium from which a slender mycelial tube pierces the cuticle, and after developing into a hypha of normal diameter, grows between the cuticle and the outer cell wall of the epidermal cells. For a few days after infection, the epidermal cells show no injury at all, but by the time the lesion appears these cells show a gradual depletion of their contents, and they eventually collapse and die. Soon the palisade, and later the mesophyll, cells exhibit the same reactions, while the fungus remains largely in the subcuticular position. It is assumed that the fungus obtains its nutrients and causes the death of the cells by secreting enzymes (and possibly toxic substances) that alter the permeability of the cell membranes and break down macromolecular components of the cell into small molecules, which then move along osmotic pressure gradients, reach the mycelium and are absorbed by it.

Once the mycelium is established in the host, it produces enormous numbers of conidia, which push outward, rupture the cuticle, and within 8–15 days of inoculation, form the olive-green, velvety scab lesions. Conidia remain attached to the conidiophores in dry weather, but upon wetting during a rain they are easily detached and may be washed down or blown away to other leaves or fruit on which they germinate and cause infection in the same way ascospores do. Additional infections by conidia continue throughout the growing season following a rain of sufficient duration. Infections, however, are more abundant in the cool, wet periods of spring and early summer and again in the fall. They are infrequent, if not entirely absent, in the dry, hot summer weather.

After infected leaves fall to the ground, the mycelium penetrates into the interior of the leaf and forms pseudothecia, which carry the fungus through the winter.

○ *Control.* Several apple varieties resistant to scab are available, but all the popular ones are moderately to highly susceptible. A number of fungi are antagonistic to the apple scab fungus, and some of them, such as *Athelia bombasina* and *Chaetomium globosum,* when applied to scab-infected apple leaves on the orchard floor decrease production of ascospores by the apple scab fungus. So far, no effective practical biological control of apple scab has been developed. Apple scab, however, can be thoroughly controlled by timely sprays with the proper fungicides.

For an effective apple scab control program, apple trees must be diligently sprayed or dusted before, during, or immediately after a rain from the time of budbreak until all the ascospores are discharged from the pseudothecia. If these primary infections from ascospores are prevented, there will be less need to spray for scab during the remainder of the season. If primary infections do develop, spraying will have to be continued throughout the season. In most areas application of fungicides for scab control is based on the phenological development of the trees, buds, and flowers, and the first application is made when buds show a slight green tip, the second when 1–2 cm green leaf shows, the third at tight bud cluster, the fourth when some pink shows, at full bloom, and the fifth at 90 percent petal fall. Sprays begin in the spring when a rainy period is sufficiently long at the existing temperature to produce an infection, and are repeated every 5–7 days, or according to rainfall, until petal fall. One must make sure that the new tissues on rapidly expanding young leaves and fruit are always covered with fungicide during an infection period. After petal fall, and depending on the success of the control program to that point, sprays are usually repeated every 10–14 days for several more times.

In recent years, considerable progress has been made in developing simple or computerized apple scab prediction systems for scheduling fungicide applications for scab control. All these systems are based on the interactions between temperature, amount and duration of rainfall, and duration of leaf wetness on the one hand, with the period required for the pathogen to initiate infection on the other. When conditions favorable to infection are predicted to prevail or have prevailed for sufficiently long periods, the recommendation to apply fungicides is made. The availability of fungicides with protectant or curative activity (or both) against apple scab makes possible the use of apple scab–predicting systems with excellent control results and with reduced numbers of fungicide applications.

Several fungicides that give excellent control of apple scab are available. Some of them are protectant, since they can protect a plant from becoming infected, but they cannot cure an infection, although some have so-called "kickback" action, that is, the ability to stop infections that may have started, and some have an eradicant action, that is, they can "burn out" young scab spots early in the season. Dodine gives excellent scab control and is one of the fungicides exhibiting eradicant action. It reduces spore formation on scab spots and inhibits germination of spores produced on such spots. Excellent scab control is also obtained with captan, ferbam, mancozeb, dichlone, thiram, sulfur, captafol, and the systemic fungicides benomyl and thiophanate-methyl. These fungicides may be used alone or in combinations since they differ in their ability to control scab and other diseases, in duration of "kickback" action, in compatibility with other fungicides, and in phytotoxicity. Captafol has recently been used as a single application treatment at the green tip stage of bud development and has been shown to protect new growth from scab until petal fall.

Several newer fungicides, most of them systemics and acting by inhibiting ergosterol biosynthesis in the fungus, show excellent protectant as well as curative control of apple scab. They include etaconazole, bitertanol, fenarinol, and triforine. As with all systemic fungicides, because of their specific mechanism of action against the fungus and the quick appearance of fungus strains resistant to the fungicide, it is recommended that they be applied only in combination with captan, mancozeb, or another broad-spectrum fungicide.

In some areas, new strains of *Venturia inaequalis* have now appeared that are resistant to dodine, benomyl, and some other systemics. These chemicals, therefore, can no longer be relied on to control the disease by themselves.

• Brown Rot of Stone Fruits

Brown rot occurs throughout the world where stone fruits are grown and where there is sufficient rainfall during blossoming and fruit ripening periods. It affects peaches, cherries, plums, apricots, and almonds with about equal severity.

Losses from brown rot result primarily by rotting of the fruit in the orchard, although serious losses may also appear during transit and marketing of the fruit. Yields may be reduced also by destruction of the flowers during the blossom blight stage of the disease. In severe infections, and in the absence of good control measures, 50–75 percent of the fruit may rot in the orchard, and the remainder may become infected before it reaches the market.

◦ *Symptoms.* The first symptoms of the disease appear on the blossoms (Figure 11-67A). Brown spots appear on petals, stamens, or pistils, and they spread rapidly, involving the entire flower and its stem. In humid weather the infected organs are covered with the grayish-brown conidia of the fungus and later shrivel and dry up, the rotting mass clinging to the twigs for some time. Twigs bearing infected flowers develop small, elliptical, sunken, brown cankers around the flower stem, which sometimes may encircle the stem and cause twig blight. In humid weather, gum and also gray tufts of conidia appear on the bark surface.

Fruit symptoms appear when the fruit approaches maturity. Small, circular, brown spots appear, which spread rapidly in all directions. Depending on the humidity, they are sooner or later covered with ash-colored tufts of conidia, which break through the skin and are either scattered or arranged in concentric rings on the fruit surface (Figure 11-67B). One large or several small rotten areas may be present on the fruit, which finally becomes completely rotted and either dries up into a mummy and remains hanging from the tree or falls to the ground, where it also forms a mummy (Figure 11-67, C, D).

Sometimes small cankers also develop on twigs or branches bearing infected fruit.

◦ *The pathogen: Monilinia (Sclerotinia) fructicola.* In addition to *M. fructicola,* two other species, *M. laxa* and *M. fructigena,* cause brown rot of stone fruits. The former occurs in the west coast and the Wisconsin–Michigan area of the United States; the latter is found exclusively in Europe, where it is as serious on apples as it is on stone fruits. With slight differences, the development of the disease caused by each species is essentially the same.

The mycelium produces chains of elliptical *Monilia*-type conidia on hyphal branches arranged in groups or tufts (sporodochia). The fungus also produces microconidia (spermatia) in culture and on mummied fruit. The micronidia are borne in chains on bottle-shaped condiophores, and they do not germinate, but seem to be involved in the fertilization of the fungus. The sexual stage (apothecium) originates from pseudosclerotia formed in mummified fruit partly or wholly buried in the soil or debris. More than 20 apothecia may form on one mummy (Figure 11-67E). Small, bulblike protrusions appear on the mummy, and they extend to form the stipe of the fruiting body. As the stipe emerges above ground, its upper portion becomes swollen and a depression appears at the tip. Subsequent growth of the sides of the swelling forms the apothecial "cup," which is usually funnel or, later, disk

FIGURE 11-67 (A-E) Brown rot of stone fruits caused by *Monilinia fructicola*. (A) Blossom blight, twig cankers and killing of the tip. (B) Plums rotting on twig and covered with tufts of conidia. (C) Mummified peach fruit hanging from the twig and producing conidia. (D) Pile of peaches discarded because of brown rot infection. (E) Funnel-shaped apothecia of *M. fructicola* produced on mummified peach fruit on the ground. (F) Monilia pod rot of cocoa caused by *Monilia roreri*. (Photos A, B courtesy Dept. Plant Pathol., Cornell Univ. Photos D–F courtesy U.S.D.A.)

shaped. The inside or upper surface of the apothecium is lined with thousands of cylindrical asci interspersed with paraphyses. Each ascus contains eight single-celled ascospores.

◦ *Development of Disease.* The pathogen overwinters as mycelium in mummified fruit on the tree and in cankers of affected twigs, or as pseudosclerotia in mummies in the ground (Figure 11-68). In the spring the mycelium in mummified fruit on the tree and in the twig cankers produces new conidia, while the pseudosclerotia in mummified fruit buried in the ground produce several apothecia, which will form asci and ascospores.

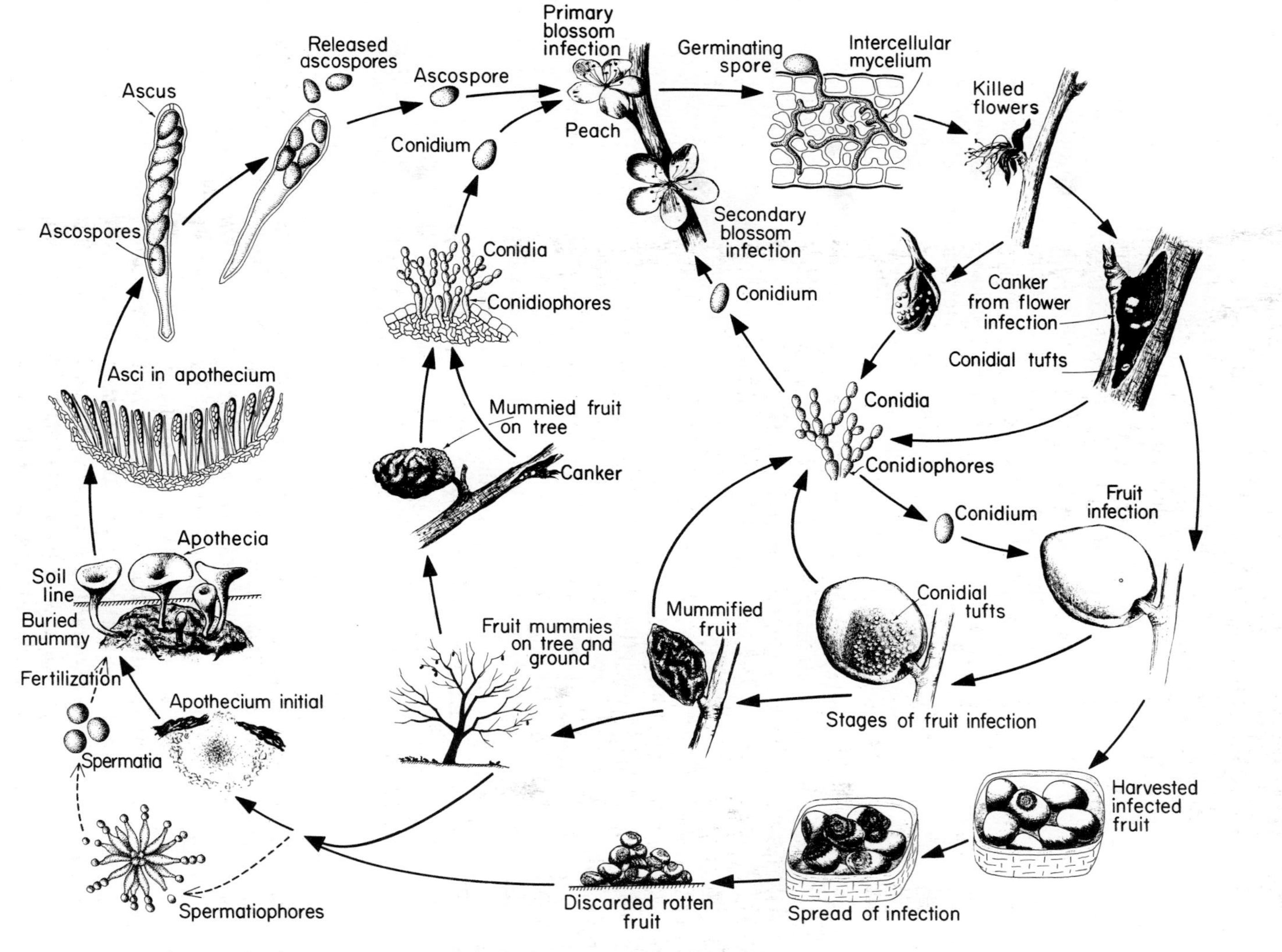

FIGURE 11-68 Disease cycle of brown rot of stone fruits caused by *Monilinia fructicola*.

Both conidia and ascospores can cause blossom infections. The conidia are windblown or may be carried by rainwater and splashes or insects to floral parts. The ascospores are forcibly discharged by the ascus, forming a whitish "cloud" over the apothecium. Air currents then carry the ascospores to the flowers. Conidia and ascospores germinate and can cause infection within a few hours.

The mycelium, especially in humid weather, produces short hyphae, which group together, push outward through the epidermis, and form numerous conidial tufts on the rotten, shriveled floral parts from which new masses of conidia are released. In the meantime, the mycelium advances rapidly down the tissues of the blossom petioles and from there into the fruit spurs and the twigs. In the twig the mycelium causes disintegration and collapse of the cells around the fruit spur, and a depressed, reddish-brown, shield-shaped canker forms. The advance of the mycelium may encircle the twig, which then becomes girdled and dies. The surface of the canker is soon covered with conidial tufts, and conidia from these serve as inoculum for fruit infection later in the season.

The brown rot fungus does not infect leaves or bark directly, and the new conidia and ascospores are short lived. The apothecia themselves soon disintegrate, and therefore no ascospores are present when the fruit becomes susceptible. The gap between the time the blossoms are infected and when the fruit on the same tree can become infected is bridged by the conidia, which are formed on twig cankers during humid weather in the summer. Also, the fruit of some early blooming stone fruits begins to ripen, and therefore becomes susceptible to infection, when some late-blooming ones are still in bloom or have just finished. In the latter case, conidia produced on the flowers of late-blooming trees can be carried to and infect the fruit of early ripening trees.

The susceptibility of the fruit to infection increases with its maturity. Conidia usually penetrate fruit through wounds made by insects, twig punctures, or hail, but on some, fruit penetration can take place through stomata or directly through the cuticle. The fungus grows intercellularly at first and through secretion of enzymes causes the maceration and browning of the infected tissues. The invasion of the fruit by the fungus is quite rapid. As the fungus advances into the fruit, it also produces conidial tufts on the already rotted area, and the conidia may be carried away and infect more fruit. The entire fruit may become completely rotten within a few days, and it either clings to and hangs from the tree, or falls to the ground. Fruit falling to the ground soon after infection usually disintegrates under the action of saprophytic fungi and bacteria. Fruit left hanging on the tree soon loses moisture and shrivels, and by the end of the season it becomes a dry, distorted mummy with a corrugated surface. The skin remains as a covering, and beneath it the remains of the fruit cells are held in place by the mycelial threads, which are closely interwoven and form a hard rind. Once the fruit has dried into a mummy, it may fall to the ground, but it is not affected by soil microorganisms and may persist there for two or more years.

Fruit infection also takes place after harvest, in storage, and in transit. Infected fruit will continue to rot after harvest, and the mycelium will attack directly healthy fruit in contact with infected ones. Healthy fruit may also be attacked by conidia at any time between harvest and use by the consumer.

◦ *Control.* Brown rot of stone fruits can be controlled best by completely controlling the blossom blight phase of the disease. This can be done by spraying two to four times with an effective fungicide from the time the blossom buds show pink until the petals fall. Captan, benomyl, thiophanate-methyl, triforine, chlorothalonil, dichlone, sulfur, and thiram are the fungicides used for brown rot control. Several newer fungicides that provide excellent control of brown rot are now being tested. They include the contact fungicides iprodione and vinclozolin and the systemic bitertand and etaconazole. Resistant strains of the brown rot fungus have developed to each of the systemic fungicides, and therefore, these chemicals are generally used in combination with one of the broad-spectrum fungicides such as captan, thiram, or sulfur.

Twigs bearing infected blossoms or cankers should be removed as early as possible to reduce the inoculum available for fruit infections later on in the season and to eliminate one of the sources of overwintering mycelium and spores that cause blossom infections the following season.

To control brown rot in ripening fruit, the fungicides are applied to the trees a few weeks before harvest, and applications continue weekly or biweekly until just before harvest. Since most infections of immature fruit and many of mature fruit orginate in wounds made by insect punctures, control of insects will also help control the disease.

To prevent infections at harvest and during storage and transit, fruit should be picked and handled with the greatest care to avoid punctures and skin abrasions on the fruit, which enable the brown rot fungus to gain entrance more easily. All fruit with brown rot spots should be discarded. Postharvest brown rot can be reduced by dipping or drenching fruit in a dichloran-benomyl or a dichloran-triforine-benomyl solution before storing and again during the sorting operation; and by hydrocooling or cooling fruit in air before refrigeration at 0–3°C. Recently, biological control of brown rot with the bacterium *Bacillus subtilis* has been reported and is being tested for prevention of pre- and postharvest decay of stone fruits.

Botrytis Diseases

Botrytis diseases are probably the most common and most widely distributed diseases of vegetables, ornamentals, fruits, and even field crops throughout the world. They are the most common diseases of greenhouse-grown crops. *Botrytis* diseases appear primarily as blossom blights and fruit rots but also as damping-off, stem cankers or rots, leaf spots and tuber, corm, bulb, and root rots (Figures 11-69 and 11-70). Under humid conditions, the fungus produces a noticeable gray-mold fruiting layer on the affected tissues that is characteristic of the *Botrytis* diseases. Some of the most serious diseases caused by *Botrytis* include gray mold of strawberry, gray mold rot of vegetables such as artichoke, bean, beet, cabbage, carrot, cucumber, eggplant, tip-end rot of bananas, lettuce, pepper, squash, tomato, onion blast and neck rot, calyx end rot of apples, blossom and twig blight of blueberries, blight or gray mold of ornamentals such as African violet, begonia, cyclamen, chrysanthemum, dahlia, geranium, hyacinth, lily, peony, rose, snapdragon, stock and tulip, bulb rot of amaryllis, corm rot and also leaf spot and stem rot of gladiolus.

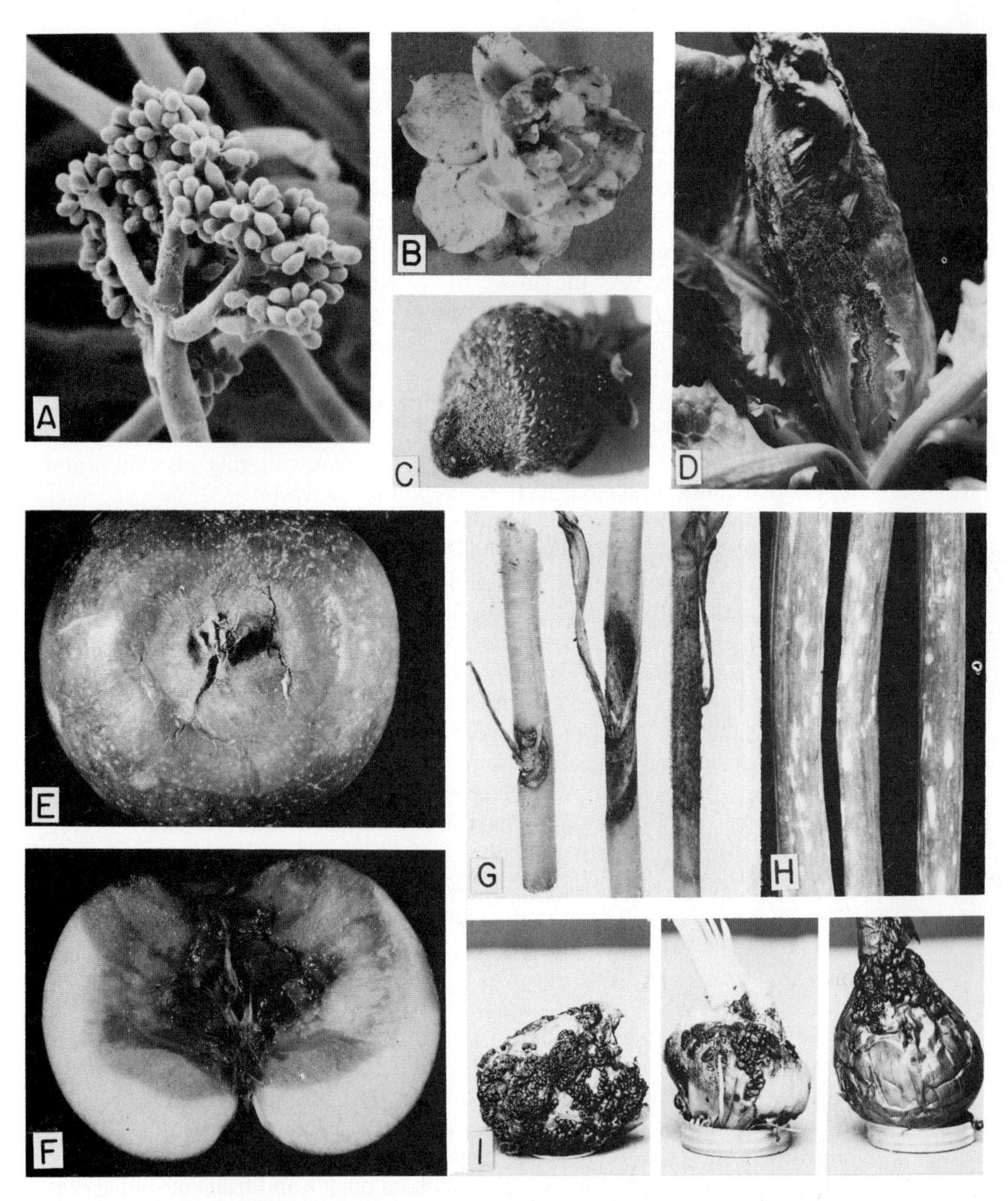

FIGURE 11-69 Scanning electron micrograph of a typical grape clusterlike conidiophore and conidia of *Botrytis* (A) and various symptoms caused by *Botrytis* (B-I). (B) Blossom blight of gardenia. (C) Strawberry gray mold. (D) Lettuce gray mold. (E, F) External (E) and internal (F) symptoms of *Botrytis* blossom end rot of apple. (G) Stem lesions; note that they originate at a dead leaf. (H): *Botrytis* blast of onion. (I) *Botrytis* neck rot of onion in storage. The black bodies on the onions are sclerotia of the fungus. (Photo A courtesy M. F. Brown and H. G. Brotzman.)

Botrytis also causes secondary soft rots of fruits and vegetables in storage, transit, and market.

In the field, blossom blights often precede and lead to fruit rots and stem rots. The fungus becomes established in flower petals, which are particularly

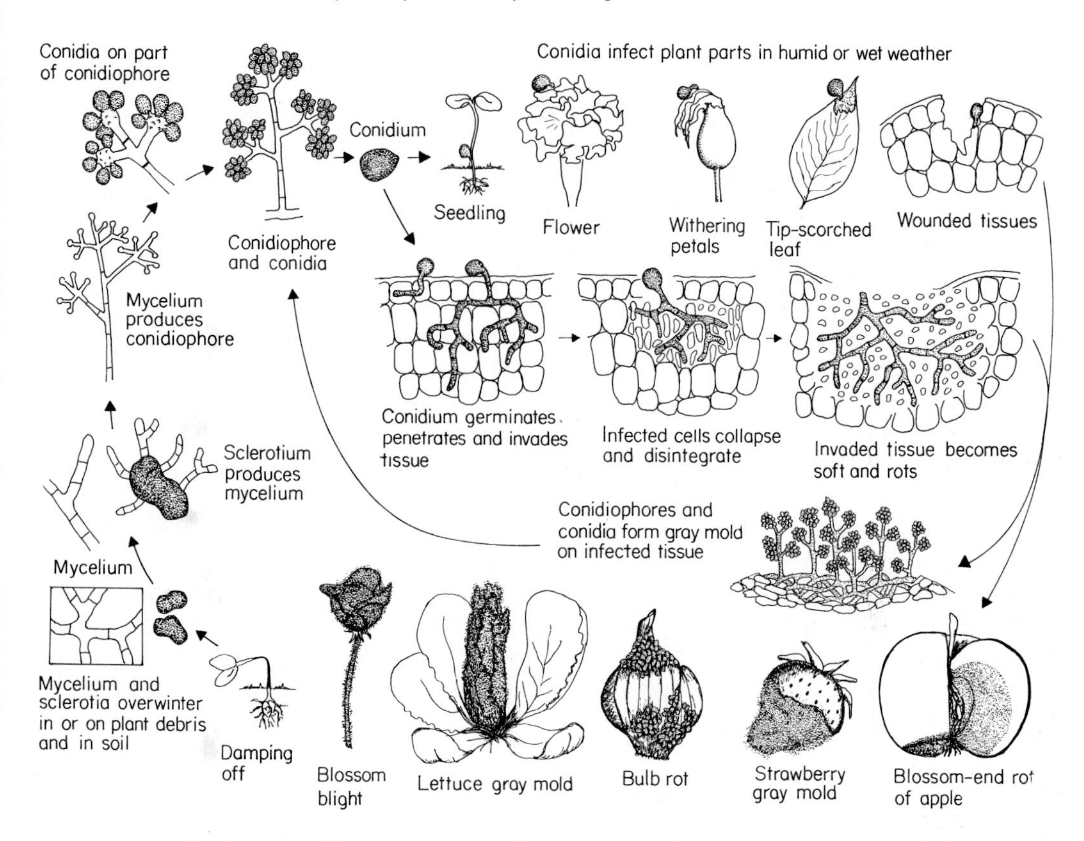

FIGURE 11-70 Development of *Botrytis* gray mold diseases.

susceptible when they begin to age, and there it produces abundant mycelium (Figure 11-69B). In cool, humid weather the mycelium produces large numbers of conidia, which may cause further infections, but the mycelium itself also grows and penetrates and invades the rest of the inflorescence, which becomes filled and covered with a whitish-gray or light brown cobwebby mold. The fungus then spreads to the pedicel, which rots and lets the buds and flowers lop over. If any fruit has developed, the fungus moves from the petals into the green or ripe fruit and causes a blossom-end rot of the fruit, which advances and may destroy part or all of the fruit and may spread to other fruits touching the diseased one. Infected fruit and succulent stems become soft and watery, and later, invaded tissues appear light brown (Figure 11-69, C, E, F). As the tissue rots, the epidermis cracks open and the fungus fruits abundantly. The tissues then become wrinkled and dry, and flat black sclerotia may appear on the surface or are sunken within the tissue.

Damping off of seedlings due to *Botrytis* occurs primarily in cold frames, where the humidity is high, but also in the field if the seed is contaminated with sclerotia of the fungus, or fungus mycelium or sclerotia are present in the soil.

Some species of *Botrytis* cause leaf spots on their hosts, for example, on gladiolus, onions, and tulip, the spots being small and yellowish at first but later becoming larger, whitish gray or tan, sunken, coalescing, and frequently involving the entire leaf (Figure 11-69 D, H). In many hosts, however, foliage infection occurs only after the fungus has grown on dying parts of the plants or in decaying matter on the soil that comes into contact with healthy leaves.

Stem lesions usually appear on succulent stems or stalks and may be either dark, sunken, elongated lesions with a definite outline or may spread through the stalk and cause it to weaken and break over at the point of infection, as happens with peony, rose, and tulip. (Figure 11-69G). Such stalks are usually susceptible to attack by the fungus along their entire length, and in wet weather the diseased parts become covered with the grayish-brown felty coat of fungus spores. Sclerotia may also be produced on infected stems.

Infection of belowground parts, such as bulbs, corms, tubers, and roots, may begin while these organs are still in the ground or upon harvest. Lesions may develop at any point of their surface, but in most hosts they are more likely to begin at the crown or at the base of these organs. Infected tissues usually appear soft and watery at first, but as the infection progresses, these areas enlarge, turn tan to brown and finally dark brown, and become spongy or corky and light in weight. Pockets of mycelium may develop between decayed bulb scales, under the husks of decayed corms, or on the surface of the lesions of such organs. Black sclerotia are often found on the surface or intermingled with the rotted tissues and mycelium (Figure 11-69I).

The pathogen, *Botrytis* sp., produces abundant gray mycelium and long, branched conidiophores with rounded apical cells bearing clusters of colorless or gray, one-celled, ovoid conidia (Figure 11-69A). The conidiophores and clusters of conidia resemble a grapelike cluster. The conidia are released readily in humid weather and are carried by air currents. The fungus frequently produces black, hard, flat, irregular sclerotia. Some species of *Botrytis* occasionally produce a *Sclerotinia* perfect stage in which ascospores are produced in an apothecium.

Botrytis overwinters in the soil as mycelium growing on decaying plant debris and as sclerotia (Figure 11-70). The fungus does not seem to infect seeds, but it can be spread with seed contaminated with sclerotia the size of the seed or with bits of plant debris infected with the fungus. The overwintering stages can also be spread by anything that moves soil or plant debris that may carry sclerotia or mycelium. The fungus requires cool (18–23°C), damp weather for best growth, sporulation, spore release and germination, and establishment of infection. The pathogen is active at low temperatures and causes considerable losses on crops kept for long periods in storage, even if the temperatures are between 0 and 10°C. Germinating spores seldom penetrate actively growing tissue directly, but they can penetrate tissues through wounds or after they have grown for a while and produced mycelium on old flower petals, dying foliage, dead bulb scales, and so on. *Botrytis* sclerotia usually germinate by producing mycelial threads that can infect directly, but in a few cases sclerotia germinated by producing apothecia and ascospores.

Control of *Botrytis* diseases is aided by the removal of infected and infested debris from the field and storage rooms, and by providing conditions for proper aeration and quick drying of plants and plant products. In greenhouses, humidity should be reduced by ventilation and heating. Storage organs such as onion bulbs can be protected by keeping them at 32–50°C for 2–4 days to remove excess moisture and then keeping them at 3°C in as dry an environment as possible. Biological control of *Botrytis* gray mold of apple fruit was obtained by spraying the flowers with a spore suspension of the antagonistic fungus *Trichoderma harzianum,* but this is not used in practice yet. Control of *Botrytis* in the field through chemical sprays has been only partially successful, especially in cool, damp weather. For *Botrytis* rot of lettuce, sprays with dichloran or zineb are recommended, while difolatan, dyrene, maneb-zinc, maneb, and chlorothalonil seem to give better control on crops such as onion and tomatoes. To control fruit rots, such as gray mold of strawberry, sprays or dusts with captan, thiram, or benomyl are recommended. Two newer contact fungicides, iprodione and vinclozolin, give excellent control of *Botrytis* and are now being tested on a wide variety of crops. *Botrytis* strains resistant to benomyl, dichloran, iprodione, and even captan have been found in various crops sprayed with these chemicals. Therefore, the use of different fungicides and fungicide combinations is recommended to reduce appearance and establishment of resistant strains.

SELECTED REFERENCES

Bove, F. J. (1970). "The Story of Ergot." Karger, Basel.

Bratley, C. O. (1937). Incidence and development of apple scab on fruit during the late summer and while in storage. *U. S., Dept. Agric. Tech. Bull.* **563,** 1–45.

Burr, T. J., ed. (1985). "Brown Rot of Stone Fruit," Workshop Proc. Am. Phytopathol. Soc., St. Paul, Minnesota, and Cornell University, Ithaca, New York.

Coley-Smith, J. R., Verhoeff, K., and Jarvis, W. R., eds. (1980). "The Biology of *Botrytis.*" Academic Press, New York.

Dickens, J. S. W., and Mantle, P. G. (1974). Ergot of cereals and grasses. *Minist. Agric. Fish Food, Advis. Leafl.* **548.**

Ellis, M. A., Madden, L. V., and Wilson, L. L. (1984). Evaluation of an electronic apple scab predictor for scheduling fungicides with curative activity. *Plant Dis.* **68,** 1055–1058.

Ezekiel, W. N. (1924). Fruit rotting *Sclerotinias.* II. The American brown-rot fungi. *Md., Agric. Exp. Stan., Bull.* **271,** 87–142.

Forsberg, J. L. (1975). "Diseases of Ornamental Plants," Spec. Publ. No. 3 Rev. University of Illinois, College of Agriculture, Urbana-Champaign.

Foster, H. H. (1937). Studies on the pathogenicity of *Physalospora obtusa. Phytopathology* **27,** 803–832.

Gould, C. J. (1954). *Botrytis* diseases of gladiolus. *Plant Dis. Rep., Suppl.* **224,** 1–33.

Heye, C. C., and Andrews, J. H. (1983). Antagonism of *Athelia bombacina* and *Chaetomium globosum* to the apple scab pathogen. *Venturia inaequalis. Phytopathology* **73,** 650–654.

James, J. R., and Sutton, T. B. (1982). Environmental factors influencing pseudothecial development and ascospore maturation of *Venturia inaequalis. Phytopathology* **72,** 1073–1080.

Jones, A. L., Lillevik, S. L., Fisher, P. D., and Stebbins, T. C. (1980). A microcomputer-based instrument to predict primary apple scab infection periods. *Plant Dis.* **64,** 69–72.

Jones, A. L. Fisher, P. D., Seem, R. C., Kroon, J. C., and Van DeMotter, P. J. (1984). Development and commercialization of an in-field microcomputer delivery system for weather-driven predictive models. *Plant Dis.* **68,** 458–463.

Keitt, G. W., and Jones, L. K. (1926). Studies of the epidemiology and control of apple scab. *Res. Bull.— Wisc., Agric. Exp. Stn.* **73,** 1–104.

Landgraf, F. A., and Zehr, E. I. (1982). Inoculum sources for *Monilinia fructicola* in South Carolina peach orchards. *Phytopathology* **72,** 185–190.

McClellan, W. D., and Hewitt, W. B. (1973). Early *Botrytis* rot of grapes: time of infection and latency of *Botrytis cinerea* in *Vitis vinifera. Phytopathology* **63,** 1151–1156.

McColloch, L. P., Cook, H. T., and Wright, W. R. (1968). Market diseases of tomatoes, peppers, and eggplants. *U. S., Dep. Agric., Agric. Handb.* **28,** 1–74.

MacHardy, W. E., and Gadoury, D. M. (1985). Forecasting the seasonal maturation of ascospores of *Venturia inaequalis. Phytopathology* **75,** 185–190.

McKeen, W. E. (1974). Mode of penetration of epidermal cell walls of *Vicia faba* by *Botrytis cinerea. Phytopathology* **64,** 461–467.

Mower, R. L., Snyder, W. C., and Hancock, J. G. (1975). Biological control of ergot by *Fusarium. Phytopathology* **65,** 5–10.

Nusbaum, C. J., and Keitt, G. W. (1938). A cytological study of host-parasite relations of *Venturia inaequais* on apple leaves. *J. Agric. Res. (Washington, D.C.)* **65,** 595–618.
Ogawa, J. M. *et al.* (1975). Monilinia life cycle on sweet cherries and its control by overhead sprinkler fungicide applications. *Plant Dis. Rep.* **59,** 876–880.
Pepin, H. S., and MacPherson, E. A. (1982). Strains of *Botrytis cinerea* resistant to benomyl and captan in the field. *Plant Dis.* **66,** 404–405.
Pierson, C. F., Ceponis, M. J., and McColloch, L. P. (1971). Market diseases of apples, pears, and quinces. *U. S., Dept. Agric., Agric. Handb.* **376,** 1–112.
Puranik, S. B., and Mathre, D. E. (1971). Biology and control of ergot on male sterile wheat and barley. *Phytopathology* **61,** 1075–1080.
Pusey, L. P., and Wilson, C. L. (1984). Postharvest biological control of stone fruit brown rot by *Bacillus subtilis. Plant Dis.* **68,** 753–757.
Riggs, R. K., Henson, L., and Chapman, R. A. (1968). Infectivity of and alkaloid production by some isolates of *Claviceps purpurea. Phytopathology* **58,** 54–55.
Roberts, J. W., and Dunegan, J. C. (1932). Peach brown rot. *U. S., Dep. Agric., Tech. Bull.* **328,** 1–59.
Rose, D. H., Fisher, D. F., and Brooks, C. (1937). Market diseases of fruits and vegetables: Peaches, plums, cherries and other stone fruits. *Misc. Publ.-U. S. Dep. Agric.* **228,** 1–26.
Savage, S. D., and Sall, M. A. (1984). Botrytis bunch rot of grapes: Influence of trellis type and canopy microclimate. *Phytopathology* **74,** 65–70.
Segall, R. H., and Newhall, A. G. (1960). Onion blast or leaf spotting caused by species of *Botrytis. Phytopathology* **50,** 76–82.
Smith, M. A., McColloch, L. P., and Friedman, B. A. (1966). Market diseases of asparagus, onions, beans, peas, carrots, celery and related vegetables. *U. S., Dep. Agric., Agric. Handb.* **303,** 1–65.
Smoot, J. J., Houck, L. G., and Johnson, H. B. (1971). Market diseases of citrus and other subtropical fruits. *U. S., Dep. Agric., Agric. Handb.* **398,** 1–101.
Tronsino, A., and Ystaas, J. (1980). Biological control of *Botrytis cinerea* on apple. *Plant Dis.* **64,** 1009.
Walker, J. C. (1926). *Botrytis* neck rot of onions. *J. Agric. Res. (Washington, D.C.)* **33,** 893–928.
Wicks, T. (1974). Tolerance of the apple scab fungus to benzimidazole fungicides. *Plant Dis. Rep.* **58,** 886–889.

Vascular Wilts Caused by Ascomycetes and Imperfect Fungi

Vascular wilts are widespread, very destructive, spectacular, and frightening plant diseases appearing as more or less rapid wilting, browning, and dying of leaves and succulent shoots of plants followed by plant death. Wilts occur as a result of the presence and activities of the pathogen in the xylem vascular tissues of the plant. Entire plants or plant parts above the point of vascular invasion by the pathogen may die within a matter of weeks in most annuals and some perennials, although in certain perennials death may not occur until several years after infection. The pathogen usually continues to spread internally through the xylem vessels as mycelium or conidia until the entire plant is killed. As long as the infected plant is alive, the fungi that cause vascular wilts remain limited to the vascular (xylem) tissues and a few surrounding cells and never come to the surface of the plant—not even to produce spores. Only when the infected plant is killed by the disease do these fungi move into other tissues and sporulate at or near the surface of the dead plant.

There are three genera of fungi that cause vascular wilts: *Ceratocystis, Fusarium,* and *Verticillium.* Each of them causes widespread and severe disease on several important crop, forest, and ornamental plants.

Ceratocystis causes vascular wilts primarily of trees, such as Dutch elm disease *(C. ulmi)* and oak wilt *(C. fagacearum).*

Fusarium causes vascular wilts primarily of annual vegetables and flowers, herbaceous perennial ornamentals, plantation crops, weeds, and of the mimosa tree (silk tree). Most of the vascular wilt-causing *Fusaria* belong to the species *Fusarium oxysporum.* Different host plants are attacked by special forms or races of the fungus. Thus, the fungus that attacks tomato is designated as *F. oxysporum* f. *lycopersici,* cucurbits *F.* o. f. *niveum,* sweet potato *F. o.* f. *batatas,* onion *F. o.* f. *cepae,* cabbage *F. o.* f. *conglutinans,* banana *F. o.* f. *cubense,* cotton *F. o.* f. *vasinfectum,* carnation *F. o.* f. *dianthii,* chrysanthemum *F. o.* f. *chrysanthemi,* and so on.

Verticillium causes vascular wilts of annual vegetables, flowers, crop plants, and weeds, and of perennial ornamentals, fruit trees, forest trees, and weeds. It exists as two species, *Verticillium albo-atrum* and *V. dahliae,* and it attacks hundreds of kinds of plants causing wilts and losses of varying severity.

All vascular wilts, regardless of which pathogen causes them, have certain characteristics in common. The leaves of infected plants or of parts of infected plants lose turgidity, become flaccid, lighter green to greenish yellow, droop and finally wilt, turn yellow, then brown, and die. Wilted leaves may be flat or curled. Young, tender shoots also wilt and die. In cross sections of infected stems and twigs, discolored brown areas appear as a complete or interrupted ring consisting of discolored vascular tissues. In the xylem vessels of infected stems and roots, mycelium and spores of the causal fungus may be present. Some of the vessels may be clogged with mycelium, spores, or polysaccharides produced by the fungus. Clogging is further increased by gels and gums formed by the accumulation and oxidation of breakdown products of plant cells attacked by fungal enzymes. The oxidation and translocation of some such breakdown products seems to also be responsible for the brown discoloration of affected vascular tissues. In newly infected young stems the number of xylem vessels formed is reduced, and their cell walls are thinner than normal. Often the parenchyma cells surrounding xylem vessels are stimulated by secretions of the pathogen to divide excessively, and this, combined with the thinner and weaker vessel walls, results in reduction of the diameter or complete collapse of the vessels. In some hosts, tyloses are produced by parenchyma cells adjoining some xylem vessels. The balloonlike tyloses protrude into the vessels and contribute to their clogging. Toxins secreted in the vessels by the wilt-causing fungi are carried upward in the water stream and affect living parenchyma cells adjacent to the xylem, thus causing some of the effects described above. Toxins may also be carried to the leaves in which they cause reduced chlorophyll synthesis along the veins (vein clearing) and reduced photosynthesis, disrupt the permeability of the leaf cell membranes and their ability to control water loss through transpiration and thereby result in leaf epinasty, wilting, interveinal necrosis, browning and death.

Of the wilt fungi, *Fusarium* is a soil inhabitant, while *Verticillium* is more of a soil invader. Both infect plants through the roots, which they penetrate directly or through wounds. Large numbers of parasitic nematodes in the soil usually increase the incidence of *Fusarium* and *Verticillium* wilts, presumably by providing effective penetration points. Once in the root, the mycelium reaches the xylem vessels, where it may produce microconidia *(Fusarium)* or conidia *(Verticillium),* and subsequently mycelium and spores spread upward through the vessels, the spores being carried by the transpiration stream. Both fungi produce only asexual spores. *Fusarium* overwinters in the soil or plant debris as thick-walled asexual spores called chlamydospores, or as mycelium and spores in plant debris. *Verticillium* overwinters in the soil as microsclerotia or as mycelium in perennial hosts and in plant debris. Both are effective saprophytes, and once introduced into a field, they become established there forever, although their populations may vary considerably depending on the susceptibility and the duration of cultivation of the host plant grown in the field. *Fusarium* and *Verticillium* spread through the soil to a small extent as mycelium growing through roots or plant debris, but primarily as mycelium, spores, or sclerotia carried in soil water, on farm equipment, transplants, tubers, seeds of some hosts, cuttings of infected plants and in some cases as wind-blown spores or sclerotia.

The two wilt-causing species of the third fungus, *Ceratocystis,* live primarily in the xylem vessels and adjacent parenchyma cells of infected elm and oak trees and in the outer layers of wood and inner layers of bark in elms and oaks killed by them. The fungus produces both ascospores and conidia and spreads from tree to tree either as spores carried by certain bark-feeding beetles or by natural root grafts. The beetles introduce the fungus into the xylem of young, vigorously growing twigs or larger branches on which they feed, and from there, as well as from the natural root grafts, the fungus spreads as mycelium or conidia into the vascular system of the tree and induces wilt. The spores of *Ceratocystis* are not soil-borne.

Vascular wilts are more or less worldwide in distribution, causing tremendous losses on most kinds of vegetables and flowers and on field crops such as cotton, alfalfa (*Fusarium* and *Verticillium*), on fruit "trees" such as banana *(Fusarium)* and stone fruits *(Verticillium),* on forest and shade trees *(Verticillium),* and on elm and oak trees, in particular *(Ceratocystis). Fusarium* wilts are much more common and destructive in the warmer temperate regions and in the tropics and subtropics, becoming less damaging or rare in colder climates except for greenhouse crops in these areas. *Verticillium* is much more common in the temperate zones and is considerably more cold-resistant than *Fusarium,* especially in its tree hosts and causes diseases in greater latitudes than the latter. *Ceratocystis* wilts are less widespread than the others, their distribution depending on the availability of the hosts (elms and oaks), of the pathogen, and of contaminated insect vectors.

Vascular wilts are among the most difficult to control. The fact that only a single infection of a plant by one spore is sufficient to introduce the pathogen into the plant, in which it then grows and spreads internally, makes prevention of infection and subsequent control with surface fungicides practically impossible. Also the fact that *Fusarium* and *Verticillium* can survive in the soil of a field saprophytically almost forever, makes control through crop rotation or other cultural practices impractical or ineffective. On the other hand, the dissemination of *Ceratocystis* spores over long distances by insect vectors feeding on trees of all sizes makes its control even more problematic.

The most effective means of controlling *Fusarium* and *Verticillium* wilts has been the use of resistant varieties. Due to the relative immobility of the pathogens and therefore slow development and distribution of any new pathogen races, varieties remain resistant for rather long periods of time. Cultural practices such as deep plowing, crop rotation, leaving the soil fallow, and flooding the field have been helpful in reducing the pathogen populations in the soil but do not eliminate it completely. Soil fumigation has been used with success in some cases, but it is too expensive and its effect does not last long enough to make its use profitable. In the greenhouse, soil sterilization gives effective control of both diseases. Control of *Ceratocystis* wilts has been attempted by efforts to control the insect vectors of the pathogen by insecticides sprayed on or injected into the trees, removal and burning of infected trees and logs to eliminate the fungus and the breeding grounds of the insect vectors and through selection of resistant trees. None of these measures has been successful against the *Ceratocystis* wilts, and the diseases keep on spreading.

An apparent breakthrough in the control of fungal vascular wilts seems to have been made by the discovery of the systemic fungicides containing thiabendazole or its derivatives, including, and particularly, benomyl in its various formulations. These chemicals, although not yet proven to give complete control of any of these wilts, when injected into elm trees before and sometimes after infection have given promising results with *Ceratocystis ulmi,* the cause of Dutch elm disease.

Fusarium Wilts

As mentioned above, *Fusarium* wilts affect and cause severe losses on most vegetables and flowers, several field crops such as cotton and tobacco, plantation crops such as banana, plantain, coffee, and sugarcane, and a few shade trees. Fusarial wilts are greatly favored by warm soil conditions and in greenhouses. Since most Fusarial wilts have very similar disease cycles and development, only *Fusarium* wilt of tomato will be described in some detail to illustrate this group of vascular wilts.

• Fusarium Wilt of Tomato

Fusarium wilt is one of the most prevalent and damaging diseases of tomato wherever tomatoes are grown intensively. The disease is most destructive in warm climates and warm, sandy soils of temperate regions. In the United States the disease is most severe in the central states and in the southern regions, whereas in the northern states it can become important only on greenhouse tomatoes.

Great losses may be caused by the disease, especially on susceptible varieties and under favorable weather conditions. *Fusarium* wilt damages plants by causing stunting of plants, which soon wilt and finally die. Occasionally entire fields of tomatoes are killed or severely damaged before a crop can be harvested. Generally, however, the disease does not cause serious losses unless soil and air temperatures are rather high during much of the season.

◦ *Symptoms.* The first symptoms appear as slight vein clearing on the outer, younger leaflets, followed by epinasty of the older leaves caused by drooping of the petioles. When plants are infected at the seedling stage, they usually wilt and die soon after appearance of the first symptoms. Older plants in the field may wilt and die suddenly if the infection is severe and if the weather is favorable for the pathogen. More commonly, however, in older plants vein clearing and leaf epinasty are followed by stunting of the plants, yellowing of the lower leaves, occasional formation of adventitious roots, wilting of leaves and young stems, defoliation, marginal necrosis of the remaining leaves, and finally death of the plant (Figure 11-71A). Often these symptoms appear on only one side of the stem and progress upward until the foliage is killed and the stem dies. While the plant is still living, no fungus mycelium or fructifications appear on its surface. Fruit may occasionally become infected, and then it rots and drops off without becoming spotted. Roots also become infected, and after an initial period of stunting, the smaller side roots rot.

In cross sections of the stem near the base of the infected plant a brown ring is evident in the area of the vascular bundles, and the upward extent of the discoloration depends on the severity of the disease.

◦ *The pathogen: Fusarium oxysporum* f. *lycopersici.* The mycelium is colorless at first, but becomes cream-colored or pale yellow with age, and under certain conditions it produces a pale pink or somewhat purplish coloration. The fungus produces three kinds of asexual spores (Figure 11-72). Microconidia, which are one or two celled, are usually the most frequently and abundantly produced spores under all conditions. They are the spores most frequently produced by the fungus inside the vessels of infected host plants. Macroconidia are the typical *"Fusarium"* spores, three to five celled, gradually pointed, and curved toward both ends. They are common on the surface of plants killed by the pathogen and usually appear in sporodochialike groups. Chlamydospores are one- or two-celled, thick-walled, round spores produced terminally or intercalary on older mycelium or in macroconidia. All three types of spores are produced in cultures of the fungus and probably in the soil, although only chlamydospores can survive in the soil for long.

◦ *Development of disease.* The pathogen is a soil inhabitant that survives between crops in infected plant debris in the soil as mycelium and in all its spore forms but most commonly, especially in the cooler temperate regions, as chlamydospores (Figure 11-72). It spreads over short distances by means of water and contaminated farm equipment, and over long distances primarily in infected transplants or in the soil carried with them. Usually, once an area becomes infested with *Fusarium,* it remains so indefinitely.

When healthy plants grow in contaminated soil, the germ tube of spores or the mycelium penetrate root tips directly, or enter the roots through wounds or at the point of formation of lateral roots. The mycelium advances through the root cortex intercellularly and, when it reaches the xylem vessels, enters them through the pits. The mycelium then remains exclusively in the vessels and travels through them, mostly upward, toward the stem and crown of the plant. While in the vessels, the mycelium branches and produces microconidia, which are detached and carried upward in the sap stream. The microconidia germinate at the point where their upward movement is stopped, the mycelium penetrates the upper wall of the vessel, and more microconidia are produced in the next vessel. The mycelium also advances laterally into the adjacent vessels penetrating them through the pits.

Presumably a combination of the processes discussed earlier—of vessel clogging by mycelium, spores, gells, gums, and tylosis and crushing of the vessels by proliferating adjacent parenchyma cells—is responsible for the breakdown of the water economy of the infected plant. When the amount of water available to the leaves is below the required minimum for their function, the stomata close and the leaves wilt and finally die, followed in death by the rest of the plant. The fungus then invades the parenchymatous tissues of the plant extensively, reaches the surface of the dead tissues, and there it

FIGURE 11-71 (A) *Fusarium* wilt of tomato. Some shoots are dead but others show only drooping of leaves. (B) *Verticillium* wilt of cotton. Healthy plant at left. (C) *Verticillium* wilt of peach tree. (D) Short brown streaks in branch of tree infected with *Verticillium*. (Photo A courtesy U.S.D.A. Photo B courtesy G. C. Papavizas.)

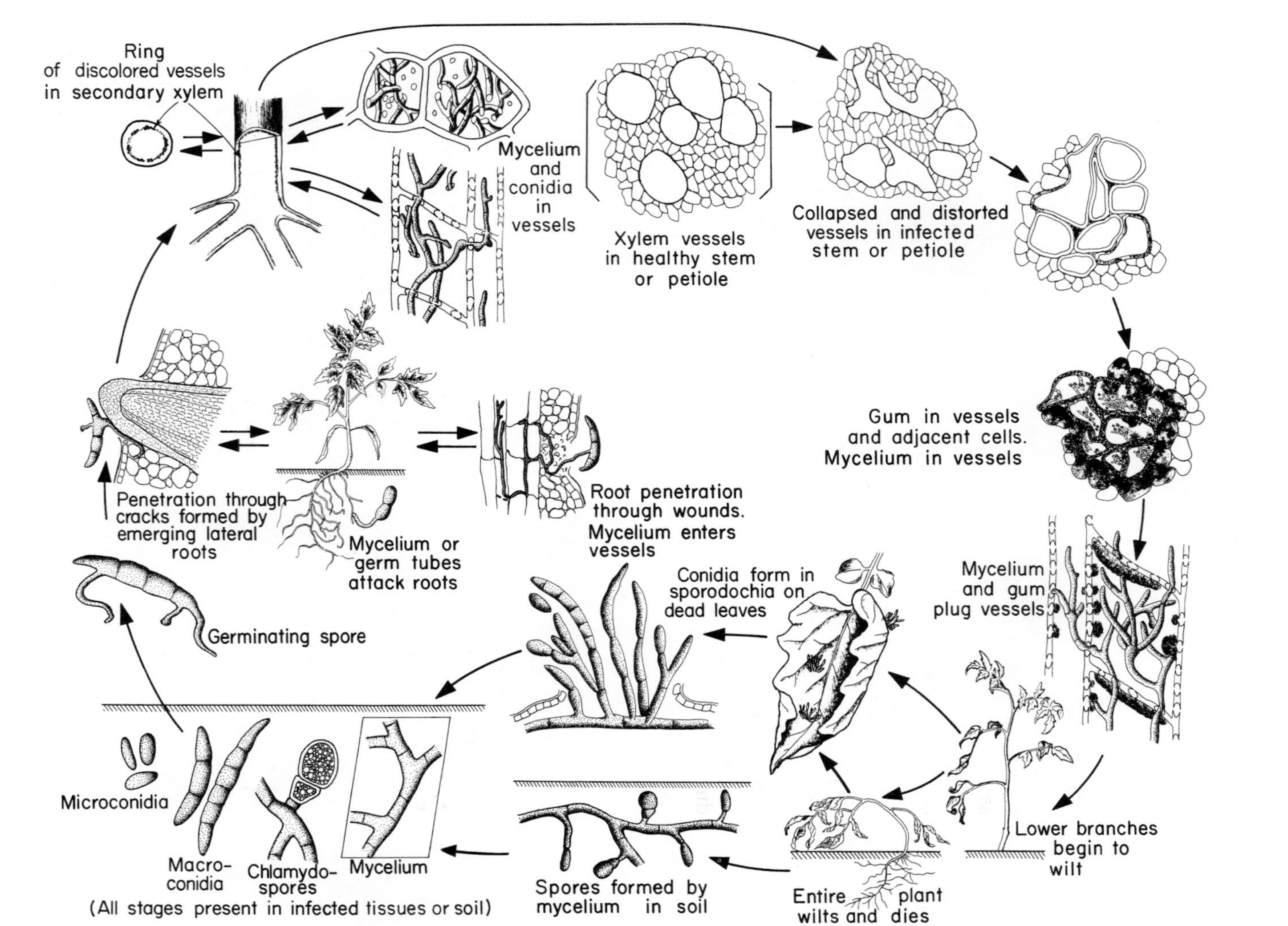

FIGURE 11-72 Disease cycle of *Fusarium* wilt of tomato caused by *Fusarium oxysporum* f. *lycopersici*.

sporulates profusely. The spores may be disseminated to new plants or areas by wind, water, and so on.

Occasionally, the fungus may reach the fruit of infected plants and penetrate or contaminate the seed. This happens primarily when the soil moisture is high and the temperature relatively low, conditions that allow plants to produce good yields although infected with the fungus. Usually, however, infected fruits decay and drop, and even if harvested, infected seeds are so light that they are eliminated in the procedures of extraction and cleaning of the seed and therefore play little role in the spread of the fungus.

◦ *Control.* Use of tomato varieties resistant to the fungus is the only practical measure for controlling the disease in the field. Several such varieties are available today. The fungus is so widespread and so persistent in soils that seedbed sterilization and crop rotation, although always sound practices, are of limited value. Soil sterilization is too expensive for field application, but should be always practiced for greenhouse-grown tomato plants. Use of healthy seed and transplants is of course mandatory, and hot-water treatment of seed suspected of being infected should precede planting.

A great deal of research has been done in the last several years regarding the possibility of biological control of Fusarial wilt of tomatoes and of many other crops. Encouraging results have been obtained by prior inoculation of plants with formae specialis of *F. oxysporum* not pathogenic to each crop, by the use of antagonistic fungi, such as *Trichoderma,* and of siderophore-producing *Pseudomonas* bacteria. Solar heating (solarization) of field soil by covering with transparent plastic film during the summer also reduces disease incidence. Although promising, none of these methods is used for practical control of Fusarial wilts so far.

***Verticillium* Wilts**

Verticillium wilts are worldwide in distribution but most important in areas of the temperate zones. *Verticillium* attacks more than 200 species of plants, most of them vegetables (such as tomato, eggplant, pepper, cantaloupe, and watermelon), flowers (such as chrysanthemum, aster, and dahlia), fruit trees (such as apricot, cherry, and peach); also strawberries, raspberries and roses, field crops (such as cotton, potato, alfalfa, peanuts, and mint); and shade and forest trees (such as maple and elm).

The symptoms of *Verticillium* wilts are almost indentical to those of *Fusarium* wilts, and on hosts affected by both, the two pathogens cannot be distinguished except by laboratory examinations. In many hosts and most areas, however, *Verticillium* induces wilt at lower temperatures than *Fusarium,* the symptoms develop more slowly, and often appear only on the lower or outer part of the plant or on only a few of its branches. In some hosts, *Verticillium* wilt develops primarily in seedlings, which usually die shortly after infection, but more common are late infections, which cause epinasty of the upper leaves followed by the appearance in leaves of irregular chlorotic patches that become necrotic. Older plants infected with *Verticillium* are usually stunted in various degrees, and their vascular tissues show characteristic discoloration. In many hosts, *Verticillium* infection results in defoliation, gradual wilting and death of successive branches, or an abrupt collapse and death of the entire plant (Figure 11-71, B–D).

Initial outbreaks of *Verticillium* wilt in a field are typically mild and local. In subsequent years, attacks become severe and widespread until the crop has to be discontinued or is replaced with resistant varieties. The increasing severity of the disease from year to year is brought on by a greater build-up in inoculum potential, by the appearance of strains of the fungus more virulent than the original, or by both.

Two species of *Verticillium, V. albo-atrum* and *V. dahliae,* are the causes of verticillium wilts in most plants. Both produce conidia that are short lived. *Verticillium dahliae* also produces microsclerotia, while *V. albo-atrum* produced microsclerotial-like dark thick-walled mycelium but not microsclerotia. *V. albo-atrum* grows best at 20–25°C, while *V. dahliae* prefers slightly higher temperatures (25–28°C) and is somewhat more common in warmer regions. There is, however, considerable variability in virulence and other characteristics of strains within each species. Although some *Verticillium* strains show host specialization, most of them show little host specialization and attack a wide range of host plants. *V. dahliae* overwinters in the soil as microsclerotia, which can survive up to fifteen years. Both, however, can overwinter as mycelium within perennial hosts, in propagative organs, or in plant debris. *Verticillium* penetrates young roots of host plants directly or through wounds. The fungus is spread by contaminated seed, by vegetative cuttings and tubers, by scions and buds, by wind and surface groundwater, and by the soil itself, which may contain up to 100 or more microsclerotia per gram; 6–50 microsclerotia per gram are sufficient to give 100 percent infection in most susceptible crops. Many fields have become contaminated with *Verticillium* for the first time by planting infected potato tubers or other crops, and it is known that solanaceous crops such as potato, eggplant, and tomato increase the inoculum level in the soil. However, *Verticillium* is often found in uncultivated areas, indicating that the fungus is native to the soils and can attack susceptible crops as soon as they are planted.

Control of *Verticillium* wilts depends on the use of disease-free plants in disease-free soil, the use of resistant varieties, and on avoiding the planting of susceptible crops where solanaceous crops have been grown repeatedly. Soil fumigation can be profitable when used to protect high-value crops, but it is too expensive on large areas.

Thermal inactivation via soil solarization is proving useful for control of *Verticillium* in regions with high summer temperatures and low rainfall, such as California and Israel.

Ceratocystis Wilts

The Ceratocystis wilts include mainly the Dutch elm disease and oak wilt.

• Dutch Elm Disease

Dutch elm disease owes its name to the fact that it was first described on elm in Holland in 1921. Since then the disease has spread throughout Europe, parts of Asia, and most of the temperate zones in North America. In the United States the disease was first found in Ohio and some states in the east coast in the early 1930s; by 1973, it had spread westward to the Pacific coast states.

Dutch elm disease is the most destructive shade tree disease in the United States today. It affects all elm species but most severely the American elm. The disease may kill branches and entire trees within a few weeks or a few years from the time of infection. Hundreds of thousands of elm trees in towns across the country die from Dutch elm disease every year. The cost of cutting down diseased and dead elm trees amounts to many millions of dollars per year. And, of course, no one can estimate the value of the natural beauty destroyed by the diease in countless communities.

◦ *Symptoms.* The first symptoms of Dutch elm disease appear as sudden or prolonged wilting of the leaves of individual branches or of the entire tree (Figure 11-73). Wilted leaves frequently curl, turn yellow, then brown, and finally fall off the tree earlier than normal (Figure 11-73). Most affected branches die immediately after defoliation. The disease usually appears first on one or several branches and then spreads to other portions of the tree. Thus, many dead branches may appear on a tree or a portion of a tree. Such trees may die gradually, branch by branch, over a period of several years or they may recover. Sometimes, however, entire trees suddenly develop disease symptoms and may die within a few weeks (Figure 11-73A, B). Usually trees that become infected in the spring or early summer die quickly, while those infected in late summer are much less seriously affected and may even recover, unless they become reinfected.

When the bark of infected twigs or branches is peeled back, brown streaking or mottling appears on the outer layer of wood. In cross section of the branch, the browning appears as a broken or continuous ring in the outer rings of the wood (Figure 11-73D).

◦ *The pathogen: Ceratocystis ulmi.* The mycelium is creamy white. While in the vessels, the mycelium produces short hyphal branches on which clusters of *Sporothrix (Cephalosporium)*-type conidia are formed (Figure 11-74). In dying or dead trees, the mycelium produces some *Sporothrix,* but mostly *Pesotum (Graphium)*-type spores on coremia developing on bark which is somewhat loose from the wood and in tunnels made in the bark by insects. The coremia consist of hyphae grouped into an erect, dark, solid stalk, and a colorless, flaring head to which the spores adhere, forming a sticky glistening, whitish at first, and later slightly yellowish, droplet.

The fungus is heterothallic and requires the contact of two sexually compatible strains for sexual reproduction. Since, frequently, only one of the mating types is found in large areas in nature, sexual reproduction is extremely rare. In the United States, for example, the fungus rarely reproduces sexually, but it does so rather frequently in Europe. When the two mating types do come in contact, perithecia develop. The perithecia are spherical and black, about 120 μm in diameter, and have a long (about 300–400 μm) neck. Perithecia form singly or in groups and in the same areas in the bark as the coremia.

Inside the perithecium, many asci develop but as the asci mature, they disintegrate, leaving the ascospores free in the perithecial cavity. The ascospores are discharged through the neck canal and accumulate in a sticky droplet.

◦ *Development of disease.* Dutch elm disease is the result of an unusual partnership between a fungus and an insect (Figure 11-74). Although the fungus alone is responsible for the disease, the insect is the indispensable

A
B
C
D
a
b
c
E
F

vector of the fungus, carrying the fungus spores from infected elm wood to healthy elm trees. The insects responsible for the spread of the disease are the European elm bark beetle *(Scolytus multistriatus)* and the native elm bark beetle *(Hylurgopinus rufipes)* (Figure 11-73E).

The fungus overwinters in the bark of dying or dead elm trees and logs as mycelium and as spore-bearing coremia. Elm bark beetles prefer to lay their eggs in the intersurface between bark and wood of trees weakened or dying by drought or disease. The adult female beetle tunnels through the bark and opens a gallery either parallel with the grain of the wood *(Scolytus)* or at an angle or perpendicular *(Hylurgopinus)*. The female lays eggs along the sides of the gallery, the eggs soon hatch, and the larvae open tunnels at right angles to the maternal gallery. If the tree was already infected with the fungus, the fungus produces mycelium and sticky, *Graphium*-type spores in the beetle tunnels. When the adult beetles emerge, they carry thousands of fungus spores on and in their bodies. *Scolytus* beetles feed in the crotches of living, vigorous elm twigs. *Hylurgopinus* beetles feed on stems 5–30 cm in diameter (Figure 11-73F). As the beetles burrow into the bark and wood, the spores are deposited in the wounded tissues of the tree, germinate, and grow rapidly into the injured bark and the wood. When the fungus reaches the large xylem vessels of the spring wood, it may produce *Sporothrix*-type spores, which are carried up by the sap stream. These spores reproduce by yeastlike budding, germinate, and start new infections. The extent of symptoms in the crown is correlated with the extent of vascular invasion. In early stages of infection, the mycelium invades primarily the vessels, and only occasionally tracheids, fibers, and the surrounding parenchyma cells. General invasion of tissue begins at the terminal or extensive dieback phase of the disease, at which time there is also considerable intercellular growth of the fungus between the parenchyma cells. Gums and tyloses are produced in the larger vessels, and sometimes isolated areas of the sapwood are blocked by a combination of gums, tyloses, and fungal growth. Infection also induces browning of the water-conducting vessels. Infected twigs and branches soon wilt and die.

Infections that take place in the spring or early summer result in invasion of the long vessels of the elm springwood through which the spores can be carried rapidly to all parts of the tree. If vascular invasion becomes general, the tree may die within a few weeks. During later infections, vascular invasion is limited to the outer, shorter vessels of the summerwood in which they move only for short distances. As a result, late infections may produce only localized infections and seldom cause serious immediate damage to the tree but may kill the tree the next year.

The elm bark beetles feed on living trees for only a few days and then fly back to dying or weakened elm wood in which they construct new galleries

FIGURE 11-73 Dutch elm disease caused by *Ceratocystis ulmi.* (A) Infected elm tree most of which shows thin, yellowish foliage in early summer. (B) The same tree dead and defoliated in late summer. (C) Twig with wilted, rolled, brown leaves. (D) Diagonal and cross sections of elm twig and branch showing a ring of brown discoloration near the surface of the twig and deeper in the wood of the branch. (E) Beetle carriers of the Dutch elm disease. Side and top views of the European (a) and the native (b) elm bark beetles, (c) larva of (a). (F) Galleries beneath the bark of dead elm trees made by the female and the larvae of the European (upper left) and the native (lower right) elm bark beetles. At upper right and lower left of (F) can be seen the bark punctures or wounds on healthy elm made by these beetles. (Photos D–F courtesy Shade Tree Lab., Univ. of Mass.)

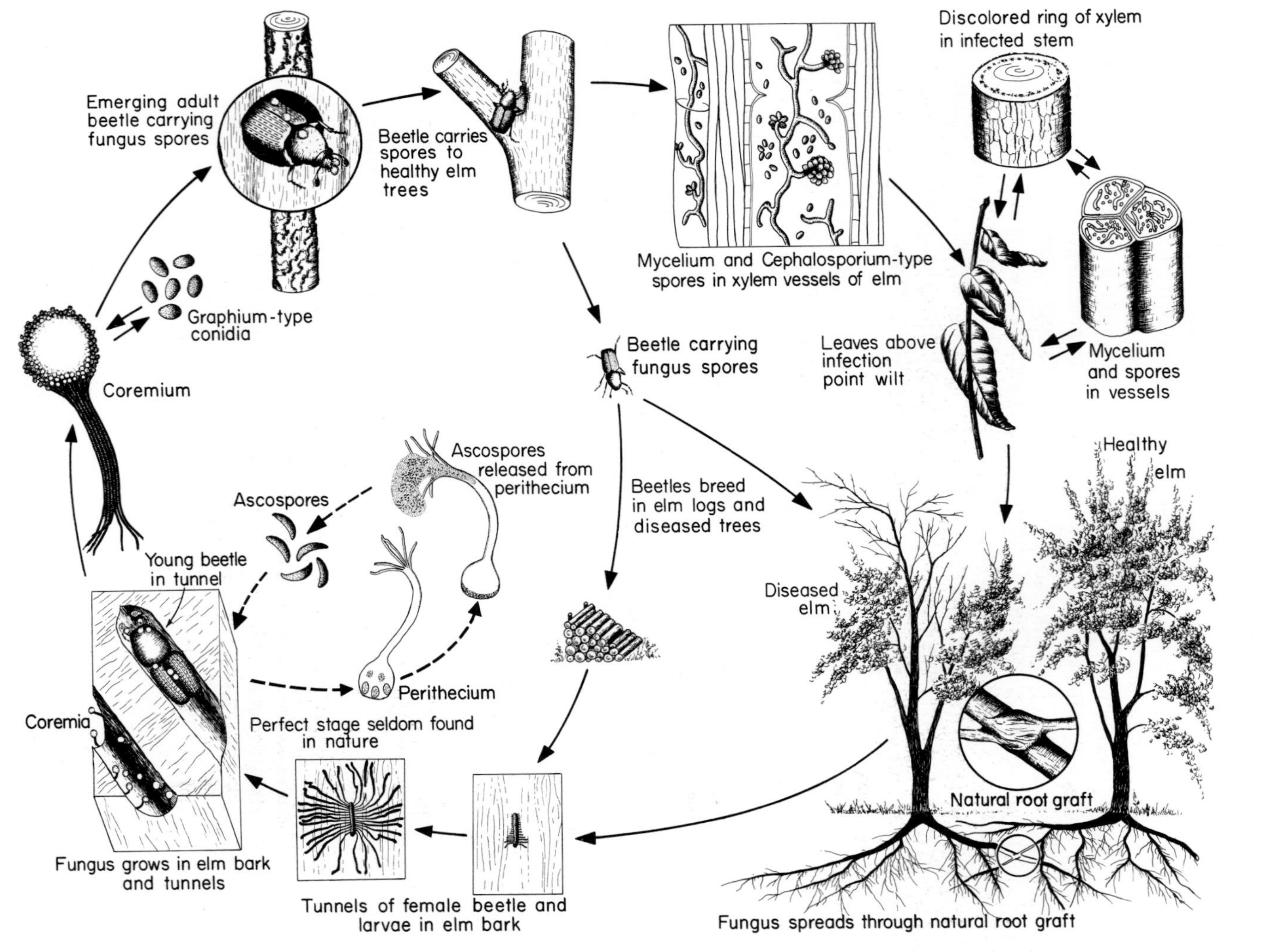

FIGURE 11-74 Disease cycle of Dutch elm disease caused by *Ceratocystis ulmi.*

and lay eggs. There are usually two generations of beetles per season. In each generation the young adult goes from dead or weakened elm trees to living, vigorous ones on which it feeds and then returns to the dead or weakened trees to lay its eggs. Therefore, once an insect becomes contaminated with fungus spores, it may carry them either to healthy or diseased wood, in both of which the fungus grows and multiplies and may contaminate all the offspring of the insect as well as any other insects that visit the infected wood.

◦ *Control.* Great efforts have been made and are being continued to find resistant clones within the susceptible American elm species and in other species. Certain Asiatic species, such as the Siberian and the Chinese elm, are resistant to Dutch elm disease but produce poor shade trees. Hybrids between various species have shown resistance in varying degrees and some of them look promising, but so far none of them has been planted widely or proved completely resistant.

For a long time, control of Dutch elm disease in the United States was attempted primarily through sanitation measures and through chemical control of the insect vectors of the fungus. Sanitation involves the removal and destruction of weakened or dead elm trees and elm logs, thus destroying the larvae contained in them or denying the insect and the fungus their overwintering habitat. Pruning out infected twigs and branches sometimes eliminates the disease. Naturally grafted roots of elm trees can be cut or killed with chemicals to prevent spread of the fungus to adjacent trees. Control of the insect vector by chemicals involves spraying the healthy elm trees while dormant and in the spring with methoxychlor and other insecticides, but spraying has been only partially effective. In some areas trap logs and pheromone traps are being tested as a means of reducing the number of insect vectors of Dutch elm disease but so far with little success.

The most promising results for Dutch elm disease control in individual trees have been obtained with trunk or root injections of healthy or diseased elm trees with solubilized benomyl or thiabendazole. These systemic fungicides, in various chemical formulations, have in some cases arrested the advance of the disease in infected trees and reduced the appearance of new infections on treated healthy trees, but they are not particularly dependable.

SELECTED REFERENCES

Armstrong, G. M. and Armstrong, J. K. (1975). Reflections on the wilt *Fusaria. Annu. Rev. Phytopathol.* **13,** 95–103.

Banfield, W. M. (1941). Distribution by the sap stream of spores of three fungi that induce vascular wilt disease of elm. *J. Agric. Res., (Washington, D.C.)* **62,** 637–681.

Banfield, W. M. (1968). Dutch elm disease recurrence and recovery in American elm. *Phytopathol. Z.* **62,** 21–60.

Beckman, C. H. (1964). Host responses to vascular infection. *Annu. Rev. Phytopathol.* **2,** 231–252.

Beckman, C. H. (1987). "The Nature of Wilt Diseases of Plants." APS Press, St. Paul, Minnesota.

Brown, M. F., and Wyllie, T. D. (1970). Utrastructure of microsclerotia of *Verticillium albo-atrum. Phytopathology* **60,** 538–542.

Chambers, L., and Corden, M. E. (1963). Semeiography of *Fusarium* wilt of tomato. *Phytopathology* **53,** 1006–1010.

Dimond, A. E. (1970). Biophysics and biochemistry of the vascular wilt syndrome. *Annu. Rev. Phytopathol.* **8,** 301–322.

Fletcher, J. T., and Martin, J. A. (1972). Spread and control of *Fusarium* wilt of carnations. *Plant Pathol.* **21,** 181–187.

Forsberg, J. L. (1975). "Diseases of Ornamental Plants," Spec. Publ. No. 3 Rev. University of Illinois College of Agriculture, Urbana-Champaign.
Frank, J. A., Webb, R. E., and Wilson, D. R. (1975). The effect of inoculum levels on field evaluations of potatoes for Verticillium wilt resistance. *Phytopathology* **65,** 225–228.
Gibbs, J. N. (1978). Intercontinental epidemiology of Dutch elm disease. *Annu. Rev. Phytopathol.* **16,** 287–307.
Holmes, F. W. (1965). Virulence to *Ceratocystis ulmi. Neth. J. Plant Pathol.* **71,** 97–112.
Jones, J. P., and Crill, P. (1973). The effect of *Verticillium* wilt on resistant, tolerant and susceptible tomato varieties. *Plant Dis. Rep.* **57,** 122–124.
Jordan, V. W. L. (1974). *Verticillium* wilt of strawberry: Cultivar reaction and effect on runner health and production. *Plant Pathol.* **23,** 8–13.
Kendrick, J. B. (1944). Fruit invasion and seed carriage of tonato *Fusarium* wilt. *Phytopathology* **34,** 1005–1006.
Kondo, E. S., and Huntley, G. D. (1973). Root-injection field trials of MBC-phosphate in 1972 for Dutch elm disease control. *Can. For. Serv. Inf. Bull.* **0-X-182,** 1–17.
Krause, C. R., and Wilson, C. L. (1972). Fine structure of *Ceratocystis ulmi* in elm wood. *Phytopathology* **62,** 1253–1256.
Mace, M. E., Bell, A. A., and Beckman, C. H., eds. (1981). "Fungal Wilt Diseases of Plants." Academic Press, New York.
Nelson, P. E., Toussoun, T. A., and Cook, R. J., eds. (1981). "Fusarium: Diseases, Biology, and Taxonomy." Pennsylvania State Press, University Park.
Papavizas, G. C. (1974). The relation of soil microorganisms to soilborne plant pathogens. *Va. Polytech. Inst. South. Coop. Ser., Bull.* **183,** 1–98.
Parker, K. G. (1959). *Verticillium* hadromycosis of deciduous tree fruits. *Plant Dis. Rep., Suppl.* **225,** 39–61.
Pegg, G. F. (1974). *Verticillium* diseases. *Rev. Plant Pathol.* **53,** 157–182.
Pomerleau, R. (1970). Pathological anatomy of the Dutch elm disease. Distribution and development of *Ceratocystis Ulmi* in elm tissues. *Can. J. Bot.* **48,** 2043–2057.
Pullman, G. S., and DeVay, J. E. (1982). Epidemiology of Verticillium wilt of cotton: A relationship between inoculum density and disease progression. *Phytopathology* **72,** 549–554.
Rudolph, B. A. (1931). *Verticillium* hadromycosis. *Hilgardia,* **5,** 201–361.
Sinclair, W. A., and Campana, R. J., eds. (1978). Dutch elm disease: Perspectives after 60 years. *Cornell Univ., Agric. Exp. Stn. Search (Agric.)* **8,** 1–52.
Stipes, R. J., and Campana, R. J., eds. (1981). "Compendium of Elm Diseases." Am. Phytopathol. Soc., St. Paul, Minnesota.
Stover, R. H. (1962). Fusarial wilt (Panama disease) of bananas and other *Musa* species. *Common. Mycol. Inst. Phytopathol. Pap.* **4,** 1–117.
Strobel, G. A., and Lanier, G. N. (1981). Dutch elm disease. *Sci. Am.* **245,** 56–66.
Strong, M. C. (1946). The effect of soil moisture and temperature on *Fusarium* wilt of tomato. *Phytopathology* **36,** 218–225.
Van Alfen, N. K., and Walton, G. S. (1974). Pressure injection of benomyl and methyl-2-benzimidazolecarbamate hydrochloride for control of Dutch elm disease. *Phytopathology* **64,** 1231–1234.
Walker, J. C., and Foster, R. E. (1946). Plant nutrition in relation to disease development. III. *Fusarium* wilt of tomato. *Am. J. Bot.* **33,** 259–264.

Root and Stem Rots Caused by Ascomycetes and Imperfect Fungi

Several Ascomycetous fungi attack primarily the roots and lower stems of plants, particularly of cereals. The most important of these are:

Cochliobolus (imperfect stage *Bipolaris=Helminthosporium*), causing root and foot rot and also blight of cereals and grasses.

Gibberella (imperfect stage *Fusarium*), causing seedling blight and foot or stalk rot of corn and small grains.

Gaeumannomyces (Ophiobolus), causing the take-all and whiteheads disease of cereals.

Another fungus, *Sclerotinia,* causes crown rots and blights of nearly all kinds of succulent plants, primarily flowers and vegetables, and also root and crown rot of turf grasses, a disease known as dollar spot.

In few, if any, of the diseases caused by the first two fungi, *Cochliobolus* and *Gibberella,* does the ascigerous stage play a role of any consequence. In almost all cases, the diseases are caused strictly by the asexual stages of these fungi, which are primarily species of *Helminthosporium* for *Cochliobolus* and species of *Fusarium* for *Gibberella.* Therefore, the diseases caused by *Cochliobolus* are discussed as diseases caused by *Helminthosporium* (see pages 359–366) and all but one *Gibberella* disease are discussed as diseases caused by *Fusarium.*

Among the Imperfect Fungi causing root and stem rots are several widely distributed and extremely destructive plant pathogens:

Dipodia maydis, causing Diplodia stalk and ear rot of corn, collar rot of peanuts, etc.

Fusarium, especially *F. solani,* causing root rot of bean, asparagus, onion, foot rot of squash, dry rot of potatoes, basal rot of iris and lily, stem rot of carnation and chrysanthemum, corm rot of gladiolus, and seed rot, damping off, and seedling blights of these and numerous other plants

Phoma, causing black leg of crucifers (Figure 11-75), root and crown rot of celery

FIGURE 11-75 Black-leg disease of cabbage caused by *Phoma lingam.*

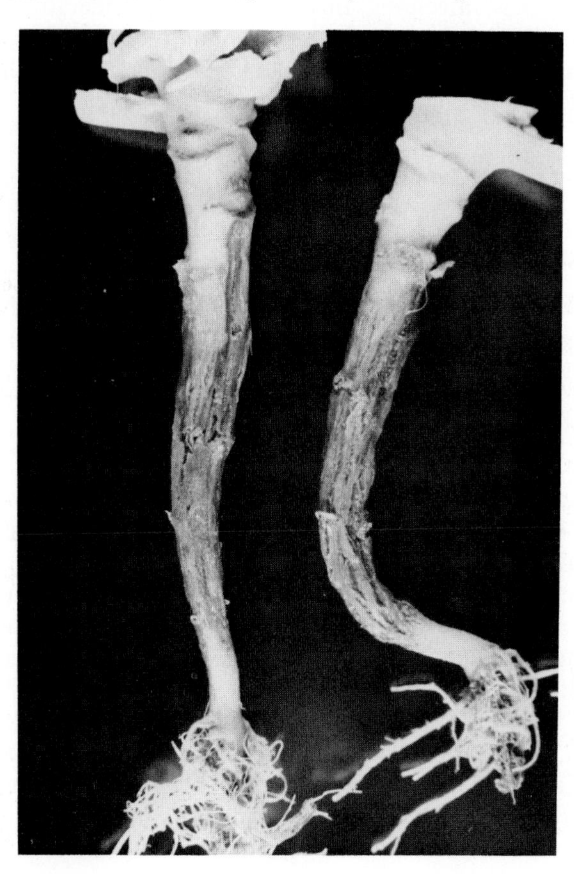

and delphinium, heart rot of beet, stem blight and fruit rot of tomato and pepper, etc.

Phymatotrichum omnivorum, causing the Texas root rot of fruit and shade trees, ornamental shrubs, cotton, alfalfa, most flowers and vegetables, and many weeds.

Thielaviopsis, causing black root rot and damping off of many vegetables and flowers, particularly bean, beet, carrot, celery, pansy, pea, poinsettia, squash, sweet pea, sweet potato, tomato, and watermelon, and of many field crops, including cotton, cowpeas, flax, peanuts, clover, soybean, and tobacco

As a general rule, the root and stem rot diseases caused by the above and by other Ascomycetes and Imperfect Fungi appear on the affected plant organs at first as water-soaked areas that later turn brown to black, although in some diseases they are frequently covered by white fungal mycelium. The roots and stems are killed more or less rapidly, and the entire plant grows poorly or is killed. The fungi that cause these diseases are nonobligate parasites that live, grow, and multiply in the soil as soil inhabitants, usually in association with dead organic matter, and they are favored by high soil moisture and high relative humidity in the air. Most of them produce conidia, and some produce ascospores occasionally or regularly. Several produce sclerotia. In all of the above fungi the fungus can overwinter as mycelium in infected plant tissues or debris, as sclerotia, or as any of the kinds of spores the fungus may produce. These same stages also serve as inoculum that can be spread and start new infections. In recent years, considerable progress has been made in biological control of several root and stem rot fungi by treating the seed with antagonistic fungi and bacteria. Such treatments, however, are still at the experimental stage.

Gibberella Diseases

• Gibberella Stalk Rot, Root Rot, Ear Rot, Kernel Rot, and Seedling Blight of Corn

Stalk rots of corn are often caused by different combinations of several species of fungi and bacteria and affect plants when they are nearly mature. The fungi most commonly responsible for stalk rots in corn include *Gibberella zeae, G. moniliforme, G. moniliforme* var. *subglutinans, Diplodia maydis,* and *Macrophomina phaseoli.* The stalk rot complex often causes losses between 10 and 30 percent.

The *Gibberella* diseases of corn are worldwide in distribution and cause serious losses. The most important phases of the diseases are stalk rot and either ear rot *(G. zeae)* or random kernel rot *(G. moniliforme).*

In stalk rot caused by *G. zeae,* lower internodes become soft and appear tan or brown on the outside while internally they may appear pink or reddish (Figure 11-76, A, B). The pith disintegrates, leaving only the vascular bundles intact. The rot may also affect the roots. Stalk rot leads to a dull gray appearance of the leaves, premature death, and stalk breakage (Figure 11-76D). Small, black, round perithecia are often produced superficially on rotting stalks. In stalk rot caused by *G. moniliforme,* the rot affects roots and lower internodes beginning soon after pollination and becomes more severe as plants mature. Salmon-colored mycelium often appears on affected stalks

FIGURE 11-76 (A–C) *Gibberella* stalk and ear rot of corn caused by *Gibberella zeae.* External (A) and internal (B) symptoms. (C) Ear rot. (D) A corn field destroyed by a combination of *Gibberella* and *Diplodia* stalk rots. (E–G) *Diplodia* stalk and ear rot of corn. (E) External discoloration of stalk (left), stalk with pycnidia appearing as black dots (middle), and internal disorganization of tissues in infected stalk (right). (F) Corn variety resistant (left) and susceptible to *Diplodia* stalk rot, the latter prematurely dead and broken. (G) *Diplodia* ear rot. (Photos courtesy Illinois Agr. Expt. Sta.)

in warm, moist weather. In the other, most common stalk rot, caused by *Diplodia,* stalk symptoms are similar (Figure 11-76), except for the presence of minute dark, subepidermal pycnidia near the nodes.

In the ear rot caused by *G. zeae,* often called red ear rot, ears develop a reddish mold that often begins at the tip of the ear (Figure 11-76C). If infection occurs early, the ears may rot completely and a pinkish to reddish mold grows between the ears and the tightly adhering husks. Perithecia may form on husks and ear shanks. Corn ears infected with *G. zeae* are toxic to humans and certain animals such as hogs. In kernel rot, caused by *G. moniliforme,* randomly scattered groups of kernels on the ear are covered with pinkish or reddish mycelium, especially near the top of the ear.

Gibberella sp. is only one of many fungi causing blight of corn seedlings. It may be carried on or in infected seed, or it may attack the seed and seedling from the soil. In either case, the germinating seed may be attacked and killed before the seedling emerges from the soil, or after emergence, in which case the seedling may be killed or become dwarfed and chlorotic and later die. Light brown to dark-colored lesions are usually evident on the tap and lateral roots and in the lower internode.

Two species of *Gibberella, G. zeae* and *G. moniliforme (fujikuroi),* are primarily responsible for the symptoms observed on corn and for those that will be described on small grains below. Both fungi produce ascospores in perithecia and *Fusarium*-type conidia *(F. graminearum* and *F. moniliforme,* respectively). Perithecia are rather rare in *G. moniliforme.* The fungi overwinter as perithecia or as mycelium or chlamydospores in infected plant debris, particularly corn stalks. In the spring, during wet, warm conditions, mature ascospores are released and are carried by wind to corn stalks or ears, which they penetrate directly or through wounds and cause infections. Conidia may also be produced on infected corn debris, but they are more commonly produced on infected plant parts in moist, warm weather and serve as the secondary inoculum. The diseases are favored by dry weather early in the season and wet weather near or after silking. Also, high plant density, high nitrogen and low potassium in the plant, and early maturity of hybrids makes them more susceptible to the diseases.

Control of *Gibberella* diseases of corn depends on the use of resistant varieties, balanced nitrogen and potassium fertilization, and lower plant density in the field.

• Gibberella Scab, Seedling Blight, and Foot Rot of Small Grains

Gibberella scab, seedling blight, and foot rot of small grains are also worldwide. They are caused by the same fungi that cause the diseases in corn and perhaps some additional species of *Gibberella* and of *Fusarium* that do not produce a perfect stage (such as *Fusarium culmorum*). Losses may be as high as 50 percent of the yield. In some areas where corn is grown extensively, this disease makes wheat and barley production unfeasible.

Scab or head blight causes severe damage to wheat and the other cereals, especially in areas with high temperature and relative humidity during the heading and blossoming period. Infected spikelets first appear water soaked, then lose their chlorophyll and become straw colored. In warm, humid weather, pinkish-red mycelium and conidia develop abundantly in the infected spikelets, and the infection spreads to adjacent spikelets or through the entire head. Purplish perithecia may also develop on the infected floral bracts. Infected kernels become shriveled and discolored with a white, pink, or light brown scaly appearance as a result of the mycelial outgrowths from the pericarp. As with corn, infected kernels of cereals are also toxic to humans, hogs, and other animals because they contain a substance that acts as an emetic.

Seedling blight appears as a light brown to reddish-brown cortical rot and blight either before or after emergence of the seedling above the soil line. In older plants, a foot rot develops appearing as a browning of the basal leaf sheaths or as a pronounced rotting of the basal part of the plant around soil level and for some distance above the soil line.

Control measures against small grain diseases caused by *Gibberella* are the same as those described for the same diseases of corn.

Fusarium Root and Stem Rots

Several *Fusarium* species, but primarily *F. solani* and its formae specialis, and also some formae specialis of *F. oxysporum* cause, instead of vascular wilts, rotting of seeds and seedlings (damping off), rotting of roots, lower stems and crowns, and rots of corms, bulbs, and tubers (Figure 11-77), A, B). The plants

FIGURE 11-77 (A) Damping off of cucumber seedlings caused by *Fusarium* sp. Healthy plant at right. (B) *Fusarium* root and stem rot of bean. (C–E) Black root rot of tobacco (C, D) and of bean (E) caused by *Thielaviopsis basicola.* Inset in (D) shows spores of the fungus. (Photo B courtesy U.S.D.A. Photo E courtesy G. C. Papavizas.)

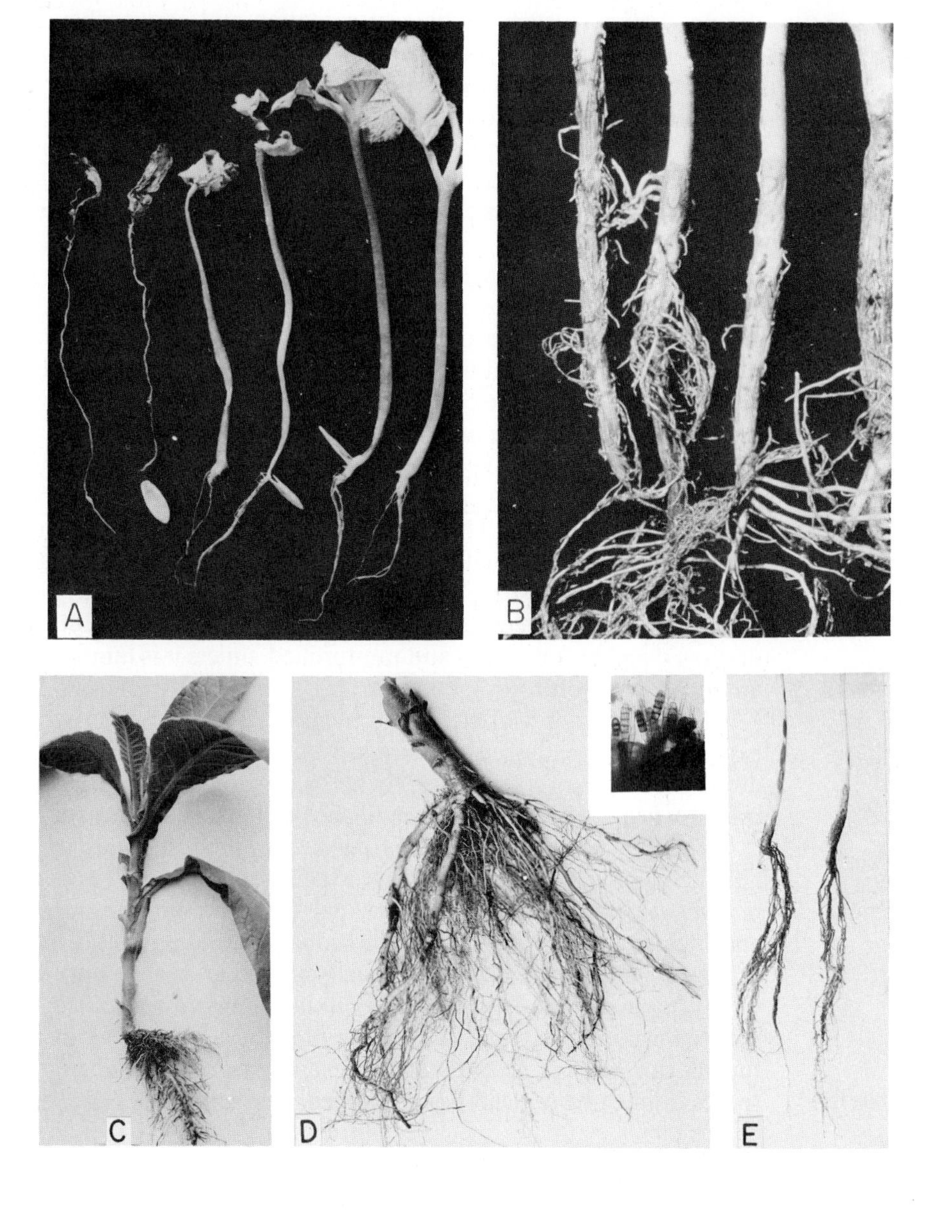

affected belong to widely separated families and may be vegetables, flowers, field crops, and weeds. These diseases are worldwide in distribution and cause severe losses by reducing stands and the growth and yield of infected plants.

In **root rots,** as those of bean, peanut, soybean, and asparagus, tap roots of young plants at first show a slightly reddish discoloration; this later becomes darker red to brown and larger, more or less covering the tap root and the stem below the soil line without a definite margin, or appearing as streaks extending up to the soil line. Longitudinal fissures appear along the main root, while the small lateral roots are killed. Plant growth is generally retarded, and in dry weather the leaves may turn yellow and even fall off. Sometimes, infected plants develop secondary roots and a large number of rootlets just below the soil line, and these roots, under favorable moisture conditions, may be sufficient to carry the plant to maturity and to production of a fairly good crop. In many cases, however, infected plants decline and die with or without wilt symptoms.

In **stem rots,** such as those of carnation and chrysanthemum, infected plants wilt and die from a stem rot at the base of the plant. Lesions develop on the stem at or below the soil line, and their edges often have pink or red discolorations. The lesions develop inward from the outside, and usually there is no internal stem discoloration, but in some plants a brown discoloration may extend in the wood of the stem for a considerable distance above the ground. In older plants the roots may have also rotted and may have sloughed off.

Rots of bulbs, corms, and tubers by species of *Fusarium* can occur in the field and in storage. They are common on plants such as onion, iris, lily, and gladiolus. The rot can start at uninjured sides of bulbs, corms, or tubers, but it often starts at wounds, at the base of these organs, at injured or diseased roots or foliage, or through the cuts formed on such tissues during harvest. Invaded bulbs and corms may or may not show outward symptoms, although usually the basal plate and fleshy scales, as well as the roots, are brown to black, sunken and decaying, and often containing mats of mycelium. The rot is generally dry and firm. The foliage turns yellow, purple, or brown and dies prematurely. Tubers usually develop small brown patches that soon enlarge, become sunken, and show concentric wrinkles that contain cavities lined with white mycelium. Eventually large parts of the tuber or entire tubers are destroyed and become hard and mummified, unless it is humid and then they are invaded by soft rotting bacteria.

Fusarium root and stem rots increase greatly in severity when plants exposed to the pathogen are stressed by low temperature, by intermittent drought or excessive soil water, by herbicides, by surface soil compaction by tractor wheels, and by subsurface tillage pans, which restrict root growth.

The fungus, *Fusarium solani,* generally produces only asexual spores, although under certain conditions a perithecial stage identified as *Nectria haematococca* is produced. The asexual spores are produced in sporodochia and include one- to two-celled microconidia, and the typical *Fusarium*-type macroconidia consisting of 3–9 (usually 4–5) cells, slightly curved, with more or less pointed ends. *Fusarium* also produces one- to two-celled, thick-walled chlamydospores that can withstand drought and low temperatures. The fungus can live on dead plant tissue and can overwinter as mycelium or spores in infected or dead tissues or seed. The spores are easily spread by air, equipment, water, and contact, and the fungus is already present in many soils as spores.

Control of *Fusarium* rots of root, stem, and corms is generally possible in the greenhouse through soil sterilization and use of healthy propagative stock. Because of rapid reestablishment of the pathogen in the soil from airborne spores, however, it may be necessary to eliminate such spores within the greenhouse by fumigation, and the soil may have to be treated with chemicals such as captafol after soil sterilization to inhibit recolonization of the steamed soil by the pathogen. There are currently no adequate control measures for these diseases in the field. Loosening compacted soil with subsoiler chisels to a depth of 25–50 cm in the last tillage before planting has been the most dependable method in reducing Fusarium root rot of bean. Rotation with nonsusceptible crops, good soil drainage, and the use of disease-free or fungicide-treated seed or other propagative stock may help reduce losses. Fertilization with the nitrate form of nitrogen also helps reduce disease, as does the use of resistant varieties when available. Treatment of propagative stock with benomyl or application of benomyl sprays on the plants in the field or greenhouse has helped reduce the incidence of Fusarium rots on some kinds of plants. Biological control of Fusarium root and stem rots has been attempted with some success by incorporating organic materials such as barley straw, lettuce, and chitin in the soil, thus favoring the increase of several fungi and bacteria antagonistic to Fusarium, or by treating seeds or transplants with spores of fungal antagonists, mycorrhizal fungi, or antagonistic *Pseudomonas* bacteria. None of the biological controls is used in practice so far.

Take-All of Wheat, Other Cereals, and Grasses

Take-all is a wide-spread and destructive disease of wheat and of the other cereals and grasses in temperate climates around the world. It is primarily a root and basal stem disease of winter wheat, particularly in areas of intensive, continuous cultivation of cereals. Losses may vary from negligible to 50 percent.

The presence of take-all in a wheat field is often suggested early in the season by the appearance of patches of poorly developed, yellowish seedlings or stunted and unthrifty plants producing few tillers. Later, as plants approach maturity, affected plants ripen prematurely and produce heads that have sterile, bleached spikelets and are known as white-heads. Infected plants are easily pulled from the soil because much of their root system has been destroyed by the fungus and the remaining few roots are short, brown-black, and brittle. The brown-black dry rot usually extends to the crown and the base of the stem up to the lower leaf bases. A dark mat of mycelium develops between the stem and the lowest leaf sheath, and the necks of fungal perithecia appear on the leaf sheath as small black raised spots. A diagnostic feature of the disease is the presence on the surface of roots of thick brown strands of runner hyphae.

The pathogen of take-all is the fungus *Gaeumannomyces (Ophiobolus) graminis,* of which var. *tritici* attacks most cereals and grasses but not oats, while var. *avenae* attacks all cereals, including oats. The fungus produces runner hyphae, which grow superficially on roots and produce short, darker, haustoriumlike feeder hyphae called **hyphopodia.** The hyphopodia grow toward the host, are flattened, and have a pore through which a hyphal peg grows into the root. The fungus produces only one kind of spores in nature, ascospores in asci in perithecia. In culture, however, it also produces **phialospores** (conidia) from **phialids** (bottle-shaped terminal hyphal cells). The perithecia are black and embedded in basal leaf sheaths, with protruding black necks.

The fungus overwinters in infected wheat and grass plant roots and stems and in host debris. The ascospores are discharged forcibly from the asci in wet weather but rarely seem to cause infection. By far the most infections are caused by mycelium coming in contact with roots of growing plants. The superficial mycelium produces feeder hyphae that penetrate root tissues directly through pegs. The fungus invades the cortex and the vascular system but does not grow systemically through the latter. Invaded roots are killed. In young plants, the fungus extends into the crown and base of the stem, while in more mature plants its spread is slower and usually remains confined to the roots. The fungus can infect plants throughout the growing season but is more active at temperatures between 12 and 18°C. Take-all is most severe in infertile, compacted, alkaline, and poorly drained soils. Its severity increases for several (3–6) years in fields cultivated continuously with wheat, but then it declines (take-all decline) and stabilizes at a lower level.

Control of take-all depends primarily on cultural practices, particularly crop rotation with nonhost plants. Other control measures include destruction of grassy weeds and volunteer wheat plants that can harbor the fungus, application of adequate potassium, phosphorus, and ammonium- but not nitrate-type nitrogen fertilizer, and the use of tolerant varieties since no highly resistant ones are available. Recently, yield losses due to take-all were reduced by 60–75 percent through seed treatment with the systemic fungicide triadimenol.

In recent years, a great deal of research has been carried out to discover and to develop biological controls of take-all. It was observed that some soils were suppressive to take-all, while others were conducive to the disease and that suppressiveness could be transferred from field to field and could be eliminated by high (60°C) temperatures and by fumigation. It was later shown that take-all decline (soil suppressiveness) was brought about by root-colonizing fluorescent pseudomonad bacteria that are antagonistic to *Gaeumannomyces* and inhibit its growth on the root surface or in the lesion formed after infection. It appears that the antagonistic bacteria are favored by wheat root exudates and multiply 5–10 times faster than other bacteria. Through their increased populations, and increased inhibitory action on the pathogen by means of their antibiotics, siderophores, and so on, the antagonistic bacteria bring about the decline of take-all. Many bacterial strains have now been found that effectively inhibit the fungus in laboratory tests. When the same bacteria are applied on the seed, however, and the seed is planted in *Gaeumannomyces*-infested soil in the greenhouse and in field plots, control is only partial and frequently fails completely. Therefore, biological control of take-all in the field is not yet possible, but it is likely that, through research, the obstacles will soon be overcome.

Sclerotinia Diseases

Fungi of the genus *Sclerotinia,* especially *S. sclerotiorum* and *S. minor,* cause destructive diseases of numerous succulent plants, particularly vegetables and flowers and some shrubs. *S. minor* attacks primarily peanuts and lettuce, while another species, *S. trifoliorum,* attacks forage legumes, and *S. homeocarpa* causes a destructive disease of turf grasses. Sclerotinia diseases are worldwide in distribution and affect plants in all stages of growth, including seedlings, mature plants, and harvested products in transit and storage.

• Sclerotinia Diseases of Vegetables and Flowers

The symptoms caused by *Sclerotinia* vary somewhat with the host or host part affected and with the environmental conditions. The *Sclerotinia* diseases are known under a variety of names. The most common of these, along with some of the host plants most seriously affected, are the following: cottony rot, white mold, or watery soft rot of bean, cabbage, carrot, eggplant, citrus, peanut, potato, stock, tobacco; stem rot and timber rot of cucumber, squash, bean, artichoke, asparagus, chrysanthemum, dahlia, delphinium, peony, potato, tomato, soybean, sweet potato; drop of lettuce, broad bean, beet, cabbage; damping off of celery, lettuce; crown rot or wilt of columbine, snapdragon; blossom blight of narcissus, camellia; pink joint of red pepper, stem canker of hollyhock, and root and crown rot of clover.

The most obvious and typical early symptom of Sclerotinia diseases is the appearance on the infected plant of a white fluffy mycelial growth in which soon afterwards develop large, compact resting bodies or sclerotia (Figure 11-78, A–D). The sclerotia are white at first but later become black and hard on the outside and may vary in size from 0.5–1 mm in *S. minor* to 2–10 mm in diameter in *S. sclerotiorum,* although they are usually more flattened and elongated than spherical.

Stems of infected succulent, herbaceous plants at first develop pale or dark brown lesions at their base. The lesions are often quickly covered by white cottony patches of fungal mycelium. In the early stages of lesion development in the stem, the foliage may show little sign of attack, and infected plants are easily overlooked until the fungus grows completely through the stem and the stem rots. Then the foliage above the lesion wilts and dies more or less quickly. In some cases the infection may begin on a leaf and then move into the stem through the leaf. The sclerotia of the fungus may be formed either internally in the pith of the stem, giving no outward signs of their presence there, or they may be formed on the outside of the stem where they are quite apparent.

Leaves and petioles of plants such as lettuce, celery, and beets suddenly collapse and die as the fungus infects the base of the stem and the lower leaves. Rapidly the fungus invades and spreads through the stem, and the entire plant dies and collapses, each leaf dropping downward until it rests on the one below. Mycelium and sclerotia usually appear on the lower surface of the outer leaves, but under moist conditions the fungus invades the plant completely and causes it to rot, producing a white, fluffy, mycelial growth over the entire plant. If dry weather follows infection, the fungus forms cankers in the stem that kill the plant without a soft rot. Attack of celery produces a characteristic pink or reddish-brown, water-soaked area at the base of the affected petioles that is often covered by the white mycelium, and the rot may spread through the stalks causing the collapse of the whole plant.

Fleshy storage organs, such as carrots, infected by *Sclerotinia* develop a white, cottony growth on their surface whether they are still in the field or in

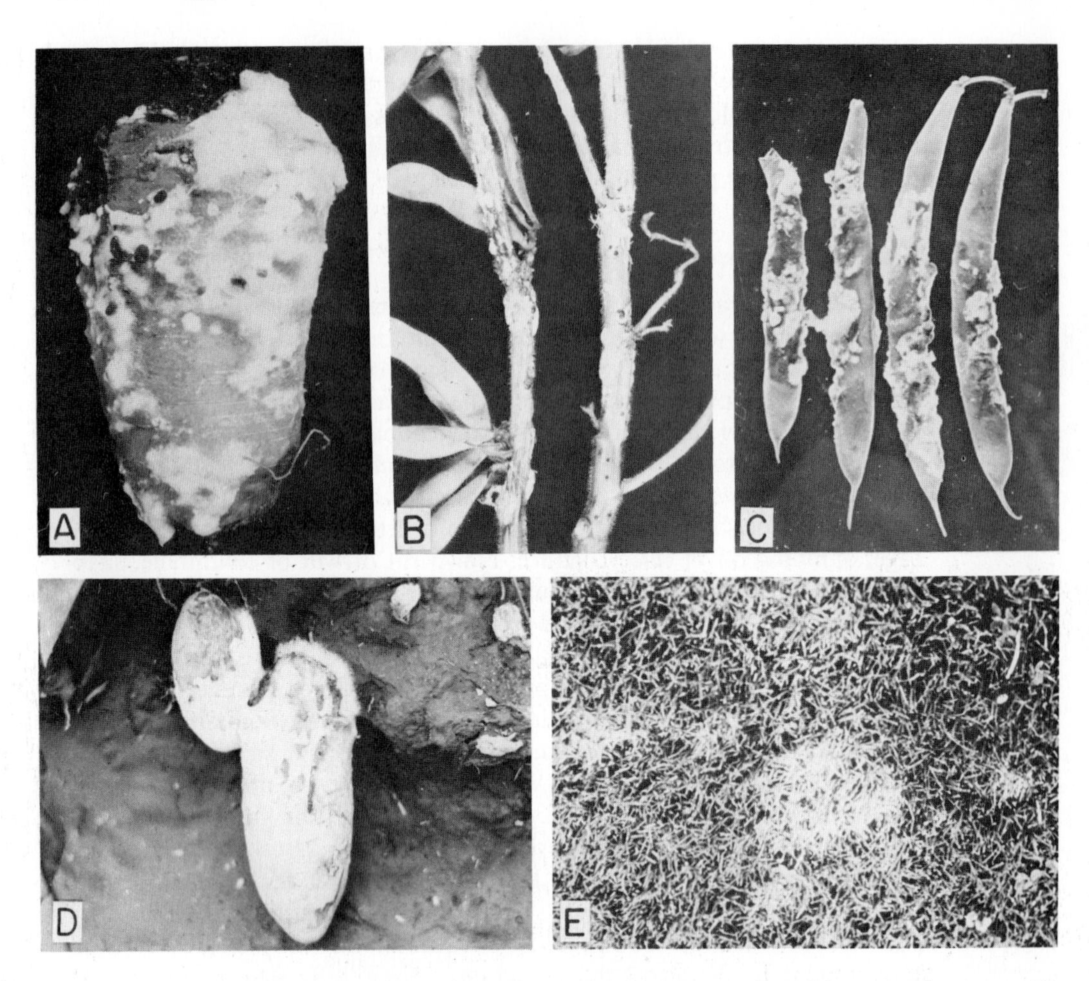

FIGURE 11-78 (A–D) *Sclerotinia* diseases of vegetables and field crops caused by *Sclerotinia sclerotiorum.* (A) Cottony rot of carrot. (B) Stem rot of soybean. (C) Watery soft rot of bean. (D) Watery soft rot of cucumber. The black bodies in photos A–C are sclerotia of the fungus. (E) Dollar spot disease of turfgrasses caused by *S. homeocarpa.* (Photos A, C, E courtesy U.S.D.A.)

storage. Black sclerotia are formed externally (Figure 11-78A) and invaded tissues appear darker than healthy ones and become soft and watery. If the disease develops after harvest in the storage house, the rot spreads to adjacent roots or whatever the storage organs are and produces pockets of rotted organs, or all the organs in the crate may become infected and collapse producing a watery soft rot, covered by fungus growth.

Fleshy fruits, such as cucumber, squash, and eggplant, and seed pods of bean, are also attacked by *Sclerotinia* through their closest point to the ground, at the point of their contact with the ground, or through their senescent flower parts. The fungus causes a wet rot that spreads from the tip of the fruit or pod to the rest of the organ, which eventually becomes completely rotted and disintegrated (Figure 11-78D). The white fungal mycelium and the black sclerotia can usually be seen both externally and within the affected pods and fruits.

Flower infection is important primarily in camellias and narcissus. Few to many small, watery, light-brown spots appear on the petals. The spots may enlarge, coalesce and involve the entire petal, and eventually the entire flower becomes dark brown and drops, but disintegration of the flowers occurs only after they have fallen and in wet weather, when the fungus produces abundant mycelium and sclerotia.

The fungus *Sclerotinia sclerotiorum* and some related species overwinter as sclerotia on or within infected tissues or as sclerotia that have fallen on the ground and as mycelium in dead or living plants (Figure 11-79). In the spring or early summer, the sclerotia germinate and produce one to many slender stalks terminating at a small, 5–15 mm in diameter, disk- or cup-shaped apothecium in which asci and ascospores are produced. Large numbers of ascospores are discharged from the apothecia into the air over a period of 2–3 weeks. The ascospores are blown away, and if they land on senescent plant parts, such as old blossoms, which provide a readily available source of food, the ascospores germinate and cause infection. In some *Sclerotinia* the sclerotia cause infection by producing mycelial strands that attack and infect young plant stems directly. Under moist conditions the latter method of infection is probably more common than the one by ascospores, although in *S. sclerotiorum* almost all infections are initiated by ascospores.

Control of *Sclerotinia* diseases depends on a number of cultural practices and on chemical sprays. In most affected crops, few varieties show

FIGURE 11-79 Development and symptoms of diseases of vegetables and flowers caused by *Sclerotinia sclerotiorum*.

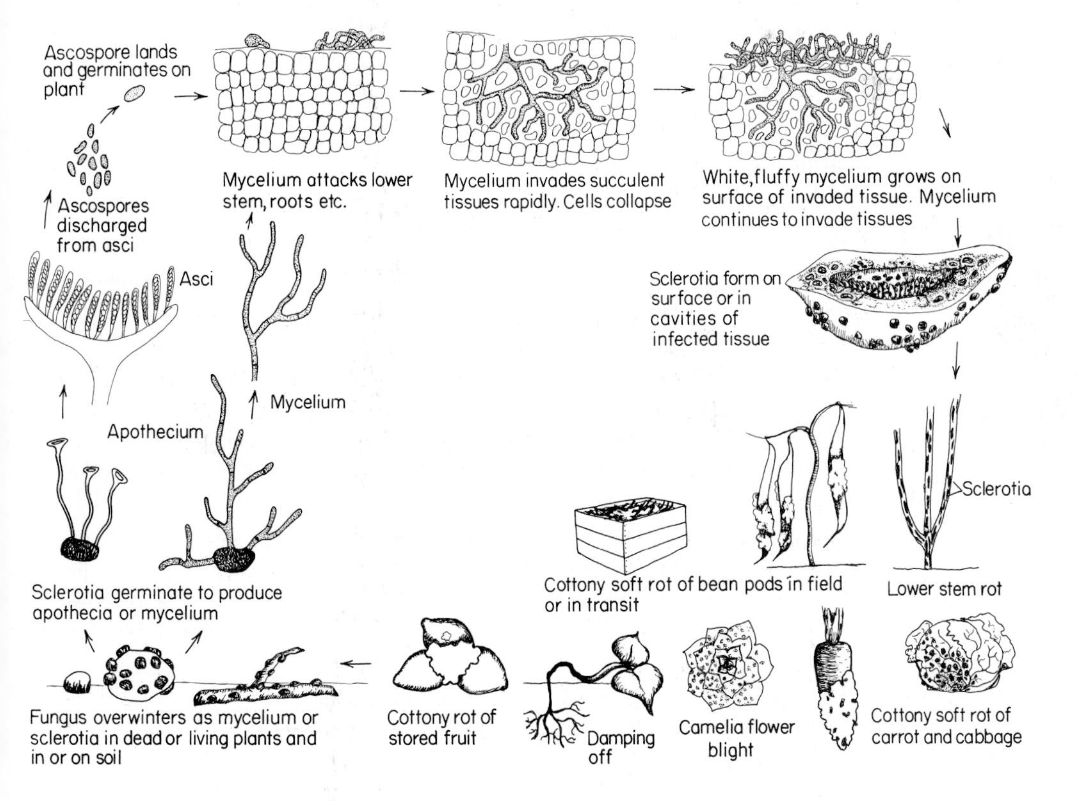

appreciable degrees of resistance to the pathogen. In the greenhouse, soil sterilization with steam eliminates the pathogen. In the field, chemical soil sterilants have been rather ineffective in controlling the disease. Since the disease is favored by high soil moisture and high air humidity and affects many cultivated and wild plants, susceptible crops should be planted only in well-drained soils, the plants should not be spaced too close together for air drainage, and the soil should be kept free of weeds between crops. If the disease has become severe on susceptible crops, infected plants should be pulled and burned to either prevent the fungus from forming sclerotia or to remove from the field as many of the sclerotia as possible. Since sclerotia remain viable in the soil for at least 3 years and since they do not all germinate or die out at the same time, infected fields should be planted to nonsusceptible crops such as corn and small grains for at least 3 years before susceptible crops are planted again. In several crops, good control of the *Sclerotinia* disease has been obtained by spraying the soil with metham sodium or spraying the plants with benomyl, dichloran, or thiophanate-methyl before and during their stage of susceptibility to the pathogen. Two newer contact fungicides, iprodione and vinclozolin, give excellent control of *Sclerotinia* and are now being tested on a wide variety of crops.

In the last several years, more than 30 species of fungi, bacteria, insects, and other organisms have been reported to parasitize or to interfere with the growth of *Sclerotinia* sp. Encouraging results of biological control of Sclerotinia diseases in some crops have been obtained by incorporating in *Sclerotinia*-infested soil the mycoparasitic fungi *Coniothyrium minitans, Gliocladium roseum, G. virens, Sporodesmium sclerotivorum,* and *Trichoderma viride.* The mycoparasites destroy existing sclerotia or inhibit the formation of new sclerotia by the fungus and, thereby, markedly reduce the fungus population in the soil. So far, however, no practical control recommendations have been developed.

Phymatotrichum Root Rot

Phymatotrichum root rot, usually called Texas root rot or cotton root rot, occurs only in the southwestern United States and Mexico. This disease probably affects more kinds of cultivated and wild dicotyledonous plants than any other. Its hosts include many fruit, forest, and shade trees, most vegetables and flowers, field crops such as cotton and alfalfa, ornamental shrubs, and many weeds. It causes its greatest losses in cotton in the area from Texas to Arizona and Mexico.

In cotton, for example, infected plants appear in patches in the field and at first show a yellowing and bronzing of the leaves. The leaves then show a slight wilting and turn brown and dry but remain attached to the plant. Below the soil line, and in some plants up to a foot or more above the soil line, the bark and cambium turn brown, resulting in a firm brown rot of the root and the lower stem. The surface of rotted roots is usually partly covered by coarse, brown, parallel strands of mycelium, and this characteristic helps diagnose the disease.

The fungus *Phymatotrichum omnivorum* produces mostly yellowish, thin-celled mycelium but also mycelium made of larger cells. The hyphae grow closely compressed together or interwoven into thick mycelial strands that have characteristic slender, tapering, crosslike side branches. Older strands are dark brown and have few side branches. *Phymatotrichum* produces short, stout, simple, or branched conidiophores that have swollen tips and bear loose heads of dry, colorless, one-celled conidia that apparently do not germinate or cause infection. The fungus also produces numerous small, brown to black sclerotia, singly or in chains, which germinate to produce mycelium. Most of the fungus mycelium and sclerotia are found in soil depths between 30 and 75 cm. Closer to the soil surface the fungus is adversely affected by the normal soil microflora.

The mycelial strands and especially the sclerotia of the fungus can survive in the soil for 5 or more years. The fungus survives best and causes considerably more damage to plants growing in alkaline, black, heavy clay soils that are poorly aerated. The fungus requires high temperature and adequate soil moisture for greatest activity, provided the soil pH is near or above neutral.

The fungus enters the plant below the soil line and then grows downward throughout the root and, on some plants, it also invades the lower stem. The fungus spreads from plant to plant through the growth of the mycelial strands and through the spread of such strands or sclerotia by farm equipment, transplants, and so on. Once introduced into an area, the fungus can survive on cultivated plants and weeds indefinitely, provided the soil and temperature conditions are favorable. The pathogen cannot stand temperatures below freezing for any appreciable time, and its narrow geographic distribution seems to be the result of its high temperature and alkalinity requirements.

Control of *Phymatotrichum* root rot depends on long rotations with grain crops, weed eradication, deep and frequent plowing to keep the soil well aerated, and the use of green manure crops, such as thickly planted corn, sorghum, or legumes, which upon decay favor the buildup of large populations of microorganisms that are antagonistic to *Phymatotrichum*. Soil fumigation with vapam, for example, is effective if applied annually and the value of the crop justifies the expense, but it has not generally proven practical because of the rapid spread of the pathogen from deeper in the soil to the root zone once the fumigant has evaporated.

SELECTED REFERENCES

Abawi, G. S., and Grogan, R. G. (1975). Sources of primary inoculum and effects of temperature and moisture on infection of beans by *Whetzelinia sclerotiorum*. *Phytopathology* **65,** 300–309.

Adams, P. B., and Tate, C. J. (1975). Factors affecting lettuce drop caused by *Sclerotinia sclerotiorum*. *Plant Dis. Rep.* **59,** 140–143.

Anderson, A. L. (1948). The development of *Gibberella zeae* head blight of wheat. *Phytopathology* **38,** 595–611.

Anonymous (1979). Symposium on *Sclerotinia (= Whetzelinia):* Taxonomy, biology and pathology. (Several papers.) *Phytopathology* **69,** 873–910.

Atanasoff, D. (1920). *Fusarium*-blight scab of wheat and other cereal. *J. Agric. Res. (Washington, D.C.)* **20,** 1–32.

Baker, K. F., and Cook, R. J. (1974). "Biological Control of Plant Pathogens." Freeman, San Francisco, California.

Bockus, W. W. (1983). Effects of fall infection by *Gaeumannomyces graminis* var. *tritici* and triadimenol seed treatment on severity of take-all in winter wheat. *Phytopathology* **73,** 540–543.

Booth, C. (1975). The present status of *Fusarium* taxonomy. *Annu. Rev. Phytopathology.* **13,** 83–93.

Bruehl, G. W., ed. (1975). "Biology and Control of Soil-Borne Plant Pathogens." Am. Phytopathol. Soc., St. Paul, Minnesota.

Burke, D. W., and Miller, D. E. (1983). Control of *Fusarium* root rot with resistant beans and cultural management. *Plant Dis.* **67,** 1312–1317.
Christensen, J. J., and Wilcoxson, R. D. (1966). "Stalk Rot of Corn," Monogr. No. 3. Am. Phytopathol. Soc., St. Paul. Minnesota.
Cook, R. J. (1980). Fusarium foot rot of wheat and its control in the Pacific Northwest. *Plant Dis.* **64,** 1061–1066.
Cook, R. J., and Baker, K. F. (1983). "The Nature and Practice of Biological Control of Plant Pathogens." Am. Phytopathol. Soc., St. Paul, Minnesota.
Dickson, J. G. (1956). "Diseases of Field Crops," 2nd ed., McGraw-Hill, New York.
Dodd, J. L. (1980). The role of plant stresses in development of corn stalk rots. *Plant Dis.* **64,** 533–537.
Garrett, S. D. (1970). "Pathogenic Root-Infecting Fungi." Cambridge Univ. Press, London and New York.
Koehler, B. (1959). Corn ear rot in Illinois. *Ill., Agric. Exp. Stn., Bull.* **639.**
Koehler, B. (1960). Corn stalk rots in Illinois. *Ill., Agric. Exp. Stn., Bull.* **658.**
Kucharek, T. A., and Kommedahl, T. (1966). Kernel infection and corn stalk rot caused by *Fusarium moniliforme. Phytopathology* **56,** 983–984.
Lumsden, R. D., and Dow, R. L. (1973). Histopathology of *Sclerotinia sclerotiorum* infections of bean. *Phytopathology* **63,** 708–715.
Lyda, S. D. (1978). Ecology of *Phymatotrichum omnivorum. Annu. Rev. Phytopathol.* **16,** 193–209.
Nelson, P. E., Toussoun, T. A., and Cook, R. J., eds. (1981). "Fusarium: Diseases, Biology and Taxonomy." Pennsylvania State Univ. Press, University Park.
Papavizas, G. C., ed. (1974). The relation of soil microorganisms to soil borne plant pathogens. *Va. Polytech. Inst. State Univ., South. Coop. Ser., Bull.* **183.**
Papendick, R. I., and Cook, R. J. (1974). Plant water stress and development of Fusarium foot rot in wheat subjected to different cultural practices. *Phytopathology* **64,** 358–363.
Percich, J. A., and Lockwood, J. L. (1975). Influence of Atrazine on the severity of Fusarium root rot in pea and corn. *Phytopathology* **65,** 154–159.
Rowe, R. C., and Farley, J. D. (1981). Strategies for controlling Fusarium crown rot and root rot in greenhouse tomatoes. *Plant Dis.* **65,** 107–112.
Rush, C. M., Gerik, T. J., and Lyda, S. D. (1984). Factors affecting symptom appearance and development of Phymatotrichum root rot of cotton. *Phytopathology* **74,** 1466–1469.
Sherf, A. F., and MacNab, A. A. (1986). "Vegetable Diseases and Their Control." Wiley-Interscience, New York.
Willets, H. J., and Wong, J. A. L. (1980). The biology of *Sclerotinia sclerotiorum. S. trifoliorum* and *S. minor* with emphasis on specific nomenclature. *Bot. Rev.* **46,** 101–165.
Williams, G. H., and Western, J. H. (1965). The biology of *Sclerotinia trifoliorum* and other species of sclerotium-forming fungi. I. Apothecium formation from sclerotia. II. The survival of sclerotia in soil. *Ann. Appl. Biol.* **56,** 253–268.

Postharvest Diseases of Plant Products Caused by Ascomycetes and Imperfect Fungi

Postharvest diseases of plant produce or plant products are those that develop during harvesting, grading, packing, and transportation of the crop to market; during its storage at shipping points or at the market; and during the various handling operations required to move the crop from the grower to the wholesale dealer, to the retail store, and finally, to the consumer. Postharvest diseases actually continue to develop while the produce is in the possession of the consumer but is stored at room temperature or under refrigeration until the moment of actual consumption or use (Figures 11-80, 11-81, 11-82). During any of these operations the product may show symptoms of diseases that had begun in the field but remained latent till later; the product may be subjected to environmental conditions or treatments that are themselves harmful to the product and impair its appearance and food value; or the

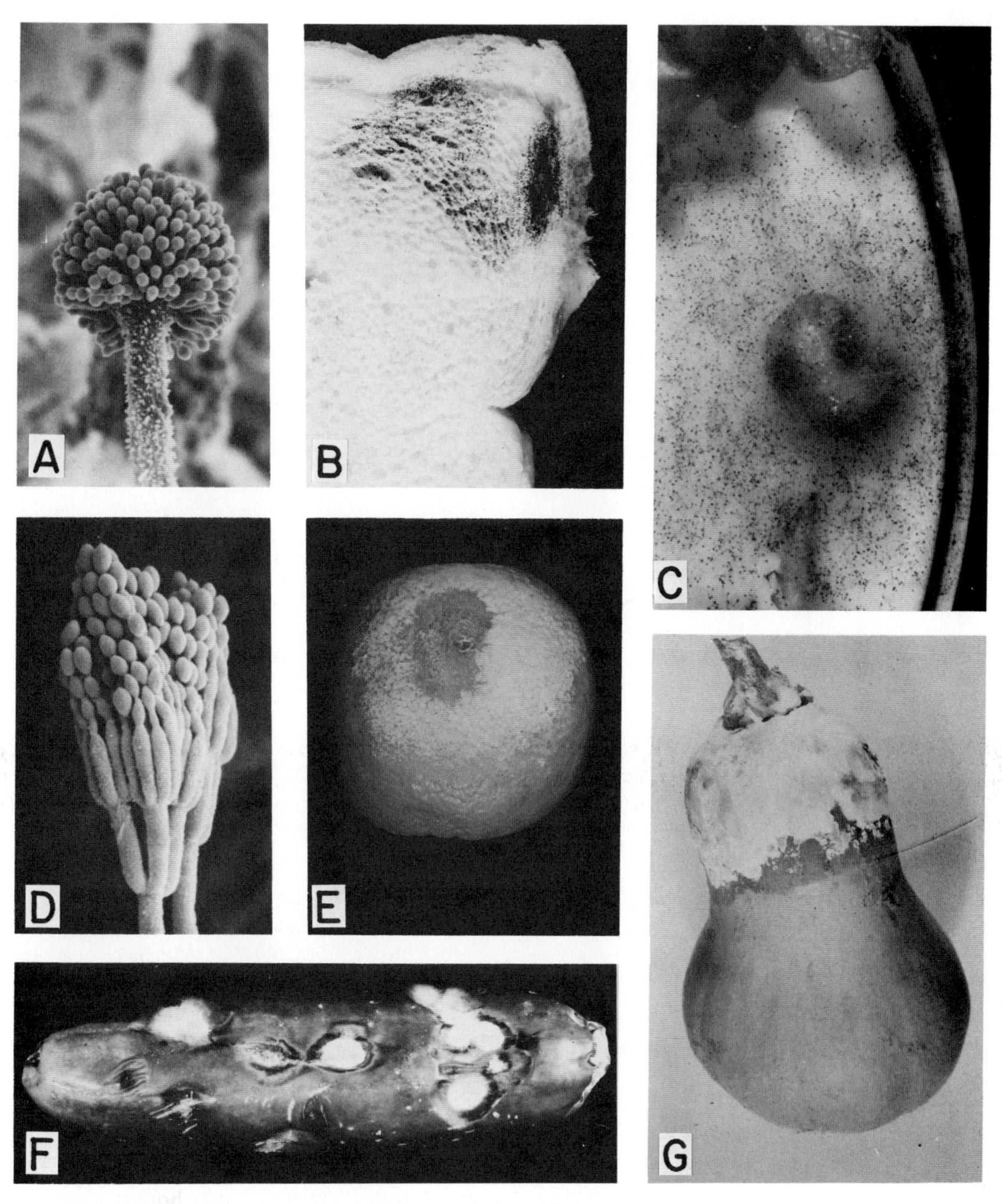

FIGURE 11-80 Some fungi causing postharvest diseases, and a few common such diseases. (A) Scanning electron micrograph of conidiophore and conidia of *Aspergillus flavus.* (B) Bread mold caused by *Aspergillus.* (C) *Rhizopus* mycelium and sporangia growing from naturally contaminated seed. (D) Scanning electron micrograph of *Penicillium.* (E) Blue mold of orange caused by *Penicillium italicum.* (F) Cucumber rot during cold storage caused by multiple infections by *Alternaria.* (G) Squash rot in the field or storage caused by *Fusarium.* (Photos A and D courtesy M. F. Brown and H. G. Brotzman.)

product may be subjected to conditions that favor its attack by decay-producing microorganisms, which usually cause a portion of it to rot and, in some cases, by secreting toxic substances, making the remainder of the product unfit for consumption or lower its nutritional and sale value.

FIGURE 11-81 *Rhizopus* soft rot of peaches (A) and *Sclerotinia* cottony soft rot of beans (B) developing during harvest and storage. (Photos courtesy U.S.D.A.)

All types of plant produce and products are susceptible to postharvest diseases. Generally, the more tender or succulent the exterior of the product, and the greater the water content of the entire product, the more susceptible it is to injury and to infection by fungi and bacteria. Thus, succulent, fleshy fruits and vegetables such as strawberries, peaches, tomatoes, cucurbits, citrus, green vegetables, bananas, onions, and potatoes, as well as cut flowers, bulbs, and corms, are affected by postharvest diseases to a greater or lesser extent. The extent of damage or loss depends on the particular product, on the disease organism or organisms involved, and on the storage conditions. Although rotting of fresh fruits and vegetables is much more common and is encountered by everyone at the retail store or at home, rotting of cereal grains

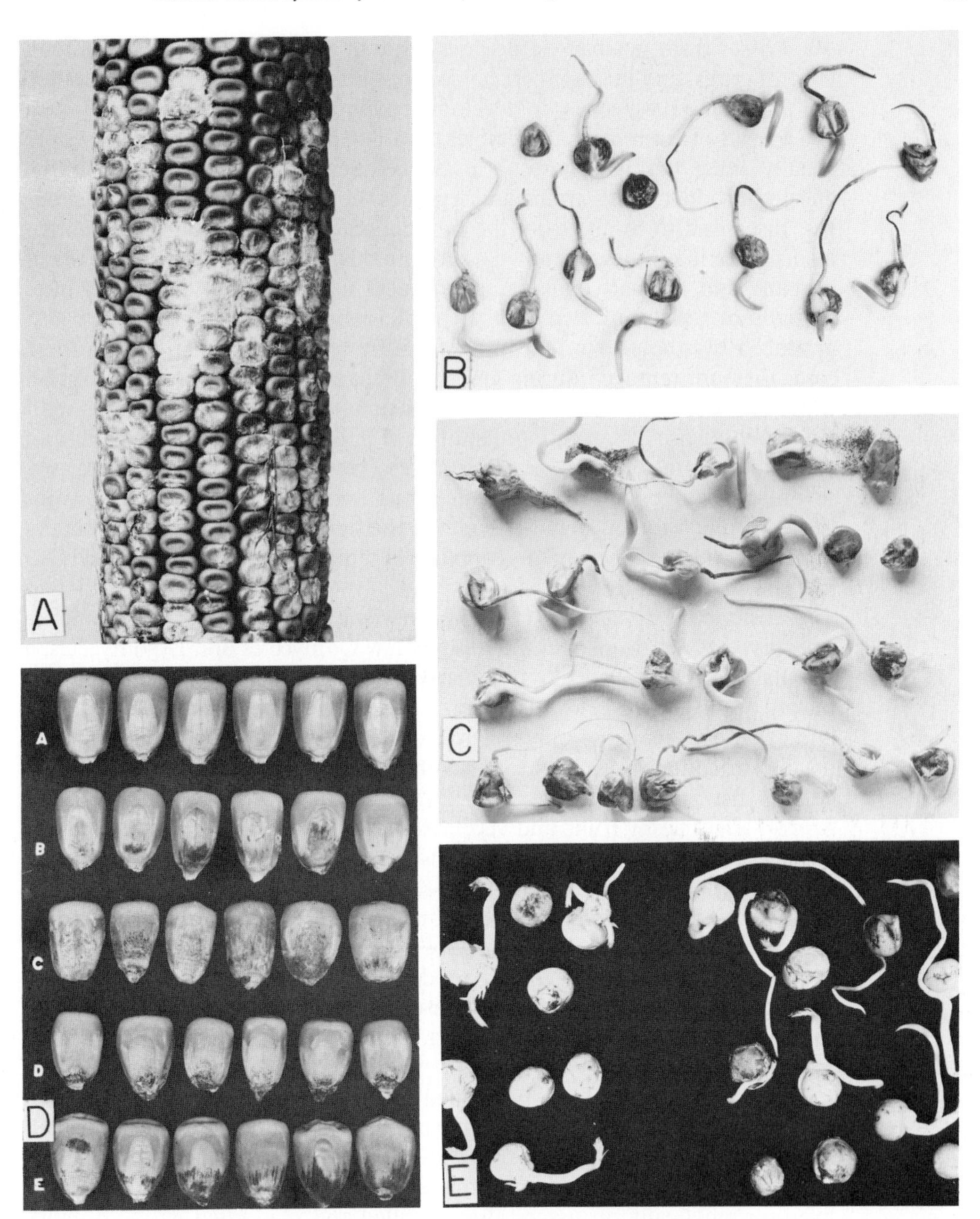

FIGURE 11-82 Some seed infections by fungi. (A) Close-up of corn ear infected with *Fusarium moniliforme.* (B) Sweet corn seedlings infected with *Fusarium* carried in corn kernels. (C) Sweet corn seedling infections after 5 days in the laboratory germinator. Top to bottom: *Rhizopus, Diplodia, Penicillium, Fusarium.* (D) Healthy corn kernels (A) and kernels infected with *Diplodia* (B), *Physalospora* (C), *Nigrospora* (D), and *Cladosporium* (E). (E) Pea seeds discolored and infected by different fungi. (Photos A and D courtesy Illinois Agr. Expt. Sta.)

and of legumes is also quite common and the losses caused by it quite large. Losses of grains and legumes, however, in spite of their magnitude, occur primarily at the large bins or warehouses of the growers, wholesalers, or manufacturers and are seldom observed by the general public. In addition, postharvest decays of hay, silage, and other feedstuffs are quite large.

Losses from postharvest diseases amount to 10–30 percent of the total yield of crops, and in some perishable crops losses greater than 30 percent are not uncommon, especially in developing countries.

Losses of fresh fruits, vegetables, and flowers to postharvest diseases are usually direct, that is, they result in reduced quality, quantity, or both, of the affected product. With grains and legumes, however, damage from postharvest diseases also results from the production by some infecting microorganisms of toxic substances known as **mycotoxins.** Mycotoxins are poisonous to humans and animals consuming products made from grains or legumes partially or totally infected with such microorganisms. Mycotoxins are also produced by some fungi that infect fresh fruits and vegetables, but in these cases they are removed during grading, preparation, or before consumption when the rotten fruits or vegetables or their rotten parts are discarded. With the increased use by large manufacturers of bulk quantities of fresh fruits and vegetables to make fruit or vegetable juices, purees, cole slaw, baby foods, and so on, quality control of individual fruits and vegetables becomes all but economically impractical, and therefore, the significance of postharvest infections and the presence of mycotoxins in bulk-prepared foods is likely to increase in the future.

Postharvest diseases are caused primarily by a relatively small number of Ascomycetes and Imperfect Fungi, by a few Oomycetes and Zygomycetes, a few Basidiomycetes, and by a few species of bacteria. The bacteria are primarily of the genera *Erwinia* and *Pseudomonas.* Of the Oomycetes, *Pythium* and *Phytophthora* cause only soft rots of fleshy fruits and vegetables that are usually in contact with or very near the soil, and they may spread to new, healthy fruit during storage. Two Zygomycetes, *Rhizopus* and sometimes *Mucor,* affect fleshy fruits and vegetables after harvest and also stored grains and legumes, as well as prepared foods such as bread, when moisture conditions are favorable (Figures 11-80, 11-81, and 11-82). Of the Basidiomycetes, *Rhizoctonia* and *Sclerotium* cause rotting of fleshy fruits and vegetables, while several fungi, for example, *Polyporus, Poria,* and *Heterobasidion (Fomes),* cause deterioration of wood and wood products. The Ascomycetes and Imperfect Fungi that cause postharvest diseases are by far the most common and most important causes of postharvest decay, and they will be discussed in some detail below.

The fungi and bacteria mentioned above as causing postharvest diseases are usually primary parasites, that is, they attack healthy, living tissue, which they disintegrate and cause to rot. They are often, however, followed into the tissue by other fungi and bacteria that act as secondary parasites, that is, they live saprophytically on tissues already killed and macerated by the primary parasites. Also, it is not uncommon for more than one of the primary parasites to attack the same tissue concurrently or in sequence. Thus, often some of the primary parasites also act as secondary ones.

Many of the postharvest diseases of fruits, vegetables, grains, and legumes are the results of incipient infections of the plants or their fruits by pathogens in the field while the plants and fruits are still developing, or after the fruits or seeds have matured in the field but before they are harvested. Symptoms from such "field infections" may be too inconspicuous to be noticed at harvest. In fleshy fruits and vegetables, field infections continue to develop after harvest, while in grains and legumes the development of field infections ceases soon after harvest. In fleshy fruits and vegetables, new infections may be caused in storage by the same or other pathogens, while in grains and legumes storage infections are usually caused by pathogens other than those causing field infections.

As with all fungal and bacterial plant diseases, postharvest diseases are greatly favored by, indeed they depend on, the presence of high moisture and high temperatures. Fleshy fruits and vegetables contain plentiful amounts of water, and since they are generally kept at high relative humidities to avoid shrinkage, they make excellent substrates for attack by pathogenic microorganisms, provided the latter can penetrate the outer protective coating of the fruit or vegetable. Wounds, cuts, and bruises, which are common in fleshy tissues, provide the most common and effective courts for penetration. But penetration through natural openings, such as lenticels, also occurs, and direct penetration through the cuticle and epidermis, especially of fruits and vegetables in contact with infected ones, is quite common. Once a fresh fruit or vegetable becomes infected, futher development of the infection and spread to adjacent fruits or vegetables depends mainly on the storage temperature. Generally, the higher the temperature the faster is the development and spread of the disease, while at lower temperatures pathogens and the diseases they cause develop more slowly or cease to develop at all.

On the other hand, grains and legumes can be, and ordinarily are, kept for long periods of time because their moisture content is low or can be reduced to as low as 12–14 percent. At such low moisture content, almost none of the fungi that cause field infections can continue to grow and to cause new infections immediately, or even later when the grains become remoistened. Other fungi, however, can infect grains and legumes whose moisture content is about or slightly lower than 14–15 percent, and the severity and spread of infection increase drastically with the slightest increase in moisture above that range. High temperatures favor the infection of grains with high moisture content, just as they do of fruits and vegetables. Frequently, however, the infection itself results in a drastic rise of the temperature of the moistened infected grain due to the heat produced as a result of respiration of the actively growing fungi and bacteria that cause the infection.

• Postharvest Decays of Fruits and Vegetables

Some of the most common Ascomycetes or Imperfect Fungi and the main postharvest diseases they cause are listed below.

◦ **Alternaria.** *Alternaria's* various species probably cause decay on most, if not all, fresh fruits and vegetables either before or after harvest. The symptoms may appear as brown or black, flat or sunken spots with definite margins, or they may appear as diffuse, large, decayed areas that are shallow or extend deep into the flesh of the fruit or vegetable. The fungus develops well at a wide range of temperatures, even in the refrigerator, although its development is slower at low temperatures (Figure 11-80F). The fungus may spread into and rot tissues internally with little or no mycelium appearing on the surface, but usually a mat of mycelium that is white at first but later turns brown to black forms on the surface of the rotted area. Some of the most serious diseases caused by *Alternaria* after harvest are *Alternaria* rot of lemons and black rot of oranges, *Alternaria* rot of tomatoes, peppers, eggplant, apples, cucumber, squash, and melons, cabbage, cherries, grapes, strawberries, tuber rot of potatoes, rot of sweet potatoes, and purple blotch of onion.

- **Botrytis.** *Botrytis* causes the "gray molds" or "gray mold rots" of fruits and vegetables, both in the field and in storage. There is practically no fresh fruit, vegetable, or bulb that is not attacked by *Botrytis* in storage. Some of them, such as strawberry, lettuce, onion, grape, and apple, are also attacked in the field near maturity or while green. The decay may start at the blossom or stem end of the fruit or at any wound, crack, or cut of storage tissues. The decay appears as a well-defined water-soaked, then brownish, area that penetrates deeply and advances rapidly into the tissue. In most hosts and under humid conditions a grayish or brownish-gray, granular, velvety mold layer develops on the surface of decaying areas. Gray molds are most severe in cool, humid environments and continue to develop, although slowly, even at 0°C. Heavy losses are caused in storage annually by the gray mold fungus on many fleshy fruits and vegetables, particularly pears, apples, strawberries, citrus, tomatoes, and onions.
- **Fusarium.** *Fusarium* causes postharvest "pink or yellow molds" on vegetables and ornamentals and especially on root crops, tubers, and bulbs, but low-lying crops such as cucurbits and tomatoes are also frequently affected. A brown rot of oranges and lemons held in storage for long periods is also caused by *Fusarium.* With most vegetables, contamination with *Fusarium* takes place in the field before or during harvest, but infection may develop in the field or in storage. Losses are particularly heavy with crops, such as potatoes, that are stored for long periods of time. Affected tissues appear fairly moist and light brown at first, but later they become darker brown and somewhat dry. As the decaying areas enlarge, they often become sunken, the skin is wrinkled, and small tufts of whitish, pink, or yellow mold appear. Similar mycelial tufts also develop in hollow places formed in decaying tissues. The infection of softer tissues such as tomatoes and cucurbits develops faster and is characterized by pink mycelium and pink, rotten tissues. (Figure 11-80G).
- **Geotrichum.** *Geotrichum* causes the "sour rots" of citrus fruits, tomatoes, carrots, and other fruits and vegetables. Sour rot is one of the messiest and most unpleasant rots of susceptible fruits and vegetables. Although it may affect tomatoes at the mature green stage, it is the ripe or overripe fruits and vegetables, especially when kept in moisture-holding plastic bags or packages, that are particularly susceptible to sour rot. The fungus is widely distributed in soils and decaying fruits and vegetables and contaminates fruits and vegetables before or during harvest. The fungus penetrates fruits, usually after harvest, at stem scars, skin cracks, cuts, and punctures of various sorts. Infected areas appear water soaked and soft and are easily punctured. The decay spreads rapidly, at first mainly inside the fruit, and eventually involves the whole fruit. Later, the skin frequently cracks over the affected area and is usually filled with a white, cheesy, or scumlike development of the fungus. Also, a thin, water-soaked layer of compact cream-colored fungal growth develops on the surface, while the whole inside becomes a sour-smelling, decayed, watery mass. Fruit flies, which are attracted to tissues affected with sour rot, further spread the pathogen. The fungus prefers high temperatures (24–30°C) and humidity but is active at temperatures as low as 2°C.

◦ **Penicillium.** *Penicillium's* various species cause the "blue mold rots" and the "green mold rots," which are also known as *Penicillium* rots. They are the most common and usually the most destructive of all postharvest diseases, affecting all kinds of citrus fruits, apples, pears, and quinces, grapes, onions, melons, figs, sweet potatoes, and many other fruits and vegetables (Figure 11-80, D, E). On some fruits, such as citrus, some infection may take place in the field, but blue molds or green molds are essentially postharvest diseases and often account for up to 90 percent of decay in transit, in storage, and in the market. *Penicillium* (Figure 11-80E) enters tissues through breaks in the skin or rind and even through lenticels. However, it can spread from infected fruit to healthy ones in contact with the infected through the uninjured skin. *Penicillium* rots at first appear as soft, watery, slightly discolored spots of varying size and on any part of the fruit. The spots are rather shallow at first but quickly become deeper, and at room temperature most of the fruit or the whole fruit decays in just a few days. Soon after decay develops, a white mold begins to grow on the surface of the skin or rind, near the center of the spot. Later, the fungus growth starts producing spores. The sporulating area has a blue, bluish-green, or olive green color and is usually surrounded by a narrow or wide band of white mycelium with a band of water-soaked tissue ahead of the mycelium. The surface growth of the fungus develops on spots of any size as long as the air is moist and warm. In cool, dry air, surface mold is rare, even when the fruits are totally decayed. Under storage conditions, small, spore-bearing tufts appear on the surface of the spots. Decaying fruit has a musty odor and under dry conditions may shrink and become mummified, while under moist conditions, when secondary fungi and yeasts also enter the fruit, it is reduced to a wet, soft mass.

Although most of the damage from blue mold and green mold rots shows up in storage and market, the occurrence of these molds is greater when the fruit is picked and handled during wet, humid weather than in cool and dry weather; when fruits are delayed in going into storage; cooled slowly in storage; stored until late in the season; or held at warm temperatures after removal from storage. The most important factor, however, that favors these rots, especially early in the storage season, is mechanical injuries to the fruit surface. Although blue mold and green mold are favored by relatively high storage temperatures, they continue to be slowly active even at temperatures near freezing. Some *Penicillium* species produce ethylene, which diffuses into the container or storage room and increases fruit respiration, affects its coloring, and accelerates its maturity and senescence, thus reducing the storage life of the healthy fruit as well.

In addition to the losses caused by the rotting of fruits and vegetables by *Penicillium,* the fungus also produces several mycotoxins, such as patulin, in the affected products, which contaminate juices and sauces made from healthy and partly rotten fruits. These mycotoxins may cause lesions or degenerations of internal organs such as intestines, kidneys, and liver. They may affect the nervous system, and some of them also cause cancerous tumors.

◦ **Sclerotinia.** *Sclerotinia* causes the "cottony rot" of citrus fruits, especially lemons, and the "watery soft rot" of bean pods (Figure 11-81B), crucifers, cucurbits, strawberries, many other fruits, and practically all vegetables except onions and potatoes. In a moist atmosphere, a characteristic soft, watery decay is produced and the affected tissues are rapidly covered with a white, cottony growth of mycelium that is the outstanding distinguishing characteristic of this decay. The degree of wetness varies with the succulence of the tissues and the humidity of the surrounding air. In moist air, succulent decaying products actually leak and may be completely liquefied, leaving a pool of juice. In dry air the water frequently evaporates as fast as it is liberated by the decay, and the tissues dry down into a mummy or parchmentlike remains. Cottony rot is a rapidly spreading, contact decay that attacks both green and mature fruits and vegetables and makes so vigorous and compact a growth as it spreads from one fruit to another that it enmeshes them in its mycelium and creates so-called "nests." Black, irregularly shaped, sclerotial bodies 2–15 mm long later develop in the fungus mat. The fungus is most active and the severity of the rot increases with temperature up to 21–25°C but, once started, rotting of tissues continues at temperatures as low as 0°C.

Control of Postharvest Decays of Fresh Fruits and Vegetables

For some postharvest diseases, control depends on effective control of the pathogens that cause the same diseases in the field so that the crop will not be contaminated with the pathogens at harvest and subsequently in storage. The crop should be harvested and handled carefully to avoid wounds, bruises, and other injuries that would serve as ports of entry for the pathogen. Harvesting and handling of the crop should be done when the weather is dry and cool to avoid further contamination and infection. The crop should be cooled as quickly as possible to prevent the establishment of new infections and the development of existing ones. All fruits or vegetables showing signs of infection should be removed from the crop that is to be stored or shipped to avoid further spread of the disease. The storage containers, warehouse, and shipping cars should be clean and disinfected with formaldehyde, copper sulfate, or other disinfectant before use. The crop should be stored and shipped at a temperature low enough to slow down development of infections and the physiological breakdown of the tissues but not so low as to cause chilling injuries, which then serve as ports of entry for fungi. The crop should be free of surface moisture when placed in storage, and there should be adequate ventilation in storage to prevent excessively high relative humidity from building up and condensing on the fruit surface. Packaging in plastic bags should be avoided. The crop should be free of insects and other pests when placed in storage and should be kept free of them while in storage to avoid creation of new wounds and development of new infections. Some crops, such as sweet potatoes and onions, can be protected from some decay fungi by "curing" at 28–32°C for 10–14 days, which helps reduce surface moisture and heal any exposed wounds by suberization or wound periderm formation. Hot-air or hot-water treatment is sometimes used to eradicate incipient infections at the surface of some fruits.

In recent years, storage and transport under low oxygen (5 percent) or increased carbon dioxide levels (5–20 percent), have been used to suppress respiration of both the host and the pathogen, thereby suppressing development of postharvest rots. These results are further improved by the addition of 10 percent carbon monoxide. Biological controls have been developed that are effective against some fungal and bacterial pathogens of postharvest diseases, but they are still in the experimental stage. Gamma rays may be useful in reducing storage rots of some crops.

Finally, postharvest decays can be controlled by the use of chemical treatments to prevent infection and suppress development of pathogens on the surface of the diseased host. The chemicals most commonly used for such treatments include diphenyl, sodium-*o*-phenylphenate, dichloran, 2-aminobutane, thiabendazole, benomyl, thiophanate-methyl, imazalil, triforine, captan, iprodione, vinclozolin, soda ash, and borax. They are usually applied as fungicidal wash treatments and are more effective when used "hot," at temperatures between 28 and 50°C, depending on the susceptibility of the crop to injury from heat. Some fungicides, such as dichloran, biphenyl, acetaldehyde vapors, and some ammonia-emitting or nitrogen trichloride-forming chemicals, are used as supplementary, volatile in-package fungistats impregnated in paper sheets during storage and transport. Fungal strains resistant to one or more of the systemic fungicides are common, and precautions must be taken to include additional, preferably broad-spectrum fungicides, in control programs.

• Postharvest Decays of Grains and Legumes

Although several Ascomycetes and Imperfect Fungi such as *Alternaria, Cladosporium, Colletotrichum, Diplodia, Fusarium,* and *Helminthosporium* attack grains and legumes in the field (Figure 11-82), they require too high a moisture content in the grain (24–25 percent) in order to grow and are, therefore, unable to grow much in grains after harvest since grains are usually stored at a moisture content of 12–14 percent. Such fungi apparently die out after a few months in storage or are so weakened that they cannot infect new seeds, but by that time they may have had time to discolor seeds, kill ovules, weaken or kill the embryos, cause shriveling of seeds, and may have produced compounds (mycotoxins) toxic to humans and animals.

Most of the decay or deterioration of grains and legumes after harvest, that is, during storage or transit, is caused by several species of the fungus *Aspergillus* (Figure 11-80). Sometimes *Penicillium* infection occurs in grains or legumes stored at low temperatures and with slightly above normal moisture content. *Aspergillus,* however, particularly *A. flavus,* often infects corn kernels and groundnuts, while still in the field, and its incidence in the field is increased by damage to kernels by insects or other agents, by stalk rots, drought, severe leaf damage, lodging, and by other stresses on the plant.

Each of the various species or groups of species of *Aspergillus* responsible for seed deterioration has rather definite lower limits of seed moisture content below which it will not grow. Each also has less well-defined optimum and upper limits of seed moisture content, these, especially the upper limit, being determined mostly by competition with associated species whose requirement for optimum moisture content coincides with the upper limit at which the former species can survive. Because of competition with field fungi or for other unknown reasons, the storage fungi do not invade grains to any appreciable extent before harvest.

Aspergillus and several of the fungi that attack grains in the field, by invading the embryos of seeds, cause a marked decrease in germination percentage of infected seeds used for planting or in malting barley. Field and storage fungi also discolor the embryos and the seeds they kill or damage and this reduces the grade and price at which the grain can be sold. Flour containing more than 20 percent discolored kernels yields bread of smaller loaf volume and of "off" flavor. In many cases nearly 100 percent of the embryos of wheat may be infected with *Aspergillus* without yet showing discoloration, and this wheat is routinely and unknowingly used to make bread, but whether such grain ever poses a health hazard is not known. Infection of grains, hay, feeds, and cotton stored in bulk or during long shipping results in increased growth and respiration of the fungi, and this causes varying degrees of heating of the material. It also produces moisture of respiration, which raises moisture in adjacent grain. Although not all spoilage of stored grains results in drastic or even detectable heating, any spoilage in progress produces heat that in some materials may raise the temperature up to 70°C or more. The fungi operate at the lower moisture contents where no free water is available, and bacteria at the higher moisture contents.

• Mycotoxins and Mycotoxicoses

One of the more important effects of postharvest decays of fruits and vegetables, and especially of seed and feed deterioration by fungi (Figure 11-82), is the induction of mycotoxicoses, that is, diseases of animals and humans caused by consumption of feeds and foods invaded by fungi that produce toxic substances called mycotoxins. Ergotism and mushroom poisoning are the best known classic examples of mycotoxicoses and have been known for a long time. The magnitude of the mycotoxin problem began to be appreciated during World War II, when in Russia and elsewhere consumption of moldy grain led to necroses of the skin, hemorrhage, liver and kidney failure, and death in numerous humans and animals. Similar symptoms also appeared in horses fed moldy hay. It was not until 1960, though, when a large number of young turkeys died in England after they were fed contaminated peanut feed that intensive research on mycotoxins established that they are a global problem. Mycotoxins pose an ever-present threat to the health of humans and animals, not only when they are present in relatively high concentrations and cause acute disease symptoms, but perhaps even more by the chronic effects on health and productivity caused by the constant presence of subacute dosages of mycotoxins in the food and feed consumed throughout the world, particularly in developing countries.

Most mycotoxicoses are caused by such common and widespread fungi as *Aspergillus, Penicillium, Fusarium,* and *Stachybotrys,* and some may result in severe illness and death. *Aspergillus* and *Penicillum* produce their toxins mostly in stored seeds, hay, or commercially processed foods and feeds, although infection of seeds usually takes place in the field. *Fusarium* produces its toxins primarily on corn and other grains infected in the field or after corn is stored in cribs. *Stachybotrys* produces its toxins while colonizing straw, hay, or other cellulose products used as fodder or bedding for animals.

The mycotoxins produced by each of these fungi may differ from each other in their chemical formula, products in which they are produced, conditions under which they are produced, their effects on various animals and humans, and in their degree of toxicity. Several different fungi, however, produce some of the same or closely related toxins. The main mycotoxins produced by the above fungi and some of their properties are listed below.

- *Aflatoxin.* Aflatoxin's name derives from the fact that it was originally found to be produced by *Aspergillus flavus* but is now known to be produced by several other species of *Aspergillus.* Aflatoxin may be produced in infected cereal seeds and most legumes, but in these it reaches a rather low (about 50 ppb) and probably nontoxic concentration. During some years, a rather high percentage (30 percent or more) of the corn harvest over large areas contains more than 100 ppb aflatoxin, which is five times that allowed in food for humans and in feed for sensitive animals such as chickens. However, in peanuts, cottonseed, fishmeal, brazil nuts, copra, and probably other seeds or nuts grown in warm and humid regions, aflatoxin is produced at high concentrations (up to 1000 ppb or more) and causes mostly chronic or occasionally acute mycotoxicoses in humans and domestic animals. Aflatoxin exists in a variety of derivatives with varying effects. Some of these toxins, when ingested with the feed by dairy cattle, are excreted in the milk in still toxic form. The symptoms of mycotoxicoses caused by aflatoxin in animals, and presumably humans, vary widely with the particular toxin and animal, dosage, age of the animal, and so on. Young ducklings and turkeys fed high dosages of aflatoxin become severly ill and die. Pregnant cows, calves, fattening pigs, mature cattle, and sheep fed low dosages of aflatoxin over long periods develop weakening, debilitation, reduced growth, nausea, refusal of feed, predisposition to other infectious diseases, and may abort. Moreover, most of the ingested aflatoxin is taken up by the liver, and in some experiments, animals given feed containing even less than the permissible (20 ppb) amount of aflatoxin almost invariably developed liver cancer.
- *Fusarium Toxins.* Two groups of toxins, zearalenones and trichothecenes, are produced by several species of *Fusarium,* primarily in molded corn. Zearalenone, also known as F-2 mycotoxin, is produced by *Fusarium roseum, F. moniliforme, F. tricinctum,* and *F. oxysporum.* It seems to be most toxic to swine, in which it causes abnormalities and degeneration of the genital system, the so-called "estrogenic syndrome." Female swine fed zearalenone-containing feed develop swollen vulvas bearing bleeding lesions and atrophying, nonfunctioning ovaries. They are susceptible to abortion, and piglets that are born are small and weak. Male swine show signs of feminization—atrophy of the testes and enlargement of the mammary glands. Trichothecins, of which the most common one is known as T-2 mycotoxin, are produced by the same and by several other species of *Fusarium.* Some trichothecenes are also produced in feed infected with *Cephalosporium, Mycothecium, Trichoderma,* and *Stachybotrys.* In nature they are most toxic when fed to swine in which they cause, among other symptoms, listlessness or inactivity, degeneration of the cells of the bone marrow, lymph nodes, and intestines, diarrhea, hemorrhagia, and death. Other animals, however, such as cows, chicks, and lambs, are also affected.

  Corn infected with *Fusarium* sp. often induces vomiting in swine, or swine refuse to eat it. Although low concentrations of T-2 toxin will induce vomiting in swine, it is likely that other mycotoxins, still unknown, are also involved in inducing vomiting and refusal of corn in swine.

- *Stachybotrys Toxins.* Stachybotryotoxin and several derivatives produced by species of *Stachybotrys* on straw, hay, other fodder, and in animal feeds, bedding, commercial feed, and wheat intended for human consumption cause a typical chronic form, and a less common acute form of a disease in horses, sheep, swine, poultry, and dogs, and also in humans. The symptoms appear as a profuse hemorrhage and necrosis in a variety of body organs such as the stomach, intestines, liver, kidney, and heart. Fumes from burning molded hay may also affect animals and humans, and handling of such hay by farm workers causes in them a toxic dermatitis and conjunctivitis.
- *Other Aspergillus Toxins and Penicillium Toxins.* In addition to aflatoxins, species of *Aspergillus* also produce other toxins in infected grains. The same or similar toxins are also produced in grains infected by species of *Penicillium.* The most important such toxins are discussed below.

  Ochratoxins cause degeneration and necrosis of the liver and kidney, along with several other symptoms, in domestic animals. Some ochratoxins can persist in the meat of animals fed contaminated feed and can be transmitted to the human food chain, possibly posing a public health problem.

  Yellowed-rice toxins, primarily citreoviridin, citrinin, luteoskyrin, and cyclochlorotine, are all produced by species of *Penicillium* growing in stored rice, barley, corn, and dried fish, and cause toxicoses associated with various diseases, such as cardiac beri-beri, nervous and circulatory disorders, and degeneration of the kidneys and liver.

  Tremorgenic toxins cause marked body tremors and excessive discharge of urine, followed by convulsive seizures that often end in death. They are produced by species of both *Aspergillus* and *Penicillium* infecting foodstuffs in storage and also in refrigerated foods, grains, and cereal products. Sheep, horses, and cows seem to be the domestic animals most commonly affected by tremorgenic toxins.

  Patulin is also a carcinogenic substance produced by *Penicillium* and *Aspergillus.* Patulin is toxic to bacteria, to some fungi, and to higher plants and animals. It is commonly found to occur naturally in foodstuffs such as fruit or juices made with fruit partly infected with *Penicillium,* in spontaneously molded bread and bakery products, and in most commercial apple products. Thus, patulin may constitute a serious health hazard for humans as well as for animals.

Control of Grain Decays

The control of postharvest deterioration and spoilage by fungi of grains, legumes, fodder, and commercial feeds depends on certain precautions and conditions that must be met before and during harvest and then during storage. Provided that the crop was healthy and of high quality when harvested, its subsequent infection and spoilage in storage will be avoided if: (1) The amount of moisture content is kept at levels below the minimum required for the growth of the common storage fungi. Some hardy *Aspergillus* species will grow and cause spoilage of starchy cereal seeds with a moisture content as low as 13.0–13.2 percent, and of soybeans with a moisture content of about 11.5–11.8 percent. Others require a minimum moisture of 14 percent or more to cause spoilage. (2) The temperature of stored grain is kept as low as possible since most storage fungi grow most rapidly at temperatures between 30 and 55°C, they grow very slowly at 12–15°C and their growth almost ceases at 5–8°C. Low temperature also slows down respiration of grain and prevents increase of moisture in grain. (3) Infestation of stored products by insects and mites is kept to a minimum through the use of fumigants. This helps keep the storage fungi from getting started and growing rapidly. (4) The stored grain should not be unripe or too old, should be clean, have good germinability, and be free of mechanical damage and broken seeds. Such grain resists infection by storage fungi that could invade otherwise weakened or cracked grain.

In addition to starting with good sound crops free of insects or fumigating to eliminate the insects, the simplest and most common solution to maintaining the grain free of storage fungi is through quick air drying and through the use of aeration systems in storage bins in which air is moved through the grain at relatively low rates of flow. The airflow removes excess moisture and heat. It can be regulated so that it brings the moisture content of the grain mass to the desired level and reduces the temperature to 8–10°C, at which insects and mites are dormant and storage fungi are almost dormant.

SELECTED REFERENCES

Anonymous (1983). Symposium on deterioration mechanisms in seeds. (Several papers.) *Phytopathology* **73,** 313-339.

Boyd, A. E. W. (1972). Potato storage diseases. *Rev. Plant Pathol.* **51,** 297–321.

Ceponis, M. J., and Butterfield, J. E. (1974). Market losses in Florida cucumbers and bell peppers in metropolitan New York. *Plant Dis. Rep.* **58,** 558–560.

Christensen, C. M. (1975). "Molds, Mushrooms, and Mycotoxins." Univ. of Minnesota Press, Minneapolis.

Christensen, C. M., and Kaufmann, H. H. (1965). Deterioration of stored grains by fungi. *Annu. Rev. Phytopathol.* **3,** 69–84.

Coursey, D. G., and Booth, R. H. (1972). The post-harvest phytopathology of perishable tropical produce. *Rev. Plant Pathol.* **51,** 751–765.

Dennis, C., ed. (1983). "Post-Harvest Pathology of Fruits and Vegetables." Academic Press, New York.

Diener, U. L., Asquith, R. L., and Dickens, J. W. (1983). Aflatoxins and *Aspergillus flavus* in corn. *Ala. Agric. Exp. Stn., Auburn Univ. South. Coop. Ser. Bull.* **279,** 1–12.

Eckert, J. W., and Ogawa, J. M. (1985). The chemical control of post-harvest diseases: Subtropical and tropical fruits. *Annu. Rev. Phytopathol.* **23,** 421–454.

Eckert, J. W., and Sommer, N. F. (1967). Control of diseases of fruits and vegetables by postharvest treatment. *Annu. Rev. Phytopathol.* **5,** 391–432.

Food and Agriculture Organization (1981). Food loss prevention in perishable crops. *Agric. Serv. Bull. (F.A.O.)* **43,** 1–72.

Goldblatt, L. A., ed. (1969). "Aflatoxin." Academic Press, New York.

Guba, E. F. (1950). Spoilage of squash in storage. *Mass., Agric. Exp. Stn., Bull.* **457,** 1–52.

Harmon, G. E., and Pfleger, F. L. (1974). Pathogenicity and infection sites of *Aspergillus* species in stored seeds. *Phytopathology* **64,** 1339–1344.

Harvey, J. M. (1978). Reduction of losses in fresh market fruits and vegetables. *Annu. Rev. Phytopathol.* **16,** 321–341.

Harvey, J. M., and Pentzer, W. T. (1960). Market diseases of grapes and other small fruits. *U. S., Dep. Agric., Agric. Handb.* **189,** 1–37.

Jackson, C. R., and Bell, D. K. (1969). Diseases of peanut (ground nut) caused by fungi. *Res. Bull.—Ga., Agric. Exp. Stn.* **56,** 1–137.

Jones, R. K. (1979). The epidemiology and management of aflatoxins and other mycotoxins. *In* "Plant Disease" (J. G. Horsfall and E. B. Cowling, eds.), Vol. 4, pp. 381–392. Academic Press, New York.

Jones, R. K. (1983). Minimizing the impact of corn aflatoxin. *Plant Dis.* **67,** 1297–1298.

McColloch, L. P., Cook, H. T., and Wright, W. R. (1968). Market diseases of tomatoes, peppers, and eggplants. *U. S., Dep. Agric., Agric. Handb.* **28,** 1–74.

Marasas, W. F. O., and van Rensburg, S. J. (1979). Mycotoxins and their medical and veterinary effects. *In* "Plant Disease" (J. G. Horsfall and E. B. Cowling, eds.), Vol. 4, pp 357–379. Academic Press, New York.

Moline, H. E., ed. (1984). "Postharvest Pathology of Fruits and Vegetables: Postharvest Losses in Perishable Crops," Univ. of Calif. Publ. NE-87 (UC Bull. No. 1914).

Pierson, C. F. (1971). Market diseases of apples, pears and quinces. *U. S., Dep. Agric., Agric. Handb.* **376,** 1–112.

Ramsey, G. B., Wiant, J. S., and Link, G. K. K. (1938). Market diseases of fruits and vegetables: Crucifers and cucurbits. *Misc. Publ.—U. S., Dep. Agric.* **292,** 1–74.

Ramsey, G. G., Friedman, B. A., and Smith, M. A. (1967). Market diseases of beets, chicory, endive, escarole, globe artichokes, lettuce, rhubarb, spinach, and sweet potatoes. *U. S., Dep. Agric., Agric. Handb.* **155,** 1–42.

Rose, D. H., Fisher, D. F., Brooks, C., and Bratley, C. O. (1937). Market diseases of fruits and vegetables: Peaches, plums, cherries and other stone fruits. *Misc. Publ.—U. S., Dep. Agric.* **228,** 1–26.

Sauer, D. B., Storey, C. L., and Walker, D. E. (1984). Fungal populations in U. S. farm-stored grain and their relationship to moisture, storage times, regions, and insect infestation. *Phytopathology* **74,** 1050–1053.

Slabaugh, W. R., and Grove, M. D. (1982). Postharvest diseases of bananas and their control. *Plant Dis.* **66,** 746–750.

Smith, M. A., McColloch, L. P., and Friedman, B. A. (1966). Market diseases of asparagus, onions, beans, peas, carrots, celery, and related vegetables. *U. S., Dep. Agric., Agric. Handb.* **303,** 1–65.

Smoot, J. J., Houck, L. G., and Johnson, H. B. (1971). Market diseases of citrus and other subtropical fruits. *U. S. Dep. Agric., Agric. Handb.* **398,** 1–115.

Sommer, N. S. (1982). Postharvest handling practices and postharvest diseases of fruit. *Plant Dis.* **66,** 357–362.

Spalding, D. H., and Reeder, W. F. (1974). Postharvest control of *Sclerotinia* rot of snap bean pods with heated and unheated chemical dips. *Plant Dis. Rep.* **58,** 59–62.

Tuite, J., and Foster, G. H. (1979). Control of storage diseases of grain. *Annu. Rev. Phytopathol.* **17,** 343–366.

Williams, R. J., and McDonald, D. (1983). Grain molds in the tropics: Problems and importance. *Annu. Rev. Phytopathol.* **21,** 153–178.

Wilson, C. L., and Pusey, P. L. (1985). Potential for biological control of postharvest plant diseases. *Plant Dis.* **69,** 375–378.

Wilson, D. M., and Nuovo, G. J. (1973). Patulin production in apples decayed by *Penicillium expansum*. *Appl. Microbiol.* **26,** 124–125.

Diseases Caused by Basidiomycetes

Basidiomycetes are fungi that produce their sexual spores, called **basidiospores,** on a club-shaped or tubular spore-producing structure called a **basidium** (Figures 11-83 and 11-84). Most fleshy fungi, including the common mushrooms, the puffballs, and the shelf fungi or conks, are Basidiomycetes. Their basidia are one-celled, club-shaped structures that bear four external basidiospores on short stalks called **sterigmata.** These Basidiomycetes belong to the class **Hymenomycetes,** which include almost all the wood-decaying fungi and root-rotting fungi (Figures 11-84 and 11-85). In the other class, called **Hemibasidiomycetes,** the basidium has cross walls that divide it into four cells, each of which produces a basidiospore. Such a basidium is often called a **promycelium.** The Hemibasidiomycetes include two very common and very destructive groups of plant pathogens, the rusts and the smuts (Figure 11-83 and 11-85).

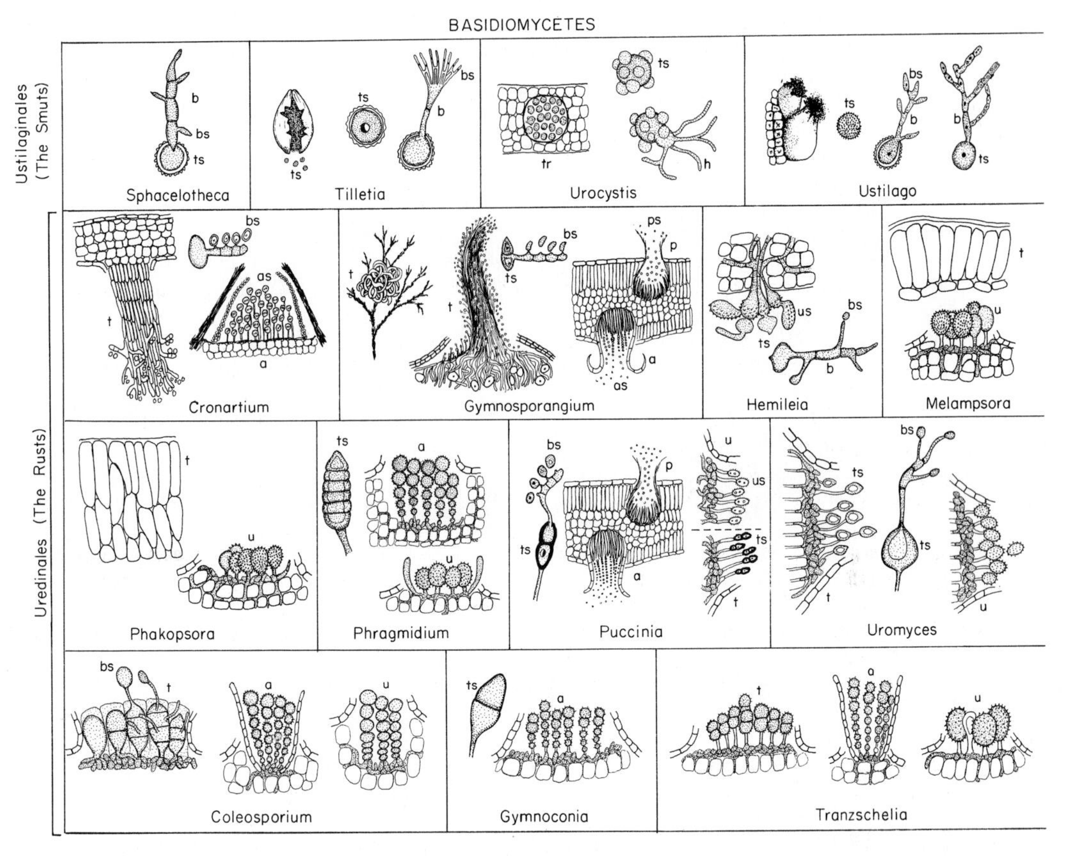

FIGURE 11-83 Basidiomycetes: Some common smut and rust fungi, a—aecium, as—aeciospore, b—basidium, bs—basidiospore, h—hypha, sg—spermogonium, s—spermatium, t—telium, tr—teliosorus, ts—teliospore, u—uredium, us—uredospore.

The Rusts

The plant rusts, caused by Basidiomycetes of the order Uredinales, are among the most destructive plant diseases. They have caused famines and ruined the economies of large areas, including entire countries. They have been most notorious for their destructiveness on grain crops, especially wheat, oats, and barley, but they also attack vegetables such as bean and asparagus, field crops such as cotton and soybeans, ornamentals such as carnation and snapdragon, and have caused tremendous losses on trees such as pine, apple, and coffee.

The rust fungi attack mostly leaves and stems and occasionally floral parts and fruits. Rust infections usually appear as numerous rusty, orange, yellow, or even white-colored spots that result in rupturing of the epidermis, in formation of swellings, and even galls. Most rust infections are strictly local spots, but some may spread internally to a more or less limited extent. There are about 4000 species of rust fungi. The most important rust fungi and the diseases they cause are listed below (Figures 11-83 and 11-85).

Puccinia, causing severe and often catastrophic diseases on numerous hosts such as the stem rust of wheat and all other small grains *(P. graminis);* yellow or stripe

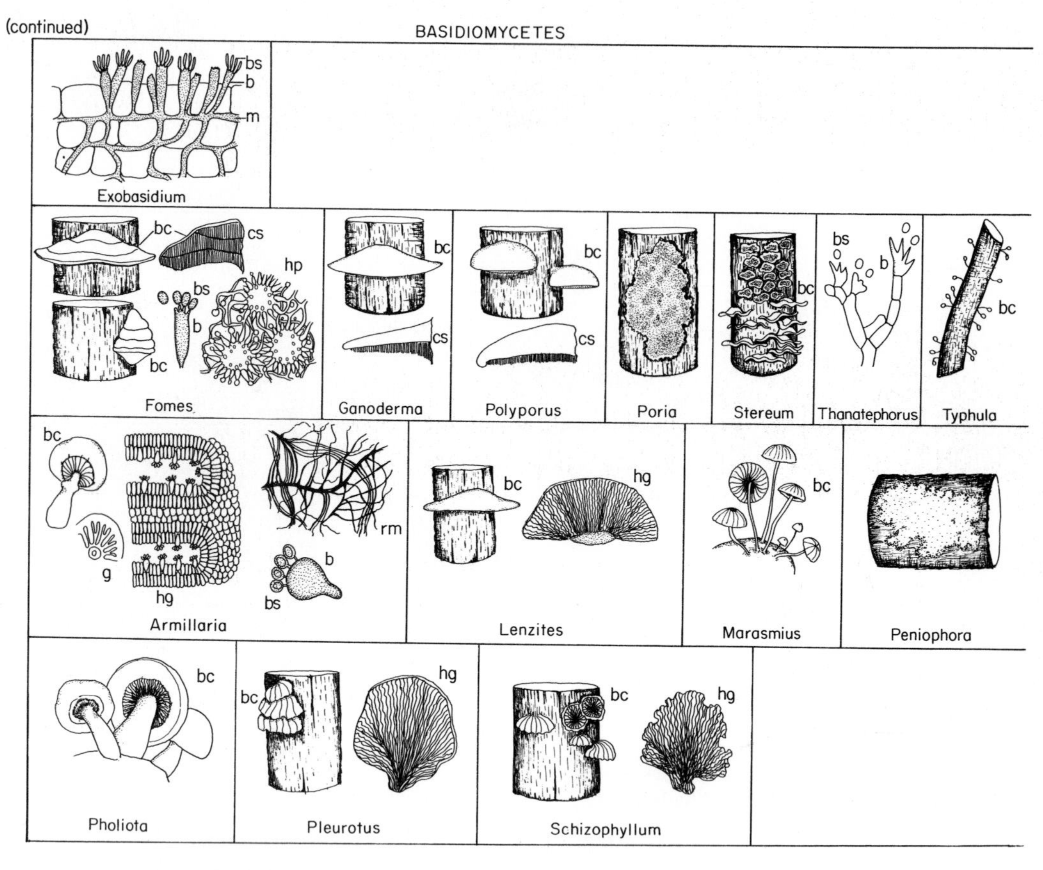

FIGURE 11-84 Basidiomycetes: some of the conk- and mushroom-forming plant pathogens. b—basidium, bc—basidiocarp, bs—basidiospore, cs—cross section, g—gill, hg—hymenial gills, hp—hymenial pores, m—mycelium, rm—rhizomycelium.

rust of wheat, barley, and rye *(P. striiformis);* leaf or brown rust of wheat and rye *(P. recondita)* (Figure 11-86); leaf or brown dwarf rust of barley *(P. hordei);* crown rust of oats *(P. coronata);* corn rust *(P. sorghi);* southern or tropical corn rust *(P. polysora);* sorghum rust *(P. purpurea);* and sugarcane rusts (*P. sacchari* and *P. kuehnii*).

Puccinia also causes severe rust diseases on field crops such as cotton *(P. stakmanii);* vegetables such as asparagus *(P. asparagi);* and flowers such as chrysanthemum *(P. chrysanthemi),* hollyhock *(P. malvacearum),* and snapdragon *(P. antirrhini).*

Gymnosporangium, causing the important cedar-apple rust *(G. juniperi-virginianae)* and hawthorn-cedar rust *(G. globosum).*

Hemileia, causing the devastating coffee leaf rust *(H. vastatrix).*

Phragmidium, causing rust on roses and yellow rust on raspberry.

Uromyces, several species causing the rusts of legumes (bean, broad bean, and pea) and one causing rust of carnation *(U. caryophyllinus).*

Cronartium, causing several severe rusts of pines, oaks, and other hosts, such as the white pine blister rust *(C. ribicola);* fusiform rust of pines and oaks (*C. quercuum* f. sp. *fusiforme*); eastern gall or pine-oak rust *(C. quercuum* f. sp. *virginianae*); pine-sweet fern blister rust *(C. comptoniae);* pine-Comandra rust *(C. comandrae);* and southern cone rust *(C. strobilinum).*

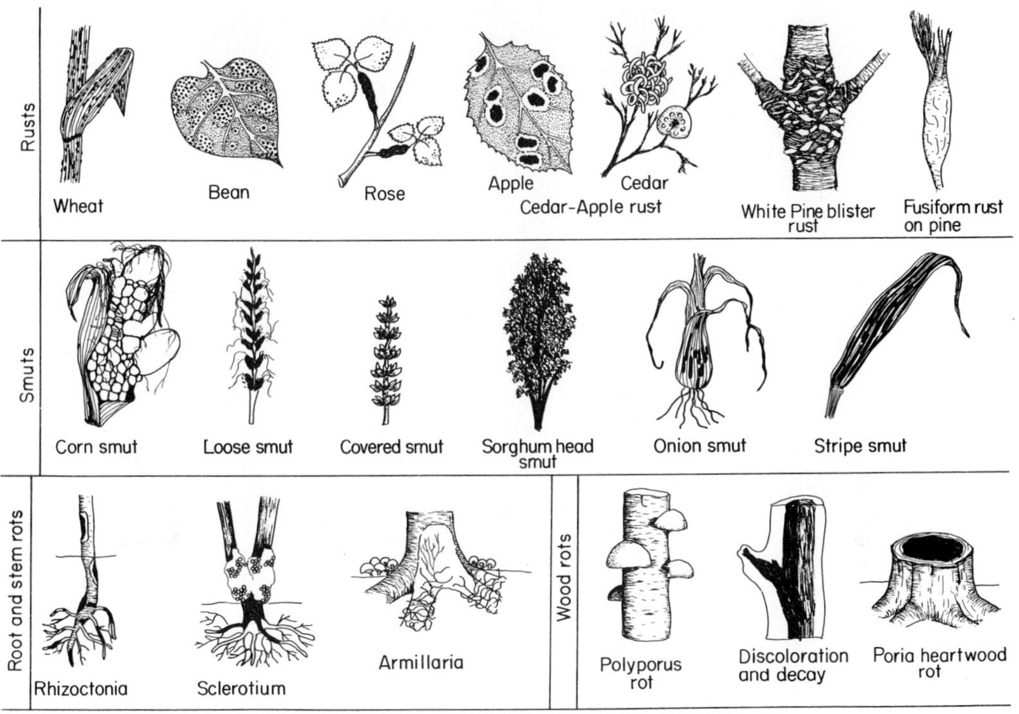

FIGURE 11-85 Common symptoms caused by Basidiomycetes.

Melampsora, causing rust of flax *(M. lini).*
Coleosporium, causing blister rust of pine needles *(C. asterinum).*
Gymnoconia, causing orange rust of blackberry and raspberry.
Phakopsora, causing the potentially catastrophic soybean rust *(P. pahyrhizi).*
Tranzschelia, causing rust of peach.

Most rust fungi are very specialized parasites and attack only certain host genera or only certain varieties. Rust fungi that are morphologically identical but attack different host genera are regarded as special forms *(formae specialis),* for example, *Puccinia graminis* f. sp. *tritici* on wheat, *P. g. f.* sp. *hordei* on barley. Within each special form of a rust there are many so-called pathogenic (physiologic) races that can attack only certain varieties within the species and can be detected and identified only by the set of differential varieties they can infect. Where sexual reproduction of the rust fungus is rare, the races are more stable and produce large populations over fairly long periods of time, but even so some of these fungi have as many races as those in which sexual reproduction is common.

The rust fungi are obligate parasites, although some of them have now been grown on special culture media in the laboratory. Most rust fungi produce five distinct fruiting structures with five different spore forms that appear in a definite sequence (Figure 11-87). Some of the spore stages parasitize one host while the others must infect and parasitize a different, alternate host. All rust fungi produce teliospores and basidiospores. Rust fungi that

FIGURE 11-86 Uredia of the leaf or brown rust on wheat leaves caused by *Puccinia recondita.*

produce only teliospores and basidiospores are called **microcyclic** or **short-cycled rusts.** Other rust fungi produce, in addition to teliospores and basidiospores, spermatia (formerly known as pycniospores), aeciopores, and uredospores (also known as urediospores or urediniospores), in that order, and are called **macrocyclic** or **long-cycled rusts** (Fig. 11-87). In some macrocyclic rusts, spermatia or uredospores or both may be absent. Although basidiospores are produced on basidia, the other spore forms are produced in specialized fruiting structures called, respectively, spermogonia, aecia, uredia (also known as uredinia), and telia (Figures 11-83 and 11-87).

Basidiospores, aeciospores, and uredospores can attack and infect host plants. The **teliospores** serve only as the sexual, overwintering stage, which

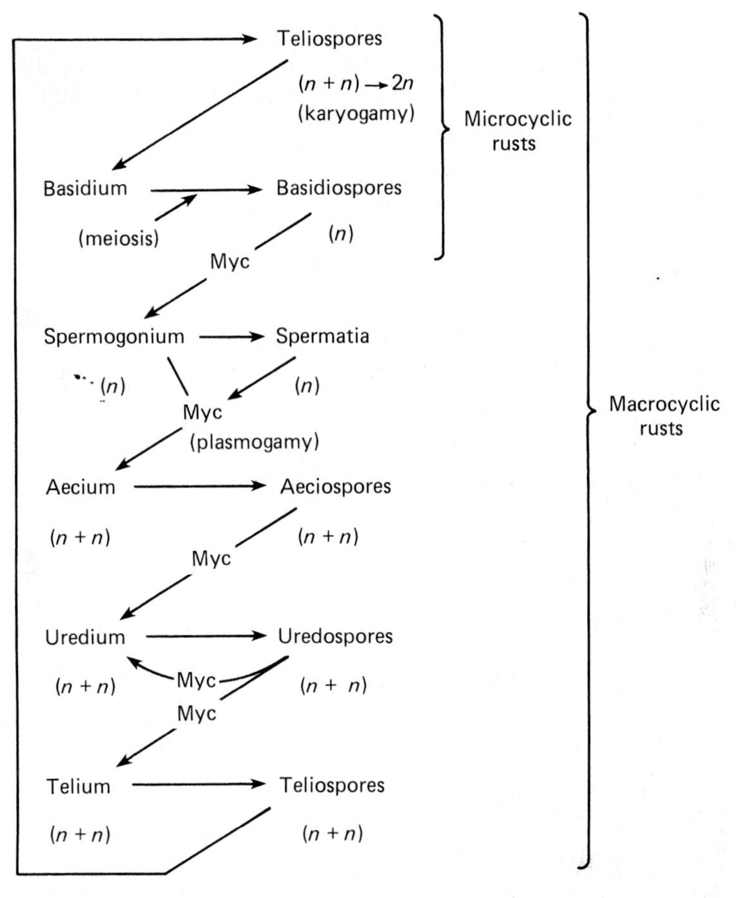

FIGURE 11-87 The kinds and sequence of spores and spore-producing structures in the rust fungi along with the nuclear condition of each. Myc= mycelium.

upon germination produce the **basidium** (promycelium). The basidium, following meiosis, produces four haploid basidiospores. The **basidiospores,** upon infection, produce haploid mycelium that forms **spermogonia** (formerly known as pycnia), containing haploid spermatia and receptive hyphae. **Spermatia** act as male gametes and are unable to infect plants; their function is the fertilization of receptive hyphae of the compatible mating type and the subsequent production of **dikaryotic mycelium** and **dikaryotic spores.** This mycelium forms **aecia** that produce **aeciospores,** which upon infection produce more dikaryotic mycelium that this time forms **uredia.** The latter produce **uredospores,** which also infect and produce more uredia and uredospores and, near host maturity, telia and teliospores. The cycle is thus completed.

Some macrocyclic rusts, for example, asparagus rust, complete their life cycles on a single host and are called **autoecious.** Others, such as stem rust of cereals, require two different or alternate hosts for completion of their full life cycle and are called **heteroecious.**

The rust fungi spread from plant to plant mostly by wind-blown spores, although insects, rain, and animals may play a role. Some of their spores are transported over long distances (several hundred miles) by strong winds and, upon landing (being scrubbed from the air by rain), can start new infections.

Control of rust diseases in some crops, such as grains, is achieved by means of resistant varieties. In some vegetable, ornamental, and fruit tree rusts, such as cedar-apple rust, the disease is controlled with chemical sprays. In others, for example, white pine blister rust, control has been attempted through removal of the alternate host and avoidance of high rust-hazard zones. With the discovery of several new systemic fungicides effective against rusts, such as triadimefon, triforine, and triadimenol, a new impetus has been given toward controlling rust diseases of annual plants as well as trees with these chemicals applied as sprays, seed dressings, soil drenches, or by injection. More recently, biological control of rust diseases has been obtained experimentally either by the application of antagonistic fungi, such as *Darluca filum* on wheat and *Tuberculina maxima* on pines, and bacteria, such as *Bacillus subtilis* on bean, on the surface of the plants, or by prior systemic inoculation of the plants with certain viruses that make the plants more resistant to rust infection. The possibility that rust diseases will be controlled in the field with any of these biological controls, however, seems quite remote at present.

Cereal Rusts

Various species or special forms of *Puccinia* attack all cultivated and wild grasses, including all small grains, corn, and sugarcane. They are among the most serious diseases of cultivated plants, resulting in losses equivalent to about 10 percent of the world grain crop per year. Rusts may debilitate and kill young plants, but more often they reduce foliage, root growth, and yield by reducing the rate of photosynthesis, increasing the rate of respiration, decreasing translocation of photosynthates from infected tissue, and instead, diverting materials into the infected tissue. The quantity of grain produced by rusted plants may be reduced greatly, and the grain produced may be of extremely poor quality since it may be devoid of starch and may consist mostly of cellulosic materials that are of low or no nutritional value to humans. Some of the most important cereal rusts are discussed below.

• Stem Rust of Wheat and Other Cereals

Stem rust of wheat is worldwide in distribution and affects wheat wherever it is grown. Similar rusts affect the other cultivated cereals and probably most wild grass genera and species.

The stem rust fungus attacks all the aboveground parts of the wheat plant and causes losses by reducing foliage and root development and the yield and quality of grain. Infected plants usually produce fewer tillers, set fewer seeds per head, and the kernels are smaller in size, generally shriveled, and of poor milling quality and food value. Under extreme situations, heavily infected plants may die. Heavy seedling infection of winter wheat may weaken the plants and make them susceptible to winter injury and to attack by other pathogens. The amount of losses caused by stem rust may vary from slight to complete destruction of wheat fields over large areas, sometimes encompassing several states. More than 1 million metric tons of wheat are lost to stem rust in North America annually, and during years of severe stem rust epidemics the losses are in the tens or hundreds of millions of tons. Losses from stem rust are at least as severe, and generally much more severe in many other wheat growing countries, particularly developing ones.

◦ *Symptoms.* The pathogen causing stem rust of wheat attacks and produces symptoms on two distinctly different kinds of host plants. The most serious, and economically important, symptoms are produced on wheat and certain related cereals (barley, oats, rye) and other grasses. Symptoms, however, although economically unimportant, are also produced on plants of common barberry *(Berberis vulgaris)* and certain other wild native species of barberry and mahonia.

The symptoms on wheat appear first as long, narrow, elliptical blisters or pustules parallel with the long axis of the stem, leaf, or leaf sheath (Figures 11-88A, B and 11-89A). In later stages of growth, blisters may appear on the neck and glumes of the wheat spike. Within a few days, the epidermis covering the pustules is ruptured irregularly and pushed back, revealing a powdery mass of brick red-colored spores, called uredospores. The pustules, called uredia, vary in size from very small to about 3 mm wide by 10 mm long. Later in the season, as the plant approaches maturity, the rusty color of the pustules turns black as the fungus produces teliospores instead of uredospores and uredia are transformed into black telia. Sometimes telia may develop independently of uredia. Although uredia and telia are rather small, either fruiting structure may exist on wheat plants in such great numbers that large parts of the plant appear to be covered with the ruptured areas that are filled with either the rust-red uredospores or the black teliospores or both.

On barberry, the symptoms appear as yellowish to orange-colored spots on the leaves and sometimes on young twigs and fruits. Within the spots, and

FIGURE 11-88 Stem rust of wheat caused by *Puccinia graminis tritici.* (A) Rust symptoms on wheat stems showing telia. (B) Close-up of infected wheat stem. (C) Barberry leaves with clusters of aecial cups of the stem rust fungus. (Photos courtesy U.S.D.A.)

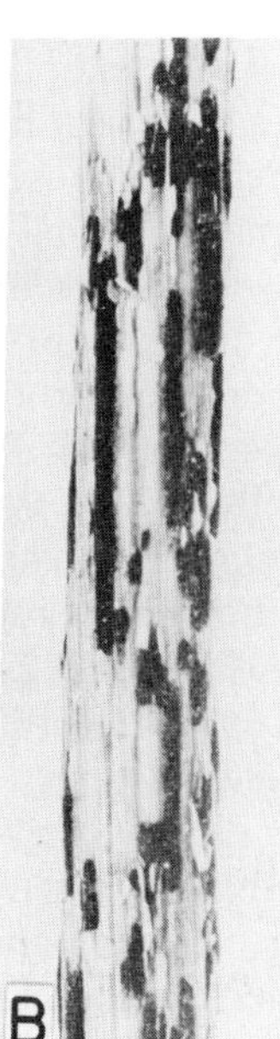

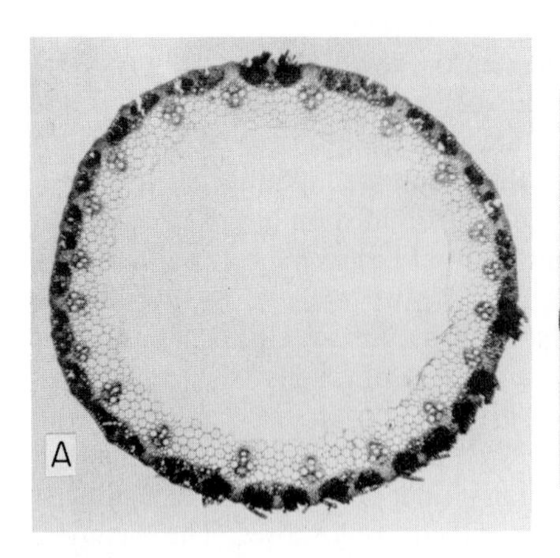

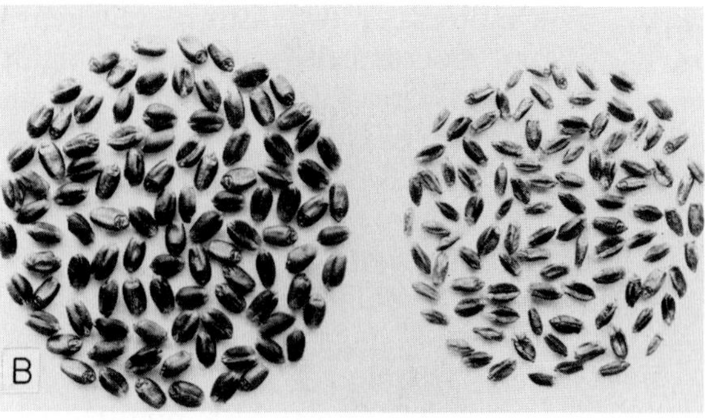

FIGURE 11-89 (A) Cross section of wheat stem showing uredia or telia some of which have ruptured the epidermis. (B) Comparison of kernels from healthy (left) and stem rust-infected wheat plant. (Photos courtesy U.S.D.A.)

in leaves generally on the upper side, appear a few minute dark-colored bodies called spermogonia (or pycnia), usually bearing a small droplet of liquid or nectar. On the lower side of the leaf, beneath the spermogonia, and occasionally on the upper surface, or next to the spermogonia on twigs, fruit, and petioles, groups of orange-yellow horn- or cuplike projections, called aecia, appear (Figure 11-88C). The infected host tissue is frequently hypertrophied. The aecial wall, called a peridium, usually protrudes at the margin of the cups, and its light, whitish color is contrasted with the orange-colored aeciospores contained in the aecia.

- *The Pathogen: Puccinia graminis.* *Puccinia graminis* is a macrocyclic, heteroecious rust fungus producing spermogonia and aecia on barberry and mahonia, and uredia and telia on wheat and other cereals and grasses.
- *Development of Disease.* In cooler, northern regions the fungus overwinters as teliospores on infested wheat debris. Teliospores germinate in the spring after dormancy is broken by alternate freezing and thawing occurring naturally. The basidiospores produced by each teliospore are forcefully ejected into the air and are carried by air currents for a few hundred meters. If the basidiospores land on young barberry leaves, they germinate and penetrate the epidermal cells directly; after that, the mycelium grows mostly intercellularly with haustoria entering the cells. Within 3 or 4 days the hyphal branches form a mat of mycelium that develops into a spermogonium (Figure 11-90). The outward pressure of the spermogonium ruptures the epidermis, and its ostiole (opening) emerges on the surface of the plant tissue. Receptive hyphae originating in the spermogonium extend beyond the ostiole, and spermatia embedded in a sticky liquid are exuded through the opening. Visiting insects become smeared with spermatia and carry them to other spermogonia. Spermatia may also be carried to compatible spermogonia by rainwater or dew running off the plant surface. When a spermatium comes in contact with a receptive hypha of a compatible spermogonium, fertilization takes place. The nucleus of the spermatium passes into the receptive hypha, but it does not fuse with the nucleus already present in the latter. Instead, it migrates through

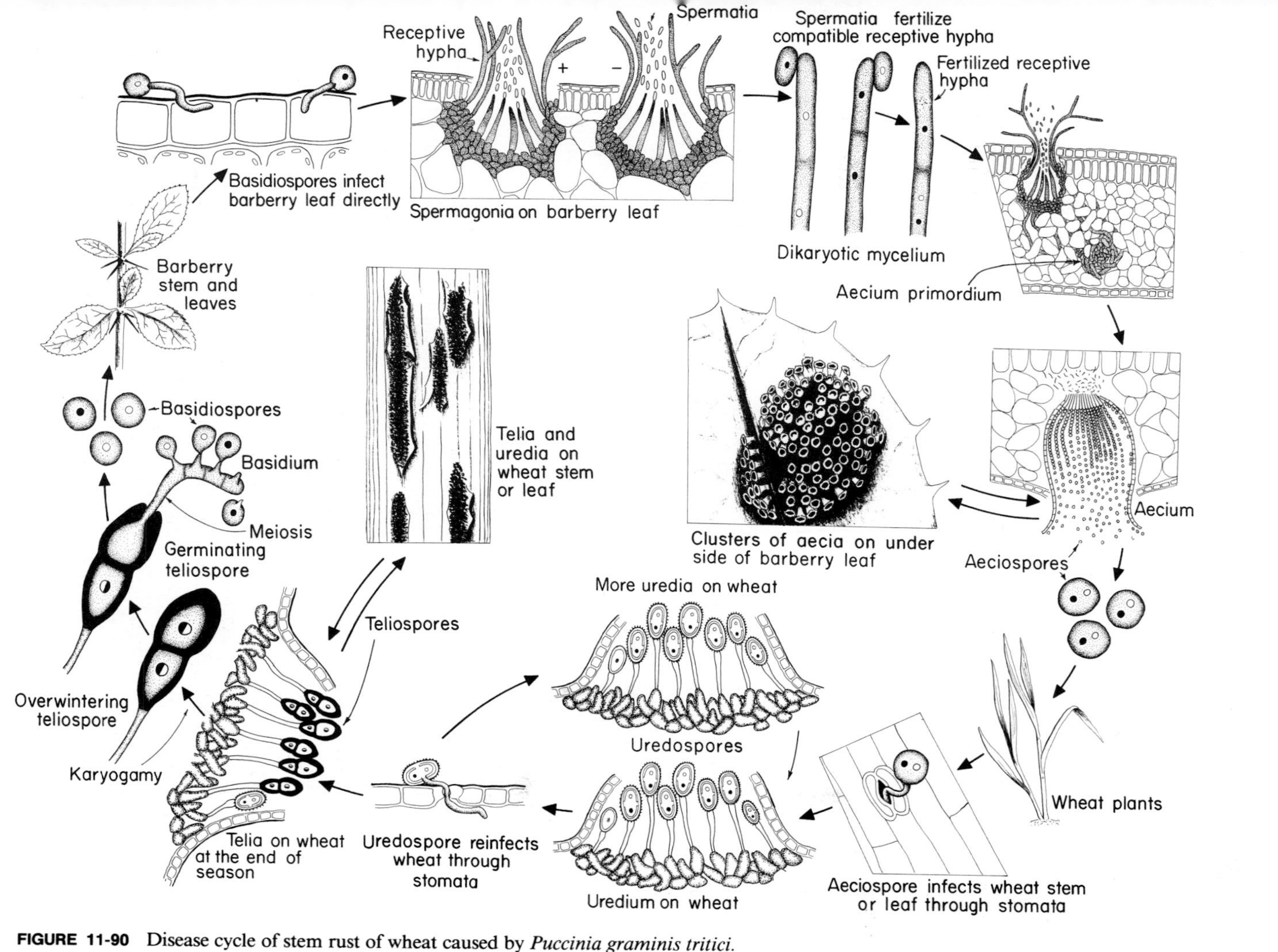

FIGURE 11-90 Disease cycle of stem rust of wheat caused by *Puccinia graminis tritici.*

the monokaryotic mycelium dividing as it progresses, to the aecial mother cells. Thus, the dikaryotic condition is reestablished, and mycelium and aeciospores formed subsequently are dikaryotic. This mycelium then grows intercellularly toward the periphery of the spermogonia present on petioles and fruit, or usually toward the lower side of the leaf bearing the spermogonium, where it forms thick mycelial mats that develop into aecia. In the meantime, the host cells surrounding the mycelium are stimulated to enlarge and, along with the increased volume of the fungus, result in a swelling of the infected area on the lower surface of the leaf.

The aecia form in groups and protrude considerably beyond the hypertrophied leaf or other tissue surface of the barberry plant. The aeciospores are produced in chains on short hyphae inside the aecium, and each spore contains two separate nuclei of opposite mating type. Aeciospores are released in late spring and are carried by wind to nearby wheat plants on which they germinate. The germ tube penetrates wheat stems, leaves, or sheaths through stomata, and after the mycelium grows intercellularly for a while, it then grows more profusely toward, but below the surface of, the wheat tissue and forms a mat of mycelium just below the epidermis. Many short hyphae arise from the mycelium, and at the tip of each forms one uredospore. The growth of the sporophores and of the uredospores exerts pressure on the epidermis, which is pushed outward and forms a pustule manifesting the presence of the uredium. Finally, the epidermis is broken irregularly and flaps back revealing several hundred thousand rust-colored uredospores, which are easily detached from the sporophores and give a powdery appearance to the uredium.

The uredospores are easily blown away by air currents. Stronger winds may carry them many miles, even hundreds of miles, from the point of their origin. The uredospores can reinfect wheat plants. When they land on wheat plants, in the presence of dew, a film of water, or relative humidities near the saturation point, they germinate and their germ tubes enter the plant through stomata. The mycelium grows intercellularly again, sends haustoria into the plant cells, and within 8–10 days from inoculation it produces a new uredium and more uredospores. Many successive infections of wheat plants by uredospores may take place within one growing season up to the time the plant reaches maturity. Most of the damage caused to wheat growth and yield results from such uredospore infections, which may literally cover the stem, leaf, leaf sheaths, and glumes with uredia.

The presence of numerous uredia on wheat plants results in an increased water loss by the plant because infected plants transpire more and because more water evaporates through the ruptured epidermis. In addition to reduced amounts of water being available to the diseased plant, the fungus itself removes much of the nutrients, and water, that would normally be used by the plant. The respiration of infected plants increases rapidly during the development of the uredia, but a few days after sporulation of the fungus, respiration drops to slightly below normal. Photosynthesis of diseased plants is reduced considerably due to the destruction of much of the photosynthetic area by the fungus and to the interference of the fungal secretions with the photosynthetic activity of the remaining green areas on the plant. The fungus also seems to interfere with normal root development and uptake of nutrients by the roots. All these effects reduce the amount of nutrients available for the production of the normal number and size of seeds on the plant. The adverse effects are further accentuated by fungus-induced earlier maturity of the plant, resulting in decreased time available for the seed to fill. The total amount of damage depends considerably on the stage of development of the wheat plant at the time rust infection becomes heavy. Thus, heavy rust infections before or at the flower stage are extremely damaging and may cause total yield loss (Figure 11-89B), whereas if heavy infections do not occur until late dough stage, the damage to yield is much smaller.

When the wheat plant approaches maturity, or when the plant fails because of overwhelming infection, the uredia produce teliospores instead of uredospores, or new telia may develop from recent uredospore infections. Teliospores do not germinate immediately and do not infect wheat but are the overwintering stage of the fungus. Teliospores also serve as the stage in which fusion of the two nuclei takes place and, after meiosis in the basidium, results in the production of new combinations of genetic characters of the fungus through genetic recombination. Several hundred races of the stem rust fungus are known to date, and new ones appear every year.

In southern regions the fungus usually overwinters as mycelium on fall-sown wheat, which becomes infected by uredospores produced on the previous year's crop. Heavy rust infections in these regions in early spring are important, not only locally, but also for the wheat crop of northern regions, since uredospores produced in the South are carried northward by the warm southern winds of spring and summer and initiate infections of wheat in successively northern regions.

○ *Control.* The most effective, and the only practical, means of control of wheat stem rust is through the use of wheat varieties resistant to infection by the pathogen. A tremendous amount of work has been and is being done for the development of wheat varieties resistant to existing races of the fungus. The best varieties of wheat that combine rust resistance and desirable agronomic characteristics are recommended annually by the agricultural experiment stations and change periodically in order to meet the existing rust races. Much effort is now directed toward development of varieties with general or nonspecific resistance and toward development of multiline cultivars.

Eradication of barberry, the alternate host of the stem rust fungus, was carried out until recently in most wheat-growing areas of the United States; this has reduced losses from stem rust by eliminating the early season infections on wheat in the areas where uredospores cannot overwinter, and by reducing the opportunity for the development of new races of the stem rust fungus through genetic recombination on barberry, thus providing for greater stability in the race population of the pathogen and contributing to the success of breeding of resistant varieties.

Several fungicides, such as sulfur, dichlone, zineb, and mixtures of zinc ion with maneb, can effectively control the stem rust of wheat. In most cases, however, 4–10 applications per season are required for complete control of the rust, and because of the low income return per acre of wheat, such a control program is not economically practical. Two applications of zinc ion–maneb mixtures, coordinated with forecasts of weather conditions favoring rust epidemics, may reduce damage from stem rust by as much as 75 percent. These chemicals have both protective and eradicative properties, and therefore even two sprays, one at trace to 5 percent rust prevalence and the second 10–14 days later, can give economically rewarding control of rust.

Certain systemic fungicides, such as fenapanil and especially triadimefon, have also been reported to give experimental control of the stem rust when applied as one or two sprays 1–3 weeks apart during the early stages of disease development.

Damage by the stem rust fungus is usually lower in fields in which heavy fertilization with nitrate forms of nitrogen and dense seeding have been avoided.

• Puccinia Rusts of Vegetables, Field Crops, and Ornamentals

The most common *Puccinia* rusts on plants other than cereals are those on asparagus, peanuts, cotton, chrysanthemum, hollyhock, and snapdragon. The asparagus rust, *Puccinia asparagi,* produces spermogonia, aecia, uredia, and telia on asparagus. In the peanut rust *(P. arachidis),* chrysanthemum rust *(P. chrysanthemi),* and snapdragon rust *(P. antirrhini)* (Figure 11-91B), spermogonia and aecia are unknown, and only uredia and occasionally telia are produced on the respective hosts. In hollyback rust *(P. malvacearum),* only telia are produced. In cotton rust *(P. stackmanii),* the disease is caused by the aecial stage, while uredia and telia develop on desert grasses of the genus *Bouteloua,* which are common through the southwestern cotton-growing areas of the United States and in Mexico.

In most of the above rusts, the symptoms appear as rust-colored uredial spots on leaves and green stems that later in the season may be replaced or supplemented by black telia. In cotton, the symptoms appear as circular, slightly elevated, orange-yellow aecia mostly on the under surface of the leaves. Depending on the severity of the infection, plants become weakened, stunted, and may even be killed.

Control of these rusts, in addition to the use of resistant varieties, depends primarily on sprays or dusts with fungicides such as polyram, maneb, zineb, or sulfur, and the systemic fungicides fenapanil, triadimefon, and others. Removal and burning of all infected plant material helps reduce or eliminate the inoculum in the area and reduces subsequent development of disease.

• Cedar-Apple Rust

Cedar-apple rust is present in North America and in Europe. It causes yellow-to-orange-colored leaf spots and occasionally fruit spots and premature defoliation on apple (Figure 11-92, A, B); and galls, often called cedar apples, that produce jellylike horns on cedar (Figure 11-92, C, D). It can cause considerable damage to both hosts when they are located near each other. Similar diseases affect hawthorn and quince.

The fungus, *Gymnosporangium juniperi-virginianae,* overwinters as dikaryotic mycelium in the galls on cedar trees. Cedar needles or axillary buds are infected in the summer by wind-borne aeciospores from apple leaves (Figure 11-93). The fungus grows little in the cedar needles during fall and winter, but the following spring or early summer, galls begin to appear as small greenish-brown swellings on the upper surface of the needle. The fungus is present in the galls as mycelium growing between the cells of cedar. The galls enlarge rapidly and, by fall, they may be 3–5 cm in diameter, turn chocolate brown, and their surface is covered with small circular depressions.

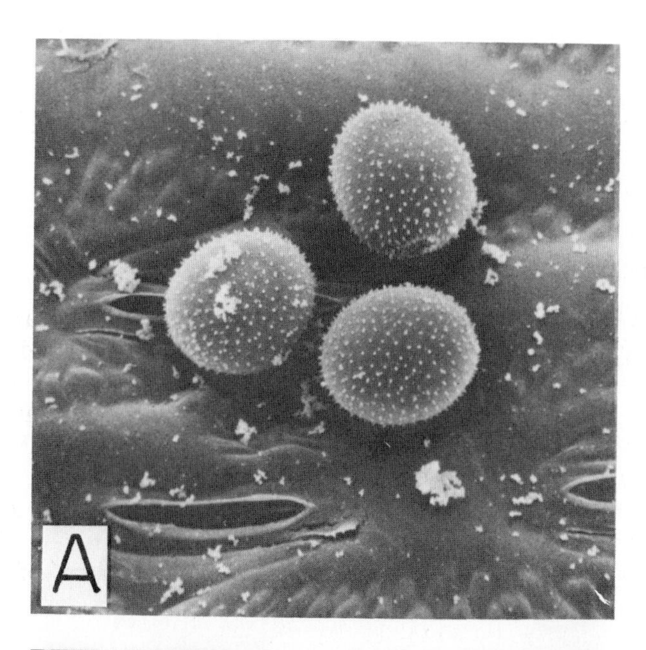

FIGURE 11-91 (A) Scanning electron micrograph of uredospores of *Puccinia sorghi* next to a stoma. (B) Snapdragon rust caused by *P. antirrhini.* (Photo A courtesy M. F. Brown and H. G. Brotzman.)

The cedar-apple rust fungus does not produce uredia or uredospores. The following spring, however, the small depressions on the galls absorb water during warm, wet weather, swell, and produce orange-brown, jellylike "horns" that are 10–20 mm long and very conspicuous (Figure 11-92, C, D). The jellylike horns are columns of teliospores that germinate in place for several weeks and produce basidiospores that can infect apple leaves. The galls eventually die but may remain attached to the tree for a year or more.

Basidiospores are wind borne and may be carried for up to 2–3 miles. Their germ tubes penetrate young apple leaves or fruit directly and produce

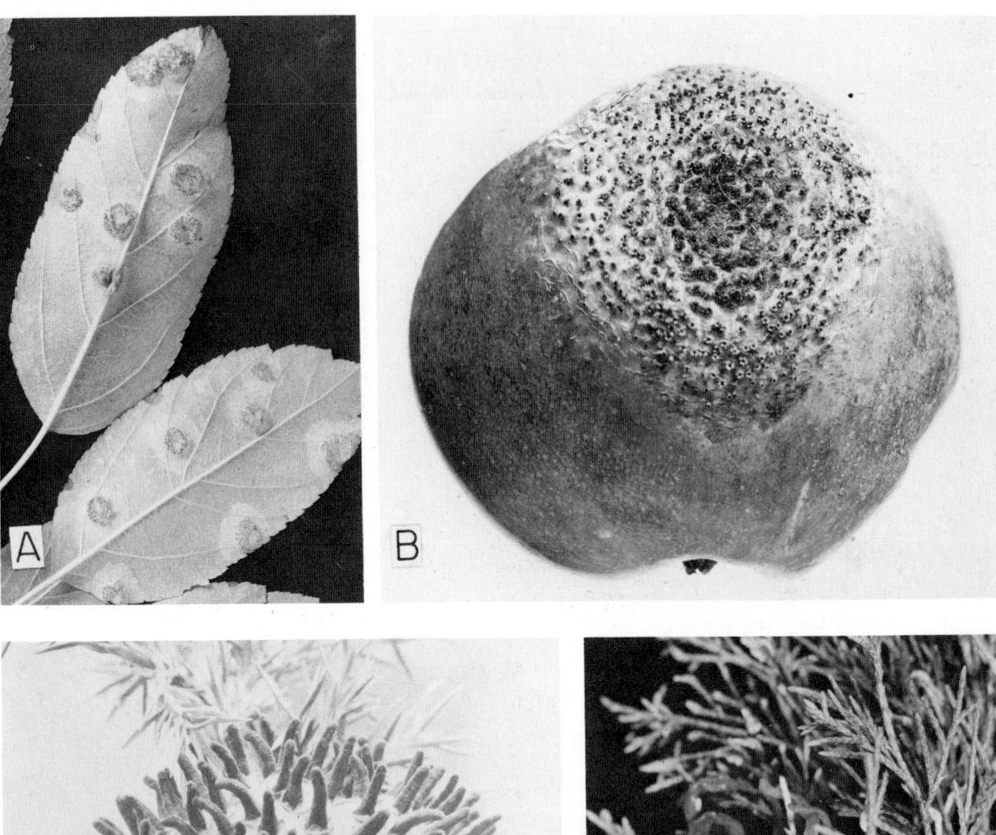

FIGURE 11-92 Cedar-apple rust caused by *Gymnosporangium juniperi-virginianae.* (A) Clusters of aecial cups on apple leaves. (B) Infected apple fruit with numerous aecial cups and a few spermogonia in center of spot. (C) Gall on cedar (cedar apple) in early spring when telial horns are just coming out. (D) Mature telial horns releasing basidiospores. (Photos A–C courtesy U.S.D.A.)

haploid mycelium that spreads through or between the apple cells. The mycelium forms orange-colored spermogonia on the upper leaf surface, and presumably after fertilization of receptive hyphae by compatible spermatia, dikaryotic mycelium ensues, which produces aecial cups in concentric rings on the lower side of leaves and on fruit. The area of the leaf where spermo-

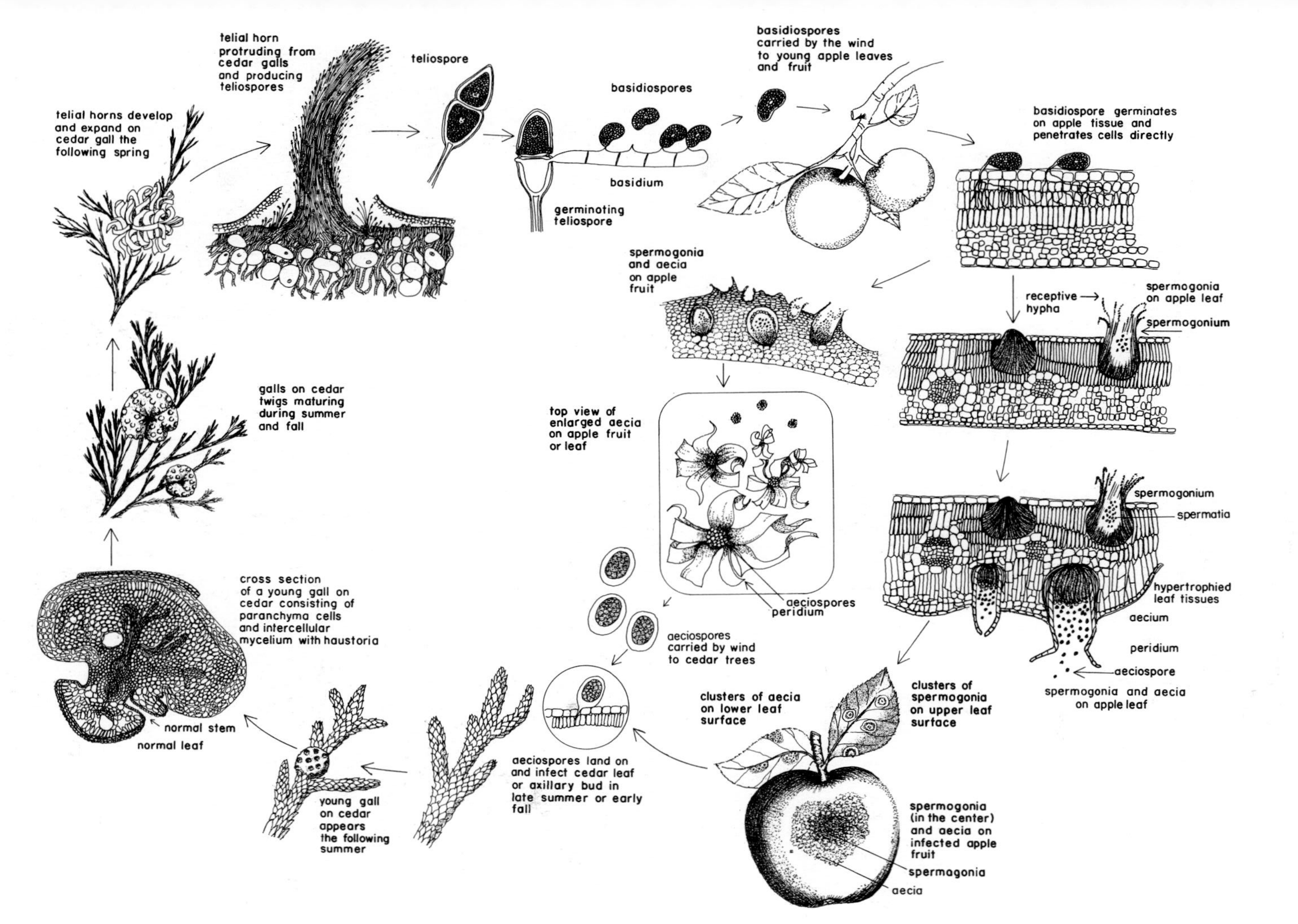

FIGURE 11-93 Disease cycle of cedar-apple rust caused by *Gymnosporangium juniperi-virginianae.*

gonia and aecia are produced is swollen, especially on the lower side where the clusters of orange-yellow aecial cups and their white peridia (cup walls) stand out conspicuously (Figure 11-92A). In the fruit, spermogonia and aecia are formed in the same areas; the spermogonia appear first in the center of the spot and the aecia subsequently in the surrounding area. Infected fruit areas are usually large and flat or depressed rather than swollen (Figure 11-92B). The aeciospores are produced in long chains. They are released in the air during dry weather in late summer and are carried by wind to cedar leaves, where they germinate and start new infections.

Control of cedar-apple rust can be effected by keeping apple and cedar trees sufficiently removed from each other so that the fungus cannot complete its life cycle. This, however, is often impossible or impractical, and therefore the disease is generally controlled on both hosts with chemical sprays with ferbam, thiram, maneb, or with the systemic triforine, which also gives good postinfection control of cedar-apple rust. Many apple varieties are also quite resistant to rust.

• Coffee Rust

Undoubtedly the most destructive disease of coffee, coffee rust damages trees and reduces yields by causing premature drop of infected leaves. Coffee rust has caused devastating losses in all coffee-producing countries of Asia and Africa. It attacks all species of coffee but is most severe on *Coffea arabica.* In 1970 the disease appeared for the first time in the Western Hemisphere, in Brazil, and has been steadily spreading into the world's most important coffee-producing countries of South and Central America, where all commercial coffees are susceptible to the rust.

The symptoms appear as orange-yellow powdery spots on the lower side of the leaves. The spots are circular and small, about 5 mm in diameter, at first, but they often coalesce and form large patches that may be ten times as large. The centers of the spots eventually become dry, turn brownish, and the leaf falls off prematurely. Infected trees produce small yields of poor quality, and repeated infections and defoliations result in the death of trees (Figure 11-94).

The fungus, *Hemileia vastatrix,* exists primarily as mycelium, uredia, and uredospores, which in the tropics, where coffee and the fungus thrive, can perpetuate themselves in infected leaves they continuously and successively infect. The fungus occasionally produces teliospores, which upon germination form basidiospores, but the latter do not infect coffee and no alternate host has so far been found. Thus, uredospores are believed to be responsible for all coffee infections. They are easily spread by wind, rain, and perhaps by insects. The spores require high humidity and probably dew for germination and infection. Under favorable conditions they can germinate and enter leaves through the stomata of the lower surface in less than 12 hours. The mycelium grows between the leaf cells and sends haustoria into the cells. Young leaves are generally more susceptible to infection than older ones and new uredia may appear on the lower side of the leaf within 10–25 days from infection, depending on the climatic conditions. Once uredia develop, premature falling of infected leaves may occur at any time; sometimes even one uredium is sufficient to cause the leaf to fall. New leaves are affected after the older ones

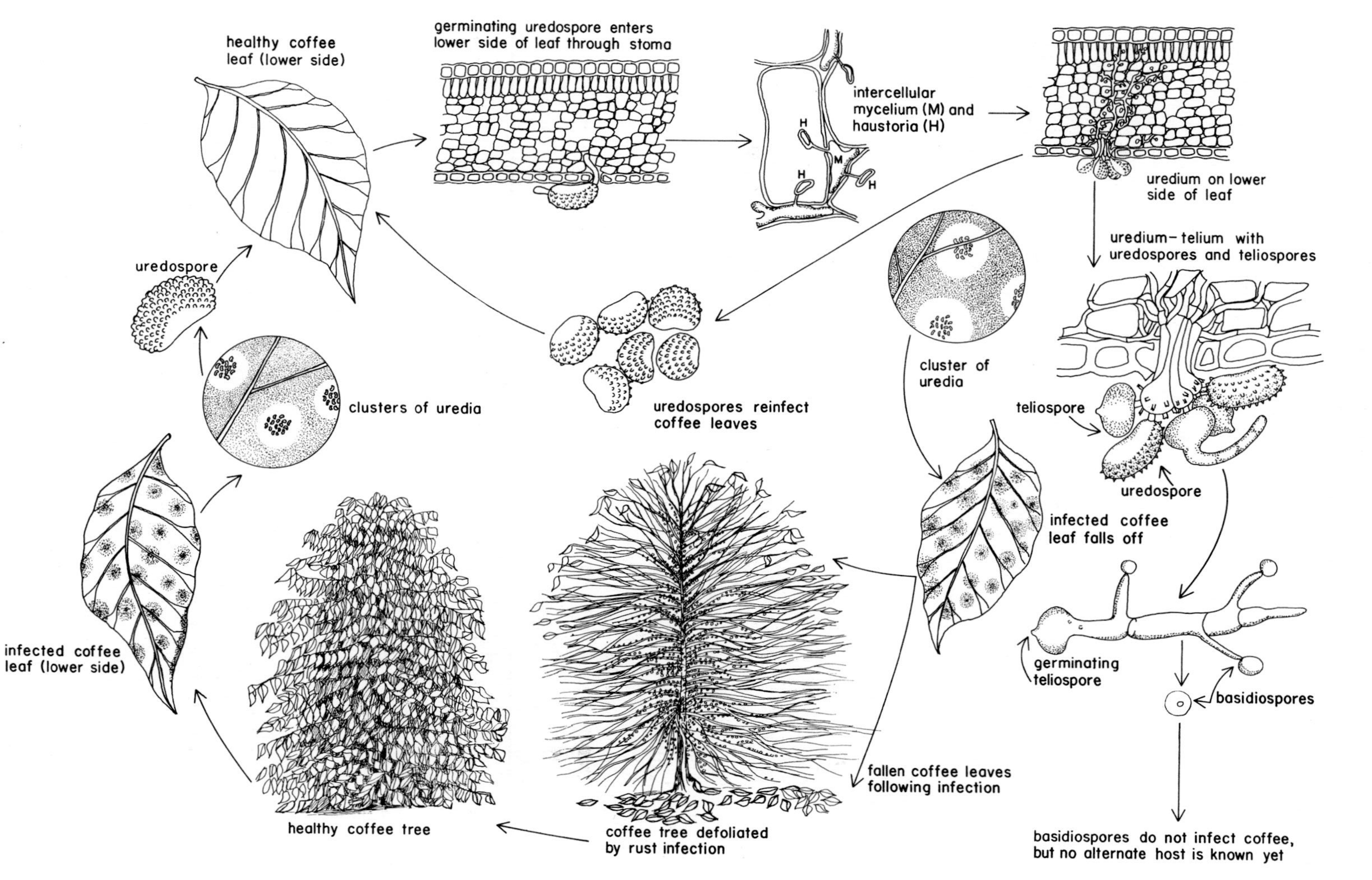

FIGURE 11-94 Disease cycle of coffee rust caused by *Hemileia vastatrix*.

have fallen. The premature shedding of leaves weakens the trees and results in reduced yields, severe dieback of twigs, and death of trees.

Control of coffee rusts is difficult, but satisfactory results can be obtained with copper fungicides such as Bordeaux mixture, copper oxychlorides, and cuprous oxide. Fungicides must be applied before and during the rainy season at 2- to 3-week intervals or less, depending on weather conditions and the severity of the attack. Recently, systemic fungicides, such as triadimefon and pyracarbolid, which have a curative effect on developing uredial pustules, have been used in alternate applications with the copper fungicides. Sufficient tree pruning, good site selection, and use of the newly found or developed resistant varieties of Arabica coffee in future plantings help minimize losses from the rust. New races of the pathogen virulent to the new resistant varieties of the host have already appeared in some regions, however.

Rusts of Forest Trees

Several species of *Cronartium* are responsible for a number of rust diseases that cause major losses in forest trees. Some *Cronartium* species attack the main stem or branches of trees and these are the most destructive; other species attack the needles or leaves and are less serious. All rusts, however, are especially destructive when they attack young trees in the nursery or in recently established plantations. The main economic host of the majority of forest tree rusts and the one to which they cause the most damage is pine. Some of these rusts have oak as their alternate host but the damage to oaks is much less severe. Other pine rusts have as their alternate hosts various wild or cultivated shrubs or weeds.

• White Pine Blister Rust

White pine blister rust is native to Asia from where it spread to Europe and, about 1900, to North America. It is one of the most important forest diseases in North America, where it causes an annual growth loss and mortality of more than 200 million cubic feet, and, if not controlled, it makes white pine growing impossible or unprofitable. White pine blister rust is caused by the fungus *Cronartium ribicola,* which produces its spermogonia and aecia on white pine (the 5-needle pines) and its uredia and telia on wild and cultivated currant and gooseberry bushes (*Ribes* sp.). Blister rust kills pines of all ages and sizes. Small pines are killed quickly, while larger pines may develop cankers that girdle and either kill the trees or retard their growth and weaken the stems, which then break at the canker. Infection of *Ribes* bushes causes relatively little loss through premature, partial defoliation and reduced fruit production.

The symptoms of blister rust on white pine stems or twigs appear first as small, discolored, spindle-shaped swellings (cankers) surrounded by a narrow band of yellow-orange bark. In the canker, small, irregular, dark brown, blisterlike spermogonia appear, which rupture, ooze droplets full of the year's crop of spermatia, and then dry. As the canker grows, the margin and the zone of spermogonia expand, and the portion formerly occupied by spermogonia is a year later the area where the aecia are produced. The aecia appear as white sacks or blisters containing orange-yellow aeciospores that push through the diseased bark. The aecial blisters soon rupture (Figure 11-95A) and the orange-yellow aeciospores are carried by the wind, sometimes for several hundred miles, some of them landing on and infecting *Ribes* leaves. After the aeciospores have been released, the blisters persist on the bark for a long time, although the bark of that area dies. Resin often flows down the stem and hardens in masses characteristic of the disease. The fungus, however, continues to spread into the surrounding healthy bark, and the sequence

FIGURE 11-95 (A) White pine blister rust caused by *Cronartium ribicola.* (B) Scanning electron micrograph of part of telium of *Cronartium quercuum* f. sp. *fusiforme* showing basidia and basidiospores. (C) Fusiform rust on pine seedlings. Masses of aeciospores can be seen on two of them. (Photo A courtesy U.S.D.A. Photo B courtesy M. F. Brown and H. G. Brotzman.)

of spore production and bark killing continues in subsequent years until the stem or branch is girdled and killed. The dead branches, called "flags," have dead, brown needles and are visible from a distance.

On currant and gooseberries, the symptoms appear on the undersides of the leaves as slightly raised, yellow-orange uredia grouped in circular or irregular spots. The uredia produce orange masses of uredospores that reinfect *Ribes*. Later, telia develop in the same or new lesions. The telia are slightly darker than the uredia and consist of brownish, hairlike structures up to 2 mm in height that bear the teliospores.

The pathogen, *Cronartium ribicola,* overwinters mostly as mycelium in infected white pines and to some extent on *Ribes*. Pines are infected only by basidiospores produced by teliospores still in the telia on the undersides of *Ribes* leaves (Figure 11-96). The basidiospores are produced only during wet, cool periods, especially during the night, and can be carried by wind and infect pines within a few hundred feet from the *Ribes* host. The basidiospores infect pine needles through stomata in late summer or early fall. Small, discolored spots may appear on the needles 4–10 weeks after infection. The mycelium grows down the conducting tissues of the needle and into the bark of the stem, which it reaches about 12–18 months after infection. Spermogonia develop on infected stems or branches in the spring and early summer 2–4 years after the needle infection, and aecia are produced in the spring, 3–6 years from inoculation. Spermatia are short lived and spread over short distances by rain or insects, while aeciospores may live for many months, may overwinter, and may be carried by wind over many miles to *Ribes* leaves. On the latter, the aeciospores germinate and infect the leaves, producing uredia and uredospores within 1–3 weeks after inoculation. Uredospores can reinfect *Ribes* plants again and again, producing many generations of uredospores in a single growing season. The uredospores can survive for many months and even through winter and can be spread by wind for a mile or more, but they can infect only *Ribes*. Finally, the same mycelium that produced the uredospores begins to produce telial columns and teliospores. The latter germinate from July to October and produce short-lived basidiospores, which if blown to nearby white pines infect the needles and complete the life cycle of the fungus.

Control of white pine blister rust can be obtained by eradication of wild and cultivated *Ribes* bushes mechanically or, better still, with herbicides such as 2,4-D and 2,4,5-T. Pruning infected branches on young trees reduces stem infections and tree mortality. Treatment of cankers of entire trees with the hyperparasite *Tuberculina maxima,* which parasitizes *Cronartium ribcola* on blister rust cankers, has also been considered as a control measure, but so far its practical value has not been demonstrated. The most promising control for blister rust seems to be the selection and breeding of resistant trees. Seed orchards with trees that have shown resistance to the disease have been established and are expected to produce millions of resistant trees in the near future.

• Fusiform Rust

Fusiform rust is one of the most important diseases on southern pines, especially loblolly and slash pines. The disease is present from Maryland to Florida and west to Texas and Arkansas, where it causes tremendous losses in

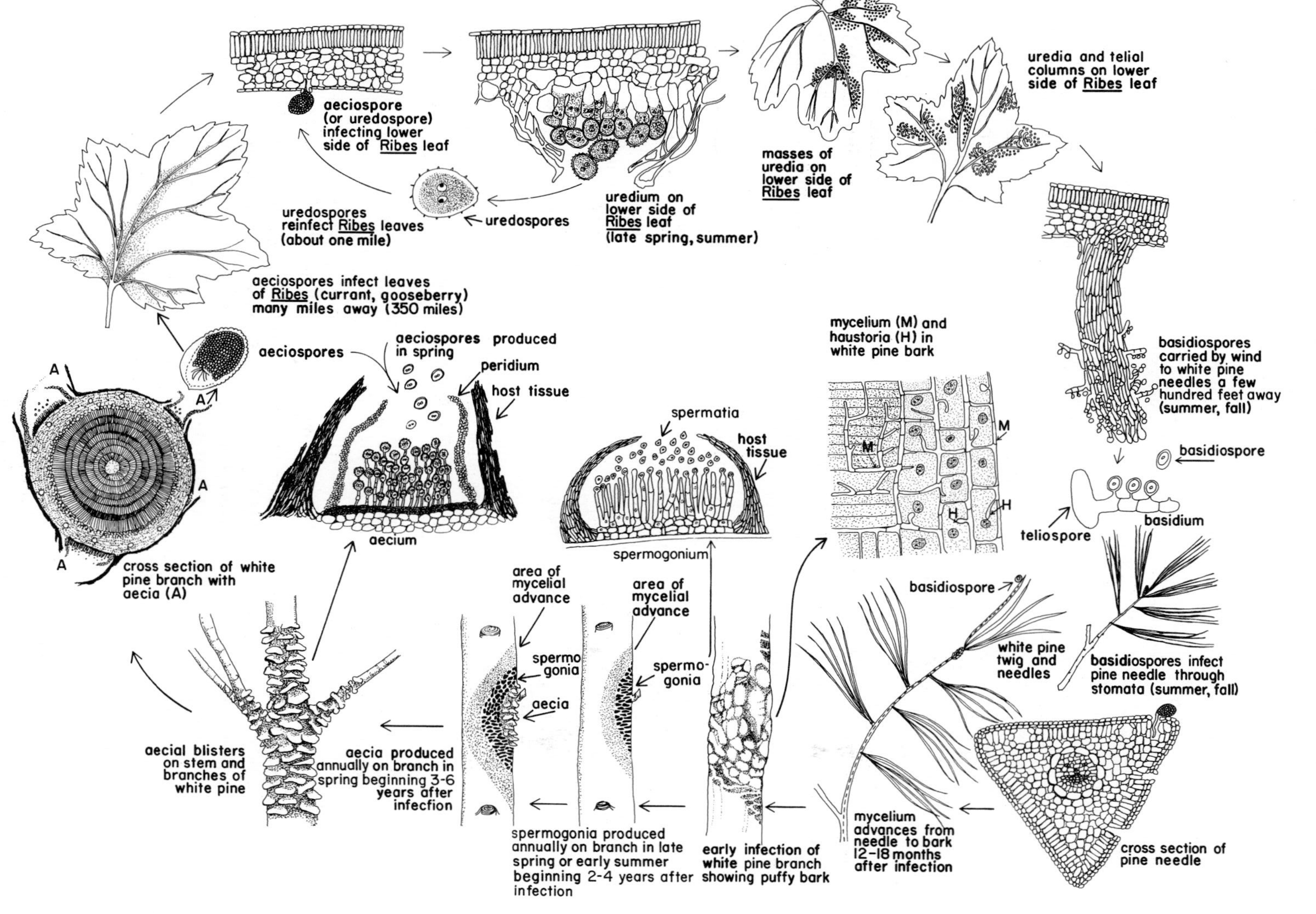

FIGURE 11-96 Disease cycle of white pine blister rust caused by *Cronartium ribicola*.

nurseries, young plantations, and seed orchards ranging from 20 to 60 percent or even more mortality of young trees. Fusiform rust is caused by *Cronarium quercuum* f. sp. *fusiforme,* which produces spermogonia and aecia on pine stems and branches and uredia and telia on oak leaves. Damage on oak is slight, mostly through occasional partial defoliation.

The symptoms on pine first appear as small, purple spots on needles and succulent shoots. These soon form small galls and later develop into spindle-shaped swellings or galls (cankers) on branches and stems of mostly young pines. These galls may elongate from 5 to 15 cm per year and often encircle the stem or branch and cause it to die (Figure 11-95C). Infection of young seedlings results in their death within a very few years, while infected young trees may branch excessively for a period and show a bushy growth. On older trees, stem or branch infections lead to weak, distorted boles or, as host tissue is killed, to sunken cankers that break easily during strong winds. Yellowish masses of spermatia and later orange-yellow aeciospores appear on the galls. On oak, the symptoms appear as orange pustules (uredia) and brown, hairlike columns (telia) on the underside of the leaves.

The fusiform rust pathogen, *Cronartium quercuum* f. sp. *fusiforme,* over-winters as mycelium in the fusiform galls. From February to April, spermogonia and spermatia form, and soon aeciospores are produced on the galls. The wind carries the aeciospores to young, expanding oak leaves, which they infect. On the oak leaves, orange uredial pustules develop in a few days and produce uredospores from February to May. Uredospores can reinfect more oak leaves and produce more uredospores. The same mycelium also produces brown telia from February to June in place of uredia or in new lesions. The teliospores germinate on the telia (Figure 11-95B), and the basidiospores produced are carried by wind to pine needles and shoots, which they infect directly. The mycelium grows first in the needles and later spreads into branches or the stem, where it induces both hyperplasia and hypertrophy and the formation of the gall.

Control of fusiform rust infections in the nursery can be obtained by frequent, twice-a-week sprays with ferbam, especially before and during cool wet weather. Some of the newer systemic fungicides, such as benodanil, triadimefon, and triadimenol, give good control of fusiform rust of seedlings when applied as sprays or as seed dressings. Several fungi antagonistic to or parasitic on the pathogen are known, for example, *Darluca filum* and *Scytalidium uredinicola,* but no practical biological control of the disease has been developed yet. All infected seedlings should be discarded. In plantations and natural stands, only limited control can be obtained against fusiform rust by either avoiding planting highly susceptible slash and loblolly pines in areas of known high rust incidence or by pruning infected branches before the fungus reaches the trunk. As with white pine blister rust, and perhaps even more so, control of fusiform rust is obtained through selection and breeding of resistant trees, with emphasis on trees possessing general rather than specific resistance.

SELECTED REFERENCES

Allen, R. F. (1930). A cytological study of heterothallism in *Puccinia graminis. J. Agric. Res. (Washington, D.C.)* **40,** 585–614.

Anonymous (1981). Stakman-Craigie symposium on rust diseases. (Several papers.) *Phytopathology* **71,** 967–1000.

Arthur, J. C., and Cummins, G. B. (1962). "Manual of the Rusts in United States and Canada." Hafner, New York.

Baker, C. J., Stavely, J. R., and Mock, N. (1985). Biocontrol of bean rust by *Bacillus subtilis* under field conditions. *Plant Dis.* **69,** 770–773.

Bliss, D. E. (1933). The pathogenicity and seasonal development of *Gymnosporangium* in Iowa. *Iowa, Agric. Exp. Stn., Res. Bull.* **166.**

Browning, J. A., and Frey, K. J. (1969). Multiline cultivars as a means of disease control. *Annu. Rev. Phytopathol.* **7,** 355–382.

Caldwell, R. M. *et al.* (1934). Effect of leaf rust *(Puccinia triticina)* on yield, physical characters, and composition of winter wheats. *J. Agric. Res. (Washington, D.C.)* **48,** 1049–1071.

Colley, R. H. (1918). Parasitism, morphology, and cytology of *Cronartium ribicola. J. Agric. Res. (Washington, D.C.)* **15,** 619–660.

Crowell, I. H. (1934). The hosts, life history, and control of the cedar-apple rust fungus *Gymnosporangium juniperi-virginianae. J. Arnold Arbor., Harv. Univ.* **15,** 163–232.

Cummins, G. B. (1959). "Illustrated Genera of Rust Fungi." Burgess, Minneapolis, Minnesota.

Dickson, J. G. (1956). "Diseases of Field Crops." McGraw-Hill, New York.

Eramus, D. S., and von Wechmar, M. B. (1983). Reduction of susceptibility of wheat to stem rust (*Puccinia graminis* f. sp. *tritici*) by brome mosaic virus. *Plant Dis.* **67,**1196–1198.

Eversmeyer, M. G., Kramer, C. L., and Browder, L. E. (1984). Presence, viability, and movement of *Puccinia recondita* and *P. graminis* inoculum in the Great Plains. *Plant Dis.* **68,** 392–395.

Flor, H. H. (1971). Current status of the gene-for-gene concept. *Annu. Rev. Phytopathol.* **9,** 275–296.

Hamilton, M., and Stakman, E. C. (1967). Time of stem rust appearance on wheat in the western Mississippi basin in relation to the development of epicemics from 1921 to 1962. *Phytopathology* **57,** 609–614.

Hart, H. (1931). Morphologic and physiologic studies on stem-rust resistance in cereals. *Minn., Agric., Exp. Stn., Tech. Bull.* **266,** 1–75.

Heath, M. C. (1981). Resistance of plants to rust infection. *Phytopathology* **71,** 971–975.

Hooker, A. L. (1967). The genetics and expression of resistance in plants to rusts of the genus *Puccinia. Annu. Rev. Phytopathol.* **5,** 183–200.

Johnson, T., Green, G. J., and Samborski, D. J. (1967). The world situation of the cereal rusts. *Annu. Rev. Phytopathol.* **5,** 183–200.

Kuchler, F., Duffy, M., Shrum, R. D., and Dowler, W. M. (1984). Potential economic consequences of the entry of an exotic fungal pest: The case of soybean rust. *Phytopathology* **74,** 916–920.

Littlefield, L. J., and Heath, M. C. (1979). "Ultrastructure of Rust Fungi." Academic Press, New York.

Lopez, A., Rajaram, S., and DeBauer, L. I. (1974). Susceptibility of triticale, rye and wheat to stem rust from these three hosts. *Phytopathology* **64,** 266–267.

McCain, J. W., and Hennen, J. F. (1984). Development of the uredinial thallus and sorus in the orange coffee rust fungus. *Hemileia vastatrix. Phytopathology* **74,** 714–721.

Pearson, R. C., Seem, R. C., and Meyer, F. W. (1980). Environmental factors influencing the discharge of basidiospores of *Gymnosporangium juniperi-virginianae. Phytopathology* **70,** 262–266.

Peterson, R. S., and Jewell, F. F. (1968). Status of American stem rusts of pine. *Annu. Rev. Phytopthol.* **6,** 23–40.

Powers, H. R., Schmidt, R. A., and Snow, G. A. (1981). Current status and management of fusiform rust on southern pines. *Annu. Rev. Phytopathol.* **19,** 353–371.

Rapilly, F. (1979). Yellow rust epidemiology. *Annu. Rev. Phytopathol.* **17,** 59–73.

Rijkenberg, F. H. J., and Truter, S. J. (1973). Haustoria and intracellular hyphae in the rusts. *Phytopathology* **63,** 281–286.

Roelfs, A. P. (1982). Effect of barberry eradication on stem rust in the United States. *Plant Dis.* **66,** 177–181.

Rowell, J. B. (1973). Control of leaf and stem rusts of wheat by seed treatment with oxycarboxin. *Plant Dis. Rep.* **57,** 567–571.

Rowell, J. B. (1981). Control of stem rust on spring wheat by triadimefon and fenapanil. *Plant Dis.* **65,** 235–236.

Schieber, E., and Zentmeyer G. A. (1984). Coffee rust in the Western Hemisphere. *Plant Dis.* **68,** 89–93.

Scott, K. J., and Chakravorty, A. K., eds. (1982). "The Rust Fungi." Academic Press, New York.
Scott, K. J., and Maclean, D. J. (1969). Culturing of rust fungi. *Annu. Rev. Phytopathol.* **7,** 123–146.
Shaw, M. (1963). the physiology and host-parasite relations of the rusts. *Annu. Rev. Phytopathol.* **1,** 259–294.
Stakman, E. C. (1914). A study in cereal rusts. Physiological races. *Stn. Bull.—Minn., Agric. Expt. Stn.* **138,** 1–56.
Stavely, J. R. (1984). Pathogenic specialization in *Uromyces phaseoli* in the United States and rust resistance in beans. *Plant Dis.* **68,** 95–99.
Subrahamanyam, P., Reddy, L. I., Gibbons, R. W., and McDonald, D. (1985). Peanut rust: A major threat to peanut production in the semiarid tropics. *Plant Dis.* **69,** 813–819.
Ward, H. M. (1882). Researches on the life history of *Hemileia vastatrix,* the fungus of the "coffee leaf disease." *Linn. Soc. J. (Bot.)* **19,** 229–335.
Zadoks, J. C. (1965). Epidemiology of wheat rusts in Europe. *FAO Plant Prot. Bull.* **13,** 97–108.

The Smuts

The plant smuts, caused by Basidiomycetes of the order Ustilaginales, occur throughout the world and, until this century, were the causes of serious grain losses that were equal to, or second only to, the losses caused by the rusts. In some respects, the smuts of cereals were dreaded by farmers even more than the rusts because many smuts attack the grain kernels themselves and replace the kernel contents with the black, dusty spore masses that resemble soot or smut. Thus the reduction in yield is conspicuous and direct, and the quality of the remaining yield is drastically reduced by the presence of the black smut spores on the surface of the healthy kernels.

In addition to the various cereals, smuts also affect sugarcane, onions, and some ornamentals such as carnation.

Most smut fungi attack the ovaries of grains and grasses and develop in them or in the fruit, that is, the kernels of grain crops, which they destroy completely (Figure 11-85). Several smuts, however, attack the leaves, stems or floral parts. Some smuts infect seeds or seedlings before they emerge from the ground, and they grow internally in the seedling until they reach the inflorescence; others cause only local infections on leaves, stems, and so on. Cells in affected tissues are either destroyed and replaced by black smut spores or they are first stimulated to divide and enlarge to produce a swelling or gall of varying size and are then destroyed and replaced by the black smut spores. The spores are present in masses called sori that may be held together only temporarily by a thin, flimsy membrane or by a more or less durable one. Smut fungi seldom kill their hosts, but in some cases infected plants may be severely stunted.

The smut fungi can be grown in culture on artificial media, but in nature they exist almost entirely as parasites on their hosts. Most smut fungi produce only two kinds of spores: teliospores and basidiospores (Figure 11-83). Their teliospores are usually formed from mycelial cells along the length of the mycelium within the smut galls, and their basidiospores either bud off laterally from the basidium cells or are produced as a cluster at the tip of a nonseptate basidium. The basidiospores of the smuts are not borne on sterigmata. When basidiospores germinate, they either unite with compatible ones while still on the basidium and then infect or their germ tubes penetrate tissues directly. Their haploid mycelium, however, cannot invade tissues extensively and does not cause typical infections until two compatible mycelia unite to produce dikaryotic mycelium. The latter, then, invades tissues inter- or intracellularly, generally without haustoria, and produces the typical symptoms and the teliospores. The smut fungi also exist in many races, which, however, are not as stable as in the rusts since each generation of smut fungi on the host plant involves meiosis, that is, genetic recombination, and this results in new races appearing constantly.

There are about 1100 species of smut fungi that attack Angiosperms in more than 75 families. The most common smut fungi and the diseases they cause are the following:

Ustilago, causing corn smut *(U. maydis),* loose smut of oats *(U. avenae),* of barley *(U. nuda)* and of wheat *(U. tritici),* semiloose smut of barley (*U. nigra),* covered smut of barley and oats *(U. hordei),* and sugarcane smut *(U. scitaminea).*

Tilletia, causing covered smut or bunt of wheat *(T. caries* and *T. foetida),* dwarf bunt of wheat *(T. contraversa),* and Karnal bunt of wheat *(T. indica).*

Sphacelotheca, causing the sorghum smuts, such as covered kernel smut *(S. sorghi),* loose kernel smut *(S. cruenta),* and head smut of sorghum and corn *(S. reiliana).*

Urocystis, causing onion smut *(U. cepulae),* and leaf or stalk smut of rye *(U. occulta).*

Neovossia, causing kernel smut of rice *(N. barclayana).*

Entyloma, causing leaf smut of rice *(E. oryzae)* and of some broad-leaved plants such as spinach *(E. ellisii).*

The smuts generally overwinter as teliospores on contaminated seed, plant debris, or in the soil. However, some smuts overwinter as mycelium inside infected kernels or as mycelium in infected plants. The teliospores cannot infect but they produce the basidiospores, which upon germination either fuse with compatible ones and then infect, or penetrate the tissue and then fuse to produce dikaryotic mycelium and the typical infection. The smut fungi have only one generation per year, each infection resulting in one crop of teliospores per growing season.

Control of smuts is primarily by resistant varieties and seed treatment. The latter may be either by chemical dusting or dip, if the fungus is present as teliospores on the seed surface or in the soil, or by hot water if the fungus is present as mycelium inside the seed. The discovery of carboxin, thiabendazole, etaconazole, and other fungicides that are absorbed and translocated systemically by seeds and seedlings allows chemical control by seed treatment of even those smuts present as mycelium inside the seeds. Soil treatments with these and other chemicals are also useful in the control of smut diseases.

General Smuts

• Corn Smut

Corn smut occurs wherever corn is grown. It is more prevalent, however, in warm and moderately dry areas, where it causes serious damage to susceptible varieties, and particularly to sweet corn.

Corn smut damages plants and reduces yields by forming galls on any of the aboveground parts of plants, including ears, tassels, stalks, and leaves. The number, size, and location of smut galls on the plant affect the amount of yield loss. Galls on the ear usually destroy it to a large extent, while large galls above the ear cause much greater reduction in yield than do galls below the ear. Losses from corn smut are highly variable from one location to another and may range from a trace up to 10 percent or more in localized areas. Some individual fields of sweet corn may show losses approaching 100 percent from corn smut. Generally, however, over large areas and with the use of resistant varieties, losses in grain yields average about 2 percent.

◦ *Symptoms.* When young corn seedlings are infected, minute galls form on the leaves and stem, and the seedling may remain stunted or may be killed. Seedling infection, because of seedling death, is seldom observed in the field.

On older plants, infections occur on the young, actively growing tissues of axillary buds, individual flowers of ear and tassel, leaves, and stalks (Figure 11-97).

Infected areas are permeated by the fungus mycelium, which stimulates the host cells to divide and enlarge, thus forming overgrowths or galls, Galls are first covered with a greenish white membrane. Later as the galls mature, they reach a size from 1 to 15 cm in diameter, their interior darkens, and they turn into a mass of powdery, dark, olive-brown spores. The silvery gray membrane then ruptures and exposes the millions of the sooty teliospores, which are released into the air. Galls on leaves frequently remain very small (about 1–2 cm in diameter); they become hard and dry and do not rupture.

◦ *The pathogen: Ustilago maydis.* Ustilago maydis produces dikaryotic mycelium, the cells of which are transformed into black, spherical, or ellipsoidal teliospores that have prominent spinelike protuberances. Teliospores germi-

FIGURE 11-97 Corn smut symptoms on young stem of corn plant (A), ear of corn (B) and on male inflorescence (C). (Photos A and B courtesy Dept. Plant Path., Cornell Univ.)

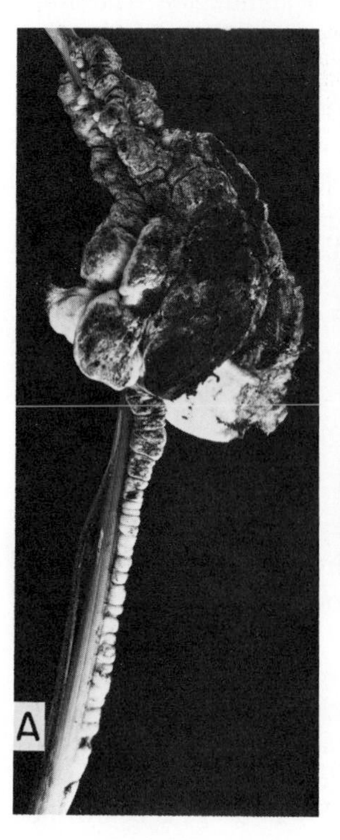

nate by producing a 4-celled basidium (promycelium), from each cell of which an ovate, hyaline, uninucleate basidiospore (sporidium) develops (Figure 11-98).

◦ *Development of Disease.* The fungus overwinters as teliospores in crop debris and in the soil, where it can remain viable for several years. In the spring and summer teliospores germinate and produce basidiospores, which are carried by air currents or are splashed by water to young, developing tissues of corn plants. The basidiospores germinate on the host surface and produce a fine hypha, which can enter epidermal cells by direct penetration. After an initial development, however, its growth stops and the hypha usually withers and sometimes dies, unless it contacts and fuses with a haploid hypha derived from a basidiospore of the compatible mating type. If fusion takes place, the resulting hypha enlarges in diameter and becomes dikaryotic. The dikaryotic hypha grows into the plant tissues mostly intercellularly (Figure 11-98).

The cells surrounding the hypha are stimulated to enlarge and divide, and galls begin to form. Hyperplasia may appear in advance of the actual invasion of the tissues by the fungus, and galls may begin to form even before the fungus actually gets there.

Galls in developed plants seem always to be the result of local infections of plant tissues. Systemic infections seldom occur, and then only in very young seedlings. Frequently, however, only a small number of the actual local infections develop into typical, large galls, the others remaining too small to be visible.

The mycelium in the gall remains intercellular during most of gall formation, but before sporulation, the enlarged corn cells are invaded by the mycelium, collapse, and die. The mycelium utilizes the cell contents for its further growth, and the gall then consists primarily of dikaryotic mycelium and cell remains. Most of the dikaryotic cells subsequently develop into teliospores and in the process seem to absorb and utilize the protoplasm of the other mycelial cells, which remain empty. Only the membrane covering the gall is not affected by the fungus, but finally the membrane breaks and the teliospores are released. Some of the released teliospores, if they land on young, meristematic corn tissues may cause new infections and new galls during the same season, but most of them fall to the ground or remain in the corn debris, where they can survive for several years.

◦ *Control.* Corn smut may be controlled to a degree through the use of corn hybrids with some resistance to the fungus. No corn varieties or hybrids completely resistant to smut are known. The pathogen, however, shows extreme variability in its pathogenicity, and new races appear constantly, making control through resistance difficult. Control through sanitation measures, such as removal of smut galls before they break open, and through crop rotation is possible only where corn in grown in small, rather isolated plots but is impractical and impossible in large corn-growing areas.

Kernel Smuts of Small Grains

- **Loose Smut of Cereals**

Loose smut of cereals is worldwide in distribution but is more abundant and serious in humid and subhumid regions.

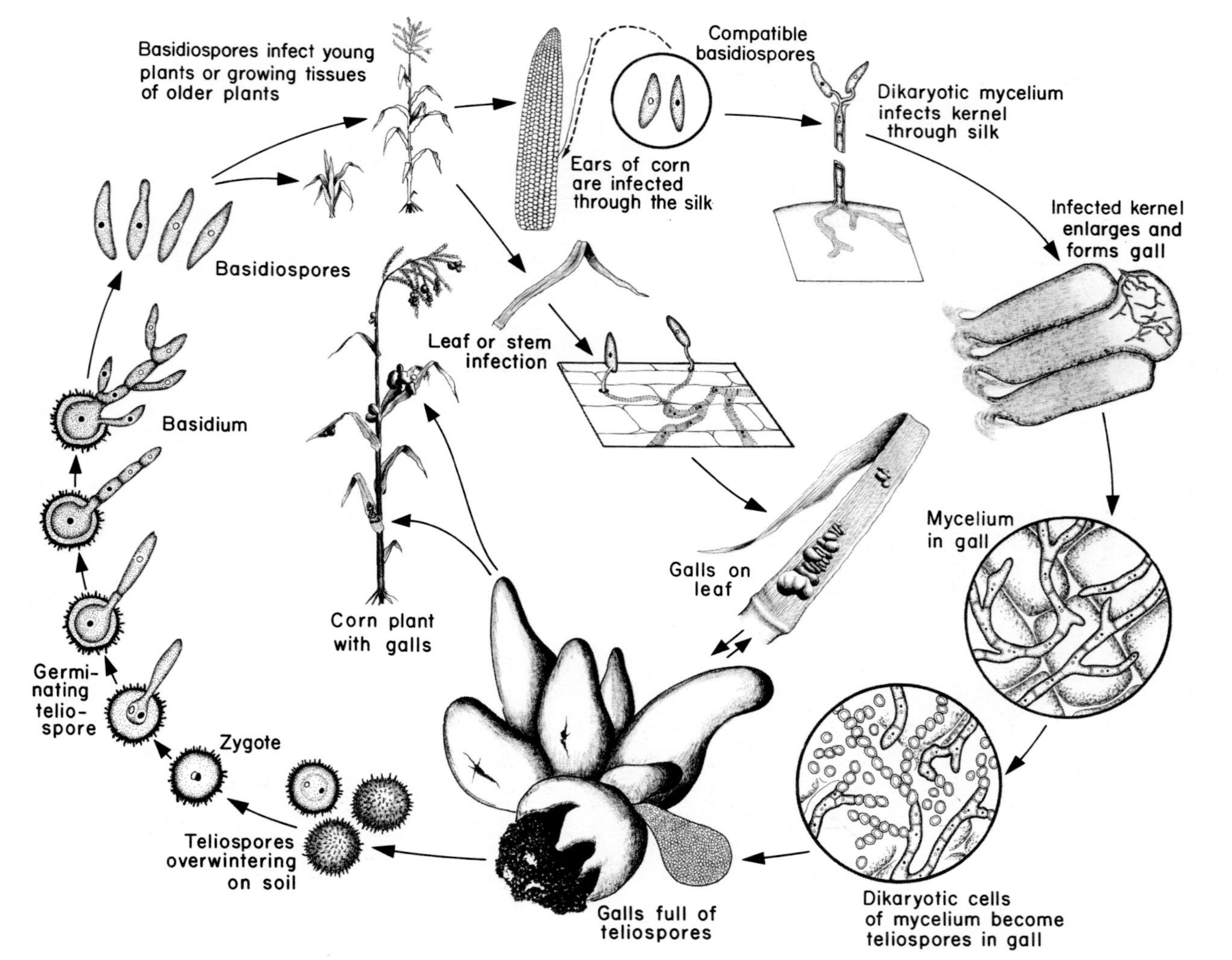

FIGURE 11-98 Disease cycle of corn smut caused by *Ustilago maydis*.

Loose smut causes damage by destroying the kernels of the infected plants and by smearing and thus reducing the quality of the grain of the noninfected plants upon harvest. Losses from loose smut may be up to 10 or 40 percent in certain localities in a given year, but the overall losses in the United States are approximately 1 percent per year.

◦ *Symptoms.* Loose smut generally does not produce discernible symptoms until the plant has headed. Smutted plants sometimes head earlier than healthy ones, and smutted heads are often elevated above those of the healthy plants. In an infected plant usually all the heads and all the spikelets and kernels of each head are smutted, although some of them may sometimes escape infection. In infected heads each spikelet is entirely transformed into a smut mass consisting of olive-green spores (Figure 11-99). This is at first covered by a delicate grayish membrane, which soon bursts and sets the powdery spores free. The spores are then blown off by the wind and leave the rachis a naked stalk.

◦ *The pathogens: Ustilago nuda, U. tritici.* The mycelium is hyaline during its growth through the plant, but it changes to brown near maturity. The mycelial cells are transformed into brown, spherical, echinulate teliospores, which germinate readily and produce a basidium consisting of one to four cells. The basidium produces no basidiospores, but its cells germinate and produce short, uninucleate hyphae that fuse in pairs and produce dikaryotic mycelium, which is capable of infection (Figure 11-100).

FIGURE 11-99 Loose smut of barley as it appears in the field (A) and on a single head of barley (B). (Photos courtesy Dept. Plant Path., Cornell Univ.)

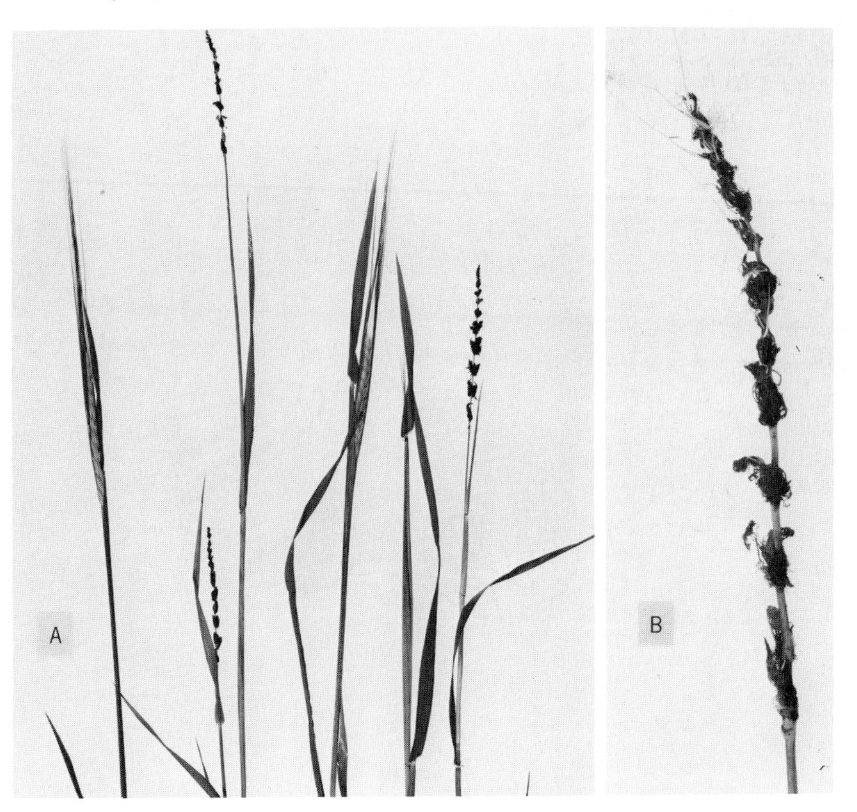

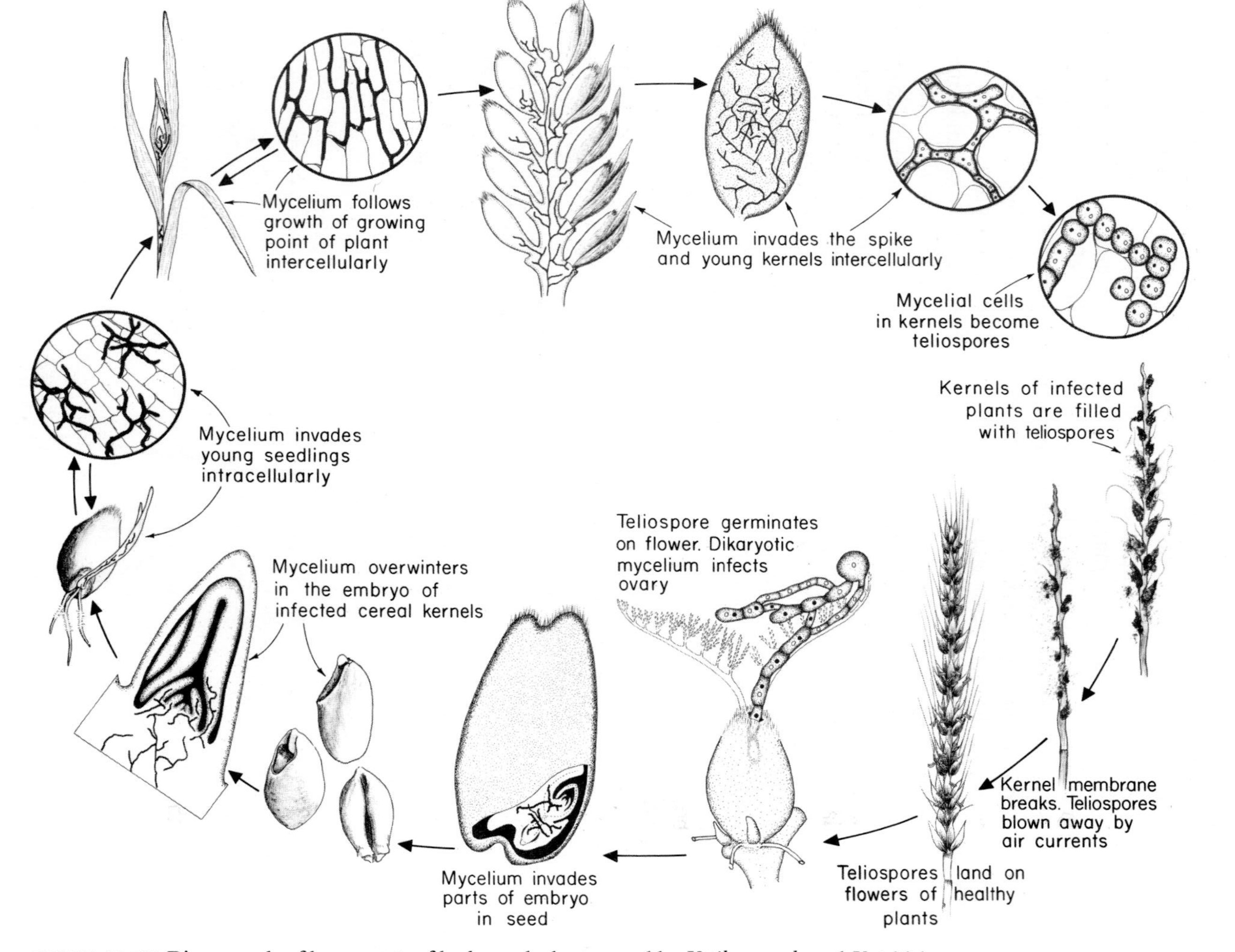

FIGURE 11-100 Disease cycle of loose smuts of barley and wheat caused by *Ustilago nuda* and *U. tritici*.

- *Development of Disease.* The pathogens overwinter as dormant mycelium in the cotyledon (sometimes called the scutellum) of infected kernels. When planted, infected kernels begin to germinate, the mycelium resumes its activity and grows intercellularly through the tissues of the embryo and the young seedling until it reaches the growing point of the plant (Figure 11-100). The mycelium then follows closely the growth of the plant and grows best just behind the growing point, while the hyphae in the tissues of the lower stem atrophy and frequently disappear. When the plant forms the head, and even before it emerges, the mycelium invades all the young spikelets, where it grows intracellularly and destroys most of the tissues of the spike, except the rachis. By this time, most of the infected plants are slightly taller than most healthy plants, probably due to stimulatory action of the pathogen. The mycelium in the infected kernels is soon transformed into teliospores, which are contained only by a delicate outer membrane of host tissue. The membranes burst open soon after maturation of the teliospores, and the spores are released and blown off by air currents to nearby healthy plants. The spore release coincides with the opening of the flowers of healthy plants. Teliospores landing on flowers germinate through formation of a basidium on which the haploid hyphae are produced. After fusion of the sexually compatible haploid hyphae, the resulting dikaryotic mycelium penetrates the flower through the stigma or through the young ovary walls and becomes established in the pericarp, integuments, and in the tissues of the embryo before the kernels become mature. The mycelium then becomes inactive and remains dormant, primarily in the scutellum, until the infected kernel germinates.
- *Control.* Control of loose smut is now obtained by treating infected seeds with carboxin and its carboxanilide derivatives before planting. These chemicals are absorbed and act systemically in the seed or in the growing plant.

  Although some barley and wheat varieties are quite resistant to loose smut, most of the commercial varieties are very susceptible to it.

  The best means of controlling loose smut is through the use of certified smut-free seed. Until the discovery of the systemic fungicides, when seed was known to be infected with loose smut mycelium, the best way of disinfecting it was by treating it with hot water. Usually small lots of seed are treated with hot water and planted in isolated fields to produce smut-free seed to be used during the next season. The hot-water treatment consists of soaking the seed, contained in half-filled burlap bags, in 20°C water for 5 hours, draining it for 1 minute, dipping it in 49°C water for about 1 minute, then in 52°C water for exactly 11 minutes, and immediately afterward in cold water for the seed to cool off. The seed is then allowed to dry so that it can be sown. Since some of the seed may be killed by the hot-water treatment, a higher seeding rate may be employed to offset the reduced germinability of the treated seed.

• Covered Smut, or Bunt, of Wheat

Covered smut, or bunt, of wheat, sometimes called stinking smut of wheat, is widely distributed in all the wheat-growing areas of the world. There are actually two kinds of bunt caused by related but different fungi: Common bunt, which now is easily controlled by treating the seed with fungicides and therefore causes few losses in most developed countries, and dwarf bunt, which still cannot be controlled and therefore continues to cause severe losses in many parts of the world including the United States, but only in parts of the Pacific Northwest.

Bunt affects plants by destroying the contents of infected kernels and replacing them with the spores of the fungus. Bunt also causes slight to severe stunting of infected plants, depending on the particular species of the bunt fungus involved. Infected plants are usually more susceptible than healthy plants to certain other diseases and to winter injury. Bunt causes losses in grain yields proportional to the number of plants affected. When bunt is not controlled, it may cause devastating losses, but even with the effective control measures practiced in the United States today, the disease continues to cause severe losses. In addition, bunt and the other smuts cause market losses by reducing the quality, and the price, of wheat contaminated with smutted kernels or smut spores because of the discoloration and the foul odor they impart to the whole wheat crop and because such wheat is reduced in grade and suitable for feed uses only. Bunt, moreover, results in explosions in combines and elevators during threshing or handling of smutted wheat because of the extreme combustibility of the oily smut spores in the presence of sparks from machinery.

◦ *Symptoms.* Plants infected with the common bunt fungi are usually a few to several centimeters shorter than healthy plants and may sometimes be only half as tall. Plants infected with the dwarf bunt fungus may be only one-fourth as tall as healthy plants and may show an increase in number of tillers. Infected plants may appear slightly bluish green to grayish green in color, but this is not easily distinguishable. Some varieties sometimes show a flecking or mottling reaction on the leaves of infected seedlings. The root system of infected plants is usually poorly developed.

Distinct bunt systems, however, appear when the heads of infected plants emerge. Their color is usually bluish green rather than the normal yellowish green, they are slimmer than healthy heads, and the glumes seem to spread apart and form a greater angle with the main axis of the head than they do in healthy plants (Figure 11-101). Infected kernels are shorter and thicker than healthy ones and are grayish brown rather than the normal golden yellow or red. When these kernels are broken after their maturation, they are found to be full of a sooty, black, powdery mass of fungus spores that give off a distinctive odor resembling that of decaying fish. During harvest of infected fields, large clouds of spores may be released in the air.

◦ *The pathogens.* *Tilletia caries* and *T. foetida* cause the common bunt, while *T. contraversa* causes dwarf bunt. The first two species are similar in their life histories and disease development, but the teliospores of *T. foetida* have smooth walls, while those of *T. caries* have highly reticulate walls. *T. contraversa's* biology is somewhat different and its spore walls have large, polygonal reticulations.

The mycelium is hyaline, and during sporulation most cells are transformed into almost spherical, brownish teliospores. The rest of the mycelial cells remain hyaline, thin walled, and sterile. Upon germination of a teliospore a basidium is produced, at the end of which 8 – 16 basidiospores develop in *T. caries* and *T. foetida,* while 14 – 30 basidiospores develop in the dwarf bunt fungus *T. contraversa.* The basidiospores are usually called primary sporidia. The primary sporidia fuse in pairs through production of lateral branches between compatible mating types and appear as H-shaped structures (Figures 11-101 and 11-102). The nucleus of each primary spori-

FIGURE 11-101 (A) Covered smut or bunt of wheat caused by *Tilletia* sp. Left two heads show abnormal spread of the glumes due to the bunt balls. Healthy head at right. (B) Germinating teliospore of *T. caries* producing the basidium (promycelium). Primary sporidia are connected by fertilization tubes and form the H-shaped structures shown at right. (C) Covered smut (a) and loose smut (b) of barley. (c) Healthy barley head. (Photos A and C courtesy U.S.D.A. Photo B courtesy M. F. Brown and H. G. Brotzman.)

dium divides, and through exchange of one of the nuclei the two fused primary sporidia become dikaryotic. When the primary sporidia germinate, they produce short hyphae on which dikaryotic secondary sporidia are formed. Upon germination the secondary sporidia produce dikaryotic mycelium, which can penetrate the plants and cause infection. After systemic development through the plant, the mycelium again forms teliospores.

◦ *Development of Disease.* The pathogens of common bunt overwinter mainly as teliospores on contaminated wheat kernels and less frequently in the soil. The teliospores of the common bunt fungi are short-lived in wet areas, losing viability within 2 years, while those of the dwarf bunt fungus may remain viable in any soil for at least 3 years and often for as long as 10 years.

When contaminated seed is sown or healthy seed is sown in bunt-infested fields, approximately the same conditions that favor germination of seeds favor germination of common bunt teliospores. Teliospores of the common bunt fungi germinate readily, and as the young seedling emerges from the kernel, the teliospore on the kernel or near the seedling also germinates through production of the basidium, primary sporidia, and secondary sporidia (Figure 11-102). The secondary sporidia then germinate, and the dikaryotic mycelium they produce penetrates the young seedling directly.

Teliospores of the dwarf bunt fungus, however, germinate slowly even under optimum conditions of temperature (3–8°C), moisture, and oxygen, requiring from 3 to 10 weeks or longer for maximum germination. Persistent

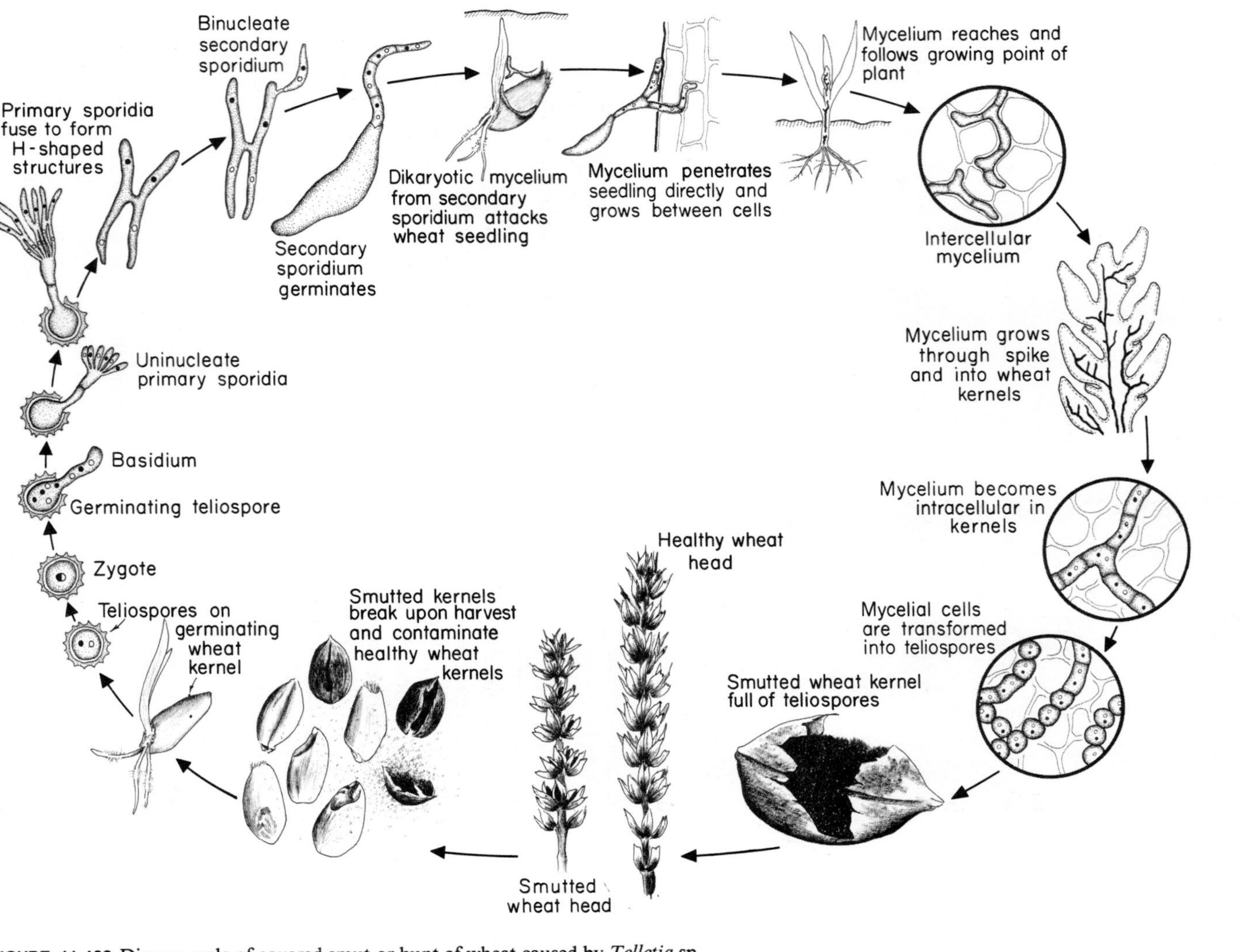

FIGURE 11-102 Disease cycle of covered smut or bunt of wheat caused by *Telletia* sp.

snow cover, providing soil surface temperatures of −2 to 2°C is consistently correlated with high dwarf bunt incidence. Dwarf bunt infections apparently originate from teliospores germinating at or near the soil surface from December through early April. Germinating secondary sporidia penetrate the tiller initials of wheat seedlings after seedling emergence. The more tiller initials formed during the infection period, the greater the incidence of bunted plants and of bunted heads per plant. Germinating seedlings and older tillers apparently are not susceptible to infection by the dwarf bunt fungus.

After penetration the mycelium grows intercellularly and invades the developing leaves and the meristematic tissue at the growing point of the plant. The mycelium remains dormant in the seedling during the winter, but when the seedling begins to grow again in the spring, the mycelium resumes its growth and grows with the growing point. When the plant forms the head of the grain, the mycelium invades all parts of it even before the head emerges out of the "boot." As the head fills and becomes mature, the mycelial threads increase in numbers and soon take over and consume the contents of the kernel cells. The mycelium, however, does not affect the tissues of the pericarp of the kernel, which form a rather sturdy covering for the smutted mass they contain. At the same time most hypha cells are transformed into teliospores.

Smutted kernels are usually kept intact while on the plant but break and release their spores upon harvest or threshing. The liberated spores contaminate the healthy kernels and are also blown away by air currents, thus contaminating the soil.

◦ *Control.* Common bunt can be controlled by using smut-free seed of a resistant variety treated with an appropriate fungicide. Contaminated seed should be cleaned to remove any unbroken, infected kernels and as many of the smut spores on the seed as possible. The seed is then treated with the fungicide hexachlorobenzene (HCB) or carboxin, although other fungicides, including thiram, chloranil, thiabendazole, and benomyl, give good control of the disease. In dwarf bunt, and in common bunt in drier areas, the spores survive in the soil for long periods and can cause infection of seedlings. In such cases, the most effective control is through the use of resistant cultivars, while seed treatment with the systemic fungicides carboxin, thiabendazole, triadimefon, triadimenol, etaconazole, and fenapanil are moderately effective.

SELECTED REFERENCES

Ben-Yephet, Y., Henis, Y., and Dinoor, A. (1975). Inheritance of tolerance to carboxin and benomyl in *Ustilago hordei. Phytopathology* **65**, 563–567.

Christensen, J. J. (1963). Corn smut caused by *Ustilago maydis. Am. Phytopathol. Soc. Monogr.* **2**, 1–41.

Churchward, J. G. (1940). The initiation of infection by bunt of wheat *(Tilletia caries). Ann. Appl. Biol.* **27**, 58–64.

Comstock, J. C., Ferreira, S. A., and Tew, T. L. (1983). Hawaii's approach to control of sugarcane smut. *Plant Dis.* **67**, 452–457.

Davis, G. N. (1936). Some of the factors influencing the infection and pathogenicity of *Ustilago zeae* (Backm.) Unger on *Zea mays* L. *Iowa, Agric. Exp. Stn., Res. Bull.* **199,** 247–278.

Dickson, J. G. (1956). "Diseases of Field Crops." McGraw-Hill, New York.

Dietrich, S. (1959). Untersuchungen zur Biologie and Bekakmpfung von *Ustilago zeae* (Beckm.) Unger. *Phytopathol. Z.* **35**, 301–322.

Fischer, G. W. (1953). "Manual of the North American Smut Fungi." Ronald Press, New York.
Fischer, G. W., and Holton, C. S. (1957). "Biology and Control of the Smut Fungi." Ronald Press, New York.
Halisky, P. M. (1965). Physiologic specialization and genetics of the smut fungi. III. *Bot. Rev.* **31**, 114–150.
Hoffmann, J. A. (1982). Bunt of wheat. *Plant Dis.* **66**, 979–986.
Holton, C. S., Hoffmann, J. A. and Duran, R. (1968). Variation in the smut fungi. *Annu. Rev. Phytopathol.* **6**, 213:242.
Joshi, L. M. *et al.* (1983). Karnal bunt: A minor disease that is now a threat to wheat. *Bot. Rev.* **49**, 309–330.
Kendrick, E. L. (1965). The reaction of varieties and hybrid selections of winter wheat to pathogenic races of *Tilletia caries* and *T. foetida. Plant Dis. Rep.* **49**, 843–846.
Mathre, D. E., ed. (1982). "Compendium of Barley Diseases." Am. Phytopathol. Soc., St. Paul, Minnesota.
Purdy, L. H., Hoffman, J. A., Meiners, J. P., and Stewart, V. R. (1963). Time of year of infection of winter wheat by the dwarf bunt fungus. *Phytopathology* **53**, 1419–1421.
Spencer, J. L., and White, H. E. (1951). Another smut of carnation. *Phytopathology* **41**, 291–299.
Thomas, P. L. (1984). Barley smuts in the Prairie Provinces of Canada. 1978–1982. *Can. J. Plant Pathol.* **6**, 78–80.
Trione, E. J. (1982). Dwarf bunt of wheat and its importance in international wheat trade. *Plant Dis.* **66**, 1083–1088.
Urech, P. A. (1972). Investigations on the corn smut caused by *Ustilago maydis. Phytopathol. Z.* **73**, 1–26.
Wiese, M. V. (1987). "Compendium of Wheat Diseases." 2nd Ed., APS Press, Am. Phytopathol. Soc., St. Paul, Minnesota.

Root and Stem Rots Caused by Basidiomycetes

Several Basidiomycetes cause serious plant losses by attacking primarily the roots and lower stems of plants (Figure 11-85). Some of these fungi, for example, *Rhizoctonia (Thanatephorus)* and *Sclerotium (Athelia, Pellicularia),* attack primarily herbaceous plants, and *Typhula* attacks only grasses. On the other hand, some other fungi, for example, *Armillaria,* some species of *Heterobasidion,* particularly *H. annosum,* and of *Poria* and *Polyporus,* attack only roots and lower stems of woody plants, primarily forest trees and certain fruit trees. Another fungus, *Marasmius,* includes species that affect turf grasses, others that cause root or crown rot of banana and sugarcane, and others that cause witches'-broom of cacao and wiry cord blights on the tops of tropical trees.

Root and Stem Rot Diseases Caused by the "Sterile Fungi" Rhizoctonia and Sclerotium

The fungi *Rhizoctonia* and *Sclerotium* are soil inhabitants and cause serious diseases on many hosts by affecting the roots, stems, tubers, corms, and other plant parts that develop in or on the ground. These two fungi are known as sterile fungi because for many years they were thought to be incapable of producing spores of any kind, either sexual or asexual. It is known now that at least some species within these two genera produce spores, some of them sexual spores and some conidia. Thus, *Rhizoctonia solani* produces basidiospores, which thus makes *Rhizoctonia* a basidiomycete and is called *Thanatephorus cucumeris.* On the other hand, the most common species of *Sclero-*

tium, S. rolfsii, also produces basidiospores and has as its perfect stage the basidiomycete *Athelia (Pellicularia) rolfsii,* while two other species, *S. bataticola* and *S. cepivorun,* have been reported to produce conidia belonging to the imperfect fungi *Macrophomina* and *Sphacellia,* respectively. However, the spores of these fungi are either produced only under special conditions in the laboratory or are extremely rare in nature and therefore of little value in diagnosing the fungus. For these reasons, these fungi continue to be considered as sterile mycelia, and since, for all practical purposes, they behave as such they continue to be referred to by their names as *Rhizoctonia* and *Sclerotium.*

• Rhizoctonia Diseases

Rhizoctonia diseases occur throughout the world and cause losses on most annual plants, including almost all vegetables and flowers, several field crops, and also on perennial plants such as turf grasses, perennial ornamentals, shrubs, and trees. The symptoms of *Rhizoctonia* diseases may vary somewhat on the different crops and even on the same host, depending on the stage of growth at which the plant becomes infected and the prevailing environmental conditions. The most common symptoms caused by *Rhizoctonia,* primarily *R. solani,* on most plants are damping off of seedlings and root rot, stem rot, or stem canker of growing and grown plants. On some hosts, however, *Rhizoctonia* also causes rotting of storage organs and foliage blights or spots, especially of foliage near the ground (Figure 11-103).

Damping off is probably the most common symptom caused by *Rhizoctonia* on most plants it affects. It occurs primarily in cold, wet soils. Very young seedlings may be killed before or soon after they emerge from the soil (Figure 11-103C). Before emergence, the fungus attacks and kills the growing tip of the seedling, which then soon dies. However, thick, fleshy seedlings such as those of legumes and the sprouts from potato tubers may show noticeable brown, dead tips and lesions before they are killed. After the seedlings have emerged, the fungus attacks their stem and makes it water soaked, soft, and incapable of supporting the seedling, which then falls over and dies. Older seedlings may also be attacked, but on these the invasion of the fungus is limited to the outer cortical tissues on which the fungus produces elongate, tan to reddish brown lesions. The lesions may increase in length and width until they finally girdle the stem and the plant may die, or as it often happens in crucifers, before the plant dies the stem may turn brownish black and may be bent or twisted without breaking, giving the disease the name wire stem (Figure 11-103B).

A **seedling stem canker** known as "soreshin" is common and destructive in cotton seedlings that have escaped the damping off or seedling blight phase of the disease and develops under conditions that are not especially favorable to the disease. The "soreshin" lesions appear as reddish brown, sunken cankers that range from narrow to completely girdling the stem near the soil line (Figure 11-103A). As soil temperature rises later in the season, affected plants may show partial recovery due to new root growth. "Soreshin" also affects tobacco and other crops in the seed bed or in the field. Dark-colored cankers or rotting develops at the base of the stem and may extend into the woody tissues and the pith as well as up the stem and into the lower leaves.

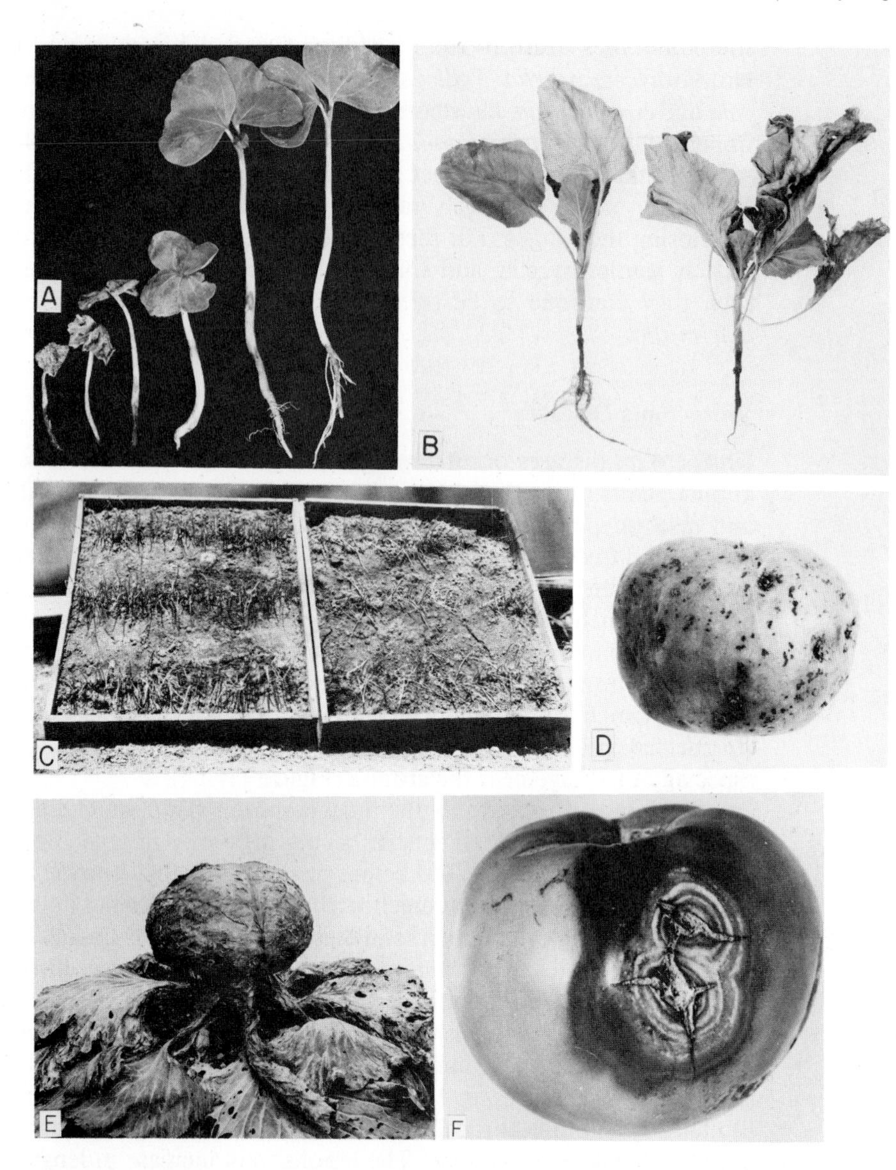

FIGURE 11-103 (A) Cotton seedling stem canker ("soreshin") caused by *Rhizoctonia.* (B) "Wire stem" of young cabbage plants caused by *Rhizoctonia.* (C) Onion seedlings damped off by *Rhizoctonia* added to flat at right. (D) Potato showing sclerotia of *Rhizoctonia.* (E) *Rhizoctonia* head rot of cabbage. (F) *Rhizoctonia* soil rot of tomato fruit. (Photos A and B courtesy G. C. Papavizas, E and F courtesy U.S.D.A.)

The invaded tissues die and collapse, the black rotted area being relatively dry.

Root lesions form even in the damping off phase of the disease, since *Rhizoctonia* frequently attacks the roots at the same time it attacks the stem. On most partly grown or mature plants, the reddish brown lesions usually appear first just below the soil line, but in cool,wet weather the lesions enlarge in all directions and may increase in size and numbers to include the whole base of the plant and most of the roots. This results in weakening, yellowing, and sometimes death of the plant. If infested soil is splashed by rain onto stems and lower branches or their junctions, stem cankers may also develop.

On **low-lying plants** such as lettuce and cabbage, the lower leaves of which touch the ground or are close to it, *Rhizoctonia* attacks these leaves at the petioles and midribs on which it produces reddish brown, slightly sunken lesions, while the entire leaf becomes dark brown and slimy. From the lower leaves the infection spreads upward to the next leaves until most or all leaves and the head may be invaded and rot, with mycelium and sclerotia permeating the tissues or nestled between the leaves (Figure 11-103E, F).

On **lawn and fine turf** grasses, *Rhizoctonia* causes the **brown patch** disease. Brown patch is particularly severe during periods of hot and humid or wet weather, especially with heavy dew periods. It appears as roughly circular areas, ranging from a few centimeters to one or more meters in diameter, in which the grass blades become water soaked and dark at first but soon become dry, wither, and turn light brown. Diseased areas appear slightly sunken, but at the border of the diseased areas, where the fungus is still active and attacking new grass blades making them look water soaked and dark, a characteristic grayish black "smoke" ring 2–5 cm wide appears in damp days or in the early morning hours. As the grass dries, the activity of the fungus slows down or stops and the ring disappears. Brown to black, hard, round sclerotia about 2 mm in diameter form in the thatch, diseased plants, and soil. In brown patch, *Rhizoctonia* usually kills only the leaf blades, and plants in the affected area begin to recover and grow again from the center outward, resulting in a doughnut-shaped diseased area.

On **fleshy, succulent stems and roots,** and on tubers, bulbs, and corms, *Rhizoctonia* causes brown rotten areas that may be superficial or may extend inward to the middle of the root or stem. The rotting tissues usually decompose and dry, forming a sunken area filled with the dried plant parts mixed with fungus mycelium and sclerotia. The lesions may begin at the top of the fleshy root resulting in a crown rot that, in the field, may cause stunting and yellowing or death of the foliage. Lesions may also develop on the sides of fleshy tissues and may reach various sizes depending on host, weather, the presence of cracks, and so on. White, cream-colored or brown mycelium may cover the lesions in wet weather, and when the tissues rot and dry, sclerotia also develop.

On **potato tubers,** *Rhizoctonia* causes characteristic signs called "black scurf," in which small, hard, black sclerotia occur on the tuber surface and are not removed by washing (Figure 11-103D), or "russeting" or "russet scab" in which the skin becomes roughened in a crisscross pattern resembling the shallow form of common potato scab.

Finally, *Rhizoctonia* causes **rots on fruits and pods** lying on or near the soil, such as cucumbers, tomatoes (Figure 11-104F), eggplant, and beans. These rots develop most frequently in wet, cool weather and appear first in the field but may continue and spread to other fruits after harvest and during transportation and storage. The lesions appear at first as rather firm, water-soaked areas in which the tissues soon collapse and form a shallow, sunken area. In moist weather, mycelium appears on the spots, white at first but turning brown with age. The affected fruits and pods also turn brown and dry, or they may be invaded by soft-rotting bacteria that cause them to become mushy or watery.

Other *Rhizoctonia* species cause somewhat different symptoms. Thus, *R. crocorum* attacks only underground parts of many vegetables and ornamentals, and the diseased plant parts show a violet or red coloration due to the purple color of the superficial growth of the fungus, which also contains many closely aggregated and darker sclerotialike bodies.

In the sheath and culm blight of rice, one of the most serious diseases of rice and sometimes important on other cereals also, different *Rhizoctonia* species cause large, irregular lesions that have a straw-colored center and a wide, reddish-brown margin. Seedling and mature plants may become blighted under favorable conditions for the pathogen.

The pathogen, Rhizoctonia sp., and particularly *R. solani,* exists primarily as a sterile mycelium that is colorless when young but turns yellowish or light brown with age. The mycelium consists of long cells and produces branches that grow at approximately right angles to the main hypha, are slightly constricted at the junction, and have a cross wall near the junction (Figure 11-104). The branching characteristics are usually the only ones available for identification of the fungus as *Rhizoctonia.* Under certain conditions the fungus produces sclerotialike tufts of short, broad, ovate- to triangular-shaped cells that function as chlamydospores, or eventually the tufts develop into rather small, loosely formed brown to black sclerotia, which are common on some hosts such as potato. As mentioned earlier, *R. solani* infrequently

FIGURE 11-104 Disease cycle of *Rhizoctonia solani (Thanatephorus cucumeris).*

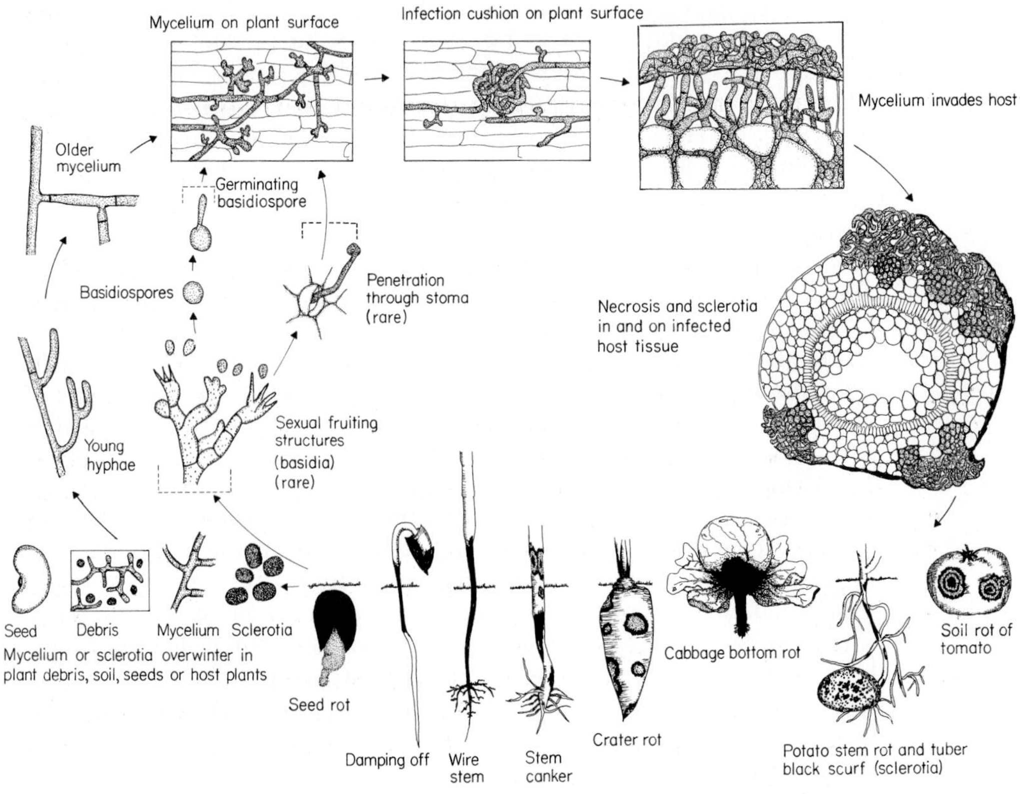

produces a basidiomycetous perfect stage known as *Thanatephorus cucumeris*. The perfect stage forms under high humidity and appears as a thin, mildewlike growth on soil, leaves, and infected stems just above the ground line. The basidia are barrel-shaped, are produced on a membranous layer of mycelium, and have four sterigmata, each bearing an ovoid basidiospore.

It has now become evident that *Rhizoctonia solani* is a "collective" species, consisting of at least four and possibly more, more or less unrelated strains. The *Rhizoctonia* strains are distinguished from each other because **anastomosis** (fusion of touching hyphae) occurs only between isolates of the same **anastomosis group.** After anastomosis, which can be detected microscopically, an occasional heterokaryon hypha may be produced, under certain conditions, from one of the anastomosing cells. In the vast majority of anastomoses, however, five to six cells on either side of the fusion cells become vacuolated and die, appearing as a clear zone at the junction of two colonies. This "killing reaction" between isolates of the same anastomosis group is the expression of somatic or vegetative incompatibility. Such somatic incompatibility limits outbreeding to a few compatible pairings. The existence of the anastomosis groups in *Rhizoctonia solani* represents genetic isolation of the populations in each group.

Although the various anastomosis groups are not entirely host specific, they show certain fairly well-defined tendencies: isolates of anastomosis group one (AG 1) cause seed and hypocotyl rot and aerial (sheath) and web blights of many plant species; isolates of AG 2 cause a canker of root crops, and many cause root diseases on crucifers; isolates of AG 3 affect mostly potato, causing stem cankers and stolon lesions and producing black sclerotia on tubers; isolates of AG 4 infect a wide variety of plant species, causing seed and hypocotyl rot on almost all angiosperms and stem lesions near the soil line on most legumes, cotton, and sugar beets. Recognition of the existence of anastomosis groups and of their lesser or greater host specificity has been important in determining the anastomosis group of the isolate that must be used for inoculations in breeding different crops for resistance to *Rhizoctonia* and of the propagules counted for making disease predictions for the various crops to that fungus.

The pathogen overwinters usually as mycelium or sclerotia in the soil, in or on infected perennial plants or propagative material such as potato tubers. In some hosts, such as bean, eggplant, pepper, and tomato, the fungus even invades and may be carried in the seed. The fungus is present in most soils and, once established in a field, remains there indefinitely. Different races of the fungus probably equivalent to the anastomosis groups, but also within each anastomosis group, exist, exhibiting different preferences as to hosts, temperature optimum, and so on. The fungus spreads with rain, irrigation, or flood water, with tools and anything else that carries contaminated soil, and with infected or contaminated propagative materials. For most races of the fungus the optimum temperature for infection is about 15–18°C, but some races are most active at much higher temperatures—up to 35°C. Disease is more severe in soils that are moderately wet than in soils that are waterlogged or dry. Infection of young plants is most severe when plant growth is slow due to adverse environmental conditions for the plant. Rapidly growing plants are likely to escape infection by *Rhizoctonia,* even when moisture and temperature may be favorable for the fungus.

Control of *Rhizoctonia* diseases, when the fungus is carried with the seed, depends on the use of disease-free seed or seed treated with hot water and chemicals. Wet, poorly drained areas should be avoided or drained better and seeds should be planted on raised beds and in soil in the best possible condition to encourage fast growth of the seedling. There should be wide spaces among plants for good aeration of soil surface and of plants. When possible, as in greenhouses and seed beds, the soil should be sterilized with steam or treated with chemicals. Drenching of soil with pentachloronitrobenzene (PCNB) helps reduce damping off in seed beds and greenhouses. When specific races of the pathogen have built up, a three-year crop rotation with another crop may be valuable. With most vegetables, no effective fungicides are available against Rhizoctonia diseases, although chlorothalonil, thiophanate methyl, iprodione, and some other chemicals are sometimes recommended as sprays on the soil before planting and once or twice on the seedlings soon after emergence. On turf grasses, along with proper drainage and removal of thatch, preventive fungicide applications are recommended, especially when temperatures stay above 21°C at night or 28°C during the day. Several fungicides, including some contact (iprodione and chlorothalonil) and systemic fungicides (carboxin, triadimefon, thiophanate methyl), seem to provide effective control.

In recent years, tremendous efforts have gone into developing alternative, more effective means of control of Rhizoctonia diseases. Such methods include mulching of fields with a 2- to 3-cm-thick layer of rice husks to control web blight of beans, or with photodegradable plastic mulch to control fruit rot of cucumber, and avoiding application of some herbicides that seem to increase Rhizoctonia diseases in certain crops.

The greatest effort, however, has gone into developing biological controls against Rhizoctonia diseases. *Rhizoctonia* is parasitized by several microorganisms such as the fungi *Trichoderma, Gliocladium,* and *Laetisaria,* several soil myxobacteria, and by mycophagous nematodes like *Aphelenchus avenae. Rhizoctonia* also often suffers by the so-called Rhizoctonia decline, which is caused by two or three infectious double-stranded RNAs. These RNAs, through anastomoses, spread from infected hypovirulent *Rhizoctonia* individuals to healthy virulent ones and reduce both their ability to cause disease and their ability to survive. Addition of these organisms to *Rhizoctonia*-infested soil, or treatment of seeds, tubers, and transplants with suspensions of spores or mycelium of the antagonistic or hypovirulent fungi, or with the myxobacteria, before planting in *Rhizoctonia*-infested soil, greatly reduces disease incidence and severity in almost all crops, for example carrot, bean, carnation, and potato, on which it has been tried. So far, however, biological controls are still at the experimental stage and are not available for use by farmers. A biological control is currently gaining acceptance for control of *Rhizoctonia* and other soil-borne fungi in container-grown plants. This control involves the use of composted hardwood bark in the soil medium. It apparently works by increasing the population of *Trichoderma* and other antagonistic microorganisms in the container and, possibly, by releasing some fungitoxic chemicals.

• Sclerotium Diseases

Sclerotium diseases appear as damping off of seedlings, stem canker, crown blight, root and crown rot, bulb and tuber rot, and fruit rots. *Sclerotium* frequently causes severe losses of fleshy fruits and vegetables during shipment and storage. Sclerotium diseases are primarily diseases of warm climates, affecting plants in countries within 38° latitude on either side of the equator. Because in the United States they are more common and severe in the southern states, they are often called "southern wilts or southern blights." Sclerotium diseases affect a wide variety of plants, including vegetables, flowers, cereals, forage plants, and weeds. Some of the most common hosts of *Sclerotium* include legumes, crucifers, cucurbits, carrot, celery, sweet corn, eggplant, lettuce, okra, onion, peppers, potato, sweet potato, tomato, amaryllis, chrysanthemum, delphinium, iris, narcissus, tulip, alfalfa, cereals, cotton, peanuts, and tobacco.

When seedlings are attacked, the fungus invades all the parts of the seedling, and the seedlings die quickly. When the fungus attacks plants that have already developed some woody tissue, it does not invade them throughout, but it grows into the cortex and slowly or quickly girdles the plants, which eventually die. Usually the infection begins on the succulent stem as a dark-brown lesion just below the soil line. The first visible symptoms appear as yellowing or wilting of the lower leaves or dying back of leaves from the tips downward. These symptoms then progress to the upper leaves. In plants with very succulent stems, such as celery, the stem may fall over, while in plants with harder stems, such as alfalfa, bean, tomato, and tobacco, the invaded stem stands upright and begins to lose its leaves or to wilt. In the meantime, the fungus grows upward in the plant covering the stem lesion with a cottony, white mass of mycelium, the upward advance of the fungus depending on the amount of moisture present. The fungus moves even more rapidly downward into the roots and finally destroys the root system. The white mycelium is always present in and on infected tissues, and from these it grows over the soil to adjacent plants, starting new infections. Invaded stem, tuber, and fruit tissues are usually pale brown and soft but not watery. The margin between healthy and diseased tissue is often darker than the other tissues. When fleshy roots or bulbs are infected, a watery rot of the outer scales or root tissues may develop or the entire root or bulb may rot and disintegrate and be replaced by debris interwoven with mycelium. If bulbs, roots, and fruits are infected late in their development, symptoms may go unnoticed at harvest, but the disease continues as a storage rot.

On all infected tissues, and even on the nearby soil, the fungus produces numerous small sclerotia of uniform size that are roundish or irregular and white when immature, becoming dark brown to black when they mature. The mature sclerotia are not connected with mycelial strands and have the size, shape, and color of mustard seed (Figure 11-105).

The fungus, *Sclerotium* sp., produces abundant white, fluffy, branched mycelium that forms numerous sclerotia but is usually sterile, that is, it does not produce spores. *Sclerotium rolfsii,* which causes the symptoms described above on most of the hosts, occasionally produces basidiospores at the margins of lesions under humid conditions. Its perfect stage is *Athelia (Pellicularia) rolfsii.* Another species, *S. bataticola,* which causes diseases in several different hosts, including ashy stem blight of bean and soybean, charcoal rot of sorghum and corn, stem rot on watermelon, and root and wood rot of citrus, produces numerous, small, black, irregular sclerotia, usually in the pith of stems or stalks of infected plants and, occasionally, also conidia in pycnidia of the *Macrophomina* type. A third *Sclerotium* species, *S. cepivorun,* which

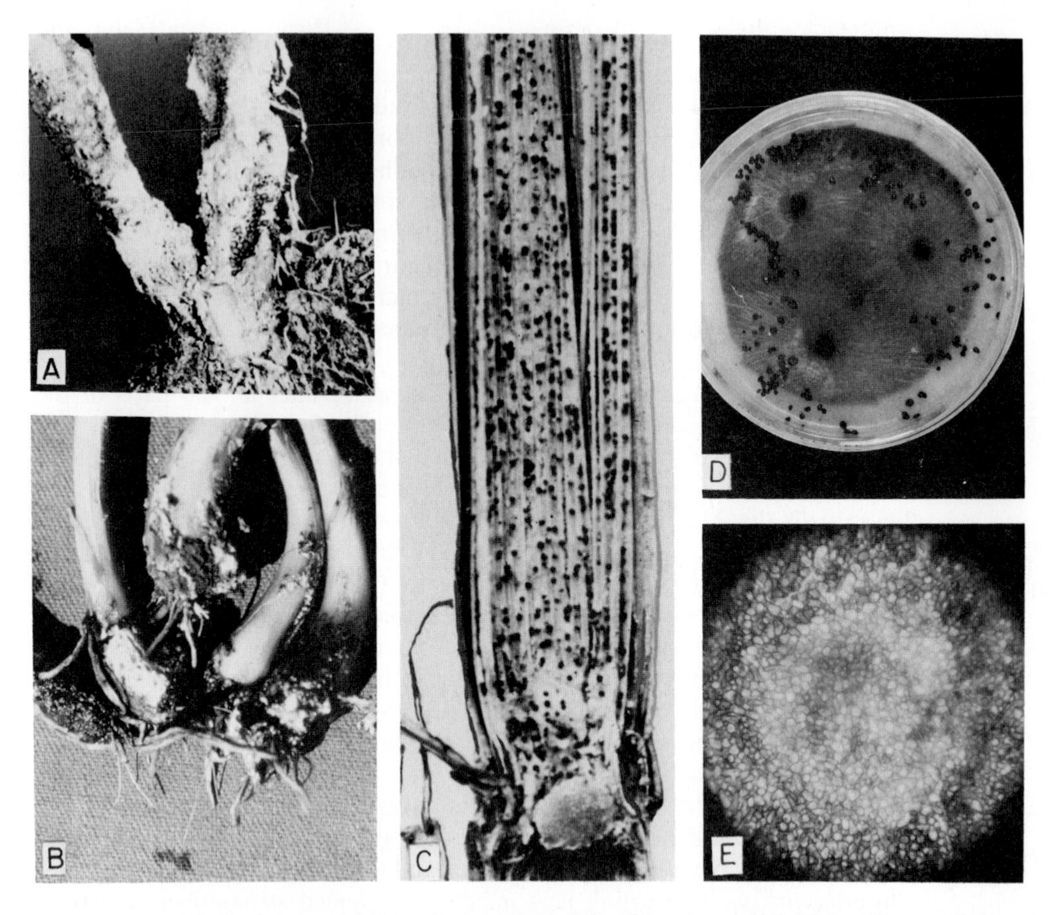

FIGURE 11-105 (A) Southern stem blight of tomato caused by *Sclerotium rolfsii.* Note white mycelium and round, uniform sclerotia. (B) Onion white rot caused by *Sclerotium cepivorum.* (C) Stem rot of rice caused by *Sclerotium oryzae.* Numerous sclerotia are showing. (D) Mycelium and sclerotia of *Sclerotium rolfsii* in culture. (E) Cross section of sclerotium showing the compact mass of mycelial cells. (Photos A and C courtesy U.S.D.A. Photo B courtesy P. B. Adams.)

causes the white rot disease of onion and garlic, in addition to sclerotia, also produces occasional conidia on sporodochia; these conidia, however, seem to be sterile.

The fungus seems to overwinter mainly as sclerotia. It is spread by moving water, infested soil, contaminated tools, infected transplant seedlings, infected vegetables and fruits, and in some hosts, as sclerotia mixed with the seed. Basidiospores and conidia may also participate in the dissemination of the species that produce them, but their role in this is not clearly established.

The fungus attacks tissues directly. However, it produces a considerable mass of mycelium and kills and disintegrates tissues by secreting oxalic acid, and also pectinolytic, cellulolytic, and other enzymes before it actually penetrates the host. Once the fungus becomes established in the plants, its subsequent advance and production of mycelium and sclerotia are quite rapid, especially during conditions of high moisture and high temperature (between 30 and 35°C). The pathogen seems to grow, survive, and attack plants best near the soil line, perhaps because of more favorable temperatures there, more plentiful supply of organic substances the fungus uses for food, and perhaps, less competition or antagonism by other soil organisms.

Control of *Sclerotium* diseases is difficult and depends partly on crop rotation with crops such as corn and wheat, which seem not to be affected by the pathogen, partly on cultural practices such as deep plowing to bury fungal sclerotia in surface debris, fertilizing with ammonium-type fertilizers, and application of calcium compounds and, in some cases, by applying fungicides such as pentachloronitrobenzene (PCNB), captafol, and dichloran to the soil before planting or in the furrow during planting.

In recent years control of *Sclerotium rolfsii* diseases has been obtained by soil solarization, that is, mulching moistened soil with transparent polyethylene sheets during the hot season, which results in increased temperature in the soil and subsequent control of soil-borne diseases. Even more promising appears to be the biological control of Sclerotium diseases primarily by use of parasitizing and antagonistic species of the fungus *Trichoderma,* of some pseudomonad bacteria, and some *Streptomyces* species, for treatment of seeds or other propagative organs of crops planted in *Sclerotium*-infested fields. So far, however, all such controls are at the experimental stage.

SELECTED REFERENCES

Anderson, N. A. (1982). The genetics and pathology of *Rhizoctonia solani. Annu. Rev. Phytopathol.* **20**, 329–347.

Aycock, R. (1966). Stem rot and other diseases caused by *Sclerotium rolfsii. N. C. Agric. Exp. Stn., Tech. Bull.* **174,** 1–202.

Baker, K. F., and Cook, R. J. (1974). "Biological Control of Plant Pathogens." Freeman, San Francisco, California.

Baker, K. F., and Snyder, W. C. (1965). "Ecology of Soil-Borne Plant Pathogens." Univ. of California Press, Berkeley and Los Angeles.

Beagle-Ristaino, J. E., andf Papavizas, G. C. (1985). Biological control of *Rhizoctonia* stem canker and black scurf of potato. *Phytopathology* **75**, 560–564.

Bolkan, H. A., and Ribeiro, W. R. C. (1985). Anastomosis groups and pathogenicity of *Rhizoctonia solani* isolates from Brazil. *Plant Dis.* **69**, 599–601.

Boosalis, M. G. (1950). Studies on the parasitism of *Rhizoctonia solani* on soybeans. *Phytopathology* **40**, 820–831.

Bruehl, G. W., ed. (1975). "Biology and Control of Soil-Borne Plant Pathogens." Am. Phytopathol. Soc., St. Paul, Minnesota.

Christou, T. (1962). Penetration and host-parasite relationships of *Rhizoctonia solani* in the bean plant. *Phytopathology* **52**, 381–389.

Coley-Smith, J. R. (1959). Studies of the biology of *Sclerotium cepivorum.* III. *Ann. Appl. Biol.* **47**, 511–518.

Coley-Smith, J. R. (1960). Studies of the biology of *Sclerotium cepivorun.* IV. *Ann. Appl. Biol.* **48**, 8–18.

Costanho, B., and Butler, E. E. (1978). Rhizoctonia decline: A degenerative disease of *Rhizoctonia solani.* II. Studies on hypovirulence and potential use in biological control. III. The association of double stranded RNA with *Rhizoctonia* decline. *Phytopathology* **68**, 1505–1519.

Elad, Y., Chet, I., Boyle, P., and Henis, V. (1983). Parasitism of *Trichoderma* sp. on *Rhizoctonia solani* and *Sclerotium rolfsii*—Scanning electron microscopy and fluorescence microscopy. *Phytopathology* **73**, 85–88.

Forsberg, J. L. (1975). "Diseases of Ornamental Plants,"Spec. Publ. No. 3 Rev. University of Illinois, College of Agriculture, Urbana-Champaign.

Garrett, S. D. (1970). "Pathogenic Root-Infecting Fungi." Cambridge Univ. Press, London and New York.

Lee, F. N., and Rush, M. C. (1983). Rice sheath blight: A major rice disease. *Plant Dis.* **67**, 829–832.

Neal, D. C. (1942). *Rhizoctonia* infection of cotton and symptoms accompanying the disease in plants beyond the seedling stage. *Phytopathology* **32**, 641–642.
Nelson, E. G., and Hoitink, H. A. J. (1983). The role of microorganisms in the suppression of *Rhizoctonia solani* in container media amended with composted hardwood bark. *Phytopathology* **73**, 274–278.
Ogoshi, A. (1975). Grouping of *Rhizoctonia solani* and their perfect stages. *Rev. Plant Prot. Res.* **8**, 93–103.
Papavizas, G. C. (1974). "The Relation of Soil Microorganisms to Soilborne Plant Pathogens," South. Coop. Ser. Bull. No. 183, Virginia Polytechnic Institute, Blacksburg.
Parmeter, J. R., Jr., ed. (1970). "*Rhizoctonia solani,* Biology and Pathology." Univ. Of California Press, Berkeley and Los Angeles.
Punja, Z. K. (1985). The biology, ecology, and control of *Sclerotium rolfsii. Annu. Rev. Phytopathol.* **23**, 97–127.
Smith, A. M. (1972). Drying and wetting sclerotia promotes biological control of *Sclerotium rolfsii. Soil Biol. Biochem.* **4**, 119–123, 125–129, 131–134.
Sumner, D. R., and Bell, D. K. (1982). Root diseases induced in corn by *Rhizoctonia solani* and *R. zeae. Phytopathology* **72**, 86–91.
Toussoun, T. A., Bega, R. V., and Nelson, P. E. (1970). "Root Diseases and Soil-Borne Pathogens." Univ. of California Press, Berkeley and Los Angeles.
Wellman, F. L. (1932). *Rhizoctonia* bottom rot and head rot of cabbage. *J. Agric. Res. (Washington, D. C.)* **45**, 461–469.

Root Rots of Trees

• Armillaria Root Rot of Fruit and Forest Trees

Armillaria root rot is worldwide in distribution and affects hundreds of species of fruit trees, vines, shrubs, and shade and forest trees, as well as other plants such as potatoes and strawberries, both in the temperate and tropical regions. The disease is often known as "shoestring root rot," "mushroom root rot," "crown rot," or "oak root fungus disease." The pathogen, *Armillaria mellea,* and probably other closely related species, is one of the most common fungi in forest soils. The most spectacular losses occur in orchards or vineyards planted in recently cleared forest lands or in forest tree plantations, particularly in stands recently thinned. Most commonly, however, the losses from Armillaria root rot are steady but inconspicuous, appearing as slow decline and death of occasional trees, with greater numbers of trees dying from this disease during periods of moisture stress or after defoliation.

The aboveground parts of affected trees show symptoms similar to those caused by other root rot diseases, that is, reduced growth, smaller, yellowish leaves, dieback of twigs and branches, and gradual or sudden death of the tree. Affected trees may be scattered at first, but soon circular areas of diseased trees appear because of the spread of the fungus from its initial infection point. Diagnostic characteristics of Armillaria root rot appear at decayed areas in the bark, at the root collar, and on the roots. White mycelial mats, their margins often veined and shaped like fans, form between the bark and wood (Figure 11-106B). The mycelium may extend for a few feet upward in the phloem and cambium of the trunk, and in some trees, such as oak, sugar maple, and hemlock, it may cause a white rot decay. In addition to the mycelial fans, another even more characteristic sign of the disease is the formation of reddish brown to black "rhizomorphs" or "shoestrings," that is, cordlike threads of mycelium 1–3 mm in diameter, consisting of a compact outer layer of black mycelium and a core of white or colorless mycelium. These often form a branched network of sorts on the roots, under the bark, or

FIGURE 11-106 (A) *Armillaria mellea* basidiospores growing at the base of infected tree. (B) Fan-shaped mycelial mats of *Armillaria* advancing on the wood surface of main roots and the trunk. (C) Rhizomorphs of *Armillaria* growing on the wood of tree killed by the fungus. (D) Close-up of rhizomorphs on wood surface of trunk. (Photo A courtesy U.S.D.A. Photo B courtesy U.S. Forest Service.)

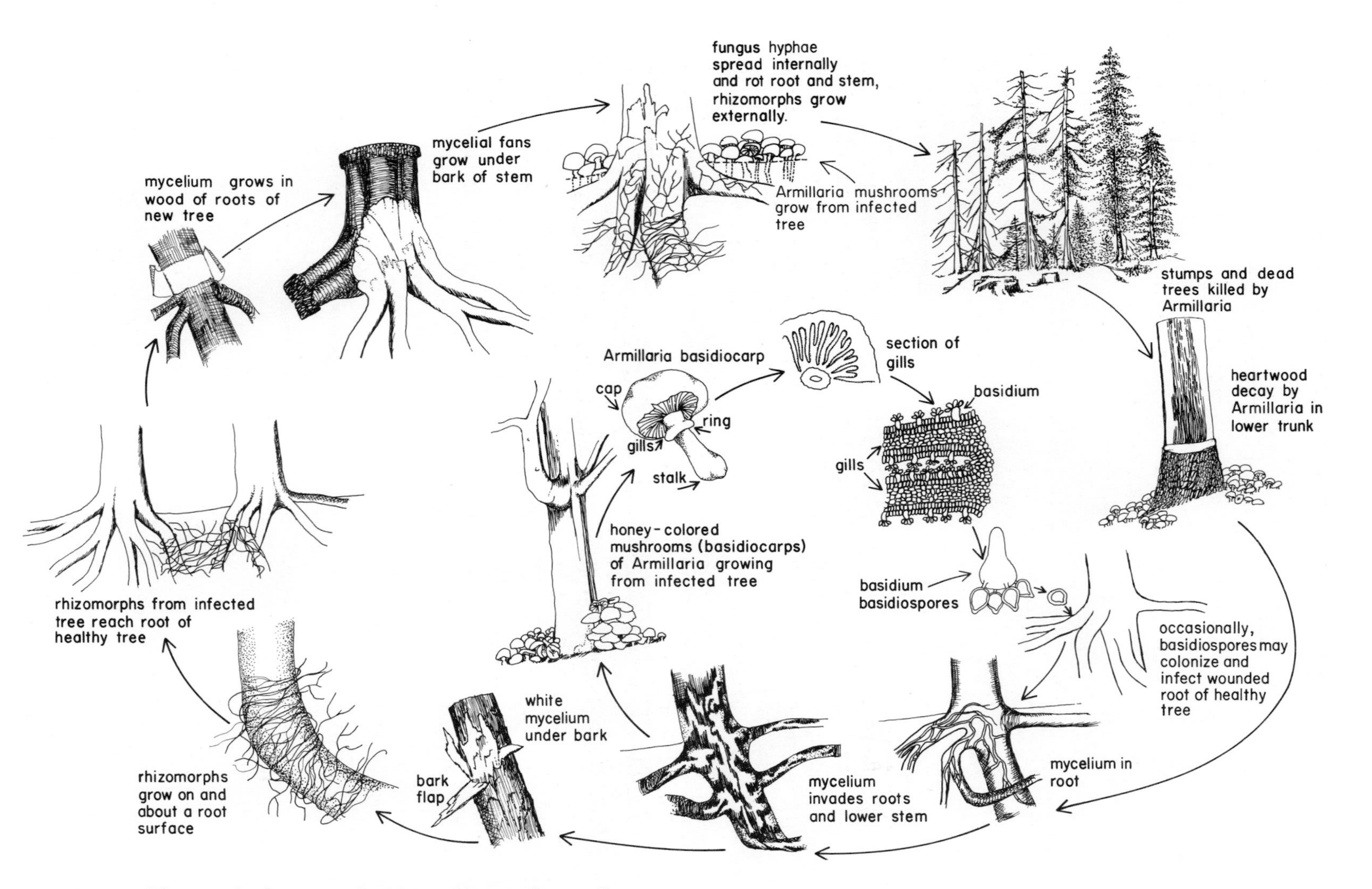

FIGURE 11-107 Disease cycle of root rots of trees caused by *Armillaria mellea.*

in severely decayed wood, with some strands spreading into the soil surrounding the roots (Figure 11-106C, D). In areas in which the mycelium has invaded the cambium, cankers form on both hardwoods and conifers, and gum or resin is exuded from the infected area and flows into the soil. As the fungus gradually girdles and kills the tree at the base, infected wood changes from firm and slightly moist to somewhat soft and dry. At the base of dead or dying trees, a few to many honey-colored, speckled mushrooms, about 7 or more centimeters tall and with a cap 5–15 cm in diameter grow from trunks, stumps or on the ground near infected roots (Figure 11-106A). These are the fruiting bodies of *A. mellea,* which appear in early fall and within their radial gills produce numerous basidia and basidiospores.

The fungus overwinters as mycelium or rhizomorphs in diseased trees or in decaying roots. The principal method of tree-to-tree spread of the fungus is through rhizomorphs or direct root contact. Rhizomorphs grow from roots of infected trees or from decaying roots or stumps through the soil to roots of adjacent healthy trees (Figure 11-107). Also, pieces of rhizomorphs in infected plant debris may be carried by cultivating equipment into new areas. The fungus can apparently spread by basidiospores, but spores generally colonize dead stumps or woody material first and then rhizomorphs radiating from these attack living roots directly or through wounds. When roots of trees are in contact with infected or decaying roots, mycelium may directly invade healthy roots appressed to diseased roots without forming rhizomorphs. In all cases, trees and roots weakened from other causes are much more easily attacked by *Armillaria* than are vigorous trees.

Control of Armillaria root rot is usually not attempted under forest conditions. Generally, however, losses can be reduced by removing the substrate such as tree stumps and roots and avoiding or delaying planting for several years, susceptible fruit or forest trees in recently reclaimed forest land originally occupied by oaks or other plants favoring buildup of large amounts of *Armillaria* inoculum. Control of the disease in orchards and occasionally in forest plantations is attempted by digging a trench around an infected tree and its neighbors to prevent growth of rhizomorphs to adjacent trees and local soil fumigation of the infested area to destroy the fungus in the soil before *Armillaria*-killed trees can be replaced.

SELECTED REFERENCES

Bliss, D. E. (1951). The destruction of *Armillaria mellea* in citrus soils. *Phytopathology* **41**, 665–683.

Gremmen, J. (1963). Biological control of the root rot fungus *Fomes annosus* by *Peniophora gigantea. Ned. Bosbouw. Tijdschr.* **35**, 356–367.

Leaphart, C. D. (1963). *Armillaria* root rot. *U. S., For. Serv., For. Pest Leafl.* **78**, 1–8.

Munnecke, D. E., Kolbezen, M. J., Wilbur, W. D., and Ohr, H. D. (1981). Interactions involved in controlling *Armillaria mellea. Plant Dis.* **65**, 384–389.

O'Reilly, H. J. (1963). *Armillaria* root rot of deciduous fruits, nuts and grapevines. *Calif., Agric. Exp. Stn., Ext. Serv., Circ.* **525,** 1–15.

Redfern, D. B. (1975). The influence of food base on rhizomorph growth and pathogenicity of *Armillaria mellea* isolates. *In* "Biology and Control of Soil-Borne Plant Pathogens" (G. W. Bruehl, ed., pp. 69–73. Am. Phytopathol. Soc., St. Paul, Minnesota.

Sinclair, W. A. (1964). Root- and butt-rot of conifers caused by *Fomes annosus.* with special reference to inoculum dispersal and control of the disease in New York. *Mem.—N.Y., Agric. Exp. Stn.* **391,** 1–54.

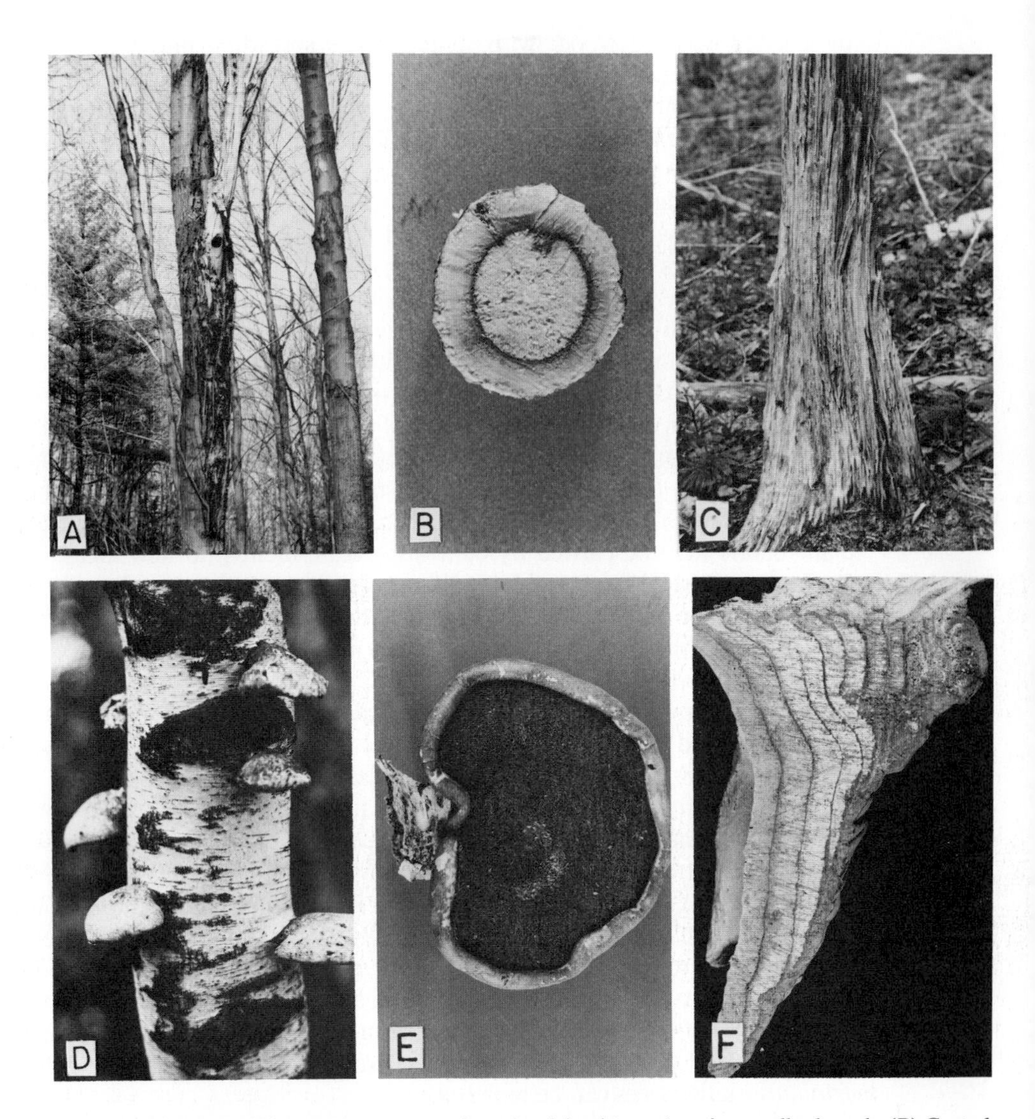

FIGURE 11-108 (A-F) (A) Large canker and rotten area of trunk originating at decaying smaller branch. (B) Central portion of apple branch rotted by the silverleaf fungus *Stereum purpureum.* (C) Remnants of tree trunk attacked by wood rot fungi. (D) Beech trunk rotted by *Polyporus betulinus* and several basidiocarps of the fungus. (E) The lower side of a basidiocarp of *Polyporus betulinus.* (F) Tangential section of a perennial basidiocarp (conk) of a *Fomes* species. *(Figure continues)*

Wargo, P. M., and Houston, D. R. (1974). Infection of defoliated sugar maple trees by *Armillaria mellea. Phytopathology* **64**, 817–822.

Wargo, P. M., and Shaw, C. G., III (1985). *Armillaria* root rot: The puzzle is being solved. *Plant Dis.* **69**, 826–832.

Wood Rots and Decays Caused by Basidiomycetes

Huge losses of timber in the living trees in the forest and in harvested wood or in wood products are caused every year by the wood-rotting Basidiomycetes

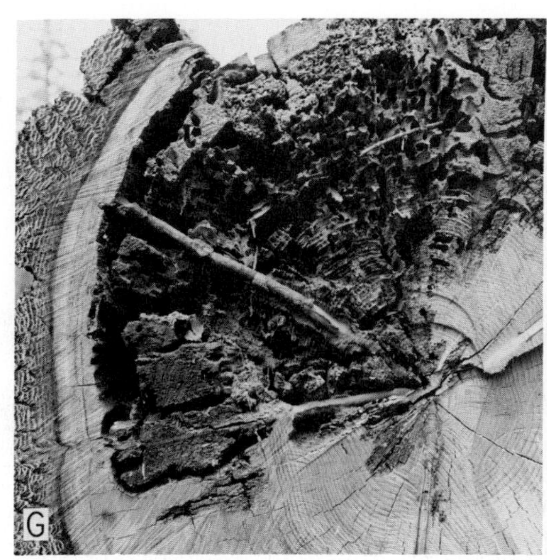

FIGURE 11-108 (G, H) (G) Close-up of wood rot in tree trunk originating at large wound. (H) Most wood of this trunk has been rotted away by wood rot fungi although there are no signs of exterior damage at the level of the cut. (Photos courtesy U.S.D.A.)

(Figures 11-84, 11-108, and 11-109). In living trees, most of the rotting is confined to the older, central wood of roots, stem, or branches sometimes referred to as heartwood (Figure 11-108G, H). Once the tree is cut, however, the outer wood, which is sometimes referred to as sapwood, is also attacked by the wood-rotting fungi, as are the wood products made from it—if moisture and certain other conditions remain favorable for the fungi. When large wounds or cuts are present on the tree, discoloration and decay may also spread into the outer wood, and the entire tree, especially if a hardwood tree, loses its economic value.

Depending on the host portion attacked, wood rots may be called root rots, root and butt rots, stem rots, or top rots. The fungi that cause these rots or wood product decays grow inside the wood cells and utilize the cell wall components for food and energy. Some of them, the brown-rot fungi, which attack preferably softwoods, can break down and utilize primarily the cell wall polysaccharides (cellulose and hemicellulose), leaving the lignin more or less unaffected. This usually results in rotten wood that is some shade of brown and, in advanced stages, has a cubical pattern of cracking and a crumbly texture (Figure 11-109E). Other wood rotters, the white-rot fungi, either decompose lignin and hemicellulose first and cellulose last or decompose all wood components simultaneously, in either case reducing the wood to a light-colored spongy mass (white rot) with white pockets or streaks separated by thin areas of firm wood (Figure 11-109F, G). The white-rot fungi are able to or preferably attack hardwoods normally resistant to brown-rot fungi.

It should be noted here that, in addition to the brown rots and white rots caused by Basidiomycetes, wood is also attacked by certain Ascomycetes and Imperfects. Some Ascomycetes, such as *Daldinia, Hypoxylon,* and *Xylaria,* cause a relatively slow white rot with variable black zone lines in and around the rotting wood, both in standing hardwood trees and in slash. In standing

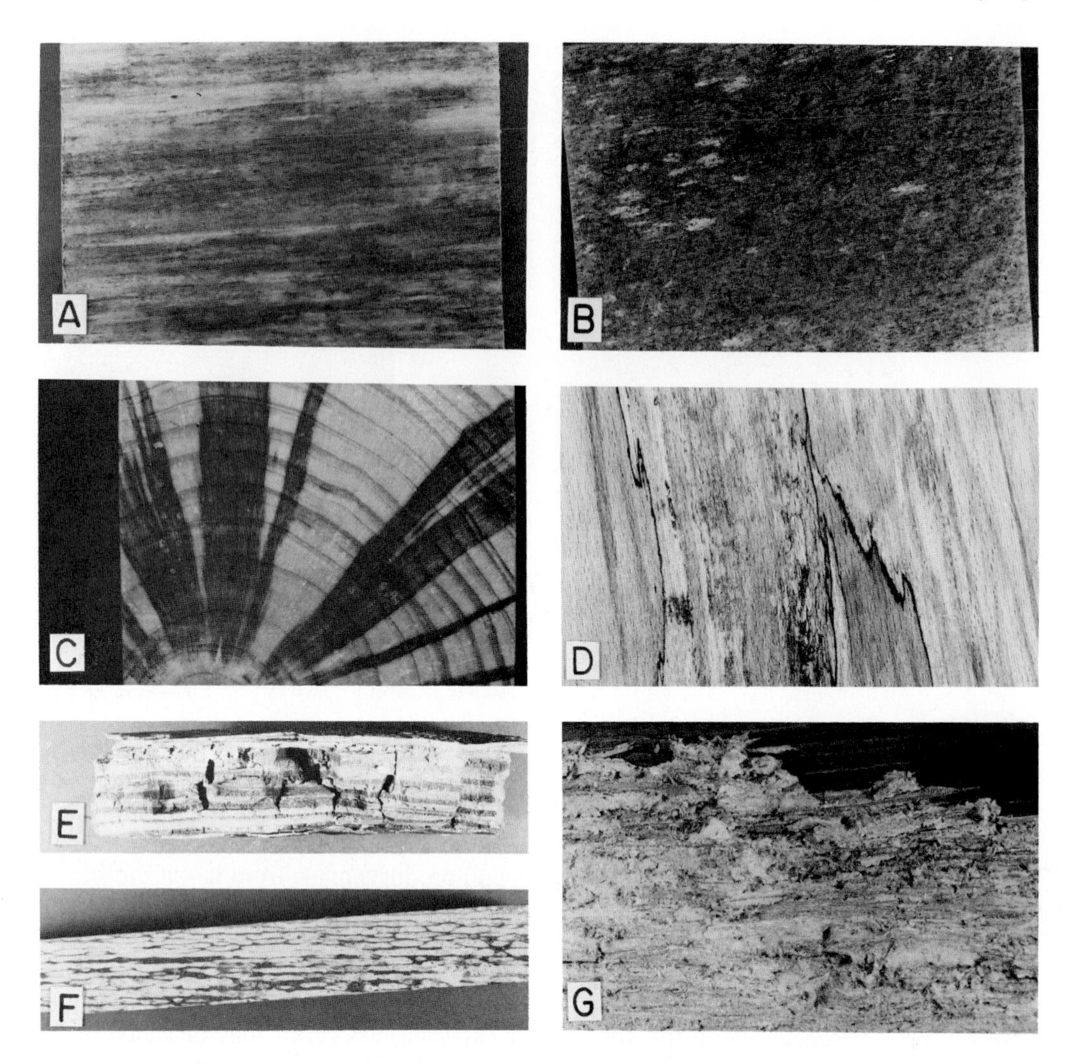

FIGURE 11-109 Wood discolorations and decays. (A, B) Surface molds or mildews on wood caused by fungi such as *Penicillium.* (C) Sapstain or blue stain caused by fungi such as *Ceratocystis.* (D) Zone lines of white rot fungus *(Fomes igniarius)* in beech. (E) Brown rot caused by *Poria incrasata,* the water conducting fungus. (F) White pocket rot caused by species of *Polyporus* and *Fomes.* (G) Advanced white rot on wood caused by *Polyporus* and other wood-rotting fungi.

trees the decay is usually associated with wounds or cankers, while in wood pieces the decay is usually at or near the surface of wood with high moisture content. Others, such as species of *Alternaria, Bisporomyces, Diplodia,* and *Paecilomyces,* cause the so-called soft rots of wood that affect the surface layers of wood pieces maintained more or less continuously at a high moisture content. The soft-rot fungi utilize both polysaccharides and lignin. They invade wood preferably through rays or vessels, from where they grow into the adjacent tracheids and invade their cell wall. Within the cell wall they produce conical or cylindrical cavities parallel to the orientation of the microfibrils and, with progressing decay, the entire secondary wall is interlaced by confluent cavities. Several types of bacteria also attack wood, primarily in wood parenchyma rays, where they break down and utilize the contents and walls of the parenchyma cells and thus increase the porosity and permeability of the wood to liquids, including fungal enzymes. Furthermore, several Ascomycetes and Imperfects result in the appearance of unsightly discolorations in the wood and thus reduce the quality but not the strength of the wood. Some of the wood-staining fungi are simply surface molds that usually grow on freshly cut surfaces of wood and impart to the wood the color of their spores, for example, *Penicillium* (green or yellow), *Aspergillus* (black, green). *Fusarium* (red), and *Rhizopus* (gray) (Figure 11-109). Other wood-staining fungi, however, usually called sap-stain or blue-stain fungi, cause discoloration of the sapwood by producing pigmented hyphae that grow mainly in the ray parenchyma but can spread throughout the sapwood and cause lines of discoloration (Figure 11-109C, D). Among the blue-stain fungi are species of *Ceratocystis, Hypoxylon, Xylaria, Graphium, Diplodia,* and *Cladosporium.*

The bulk of wood rotting, however, is carried out by Basidiomycetes, and the most important fungi that rot wood in standing trees or in wood products are the following:

Fomes (Heterobasidion), causing root and butt rot of conifers *(H. annosum),* mottled white root and butt rot of hardwoods *(F. aplanatus),* white heart rot of deciduous trees *(F. connatus),* spongy white sap and heart rot of birch and aspen *(F. fomentarius),* white heart and sap rot of many living deciduous trees *(F. igniarius),* red heart rot of conifers *(F. pini),* brown cubical sap and heart rot of conifers and hardwoods *(F. pinicola).*

Polyporus, many species causing rot of dead trees or logs *(P. adustus, P. gilvus, P. hirsutus, P. pargamenus, P. picipes,* and *P. versicolor,* all of them decaying primarily dead hardwood trees and logs). Many other species attack living as well as dead trees. Some of them attack only or mainly conifers and cause brown cubical root and butt rot *(P. schweinitzii),* red root and butt rot *(P. tomentosus),* white pocket rot of roots and butt *(P. circinatus),* red heart and sap rot of trunks *(P. anceps).* Others attack only or mainly deciduous hardwoods and cause a white heart rot *(P. squamosus* and *P. obtusus),* a brown cubical rot of the trunk *(P. sulphureus),* a spongy white rot *(P. hispidus),* a yellow cubical sap and heart rot of birch *(P. betulinus),* a canker and decay of maple *(P. glomeratus),* a root and butt rot *(P. lucidus),* or a trunk canker and localized decay *(P. hispidus).*

Poria, causing a root rot of most conifers *(P. weirii),* a yellow, feathery root and butt rot of balsam fir *(P. subacida),* and a butt swelling and advanced decay on birch *(P. obliqua),* while another species, *P. incrasata,* causes the common brown cubical rot in buildings and stored lumber, and by means of rhizomorphs can transport water for distances up to 5 meters or more.

Ganoderma, similar to *Fomes,* causing white mottled rot in hardwoods.

Stereum, causing "silver leaf" of fruit trees as a result of decay of the interior of the tree trunk and branches *(S. purpureum).* Other species cause a white pocket rot of oaks *(S. frustulosum),* a white mottled heart rot of sprout oaks *(S. gausapatum),* and a red top rot of balsam fir *(S. sanguinolentum).*

Peniophora, causing decay in coniferous logs and pulpwood.

Lenzites, causing a brown cubical rot on coniferous logs, posts, and poles *(L. sepiaria),* and decay of hardwood slash *(L. betulina).*

Pholiota, causing brown rot in hardwoods.

Pleurotus, causing white rot in hardwoods.

Schizophyllum, also causing white rot in hardwoods.

The development of wood rots varies, of course, with the particular fungus involved and the host tree attacked. There are, however, many similarities (Figure 11-110). Thus, the wood rot fungi enter trees as germinating

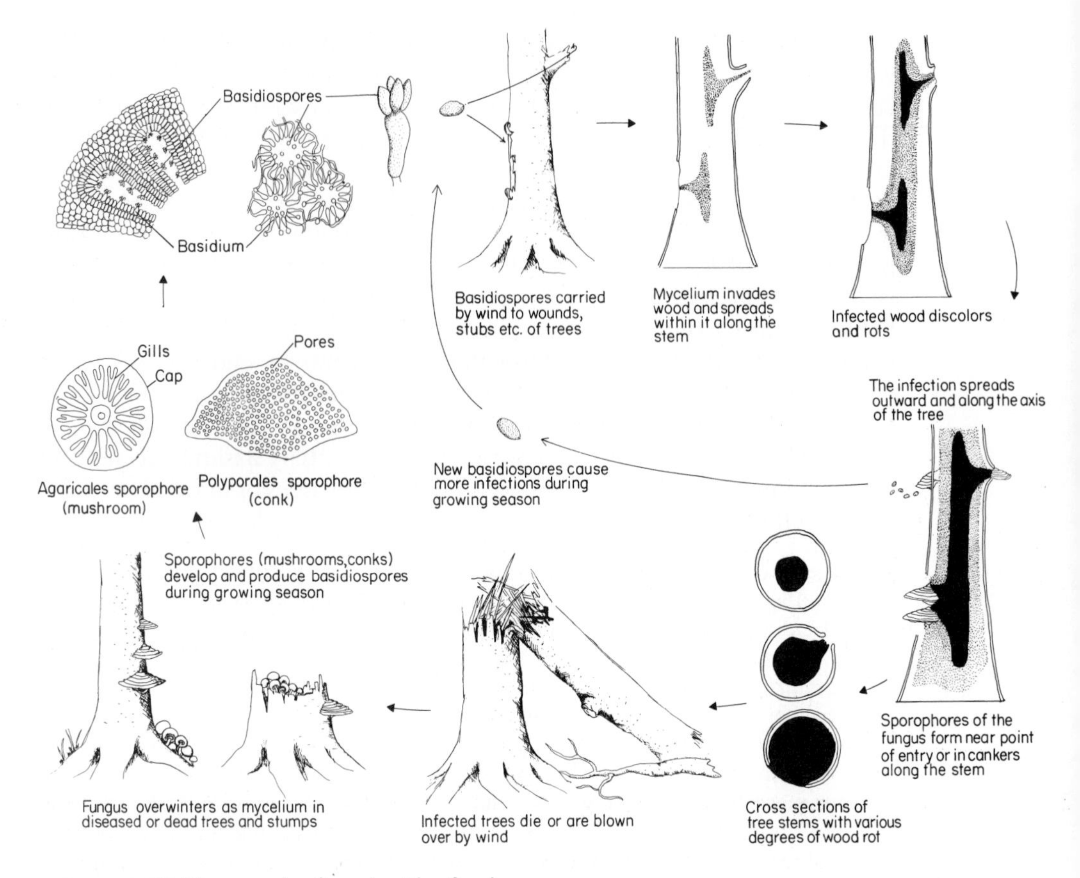

FIGURE 11-110 Disease cycle of wood-rotting fungi.

basidiospores or mycelium through wounds, dead branches, branch stubs, tree stumps or damaged roots and spread from there to the heartwood or sapwood of the tree or tree sprout. Wounds caused by fire and by cutting and thinning operations are the most common points of entry of these fungi. The fungi develop in the wood and spread upward or downward, or both, in a cylinder much faster than they do radially. In some wood rots, especially those of hardwoods originating from wounds or branch stubs, the rotten cylinder is only a few inches in diameter, forms a column no larger than the diameter of the tree at the time of injury, and may extend to one or a few meters above and below the area where the fungus entered the tree or where its fruiting bodies (conks) appear. In other wood rots, particularly those of conifers, the rotten cylinder enlarges steadily until the tree is killed or blown over by heavy winds and it may extend upward over much of the height of the tree.

The process of discoloration and decay in the wood of living trees appears to be quite complex, involving a number of successive or overlapping events. First, there must be an injury to the tree that exposes the wood as a result of a dead or broken branch, animal damage, fire burn, or mechanical scraping. The injured cells and those around them undergo chemical changes such as oxidation and become discolored. As long as the wound is open, discoloration advances toward the pith and around the tree, but if the wound is small and occurs early in the season, a new growth ring forms and its cells act as a barrier to the discoloration process. The discoloration moves up and down with the cylinder of barrier cells but not outward into the new and subsequent growth rings. Of course, many microorganisms are likely to land or be brought to the surface of a tree wound, and many of them begin to grow on the moist surface. Among these, however, only some bacteria and some Ascomycetes or Imperfect Fungi manage to survive on the discolored wood of the wound. These microorganisms do not cause wood decay, but they increase the discoloration and wetness of the wood and erode parts of the cell walls. Such wood is called wetwood, redheart, or blackheart. Finally, however, the wood-rotting Basidiomycetes become active and begin to disintegrate and digest the cell wall components. These wood-rotters attack only the tissues that have already been altered first by the chemical processes and then by the bacteria and the Ascomycetes and Imperfects. So, the wood rotting Basidiomycetes also remain confined to the discolored column within the new growth, being unable to attack the latter. The decay in the discolored column continues until the wood is completely decomposed, but the influx of new microorganisms through the wound continues, even after the first decay fungus has caused the tissue to rot, and stops only when all tissues are completely digested. It should be noted that this process of discoloration and decay may take 50–100 years to develop. It is most common and rapid in older, larger trees, and the older the trees the more likely they are to contain decay columns. The discoloration and decay process starting at a particular wound need not, of course, go through to completion. Quick healing of the wound, antagonisms among the microorganisms involved, natural wood resistance, and other factors may stop the process at any stage. On the other hand, a large tree is likely to be injured many times over its long period of life. The events described above may be repeated many times after each new wound is formed, and thus more and more of the wood may be involved in the more or less continuous process of discoloration and decay. The end result is the formation of a single large column or multiple columns of discolored and decayed wood.

The sporophores or conks of the wood-rotting Basidiomycetes appear near the point of entry of the fungus, near the base of the tree, in cankers or swollen knots along the stem of living trees, or along the length of the tree stem after its death. The sporophores of most wood rotting fungi, such as *Polyporus* and *Poria,* are formed annually and do not last for more than a year, but those of *Fomes* are perennial, adding a layer of tissue with vertical tubes and pores each year for 50 years or more. The sporophores produce basidiospores during part or most of the growing season, and the spores are carried by wind, rain, or animals to the nearby trees.

Control of wood rots and decays is impossible in the forest, but losses can be reduced: (1) by management practices that reduce or eliminate the chance of introducing the fungi into healthy stands, (2) by conducting logging and thinning operations in a way that minimizes breakage of branches or other wounds of remaining trees and conducting them during the dry season to avoid mechanical damage to the root systems of the residual trees, and (3) by harvesting trees before the age of extreme susceptibility to wood rot fungi. Damage caused by wood-rotting fungi in shade and fruit trees can be prevented or minimized by avoiding or preventing wounds; by pruning dead and dying branches with a flush cut as close to the main stem as possible but without cutting the collarlike part of the stem that surrounds the base of the branch; by cleaning wounds by cutting the torn bark and shaping the wound like a vertical ellipse; and by keeping the trees in good vigor through adequate irrigation and proper fertilization. Treating large cuts or wounds with a wound dressing or tree paint has been routinely practiced in the past, but its usefulness in preventing wood discoloration and decay is questionable.

In recent years considerable success in controlling wood rot and decays has been obtained through treatment of tree wounds and stumps with antagonistic fungi such as *Trichoderma.* In the case of *Heterobasidion (Fomes) annosum* root and butt rot of forest trees, commercial control of the disease is obtained by applying to freshly cut stumps conidia of the antagonistic fungus *Peniophora (Phlebia) gigantea.* This is often accomplished by mixing the spores with the oil that constantly lubricates the chain of the chain saw used to cut the trees.

Control of discoloration or decays in lumber and wood products is usually accomplished by drying the wood or by treating it with an organic mercuric or a chlorophenate fungicide or with a mixture of the two. Wood that is likely to be in contact with soil or other moist surfaces should be treated with one of several wood preservatives such as creosote, pentachlorophenol, copper naphthanate, and zinc chromate.

SELECTED REFERENCES

Blanchette, R. A. (1984). Selective delignification of eastern hemlock by *Ganoderma tsugae. Phytopathology* **74**, 153–160.

Boyce, J. S. (1961). "Forest Pathology." McGraw-Hill, New York.

Cowling, E. B. (1961). Comparative biochemistry of the decay of sweetgum sapwood by white-rot and brown-rot fungi. *USDA For. Serv., Tech. Bull.* **1258,** 1–79.

Eslyn, W. E., Kirk, T. K., and Effland, M. J. (1975). Changes in the chemical composition of wood caused by six soft-rot fungi. *Phytopathology* **65**, 473–476.

Greaves, H. (1969). Micromorphology of the bacterial attack of wood. *Wood Sci. Technol.* **3**, 150–166.

Jacobi, W. R. *et al.* (1980). Disease losses in North Carolina forests. I. Losses in softwoods. 1973-74. II. Losses in hardwoods. 1973-74. III. Rationale and recommendations for future cooperative survey efforts. *Plant Dis.* **64**, 573–576, 576–578, 579–581.

Levy, J. F. (1965). The soft rot fungi: Their mode of action and significance in the degradation of wood. *Adv. Bot. Res.* **31**, 323–357.

Liese, W. (1970). Ultrastructural aspects of woody tissue disintegration. *Annu. Rev. Phytopathol.* **8**, 231–258.

Merrill, W. (1970). Spore germination and host penetration by heartrotting Hymenomycetes. *Annu. Rev. Phytopathol.* **8**, 281–300.

Peace, T. R. (1962). "Pathology of Trees and Shrubs." Oxford Univ. Press, London and New York.

Rishbeth, J. (1963). Stump protection against *Fomes annosus.* III. Inoculation with *Peniophora gigantea. Ann. Appl. Biol.* **52**, 63–77.

Setliff, E. C., and Wade, E. K. (1973). *Stereum purpureum* associated with sudden decline and death of apple trees in Wisconsin. *Plant Dis. Rep.* **57**, 473–474.

Shigo, A. L. (1967). Successions of organisms in discoloration and decay of wood. *Int. Rev. For. Res.* **2**, 237–299.

Shigo, A. L. (1979). Tree decay: An expanded concept. *U. S. Dep. Agric., For. Serv., Inf. Bull.* **419,** 1–73.

Shigo, A. L. (1982). Tree decay in our urban forests: What can be done about it? *Plant Dis.* **66**, 763–768.

Shigo, A. L. (1984). Compartmentalization: A conceptual framework for understanding how trees grow and defend themselves. *Annu. Rev. Phytopathol.* **22**, 189–214.

Shigo, A. L. (1985). Compartmentalization of decay in trees. *Sci. Am.* **252,** 96–103.
Shigo, A. L., and Hillis, W. E. (1973). Heartwood, discolored wood, and microorganisms in living trees. *Annu. Rev. Phytopathol.* **11**, 197–222.
Shigo, A. L., and Larson, E. H. (1969). A photo guide to the patterns of discoloration and decay in living northern hardwood trees. *USDA For. Serv. Res. Pap. NE* **127,** 1–100.
Smith, K. T., Blanchard, R. D., and Shortle, W. C. (1981). Postulated mechanism of biological control of decay fungi in red maple wounds treated with *Trichoderma harzianum. Phytopathology* **71**, 496–498.
Smith, W. H. (1970). "Tree Pathology: A Short Introduction." Academic Press, New York.

Mycorrhizae and Plant Growth

The feeder roots of most flowering plants growing in nature are generally infected by symbiotic fungi that do not cause root disease but, instead, are beneficial to their plant hosts. The infected feeder roots are transformed into unique morphological structures called mycorrhizae, that is, "fungus roots." Mycorrhizae, known for many years to be common in forest trees, are now considered to be the normal feeder roots for most plants including cereals, vegetables, ornamentals, and of course, trees.

There are three types of mycorrhizae, distinguished by the way the hyphae of the fungi are arranged within the cortical tissues of the root.

- *Ectomycorrhizae.* Ectomycorrhizae roots are usually swollen and, in some host-fungus combinations, appear considerably more forked than nonmycorrhizal roots. Ectomycorrhizae are formed primarily on forest trees mostly by mushroom- and puffball-producing basidiomycetes and by some ascomycetes. Spores of ectomycorrhizal fungi are produced aboveground and are wind disseminated. The hyphae of ectomycorrhizal fungi usually produce a tightly interwoven "fungus mantle" around the outside of the feeder roots, the mantle varying in thickness from 1 or 2 hyphal diameters to as many as 30–40. These fungi also enter the roots, but they only grow around the cortical cells, replacing part of the middle lamella between the cells, and forming the so-called Hartig net. Ectomycorrhizae appear white, brown, yellow, or black, depending on the color of the fungus growing on the root.
- *Endomycorrhizae.* Endomycorrhizae roots externally appear similar to nonmycorrhizal roots in shape and color, but internally the fungus hyphae grow into the cortical cells of the feeder root either by forming specialized feeding hyphae (haustoria), called arbuscules, or by forming large swollen food-storing hyphal swellings, called vesicles. Most endomycorrhizae contain both vesicles and arbuscules and are, therefore, called **"vesicular-arbuscular" (VA)** mycorrhizae (Figure 11-111). Endomycorrhizae are not surrounded by a dense fungal mantle but by a loose mycelial growth on the root surface from which hyphae and large pearl-covered zygospores or chlamydospores are produced underground. Endomycorrhizae are produced on most cultivated plants and on some forest trees mostly by zygomycetes, primarily of the genus *Glomus,* but also other fungi, such as *Acaulospora.* Endomycorrhizae are also produced by some basidiomycetes.
- *Ectendomycorrhizae.* Ectendomycorrhizae are intermediate between the other two. They are caused by fungi of unknown identity that grow into and also around the cortical cells of the root and may or may not have a fungus mantle on the surface of the feeder roots.

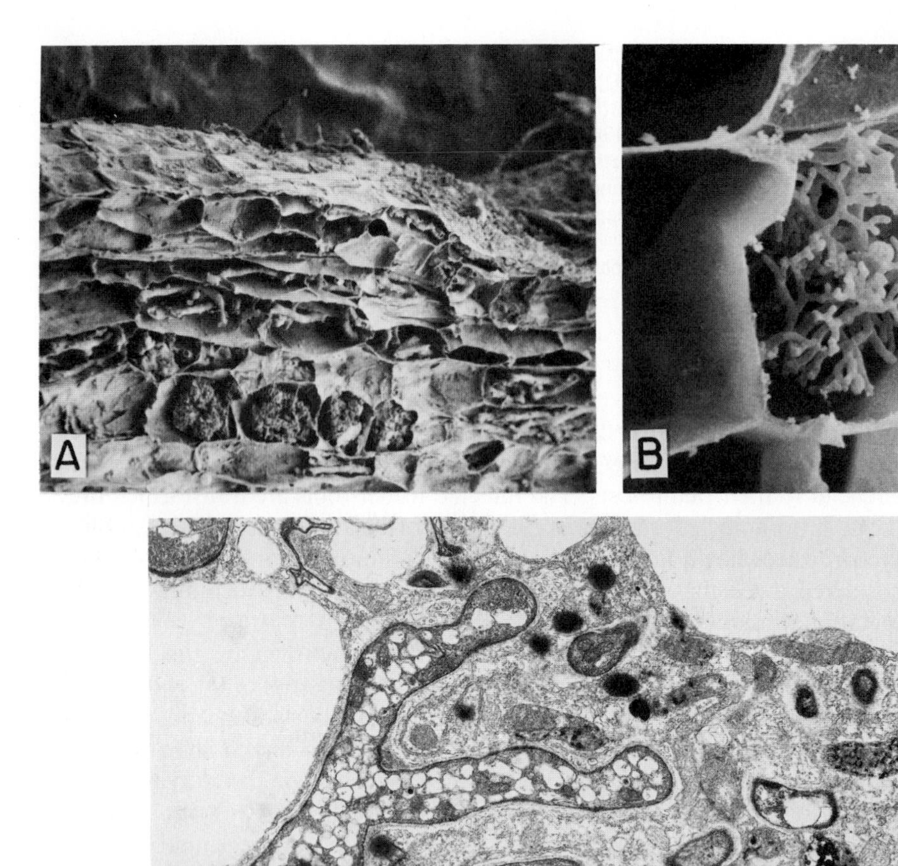

FIGURE 11-111 Vesicular-arbuscular mycorrhizae (endomycorrhizae) on yellow poplar *(Liriodendron tulipifera)* produced by *Glomus mosseae.* (A) Scanning electron micrograph of interior of mycorrhizal root showing coiled intracellular hyphae in outer cortical cells and three inner cortical cells which contain arbuscules. Some external mycelium of the fungus can be seen on the outside of the epidermis (top center). (B) Scanning electron micrograph of arbuscular morphology in a sample treated to remove host cytoplasm which surrounds the structure. This is a mature, viable arbuscule prior to the initiation of degenerative processes which lead to breakdown of this part of the endophyte. (C) Transmission electron micrograph of a similar arbuscule in a cortical cell. (Photos courtesy M. F. Brown and D. A. Kinden.)

Mycorrhizae apparently improve plant growth by increasing the absorbing surface of the root system; by selectively absorbing and accumulating certain nutrients, especially phosphorus; by solubilizing and making available to the plant some normally nonsoluble minerals; by somehow keeping feeder roots functional longer; and by making feeder roots more resistant to infection by certain soil fungi such as *Phytophthora, Pythium,* and *Fusarium.* It should be kept in mind, however, that there may be many different host–fungus mycorrhizal associations, and each combination may have different effects on the growth of the plant. Some mycorrhizal fungi have a broad host range while others are more specific. Also, some mycorrhizal fungi are more beneficial to a certain host than other fungi, and some hosts need and profit from association with a certain mycorrhizal fungus much more than do other hosts. Mycorrhizal fungi also need the host in order to grow and reproduce; in the absence of hosts the fungi remain in a dormant condition as spores or resistant hyphae.

The symbiosis between the host plant and the mycorrhizal fungus is generally viewed as providing equal benefits to both partners. Yet, it is quite probable that under certain nutritional conditions, one of the two partners may dominate and benefit more than the other. It has been suggested that the fungus is most aggressive in its invasion of root tissues when the host is growing at suboptimal nutritional levels (host defenses weak?) and the symbiotic relationship is terminated when nitrogen supply in the host reaches its optimum (host defenses at their best?). If nitrogen supply is again reduced to deficiency levels, the fungus partner begins to dominate and forms in abundance while plant growth is suppressed.

As far as is known, mycorrhizae do not cause disease, but absence of mycorrhizae in certain fields results in plant stunting and poor growth, which can be avoided if the appropriate fungi are added to the plants. Also, soil fumigation often results in the eradication of mycorrhizal fungi, and this in turn causes plants to remain smaller than plants growing in nonfumigated soil.

SELECTED REFERENCES

Anonymous (1982). Symposium on aspects of vesicular-arbuscular mycorrhizae and plant disease research. (Several papers.) *Phytopathology* **72**, 1101–1132.

Bethlenfalvay, G. J., Brown, M. S., and Pacovsky, R. S. (1982). Parasitic and mutualistic associations between a mycorrhizal fungus and soybean. I. Development of the host plant. II. Development of the endophyte. *Phytopathology* **72**, 889–893, 894–897.

Gerdemann, J. W. (1968). Vesicular-arbuscular mycorrhiza and plant growth. *Annu. Rev. Phytopathol.* **6**, 397–418.

Hackskaylo, E. (1971). Mycorrhizae. *Misc. Publ.—U. S. Dep. Agric.* **1189,** 1–255.

Kleinschmidt, G. D., and Gerdemann, J. W. (1972). Stunting of citrus seedlings in fumigated nursery soils related to the absence of endomycorrhizae. *Phytopathology* **62**, 1447–1453.

Marx, D. H. (1972). Mycorrhizae as biological deterrents to pathogenic root infections. *Annu. Rev. Phytopathol.* **10**, 429–454.

Mosse, B. (1973). Advances in the study of vesicular-arbuscular mycorrhizae. *Annu. Rev. Phytopathol.* **11**, 171–196.

Schenk, N. C. (1981). Can mycorrhizae control root disease? *Plant Dis.* **65**, 230–234.

Slankis, V. (1974). Soil factors influencing formation of mycorrhizae. *Annu. Rev. Phytopathol.* **12**, 437–457.

Wilcox, H. E. (1983). Fungal parasitism of woody plant roots from mycorrhizal relationships to plant disease. *Annu. Rev. Phytopathol.* **21**, 221–242.

12 PLANT DISEASES CAUSED BY PROKARYOTES

Introduction

Prokaryotes are generally single-celled microorganisms that have a cell membrane or a cell membrane and a cell wall surrounding the cytoplasm, the latter containing small (70S) ribosomes and genetic material (DNA) not bound by a membrane, that is, not organized into a nucleus. The cells of all other organisms (eukaryotes) contain membrane-bound organelles (nuclei, mitochondria and—in plants only—chloroplasts). Eukaryotes also have two types of ribosomes, larger ones (80S) in the cytoplasm and smaller ones (70S) in mitochondria and chloroplasts. In fact, cellular organelles and prokaryotes have much in common. For example, antibiotics that affect bacteria often inhibit mitochondria or chloroplasts but do not interfere with the other functions of eukaryotic plant cells.

Two kinds of prokaryotes cause disease in plants: **bacteria,** which have a cell membrane and a rigid cell wall and, often, one or more flagella, and the **mollicutes,** or **mycoplasmalike organisms** (MLO), which lack a cell wall and have only a typical single-unit membrane. Plant pathogenic bacteria have been known since 1882; they are by far the largest group of plant pathogenic prokaryotes, cause a variety of plant disease symptoms, and are the best understood prokaryotic pathogens of plants. Even so, some types of phytopathogenic bacteria, for example, the vascular fastidious bacteria, which for several years were thought to be rickettsialike organisms, were only discovered in 1972, and their properties and relationships to the other plant pathogenic bacteria are poorly understood. A general classification of the plant pathogenic prokaryotes is shown below:

Kingdom: PROKARYOTAE: Organisms with genetic material (DNA) not organized into a nucleus, i.e., not surrounded by a membrane.
- BACTERIA: Have cell membrane and cell wall.
  - Part I: Gram-negative aerobic rods and cocci.
    - Family: Pseudomonadaceae
      - Genus: *Pseudomonas,* rod-shaped, one or several polar flagella, colonies white or yellow. DNA (G + C) content 58–70 mol%.
        - *Xanthomonas,* rod-shaped, one polar flagellum, colonies yellow. DNA (G + C) content 63–71 mol%.
          - *Xylella,* rod shaped, under some cultural conditions filamentous. Nonmotile, aflagellate, nonpigmented. Nutritionally fastidious, requiring specialized media. Habitat is xylem of plant tissue. (G + C) content of DNA is 51–53 mol%.
    - Family: Rhizobiaceae
      - Genus: *Agrobacterium,* rod-shaped, sparse lateral flagella, colonies white, rarely yellow.

Part II: Gram-negative facultative anaerobic rods.
Family: Enterobacteriaceae
Genus: *Erwinia,* peritrichous flagella; colonies white or yellow.
Part III: Irregular, Gram-positive, nonsporing rods.
Genus: *Clavibacter.* Contains most important phytopathogenic bacteria formerly classified as *Corynebacterium.* Nonmotile, pleomorphic rods, often arranged in V-formations. Obligate aerobic. DNA ratio of (G + C) 70 ± 5 mol%. Some species are fastidious, xylem-limited, and grow slowly and only on specialized media.

A few phytopathogenic former *Corynebacterium* species are still listed as *Corynebacterium* but they, too, are expected to be transferred to other genera.

Part IV: Actinomycetes: Bacteria forming branching filaments.
Genus: *Streptomyces,* Gram-positive; aerial mycelium with chains of nonmotile conidia. DNA (G + C) 72.2 mol%.
Part V: MOLLICUTES: Prokaryotes that have a cell membrane but no cell wall.
Family: Mycoplasmataceae (The plant mycoplasmalike organisms?)
Family: Spiroplasmataceae
Genus: *Spiroplasma,* helical, motile but lacking flagella.

The taxonomy of the plant pathogenic fastidious phloem-limited bacteria is still unknown, and even the taxonomy of the plant pathogenic mycoplasmalike organisms (MLO), and of the spiroplasmas, is still tentative.

A. PLANT DISEASES CAUSED BY BACTERIA

Bacteria are simple microorganisms, usually consisting of single prokaryotic cells. About 1600 bacterial species are known. The great majority of bacteria are strictly saprophytic and as such are beneficial to humans because they help decompose the enormous quantities of organic matter produced yearly by humans and their factories as waste products or as a result of the death of plants and animals. Several species cause diseases in humans, including tuberculosis, pneumonia, and typhoid fever, and a similar number cause diseases in animals, such as brucellosis and anthrax. About 80 species of bacteria, many of them consisting of numerous **pathovars,** that is, strains differing only in the plant species they infect, have been found to cause diseases in plants. Most pathogenic bacteria are facultative saprophytes and can be grown artificially on nutrient media, but the fastidious vascular bacteria are difficult to grow in culture, and some of them have yet to be grown in culture.

Bacteria may be rod shaped, spherical, ellipsoidal, spiral, comma shaped, or filamentous (threadlike). Some bacteria can move through liquid media by means of flagella, while others have no flagella and cannot move themselves. Some can transform themselves into spores, and certain filamentous forms can produce spores, called conidia, at the end of the filament. Other bacteria, however, do not produce any spores. The vegetative stages of most types of bacteria reproduce by simple fission. Bacteria multiply with astonishing rapidity, and their significance as pathogens stems primarily from the fact that they can produce tremendous numbers of cells in a short period of time. Bacterial diseases of plants occur in every place that is reasonably moist or warm, they affect almost all kinds of plants, and under favorable environmental conditions, they may be extremely destructive.

Characteristics of Plant-Pathogenic Bacteria

Morphology

Almost all plant-pathogenic bacteria are rod-shaped (Figures 12-1 and 12-2), the only exception being *Streptomyces*, which is filamentous. The rod-shaped bacteria are more or less short and cylindrical, and in young cultures they range from 0.6 to 3.5 μm in length and from 0.5 to 1.0 μm in diameter. In older cultures or at high temperatures, the rods of some species are much longer and they may even appear filamentous. Sometimes deviations from the rod shape in the form of a club, a Y or V shape, and other branched forms occur, and some bacteria may occasionally occur in pairs or in short chains.

The cell walls of bacteria of most species are enveloped by a viscous, gummy material, which may be thin (when it is called a **slime layer**) or may be thick, forming a relatively large mass around the cell (when it is called a **capsule**). Most plant-pathogenic bacteria are equipped with delicate, thread-like **flagella**, which are usually considerably longer than the cells by which they are produced. In some bacterial species each bacterium has only one

FIGURE 12-1 Electron micrographs of some of the most important genera of plant-pathogenic bacteria. (A) *Agrobacterium*. (B) *Erwinia*. (C) *Pseudomonas*. (D) *Xanthomonas*. (Photo A courtesy of R. E. Wheeler and S. M. Alcorn, B-D courtesy R. N. Goodman and P. Y. Huang.)

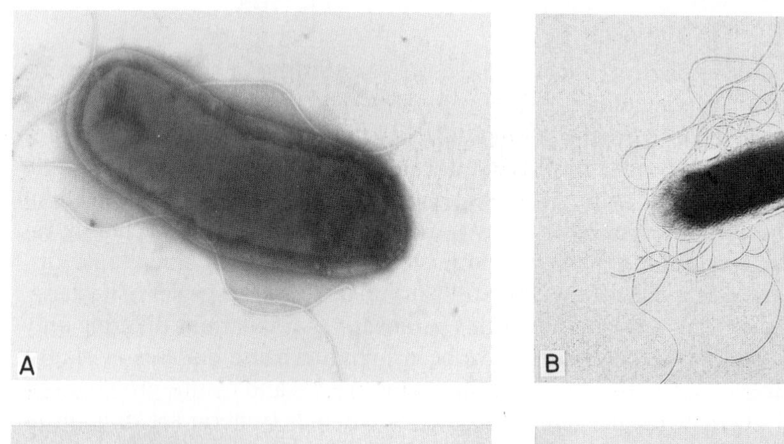

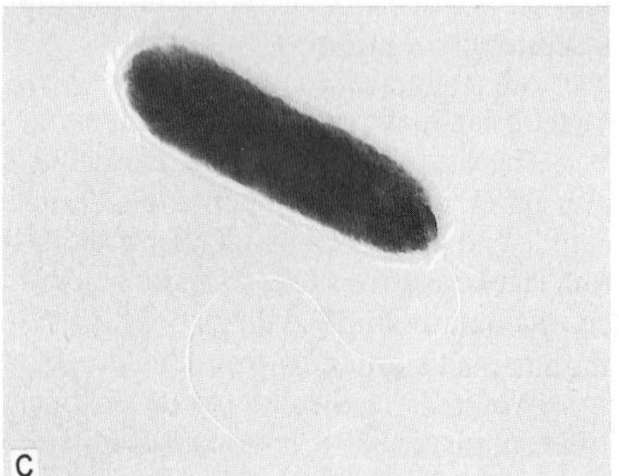

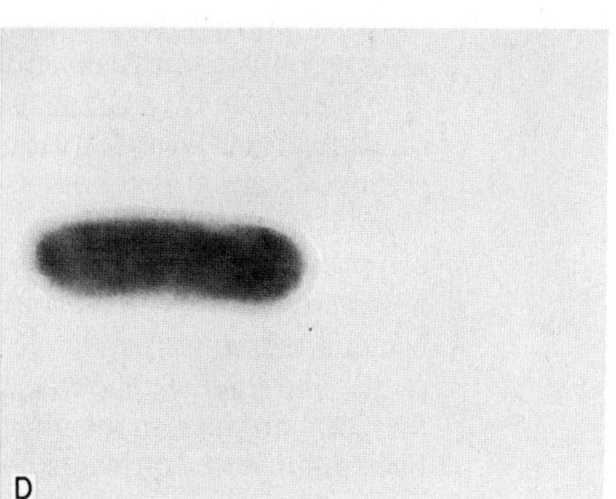

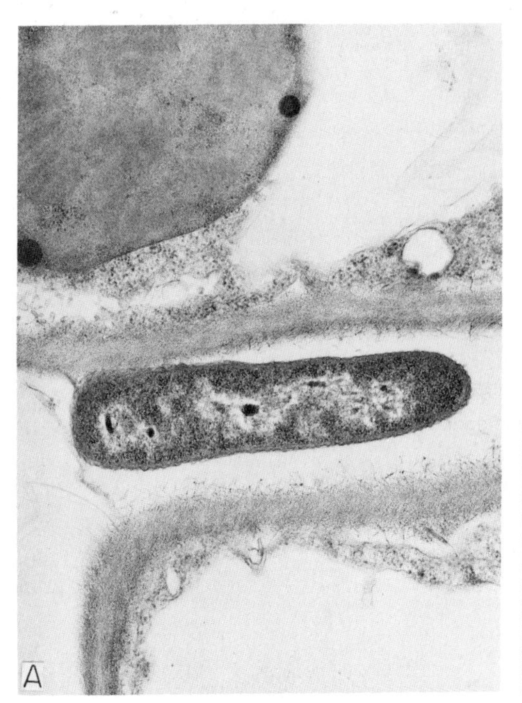

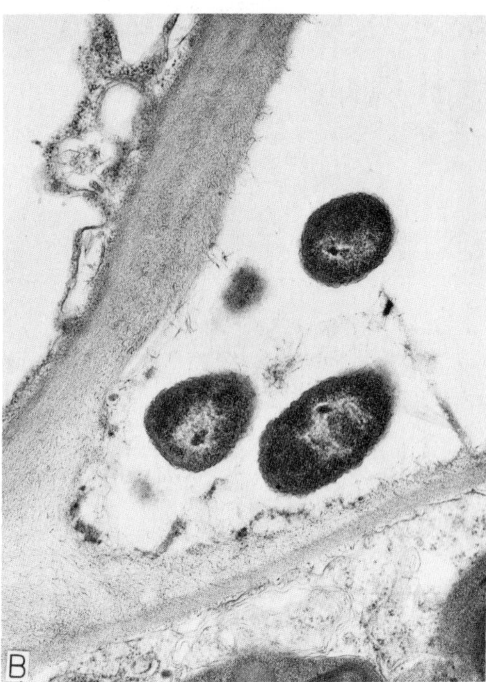

FIGURE 12-2 Electron micrographs of longitudinal (A) and cross sections (B) of bacteria *(Pseudomonas tabaci)* in the intercellular spaces of tobacco leaf mesophyll cells. (Photos courtesy D. J. Politis and R. N. Goodman.)

flagellum, others have a tuft of flagella at one end of the cell (**polar** flagella); some have a single flagellum or a tuft of flagella at each end, and still others have **peritrichous** flagella, that is, distributed over the entire surface of the cell.

In the filamentous *Streptomyces* species, the cells consist of nonseptate branched threads, which usually have a spiral formation and produce conidia in chains on aerial hyphae (Figure 12-3).

Single bacteria appear hyaline or yellowish-white under the compound microscope and are very difficult to observe in detail. When a single bacterium is allowed to grow (multiply) on the surface or within a solid medium, its progeny soon produces a visible mass called a **colony.** Colonies of different species may vary in size, shape, form of edges, elevation, and color, and are sometimes characteristic of a given species. Colonies may be a fraction of a millimeter to several centimeters in diameter, and they are circular, oval, or irregular. Their edges may be smooth, wavy, or angular, and their elevation may be flat, raised, dome shaped, or wrinkled. Colonies of most species are whitish or grayish, but some are yellow, red, or other colors. Some produce diffusible pigments into the agar.

Bacterial cells have thin, relatively tough, and somewhat rigid cell walls, which seem to be quite distinct from the inner cytoplasmic membrane but which sometimes appear to intergrade and merge with the outer slime layer or capsule. The cell wall contains the cell contents and allows the inward passage of nutrients and the outward passage of waste matter, digestive enzymes, and other products given off by the bacterial cell.

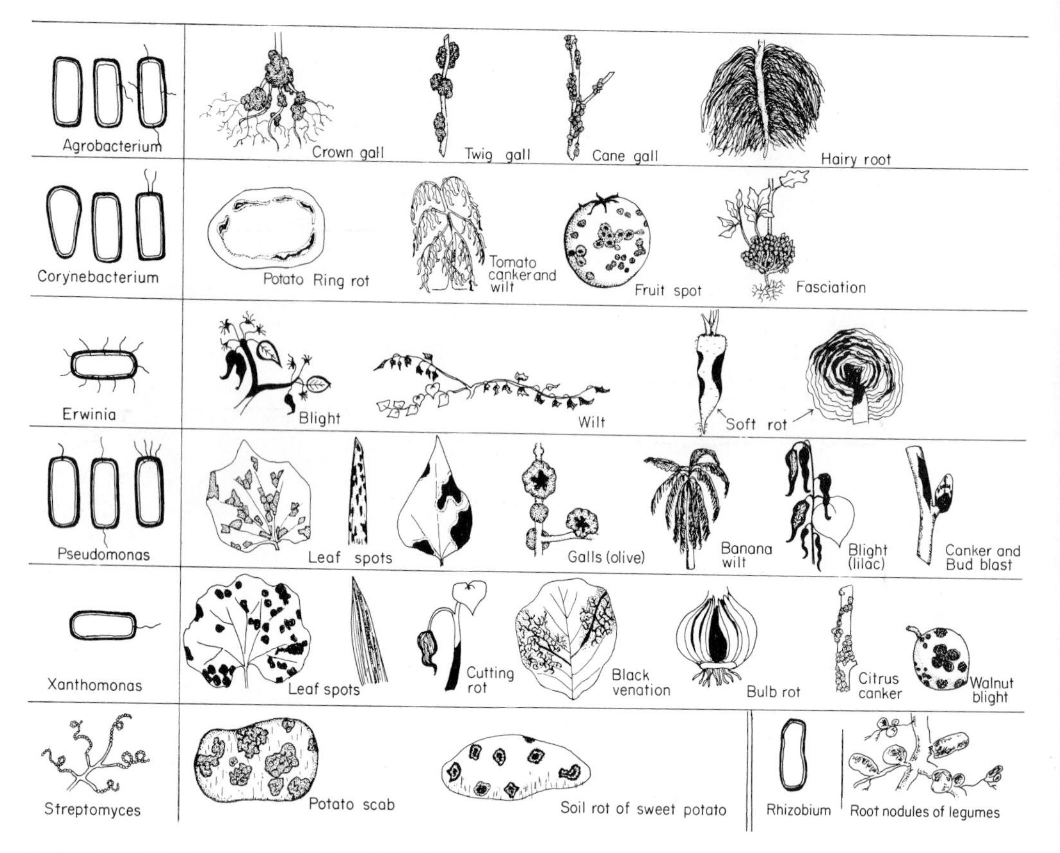

FIGURE 12-3 Genera of bacteria and the kinds of symptoms they cause.

All the material inside the cell wall constitutes the **protoplast.** The protoplast consists of the cytoplasmic or protoplast membrane, which determines the degree of selective permeability of the various substances into and out of the cell; the **cytoplasm,** which is the complex mixture of proteins, lipids, carbohydrates, many other organic compounds, and minerals and water; and the nuclear material, which consists of a large circular **chromosome** composed of DNA that makes up the main body of the genetic material of a bacterium and appears as a spherical, ellipsoidal, or dumbbell-shaped body within the cytoplasm. Often, bacteria also have single or multiple copies of additional smaller circular chromosomes called **plasmids** that can move or be moved between bacteria or between bacteria and plants, as, for example, in the crown gall disease.

Reproduction

Rod-shaped phytopathogenic bacteria reproduce by the asexual process known as **binary fission, or fission.** This occurs by the inward growth of the cytoplasmic membrane toward the center of the cell forming a transverse membranous partition dividing the cytoplasm into two approximately equal parts. Two layers of cell wall material, continuous with the outer cell wall, are then secreted or synthesized between the two layers of membrane. When the formation of these cell walls is completed, the two layers separate, splitting the two cells apart.

While the cell wall and the cytoplasm are undergoing fission, the nuclear material becomes organized in a circular chromosomelike structure that duplicates itself and becomes distributed equally between the two cells formed from the dividing one. Plasmids also duplicate themselves and distribute themselves equally in the two cells.

Bacteria reproduce at an astonishingly rapid rate. Under favorable conditions bacteria may divide every 20 minutes, one bacterium becoming two, two becoming four, four becoming eight, and so forth. At this rate one bacterium conceivably could produce one million bacteria in 10 hours. But because of the diminution of the food supply, the accumulation of metabolic wastes, and other limiting factors, reproduction slows down and may finally come to a stop. Bacteria do reach tremendous numbers in a short time, however, and cause great chemical changes in their environment. It is these changes caused by large populations of bacteria that make them of such a great significance in the world of life in general and in the development of bacterial diseases of plants in particular.

Ecology and Spread

Almost all plant-pathogenic bacteria develop mostly in the host plant as parasites and partly in plant debris or in the soil as saprophytes. There are great differences among species, however, in the degree of their development in the one or the other environment.

Some bacterial pathogens, such as *Erwinia amylovora,* which causes fire blight, produce their populations in the plant host, while in the soil their numbers decline rapidly and usually do not contribute to the propagation of the disease from season to season. These pathogens have developed sustained plant-to-plant infection cycles, often via insect vectors, and either because of the perennial nature of the host or the association of the bacteria with its vegetative propagating organs or seed, they have lost the requirements for survival in the soil.

Some other bacterial pathogens, such as *Agrobacterium tumefaciens,* which causes crown gall, *Pseudomonas solanacearum,* which causes the bacterial wilt of solanaceous crops, and particularly *Streptomyces scabies,* which causes the common scab of potato, are rather typical **soil inhabitants** since they build up their populations within the host plants, but these populations only gradually decline when they are released into the soil. If susceptible hosts are grown in such soil in successive years, sufficiently high numbers of bacteria could be released to cause a net increase of bacterial populations in the soil from season to season. Most plant pathogenic bacteria, however, can be considered soil invaders that enter the soil in host tissue or as free cells, and because they have poor ability to compete as saprophytes, they persist in the soil either as long as the host tissue resists decomposition by the saprophytes, or for varying durations afterwards, depending on the bacterial species and on the soil temperature and moisture conditions.

When in the soil, bacteria live mostly on plant material and less often freely or saprophytically, or in their natural bacterial ooze, which protects them from various adverse factors. Bacteria may also survive in or on seeds, other plant parts, or insects found in the soil. On the plants bacteria may survive epiphytically, in buds, on wounds, in their exudate, or inside the various tissues or organs that they infect (Figure 2-9).

The dissemination of plant pathogenic bacteria from one plant to another or to other parts of the same plant is carried out primarily by water, insects, other animals, and humans (Figure 2-8). Even bacteria possessing flagella can move only very short distances on their own power. Rain, by its washing or spattering effect, carries and distributes bacteria from one plant to another, from one plant part to another, and from the soil to the lower parts of plants. Water also separates and carries bacteria on or in the soil to other areas where host plants may be present. Insects not only carry bacteria to plants, but they inoculate the plants with the bacteria by introducing them into the particular sites in plants where they can almost surely develop. In some cases bacterial plant pathogens also persist in the insect and depend on them for their survival and spread. In other cases, insects are important but not essential in the dissemination of certain bacterial plant pathogens. Birds, rabbits, and other animals visiting or moving among plants may also carry bacteria on their bodies. Humans help spread bacteria locally by handling plants and by their cultural practices, and over long distances by transportation of infected plants or plant parts to new areas or by introduction of such plants from other areas. In cases in which bacteria infect the seeds of their host plants, they can be carried in or on them for short or long distances by any of the agencies of seed dispersal.

Identification of Bacteria

The main characteristics of the plant-pathogenic genera of bacteria (Figure 12-3) are:

Agrobacterium. The bacteria are rod shaped, 0.8 by 1.5–3 μm. They are motile by means of 1–4 peritrichous flagella; when only one flagellum is present it is more often lateral than polar. When growing on carbohydrate-containing media the bacteria produce abundant polysaccharide slime. The colonies are nonpigmented and usually smooth. These bacteria are rhizosphere and soil inhabitants.

Clavibacter (Corynebacterium). Straight to slightly curved rods, 0.5–0.9 by 1.5–4 μm. Sometimes they have irregularly stained segments or granules and club-shaped swellings. The bacteria are generally nonmotile, but some species are motile by means of one or two polar flagella. Gram-positive.

Erwinia. Straight rods, 0.5–1.0 by 1.0–3.0 μm. Motile by means of several to many peritrichous flagella. *Erwinias* are the only plant pathogenic bacteria that are facultative anaerobes. Some *Erwinias* do not produce pectic enzymes and cause necrotic or wilt diseases (the *"amylovora"* group), while other *Erwinias* have strong pectolytic activity and cause soft rots in plants (the *"carotovora"* group).

Pseudomonas. Straight to curved rods, 0.5–1 by 1.5–4 μm. Motile by means of one or many polar flagella. Many species are common inhabitants of soil or of fresh water and marine environments. Most pathogenic *Pseudomonas* species infect plants; few infect animals or humans. Some plant pathogenic *Pseudomonas* species, e.g., *Ps. syringae,* are called fluorescent pseudomonads because, on a medium of low iron content, they produce yellow-green, diffusible, fluorescent pigments. Others, e.g., *Ps. solanacearum,* do not produce fluorescent pigments and make up the nonfluorescent pseudomonads.

Xanthomonas. Straight rods, 0.4–1.0 by 1.2–3 μm. Motile by means of a polar flagellum. Growth on agar media usually yellow. Most are slow growing. All species are plant pathogens and are found only in association with plants or plant materials.

Streptomyces. Slender, branched hyphae without cross walls, 0.5–2 μm in diameter. At maturity the aerial mycelium forms chains of three to many spores. On nutrient media, colonies are small (1–10 mm in diameter) at first with a rather smooth surface but later with a weft of aerial mycelium that may appear granular, powdery, or velvety. The many species and strains of the organism produce a wide variety of pigments that color the mycelium and the substrate; they also produce one or more antibiotics active against bacteria, fungi, algae, viruses, protozoa, or tumor tissues. All species are soil inhabitants. Gram-positive.

Xylella. Mostly single, straight rods, 0.3 by 1–4 μm, producing long filamentous strands under some cultural conditions. Colonies small, with smooth or finely undulated margins. Gram-negative, nonmotile, aflagellate, strictly aerobic, nonpigmented. Nutritionally fastidious, requiring specialized media. Habitat is xylem of plant tissue.

Each bacterial genus, of course, consists of a number of bacterial species (3 in *Agrobacterium,* 5 in *Clavibacter,* 21 in *Erwinia,* 17 in *Pseudomonas,* 5 in *Xanthomonas,* and 2 in *Streptomyces*), the names of which were approved and listed in 1980 by an international committee on bacterial taxonomy. Unfortunately, because not all the taxonomic characteristics of plant pathogenic bacteria were available at the time, many phytopathogenic bacteria formerly thought to be distinct species were downgraded to **subspecies** or **pathovars** of a few recognized species, that is, they were designated as belonging to that one species but distinguishable from each other by their infecting different host plants. According to this scheme, 40 previously distinct *Pseudomonas* species are now listed as pathovars of *Pseudomonas syringae,* while more than 100 previously distinct *Xanthomonas* species are now listed as pathovars of *Xanthomonas campestris.*

Differential media on which the above genera can be separated have been developed.

The genus *Streptomyces* can be easily distinguished from the other bacterial genera because of its much-branched, well-developed mycelium and curled chains of conidia. Identification of bacteria belonging to the rod-shaped genera, however, is a much more complex and difficult process, and it can be made by taking into consideration not only visible characteristics such as size, shape, structure, and color, but also such obscure properties as chemical composition, antigenic reactivity, nutritional versatility, enzymatic action, pathogenicity to plants, susceptibility to certain viruses (bacteriophages), and growth on selective media.

The shape and size of bacteria of a given species in culture can vary with the age of the culture, the composition and pH of the medium, temperature, and staining method. Under given conditions, however, the predominating form, size, and arrangement of cells in a pure culture are quite apparent, and they are important and reliable characteristics. The presence, number, and arrangement of flagella on the bacterial cell are also determined, usually after the flagella have been stained with specific stains.

The chemical composition of certain substances in bacterial cells can be detected with specific staining techniques. Information about the presence or absence of such substances is used for identification of bacteria. Gram's staining reaction differentiates bacteria into gram-positive and gram-negative. In this reaction bacteria are treated with a crystal violet solution for 30 seconds, rinsed gently, treated with iodine solution, and rinsed again with water and then alcohol. Gram-positive bacteria retain the violet-iodine stain combination because it forms a complex with certain components of their cell wall and cytoplasm. Gram-negative bacteria have no affinity for the stain combination, which is therefore removed by the alcohol rinse, and the bacteria remain as nearly invisible as before. Unfortunately, of the rod-shaped phytopathogenic bacteria, only the genus *Clavibacter* is gram-positive. *Agrobacterium, Erwinia, Pseudomonas, Xanthomonas,* and *Xylella* are gram-negative.

The nutritional spectrum of bacterial cells is studied by recording the substances that the bacteria can or cannot use for food. Extracellular hydrolases, that is, enzymes produced when the bacteria grow on certain media, are important determinative tools.

Phytopathogenic bacteria are also tested on various species and varieties of host plants for their pathogenicity on them. This test sometimes, and for practical purposes, may be sufficient for tentative identification of the bacterium.

Serological methods, especially those employing antibodies labeled with a fluorescent compound (immunofluorescent staining), are used for quick and fairly accurate identification of bacteria and have gained popularity in recent years. The use of serological methods is becoming widespread in plant pathology as the availability of species-specific and pathovar-specific antisera increases.

In a few cases, in which specific bacteriophages are available, bacterial species and strains can be identified by the bacteriophages (viruses) that infect them.

Recently, a group of compounds called bacteriocins have been used to differentiate or "type" bacterial isolates by their sensitivity patterns to these compounds or by their production of bacteriocins. **Bacteriocins** are antibacterial substances produced by certain bacteriocinogenic strains of many bacterial species. They are present in cultures of such strains in small amounts, presumably as a result of spontaneous lysis of cells. Bacteriocins are highly specific proteinaceous substances that inhibit and lyse only certain indicator strains of bacteria. Bacteriocins resemble bacteriophage in many respects but differ from them mainly in that they do not reproduce in bacterial host cells. Their production is genetically controlled by extrachromosomal DNA (plasmids) that replicate with the bacterial chromosome and are maintained as long as the bacteriocinogenic strain exists.

An excellent method for isolation and identification of bacteria obtained from plant tissues (Figure 12-4) or soil is through the use of selective nutrient media. Selective media contain nutrients that promote the growth of a particular type of bacterium while at the same time contain substances that inhibit the growth of other types of bacteria. Considerable progress toward perfecting such selective media has been made, and the available selective media for plant pathogenic bacteria are quite satisfactory for routine use in identification of bacterial genera and of several species and even pathovars.

Control of bacterial spots and blights, in addition of the use of resistant varieties, crop rotation, and sanitation, can be obtained to some extent by spraying several times during the period of plant susceptibility with chemicals such as Bordeaux mixture, other copper compounds, zineb, antibiotics such as streptomycin and tetracyclines, and in trees, by injecting antibiotics into the trunks.

• Wildfire of Tobacco

Wildfire of tobacco occurs in all parts of the world where tobacco is grown. In some regions it occurs year after year and is very destructive, whereas in others it appears sporadically and its destructiveness varies. It has been reported to attack other plants; however, it seems to be economically important only on tobacco and soybean.

Wildfire causes losses in both seedbed and field. Affected seedlings may be killed. In tobacco plants already in the field, wildfire causes large, irregular, dead areas on the leaves, which may fall off or become commercially worthless.

◦ *Symptoms.* The first symptoms of wildfire of tobacco appear usually on the leaves of young plants in seedbeds, although plants of any age can be attacked. The leaves of poorly growing seedlings show an advancing wet rot at the margins and tips, with a water-soaked zone separating the rotting and the healthy tissues. The whole leaf or only parts of it may rot and fall off. Some seedlings are killed in the seedbed while others may die after they are transplanted.

The most common symptoms appear on leaves of plants in the field and consist of round, yellowish green spots about 0.5–1.0 cm in diameter. Within a day or so the centers of the spots turn brown and are surrounded by yellowish green haloes (Figure 12-5A). As the disease advances, the brown spots and the chlorotic haloes enlarge. In a few days the brown spots may be 2–3 cm in diameter, although they are not always circular. Adjacent spots

FIGURE 12-5 Wildfire lesions with chlorotic "haloes" on young tobacco leaf (A) and symptoms of wildfire on young tobacco plants (B). Healthy plant at right. (Photo B courtesy G. C. Papavizas.)

usually coalesce and form large, irregular, dead areas, which may involve a large portion of the leaf (Figure 12-5B). In dry weather, these diseased areas dry up and remain intact. But in wet weather they fall off and give a distorted, ragged, and torn appearance to the leaves, which thus become worthless. Spots appear less frequently on flowers, seed capsules, petioles, and stems.

- *The Pathogen: Pseudomonas syringae pv. tabaci.* This bacterium produces a potent toxin, called tabtoxin or wildfire toxin, in the host plants and on many nutrient media. Only 0.05 μg of this toxin can produce a yellow lesion on a tobacco leaf in the absence of bacteria.
- *Development of Disease.* The wildfire bacterium overwinters in plant debris in the soil, in dried or cured diseased tobacco leaves, on seed from infected seed capsules, on seedbed covers, and in the roots of many weeds and crop plants. From these sources the bacteria are carried to the leaves by rain splashes or by wind during wet weather (Figure 12-6). They may also be spread by contaminated tools and hands during handling of the plants.

Very high humidity or a film of moisture on the plants must be present for infections to occur and hence for development of epidemics. Water-soaked areas present in the leaves during long rainy periods or during rains accompanied by strong winds are excellent infection courts for the bacterium and result in extensive lesions within 2–3 days. The bacteria enter the leaf through the large stomata and hydathodes and through wounds caused by insects and other factors. Certain insects such as flea beetles, aphids, and white flies also act as vectors of this pathogen.

Once inside the leaf tissues the bacteria multiply intercellularly (Figure 12-2) at a rapid rate. At the same time they secrete the wildfire toxin, which spreads radially from the point of infection and results in the formation of the chlorotic halo. This consists of a rather broad zone of cells that is free of bacteria and surrounds the bacteria-containing spot. Variants of the bacterium that do not produce tabtoxin produce a similar disease without haloes, known as angular leafspot or blackfire.

In wet weather the bacteria continue to spread intercellularly, and through the toxin and enzymes they secrete cause the breakdown, collapse, and death of the parenchymatous cells in the leaf tissues they invade. Collapsed cells are invaded by the wildfire bacteria and also by saprophytic bacteria and fungi, which further disintegrate the tissues. Dead, disintegrated areas of the leaf are loosely held together, and during humid weather, they are easily detached from the healthy tissues and fall to the ground or are carried by air currents to other plants.

- *Control.* Whenever possible, only resistant varieties should be planted. With susceptible varieties, it is important that control practices begin in the seedbed, since the disease often starts there. Only healthy seed should be used, and if it is suspected of being contaminated with bacteria it should be disinfested by soaking it in a formaldehyde solution for 10 minutes. The seedbed soil should be sterilized, preferably with steam, before planting or with a chemical, such as Vapam, Mylone, or methyl bromide, in the fall. After seedlings emerge, and if wildfire has been present in the area during the previous year, seedbeds should be sprayed with a neutral copper fungicide and streptomycin. The streptomycin sprays should be continued at weekly intervals until plants are transplanted. If isolated spots of wildfire appear, the infected plants plus all healthy plants in a 25-cm band around them should be

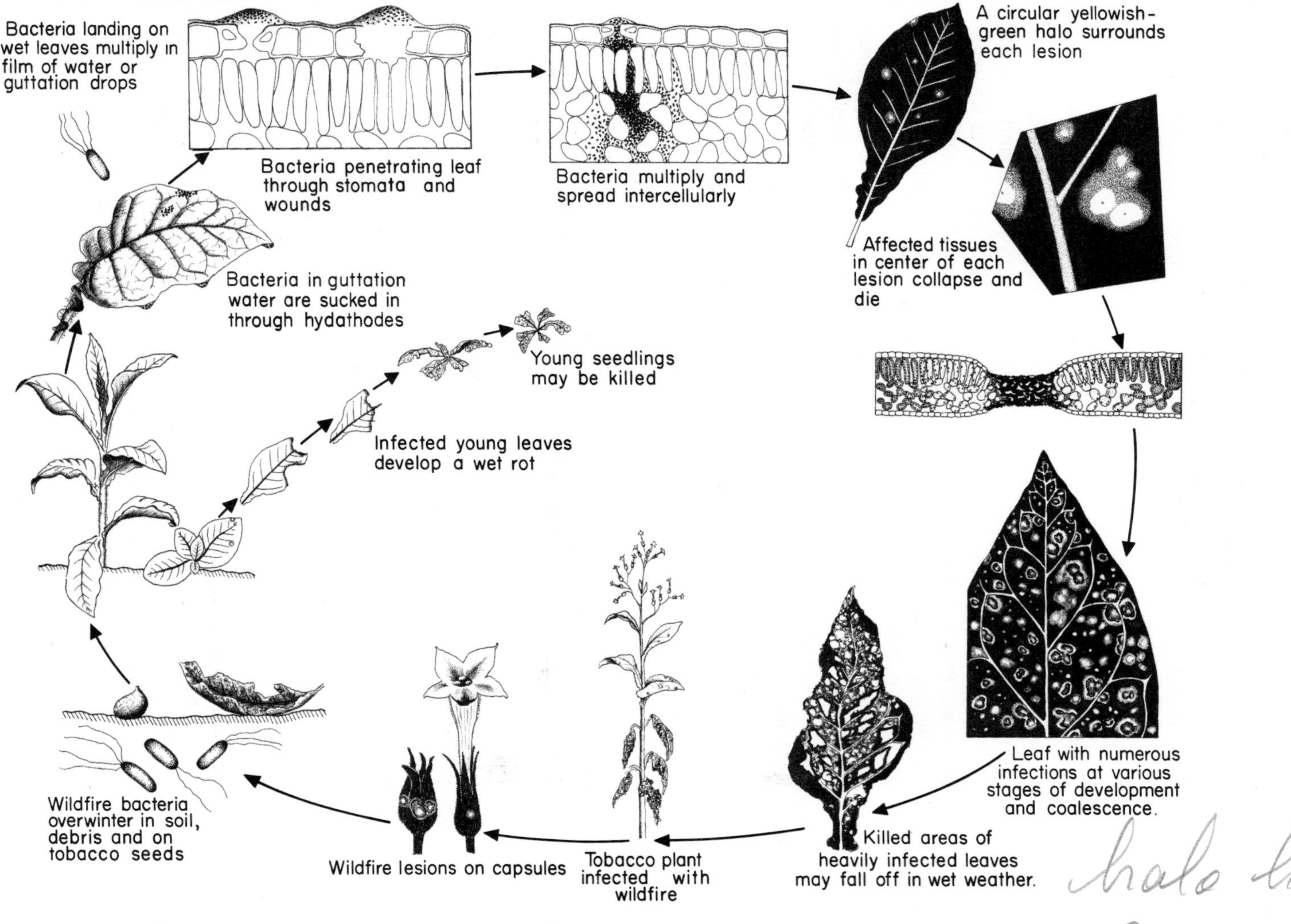

FIGURE 12-6 Disease cycle of a bacterial leaf blight, e.g. wildfire of tobacco caused by *Pseudomonas syringae* pv. *tabaci*.

destroyed by drenching with formaldehyde. Only healthy seedlings should be transplanted into the field, and they should be planted only in fields that did not have a diseased crop during the previous year. Overfertilization, especially with nitrogen, should be avoided, since rapidly growing, succulent plants are much more susceptible to the disease than those that have made slow, normal growth.

• Bacterial Blights of Bean

Common blight is caused by *Xanthomonas campestris* pv. *phaseoli,* halo blight is caused by *Pseudomonas syringae* pv. *phaseolicola,* and bacterial brown spot is caused by *P. syringae* pv. *syringae.* All diseases occur wherever beans are grown and cause similar symptoms. In the field, the diseases are usually impossible to distinguish from one another, and all affect the leaves, pods, stems, and seeds in a similar way.

The symptoms appear first on the lower sides of the leaves as small, water-soaked spots. The spots enlarge, coalesce, and form larger areas that later become necrotic. The bacteria may also enter the vascular tissues of the leaf and spread into the stem. In halo blight, a halolike zone of greenish yellow tissue 10 mm or more in width forms outside the water-soaked area, giving the leaves a yellowish appearance (Figure 12-7A.). In common blight and in bacterial brown spot the infected area, which is surrounded by a much narrower zone of bright, lemon-yellow tissue, turns brown, becomes rapidly necrotic, and through coalescence of several small spots, may produce large dead areas of various shapes. All diseases produce identical symptoms on the stems, pods, and seeds, but when a bacterial exudate is produced on them, it is yellow in common blight *(Xanthomonas)* and light cream or silver-colored in halo blight and in bacterial brown spot *(Pseudomonas).*

The symptoms on the stem appear as water-soaked, sometimes sunken lesions that gradually enlarge longitudinally and turn brown, often splitting at the surface and exuding a bacterial exudate. Such lesions are most common in the vicinity of the first node, where they girdle the stem, usually at about the time the pods are half mature. The weighted plant thus breaks at the lesion, and this symptom is called girdle stem or joint rot. On the pods, small water-soaked spots also develop that may enlarge, coalesce, and turn brownish or reddish with age (Figure 12-7B). Often the vascular systems of the pod sutures become infected causing the adjoining tissues to become water soaked and resulting in the infection of the seed through its connection (funiculus) with the pod. Seeds may rot or shrivel if infected quite young or they may show various degrees of shriveling and discoloration depending on the timing and degree of infection. Similar symptoms are caused on pea and soybean by two different species of *Pseudomonas* (Figure 12-7C,D).

In all three bacterial blights the bacteria overwinter in infected seed and infected bean stems. From the seed, the bacteria infect the cotyledons and from these they either spread to the leaves later on or they enter the vascular system and cause systemic infection producing stem and leaf lesions. Internally, the bacteria move between cells, but the latter collapse, are invaded and digested, and cavities form. When in the xylem, the bacteria multiply rapidly and move up or down in the xylem and out into the parenchyma. They may ooze out through stomata or splits in the tissue and may reenter stems or leaves through stomata or wounds.

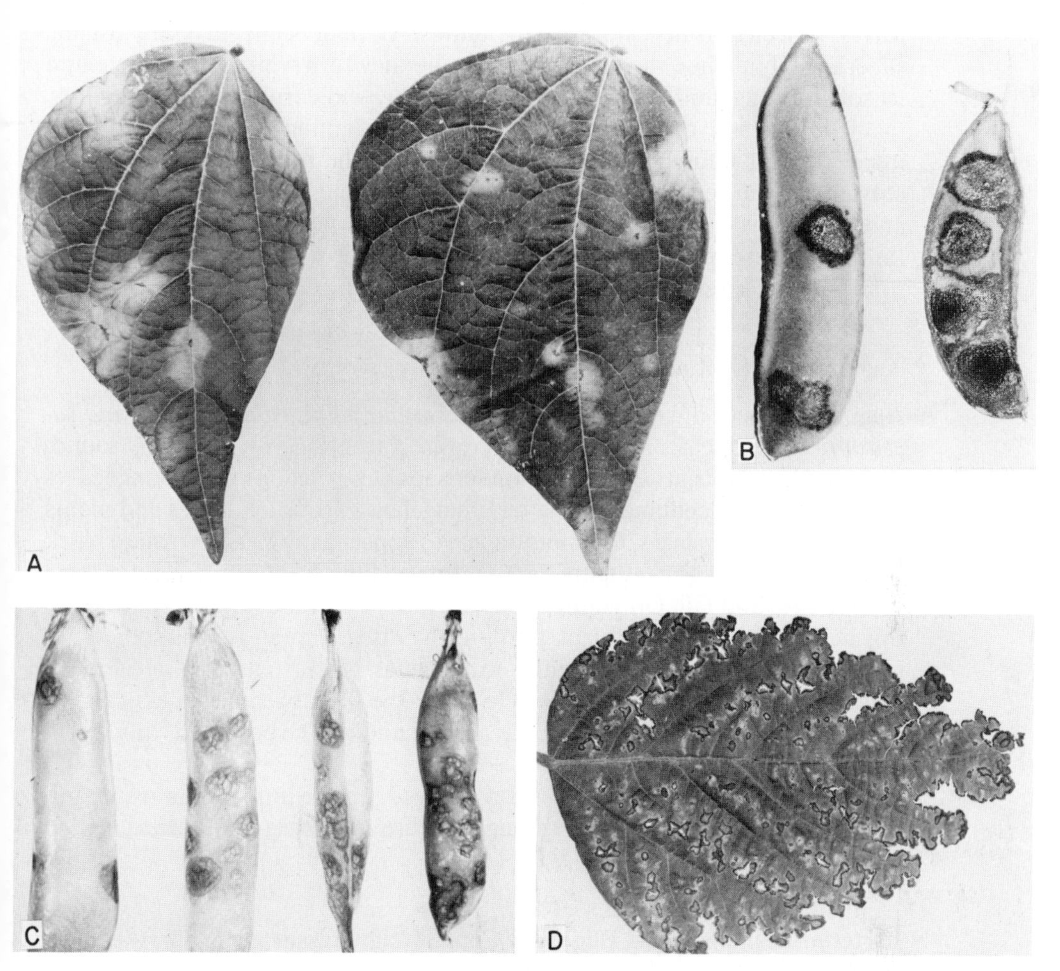

FIGURE 12-7 (A) Early foliar symptoms of halo blight of beans caused by *Pseudomonas syringae* pv. *phaseolicola*. (B) Bacterial blight symptoms on lima bean pods caused by *Xanthomonas campestris* pv. *phaseoli*. (C) Symptoms on pea pods caused by *P. pisi*. In (D) advanced foliar symptoms of bacterial blight of soybean caused by *P. syringae* pv. *glycines*. (Photos A–C courtesy U.S.D.A.)

Control of the bacterial bean blights is through the use of disease-free seed, three-year crop rotation, and sprays with copper fungicides.

Bacterial blight of soybeans caused by *Pseudomonas syringae* pv. *glycinea* and bacterial pustule of soybeans caused by *Xanthomonas campestris* pv. *glycines* are similar in almost all aspects to the bacterial blights of bean.

• Angular Leaf Spot of Cucumber

Angular leaf spot of cucumber is caused by *Pseudomonas syringae* pv. *lachrymans*. It affects the leaves, stems and fruit of cucumber, cantaloupe, squash, and some other cucurbits in North America, Europe, and probably elsewhere. At first, small and circular spots that soon become large, angular to irregular, water-soaked areas develop on the leaves. In wet weather, droplets of bacterial ooze exude from the spots on the lower leaf sides and later dry into a whitish crust. Later, the infected areas turn gray, die, and shrink, often tearing away from the healthy tissue, falling off, and leaving large, irregular holes in the leaves. Infected fruits show small, almost circular spots that are usually superficial, but when the affected tissues die they turn white, crack open, and let soft-rot fungi and bacteria enter and rot the whole fruit.

The bacteria overwinter primarily on contaminated seed and also in infected plant refuse. From the seed or debris the bacteria are splashed to cotyledons and leaves, which they penetrate through stomata and wounds, and may move systemically to other parts of the plant. Control is obtained through the use of clean or treated seed, resistant varieties, crop rotation, and somewhat by spraying with fixed copper-containing bactericides.

• Angular Leaf Spot of Cotton

Angular leaf spot of cotton is caused by *Xanthomonas campestris* pv. *malvacearum.* The disease is present wherever cotton is grown. Small, round, water-soaked spots appear on the undersides of cotyledons and young leaves and on stems of seedlings soon after emergence. Most such leaves and plants are killed. In later stages, the spots on leaves appear as angular, brown to black lesions of varying sizes (Figure 12-8A). Infected leaves of some varieties turn yellow, curl, and fall. On young stems the lesions become long and black, and this has given the name "black arm" to the disease. Stem lesions sometimes girdle and kill the stems. Angular to irregular black spots also develop on young cotton bolls (Figure 12-8B). On these, the spots become sunken, and in hot, humid weather the bacteria may invade and rot the bolls and cause them to drop or to become distorted.

The bacteria overwinter in or on the seed, on the lint, and on undecomposed plant debris. Control is through the use of disease-free or treated seed and of resistant varieties.

• Bacterial Leaf Spots and Blights of Cereals and Grasses

Several *Pseudomonas* and *Xanthomonas* species and pathovars attack each of the cultivated cereals and wild grasses, and some of them cause severe losses to their respective hosts. The most common bacterial diseases of these crops are bacterial stripe of sorghum and corn *(P. andropogonis),* leaf blight of all cereals *(P. avenae),* red stripe and top rot of sugarcane *(P. rubrilineans),* basal glume rot of cereals (*P. syringae* pv. *atrofaciens*), halo blight of oats and other cereals (*P. syringae* pv. *coronafaciens*), bacterial blight or stripe of several

FIGURE 12-8 Angular leaf spot symptoms on leaves (A) and bolls (B) of cotton caused by *Xanthomonas campestris* pv. *malvacearum.* (Photo A courtesy G. C. Papavizas. Photo B courtesy U.S.D.A.)

FIGURE 12-9 Bacterial blight or stripe of barley caused by *Xanthomonas campestris* pv. *translucens.* (Photo courtesy U.S.D.A.)

cereals and streak of sorghum and maize (*X. campestris* pv. *translucens*), leaf scald of sugarcane *(X. albilineans),* and a few other minor diseases.

Most bacterial leaf spots and blights of cereals are probably worldwide in distribution. They cause more or less similar diseases on one or more of the cereals and grasses. Most bacterial diseases of cereals only occasionally cause reduction in yield, but some of them are of major importance. The symptoms appear on leaf blades and sheaths as small, linear, water-soaked areas that soon elongate and coalesce into irregular, narrow, yellowish, or brownish glossy stripes (Figure 12-9). Droplets of white exudate are common on the stripes. Severe infections cause leaves to turn yellow and die from the tip downward and, along with the lesions on the leaf sheath and floral bracts, retard spike elongation, and cause blighting. Small lesions also form on the kernels. The diseases are favored by and develop mainly in rainy, damp weather. The bacteria overwinter on the seed and in crop residue and spread by rain, direct contact, and insects. The main control measures are use of disease-free or treated seed and crop rotation.

• Bacterial Spot of Tomato and Pepper

Bacterial spot of tomato and pepper is caused by *Xanthomonas campestris* pv. *vesicatoria* and is widespread. It causes considerable injury to the leaves and stems, especially of seedlings, but the disease is most noticeable by its effect on the fruit. On the leaves, the symptoms appear as small (about 3 mm), irregular, purplish gray spots with a black center and a narrow yellow halo. Numerous spots may cause defoliation or make the leaves appear ragged. Infection of

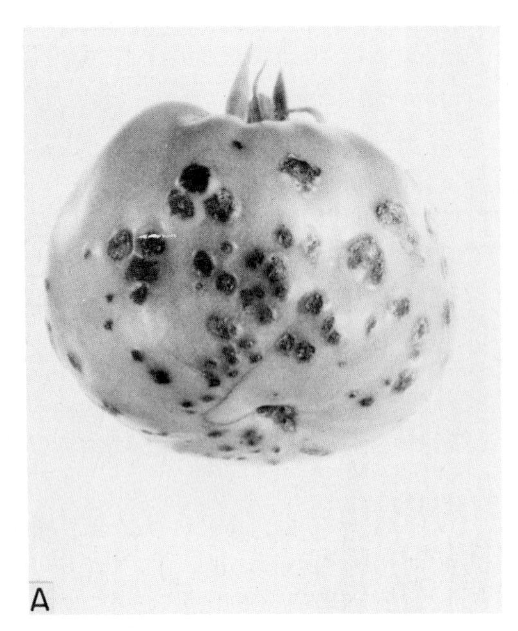

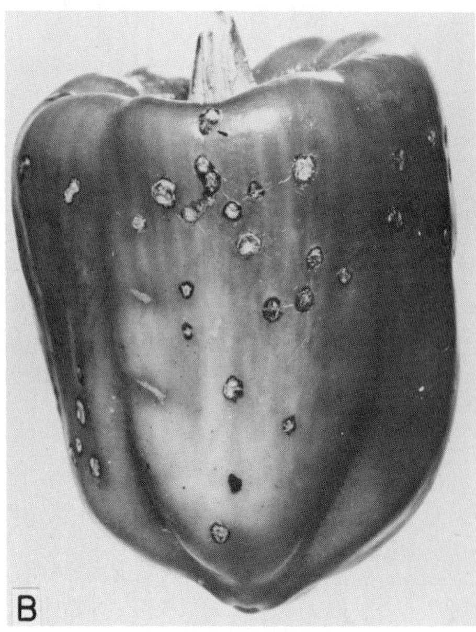

FIGURE 12-10 Bacterial spot of tomato (A) and pepper (B) caused by *Xanthomonas campestris* pv. *vesicatoria*. (Photo courtesy U.S.D.A.)

flower parts usually results in serious blossom drop. On green fruit, small, water-soaked spots appear that are slightly raised, have greenish-white halos, and enlarge to about 3–6 mm in diameter (Figure 12-10). Soon afterward, the halos disappear and the spots become brown to dark, slightly sunken, with a rough, scabby surface and the fruit epidermis rolled back. The bacteria overwinter on seed contaminated during extraction, in infected plant debris in the soil, and on weeds and other hosts. It is spread by rain, wind, or contact. It penetrates leaves through stomata and wounds, and fruits through wounds. Control of the disease depends on the use of bacteria-free seed and seedlings, resistant varieties, crop rotations, and sprays with premixed maneb and fixed copper fungicides. The disease, however, after it appears in the field can be controlled with copper-maneb premixed fungicides only under reasonably dry weather.

A disease called bacterial speck of tomato is similar to bacterial spot but is caused by the bacterium *Pseudomonas syringae* pv. *tomato*. Bacterial speck has become economically important throughout the world since the mid-1970s. The lesions on leaves, stems, and fruit are similar but smaller than those of bacterial spot, although they often coalesce and appear as scabby areas that, on the fruit, may cover one-fourth or more of its surface. Bacterial speck is favored by cool moist weather. Control is the same as for bacterial spot.

• Bacterial Spot of Stone Fruits

Bacterial spot of stone fruits is caused by *Xanthomonas campestris* pv. *pruni*. It is present in most areas where stone fruits are grown and may cause serious losses by directly reducing the marketability of the fruit and by devitalizing the trees by causing leaf spotting and defoliation, and lesions on twigs. The disease is most severe on peach, plum, and apricot, but it affects all stone fruits.

Symptoms appear on the leaves as small, circular-to-irregular, water-soaked spots that soon enlarge somewhat to about 1–5 mm in diameter, become more angular, and turn purple or brown. Often cracks develop around the spots, and the affected areas break away from the surrounding healthy tissue, drop out, and give a "shot-hole" effect to the leaves (Figure 12-11). Several spots may coalesce and may involve large areas of the leaf. Severely affected leaves turn yellow and drop. On the fruit, small, circular, brown, slightly depressed spots appear, usually on a localized area of the fruit. Pitting and cracking occur in the vicinity of the fruit spots and, after rainy weather, gum may exude from the injured areas. On the twigs, dark purplish to black, slightly sunken, circular-to-elliptical lesions form usually around buds in the spring or on green shoots later in the summer.

The bacteria overwinter in twig lesions and in the buds. In the spring they ooze out and are spread by rain splashes and insects to young leaves, fruits, and twigs, which they infect through natural openings, leaf scars, and wounds. The disease is more severe on weakened trees than on vigorous ones, and therefore, keeping trees in good vigor helps them resist the disease. Chemical sprays have not been effective so far, but recent application of antibiotics by injection into trees after the fruit has been harvested has given promising control results during the following season.

FIGURE 12-11 Bacterial leaf spot and shot-hole on stone fruits caused by *Xanthomonas campestris* pv. *pruni.* (A) On ornamental cherry *(Prunus tomentosa)* leaves where characteristic broad, light green haloes form around the infected area before all affected tissue falls off. (B) On peach. (C) On plum. The shot-hole effect is particularly obvious on the plum leaves.

SELECTED REFERENCES

Brinkerhoff, L. A. (1970). Variation in *Xanthomonas malvacearum* and its relation to control. *Annu. Rev. Phytopathol.* **8,** 85–110.

Brinkerhoff, L. A., and Fink, G. B. (1964). Survival and infectivity of *Xanthomonas malvacearum* in cotton plant debris and soil. *Phytopathology* **54,** 1198–1201.

Brinkerhoff, L. A., Verhalen, L. M., Johnson, W. M., Essenberg, M., and Richardson, P. E. (1984). Development of immunity to bacterial blight of cotton and its implications for other diseases. *Plant Dis.* **68,** 168–173.

Burr, T. J., and Katz, B. H. (1984). Overwintering and distribution pattern of *Pseudomonas syringae* pv. *papulans* and pv. *syringae* in apple buds. *Plant Dis.* **68,** 383–385.

Clayton, E. E. (1936). Water soaking of leaves in relation to development of the wildfire disease of tobacco. *J. Agric. Res. (Washington, D.C.)* **52,** 239–269.

Daft, G. C., and Leben, C. (1972). Bacterial blight of soybeans: Epidemiology of blight outbreaks. *Phytopathology* **63,** 57–62.

Daft, G. C., and Leben, C. (1972). Seedling infection after emergence. *Phytopathology* **63,** 1167–1170.

Davis, R. G., and Sandridge, T. L. (1977). Epidemiology of bacterial blight of cotton. *Miss., Agric. For. Exp. Stn., Tech. Bull.* **88.**

Dunegan, J. C. (1932). The bacterial spot disease of the peach and other stone fruits. *U.S., Dep. Agric., Tech. Bull.* **273**

Fahy, P. C., and Persley, G. J. (1983). "Plant Bacterial Diseases: A Diagnostic Guide." Academic Press, New York.

Feliciano, A., and Daines, R. H. (1970). Factors influencing ingress of *Xanthomonas pruni* through peach leaf scars and subsequent development of spring cankers. *Phytopathology* **60,** 1720–1726.

Getz, S., Stephens, C. T., and Fulbright, D. W. (1983). Influence of developmental stage on susceptibility of tomato fruit to *Pseudomonas syringae* pv. *tomato. Phytopathology* **73,** 36–38, 39–43.

Goode, M. J., and Sasser, M. (1980). Prevention—the key to controlling bacterial spot and bacterial speck of tomato. *Plant Dis.* **64,** 831–834.

Higgins, B. B. (1922). The bacterial spot of pepper. *Phytopathology* **12,** 501–516.

Hirano, S. S., and Upper, C. D. (1983). Ecology and epidemiology of foliar bacterial plant pathogens. *Annu. Rev. Phytopathol.* **21,** 243–269.

Kritzman, G., and Zutra, D. (1983). Systemic movement of *Pseudomonas syringae* pv. *lachrymans* in the stem, leaves, fruits, and seeds of cucumber. *Can. J. Plant Pathol.* **5,** 273–279.

Leben, C. (1983). Chemicals plus heat as seed treatments for control of angular leaf spot of cucumber seedlings. *Plant Dis.* **67,** 991–993.

McCarter, S. M., Jones, J. B., Gitaitis, R. D., and Smitley, D. R. (1983). Survival of *Pseudomonas syringae* pv. *tomato* in association with tomato seed, soil, host tissue, and epiphytic weed hosts in Georgia. *Phytopathology* **73,** 1393–1398.

Park, E. W., and Lim, S. M. (1985). Overwintering of *Pseudomonas syringae* pv. *glycinea* in the field. *Phytopathology* **75,** 520–524.

Reddy, A. P. K., *et al.* (1979). Relationship of bacterial leaf blight severity to grain yield of rice. *Phytopathology* **69,** 967–969, 970–973.

Sands, D. C., and Walton, G. S. (1975). Tetracycline injections for control of eastern X-disease and bacterial spot of peach. *Plant Dis. Rep.* **59,** 573–576.

Webster, D. M., Atkin, J. D., and Cross, J. E. (1983). Bacterial blights of snap beans and their control. *Plant Dis.* **67,** 935–940.

Wiles, A. B., and Walker, J. C. (1952). Epidemiology and control of angular leaf spot of cucumber. *Phytopathology* **42,** 105–108.

Williams, P. H., and Keen, N. T. (1967). Histology of infection by *Pseudomonas lacrymans. Phytopathology* **52,** 254–256.

Zapata, M., Freytag, G. F., and Wilkinson, R. E. (1985). Evaluation of bacterial blight resistance in beans. *Phytopathology* **75,** 1032–1039.

Bacterial Vascular Wilts

Vascular wilts caused by bacteria affect mostly herbaceous plants such as several vegetables, field crops, ornamentals, and tropical plants.

The bacterial pathogens that cause vascular wilts and the most important diseases they cause are listed below:

Clavibacter (Corynebacterium), causing bacterial wilt of alfalfa (*C. michiganense* subsp. *insidiosum*) and bean (*C. flaccumfaciens*) ring rot of potato (*C. michiganense* subsp. *sepedonicum*), and bacterial canker and wilt of tomato (*C. michiganense* subsp. *michiganense*).

Erwinia, causing bacterial wilt of cucurbits (*E. tracheiphila*), Stewart's wilt of corn (*E. stewartii*), and fire blight of pome fruits (*E. amylovora*).

Pseudomonas, causing the southern bacterial wilt of solanaceous crop and the Moko disease of banana (*P. solanacearum*), and bacterial wilt of carnation (*P. caryophylli*).

Xanthomonas, causing black rot or black vein of crucifers (*X. campestris* pv. *campestris*), and gumming disease of sugarcane *(X. vascularum).*

In vascular wilts, the bacteria enter, multiply in, and move through the xylem vessels of the host plants (Figure 12-12). In the process, they interfere with the translocation of water and nutrients, and this results in the drooping, wilting, and death of the aboveground parts of the plants. In these respects bacterial vascular wilts are similar to the fungal vascular wilts caused by *Ceratocystis, Fusarium,* and *Verticillium.* However, while in the fungal wilts the fungi remain almost exclusively in the vascular tissues until the death of the plant, in the bacterial wilts the bacteria often destroy (dissolve) parts of cell walls of xylem vessels or cause them to rupture quite early in disease development. Subsequently, they spread and multiply in adjacent parenchyma tissues, at various points along the vessels, kill and dissolve the cells, and cause the formation of pockets or cavities full of bacteria, gums, and cellular debris. In some bacterial vascular wilts, for example, those of corn and sugarcane, the bacteria, once they reach the leaves, move out of the vascular bundles, spread throughout the intercellular spaces of the leaf, and may ooze out through the stomata or cracks onto the leaf surface. Similarly, in some cases, as in the bacterial wilt of carnation, the bacteria ooze to the surface of stems through cracks formed over the bacterial pockets or cavities. More commonly, however, the wilt bacteria, although they may not be confined entirely to the vascular elements, do not spread extensively through the rest of the plant tissues and do not reach the plant surface until the plant is overcome and killed by the disease.

Bacterial vascular wilts can sometimes be determined by cutting an infected stem with a sharp razor blade and then pulling the two parts apart slowly, in which case a thin bridge of a sticky substance can be seen between the cut surfaces while they are being separated; or better still, by placing small pieces of infected stem, petiole, or leaf in a drop of water and observing it under the microscope, in which case masses of bacteria will be seen flowing out from the cut ends of the vascular bundles.

The mechanisms by which bacteria induce vascular wilt in plants seem to be the same as those operating in the fungal vascular wilts. Thus, the bacterial cells themselves along with their polysaccharides seem to cause occlusion of some vessels. The bacteria also secrete enzymes, such as pectinases and cellulases, that break down cell wall substances and that, when carried in the transpiration stream, collect at vessel ends, form gels and gums that help clog the vessel pores, and thus block movement of water. These enzymes also cause softening and weakening of the cell walls, which then collapse, and the tissues droop and wilt. Phenoloxidases secreted by the bacteria or released by

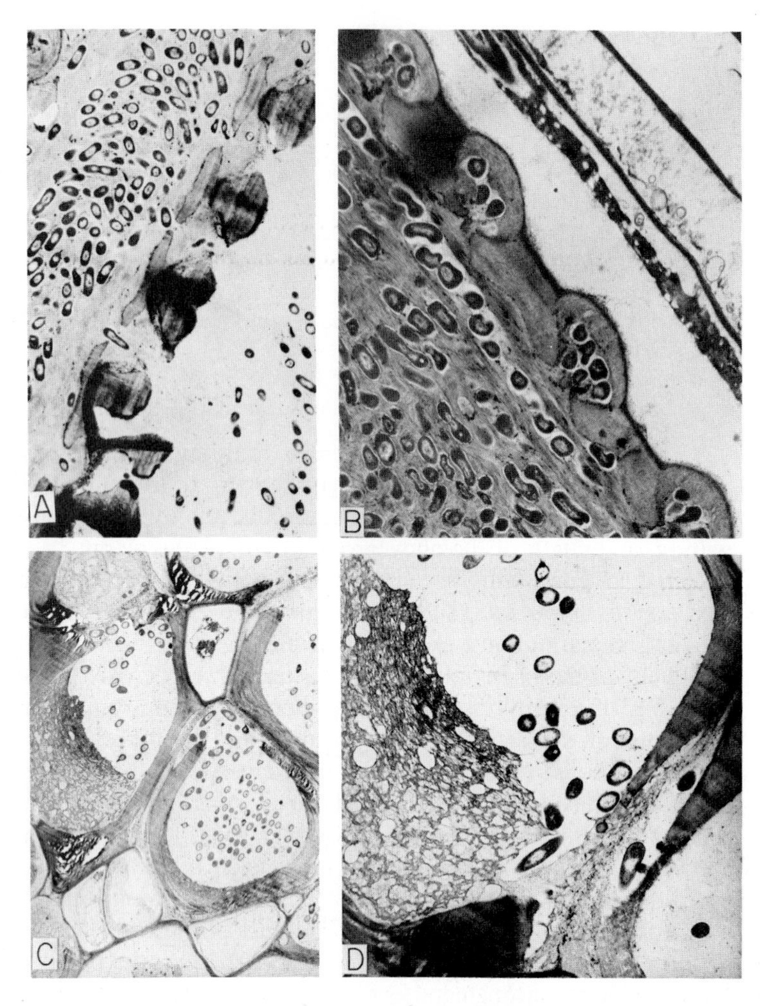

FIGURE 12-12 Histopathology of cabbage leaf veins infected with black rot caused by *Xanthomonas campestris* pv. *campestris*. (A) Uneven distribution of bacteria in xylem vessels and passage of bacteria between adjacent vessels. (B) Bacteria in xylem vessel and in bulges in interspiral regions toward xylem parenchyma cell. (C) Vascular bundle showing bacteria-containing and apparently bacteria-free vessels completely or partially occluded with plugging material. (D) A few bacteria and a mass of plugging material in invaded vessel. (Photos courtesy F. M. Wallis, Univ. of Natal, S. Africa, from Wallis *et al.*, *Physiol. Plant Pathol.* **3,** 371–378.)

the disrupted plant cells cause oxidation of phenolics to quinones, which then polymerize to form melanoid substances. The latter impart a brown coloration to any cell wall or substance to which they become adsorbed. Growth regulators secreted by bacterial pathogens may cause hyperplasia of xylem parenchyma cells with subsequent crushing of xylem vessels, formation of tyloses, and so on. Whether wilt bacteria produce toxins is not known, but many of their secretions certainly have a detrimental effect on plant growth and development.

The wilt bacteria overwinter in plant debris in the soil, in the seed, in vegetative propagative material, or in some cases, in their insect vectors. They enter the plants through wounds that expose open vascular elements and multiply and spread in the latter. They spread from plant to plant through the soil, through handling and tools, through direct contact of plants, or through insect vectors. Nematode infections, by injuring roots, seem to facilitate infection by wilt bacteria in at least some of the vascular wilts. Control of bacterial vascular wilts is difficult and depends primarily on the use of crop rotation, resistant varieties, the use of bacteria-free seed or other propagative material, control of the insect vectors of the bacteria when such vectors exist, and through removal of infected plant debris and proper sanitation.

• Bacterial Wilt of Cucurbits

Bacterial wilt of cucurbits is found in all the United States, although it is most severe in the eastern half of the country. It also occurs in central and northern Europe, South Africa, and Japan. It affects many cultivated and wild species of plants of the family Cucurbitaceae. Cucumber seems to be the most susceptible host to the disease, followed in susceptibility by muskmelon, squash, and pumpkin. Watermelon is extremely resistant or immune to bacterial wilt.

Bacterial wilt affects plants by causing sudden wilting of foliage and vines and finally death of the plants. It also causes a slime rot of squash fruit in storage. The severity of the disease varies widely in different seasons and localities from an occasional wilted plant up to a destruction of 75 to 95 percent of the crop.

◦ *Symptoms.* The first symptoms of bacterial wilt appear as drooping of one or more leaves of a vine; this is soon followed by drooping and wilting of all the leaves of that vine and quickly afterward by wilting of all leaves and collapse of all vines of the infected plant (Figure 12-13). Wilted leaves shrivel and dry up; affected stems first become soft and pale but later they, too, shrivel and become hard and dry. Symptoms in less susceptible plants or under unfavorable conditions develop slowly and may be accompanied by reduced growth and, occasionally by excessive blossoming and branching of the infected plants. When infected stems are cut and pressed between the fingers, droplets of white bacterial ooze appear on the cut surface. The viscid sap sticks to the finger or to the cut sections, and if they are gently pulled apart the ooze forms delicate threads that may be extended for several centimeters. The stickiness and somewhat milky appearance of the sap of infected plants are frequently used as diagnostic characteristics of the disease, although they are not entirely dependable.

The slime of stored squash progresses internally and may cause the spoilage of every portion of the fruit while the exterior surface of the fruit may appear perfectly sound. Usually, however, as the internal rot progresses there appear on the surface dark spots or blotches that coalesce and enlarge. The disease develops over several months in storage. Infected squash fruits are further invaded by soft-rot microorganisms and are completely destroyed.

◦ *The Pathogen: Erwinia tracheiphila.* *Erwinia tracheiphila* is very sensitive to drying and does not survive in infected, dried, plant tissue for more than a few weeks. It survives, instead, in the intestines of striped cucumber beetles *(Acalymma vittata)* and spotted cucumber beetles *(Diabrotica undecimpunctata),* on which the cucurbit wilt bacteria are completely dependent for dissemination, inoculation, and overwintering (Figure 12-14).

Development of Disease. The cucurbit wilt bacteria hibernate in the digestive tracts of a relatively small number of overwintering striped cucumber beetles and spotted cucumber beetles. In the spring these insects feed on the leaves of cucurbit plants, on which they cause deep wounds. The bacteria are deposited in these wounds with the feces of the insects. Swimming through the droplets of sap present in the wounds, the bacteria enter the xylem vessels, where they multiply rapidly and spread to all parts of the plant (Figure 12-14). Penetration through stomata does not take place.

As the bacteria multiply in the xylem they cause a mechanical obstruction of the vessels and so reduce the efficiency of the water-conducting system of diseased plants. Furthermore, gum deposits are commonly found in the xylem elements of infected plants and in some wilting plants tyloses are also present. In some instances the presence of gums or tyloses, or both, appears to be as important in plugging the transpiration stream as the polysaccharides and the bacteria. When wilt symptoms begin to appear, the transpiration rate of infected plants is lower than that of healthy ones and steadily decreases as wilting proceeds. Stems of wilted plants allow less than one-fifth the normal water flow, indicating that an extensive plugging of the vessels is the primary cause of wilting.

Spread of the bacteria from one plant to another is achieved primarily through the striped and the spotted cucumber beetles and to a smaller extent through other insects, such as grasshoppers. When these feed on infected plants, their mouthparts become contaminated with the wilt bacteria. Later

FIGURE 12-13 Bacterial wilt of cucumber caused by *Erwinia tracheiphila*. (Photo courtesy Department of Plant Pathology, Cornell University.)

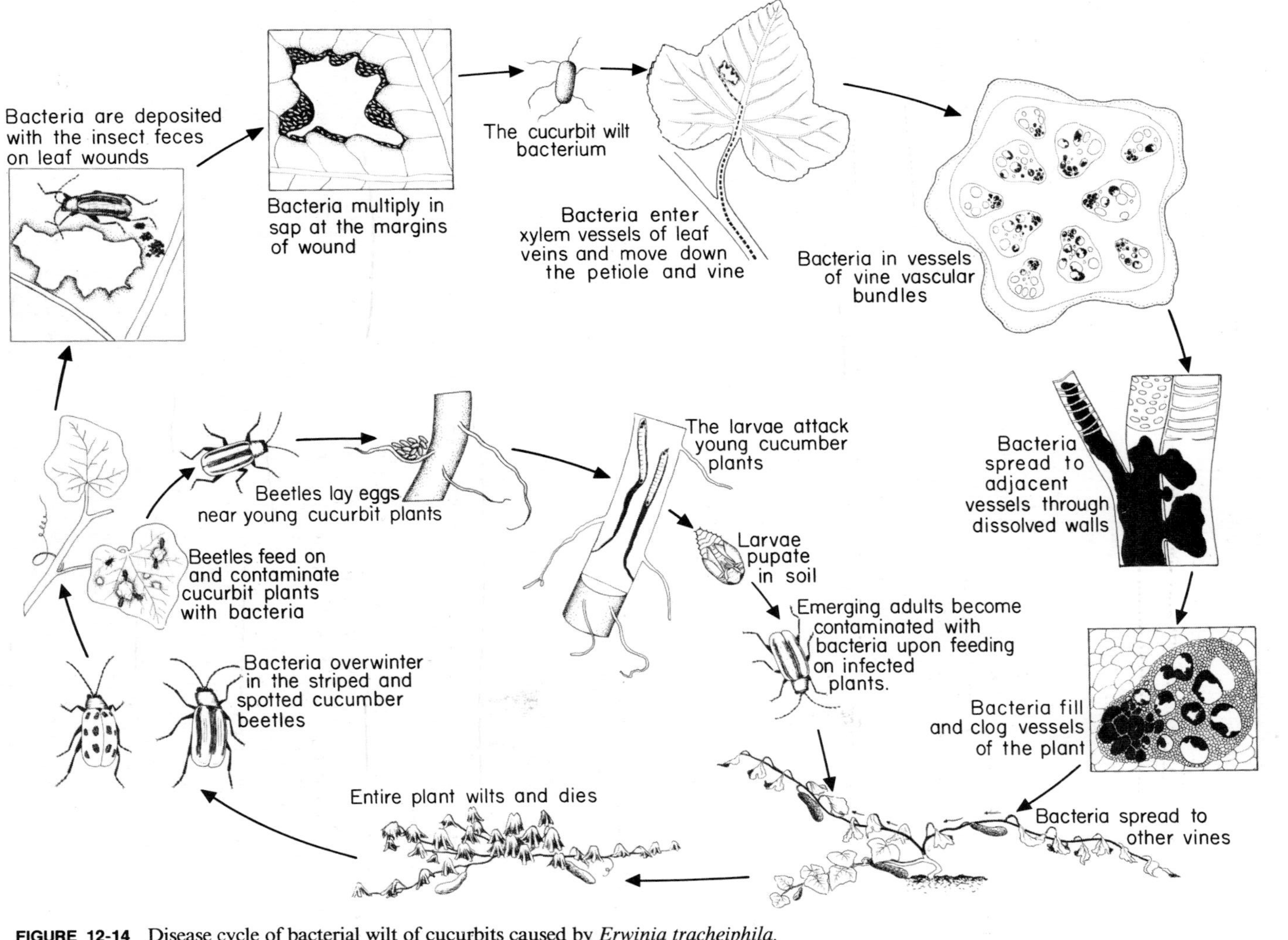

FIGURE 12-14 Disease cycle of bacterial wilt of cucurbits caused by *Erwinia tracheiphila*.

the beetles move on to healthy plants and carry with them bacteria, which they place in the new wounds they make. Each contaminated beetle can infect at least three or four healthy plants after one feeding on a wilted plant, although some beetles are capable of spreading infection for more than three weeks after one wilt feeding. Only a rather small percentage of beetles, however, become carriers of bacteria. Infections take place only when a film of water is present on the tissues and allows the pathogen to reach the wound and move into the xylem vessels. The first wilt symptoms appear 6 or 7 days after infection, and the plant is usually completely wilted by the fifteenth day. The bacteria present in the vessels of infected plants die within one or two months after the dead plants dry up. The bacteria are also incapable of surviving the winter in the soil and in or on seeds from infected plants.

Fruit infection of squash plants usually takes place through infected vines, but it is also possible through the blossoms and the rind of the squash by beetles that feed on the blossoms and the fruits during the growing season.

The disease is strongly influenced by certain environmental factors. Thus, the greater the number of cucumber beetles in an area, the younger and more succulent the plants, and the more humid the weather, the greater the number of plants that will become diseased and the more severe the symptoms.

◦ *Control.* Bacterial wilt of cucurbits can be controlled best by controlling the cucumber beetles with insecticides, such as carbaryl (Sevin), methoxychlor, and rotenone. Control of the early beetles is most important in limiting or eliminating the primary infections of plants and the multiplication and secondary spread of the pathogen.

To avoid squash rot in storage, only fruit from healthy plants should be picked and it should be stored in a clean, fumigated warehouse.

Several varieties within each cucurbit species are resistant to bacterial wilt. These should be preferred to more susceptible ones.

• Fire Blight of Pear and Apple

Fire blight is the most destructive disease of pear in the eastern half of the United States and also causes damage to pear and apple orchards in other parts of the United States, in Canada, New Zealand, Japan, and since 1957, Europe. It has been reported from many other parts of the world.

Fire blight is most destructive on pear, making commercial pear growing under certain conditions impossible. Certain apple and quince varieties are very susceptible to the disease and may be damaged as severely as pear trees. Many other plant species in the rose family (Rosaceae) and some nonrosaceous hosts are affected by fire blight, including several of the stone fruits and many cultivated and wild ornamental species. Although most of these other species can serve as hosts for overwintering of the pathogen and may be affected to varying degrees, only those in the pome-fruit group are affected seriously.

Fire blight damages susceptible hosts by killing flowers and twigs (Figure 12-15), and by girdling of large branches and trunks resulting in the death of the trees. Young trees in the nursery or in the orchard may be killed to the ground by a single infection in one season (Figure 12-16).

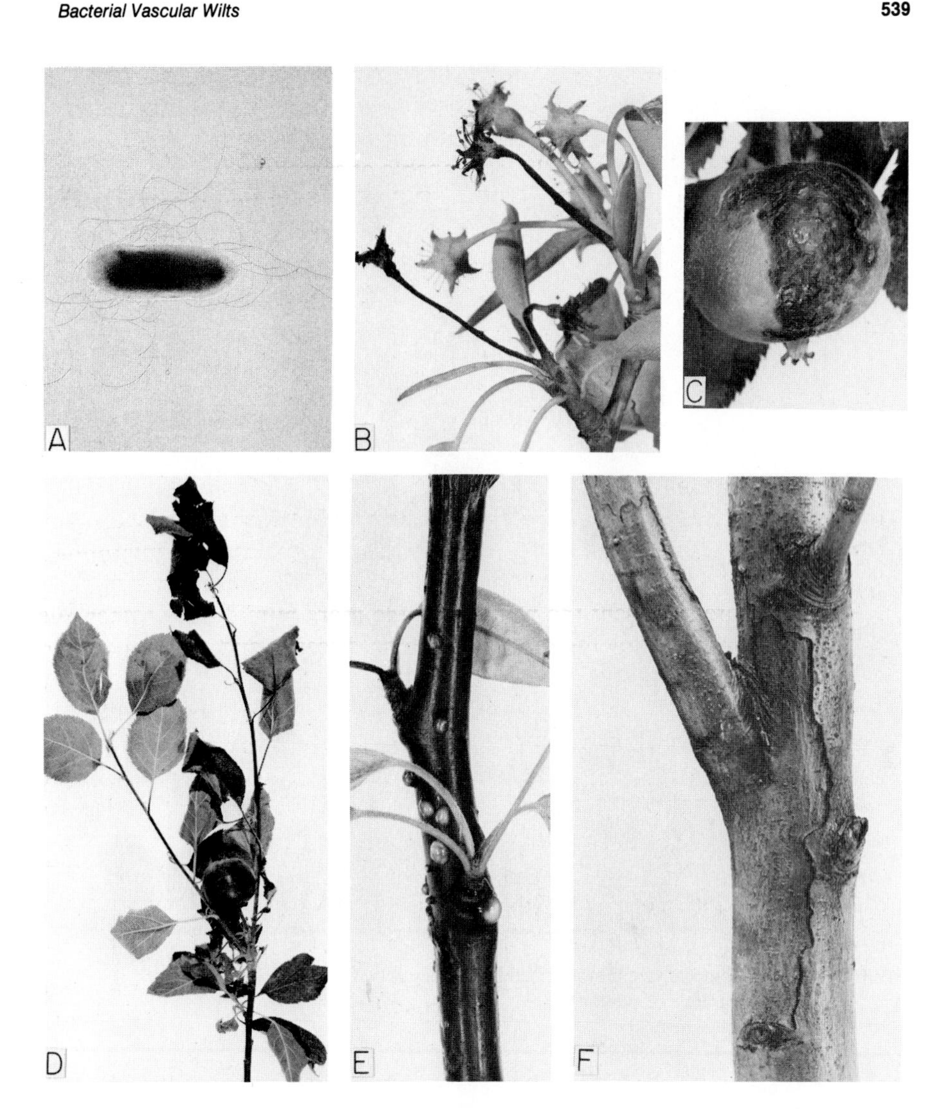

FIGURE 12-15 *Erwinia amylovora* bacterium (A) and fire blight symptoms on pear blossoms (B), fruit (C), and young twig (D). Droplets of bacterial ooze running down the surface of infected pear twig (E). A fire blight canker is shown in (F). (Photo A courtesy R. N. Goodman and P. Y. Huang. Photos B-F courtesy Department of Plant Pathology, Cornell University.)

◦ *Symptoms.* The first symptoms of fire blight appear usually on the flowers, which become water soaked, then shrivel rapidly, turn brownish to black in color, and may fall or remain hanging in the tree (Figure 12-15). Soon the symptoms spread to the leaves on the same spur or on nearby twigs, starting as brown-black blotches along the midrib and main veins or along the margins and between the veins. As the blackening progresses, the leaves curl and shrivel, hang downward, and usually cling to the curled, blighted twigs.

Terminal twigs and watersprouts ("suckers") are usually infected directly and wilt from the tip downward. Their bark turns brownish black and is soft at first but later shrinks and hardens. The tip of the twig is hooked, and the leaves turn black and cling to the twig. From fruit spurs and terminals the symptoms progress down to the supporting branches, where they form

FIGURE 12-16 Young pear tree almost killed by fire blight two months after the first appearance of symptoms.

cankers. The bark of the branch around the infected twig appears water soaked at first, later becoming darker, sunken, and dry. If the canker enlarges and encircles the branch, the part of the branch above the infection dies. If the infection stops short of girdling the branch, it becomes a dormant or inactive canker, with sunken and sometimes cracked margins (Figure 12-15).

Fruit infection usually takes place through the pedicel, but direct infection is not uncommon. Small immature fruit become water soaked, then turn brown, shrivel, mummify, and finally turn black. Dead fruit may also cling to the tree for several months after infection.

Under humid conditions, droplets of a milky colored, sticky ooze may appear on the surface of any recently infected part (Figure 12-15E). The ooze usually turns brown soon after exposure to the air. The droplets may coalesce to form large drops, which may run off and form a layer on parts of the plant surface.

○ *The Pathogen: Erwinia amylovora.* *Erwinia amylovora* is a rod-shaped bacterium and has peritrichous flagella (Figure 12-15A). It does not produce pectolytic enzymes, nor yellow pigments.

◦ *Development of Disease.* The bacteria overwinter at the margins of cankers formed during the previous season, on cankers on other hosts, and possibly in buds and apparently healthy wood tissue. They survive most often in large branches and seldom in twigs less than 1 cm in diameter. In the spring, the bacteria in these "holdover" cankers become active again, multiply, and spread into the adjoining healthy bark. During humid or wet weather, water is absorbed by these bacterial masses, which increase in volume beyond the capacity of the tissues, so that parts of them exude through lenticels and cracks to the surface of the tissue. This gummy exudation, called bacterial ooze or exudate, consists of plant sap, millions of bacteria, and bacterial by-products. The ooze usually appears first about the time when the pear blossoms are opening. Various insects, such as bees, flies, and ants, are attracted to the sweet, sticky exudate and become smeared with it. When they visit flowers afterward, they leave some of the bacteria-containing exudate in the nectar and possibly the pistil of the flower. In some cases bacteria may also be carried from oozing cankers to flowers by splashing rain (Figure 12-17). When the ooze dries, it often forms aerial strands that can be spread by wind and serve as inoculum.

The bacteria multiply rapidly in the nectar, reach the nectarthodes, and penetrate into the tissues of the flower. Bees visiting an infected flower carry bacteria from its nectar to all the succeeding blossoms that they visit. Once inside the flower, the bacteria multiply quickly. Through substances they secrete, they cause plasmolysis and then death and collapse of nearby parenchyma cells. In the meantime, some of the components of the middle lamella and of the cell walls break down. The bacteria move quickly, primarily through the intercellular spaces but also through the macerated middle lamella. Sometimes the delicate walls of the flower cells are disrupted, and invasion of the protoplasts follows. Collapse of several layers of plasmolyzed parenchyma cells in some cases results in fairly large-sized cavities filled with bacteria. From the flower the bacteria move down the pedicel into the bark of the fruit spur. Infection of the spur results in the death of all flowers, leaves, and fruit on it (Figure 12-17).

Penetration and invasion of leaves, when it happens, is similar to that of flowers. Although stomata and hydathodes may serve as ports of entry for the bacteria, it seems that most leaf infections take place through wounds made by insects, hail storms, and so on. The bacteria seem to develop better and faster in the spongy mesophyll than in the palisade parenchyma. From the vein parenchyma the bacteria pass into the petiole and may reach the stem through the petiole.

It has been shown in recent years that as *E. amylovora* bacteria enter tissues through wounds, and perhaps through injured delicate flower and leaf tissues, the bacteria initially colonize and move through vessels, colonizing other tissues only later in the infection process. In contrast to other bacterial wilts, however, *E. amylovora* bacteria move through the vessels and then rapidly invade other tissues, killing cells and causing blight and canker symptoms in the process.

Young, tender twigs may be infected by bacteria through their lenticels, through wounds made by various agents, and through insects. They may also be infected through flower and leaf infections. In the twig the bacteria travel intercellularly, or if they enter injured xylem vessels, the bacteria may move

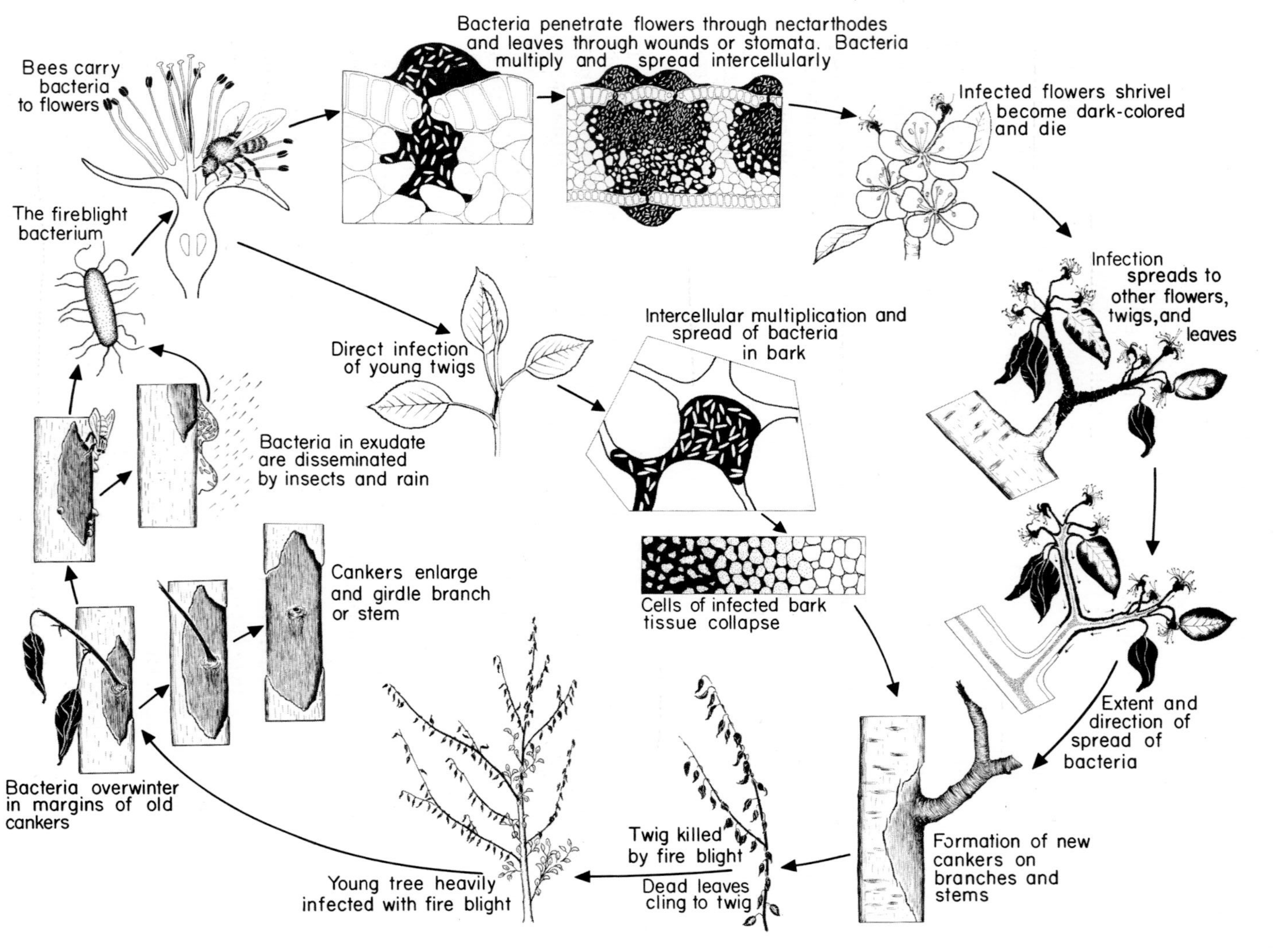

FIGURE 12-17 Disease cycle of the fire blight of pear and apple caused by *Erwinia amylovora*.

over short distances through the xylem. They soon cause collapse and breakdown of cortical cells or nearby xylem parenchyma cells, forming large cavities. In young twigs the bacteria may reach the phloem, in which they then are carried upward to the tip of the twig and to the leaves. Invasion of large twigs and branches is restricted primarily to the cortex. Progress of the infection depends on the succulence of the tissues and on the prevailing temperature and humidity. Under conditions adverse to the development of the pathogen, the host may form cork layers around the infected area and may limit the expansion of the canker. In susceptible varieties and during warm, humid weather the bacteria may progress from spurs or shoots into the second-year, third-year, and older growth, killing the bark all along the way.

Control. Several measures need be taken for a successful fire blight control program.

During the winter all blighted twigs, branches, cankers, and even whole trees, if necessary, should be cut out about 10 cm below the last point of visible infection and burned. Cutting of blighted twigs, suckers, and root sprouts in the summer can reduce the inoculum and prevent the production of large cankers on the branches supporting them. But bacteria are in a very active state in the summer, and precautions should be taken not to spread them to new branches or trees. Cutting should be done about 30 cm below the point of visible infection. The tools should be disinfested after each cut by being wiped with a sponge soaked in 10 percent commercial sodium hypochlorite solution. The latter mixture can also be used to disinfect large cuts made by the removal of branches and cankers.

Since fire blight development is greatly favored by the presence of young, succulent tissues, certain cultural practices that favor moderate growth of trees are recommended. These practices include growing trees in sod, balanced fertilization, especially avoiding the overstimulation of growth by heavy nitrogen applications, and limited pruning. Also a good insect control program should be followed in the post-blossom period to reduce or eliminate spread of bacteria by insects to succulent twigs.

No pear or apple varieties are immune to fire blight when conditions are favorable and the pathogen is abundant, but there is a marked difference between the susceptibility of the varieties available. In areas where fire blight is destructive, varieties for new plantings should be chosen from those most resistant to fire blight.

Satisfactory control of fire blight with chemicals can be obtained only in combination with the above-mentioned measures. Dormant sprays with copper sulfate (4 pounds to 100 gallons of water) before bud break, or with Bordeaux mixture (12:12:100) containing 2 percent miscible-type oil in the delayed dormant period offer some, but not much, protection from fire blight to apple trees. Bordeaux (2:6:100) and streptomycin at 100 parts per million (ppm) are the only effective blossom sprays. Bordeaux should be applied during quick drying conditions to avoid possible russeting of fruit. Streptomycin acts systemically to a limited extent and should be applied either when maximum temperatures are above 18°C or during the night, both conditions favoring absorption of streptomycin by the tissues. One to four streptomycin applications may be necessary for satisfactory control of blossom blight. Bordeaux or streptomycin is sometimes used to control twig blight on bearing and nonbearing trees, but none of them gives good control of this phase of the disease. In many areas, however, streptomycin-resistant strains of the fire blight bacterium are encountered, and that antibiotic is no longer effective in controlling the disease in these areas. Oxytetracycline has been used effectively in orchards where streptomycin is no longer effective.

It has been shown that, after the primary inoculum from fire blight cankers reaches new blossoms, the bacteria must multiply at a rapid rate for several days before they can invade and infect the tree. Such rapid multiplication occurs only when periods of high humidity or abundant rainfall coincide with warm temperatures. This observation has been used to develop a fire blight forecasting model for California and other Pacific Coast states. According to this model, on a graph in which temperature (°C) forms the *Y* axis and the dates from March 1 to May 1 form the *X* axis, a line is drawn from 16.7°C on March 1 to 14.4°C on May 1. As long as the mean temperature in an orchard is below the temperature line, no fire blight is likely to develop and no sprays are required. If, however, the mean temperature in the orchard exceeds the temperature line, epiphytic populations of *Erwinia amylovora* are increasing, and if such temperatures are accompanied or soon followed by rain (or irrigation), a severe outbreak of fire blight infections is likely to occur. Therefore, growers must begin applying bactericidal sprays as soon as the mean temperature exceeds the temperature line.

• Ring Rot of Potato

Ring rot of potato is caused by *Clavibacter (Corynebacterium) michiganense* subsp. *sepedonicum*. The disease occurs and used to cause severe losses in North America and continental Europe. Through strict inspection and quarantine of potato seed tubers, the disease has almost been eliminated from seed lots, but occasional outbreaks still occur by contamination from handling or transportation equipment at the seed producers or on the farm. Infected plants usually do not show aboveground symptoms until they are fully grown, or the symtoms may occur so late in the season that they are often overlooked or masked by senescence, late blight, or other diseases. In years with cool springs and warm summers, however, one or more of the stems in a hill may appear more or less stunted while the rest of the plant appears normal. The interveinal areas of leaflets of affected stems turn yellowish and their margins roll upward and become necrotic. The yellowing of the leaves is accompanied by a progressive wilting, which continues until all the leaves of the stem wilt and the stem then dies. Wilted stems do not usually show much internal discoloration, but if the stem is cut at the base and is squeezed, a creamy exudate oozes out of the vascular bundles.

The characteristic symptoms of the disease appear in tubers either before or after harvest and may be present in only some of the tubers of a plant. Infection and the symptoms begin to develop at the stem end of the tuber and progress through the vascular tissue. When cut through, infected tubers show at first a ring of light yellow vascular discoloration and some bacterial ooze that may be increased by squeezing the tuber (Figure 12-18). As the disease advances, a creamy yellow or light brown crumbly or cheezy rot develops in the region of the vascular ring, and if the tuber is squeezed, a soft, pulpy exudate oozes out from the diseased areas while a more or less continuous ring of cavities is formed by the rotting of tissues in the vascular area.

FIGURE 12-18 Potato tubers showing external (A) and internal (B) symptoms of potato ring rot caused by *Clavibacter michiganense* subsp. *sepedonicum.*

Secondary, soft-rotting bacteria often invade infected tubers, and these bacteria may cause complete rot of the tuber.

The characteristic morphology of *Clavibacter* cells and its gram-positive reaction, taken together with the host and the symptoms, are the primary diagnostic tools for this disease.

The ring rot bacteria overwinter mostly in infected tubers and as dried slime on machinery, crates, sacks, and so on. They do not overwinter in the soil. The bacteria, however, are easily spread by knives used to cut potato seed pieces, and a knife used to cut an infected tuber may infect the next 20 healthy seed pieces cut with it. The bacteria enter plants only through wounds and invade the xylem vessels in which they multiply profusely and may cause plugging. The bacteria also move out of the vessels into the surrounding parenchyma tissues, where they cause cavities, and then again into new vessels. The bacteria also invade the roots and cause the deterioration of the young feeder roots, which contributes to the above-ground symptoms of the plants late in the season.

Potato ring rot is controlled through the use of healthy seed tubers wherever available. The bacterium has not been reported to overwinter in soil. If a grower had ring rot in his potato crop the previous year, however, since the bacteria can also overwinter as dried slime on containers or tools, thorough disinfestation of warehouses, crates, and equipment with ethylene oxide, copper sulfate, or formaldehyde must be carried out. Knives used to cut seed tubers should be constantly disinfested by sodium hypochlorite or by boiling water.

• Bacterial Canker and Wilt of Tomato

Bacterial canker and wilt of tomato is caused by *Clavibacter (Corynebacterium) michiganense* subsp. *michiganense.* It has been reported from many parts of the world and causes considerable losses, particularly in field-grown tomatoes. A similar disease also affects peppers. The disease appears as spotting of leaves, stem, and fruit and as wilting of the leaves and shoots. In advanced stages of the disease, the whole plant wilts and collapses. Very small, indiscernible cankers may occur on stems and leaf veins.

The first noticeable symptoms are spotting or wilting of leaflets at the outer and lower parts of the plant. Leaf spotting occurs during wet weather and appears initially as white blisterlike spots, which become brown with age and may coalesce. Wilting leaves curl upward and inward and later turn brown and wither but do not fall off. Often only the leaflets on one side of the leaf are affected or only on one side of the plant. The wilt may develop gradually from one leaflet to the next or it may become general and destroy much of the foliage. In the meantime, light-colored streaks appear on the stems, shoots, and leaf stalks, usually at the joints of petioles and stems. Later, cracks may appear in the streaks and these cracks form the cankers (Figure 12-19A). Through them, in humid or wet weather, slimy masses of bacteria ooze to the surface of the stem, from which they are spread to leaves and fruits and cause secondary infections. The symptoms on the fruit appear as small, shallow, water-soaked, white spots, the centers of which later become slightly raised, tan colored, and rough. The final, bird's-eye-like appearance of the spots, which have brownish centers about 3 mm in diameter and white halos around them, is quite characteristic of the disease (Figure 12-19B).

In longitudinal sections of infected stems a creamy white, yellow, or reddish-brown line can be seen just inside the woody tissue and along the phloem. The vascular tissues show a brown discoloration, and large cavities are present in the pith and in the cortex and extend to the outer surface of the stem where they form the cankers. The discoloration of the vascular tissues extends all the way to the fruits, both outward toward the surface and inward toward the seeds, and small dark cavities may develop in the centers of such fruits.

The bacteria overwinter in or on seeds and, in some areas, in plant refuse in the soil. Primary infections may result from spread of the bacteria from the seed to cotyledons or leaves (Figure 12-19C), but most infections result from penetration of bacteria through wounds of roots, stems, leaves, and fruits. The bacteria are spread to them through handling during transplanting, by soil water, by wind-blown rain, and by cultural practices such as tying and suckering pole-type tomatoes. Once inside the plant, these bacteria enter the

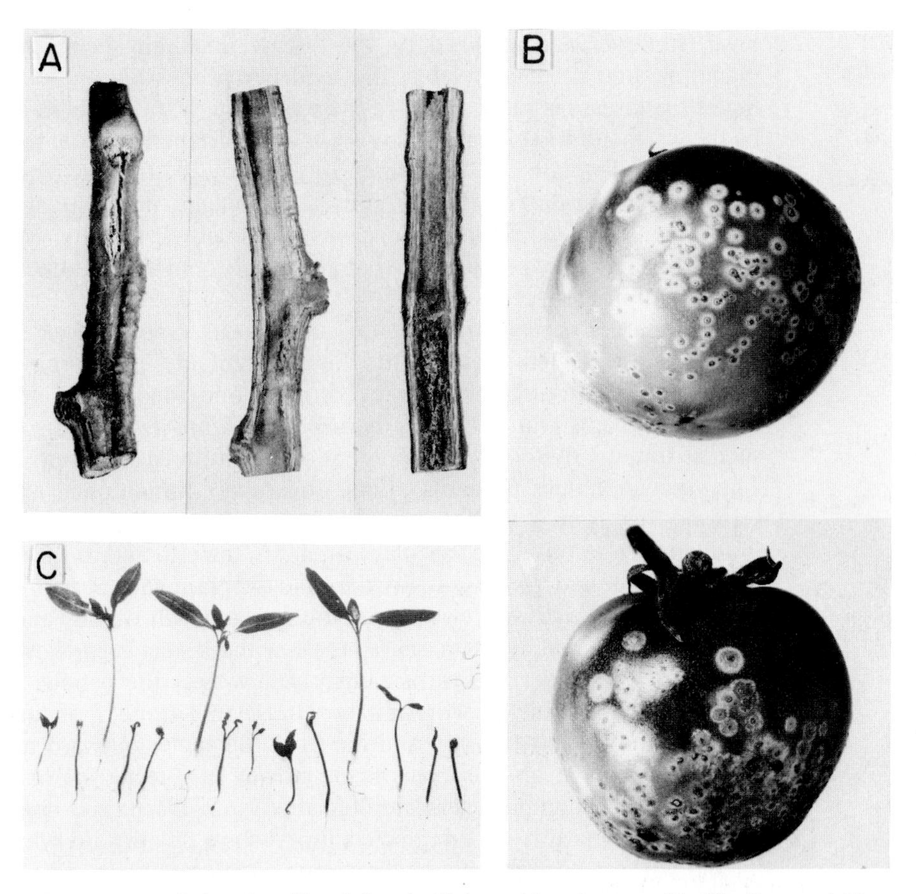

FIGURE 12-19 Tomato stems, fruit and seedlings infected with bacterial canker caused by *Clavibacter michiganense* subsp. *michiganense.* (A) Stems showing open canker (left) and split lengthwise to show discoloration and decay of vascular tissues and pith. (B) Tomatoes showing bird's-eye-like spots with dark rough centers and white haloes at the margins. (C) Three healthy and several diseased tomato seedlings infected through the seed. (Photos A and B courtesy U.S.D.A.)

vascular system and move and multiply primarily in the spiral xylem vessels and move through them and out of them into the phloem, pith, and cortex, where they form the large cavities that result in the cankers.

The disease is controlled through use of bacteria-free seed, protective application of copper or streptomycin in the seed bed, and soil sterilization of the seedbeds. Also, acid treatment of the seed reduces but does not eliminate seed-borne bacteria. Direct seeding of processing tomatoes often gives healthy crops because it avoids the secondary spread that could occur with transplanting.

• Southern Bacterial Wilt of Solanaceous Plants and Moko Disease of Banana

Southern bacterial wilt of solanaceous plants and Moko disease of banana are caused by *Pseudomonas solanacearum.* The disease is present in the tropics and in the warmer climates throughout the world. It causes its most severe losses on banana in the tropics, and it is frequently severe on tobacco, tomato, potato, and eggplant in some warm areas outside the tropics. Many other hosts, however, are attacked by the disease, including peanuts, soybeans, plantains, and other cultivated and wild herbaceous plants. In the United States the disease is most severe in the southeastern states where it is favored by the warm, humid climate and it is known as Granville wilt of tobacco or as brown rot of potato. At least three races of the pathogen are involved in causing the diseases on the various hosts, one of them attacking all the solanaceous and many nonsolanaceous crops as well as some bananas, another attacking only plants in the banana family, and a third attacking potato and sometimes tobacco.

Symptoms of bacterial wilt on solanaceous crops appear as a rather sudden wilt. Infected young plants die rapidly. Older plants may first show leaf drooping and discoloration, leaf drop, or one-sided wilting and stunting before the plants wilt permanently and die (Figure 12-20). In some plants, such as tomato, excessive development of adventitious roots may take place. The vascular tissues of stems, roots, and tubers turn brown, and in cross sections they ooze a whitish bacterial exudate. Bacterial pockets are commonly present around the vascular bundles in the pith and in the cortex, and roots often rot and disintegrate by the time the plant wilts permanently.

In the Moko disease of banana, young plants wilt rapidly and die, their central leaves breaking at a sharp angle without turning yellow. In older plants, first the inner leaf turns a dirty yellow near the petiole, the petiole breaks down, and the leaf wilts and dies. In the meantime, more and more of the surrounding leaves droop and die from the center outward until all the leaves bend down and dry out. Fruit growth in infected plants, if it had started, stops. Banana fingers are deformed, turn black, and shrivel. If the fruit was near maturity when infected, it may show no outward symptoms but the pulp of some fingers may be discolored and decaying. In cross section, an infected banana pseudostem shows many discolored, greenish-yellow to reddish-brown or almost black vascular bundles, particularly in the inner leaf

FIGURE 12-20 Symptoms of southern bacterial wilt on tomato (A) and tobacco (B) caused by *Pseudomonas solanacearum.* (Photo A courtesy U.S.D.A. Photo B courtesy G. C. Papavizas.)

sheaths and in the fruit stalk. Pockets of bacteria and decay may be present in the pseudostem, in the rhizome, and most strikingly in individual bananas that become filled with a dark, gummy substance. The pulp of such bananas finally dries out into a gray, crumbly, starchy residue that pours out when the peel splits open.

The *P. solanacearum* bacteria overwinter in diseased plants or plant debris, vegetative propagative organs, such as potato tubers and banana rhizomes, on the seeds of some crops, for example soybean and peanut, in wild host plants, and probably in the soil. Injured or decaying infected tissues release bacteria in the soil. The bacteria are spread through the soil water, infected or contaminated seeds, rhizomes, and transplants by contaminated knives used for cutting tubers and rhizomes or for pruning suckers and, in some instances, by insects. The bacteria enter plants through wounds made in roots by cultivating equipment, nematodes, and insects and at cracks where secondary roots emerge. The bacteria reach the large xylem vessels and through them spread into the plant. Along the vessels they escape into the intercellular spaces of the parenchyma cells in the cortex and pith, dissolve the cell walls, and create cavities filled with slimy masses of bacteria and cellular debris.

Control of bacterial wilt of solanaceous plants and banana depends mostly on the use of resistant varieties, when available, and proper crop rotation or fallow. Only bacteria-free rhizomes, transplants, and tubers should be used, and tools, such as knives, should be disinfected by dipping for 10 seconds or more in a 10 percent formaldehyde solution when moving from one banana plant to another. Diseased banana plants and rhizomes should be cut up and burned as should plants around them that may be infected but do not yet show symptoms. Infested banana soils can be reclaimed by keeping them fallow for about a year and by frequent disking during the dry season to accelerate desiccation of plant material and apparently the death of the wilt bacteria. Experimental biological control of the disease through treatment of propagative organs with antagonistic bacteria has been obtained and looks promising.

• Black Rot or Black Vein of Crucifers

Black rot or black vein of crucifers is caused by *Xanthomonas campestris* pv. *campestris*. The disease is present throughout the world. If affects all members of the cabbage family and sometimes causes severe losses on these crops. The disease affects plants of any age and primarily the aboveground parts of plants, but in hosts like turnip and radish that have fleshy roots, these organs may also be affected and may develop a dry rot. Infection of young seedlings causes dwarfing, one-sided growth, and drop of the lower leaves. The first symptoms, however, usually appear in the field as large, often V-shaped, chlorotic blotches at the margins of the leaves (Figure 12-21A) that progress toward the midrib of the leaf, while some of the veins and veinlets within the chlorotic area turn black. The affected area later turns brown and dry. In the meantime, the discoloration of the veins advances to the stem and from there upward and downward to other leaves and roots. When leaves become invaded systemically from bacteria moving upward through the midvein, chlorotic areas may appear anywhere on the leaves. Infected leaves may fall off

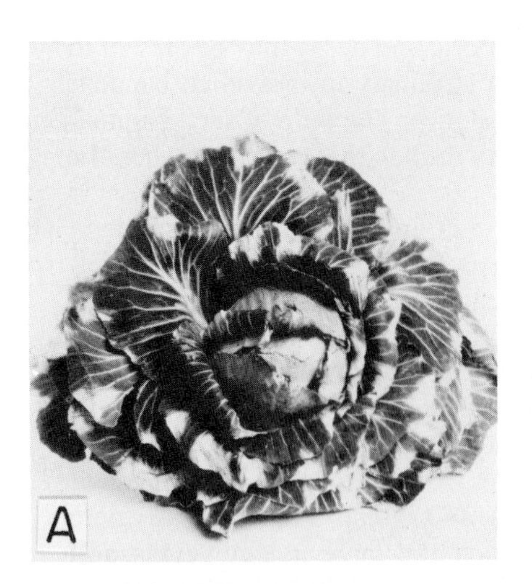

FIGURE 12-21 Cabbage black rot caused by *Xanthomonas campestris* pv. *campestris*. (A) Early stages of infection on margins of leaves. (B) Advanced stages of black rot with many leaves wilting or already fallen off.

prematurely one after the other (Figure 12-21B). The stem and the stalks of infected leaves appear healthy from the outside, but in cross section they show browning or blackening of the vascular tissues and often small yellow slime droplets of bacteria. Sometimes, cavities full of bacteria form in the pith and cortex. Cabbage and cauliflower heads are also invaded and discolored, as are the fleshy roots of turnip, radish, and so on. Infected areas are subsequently invaded by soft-rotting bacteria, which destroy the tissue, and a repulsive odor is given off.

The black rot bacteria overwinter in infected plant debris and on or in the seed. If the bacteria come in contact with or are splashed to cotyledons or young leaves, they infect them through stomata, hydathodes, or wounds and spread through them intercellularly until they reach the open ends of outer vessels, which they invade. The bacteria then multiply in the vessels and spread in them throughout the plant (Figure 12-12), reaching even the seeds. At the same time, however, disintegration of the xylem occurs in places, and the bacteria spread into the intercellular spaces of the surrounding parenchyma. These cells sooner or later are killed and disintegrated, and cavities are formed. In leaf infections, the bacteria reach the surface of the leaves through hydathodes or wounds and are subsequently spread by rain splashes and wind, or are carried by equipment to other leaves, which they invade through hydathodes, wounds, or insect injuries. In wet, warm weather infection develops rapidly, and visible symptoms may appear within hours.

Control of black rot is difficult and depends on the use of bacteria-free seed and transplants planted in soil in which no black rot was present the previous two or three years. So, crop rotation is necessary. Seed treatment with hot water (50°C for 30 minutes) and, more recently, with an antibiotic such as tetracycline or streptomycin followed by a water rinse and a 30-minute soak in 0.5 percent sodium hypochlorite, helps ensure bacteria-free seed. Sprays with copper fungicides (Kocide) at 10-day intervals help reduce spread of the disease.

Gall bacteria overwinter in galls and in the soil. They are spread by contaminated tools such as budding or pruning knives, by soil water, and rain splashes. Gall bacteria are controlled by treatment with antagonistic bacteria, avoiding wounding susceptible plants, using only disease-free rootstocks and scions, soil sterilization in greenhouses, crop rotation when possible, and in olive knot, by sprays with Bordeaux mixture.

• Crown Gall

Crown gall is worldwide in distribution. If affects many woody and herbaceous plants belonging to 140 genera of more than 60 families. In nature it is found mostly on pome and stone fruit trees, brambles, and grapes.

Crown gall is characterized by the formation of tumors or galls of varying size and form, most frequently just below the soil surface, that is, at the crown of the plant, from where it took its name. It is common on the roots and shoots of various nursery plants, which are thus unsalable because crown gall is likely to continue on the plants when they are removed to orchards and gardens. Plants with tumors at their crowns or on their main roots grow poorly and their yields are reduced. Severely infected plants or vines may die.

Crown gall tumors have certain similarities to human and animal cancers and, therefore, the cause and mechanism of their formation have been extensively studied. In spite of the apparent similarities to cancer, however, there are many and basic differences between crown gall of plants and malignant tumors of humans and animals.

Recently, however, and as a result of the above extensive studies, crown gall and the causal bacterium, *Agrobacterium tumefaciens,* have received extraordinary attention because of the discovery that the bacterium modifies the genetic material of host cells by causing part of its Ti plasmid DNA (known as T-DNA) to be transferred and incorporated into the host cell DNA and to be expressed along with the normal host DNA. It soon became possible to extract the Ti plasmid, introduce new genes (stretches of DNA) from one kind of plant (such as bean) along the length of the plasmid, reintroduce the plasmid into *A. tumefaciens* bacteria, and by allowing them to infect another kind of plant, such as sunflower, introduce the plasmid and the new (bean) gene into the second kind of plant (sunflower). This procedure is now used widely, and already several types of genes have been transferred from one kind of plant to another by such recombinant DNA technology (genetic engineering) using *A. tumefaciens* as the vehicle. More recently, it has been possible to inoculate plant protoplasts directly with intact or engineered Ti plasmids in the absence of the bacterium. It is also possible now to disarm the Ti plasmid, that is, remove the genes responsible for inducing tumors, without affecting the ability of the plasmids to carry genes into plant cells and to cause the incorporation and expression of such genes by plant cells that are no longer tumorous. *Agrobacterium tumefaciens* until now has been a natural "genetic engineer" capable of modifying the genetic material of its hosts through introduction of some of its own genetic material into their chromosomes. Now it is being used by plant biologists as the main tool (vehicle) for transferring all kinds of genes between related and unrelated plants, and even between other organisms, for example insects or viruses, and plants.

- *Symptoms.* Crown gall first appears as small overgrowths on the stem and roots, particularly near the soil line. In early stages of their development the tumors are more or less spherical, white or flesh-colored, and quite soft. Since they originate in a wound, at first they cannot be distinguished from callus. However, they usually develop more rapidly than callus. As the tumors enlarge, their surfaces become more or less convoluted. Later on, the outer tissues become dark brown or black, due to the death and decay of the peripheral cells (Figure 12-25). Sometimes there is no distinct line of demarcation between the tumor and the plant proper, the tumor appearing as an irregular swelling of the tissues and surrounding the stem or root. Almost as often, however, the tumor lies outside but close to the outer surface of the host, being connected only by a narrow neck of tissue. Some tumors are spongy throughout and may crumble or become detached from the plant. Others become much more woody and harder, looking knobby or knotty, and reaching sizes up to 30 cm in diameter. Some tumors rot partially or completely from the surface toward the center in the fall and develop again in the same places during the next growing season, or part of the tumor may rot while new tumor centers appear in other parts of the same overgrowth.

  Tumors are most common on the roots and stem near the soil line, but they can also appear on vines up to 150 cm from the ground, on branches of trees, on petioles, and on leaf veins. Several galls may occur on the same root or stem, continuous or in bunches.

  In addition to forming galls, affected plants may become stunted; they produce small, chlorotic leaves and in general are more susceptible to adverse environmental conditions, especially to winter injury.

- *The Pathogen: Agrobacterium tumefaciens.* *Agrobacterium tumefaciens* (Figure 12-25, A) is rod-shaped, with few peritrichous flagella. Virulent bacteria carry one to several large plasmids (small chromosomes composed of circular double-stranded DNA) with molecular weights between 100 and 140 million. One of these plasmids carries the genes for tumor induction and is called the Ti (tumor-inducing)-plasmid. Bacteria that lack Ti-plasmid or lose their Ti-plasmid upon heat treatment are not virulent. The Ti-plasmid also carries the genes that determine the host range of the bacterium and the kinds of symptoms that will be produced. The most characteristic property of this bacterium is its ability to introduce part of its Ti-plasmid (T-DNA) into plant cells and to transform normal plant cells to tumor cells in short periods of time. Once the transformation to tumor cells has been completed, these cells become independent of the bacterium and continue to grow and divide abnormally, even in the absence of the bacteria. Transformed cells can grow in culture on nutrient media containing none of the hormones required for growth by normal (untransformed) cells. Transformed cells also synthesize specific chemicals called "opines," which can be utilized only by bacteria that contain an appropriate Ti plasmid. This property makes the bacterium a genetic parasite since a piece of its DNA parasitizes the genetic machinery of the host cell and redirects the metabolic activities of the host cell to produce substances used as nutrients only by the parasite.
- *Development of Disease.* The bacterium overwinters in infested soils, where it can live as a saprophyte for several years. When host plants are growing in such infested soils, the bacterium enters the roots or stems near the ground through fairly recent wounds made by cultural practices, grafting, insects, and

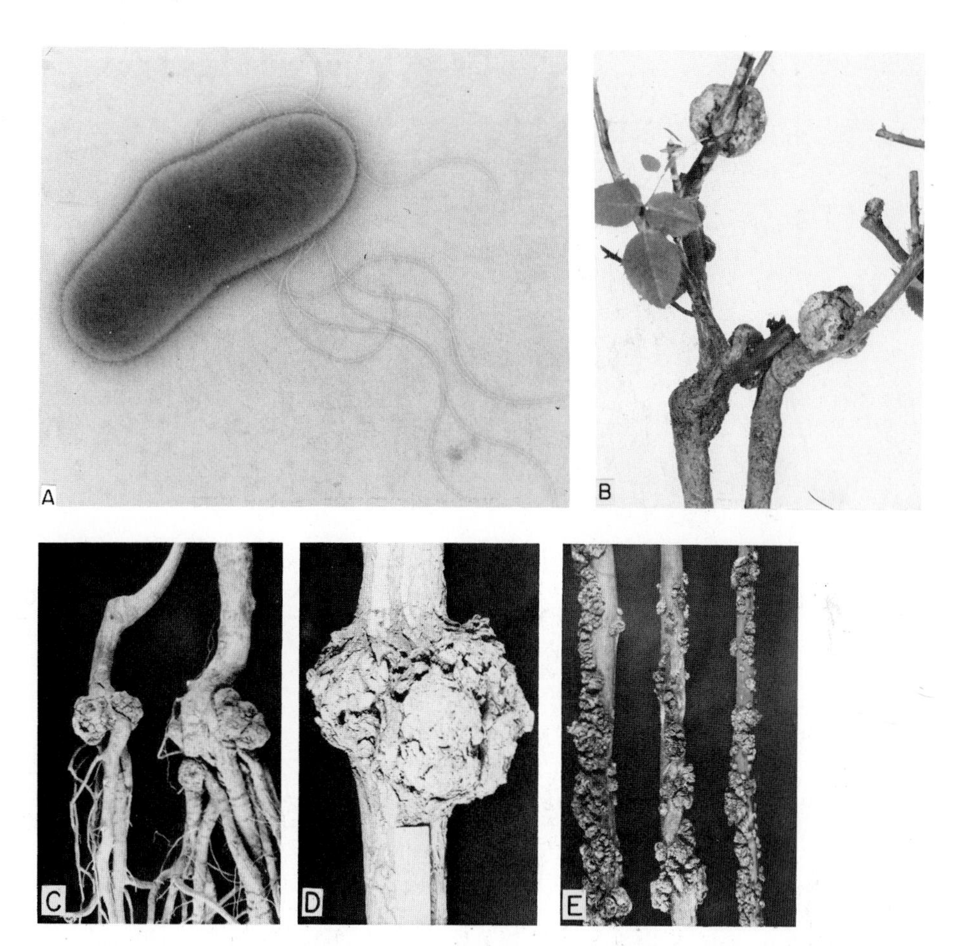

FIGURE 12-25 *Agrobacterium tumefaciens* bacterium (A), and crown gall tumors on rose stems (B), peach root (C) and willow branch (D). E shows cane gall on raspberry caused by *A. rubi.* (Photo A courtesy R. E. Wheeler and S. M. Alcorn. Photo B courtesy Department of Plant Pathology, Cornell University.)

so on. Once inside the tissue the bacteria occur primarily intercellularly and stimulate the surrounding cells to divide (Figure 12-26). One or more groups or whorls of hyperplastic cells appear in the cortex or in the cambial layer depending on the depth of the wound. These cells may contain one to several nuclei. They divide at a very fast rate, producing cells that show no differentiation or orientation, and 10–14 days after inoculation a small swelling can be seen with the naked eye. As the irregular division and enlargement of the cells continue unchecked, the swelling enlarges, developing into a young tumor. Bacteria are absent from the center of the tumors but can be found intercellularly in their periphery. By this time certain cells have differentiated into vessels or tracheids, which, however, are unorganized and with little or no connection with the vascular system of the host plant. As the tumor cells increase in number and size, they exert pressure on the surrounding and underlying normal tissues, which may become distorted or crushed. Crushing of xylem vessels by tumors sometimes reduces the amount of water reaching the upper parts of a plant to as little as 20 percent of normal.

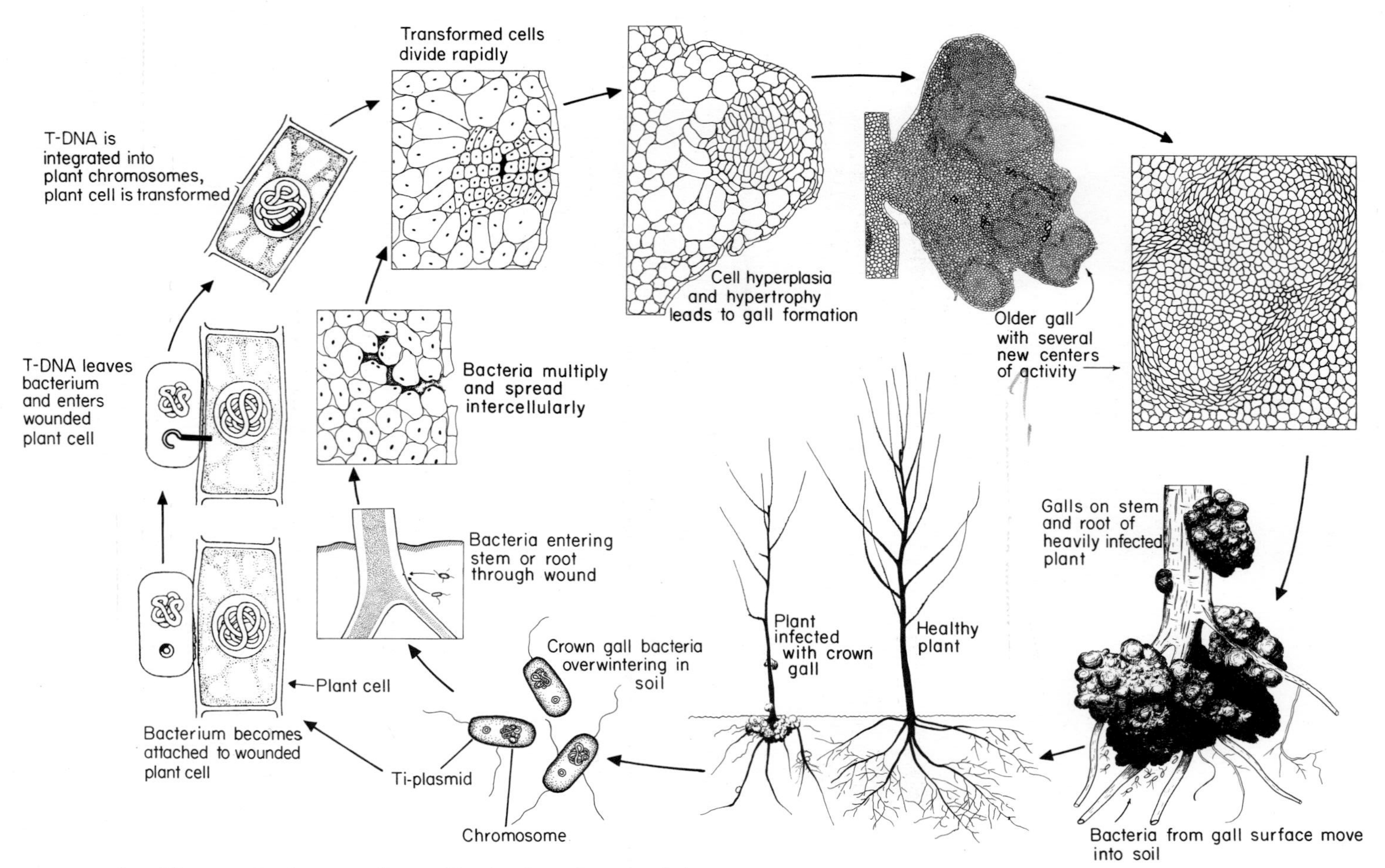

FIGURE 12-26 Disease cycle of crown gall caused by *Agrobacterium tumefaciens.*

The smooth and soft young tumors are not protected by epidermis and, therefore, are easily injured and attacked by insects and saprophytic microorganisms. These secondary invaders cause decay and discoloration of the peripheral cell layers of the tumors, which turn brown to black. Breakdown of the peripheral tumor tissues releases crown gall bacteria into the soil, where they can be carried in the water and infect new plants.

As the tumors enlarge, they sometimes become woody and hard. The incomplete and disarranged vascular bundles that may be present in the tumor itself are ineffective. When tumors are unable to obtain as much water and nourishment as is required to carry them beyond a certain point in growth, their enlargement stops, decay sets in, and the necrotic tissues are sloughed off. In some cases the tumor regresses and no new one appears. More often, however, some portion of the tumor remains alive and forms additional tumor tissue during the same or the following season.

When very young and expanding tissues are infected, in addition to the primary tumor that develops at the point of infection, secondary tumors appear. These usually form below, but often above, the primary tumor and at varying distances from it. Sometimes the secondary tumors develop at the scars of fallen leaves or at wounds made by various agents. At other times secondary tumors develop on apparently unwounded parts of the stem, on the petiole, and even on leaf midribs or larger veins several internodes above the primary tumor. Their starting point seems to be in the xylem of the vascular bundles. They are free from bacteria, since no bacteria can be isolated when these tumors are plated on nutrient media. When fragments of such bacteria-free tumors are grafted on healthy plants, they develop into large tumors similar in appearance and structure to the primary tumors but remain completely devoid of bacteria. This indicates that the bacteria are important only at the beginning of the disease, presumably by having an irritant effect on the plant cells. Once the cells have been triggered to malignancy, they produce their own irritants and their uncontrolled growth becomes autonomous.

Although the nature of the irritant and the mechanism of transformation of normal plant cells to tumor cells have been the objects of intensive studies, our knowledge on these topics is still incomplete. A great deal of information, however, has been gained in the last 10 years. As mentioned above, when Ti-plasmid-carrying virulent bacteria infect cells near a recent wound, they insert the T-DNA part of the Ti-plasmid into such cells, which having been somehow conditioned by their response to the nearby wound, become receptive and take in several copies of the T-DNA. The T-DNA copies then become incorporated in several places along the various chromosomes of the plant. It is known that such cells express the genes present on the T-DNA, that is, the cells become transformed. Transformed cells produce opines, which can be utilized only by the T-DNA-carrying bacteria, elevated amounts of indole-acetic acid (the plant hormone) and variable amounts of cytokinins and various enzymes. It is still not clear, however, how these changes are brought about and how they are related, if at all, to the uncontrollable growth of transformed cells.

◦ *Control.* Crown gall control begins with mandatory inspection of nursery stock and rejection of infected trees. Susceptible nursery stock should not be planted in fields known to be infected with the pathogen. Instead, infested fields should be planted with corn or other grain crops for several years before they are planted with nursery stock. Since the bacterium enters only through relatively fresh wounds, wounding of the crowns and roots during cultivation should be avoided, and root-chewing insects in the nursery should be controlled to reduce crown gall incidence. Nursery stock should be budded rather than grafted because of the much greater incidence of galls on graft than on bud unions. Growers should purchase and plant only crown gall–free trees. Some control of existing galls can be obtained by painting them with a commercially available mixture of aromatic hydrocarbons, which selectively kill gall tissue, but this is not practiced widely.

Excellent biological control of crown gall is obtained by soaking germinated seeds or dipping nursery seedlings or rootstocks in a suspension of a particular strain (No. 84) of *Agrobacterium radiobacter*. This strain of bacteria is antagonistic to most strains of *A. tumefaciens*. Some control is also obtained by treating nongerminated seeds with the antagonist or by drenching the soil with a suspension of the antagonistic bacterium. It is postulated that the antagonist controls crown gall initiation by establishing itself on the surface of the plant tissues, where it produces the bacteriocin agrocin 84. This bacteriocin is inhibitory to most virulent *A. tumefaciens* strains. Unfortunately, some strains of *A. tumefaciens* are resistant to agrocin 84, so this control method fails in some places.

SELECTED REFERENCES

Alconero, R. (1980). Crown gall of peaches from Maryland, South Carolina, and Tennessee and problems with biological control. *Plant Dis.* **64,** 835–838.

Amasino, R. M., and Miller, C. O. (1982). Hormonal control of tobacco crown gall tumor morphology. *Plant Physiol.* **69,** 389–392.

Anderson, A. R., and Moore, L. (1979). Host specificity in the genus *Agrobacterium. Phytopathology* **69,** 320–323.

Banfield, W. M. (1934). Life history of the crown gall organism in relation to its pathogenesis on the red raspberry. *J. Agric. Res.* **48,** 761–787.

Barton, K. A., *et al.* (1983). Regeneration of intact tobacco plants containing full length copies of genetically engineered T-DNA, and transmission of T-DNA to R1 progeny. *Cell (Cambridge, Mass.)* **32,** 1033–1043.

Chilton, M. D. (1983). A vector for introducing new genes into plants. *Sci. Am.* **248.**

Davey, M. R. *et al.* (1980). Transformation of petunia protoplasts by isolated *Agrobacterium* plasmids. *Plant Sci. Lett.* **18,** 307–313.

DeCleene, M., and DeLey, J. (1976). The host range of crown gall. *Bot. Rev.* **42,** 389–466.

Hedgecock, G. G. (1910). Field studies of the crown gall of the grape. *U. S., Dep. Agric., Bull.* **183,** 1–40.

Herrera-Estella, M. *et al.* (1983). Expression of chimeric genes transferred into plant cells using a Ti-plasmid-derived vector. *Nature (London)* **303,** 209–213.

Horsch, R. B. *et al.* (1985). A simple and general method for transferring genes into plants. *Science* **227,** 1229–1231.

Kahl, G., and Schell, J. S., eds. (1982). "Molecular Biology of Plant Tumors." Academic Press, New York.

Kerr, A. (1980). Biological control of crown gall through production of agrocin 84. *Plant Dis.* **64,** 25–30.

Lelliott, R. A. (1971). A survey of crown gall in rootstock beds of apple, cherry, plum and quince in England. *Plant Pathol.* **20,** 59–63.

Miller, H. N. (1975). Leaf, stem, crown and root galls induced in chrysanthemum by *Agrobacterium tumefaciens. Phytopathology* **65,** 805–811.

Muncie, J. H. (1926). A study of crown gall caused by *Pseudomonas tumefaciens* on rosaceous hosts. *Iowa State Coll. J. Sci.* **1,** 67–117.

Murai, N. *et al.* (1983). Phaseolin gene from bean is expressed after transfer to sunflower via tumor-inducing plasmid vectors. *Science* **222,** 474–481.

Nester, E. W., and Kosuge, T. (1981). Plasmids specifying plant hyperplasias. *Annu. Rev. Microbiol.* **35,** 531–565.
Nester, E. W., Gordon, M. P., Amasino, R. M., and Yanofsky, M. F. (1984). Crown gall: A molecular and physiological analysis. *Annu. Rev. Plant Physiol.* **35,** 387–413.
Riker, A. J., and Keitt, G. W. (1926). Studies on crown gall and wound overgrowth on apple nursery stock. *Phytopathology* **16,** 765–808.
Riker, A. J., *et al.* (1946). Some comparisons of bacterial plant galls and of their causal agents. *Bot. Rev.* **12,** 57–82.
Smith, E. F., Brown, N. A., and Townsend, C. O. (1911). Crown gall of plants: Its cause and remedy. *U. S., Dep. Agric., Bull.* **213,** 1–215.
Thomashow, M. F. *et al.* (1980). Host range of *Agrobacterium tumefaciens* is determined by the Ti-plasmid. *Nature (London)* **283,** 794–796.

Bacterial Cankers

Relatively few canker diseases of plants are caused by bacteria, but some of them are so widespread and devastating that great losses are caused by them or great efforts are required to protect the plants from them. The bacteria and the most important cankers they cause are the following:

Pseudomonas, causing the bacterial canker of stone fruit and pome fruit trees *(P. syringae* pv. *syringae* and *P. syringae* pv. *morsprunorum).*
Xanthomonas, causing the bacterial canker of citrus *(X. campestris* pv. *citri).*

In all bacterial cankers, the canker symptoms on stems, branches, or twigs are only part of the disease syndrome, and direct symptoms on fruits, leaves, buds, or blossoms may be at least as important in the overall effect of the disease on the plant as are cankers. Also, bacterial cankers are not always sunken and soft, as is the case with fungal cankers, but they may also appear as splits in the stem, as necrotic areas within the woody cylinder, or as scabby excrescences on the surface of the tissue. In some bacterial cankers, soft decayed tissue and bacterial cavities that ooze either a slimy exudate or a dark gummy substance may be present in the stem, but during much of the year populations of bacteria in woody cankers are low and their isolation from them is erratic.

The canker bacteria overwinter in perennial cankers, in buds, in plant refuse and, in tomatoes, in or on the seed. They are spread by rain splashes or runoff water, wind-blown rain, handling of plants, on contaminated tools, and on infected plant material. The bacteria enter tissues primarily through wounds, but in young plants they may also enter through natural openings. Control of bacterial cankers is through proper sanitation and eradication practices, through the use of bacteria-free seeds or budwood, and somewhat through several sprays with Bordeaux mixture, other copper formulations, or antibiotics.

• Bacterial Canker and Gummosis of Stone Fruit Trees

Bacterial canker and gummosis of stone fruit trees apparently occurs in all major fruit growing areas of the world. The disease affects primarily stone fruit trees. The same pathogen also affects pear, citrus, lilac, rose, and many other annuals and perennial ornamentals, some vegetables, and some small grains. The disease is also known as bud blast, blossom blast, dieback, spur blight, and twig blight.

Bacterial canker and gummosis is one of the most important diseases of stone fruit trees in many fruit-growing areas. Exact losses are difficult to assess because of serious damage to trees as well as reduction of yields. The disease causes cankers on branches and main trunks, kills young trees, and reduces the yield of or kills older ones. Tree losses from 10 to 75 percent have been observed in young orchards. Bacterial canker and gummosis also kills buds and flowers of trees, usually resulting in yield losses of 10–20 percent but sometimes up to 80 percent. Leaves and fruits are also attacked, resulting in weaker plants and in low quality or unsalable fruit.

◦ *Symptoms.* The most characteristic symptom of the disease, although not always the most common or the most destructive on all hosts, is the formation of cankers accompanied by gum exudation (Figures 12-27 and 12-28). Cankers usually develop at the base of an infected spur. They then spread mostly upward and to a lesser extent down and to the sides. Infected areas are slightly sunken and darker brown in color than the surrounding healthy bark. The color of the cortical tissues of the cankered area varies from bright orange to brown. Narrow brown streaks extend into the healthy tissue above and below the canker. Cankers are first noticed in late winter or early spring. As the trees break dormancy in the spring, gum is produced by the tissues surrounding most cankers, breaks through the bark, and runs down on the

FIGURE 12-27 *Pseudomonas syringae* pv. *syringae* bacteria (A) and cankers on cherry trunk (B), branch (C) and twigs (D, E). E shows the same twigs as in D but with the bark removed. (Photo A courtesy H. R. Cameron.)

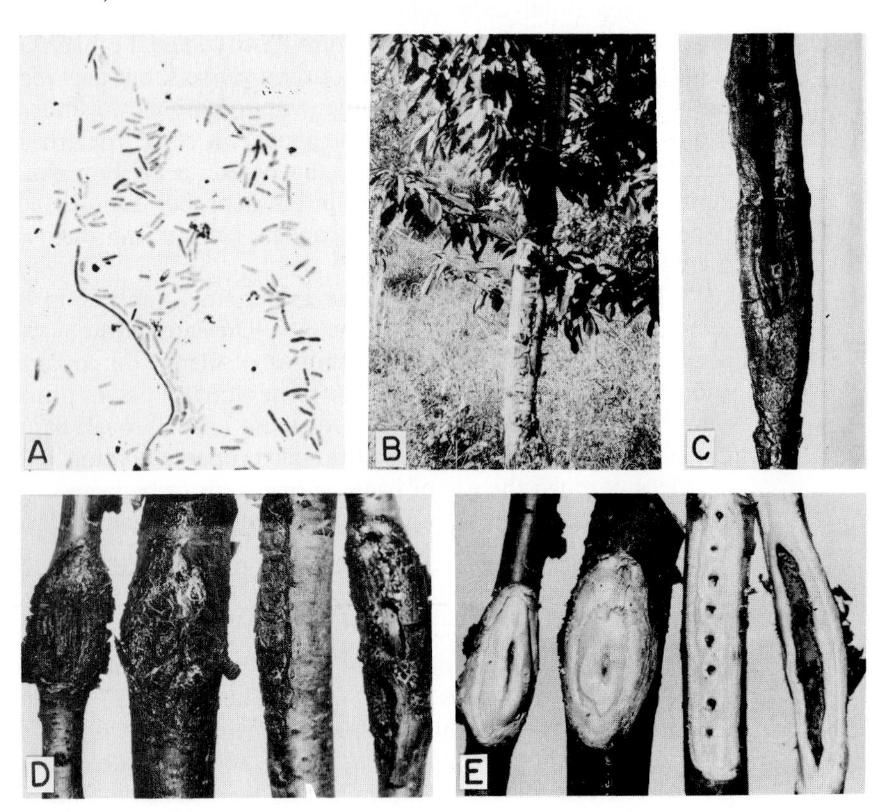

FIGURE 12-28 Young cherry tree killed by girdling of the trunk by a *Pseudomonas syringae* pv. *syringae* canker (A), side view of exposed canker (B), and brown streaks of bacteria extending into healthy tissue above rapidly advancing canker (C). (D) Leaf symptoms of *P. syringae* pv. *syringae* on cherry. From left to right: Small water-soaked spots, large angular necrotic spots, and shot-hole or tattered appearance. (Photos courtesy H. R. Cameron.)

surface of the limbs. Cankers in which gum is not produced are similar, but usually are softer, moister, sunken, and may have a sour smell. When the trunk or branch of a tree is girdled by a canker, the leaves above the girdle show an inward curling and drooping, then a light green color, and then yellow. Within a few weeks the branch or entire tree above the canker is dead (Figures 12-27 and 12-28).

Dormant bud blast is especially serious on cherry, apricot, and pear. In some areas great numbers of buds are killed. Isolated buds are often killed or fail to develop on year-old twigs. When sectioned, infected buds show brown areas at the bud scales extending across the base of the bud. The entire bud eventually dies (Figure 12-29). Both flower and leaf buds are equally affected. The damage to buds becomes most obvious in the orchard during full bloom when the light bloom of infected trees is most conspicuous (Figure 12-29E).

Infection of flowers occurs under favorable weather conditions, and it can be very severe. Infected flowers appear water soaked, turn brown, wilt, and hang on the twig. From the flower, infections may spread into the twig and cause twig blight, or they may spread into the spur and cause canker formation.

Leaf infections appear as water-soaked spots about 1 – 3 mm in diameter. As the leaves mature the spots become brown, dry, and brittle. Eventually

FIGURE 12-29 Healthy bud (A), base and scales of bud infected (B), and entire bud killed by *Pseudomonas syringae* pv. *syringae* (C). (D) Healthy cherry tree. (E) Cherry tree in same orchard with most of the lower buds killed by *P. syringae* pv. *syringae*. (Photos courtesy H. R. Cameron.)

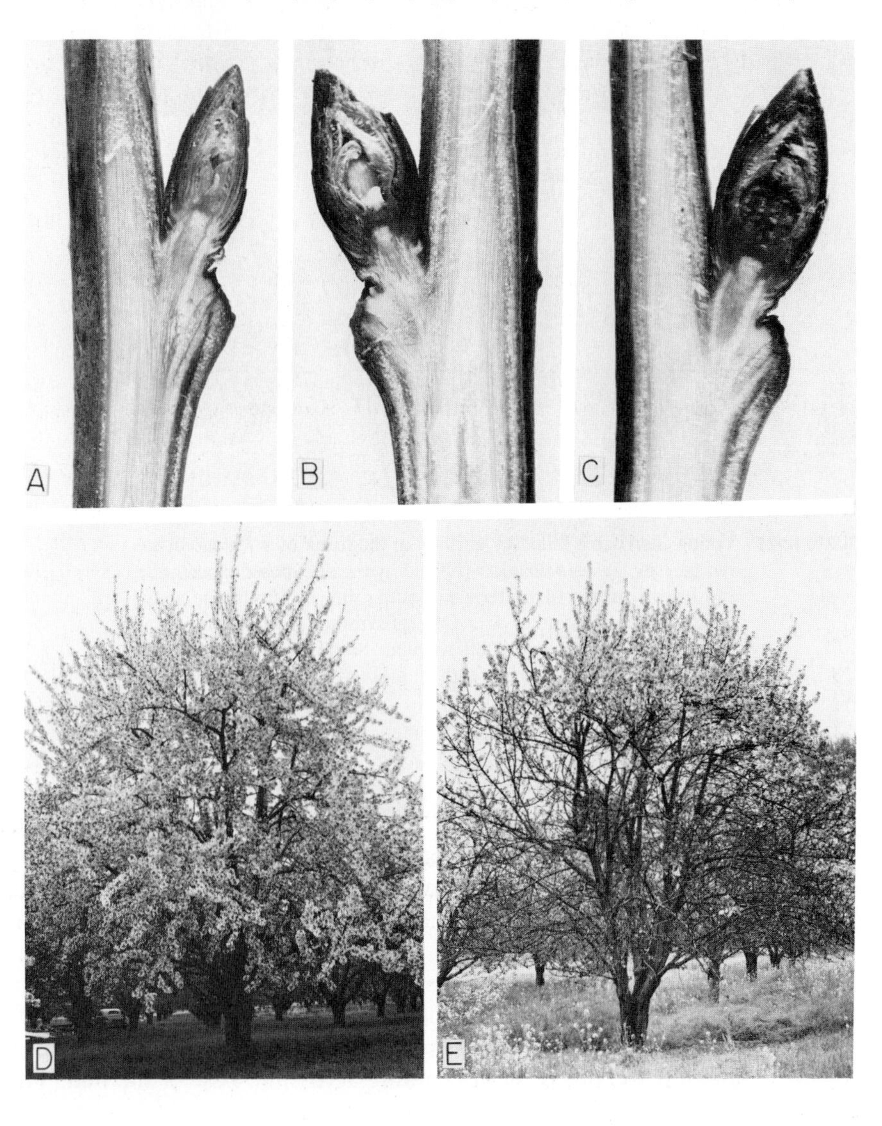

infected areas fall out and the leaves have a shot-hole or tattered appearance (Figure 12-28D).

When the fruit is infected, flat, superficial, dark brown spots develop. The spots are 2–3 mm deep, depressed, and may have underlying gum pockets on cherry, while on peach they may be 2–10 mm in both diameter and depth. The underlying tissue is dark brown to black and sometimes spongy.

◦ *The Pathogen.* The pathogens are *Pseudomonas syringae* pv. *syringae* (Figure 12-27A) and the related, more specialized *Pseudomonas syringae* pv. *morsprunorum,* which is restricted predominantly to cherry and plum. Most strains of *P. syringae* pv. *syringae* produce the phytotoxin syringomycin, which appears to play a role in the virulence of the pathogen. Many of the syringomycin-producing strains contain plasmids, but it is not known whether the two are related. The bacteria of many *Pseudomonas syringae* strains are ice-nucleation-active, that is, they serve as nuclei for ice formation, and therefore cause frost injury to plants, at relatively high freezing temperatures. The same bacteria also produce bacteriocins toxic against non-ice-nucleation-active strains, thus assuring a competitive advantage for themselves.

◦ *Development of Disease.* The bacteria overwinter in active cankers, in infected buds and leaves, epiphytically on buds and limbs of infected or healthy trees, and possibly on weeds and on nonsusceptible hosts (Figures 12-27 and 12-30).

Infection of limbs usually takes place during the fall and winter months. Bacteria enter the limbs through the bases of infected buds or spurs and also through pruning cuts, leaf scars, and injuries caused by various agents. The bacteria move intercellularly and advance into the bark and into the medullary rays of the phloem and xylem. In advanced stages of infection, the bacteria invade and break down parenchyma cells. Lysogenic cavities full of bacteria develop. Xylem vessels are sometimes invaded by bacteria, but the bacteria do not seem to move far through the vessels.

Cankers develop rather rapidly in the fall after the trees have gone into dormancy but before the onset of low winter temperatures. During the cold winter periods canker development is slow. Cankers develop most rapidly in the period between the end of the cold weather and the beginning of rapid tree growth in the spring. Cankers on the south side of trees are usually larger, due to warming by the sun during the dormant season. Cankers appear either brown, with well defined margins, or watery and gum soaked with brownish strands. The advance of the canker is checked by the advent of higher temperatures and the beginning of active growth in the spring, when the host usually forms callus tissue around the canker and the canker becomes inactive. Some cankers are permanently inactivated, but others, in which the encirclement of the canker by the callus tissue is incomplete, become active again the following year and continue to spread in succeeding years. Infections during the active growing season are seldom of any consequence and apparently are isolated very quickly by callus tissue. The ability to wall infection seems to be correlated with varietal resistance but is also affected by the age and succulence of the plant, the temperature and rainfall during a season, and the type of rootstock on which the tree is growing.

Infections of buds seem to originate at the base of the outside bud scales and then spread throughout the base of the bud, killing the tissues across the

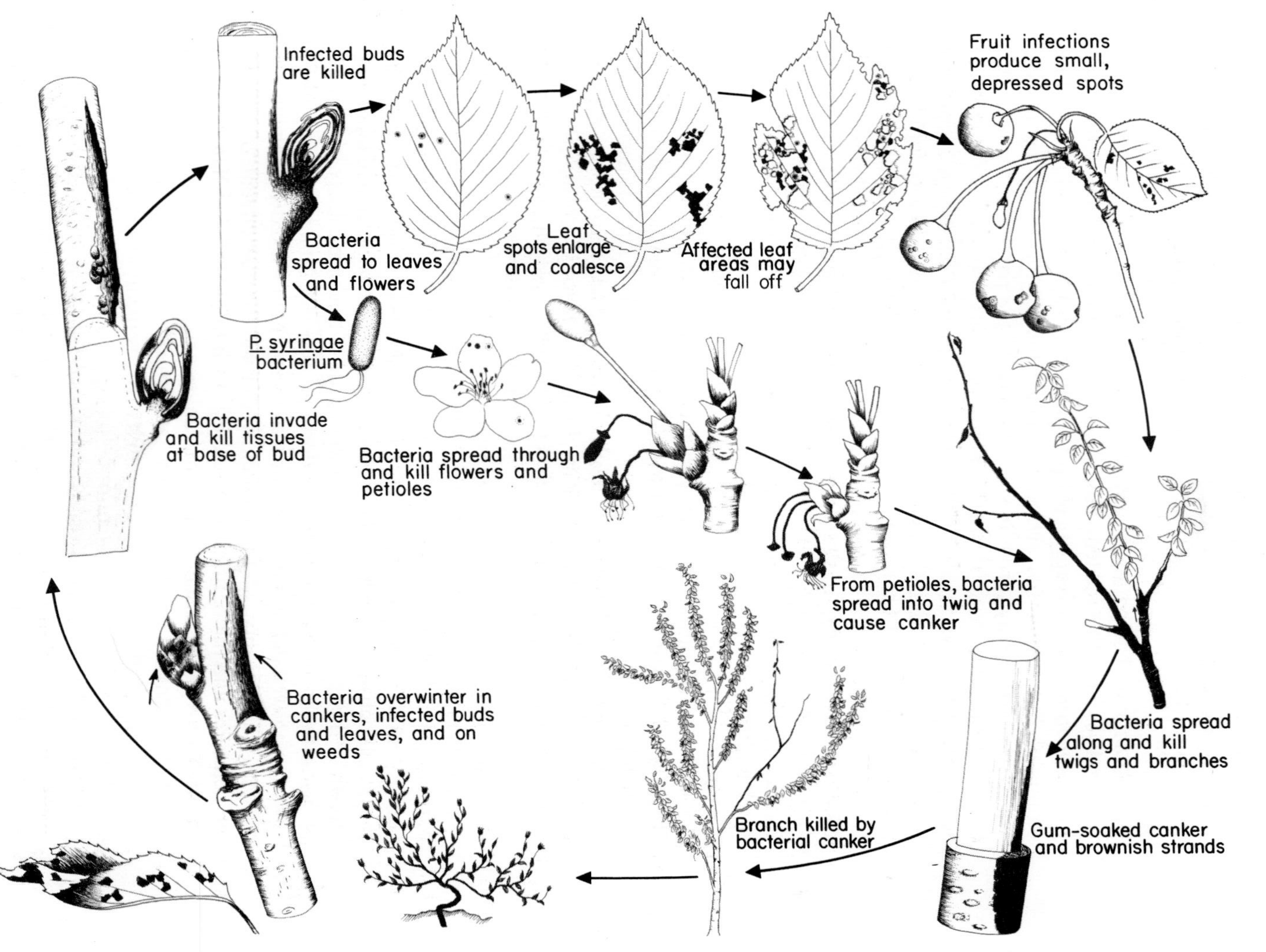

FIGURE 12-30 Disease cycle of bacterial canker and gummosis of stone fruits caused by *Pseudomonas syringae* pv. *syringae*.

base and separating the growing point from the rest of the plant (Figures 12-29 and 12-30). Most bud infections are initiated between November and February, but symptoms are first observed in mid-February and continue to develop at least through March. The bacteria sometimes spread downward and kill stem tissues around the base of the bud. Infection of buds, blossoms, and young leaves seems to be favored by frost injury to the tissues of these organs.

Flower infection is rare, but whenever it occurs it seems to take place through natural openings and through wounds made by insects or wind-blown rain. Under very humid conditions the bacteria spread through the floral parts quickly and may advance into the spur and twig, where they initiate canker formation.

Leaf infections appear on young, succulent leaves. They occur most frequently in areas with cool, wet springs and during periods of high winds and continued moisture. Infection takes place through stomata. The bacteria spread intercellularly and cause collapse and death of the cells, resulting in small angular leaf spots. During wet weather bacteria ooze out of the spots and are spread to other leaves by direct contact, by visiting insects, by rain, and so on. As leaves mature, however, they become less susceptible, and leaf infections late in the season are rare. While most of the bacteria inside or on the surface of infected leaves are dead in the fall, sufficient numbers survive to initiate new bud and stem infections.

◦ *Control.* Although no complete control of bacterial canker and gummosis of fruit trees can be obtained as yet by any single method, certain cultural practices and control measures help keep down the number and severity of infections.

Only healthy budwood should be used for propagation. Susceptible varieties should be propagated on rootstocks resistant to the disease and should be grafted as high as possible. Only healthy nursery trees should be planted in the orchard.

Chemical control of the canker phase of the disease both in the nursery and in the orchard is based on sprays with fixed copper or Bordeaux mixture (10:15:100) in the fall and in the spring (at 6:9:100 strength) before blossoming. Streptomycin applied in the spring is more effective in reducing leafspot than is Bordeaux mixture, but it does not seem to control canker initiation and development.

Cankers on trunks and large branches can be controlled by cauterization with a hand-held propane burner. The flame is aimed at the canker and especially its margins for 5–20 seconds until the underlying tissue begins to crackle and char. The treatment is carried out in early to mid-spring and, if necessary, should be repeated 2–3 weeks later.

• Citrus Canker

Citrus canker is one of the most feared of citrus diseases, affecting all types of important citrus crops. It causes necrotic lesions on fruit, leaves, and twigs. Losses are caused by reduced fruit quality and quantity and premature fruit drop. The disease is endemic in Japan and southeast Asia, from where it has spread to all other citrus-producing continents except Europe. In the United States, citrus canker was introduced into Florida in 1910, with infected nursery trees from Japan, and spread to all the Gulf states and beyond. It took 10 years, the destruction by burning of more than a quarter million bearing trees and more than three million nursery trees, many millions of dollars in expenses, and untold inconvenience and heartaches before citrus canker was practically eradicated from Florida. It took 30 more years (until 1949) to eliminate it entirely from the United States. Unfortunately, citrus canker reappeared in Florida in August 1984. Immediately, a new series of eradicating measures went into effect, resulting in the destruction of at least 17 million nursery and young orchard trees by the end of 1985, with the likelihood of additional eradications in the years to come. Citrus canker has also been eradicated from South Africa, Australia, and New Zealand. In South America, however, citrus canker was found in Brazil in 1957, it subsequently spread to several more countries, and in spite of attempts to eradicate it, the disease has apparently become permanently established there. All citrus-producing countries without canker maintain a strict prohibition on import of citrus plants and fruit from non-canker-free countries.

◦ *Symptoms.* Quite similar canker lesions are produced on young leaves, twigs, and fruit (Figure 12-31). The lesions at first appear as small, slightly raised, water-soaked, round spots, darker green than the surrounding tissue. Later, the lesions become grayish white, rupture, and appear spongy with craterlike centers. The margins of the lesions are sharply defined and are often surrounded by a yellowish halo. The size of the lesions varies with the species of citrus plant infected and may be from 1 to 6 mm in diameter on leaves and up to 1–2 cm in diameter or in length on fruits and twigs. Severe infections of

FIGURE 12-31 Fruit symptoms of the citrus canker disease as they appear on grapefruit. (Photo courtesy U.S.D.A.)

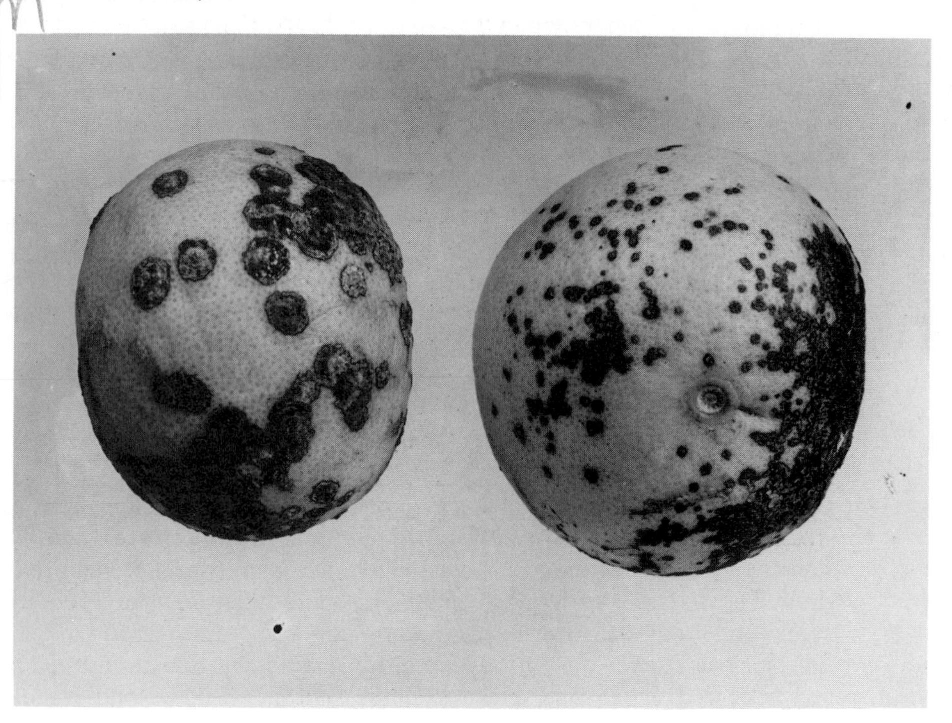

leaves, twigs, and branches debilitate the tree, while severely infected fruit appear heavily scabbed and somewhat deformed.

- *The Pathogen.* The pathogen is the bacterium *Xanthomonas campestris* pv. *citri.* At least three strains of *X. c.* pv. *citri* can be recognized by their pathogenicity to different species of *Citrus:* Strain A, native to Asia, is the most aggressive and affects primarily grapefruit, Mexican lime, sweet orange, and lemon; strain B, present in South America, affects primarily Mexican lime and lemon; and strain C, found in Brazil, affects only Mexican lime. A possibly fourth strain of *X. c.* pv. *citri* exists in Mexico, where it causes a slightly different canker on leaves and twigs of Mexican lime, and the disease is currently referred to as "citrus bacteriosis" to distinguish it from citrus canker.
- *Development of Disease.* The bacteria overwinter in developing canker lesions. During warm, rainy weather they ooze out of canker lesions and, if splashed onto young tissues, the bacteria enter them through stomata. If splashed onto older tissues, the bacteria can penetrate and cause infection only through injuries. Several cycles of infection can occur on fruit, and therefore, fruits often have lesions of many sizes. Free moisture is required for spread of the bacterium at temperatures between 20 and 35°C. Strong winds seem to greatly favor the spread of the bacteria. Citrus canker seems to be much more severe in areas in which the periods of high rainfall coincide with the period of high mean temperature (such as in Florida and other Gulf Coast states), while it is not important in areas where high temperatures are accompanied by low rainfall (such as in the southwestern United States).
- *Control.* In canker-free citrus producing areas, strict quarantine measures are practiced aimed at excluding the pathogen. When the canker bacterium is introduced into such as area (as it was in Florida in 1984), every effort must be made to eradicate the bacterium. This is attempted by burning all infected and suspected trees by flame throwers and by disinfesting of workers' tools and clothes to prevent the spread of the pathogen. In areas where the citrus canker bacterium is endemic, six or seven sprays of copper are required for even partial control of the disease on susceptible trees, such as navel oranges. Generally, control on susceptible trees has not been adequate for commercial production. Because of the presence of this bacterium, 85 percent of citrus production in Japan is of the moderately resistant Unshiu (Satsuma) orange. On such trees, satisfactory control of citrus canker is obtained by using windbreaks, pruning diseased summer and autumn shoots, forecasting of impending epidemics, and applying copper sprays.

SELECTED REFERENCES

Agrios, G. N. (1972). A severe new canker disease of peach in Greece. *Phytopathol. Mediterr.* **11,** 91–96.

Cameron, H. R. (1962). Diseases of deciduous fruit trees incited by *Pseudomonas syringae* van Hall. *Oreg., Agric. Ext. Stn., Tech. Bull.* **66,** 1–64.

Crosse, J. E. (1954). Bacterial canker, leaf spot, and shoot wilt of cherry and plum. *Annu. Rep. East Malling Res. Stn., Kent* Sect. IV, pp. 202–207.

Crosse, J. E. (1966). Epidemiological relations of the Pseudomonad pathogens of deciduous fruit trees. *Annu. Rev. Phytopathol.* **4,** 291–310.

Danos, E., Berger, R. D., and Stall, R. E. (1984). Temporal and spacial spread of citrus canker within groves. *Phytopathology* **74,** 904–908.

Endert, E., and Ritchie, D. F. (1984). Overwintering and survival of *Pseudomonas syringae* pv. *syringae* and symptom development in peach trees. *Plant Dis.* **68,** 468–470.

Gross, D. C., Cody, Y. S., Proebsting, E. L., Radamaker, G. K., and Spotts, R. A. (1984). Ecotypes and pathogenicity of ice-nucleation-active *Pseudomonas syringae* isolates from deciduous fruit tree orchards. *Phytopathology* **74,** 241–248.

Hawkins, J. E. (1976). A cauterization method for the control of cankers caused by *Pseudomonas syringae* in stone fruit trees. *Plant Dis. Rep.* **60,** 60–61.

Jones, A. L. (1971). Bacterial canker of sweet cherry in Michigan. *Plant Dis. Rep.* **55,** 961–965.

Kuhara, S. (1978). Present epidemic status and control of the citrus canker disease. *Xanthomonas citri,* in Japan. *Rev. Plant Prot. Res.* **11,** 132–142.

Latorre, B. A. *et al.* (1985). Isolation of *Pseudomonas syringae* pv. *syringae* from cankers and effect of free moisture on its epiphytic populations on sweet cherry trees. *Plant Dis.* **69,** 409–412.

Olson, B. D., and Jones, A. L. (1983). Reductions of *Pseudomonas syringae* pv. *morsprunorum* on Montmorency sour cherry with copper and dynamics of the copper residues. *Phytopathology* **73,** 1520–1525.

Peltier, G. L. (1920). Influence of temperature and humidity on the growth of *Pseudomonas citri* and its host plants and on infection and development of the disease. *J. Agric. Res. (Washington, D. C.)* **20,** 447–506.

Rodriguez, G. S., Garzal, J. G., Sarh, I. C., Stapleton, J. J., and Civerolo, E. L. (1985). Citrus bacteriosis in Mexico. *Plant Dis.* **69,** 808–810.

Stall, R. E., and Seymour, C. P. (1983). Canker, a threat to citrus in the Gulf-Coast states. *Plant Dis.* **67,** 581–585.

Wilson, E. E. (1933). Bacterial canker of stone fruit trees in California. *Hilgardia* **8,** 83–123.

Wormald, H. (1931). Bacterial diseases of stone fruit trees in Britain, III. The symptoms of bacterial canker in plum trees. *J. Pomol.* **9,** 239–256.

Bacterial Scabs

The group of diseases known as bacterial scabs includes mainly diseases that affect below-ground parts of plants and whose symptoms consist of more or less localized scabby lesions affecting primarily the outer tissues of these parts. The scab bacteria and the diseases they cause are:

Streptomyces, causing the common scab of potato and of other below-ground crops *(S. scabies)*, and the soil rot or pox of sweet potato *(S. ipomoeae)*.

The scab bacteria survive in infected plant debris and in the soil and penetrate tissues through natural openings or wounds. In the tissues, the bacteria grow mostly in the intercellular spaces of parenchyma cells, but these cells are sooner or later invaded by the bacteria and break down. In typical scabs, healthy cells below and around the lesion divide and form layers of corky cells. These cells push the infected tissues outward and give them the scabby appearance. Scab lesions often serve as points of entry for secondary parasitic or saprophytic organisms that may cause the tissues to rot.

Common Scab of Potato Common scab of potato is caused by *Streptomyces scabies* and occurs throughout the world. It is most prevalent and important in neutral or slightly alkaline and light sandy soils, especially during relatively dry years. The same pathogen also affects garden beets, sugar beets, radish, and other crops. The disease, by its usually superficial blemishes on tubers and roots reduces the value rather than the yield of the crop, although severe root infection may reduce yields, and deep scabs increase the waste in peeling.

The symptoms of common scab of potato are observed mostly on tubers. At first they consist of small, brownish, and slightly raised spots, but later they may enlarge, coalesce, and become very corky. Frequently, the lesions extend below the tuber surface and, when the corky tissue is removed, 3–4 mm deep pits are present in the tuber. Sometimes the lesions appear as small russeted areas and are so numerous that they almost cover the tuber surface, or they may appear as slight protuberances with depressed centers covered with a small amount of corky tissue (Figure 12-32).

The pathogen, *S. scabies,* is a hardy saprophyte that can survive indefinitely either in its vegetative mycelioid form or as spores in most soils except the most acidic ones. The vegetative form consists of slender (about 1 μm thick), branched mycelium with few or no cross walls. The spores are cylindrical or ellipsoid, about 0.6 by 1.5 μm, and are produced on specialized spiral hyphae that develop cross walls from the tip toward their base, and as the cross walls constrict, spores are pinched off at the tip and eventually break away from the hypha. The spores germinate by means of one or two germ tubes, which develop into the mycelioid form (Figure 12-33).

The pathogen is spread through soil water, wind-blown soil, and on infected potato seed tubers. It penetrates tissues through lenticels, wounds, stomata, and in young tubers, directly. Young tubers are more susceptible to infection than older ones. After penetration the pathogen apparently grows between or through a few layers of cells, the cells die, and the pathogen then lives off them as a saprophyte. In the meantime, however, the pathogen apparently secretes a substance that stimulates the living cells surrounding the lesion to divide rapidly and to produce several layers of cork cells that isolate the pathogen and several plant cells. As the cells that are cut off by the cork layer die, the pathogen subsists on them. Usually, several such groups of cork cell layers are produced, and as they are pushed outward and sloughed off, the pathogen grows and multiplies in the additional dead cells, and thereby large scab lesions develop. The depth of the lesion seems to depend on the variety, soil conditions, and on the invasion of scab lesions by other organisms, including insects. The latter apparently break down the cork layers and allow the pathogen to invade the tuber in great depth.

The severity of common scab of potato increases as the pH of the soil increases from 5.2 to 8.0 and decreases beyond these limits. The disease develops most rapidly at soil temperatures of about 20–22°C, but it can occur between 11 and 30°C. Potato scab incidence is greatly reduced by high soil moisture during the period of tuber initiation and for several weeks afterwards. Potato scab is also lower in fields after certain crop rotations and the

FIGURE 12-32 Early (A) and advanced (B) symptoms of common scab of potato caused by *Streptomyces scabies.* (Photos courtesy U.S.D.A.)

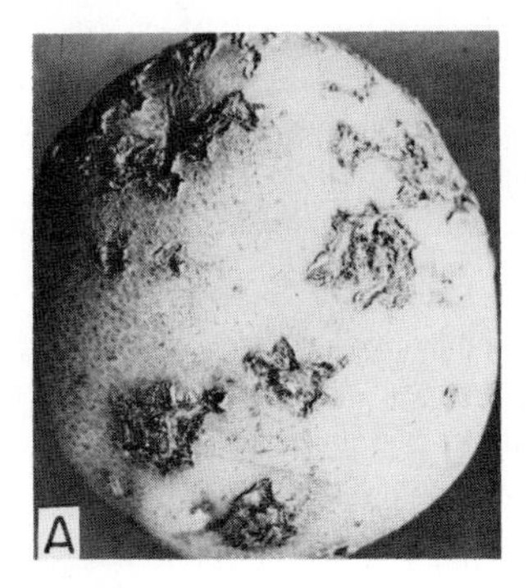

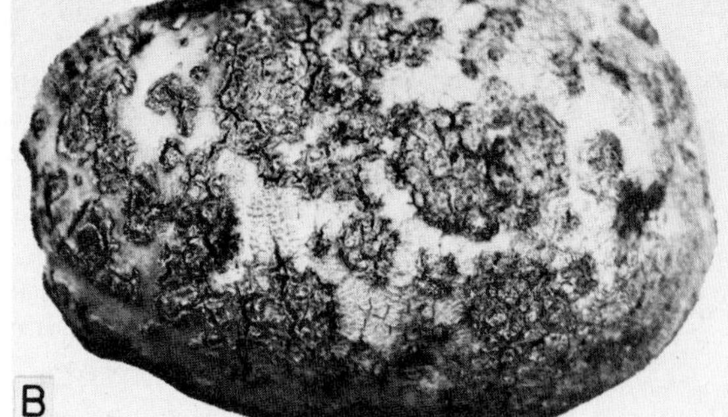

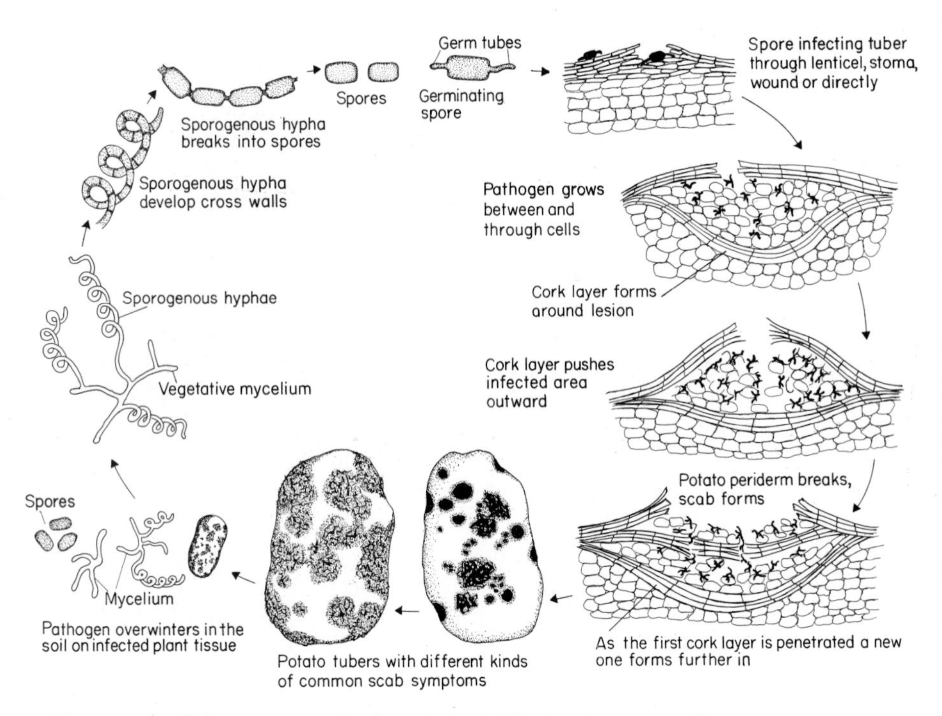

FIGURE 12-33 Disease cycle of the common scab of potato caused by *Streptomyces scabies.*

plowing under of certain green manure crops, probably as a result of inhibition of the pathogen by antagonistic microorganisms.

Control of the common scab of potato is through the use of certified scab-free seed potatoes or through seed treatment with pentachloronitrobenzene (PCNB) or with maneb–zinc dust. If the field is already infested with the pathogen, a fair degree of disease control may be obtained by using certain crop rotations, bringing and holding the soil pH to about 5.3 with sulfur, irrigating for about 6 weeks during the early stages of tuber development, and using resistant or tolerant potato varieties.

SELECTED REFERENCES

Jones, A. P. (1931). The histogeny of potato scab. *Ann. Appl. Biol.* **18,** 313–333.

KenKnight, G. (1941). Studies of soil *Actinomyces* in relation to potato scab and its control. *Mich., Agric. Exp. Stn., Tech. Bull.* **178,** 1–48.

Labruyere, R. E. (1971). Common scab and its control in seed-potato crops. *Agric. Res. Rep. (Wageningen)* **767,** 1–71.

Lapwood, D. H. (1973). Irrigation as a practical means to control potato common scab *(Streptomyces scabies):* Final experiment and conclusions. *Plant Pathol.* **22,** 35–41.

Lapwood, D. H., and Adams, M. J. (1975). Mechanism of control of common scab by irrigation. *In* "Biology and Control of Soil-Borne Plant Pathogens" (G. W. Bruehl, ed.), pp. 123–129. Am. Phytopathol. Soc., St. Paul, Minnesota.

Levick, D. R., Evans, T. A., Stephens, C. T., and Lacy, M. L. (1985). Etiology of radish scab and its control through irrigation. *Phytopathology* **75,** 568–572.

Person, L. H. (1946). The soil rot of sweet potatoes and its control with sulfur. *Phytopathology* **36,** 869–875.

Root Nodules of Legumes

Root nodules are well-organized structures produced on the roots of most legume plants after inoculation with certain species of bacteria of the genus *Rhizobium*. Root nodules, although they are the result of infection of legumes by bacteria, are considered a condition of symbiosis rather than of disease. The infecting bacteria fix (trap) atmospheric nitrogen and make it available to the plant in a utilizable organic form, and the plant profits from this nitrogen more than it loses as sugars and other nutrients to the bacteria. Unfortunately, not all root nodule bacteria are beneficial to the legume host. Some nodule bacteria are apparently strictly parasites since they form nodules on the roots but fail to fix nitrogen. Therefore, the number of root nodules does not always indicate their value to the plant unless the strain of bacteria is known to be effective in fixing nitrogen. As a result legume seeds are routinely inoculated commercially with appropriate strains of root nodule bacteria to improve plant growth and yields.

Nodules are produced on taproots as well as lateral roots of legumes and may vary in size from 1 mm to 2–3 cm (Figure 12-34A). Nodules may be round or cylindrical and as large or larger than the root diameter on which they form. Their number and size vary with the plant, bacterial strain, age of infection, and so on. On herbaceous plants nodules are fragile and short-lived while on woody plants they may persist for several years. Each nodule consists of an epidermal layer, a cortical layer, and the bacteria-containing central tissue, each of them consisting of several layers of cells (Figure 12-34B). Vascular bundles are present in the cortical layer just outside the central tissue. In elongated nodules, the tip of the nodule farthest away from the root consists of a zone of meristematic cells through which the nodule grows. In rounded nodules, the meristematic region is laid around the nodule except at the neck.

◦ *The Organism.* The root nodule bacteria, *Rhizobium* sp., vary in size and shape with age, the typical bacteria being rod-shaped (1.2–3.0 by 0.5–0.9 μm) or irregular, club-shaped forms. They have no flagella, and they are gram-negative. The bacteria survive in roots of susceptible legumes and, for varying periods of time, in the soil. Continued growth of the same legume in the soil tends to build up the population of nodule bacteria affecting that legume. Not all nodule bacteria affect all legumes. For example, the bacteria that grow on alfalfa and sweet clover do not grow on clovers, beans, peas, soybeans, and vice versa. Strains of nodule bacteria often show definite varietal preferences; for example, some soybean bacteria work better on one or two soybean varieties than on others.

◦ *Development of Nodules.* The bacteria penetrate root hairs or young epidermal cells directly. Within the cell the bacteria become embedded in a double-walled, tubular, mucoid sheath called an infection thread. The infection thread, which contains the bacteria, penetrates into the cortical parenchyma cells and branches along the way, with terminal and lateral vesicles forming on the strands. These vesicles soon break and release the bacteria mostly within the cells (Figure 12-34C). The released bacteria then enlarge and become enclosed in a membrane envelope (Figure 12-34D). These membrane-enclosed bacteria are called bacteroids. In the meantime, the cortical

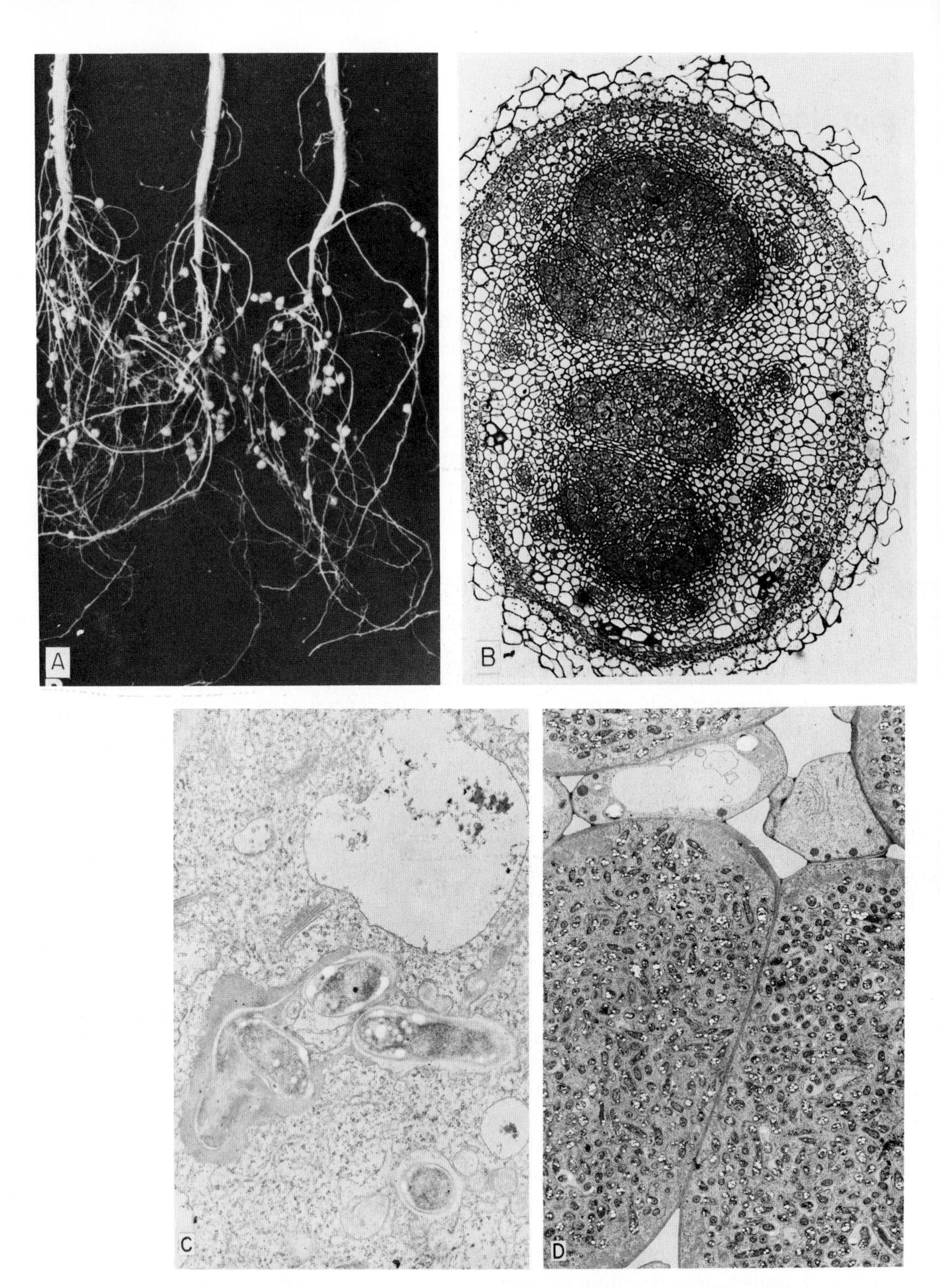

FIGURE 12-34 (A) "Healthy" soybean roots bearing numerous bacterial nodules. (B) Cross section of a developing soybean nodule, 12 days after inoculation. There are at least three central areas containing bacteroids apparently as a result of several closely adjacent infections. (C, D) Electron micrographs of sections of a soybean root nodule. (C) Area of an infection thread where bacteria are apparently being released. (D) Infected and uninfected cells in a young nitrogen-fixing nodule. Membrane envelopes are visible around some bacteria. The electron-lucent granules in the bacteria consist of poly-beta-hydroxybutyrate. (Photo A courtesy U.S.D.A. Photos B–D courtesy B. K. Bassett and R. N. Goodman.)

parenchyma cells along the path of bacterial invasion begin to divide, and the invaded cells increase in size as the bacteroids appear. The increased meristematic activity and cell enlargement of cortical cells result in the formation of the nodule, which grows outward from the root cortex. At the same time differentiation of vascular tissues, both xylem and phloem, takes place in the nodule. The vascular tissues of the nodule are not connected directly with those of the root.

While the outermost tip or layer of the nodule remains meristematic and continues to grow and thus to increase the size of the nodule up to a certain point, many of the cortical cells behind the meristematic zone and in all the central tissue of the nodule are uniformly enlarged and infected with several bacteroids. In the most recently infected cells, each bacteroid is enclosed in a membrane envelope, while in earlier infected cells several bacteroids may be enclosed in a membrane envelope. In cells that have been infected even longer than the latter, the bacteroids lack a membrane envelope and the host cellular membrane system also has deteriorated. It appears that the membraneless bacteroids, which occur in the advanced stages of infection and which increase in numbers while the nodule is still growing, lack the ability to fix nitrogen. Therefore, the efficiency of root nodules in nitrogen fixation is proportional to the number of enveloped bacteroids they contain and not necessarily to the size of the nodules. As the nodules age, first cortical cells in the earliest-infected areas and then in the entire central area of the nodule disintegrate and collapse. The bacteroids, which have by now lost their membrane envelope, either disintegrate or become intercellular bacteria and are finally released into the soil as the nodule cortex and epidermis disintegrate.

SELECTED REFERENCES

Bieberdorf, F. W. (1938). The cytology and histology of the root nodules of some Leguminosae. *J. Agron.* **30,** 375–389.

Erdman, L. W. (1967). Legume inoculation: What it is—what it does. *Farmers' Bull.* **2003,** 1–10.

Long, S. R. (1984). Genetics of *Rhizobium* nodulation. *In* "Plant-Microbe Interactions: Molecular and Genetic Perspectives" (T. Kosuge and E. E. Nester, eds.). Vol. 1, pp. 265–306. Macmillan, New York.

Tu, J. C. (1975). Rhizobial root nodules of soybeans as revealed by scanning and transmission electron microscopy. *Phytopathology* **65,** 447–454.

Vance, C. P., and Johnson, L. E. B. (1981). Nodulation: A plant disease perspective. *Plant Dis.* **65,** 118–124.

Plant Diseases Caused by Fastidious Vascular Bacteria

The fastidious vascular bacteria (previously known as rickettsialike organisms, or RLO) that cause plant diseases are only now beginning to be named and classified. Earlier, many believed them to be more or less obligate parasites closely related to rickettsiae and therefore called them rickettsialike organisms. Now, however, it has been shown that these organisms are not related to rickettsiae and they are considered to be parasitic bacteria that simply cannot be grown on simple culture media in the absence of host cells.

The fastidious phloem-limited bacteria were first observed in plants in 1972, in the phloem of clover and periwinkle plants affected with the clover club leaf disease. The following year, fastidous xylem-limited bacteria were observed in the xylem vessels of grape plants affected with Pierce's disease and of alfalfa affected with alfalfa dwarf, of peach affected with the phony peach disease, and of sugarcane affected with ratoon stunting (Figure 12-35). Subsequently, similar organisms were observed in the phloem of citrus trees affected with the greening disease, while more frequently such organisms were observed in the xylem of plants affected with one of more than twenty other diseases, for example, plum leaf scald, almond leaf scorch, and elm leaf scorch.

The fastidious vascular bacteria are generally rod-shaped cells 0.2–0.5 μm in diameter by 1–4 μm in length. They are bounded by a cell membrane and a cell wall, although in the phloem-inhabiting bacteria the cell wall appears more as a second membrane than as a cell wall. They have no flagella. The outer layer of the cell wall is usually undulating or rippled (Figure 12-36). Nearly all fastidious vascular bacteria are Gram-negative. Several such xylem-limited bacteria have been placed in the recently created genus *Xylella*. Only the xylem-inhabiting bacteria causing sugarcane ratoon stunting and Bermudagrass stunting are Gram-positive, and they are now classified as members of the genus *Clavibacter*, which contains most of the species formerly included in *Corynebacterium*. None of the phloem-inhabiting fastidious bacteria (those causing clover club leaf and citrus greening) has been grown in culture so far, but all the xylem-inhabiting fastidious bacteria can be grown in culture on more or less complex nutrient media on which they grow slowly and produce tiny (1–2 mm) colonies. All fastidious vascular bacteria are unable to grow on conventional bacteriological media.

All Gram-negative xylem-inhabiting fastidious bacteria are transmitted by xylem-feeding insects, such as sharpshooter leafhoppers (Cicadellinae) and spittlebugs (Cercopidae). The vectors can acquire and transmit the bacteria in less than 2 hours. Carrier adult insects can transmit the bacteria for life but do not pass them on to progeny. So far, no insect vector is known for the Gram-positive xylem-inhabiting fastidious bacteria, but at least one of them, the cause of sugarcane ratoon stunting, can be transmitted mechanically by cutting implements during harvest. Little definitive information exists about vectors of phloem-inhabiting fastidious bacteria, but it appears that leafhoppers and psyllid insects are the vectors of clover club leaf and citrus greening bacteria, respectively. The clover club leaf bacterium is known to multiply in its leafhopper vector and to be passed from the mother to the progeny insects through the eggs (transovarial transmission).

The symptoms of diseases caused by fastidious xylem-inhabiting bacteria often consist of marginal necrosis of leaves, stunting, and general decline and reduced yields. Such symptoms are probably caused by plugging of the xylem by bacterial cells and by a matrix material partly of bacterial and partly of plant origin. In some diseases, however, such as phony peach, no marginal leaf necrosis occurs, and in others, such as sugarcane ratoon stunting, the only diagnostic symptom is an internal discoloration of the stalk. On the other hand, the symptoms of diseases caused by fastidious phloem-inhabiting bacte-

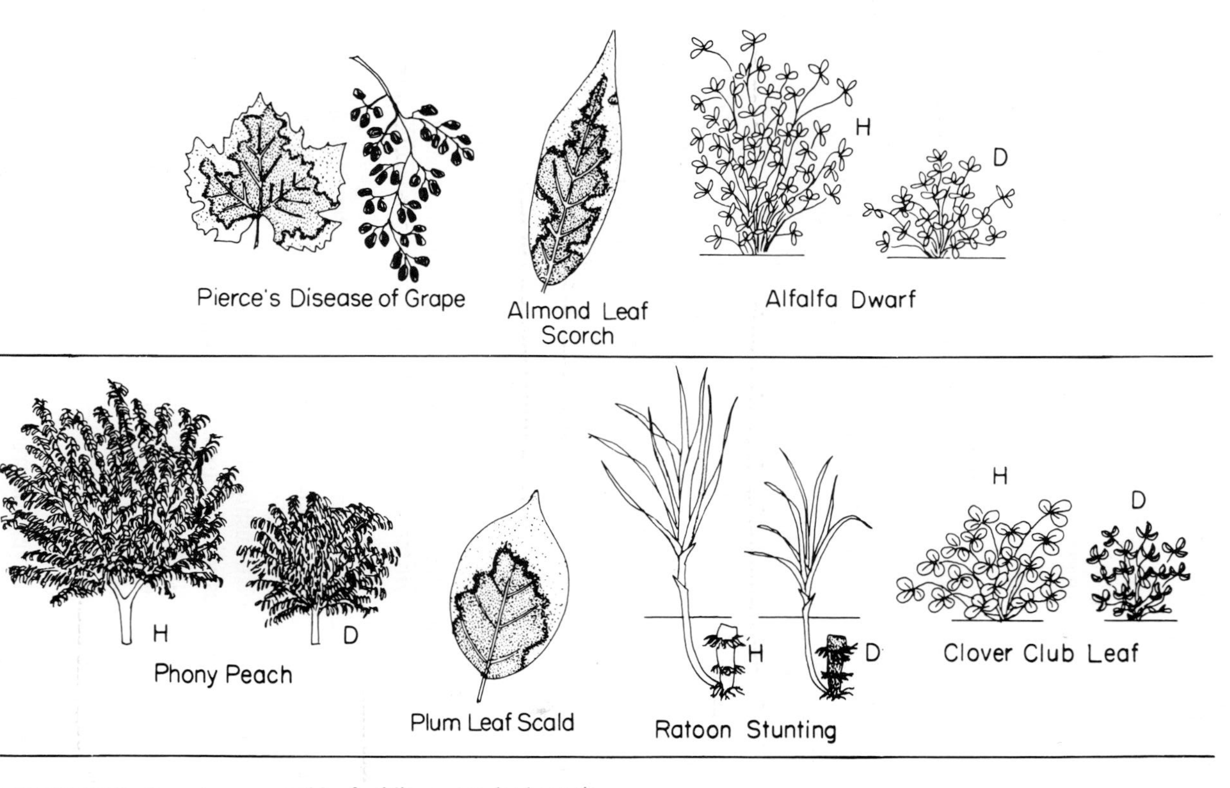

FIGURE 12-35 Symptoms caused by fastidious vascular bacteria.

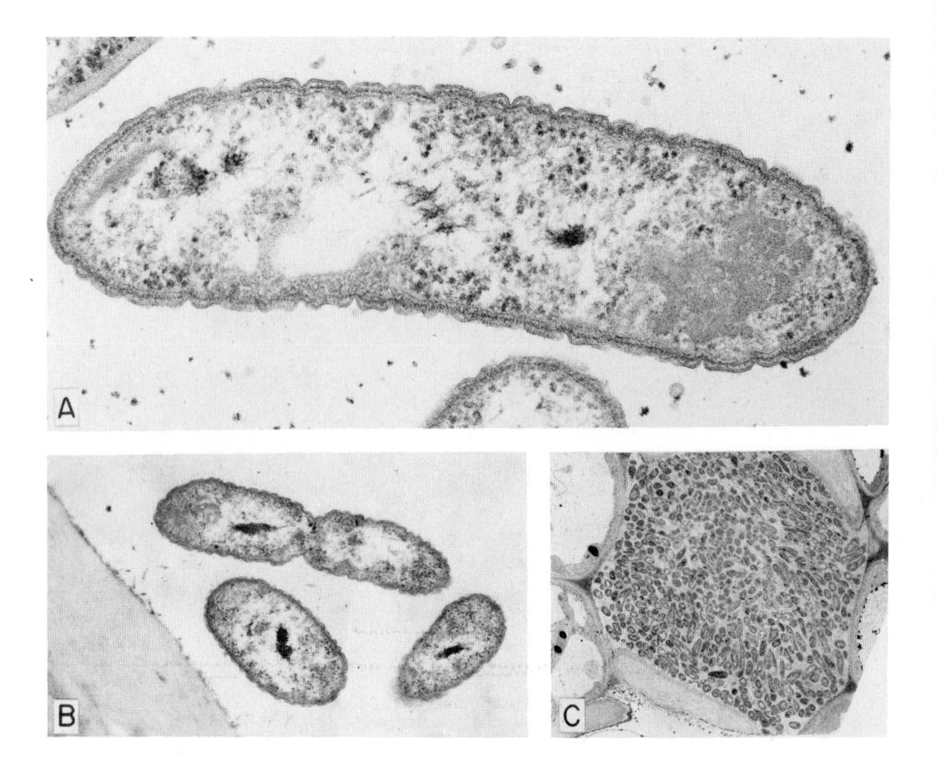

FIGURE 12-36 Morphology, multiplication, and distribution of plant pathogenic fastidious vascular bacteria, the causal agent of Pierce's disease of grapes. (A) Typical cell showing rippled cell wall. (B) Fastidious bacteria in xylem vessel, one of them undergoing binary fission. (C) Fastidious bacteria in a tracheary element of a leaf vein. (Photos courtesy H. H. Mollenhauer, from Mollenhauer and Hopkins, *J. Bacteriol.* **119,** 612–618.)

ria often consist of leaf stunting and clubbing, and in some cases may appear as shoot proliferation and witches'-brooms and as greening of floral parts. In these diseases symptoms are often mild and sometimes are followed by spontaneous recovery.

The fastidious xylem-inhabiting bacteria are serologically related to each other but show no serological relationship to any of the typical plant pathogenic bacteria *(Agrobacterium, Erwinia,* and so on). There is no information on serological relationships for the phloem-inhabiting bacteria.

Although fastidious vascular bacteria are sensitive to several antibiotics such as tetracyclines and penicillin, chemotherapy of infected plants in the field has proved difficult. For example, symptoms of Pierce's disease of grape were suppressed by weekly or biweekly drenches of the plants with tetracyclines, which would be impractical, and none of the other diseases has been controlled by antibiotics so far. Injection of Antibiotics into trunks of affected trees is more effective, but must be repeated annually.

Fastidious vascular bacteria are sensitive to high temperatures. Heat treatment of entire plants or of propagative organs by immersing them in water kept at 45–50°C for 2–3 hours; or keeping them in aerated steam at 50–53°C for 4 hours; or in hot air at 50–58°C for several (4–8) hours, has cured grapevines from Pierce's disease and sugarcane from ratoon stunting disease.

Among the most important plant diseases caused by fastidious xylem-limited, Gram-negative bacteria are Pierce's disease of grape, alfalfa dwarf, phony peach disease, almond leaf scorch, and plum leaf scald. They are all caused by the bacterium *Xylella fastidiosa,* which also causes leaf scorch diseases on elm, sycamore, oak and mulberry. The also very important ratoon stunting disease of sugarcane is caused by the xylem-limited, Gram-positive bacterium *Clavibacter xyli* subsp. *xyli,* while another subspecies causes the Bermudagrass stunting disease. Phloem-limited bacteria are so far known to cause the very important citrus greening disease and some diseases of clover and periwinkle.

• Pierce's Disease of Grape

Pierce's disease of grape is present in the southern United States from California to Florida, where it kills grapevines and makes large areas unfit for grape culture. In some areas the disease is endemic and no grapes can be grown, while in others it breaks out as infrequent epidemics. Many other annual and perennial kinds of plants of some 28 families, including grasses, herbs, shrubs, and trees are affected by the disease.

Infected grapevines may die within a few months or may live for several years after the onset of infection. Some varieties survive infections longer than others, older plants survive infections longer than young vigorous ones, and vines in colder regions survive infections longer than vines in hot areas.

In grapes the symptoms appear first as a sudden drying, that is, scalding, of part or most of the margin area of the leaf while the leaf is still green (Figure 12-37). Scalded areas advance toward the central area of the leaf and later turn

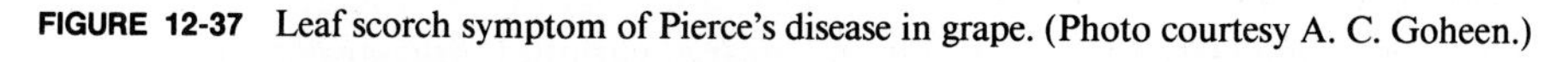

FIGURE 12-37 Leaf scorch symptom of Pierce's disease in grape. (Photo courtesy A. C. Goheen.)

brown. In late season affected leaves usually drop, leaving the petioles attached to the canes. Grape clusters on vines with leaf symptoms stop growth, wilt, and dry up. Infected canes mature irregularly, forming patches of brown bark among areas of cortex that remain immature and green. During the following season(s), infected plants show delayed spring growth, dwarfed vines, and greenish vein banding in the first few leaves, but later in the season leaves and fruit show the same symptoms as during the first season. Decline of the top is followed by dieback of the root system. Internally, the current-season wood of all parts of infected vines shows yellow to brown streaks that are readily seen in longitudinal and cross sections. In the same wood, gum forms in vessels and other types of cells and tyloses develop in vessels of all sizes. Both gum and tyloses cause plugging of vessels sufficient to account for many of the external symptoms of diseased plants.

The pathogen is *Xylella fastidiosa*, a fastidious xylem-inhabiting bacterium measuring 0.3–0.4 by 1.0–3.2 μm and having a typically rippled cell wall (Figure 12-36). This pathogen is found in xylem vessels of grapevines in large numbers and can be cultured on special nutrient media. The pathogen apparently produces one or more phytotoxins that seem to play a role in the production of marginal leaf necrosis in infected host plants.

The pathogen is transmitted by grafting and by many species of leafhoppers. The leafhopper vectors acquire the pathogen after feeding on infected hosts for about 2 hours and may continue to transmit it to healthy hosts for the rest of their lives. The presence of the pathogen in a plant is determined by the symptoms of the disease, if present, or by transmission to indicator plants either by grafting or by insect vectors. The grape varieties most commonly used as indicators are Carignane, Emperor, and Palomino. More recently, serological tests, such as ELISA and immuno-fluorescent staining have been used successfully for detection and identification of the pathogen.

There is no practical control of Pierce's disease of grape and alfalfa dwarf in the field. Control of the insect vector and roguing of infected plants have not been effective. All commercial grape varieties are susceptible to the disease, but some grape and alfalfa varieties carry some resistance. Drench treatments with 4 liters of 50 and 100 ppm tetracycline solutions applied around the base of infected plants twice weekly, weekly, or biweekly inhibited symptom development on most plants for one or two years, but such treatments are not commercially feasible. Individual plants can be freed of the pathogen by immersing the entire plant in water at 45°C for 3 hours, 50°C for 20 minutes, or 55°C for 10 minutes. Such treatments are also of little help to the grower. The best defense is to plant in areas remote from natural reservoirs of the pathogen.

SELECTED REFERENCES

Anonymous (1983). Discussion of fastidious prokaryotes as plant pathogens. (Several papers.) *Phytopathology* **73,** 341–360.

Brlansky, R. H., Lee, R. F., Timmer, L. W., Purcifull, D. E., and Raju, B. C. (1982). Immunofluorescent detection of xylem-limited bacteria *in situ. Phytopathology* **72,** 1444–1448.

Davis, J. M., Purcell, A. H., and Thomson, S. V. (1978). Pierce's disease of grapevines: Isolation of the bacterium. *Science* **199,** 75–77.

Davis, M. J., Whitcomb, R. F., and Gillaspie, A. G., Jr. (1983). Fastidious bacteria of plant vascular tissue and invertebrates (including the so-called rickettsia-like bacteria). *In* "Phytopathogenic Bacteria" (M. P. Starr, ed.), pp. 2172–2188. Springer-Verlag, Berlin and New York.

Davis, M. J., Gillaspie, A. G., Vidaver, A. K., and Harris, R. W. (1984). *Clavibacter:* A new genus containing some phytopathogenic coryneform bacteria, including *Clavibacter xyli* subsp. xyli sp. nov. and *Clavibacter xyli* subsp. *cynodontis* subsp. nov., pathogens that cause ratoon stunting disease of sugar cane and Bermudagrass stunting disease. *Int. J. Syst. Bacteriol.* **34,** 107–117.

Esau, K. (1948). Anatomic effects of the viruses of Pierce's disease and phony peach. *Hilgardia* **18,** 423–482.

Garnier, M., and Bove, J. M. (1983). Transmission of an organism associated with citrus greening disease from sweet orange to periwinkle by dodder. *Phytopathology* **73,** 1358–1363.

Goheen, A. C., Nyland, G., and Lowe, S. K. (1973). Association of a rickettsialike organism with Pierce's disease of grapevines and alfalfa dwarf and heat therapy of the diseases in grapevines. *Phytopathology* **63,** 341–345.

Hopkins, D. L. (1981). Seasonal concentration of the Pierce's disease bacterium in grapevine stems, petioles, and leaf veins. *Phytopathology* **71,** 415–418.

Hopkins, D. L. (1984). Variability of virulence in grapevine among isolates of the Pierce's disease bacterium. *Phytopathology* **74,** 1395–1398.

Hopkins, D. L. (1985). Physiological and pathological characteristics of virulent and avirulent strains of the bacterium that causes Pierce's disease of grapevines. *Phytopathology* **75,** 713–717.

Hopkins, D. L., and Mollenhauer, H. H. (1973). Rickettsia-like bacterium associated with Pierce's disease of grapes. *Science* **179,** 298–300.

Hopkins, D. L., Mortensen, J. A., and Adlerz, W. C. (1973). Protection of grapevines from Pierce's disease with tetracycline antibiotics. *Phytopathology* **63,** 443.

Hopkins, D. L., Mollenhauer, H. H., and French, W. J. (1973). Occurrence of a rickettsia-like bacterium in the xylem of peach trees with phony disease. *Phytopathology* **63,** 1422–1423.

Lafleche, D., and Bove, J. M. (1970). Structures de type mycoplasme dans les feuilles d'orangers atteints de la maladie du "greening." *C. R. Hebd. Seances Acad. Sci., Ser. D* **270,** 1915–1917.

Lee, R. F., Raju, B. C., Nyland, G., and Goheen, A. C. (1982). Phytotoxin(s) produced in culture by the Pierce's disease bacterium. *Phytopathology* **72,** 886–888.

McCoy, R. E. (1982). Chronic and insidious disease: The fastidious vascular pathogens. *In* "Phytopathogenic Prokaryotes" (M. S. Mount and G. H. Lacy, eds.), Vol. 1, pp. 475–489. Academic Press, New York.

Mircetich, S. M., Lowe, S. K., Moller, W. J., and Nyland, G. (1976). Etiology of almond leaf scorch disease and transmission of the causal agent. *Phytopathology* **66,** 17–24.

Nyland, G., Goheen, A. C., Lowe, S. K., and Kirkpatrick, H. (1973). The ultrastructure of a rickettsialike organism from a peach tree affected with phony disease. *Phytopathology* **63,** 1275–1278.

Pierce, N. B. (1892). The California vine disease. *USDA Div. Veg. Pathol. Bull.* **2,** 1–222.

Purcell, A. H. (1982). Insect vector relationships with procaryotic plant pathogens. *Annu. Rev. Phytopathol.* **20,** 397–417.

Raju, B. C., Wells, J. M., Nyland, G., Brlansky, R. H., and Lowe, S. K. (1982). Plum leaf scald: Isolation, culture and pathogenicity of the causal agent. *Phytopathology* **72,** 1460–1466.

Raju, B. C., Goheen, A. C., and Frazier, N. W. (1983). Occurrence of Pierce's disease bacteria in plants and vectors in California. *Phytopathology* **73,** 1309–1313.

Raju, B. C., and Wells, J. M. (1986). Diseases caused by fastidious xylem-limited bacteria and strategies for management. *Plant Dis.* **70,** 182–186.

Teakle, D. S., Smith, P. M., and Steindl, D. R. L. (1973). Association of a small coryneform bacterium with the ratoon stunting disease of sugarcane. *Aust. J. Agric. Res.* **24,** 869–874.

Wells, J. M., Raju, B. C., Thompson, J. M., and Lowe, S. K. (1981). Etiology of phony peach and plum leaf scald diseases. *Phytopathology* **71,** 1156–1161.

Wells, J. M., Hung, H. -Y., Weisberg, W. G., Mandelco-Paul, L., and Brenner, D. J. (1987). *Xylella fastidiosa,* gen. nov., sp. nov: Gram-negative, xylem-limited, fastidious plant bacteria related to *Xanthemonas,* spp. *Int. J. Syst. Bacteriol.* **37,** 136–143.

Wells, J. M., Raju, B. C., and Nyland, G. (1983). Isolation, culture, and pathogenicity of the bacterium causing phony disease of peach. *Phytopathology* **73,** 859–862.
Windsor, I. M., and Black, L. M. (1972). Clover club leaf: A possible rickettsial disease of plants. *Phytopathology* **62,** 1112.
Windsor, I. M., and Black, L. M. (1973). Evidence that clover club leaf is caused by a rickettsia-like organism. *Phytopathology* **63,** 1139–1148.
Worley, J. F., and Gillaspie, A. G., Jr. (1975). Electron microscopy in situ of the bacterium associated with ratoon stunting disease in sudangrass. *Phytopathology* **65,** 287–295.

B. PLANT DISEASES CAUSED BY MYCOPLASMALIKE ORGANISMS

Introduction

In 1967, wall-less microorganisms resembling mycoplasmas were seen with the electron microscope in the phloem of plants infected with one of several yellows-type diseases. Such diseases, up to that moment, were thought to be caused by viruses. That same year, similar microorganisms were seen in the insect vectors of these diseases. Furthermore, it was shown that these microorganisms were susceptible to tetracycline but not to penicillin antibiotics and that the symptoms of infected plants could be suppressed, at least temporarily, by treatment with antibiotics.

Since then, more than 200 distinct plant diseases affecting several hundred genera of plants have been shown to be caused by mycoplasmalike organisms. Among them are some very destructive diseases, especially of trees, for example, pear decline, coconut lethal yellowing, X-disease of peach, and apple proliferation, but also of herbaceous annual and perennial plants such as aster yellows of vegetables and ornamentals, and stolbur. Furthermore, several diseases, such as citrus stubborn and corn stunt, were shown to be caused by helical mycoplasmas known as spiroplasmas. The main characteristics of yellows-type diseases are a more or less gradual, uniform yellowing or reddening of the leaves, smaller leaves, shortening of the internodes and stunting of the plant, excessive proliferation of shoots and formation of witches'-brooms, greening or sterility of flowers and reduced yields, and finally, a more or less rapid dieback, decline, and death of the plant (Figure 12-38).

Although mycoplasmalike organisms have been observed in the phloem of diseased plants, in sap extracted from such plants, and in the insect vectors of some of them, the true nature of mycoplasmalike organisms and their taxonomic position among the lower organisms is still uncertain. Morphologically, the organisms observed in plants resemble the typical mycoplasmas found in animals and humans and those living saprophytically, but the mycoplasmalike organisms of plants cannot be grown on artificial nutrient media. Also, so far, no plant disease has been reproduced on healthy plants inoculated directly with mycoplasmalike organisms obtained from diseased plants. The pathogens of at least two diseases, citrus stubborn and corn stunt,

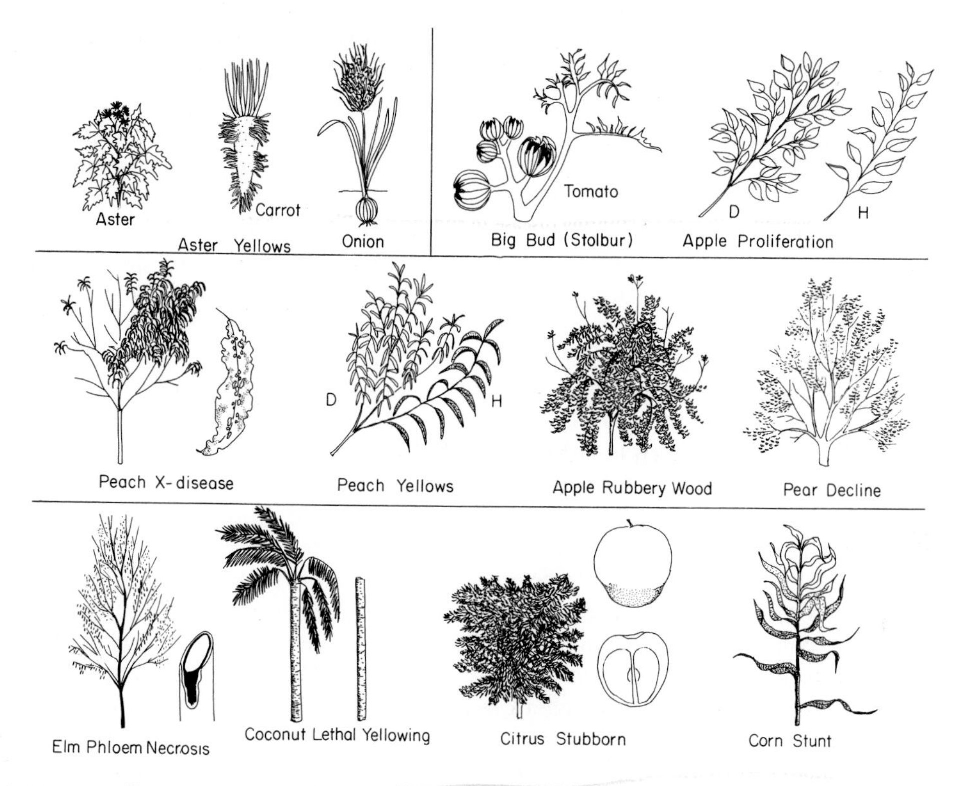

FIGURE 12-38 Symptoms caused by mycoplasmalike organisms.

have been grown on artificial nutrient media and have even reproduced the disease in plants when inoculated by insects injected with the organism from culture; however, these pathogens differ from all other plants mycoplasmalike organisms and from the true mycoplasmas in that they have a helical structure, they are motile, and in some other characteristics. At present it is believed that most of the plant mycoplasmalike organisms will be proved to be similar to the true mycoplasmas, belonging to a new taxon rather than to either of the two mycoplasma genera, that is, *Mycoplasma* and *Acholeplasma,* while the citrus stubborn and the corn stunt organisms and any others like them are placed in a newly created genus of mycoplasmas called *Spiroplasma.*

Properties of Mycoplasmas

Since, until recently, only mycoplasmas that infect animals and humans or are saprophytic were known, all information regarding mycoplasmas has been obtained from the study of these types only.

True Mycoplasmas

Mycoplasmas are prokaryotic organisms, that is, organisms without an organized and bounded nucleus.

The mycoplasmas make up the class of Mollicutes, which has one order, Mycoplasmatales. The order has three families: Mycoplasmataceae, composed of one genus, *Mycoplasma,* Acholeplasmataceae, also composed of one genus, *Acholeplasma,* and Spiroplasmataceae, also composed of one genus, *Spiroplasma. Mycoplasma* differs from *Acholeplasma* in that its species require sterol for growth and are sensitive to digitonin, while species of *Acholeplasma* do not require sterol for growth and are resistant to digotonin. Also, species of *Mycoplasma* have only half as much DNA (5×10^8 daltons) as do species of *Acholeplasma* (10^9 daltons), the amount of DNA in *Acholeplasma* being about half, or at most equal to, that of the smallest bacteria (1.5×10^9 daltons). The genome of *Spiroplasma* is 10^9 daltons.

Mycoplasmas lack a true cell wall and the ability to synthesize the substances required to form a cell wall. Mycoplasmas, therefore, are bounded only by a single triple-layered "unit" membrane. They are small, sometimes ultramicroscopic cells containing cytoplasm, randomly distributed ribosomes and strands of nuclear material. They measure from 175 to 250 nm in diameter during reproduction but grow into various sizes and shapes later on. The shapes range from coccoid or slightly ovoid to filamentous. Sometimes they produce branched mycelioid structures. The size of fully developed coccoid mycoplasmas may vary from one to a few micrometers, while slender branched filamentous forms may range in length from a few to 150 μm. Mycoplasmas seem capable of reproducing by budding and by binary transverse fission of coccoid and filamentous cells. Mycoplasmas have no flagella, produce no spores, and are gram-negative. Nearly all mycoplasmas parasitic to humans and animals and all saprophytic ones can be grown on more or less complex artificial nutrient media in which they produce minute colonies that usually have a characteristic "fried egg" appearance. Mycoplasmas have been isolated mostly from healthy and/or diseased animals and humans suffering from diseases of the respiratory and urogenital tracts; they have been associated with some arthritic and nervous disorders of animals; and some have been found to exist as saprophytes. Most mycoplasmas are completely resistant to penicillin, but they are sensitive to tetracycline, chloramphenicol, some to erythromycin, and to certain other antibiotics.

Mycoplasmalike Organisms of Plants

The organisms observed in plants and insect vectors, with the exception of spiroplasmas, resemble the mycoplasmas of the genera *Mycoplasma* or *Acholeplasma* in all morphological aspects. They lack cell wall, they are bounded by a triple-layered "unit" membrane, and have cytoplasm, ribosomes, and strands of nuclear material. Their shape is usually spheroidal to ovoid or irregularly tubular to filamentous, and their sizes comparable to those of the typical mycoplasmas (Figure 12-39).

Plant mycoplasmalike organisms are generally present in the sap of a small number of phloem sieve tubes (Figure 12-40). Most plant mycoplasmalike organisms are transmitted from plant to plant by leafhoppers (Figure 12-41), but some are transmitted by psyllids and planthoppers (see Fig. 14-15). Plant mycoplasmalike bodies also grow in the alimentary canal, hemolymph, salivary glands, and intracellularly in various body organs of their insect vectors.

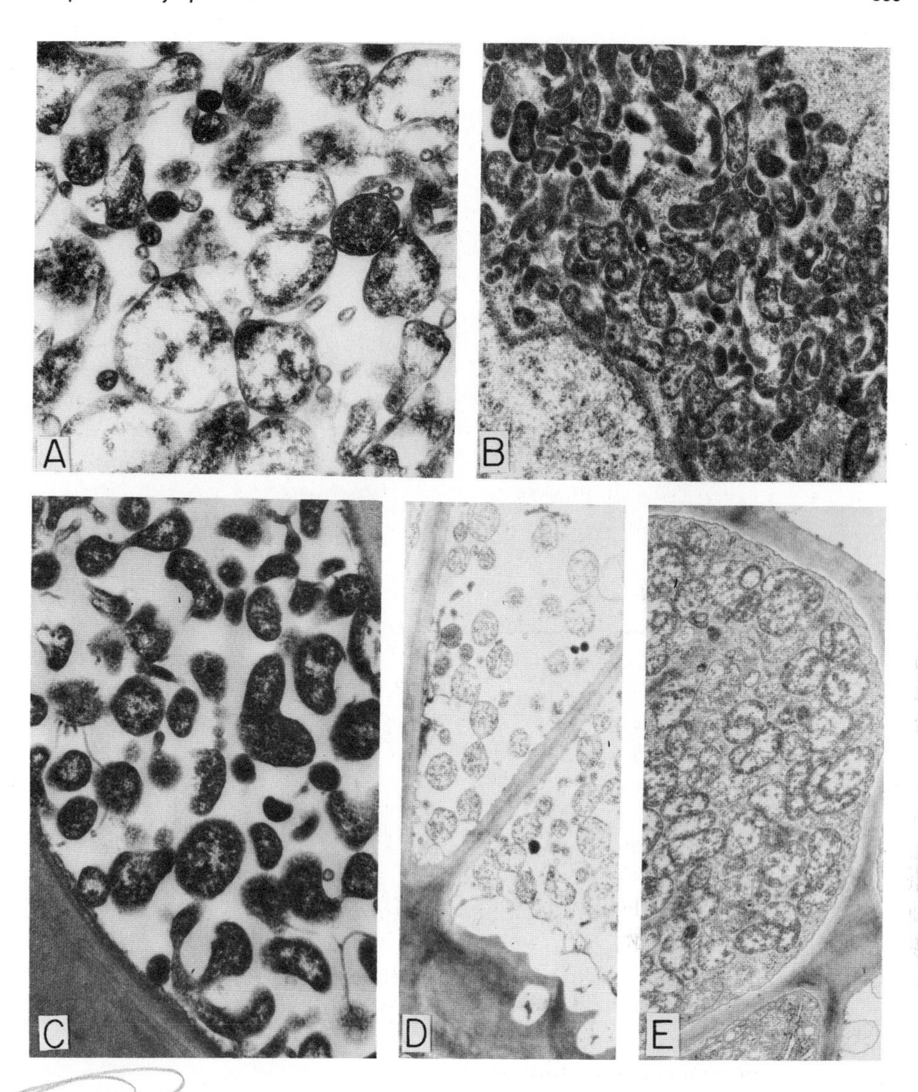

FIGURE 12-39 Aster yellows mycoplasma. (A) Typical large mycoplasmalike bodies bound by a unit membrane and containing strands resembling DNA. The smaller particles contain ribosomes. (B) Mycoplasmalike bodies in cytoplasm of infected phloem parenchyma cell. (C, D) Several polymorphic mycoplasmas (C) and some apparently undergoing binary fission or budding (C, D). (E) Invagination of some mycoplasmalike bodies by others indicating the extreme pliability of the organisms. (Photos courtesy J. F. Worley.)

The insect vectors can acquire the pathogen after feeding on infected plants for several hours or days. The insect may also become a vector if it is injected with extracts from infected plants or vectors. More insects become vectors when feeding on young leaves and stems of infected plants than on older ones. The vector cannot transmit the mycoplasma immediately after feeding on the infected plant, but it begins to transmit it after an incubation period of 10–45 days, depending on the temperature; the shortest incubation period occurs at about 30°C, the longest at about 10°C. The incubation period in insects, however, can be shortened by injecting them with high doses of extracts from infective insects.

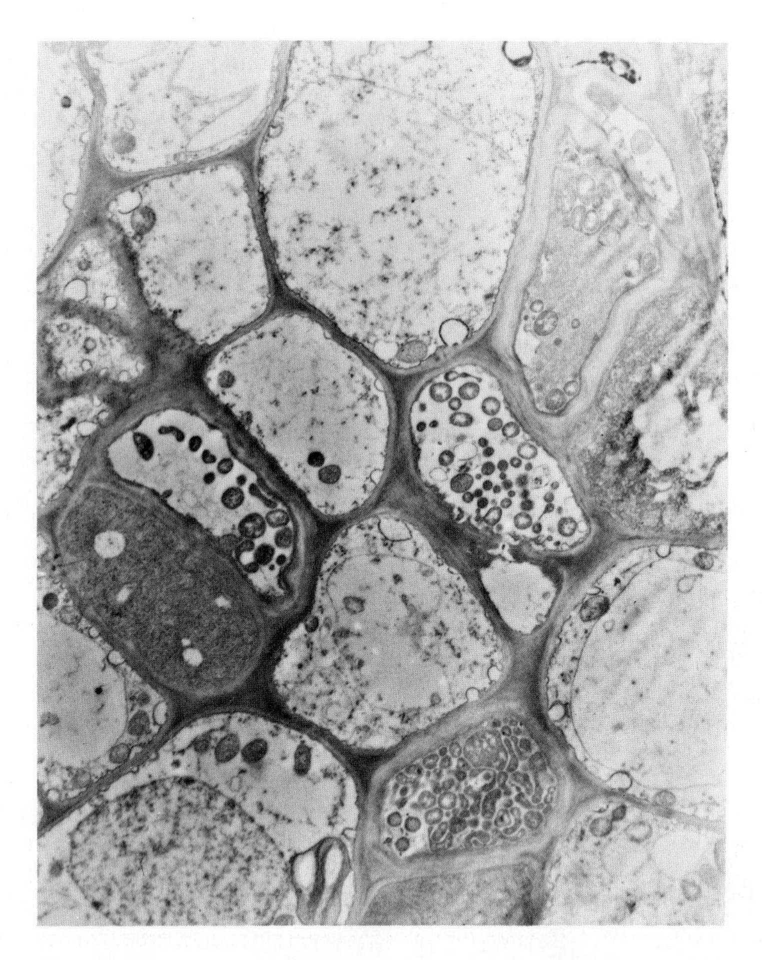

FIGURE 12-40 Distribution of the peach yellows mycoplasma within vascular tissues of diseased cherry showing symptoms. (Photo courtesy J. F. Worley.)

The incubation period is required for the multiplication and distribution of the mycoplasma within the insect (Figure 12-41). If the mycoplasma is acquired from the plant, it multiplies first in the intestinal cells of the vector; it then passes into the hemolymph and internal organs are infected. Eventually the brain and the salivary glands are invaded. When the concentration of the mycoplasma in the salivary glands reaches a certain level, the insect begins to transmit the pathogen to new plants and continues to do so more or less efficiently for the rest of its life. Insect vectors usually are not affected adversely by the mycoplasmas, but in some cases they show severe pathological effects. Mycoplasmas can usually be acquired as readily or better by nymphs as by adult leafhoppers and survive through subsequent molts, but are not passed from the adults to the eggs and to the next generation, which, therefore, must feed on infected plants in order to become infective vectors.

In spite of countless attempts by numerous investigators to culture plant mycoplasmalike organisms on artificial nutrient media, including the media on which all typical mycoplasmas grow, this has not yet been possible. Mycoplasmalike organisms, however, have been extracted from their host

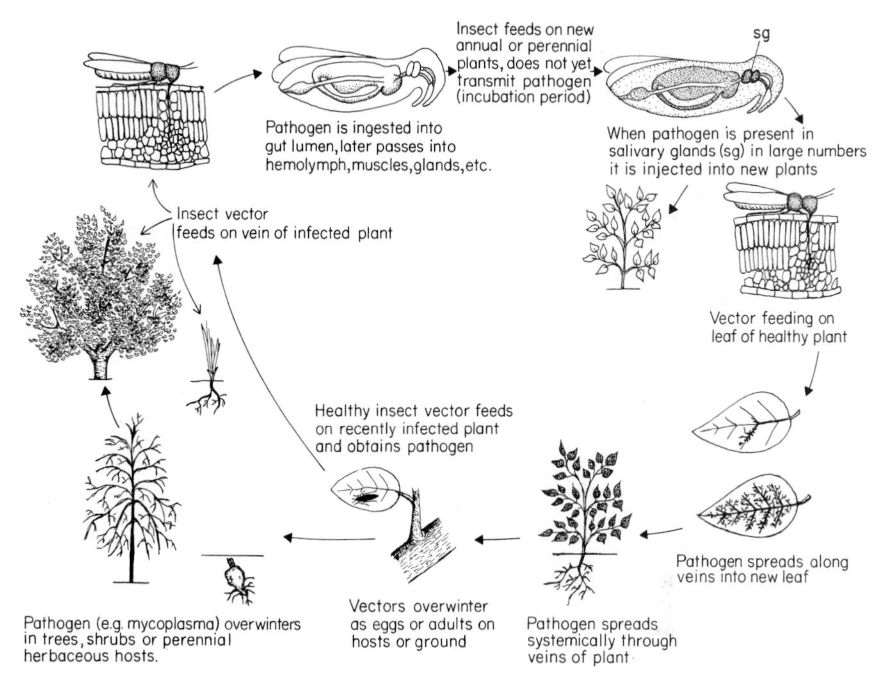

FIGURE 12-41 Sequence of events in the overwintering, acquisition, and transmission of viruses, mycoplasmas, and fastidious bacteria by leafhoppers.

plants and from their vectors in more or less pure form, and for several of them antisera, including monoclonal antibodies, have been prepared. Specific antibodies have become extremely useful in the detection and identification of the pathogen in suspected hosts and in controlling these diseases through production of pathogen-free propagating stock. Serological techniques are currently replacing other methods used to detect mycoplasmalike infections, such as indexing to sensitive hosts and fluorescent staining with either the DNA-specific stain 4,6-diamidino-2-phenylindole (DAPI) or the callose-specific stain aniline blue, or staining with the so-called "Dienes' stain."

Mycoplasmalike organisms are sensitive to antibiotics, particularly those of the tetracycline group. When infected plants are immersed periodically into tetracycline solutions, the symptoms, if already present, recede or disappear and, if not yet present, their appearance is delayed. Foliar application of tetracyclines on infected plants is ineffective, as are soil drench applications. In trees, application of antibiotics is most successful by direct injection into the trunk by pressure or by gravity flow and results in the alleviation or disappearance of symptoms for many months. None of these treatments has, so far, cured any plants from the disease: The symptoms reappear soon after treatment stops. Generally, treatment of plants during the early phases of the disease is much more effective than treatment of plants in advanced stages of the disease. Infected growing plants or dormant propagative organs can be totally freed of mycoplasmalike organisms by heat treatment. This can be applied as hot air in growth chambers at 30°–37°C for several days, weeks or months, or as hot water in which dormant organs are immersed at 30–50°C for as short as 10 minutes at the higher temperatures and as long as 72 hours at the lower temperatures.

Spiroplasmas

Spiroplasmas are helical mycoplasmas and have so far been found to cause the stubborn disease in citrus plants and the brittle root disease in horseradish *(Spiroplasma citri),* stunt disease in corn plants, and a disease in periwinkle. *Spiroplasma citri* has also been found in many other dicots, such as crucifers, lettuce, and peach, and both *S. citri* and the corn stunt spiroplasma also infect their respective leafhopper vectors. Furthermore, several kinds of spiroplasmas have been shown to infect honeybees and several other insects, and several more live saprophytically on flowers and other plant surfaces and, possibly, internally in plants.

Spiroplasmas are pleomorphic cells that vary in shape from spherical or slightly ovoid, 100–250 nm or larger in diameter, to helical and branched nonhelical filaments that are about 120 nm in diameter and 2–4 μm long during active growth and considerably longer (up to 15 μm) in later stages of

FIGURE 12-42 Corn stunt spiroplasmas isolated from infected corn plants and grown on nutrient media. (A) Electron micrograph of a spiroplasma showing typical helical morphology (scale bar, 0.5 μm). (B) Living spiroplasmas from liquid cultures observed by dark field microscopy. (C) Colonies of corn stunt spiroplasma on agar plates 14 days after inoculation (scale bar, 0.05mm). (Photos courtesy T. A. Chen, from Chen and Liao, *Science* **188,** 1015–1017. Copyright © 1975 by the American Association for the Advancement of Science.)

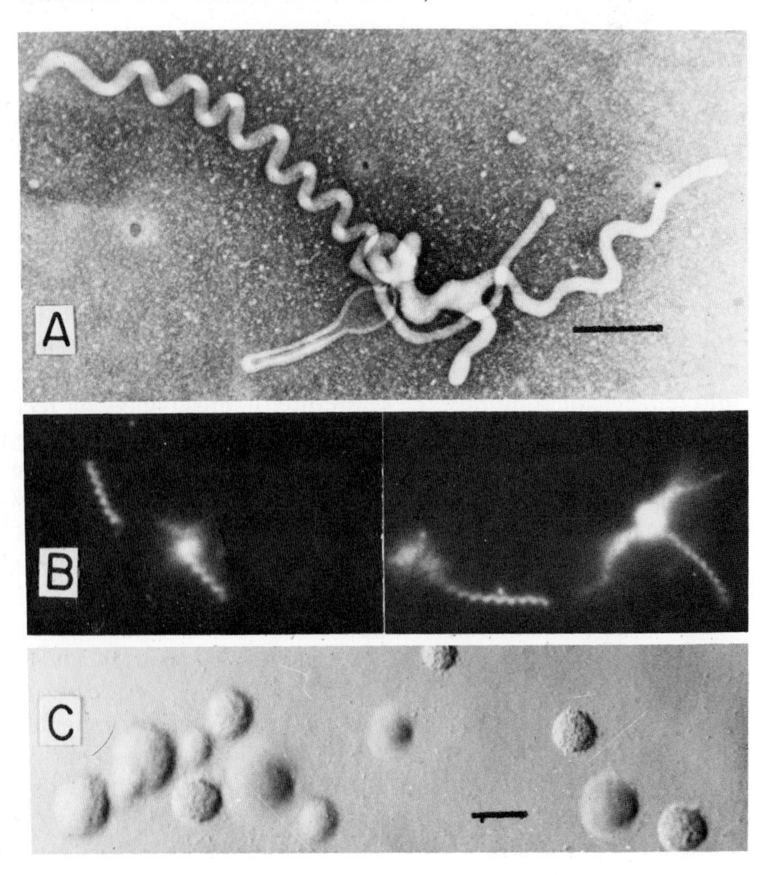

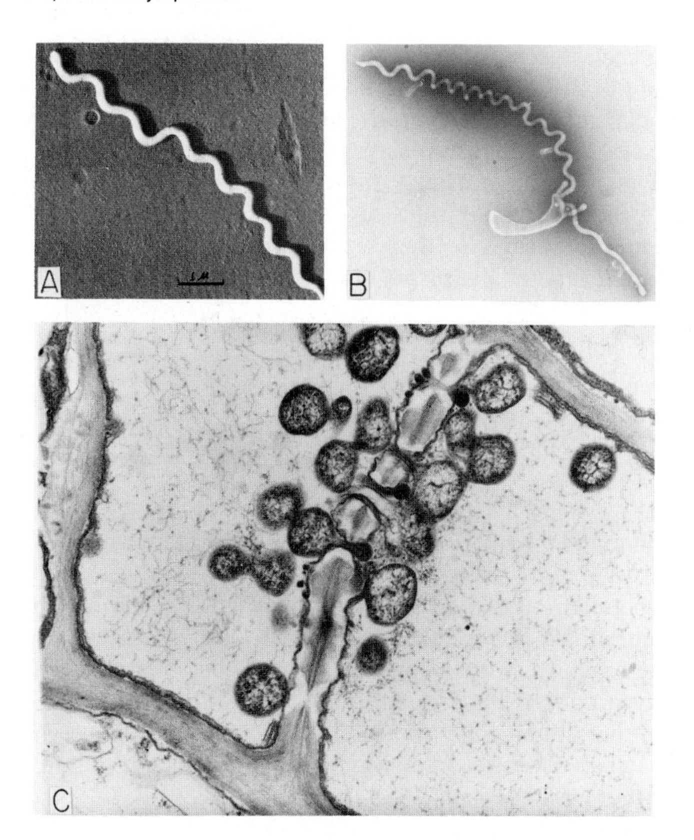

FIGURE 12-43 (A) A replicative form of *Spiroplasma citri* isolated from stubborn-infected citrus. Platinum shadowing from a log-phase culture fixed with glutaraldehyde. (B) *Spiroplasma citri* obtained from its leafhopper vector, *Circulifer tenellus,* and grown in broth culture. Note presence of bleb. (C) *Spiroplasma citri* in sieve plate in midvein of sweet orange leaf. (Photos courtesy E. C. Calavan.)

their growth. Unlike the mycoplasmalike organisms described above, spiroplasmas can be obtained from their host plants or their insect vectors and cultured on nutrient media (Figures 12-42 and 12-43). They produce mostly helical forms in liquid media. They multiply by fission. They lack a true cell wall and are bounded by a single triple-layered "unit" membrane. The helical filaments are motile, moving by a slow undulation of the filament and probably by a rapid rotary or "screw" motion of the helix. There are no flagella. Colonies of spiroplasmas on agar have a diameter of about 0.2 mm; some have a typical "fried egg" appearance, but others are granular (Figure 12-42C). Spiroplasmas require sterol for growth. They are resistant to penicillin but inhibited by erythromycin, tetracycline, neomycin, and amphotericin.

The amount of DNA in spiroplasmas is equal to that of *Acholeplasma* and of the smallest bacteria. The citrus spiroplasma has been shown to be attacked by at least three different kinds of viruses.

The known plant spiroplasmas, such as *Spiroplasma citri* and the corn stunt spiroplasma, have been obtained from their respective hosts and vectors, and have been grown as pure cultures on nutrient media. Furthermore they have been injected into or fed to their insect vectors, which then, upon feeding on the host plants, transmitted the organisms to the plants. The hosts thus inoculated developed typical symptoms of the disease. The pathogens could then be recovered from such plants again and grown and observed in culture. Thus, the spiroplasmas are definitively the causes of their respective diseases. Although some of these organisms are serologically related, they are by no means identical; also they seem to be distinct from each other in the hosts they infect and in certain nutrients required for growth of each in culture.

Other Organisms That Resemble Mycoplasmas: L-Forms of Bacteria

In addition to the three types of organisms described above, the bacteria often produce variants that fail to produce cell walls. The progeny of such variants comprise populations of wall-less bacteria, called L-form or L-phase bacteria, that are morphologically indistinguishable from mycoplasmas and the mycoplasmalike organisms observed in plants. L-form, that is, wall-less, bacteria are usually produced under laboratory conditions when penicillin or other substances that inhibit cell wall production are added to the culture medium. They can apparently also develop in living organisms during treatment with certain antibiotics.

L-form bacteria are either unstable and revert to the original bacterial form when the substance inhibiting bacterial cell wall formation is removed from the medium, or they are stable, that is, they are unable to revert to the original bacterium. L-form bacteria can be cultured on the same simple nutrient media as the original bacteria, but they usually lose any pathogenicity the original bacteria may have had. It is still uncertain, however, whether the L-form bacteria might not play a role in the persistence of disease agents during antibiotic treatments, in the recurrence of disease, and in latency by exhibiting a high degree of resistance to antibiotics acting on cell wall synthesis and by reverting to the original pathogenic bacteria at the termination of the antibiotic treatment. It is also conceivable that, *in vivo,* the L-form bacteria may themselves induce disease without reversion to the bacterial parents. Usually, however, L-form bacteria become more permeable, and thereby more sensitive, to antibiotics that affect other cell functions besides cell wall synthesis.

Although the ability to form L-forms is now accepted as a general property of bacteria, the only plant pathogenic bacteria reported to produce L-forms are *Agrobacterium tumefaciens,* the cause of crown gall disease, and *Erwinia carotovora* pv. *atroseptica,* the cause of the black leg disease of potato. Moreover, the L-forms of the first bacterium retained the pathogenicity of the parent bacteria, produced tumors identical to those produced by the bacteria, and could be reisolated and cultured from such tumors.

Since L-form bacteria are morphologically indistinguishable from mycoplasmas and plant mycoplasmalike organisms, their diagnosis depends on their ability to grow on simple nutrient media and to revert to the original

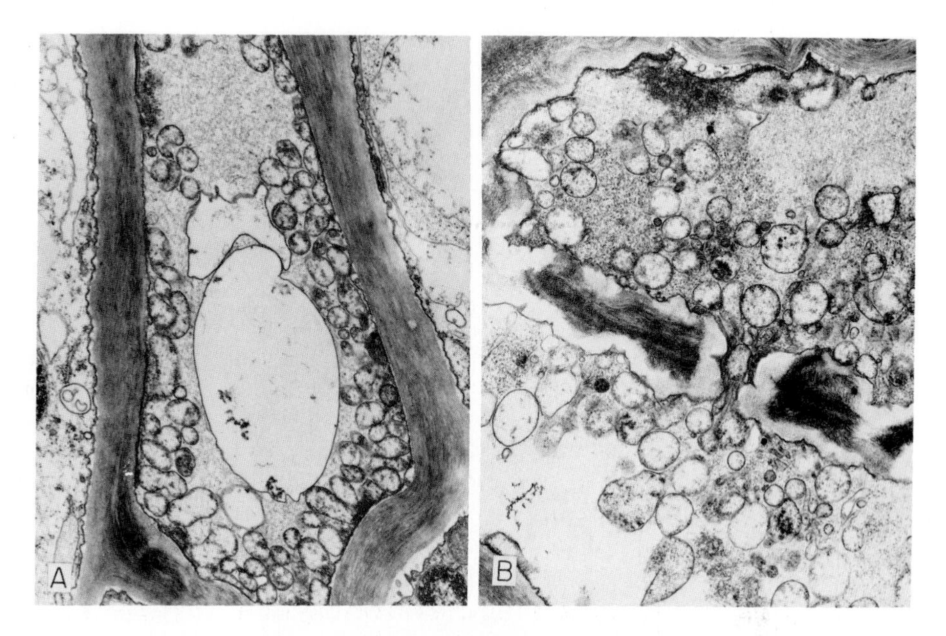

FIGURE 12-46 (A) Mycoplasmalike organism in sieve element of young inflorescence of coconut palm infected with lethal yellowing. (B) Lethal yellowing mycoplasmas passing through a sieve-plate pore lined with callose. (Photos courtesy M. V. Parthasarathy, from *Phytopathology* **64,** 667–674.)

So far control of lethal yellowing has been attempted primarily by sanitation measures, that is, removal and burning of diseased palms as soon as symptoms appear, to reduce the source of inoculum from which the vector(s) can transmit the pathogen to healthy trees. Among the various coconut palms, only certain Malayan dwarf varieties appear to be resistant or immune to lethal yellowing, and thousands of such trees and hybrids of Malayan dwarf with susceptible palms are now planted to replace the other coconut palms wherever lethal yellowing exists.

Very encouraging results of lethal yellowing control have been obtained by treating infected trees with solutions of tetracycline antibiotics. When 0.5–20 g of oxytetracycline hydrochloride is injected into trunks of diseased palms through either gravity flow or pressurized trunk injection, symptom expression is arrested or slowed down for several months, and healthy new inflorescences and leaves grow in treated palms 3–4 months after initial treatment. Palms respond much better when treated with tetracycline in the early or preyellowing stages of disease development than in more advanced stages of the disease. The higher dosages of antibiotic (6–20 g per tree) are more effective in inducing remission in intermediate to advanced stages of the disease and their effect is longer lasting than that of lower dosages. Remission lasts from 4 to 7 months and retreatment is recommended at 4-month intervals.

• Elm Phloem Necrosis (Elm Yellows)

Elm phloem necrosis occurs in about 15 central and southern states and was recently found in Pennsylvania, New York, and Massachusetts. Elm phloem necrosis epidemics have killed thousands of trees in each of numerous communities.

FIGURE 12-47 Elm phloem necrosis. (A) Diseased tree. (B) Discolored inner bark of diseased tree. (Photos courtesy W. A. Sinclair.)

FIGURE 12-48 Peach tree showing symptoms of X-disease. (Photo courtesy D. Sands.)

The symptoms consist of a general decline of the tree in which the leaves droop and curl, turn bright yellow, then brown, and finally fall (Figure 12-47A, B). Some trees are killed within a few weeks, and most trees that show symptoms in June or July die in a single growing season. If the tree was infected late, it may live through the winter, but then in the spring it produces a thin crop of small leaves and dies soon after. In later stages of the disease the inner layers of peeled bark (phloem) at the base of the stem shows a butter-scotch-brown color and have a faint odor of wintergreen. The latter characteristics are often used for a quick diagnosis of the disease. The discoloration of the phloem is apparently the result of rapid deposition of callose within the sieve tubes and then collapse of sieve elements and companion cells. The cambium produces replacement phloem but its cells become quickly necrotic also.

The pathogen is a mycoplasmalike organism present in the phloem of infected trees. It is transmitted from diseased to healthy trees by the leafhopper *Scaphoideus luteolus.*

Injection of tetracyclines into recently infected trees causes remission of symptoms for several months and up to three years. Severely diseased or dead trees should be removed and burned.

• Peach X-Disease

X-disease, including western X-disease, occurs in the northwestern and northeastern parts of the United States and the adjacent parts of Canada. It also occurs in Michigan and several other states. Where present, X-disease is one of the most important diseases of peach. Affected trees become commercially worthless in 2–4 years. Young peach trees are rendered useless within one year of inoculation. X-disease of peach also attacks sweet and sour cherries, nectarines, and chokecherries.

The symptoms of X-disease of peach appear as a slight mottle and reddish purple spots on the leaves of some or all branches. The spots die and fall out, giving a "shot-hole" appearance to the leaf. The leaves take on a reddish coloration and roll upward. Later, most leaves on affected branches drop except the ones at the tips (Figure 12-48). The fruits on affected branches usually shrivel and drop soon after the symptoms appear on the leaves. Any fruits remaining on the trees ripen prematurely, have an unpleasant taste, and are unsalable. No seeds develop in the pits of affected fruit. Fruits on healthy looking parts of infected trees show no signs of the disease.

The pathogen is a mycoplasmalike organism found in the phloem sieve tube elements of diseased trees. The pathogen is transmitted by several species of leafhoppers of the genera *Colladonus, Scaphytopius,* and, of course, by budding and grafting. Peach trees inoculated early in the season develop symptoms in less than two months, while later inoculations may not produce symptoms until the following season. The insect vector can transmit the pathogen within trees of the same species and between trees of different species, for example, from chokecherry to peach. It appears, however, that in the northeastern United States either the vector strain or the pathogen strain is different in that the pathogen is transmitted from chokechery to peach or to chokecherry but not from peach to peach. This property has allowed successful control of X-disease on peach by eradicating chokecherry from the vicinity of peach orchards in a zone within about 200 meters from the orchard. Additional controls include the use of disease-free scion wood and rootstocks and removal of diseased trees. Injections of tetracyclines into diseased trees results in temporary remission of X-disease symptoms and in reduced transmission of the disease by leafhoppers that obtain the inoculum from treated trees.

• Pear Decline

The pear decline occurs in the Pacific coast and some east coast states, in Europe, and probably in other continents where similar declinelike disorders of pear have been observed but whose relationship to pear decline has not been established. Pear decline causes either a slow, progressive weakening and final death of trees or a quick, sudden wilting and death of trees. The disease can be extremely catastrophic, having killed more than 1,100,000 trees in California between 1959 and 1962. Pear decline affects all pear varieties when

FIGURE 12-49 (A-C) Pear decline. (A) "Bartlett" pear tree on *Pyrus serotina* rootstock showing symptoms of chronic pear decline. (B) As in (A) but chronic decline is more advanced—poor tree growth; few, small fruits. (C) "Bartlett" tree recovering after injection with 1 g oxytetracycline hydrochloride the previous September. Note new enlongated shoots and denser leaf growth. (D) Mycoplasma-like organisms in phloem sieve tube of leaf of pear tree affected by pear decline. Arrows point to "unit" membrane. (Photos A-C courtesy G. Nyland, D courtesy Hibino and Schneider, from *Phytopathology* **60,** 499–501.)

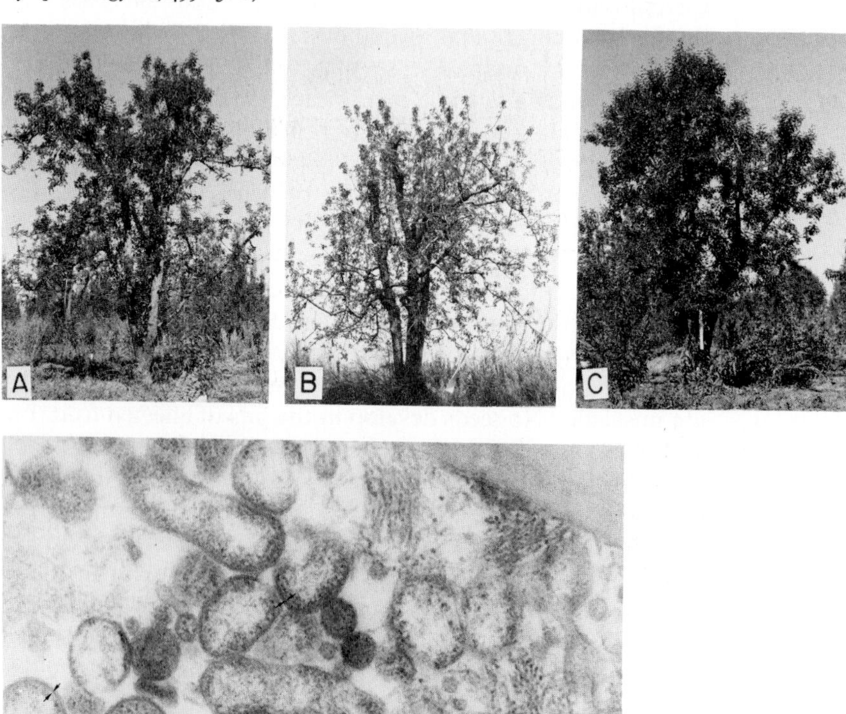

they are grafted on certain rootstocks. Although oriental rootstocks such as *Pyrus serotina* and *P. ussuriensis* are affected the most, pear decline has also been observed on pear varieties grafted on *P. communis, P. betulaefolia,* and on quince.

The symptoms of pear decline in the "slow decline" syndrome appear as a progressive weakening of the trees, which, however, may continue to live for many years but eventually, despite occasional apparent improvement, are killed by the disease. During this period there is little twig growth, the leaves are few, small, pale green, and leathery and roll slightly upward (Figure 12-49). Such leaves often turn reddish in late summer and drop off prematurely in the fall. In the early stages of the disease the trees produce abundant blossoms, but as the disease progresses the trees produce fewer blossoms, set fewer fruit, and the fruits are small. By this time, although starch accumulates above the graft union, it is almost absent below the union and most of the feeder roots of the tree are dead.

In the "quick decline" syndrome the trees wilt suddenly and die within a few weeks. Quick decline is more common in trees grafted on oriental rootstocks, while trees grafted on other rootstocks usually develop the slow decline syndrome.

Both types of pear decline can be detected and verified by a microscopic observation of the phloem below the graft union: In diseased trees the current season's ring of phloem immediately below the graft union degenerates and the degeneration becomes more pronounced as the season progresses. Also, at the graft union of diseased trees, narrow, small sieve tube elements (replacement phloem) are produced rather than normal ones. Pear decline can be transmitted by budding or grafting, although only about one-third of the buds seems to transmit the disease. Pear decline is also transmitted naturally by pear psylla *(Psylla pyricola),* large numbers of which are responsible for the outbreaks of the disease.

The pathogen is a mycoplasmalike organism present in the phloem sieve elements. The bodies are mostly spherical to oblong particles of about 50–800 nm; relatively few are elongated (Figure 12-49D). In pear psylla, the pathogens are present in various organs of the insect but seem to be almost five times more common in the cephalic part of the foregut than they are in the salivary glands.

A certain degree of control of pear decline is obtained by growing pear varieties on resistant rootstocks such as *Pyrus communis* and by avoiding the highly sensitive oriental rootstocks. Control of the pear psylla vector has not been successful. However, injection of 6–8 liters of a tetracycline solution (100 mg/liter) in the trunk of infected trees soon after harvest prevents leaf curl symptoms in the fall of the current season and greatly stimulates shoot and spur growth of treated trees the following season. When two or three such treatments are made in the fall, previously severely diseased trees are restored to normal or near normal condition (Figure 12-49C). Antibiotic treatments must continue annually or biannually, however, or the disease will reappear.

• Citrus Stubborn Disease

Citrus stubborn is present in all the Mediterranean countries, the southwestern United States, Brazil, Australia, and possibly in South Africa. In some

Mediterranean countries and in California, stubborn is regarded as the greatest threat to production of sweet oranges and grapefruit. Because of the slow development of symptoms and the long survival of affected trees, spread of stubborn is insidious and its detection difficult. However, yields are reduced drastically; the trees produce fewer fruits and many of those are too small to be marketable. In California, approximately two million orange, grapefruit, and tangelo trees are so severely affected that they are practically worthless, and many more trees are infected in one or several branches but are not yet severely damaged.

The symptoms of stubborn disease appear on leaves, fruit, and stems of all commercial varieties regardless of rootstock (Figure 12-50). Symptoms, however, vary a great deal, and frequently only a few are expressed at one time on an entire tree or parts of a tree. In general, affected trees show a bunchy upright growth of twigs and branches, with short internodes and an excessive number of shoots; multiple buds and sprouts are common. Some of the affected twigs die back. The bark is thickened and sometimes pinholed.

FIGURE 12-50 (A) Six-year-old sweet orange trees. Healthy at left, stubborn diseased at right showing extreme dwarfing. (B) Tangerine showing stylar-end greening and acorn shape (left) and healthy fruit. (C) Leaf symptoms of stubborn on sweet orange. (D) Four stubborn and one normal fruit of sweet orange. (Photos courtesy E. C. Calavan.)

The trees show slight to severe stunting and often appear flat topped. The leaves are small or misshapen or both, often mottled or chlorotic. Excessive winter defoliation is common in infected trees. Affected trees bloom at all seasons, especially in the winter, but produce fewer fruits. Some of the fruits are very small, lopsided, or otherwise deformed, frequently resembling acorns. Such fruits have normally thick rind from the stem end to the fruit equator, and abnormally thin rind from there to the stylar end. The rind is often dense or cheesy. Some fruits show greening of the stylar end or inverted development of ripe coloration in which, normally, color appears first at the stylar end. Affected fruit tends to drop prematurely and an excessive number of them become mummified. Fruits are usually sour or bitter and have an unpleasant odor and flavor. Also, fruits from affected trees or parts of trees tend to have many poorly developed, discolored, and aborted seeds.

The pathogen is *Spiroplasma citri* (Figure 12-43). It is found in the sieve tubes of stubborn-diseased citrus phloem from which it can be obtained and cultured readily on artificial media. *Spiroplasma citri* has also been found in or transmitted to plants of more than 20 dicotyledonous families and some monocots, including most crucifers, several stone fruits, such as peach and cherry, and onion. Some infected hosts, such as pea, bean, and periwinkle, become wilted and die, while most others remain symptomless.

Spiroplasma citri was the first mycoplasmalike pathogen of a plant disease to be cultured. Within phloem sieve tubes the pathogen is present mostly as what appear to be spherical, ovoid, or elongated forms and occasionally as helical filaments. In liquid cultures, the pathogen appears primarily as motile helical filaments that are sometimes connected to irregularly shaped main bodies. The pathogen may lose its helical structure and motility in older cultures or on solid agar media, and then appears mostly as irregular filaments and blebs. The pathogen is gram-positive, has a layer of surface projections on the cytoplasmic membrane, and is usually found infected with one of three distinct kinds of virus. The pathogen has a sharp optimum temperature for growth at about 30–32°C, while little growth occurs at 20°C and none at 37°C. The pathogen is insensitive to penicillin but is highly sensitive to tetracycline and less so to amphotericin, neomycin, and digitonin.

Citrus stubborn disease is transmitted with moderate frequency by budding and grafting. It is also known to spread naturally in citrus orchards by the leafhoppers *Circulifer (Neoaliturus) tenellus, Scaphytopius nitridus,* and *S. acutus delongi,* while *Macrosteles fascifrons* may be a common vector of *S. citri* in horseradish and other herbaceous hosts.

Stubborn disease can be detected and diagnosed by the symptoms it causes on trees in the field; by indexing on seedlings of several varieties of sweet orange, tangelo, grapefruit, and other citrus, which usually develop symptoms within 2–8 months in the greenhouse and within 15–24 months in the field; and by remission of stubborn disease symptoms of diseased trees after injection of erythromycin, tylosin, or tetracycline into the trunks of the trees. More recently, serolocial techniques, particularly ELISA, are used almost exclusively for detection of *S. citri* in trees either for elimination of the trees by roguing or for treatment with antibiotics.

Control of citrus stubborn depends on the use of spiroplasma-free scionwood and rootstocks, detection by ELISA or through indexing, and removal of infected trees. Young citrus trees could be protected, or showed remission of symptoms, if their roots were immersed for prolonged periods in tetracycline antibiotics, but injection or spraying of infected trees with the antibiotics failed to control the disease.

• Corn Stunt Disease

Corn stunt occurs in the southern United States, Central America, and northern South America. The disease causes severe losses in most areas where it occurs, although disease severity varies with the variety and the stage of host development at the time of infection.

The symptoms consist at first of faint yellowish streaks in the youngest leaves. As the plant matures the yellowing becomes more apparent and more general over the leaves; soon much of the leaf area turns red to purple, especially on the upper leaves. Infected plants remain stunted due to shorter stem internodes, particularly in the part of the plant produced after infection, which gives the plants a somewhat bunchy appearance at the top (Figure 12-51). Infected plants often have more ears than do healthy plants, but the

FIGURE 12-51 (A) Corn stunt disease in corn. Leaves show chlorotic streaks, plant is stunted, proliferation begins at nodes, and tassel is sterile. (B) Portions of corn stunt spiroplasmas as seen in a section of phloem tissue from a corn stunt-infected plant. (Photos courtesy R. E. Davis, B from *Phytopathology* **63,** 403–408.)

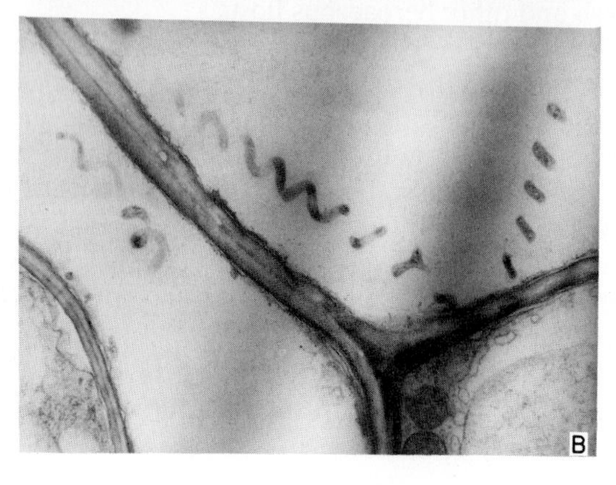

ears are smaller and bear little or no seed. Tassels of infected plants are usually sterile. There is also a proliferation of sucker shoots and, in severe infections, of roots.

The corn stunt pathogen was the first spiroplasma discovered. Its morphology is very similar to the one causing the stubborn disease of citrus (Figure 12-51B and Figure 12-42). The corn stunt spiroplasma has been grown on artificial nutrient media and its pathogenicity proven either by injecting leafhopper vectors with, or allowing them to feed on, pure cultures of the spiroplasma and then allowing them to feed on healthy corn seedlings. The inoculated plants developed typical corn stunt symptoms, and the spiroplasma was reisolated and cultured from such plants.

Corn stunt is transmitted in nature by the leafhoppers *Dalbulus elimatus, D. maidis,* and others. The leafhoppers must feed on diseased plants for several days before they can acquire the spiroplasma, and an incubation period of 2–3 weeks from the start of the feeding must elapse before the insects can transmit the spiroplasma to healthy plants. A feeding period of a few minutes to a few days may be required for the insects to inoculate the healthy plants with the spiroplasma. Plants show corn stunt symptoms 4–6 weeks after inoculation.

Where the corn stunt spiroplasma overwinters is not known with certainty, although it was previously believed to overwinter in Johnson-grass and possibly other perennial plants. In the tropics, it perpetuates itself in continuous croppings of corn.

Control of corn stunt depends on the planting of corn hybrids resistant to corn stunt.

SELECTED REFERENCES

Bove, J. M. (1984). Wall-less prokaryotes of plants. *Annu. Rev. Phytopathol.* **22,** 361–396.

Chen, T. A., and Liao, C. H. (1975). Corn stunt spiroplasma: Isolation, cultivation, and proof of pathogenicity. *Science* **188,** 1015–1017.

Chen, T. A., Wells, J. M., and Liao, C. H. (1982). Cultivation in vitro: Spiroplasmas, plant mycoplasmas, and other fastidious walled prokaryotes. *In* "Phytopathogenic Prokaryotes" (M. S. Mount and G. H. Lacey, eds.), Vol. 2, pp. 417–446. Academic Press, New York.

Daniels, M. J. (1983). Mechanisms of *Spiroplasma* pathogenicity. *Annu. Rev. Phytopathol.* **21,** 29–43.

Daniels, M. J., and Markham, P. G., eds. (1982). "Plant and Insect Mycoplasma Techniques." Croom Helm, London.

Davis, R. E., and Worley, J. F. (1973). Spiroplasma: Motile, helical microorganism associated with corn stunt disease. *Phytopathology* **63,** 403–408.

Davis R. E., Worley, J. F., Whitcomb, R. F., Ishijima T., and Steere, R. L. (1972). Helical filaments produced by a mycoplasma-like organism associated with corn stunt disease. *Science* **176,** 521–523.

Deeley, J., Stevens, W. A., and Fox, R. T. V. (1979). Use of Dienes stain to detect plant diseases induced by mycoplasmalike organisms. *Phytopathology* **69,** 1169–1171.

Doi, Y. *et al.* (1967). Mycoplasma- or PLT group-like microorganisms found in the phloem elements of plants infected with mulberry dwarf, potato witches'-broom, aster yellows, or paulownia witches'-broom. *Ann. Phytopathol. Soc. Jpn.* **33,** 259–266.

Fudl-Allah, A. A., and Calavan, E. C. (1974). Cellular morphology and reproduction of the mycoplasma-like organism associated with citrus stubborn disease. *Phytopathology* **64,** 1309–1313.

Hervey, G. E. R., and Schroeder, W. T. (1949). The yellows disease of carrot. *Bull.—N. Y., Agric. Exp. Stn. Ithaca* **737,** 1–29.

Hibino, H., and Schneider, H. (1970). Mycoplasmalike bodies in sieve tubes of pear trees affected with pear decline. *Phytopathology* **60,** 449–501.

Hibino, H., Kaloostian, G. H., and Schneider, H. (1971). Mycoplasma-like bodies in the pear psylla vector of pear decline. *Virology* **43,** 34–40.

Ishiie, T. *et al.* (1967). Suppressive effects of antibiotics of tetracycline group on symptom development of mulberry dwarf disease. *Ann. Phytopathol. Soc. Jpn.* **33,** 267–275.

Kloepper, J. W., Garrott, D. G., and Oldfield, G. N. (1982). Quantification of plant pathogenic spiroplasmas from infected plants. *Phytopathology* **72,** 577–581.

Klotz, L. J., Calavan, E. C., and Weathers, L. G. (1972). Virus and viruslike diseases of citrus. *Calif., Agric. Exp. Stn., Ext. Serv. Circ.* **559,** 1–42.

Kunkel, L. O. (1926). Studies on aster yellows. *Am. J. Bot.* **13,** 646–705.

Lin, C. P., and Chen, T. A. (1985). Monoclonal antibodies against the aster yellows agent. *Science* **227,** 1233–1235.

Lin, C. P., and Chen, T. A. (1985). Production of monoclonal antibodies against *Spiroplasma citri. Phytopathology* **75,** 848–851.

Liu, H. Y. *et al.* (1983). Transmission of *Spiroplasma citri* by *Citrulifer tenellus. Phytopathology* **73,** 582–585, 585–590.

McCoy, R. E. (1982). Use of tetracycline antibiotics to control yellows diseases. *Plant Dis.* **66,** 539–542.

McCoy, R. E. (1983). Wall-free prokaryotes of plants and invertebrates. *In* "Phytopathogenic Bacteria" (M. P. Starr, ed.), pp. 2238–2246. Springer-Verlag, Berlin and New York.

Maramorosch, K., ed. (1973). Mycoplasma and mycoplasma-like agents of human, animal, and plant diseases. *Ann. N. Y. Acad. Sci.* **225,** 1–532.

Maramorosch, K. (1981). Control of vector-borne mycoplasmas. *In* "Pathogens, Vectors, and Plant Diseases: Approaches to Control" (K. F. Harris and K. Maramorosch, eds.), pp. 265–295. Academic Press, New York.

Maramorosch, K., and Raychaudhuri, S. P. (1981). "Mycoplasma Diseases of Trees and Shrubs." Academic Press, New York.

Matteoni, J. A., and Sinclair, W. A. (1985). Role of the mycoplasmal disease, ash yellows, in decline of white ash in New York State. *Phytopathology* **75,** 355–360.

Mullin, R. S., and Roberts, D. A. (1972). Lethal yellowing of coconut palms. *Fla., Coop. Ext. Serv., Inst. Food Agric. Sci., Circ.* **358,** 1–4.

Nasu, S., Jensen, D. D., and Richardson, J. (1970). Electronmicroscopy of mycoplasmalike bodies associated with insect and plant hosts of peach western X-disease. *Virology* **41,** 186–192.

Nichols, C. W. *et al.* (1960). Pear decline in California. *Calif., Dep. Agric., Bull.* **49,** 186–192.

Nienhaus, F., and Sikora, R. A. (1979). Mycoplasmas, spiroplasmas, and rickettsia-like organisms as plant pathogens. *Annu. Rev. Phytopathol.* **17,** 37–58.

Ploaie, P. G. (1981). Mycoplasmalike organisms and plant diseases in Europe. *In* "Plant Diseases and Vectors: Ecology and Epidemiology" (K. Maramorosch and K. F. Harris, eds.), pp. 61–104. Academic Press, New York.

Posnette, A. F., ed. (1969). Virus Diseases of Apples and Pears. *Commonw. Bur. Hortic. Plant. Crops (G. B.) Tech. Commun.,* Suppl.

Purcell, A. H. (1982). Insect vector relationships with prokaryotic plant pathogens. *Annu. Rev. Phytopathol.* **20,** 397–417.

Raju, B. C. (1981). Association of a spiroplasma with brittle root of horseradish. *Phytopathology* **71,** 1067–1072.

Raju, B. C., Purcell, A. H., and Nyland, G. (1984). Spiroplasmas from plants with aster yellows disease and X-disease: Isolation and transmission by leafhoppers. *Phytopathology* **74,** 925–931.

Saglio, P. *et al.* (1971). Isolement et culture *in vitro* des mycoplasmes associés au "stubborn" des agrumes et leur observations au microscope électronique. *C. R. Hebd. Seances Acad. Sci., Ser. D* **272,** 1387–1390.

Schaper, U., and Converse, R. H. (1985). Detection of mycoplasmalike organisms in infected blueberry cultivars by the DAPI technique. *Plant Dis.* **69,** 193–196.

Sinha, R. C. (1983). Relative concentration of mycoplasma-like organisms in plants at various times after infection with aster yellows. *Can. J. Plant Pathol.* **5,** 7–10.

Sinha, R. C., and Benhamou, N. (1983). Detection of mycoplasmalike organism antigens from aster yellows diseased plants by two serological procedures. *Phytopathology* **73,** 1199–1202.

Sinha, R. C., and Chiykowski, L. N. (1967). Initial and subsequent sites of aster yellows virus infection in a leafhopper vector. *Virology* **33,** 702–708.

Sinha, R. C., and Chiykowski, L. N. (1984). Purification and serological detection of mycoplasmalike organisms from plants affected by peach eastern X-disease. *Can. J. Plant Pathol.* **6,** 200–205.

Sinha, R. F. (1979). Purification and serology of mycoloplasmalike organisms from aster yellows-infected plants. *Can. J. Plant Pathol.* **1,** 65–70.

Thomas, D. L. (1974). Mycoplasmalike bodies associated with declining palms in south Florida. *Proc. Am. Phytopathol. Soc.* **1,** 97.

Tully, J. G., and Whitcomb, R. F. (1983). The genus *Spiroplasma.* In "Phytopathogenic Bacteria" (M. P. Starr, ed.), pp. 2271–2284. Springer-Verlag, Berlin and New York.

Whitcomb, R. F., and Tully, J. G., eds. (1979). "The Mycoplasmas," Vol. 3. Academic Press, New York.

Williamson, D. L., and Whitcomb, R. F. (1975). Plant mycoplasmas: A cultivable spiroplasma causes corn stunt disease. *Science* **188,** 1018–1020.

13 PLANT DISEASES CAUSED BY PARASITIC HIGHER PLANTS

Introduction

More than 2500 species of higher plants are known to live parasitically on other plants. These parasitic plants produce flowers and seeds similar to those produced by the plants they parasitize. They belong to several widely separated botanical families and vary greatly in their dependence on their host plants. Some, for example mistletoes, have chlorophyll but no roots and depend on their hosts for water and all minerals, although they can produce carbohydrates in their green leaves and stems. Others, however, having little or no chlorophyll and no true roots, depend entirely on their hosts for their existence (for example, dodder).

Relatively few of the known parasitic higher plants cause important diseases on agricultural crops or forest trees. The most common and serious parasites belong to the following botanical families and genera:

Cuscutaceae
- Genus: *Cuscuta,* the dodders

Viscaceae
- Genus: *Arceuthobium,* the dwarf mistletoes of conifers
- *Phoradendron,* the American true mistletoes of broadleaved trees
- *Viscum,* the European true mistletoes

Orobanchaceae
- Genus: *Orobanche,* the broomrapes of tobacco

Scrophulariaceae
- Genus: *Striga,* the witchweeds of many monocotyledonous plants

• Dodder

Dodder is widely distributed in Europe and North America. In the United States, it is most serious in the southern half of the country and in the north central states, but crops like alfalfa and clover raised for seed may be destroyed by dodder wherever they are grown. Other crops that suffer losses from dodder include onions, sugar beets, several ornamentals, and potatoes.

Dodder affects the growth and yield of infected plants and causes losses ranging from slight to complete destruction of the crop in the infested areas. Names such as strangleweed, pull-down, hellbind, devil's hair, and hailweed, by which dodder is referred to in different areas, are descriptive of the ways in which dodder affects its host plants.

FIGURE 13-1 (A) Dodder on alfalfa. (B) Patches of dodder in a heavy infestation of an alfalfa field. (Photos courtesy U.S.D.A.)

Dodder may also serve as a bridge for transmission of viruses from virus-infected to virus-free plants as long as both plants are infected by the same dodder plant.

- *Symptoms.* Orange or yellow vine strands grow and entwine around the stems and the other aboveground parts of the plants. Dodder forms dense tangles of leafless strands on and through the crowns of the host plants (Figure 13-1A). The growing tips reach out and attack adjacent plants, until a gradually enlarging circle of infestation, up to 10 feet in diameter, is formed by a single dodder plant. Dodder-infested areas appear as patches in the field (Figure 13-1B), which continue to enlarge during the growth season and, in perennial plants such as alfalfa, become larger every year. During late spring and in the summer, dodder produces massed clusters of white, pink, or yellowish flowers, which soon form seed. The infected host plants become weakened by the parasite, their vigor declines, and they produce poor yields. Many are smothered and may be killed by the parasite. As the infection spreads several patches coalesce, and large areas may be formed that are easily seen by the yellowish color of the parasitic vine that covers them.
- *The Pathogen: Cuscuta sp.* Three species of dodder, largeseed *(Cuscuta indicora),* smallseed *(C. planiflora),* and field dodder *(C. campestris),* are important in the United States. The first two show preference for legumes, but the third attacks many other broadleaf plants as well as legumes.

  Dodder is a slender, twining plant (Figure 13-2). The stem is tough, curling, threadlike, and leafless, bearing only minute scales in place of leaves. The stem is usually yellowish or orange in color, sometimes tinged with red or purple; sometimes it is almost white. Tiny flowers massed in clusters occur on the stem from early June until frost. Gray to reddish-brown seeds are produced in abundance by the flowers and mature within a few weeks from bloom.
- *Development of Disease.* Dodder seed overwinters in infested fields or mixed with the seed of crop plants. During the growing season the seed germinates and produces a slender yellowish shoot but no roots (Figure 13-2). This leafless shoot rotates as though in search of a host. If no contact with a susceptible plant is made, the stem falls to the ground, where it lies dormant for a few weeks and then dies.

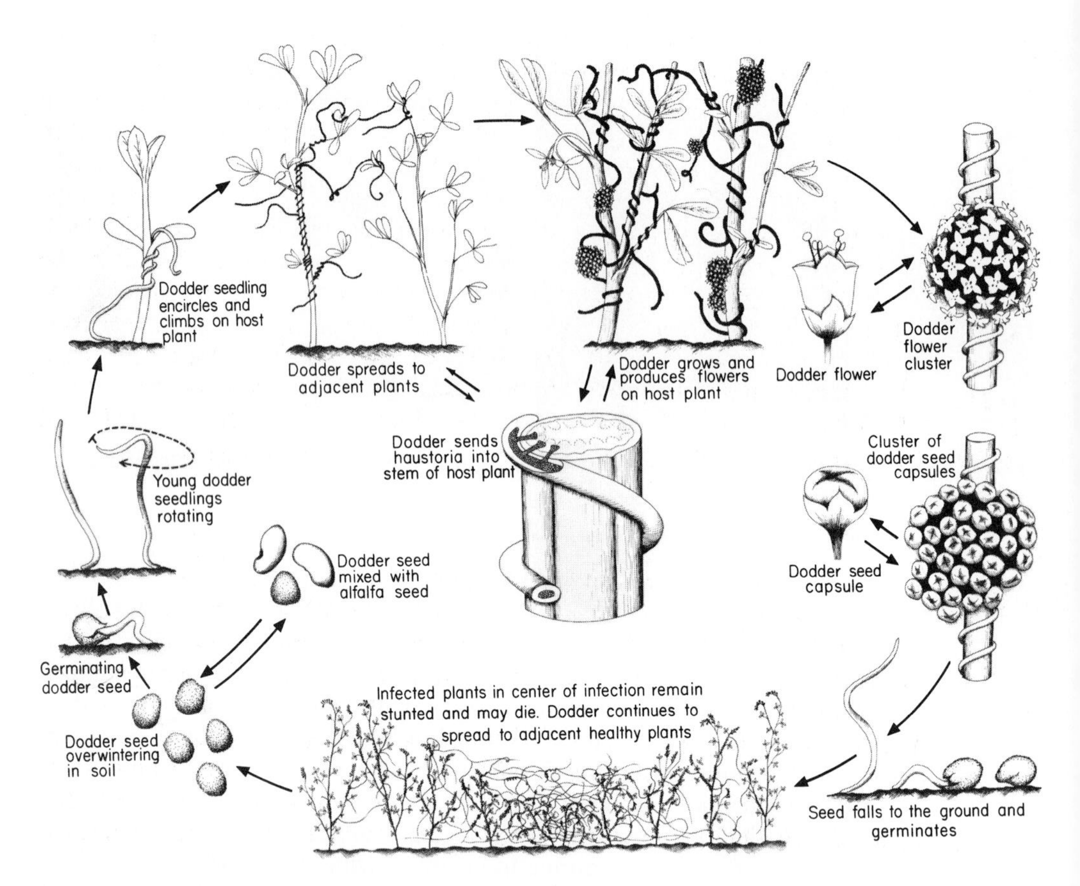

FIGURE 13-2 Disease cycle of dodder *(Cuscuta* sp.) on alfalfa.

If the dodder stem comes in contact with a susceptible host, however, the stem immediately encircles the host plant, sends haustoria into it, and begins to climb the plant. The haustoria penetrate the stem or leaf and reach into the fibrovascular tissues. Foodstuffs and water are absorbed by the haustoria and are transported to the dodder stem, where they are utilized for further growth and reproduction.

Soon after contact with the host is established, the base of the dodder shrivels and dries, so that it loses all connection with the ground and becomes completely dependent on the host for nutrients and water. The dodder continues to grow and expand at the expense of the host and the twisting tips of dodder reach out and attack adjacent plants. Thus the infection spreads from plant to plant, and patches of infected plants are formed. The growth of infected plants is suppressed and they may finally die.

In the meantime, the dodder plant has developed flowers and produced seeds. The seeds fall to the ground where they either germinate immediately or remain dormant until next season. The seed may be spread to new nearby areas by animals, water, and equipment, and over long distances it is distributed mixed with the crop seed.

Control. Dodder is best controlled by preventing its introduction into a field by the use of dodder-free seed, by cleaning equipment thoroughly before moving it from dodder-infested fields to new areas, and by limiting the movement of domestic animals from infested to dodder-free fields.

If dodder is already present in the field, scattered patches may be sprayed early in the season with contact herbicides such as diesel oil fortified with DNBP (4,6-dinitro-*o-sec*-butylphenol), PCP (pentachlorophenol), or 2,4-D. Such treatment, or cutting or burning of patches, kill both the dodder and the host plants but prevent dodder from spreading and from producing seed. When dodder infestations are already widespread in a field, dodder can be controlled by frequent tillage, flaming, and use of soil herbicides such as chloropropham, DCPA, dichlobenil, dinoseb, or pronamide. These chemicals kill the dodder plant upon its germination from the seed but before it becomes attached to the host. Another herbicide, glyphosate, was shown to be effective even when it was applied after dodder was attached to alfalfa and controlled the pathogen without causing apparent injury to alfalfa.

• Witchweed

Witchweed was known as a serious parasitic weed in Africa, Asia, and Australia before 1900. In 1956 the weed was discovered for the first time in America, in North and South Carolina. Because of effective federal and state quarantines, the spread of the parasite has been largely limited to the area of the original infestations.

Witchweed parasitizes important economic plants such as corn, sugarcane, rice, tobacco, and some small grains. Witchweed causes its host plants to become stunted and chlorotic. Heavily infected plants usually wilt and die. Losses vary with the degree of infestation in a field and may range from slight to 100 percent.

◦ *Symptoms.* Affected plants develop symptoms resembling those produced by acute drought. The plants remain stunted, wilt, and turn yellowish. Death may follow these symptoms if the plants are heavily parasitized. Infected roots of host plants bear a large number of witchweed tentacles or haustoria, which are attached to the root and feed upon it. One to several witchweed plants may be growing above ground next to the infected plants, although roots of many more witchweed plants, which do not survive to reach the surface, may parasitize the roots of the same host (Figure 13-3).

◦ *The Pathogen: Striga asiatica.* Witchweed is a small, pretty plant with bright green, slightly hairy stem and leaves. The weed grows 15–30 cm high. It produces multiple branches both near the ground and higher on the plant. The leaves are rather long and narrow in opposite pairs (Figure 13-4).

The flowers are small and usually brick red or scarlet, although some may be yellowish-red, yellowish, or almost white, always having yellow centers. Flowers appear just above the leaf attachment to the stem and are produced throughout the season. After pollination seed pods or capsules develop, each containing more than a thousand tiny brown seeds. A single plant may produce from 50,000 to 500,000 seeds.

The root of witchweed is watery white in color and round in cross section. It has no root hairs, for it obtains all nutrients from the host plant through haustoria.

The life cycle of the parasite, from the time a seed germinates until the developing plant releases its first seeds, takes 90–120 days. Although after

FIGURE 13-3 Witchweeds parasitizing roots of corn plant. (Photo courtesy U.S.D.A.)

emergence the plant turns green and can probably manufacture some of its own food, it appears that it still continues to depend upon the host, not only for all its water and minerals but for organic substances as well.

Development of Disease. The parasite overwinters as seeds, most of which generally require a rest period of 15–18 months before germination, although some can germinate without any dormancy. Seeds within a few millimeters from host roots germinate and grow toward these roots, in response to stimulants contained in the exudates of the host roots. As soon as the witchweed rootlet comes in contact with the host root, its tip swells into a conical or bulb-shaped haustorium, which presses against the host root. The haustorium dissolves host cells through enzymic secretions and penetrates the host roots within 8–24 hours. The haustorium advances into the roots through dissolution of host cell walls. Finally, its leading cells, usually tracheids, reach the vessels of the host roots (Figure 13-4). The tracheids eventually dissolve the vessel walls or force their way into the vessel, from which they absorb water and nutrients. Although xylem vessels are present in the haustorium, no typical phloem cells develop, but cells in the "nucleus" of the haustorium seem to connect the phloem of host and parasite. It has been shown that the chlorophyll of witchweed plants is functional, but still manufactured foodstuffs move from the host plant into the parasite even when the latter is fully developed.

From the initial rootlet the weed produces more roots, which move parallel to the roots of the host plant and send more haustoria into them. Furthermore, several hundred separate witchweed plants may parasitize the roots of a single host plant at once, although relatively few of these survive to reach the surface because the host plant cannot support so many.

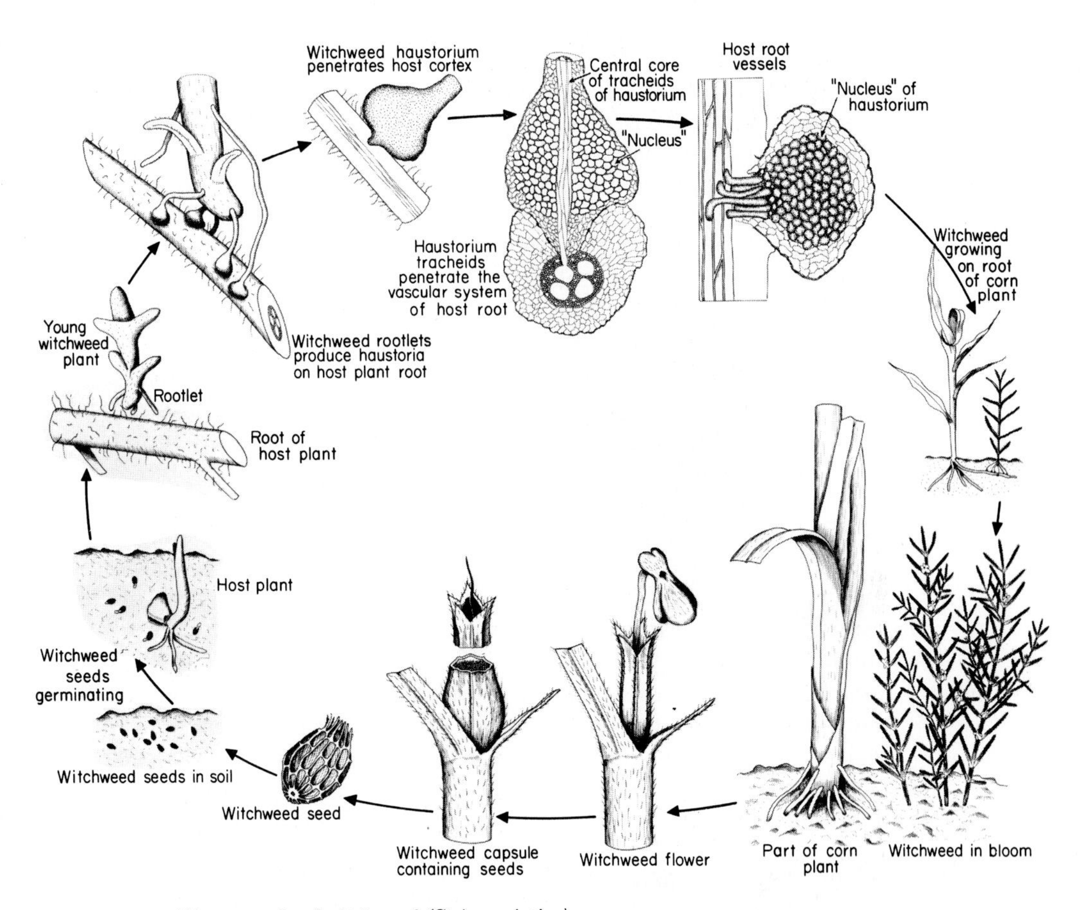

FIGURE 13-4 Disease cycle of witchweed *(Striga asiatica)* on corn.

The disease spreads in the field in a circular pattern. The circle of infected plants, however, increases year after year as the witchweed seeds spread in increasingly larger areas. The seeds are spread by wind, by water, by contaminated tools and equipment, or by contaminated soil carried on farm machinery.

○ *Control.* Witchweed can be controlled by preventing its movement from infested areas into uninfested ones on transplants, agricultural products and machinery, or in any other way. Catch crops, consisting of host plants, may be planted to force the germination of witchweed seed, and the witchweed plants then can be destroyed by plowing under or by the use of weed killers such as 2,4-D. Trap crops, consisting mostly of nonhost legumes, may be used to stimulate germination of witchweed seeds, which, however, cannot infect the trap plants and therefore starve to death. A relatively new approach to witchweed control is the injection of ethylene into fallow soils, which stimulates germination of witchweed seed resulting in subsequent death since no host is present. Several other herbicides, such as paraquat, trifluralin, and oxyfluorfen, also control witchweed directly as well as eliminate many of the host grasses. Usually, a combination of the above methods is required to prevent witchweed plants from flowering and seeding, but with appropriate timing and perseverance an area can be completely freed from the parasite.

• Broomrapes

Broomrapes are widely distributed in Europe, the United States, Africa, and Asia. They attack several hundred species of herbaceous crop plants including tobacco, potato, tomato, hemp, clover, and alfalfa. In some areas of the world, broomrapes cause losses varying from 10 to 70 percent of the crop.

Plants affected by broomrapes usually occur in small patches and may be stunted in various degrees, depending on how early in their lives and by how many broomrapes the host plants were infected.

The broomrape pathogen, *Orobanche* sp., is a whitish to yellowish annual plant that may be 15–50 cm tall. It has a fleshy stem and scalelike leaves and produces numerous pretty, white, yellow-white, or slightly purple, snapdragonlike flowers arising singly along the stem. The broomrapes produce ovoid seed pods about 5 mm long that contain several hundred minute seeds.

Broomrapes overwinter as seeds, which may survive in the soil for more than 10 years. Seeds germinate only when roots of certain plants grow near them, although not all these plants are susceptible to the pathogen. Upon germination the seed produces a radicle, which grows toward the root of the host plant, becomes attached to it, and produces a shallow disk- or cuplike appressorium that surrounds the root, penetrates the host with a mass of undifferentiated, polymorphic cells that extend to and, occasionally, into the xylem, and absorbs nutrients and water from it. Some of these cells differentiate into parasite xylem vessel elements and connect the host xylem with the main vascular system of the parasite. Other polymorphic cells become attached to phloem cells and apparently obtain nutrients from them, which they transport back to the parasite. Soon the parasite begins to develop a stem, which appears above the soil line and looks like an asparagus shoot. Meanwhile, the original root produces secondary roots that grow outward until they come in contact with other host roots to which they become attached and subsequently infect. From these points of contact, new roots and stems of the parasite are produced and result in the appearance of the typical clusters of broomrape plants arising from the soil around infected host plants. Several such broomrapes may be growing concurrently on the roots of the same host plant. The broomrape stems continue to grow and produce flowers and seeds, which mature and are scattered over the ground in less than two months from the emergence of the stems.

Control of broomrapes depends on preventing the introduction of its seeds in new areas, planting nonsusceptible crops in infested fields, frequent weeding and removal of broomrapes before they produce new seed, and where feasible, fumigating soil with methyl bromide. It was reported recently that broomrapes were effectively controlled after treatment with the herbicide glyphosate.

• Dwarf Mistletoes of Conifers

Dwarf mistletoes occur in all parts of the world where conifer trees grow. In the United States they are more prevalent and most serious in the western half of the country, especially in the states along the Pacific coast and in the Southwest, but they also cause appreciable losses in the northeastern states.

The damage caused by dwarf mistletoes in coniferous forests is extensive although not always spectacular. Trees of any age may be retarded, deformed,

FIGURE 13-5 Dwarf mistletoe on ponderosa pine. (A) Female plant (left) and a male plant on a side limb. (B) Swellings and distortion of pine branches parasitized by dwarf mistletoe. (Photos courtesy U. S. Forest Service.)

or killed. The height growth of trees may be reduced by 50 to 80 percent. Timber quality is reduced by numerous large knots and by abnormally grained, spongy wood. Seedlings and saplings, as well as trees of certain species, are frequently killed by dwarf mistletoe infections.

- *Symptoms.* Simple or branched shoots of dwarf mistletoe plants occur in tufts or scattered along the twigs of the host (Figure 13-5). If the shoots have dropped off, small basal "cups" appear on the bark. Infected twigs and branches develop swellings and cankers on the infected areas. Cross sections at the swellings of infected branches reveal yellowish, wedge-shaped haustoria of the parasite, which grow into the bark, cambium, and xylem of the branch. Large swellings or flattened cankers may also develop on the trunks of some infected trees. Infected branches often produce witches'-brooms. Tree stands with light or moderate infections are difficult to distinguish from healthy stands except for the presence of cankers, swellings, and brooms. Heavily infected stands, however, contain deformed, stunted, dying, and dead trees, or trees broken off at trunk cankers.
- *The Pathogen: Arceuthobium sp.* Some species produce shoots up to 10 cm long, while others are usually no more than 1.5 cm.

The mistletoe stems are yellowish to brownish-green or olive green. The shoots may be simple or branched and they are jointed. The leaves are inconspicuous, scalelike, in opposite pairs, and of the same color as the stem. Mistletoe plants also produce a complex, ramifying system of haustoria, which consists of a longitudinal system of strands, external to and more or less parallel to the host cambium, and a radial system of "sinkers" produced by the former and oriented radially into the phloem and xylem.

The plants are either male or female and produce flowers when they are 4–6 years old. After flowering the male shoots die; the female shoots die after the seeds are discharged. Fruits mature 5–16 months after pollination of the flowers. The fruit at maturity is a turgid, elliptical berry. On ripening, the fruit develops considerable internal pressure and, when disturbed, expels the seed upward or obliquely at lateral distances up to 15 meters. The seed is covered with a sticky substance and adheres to whatever it comes in contact with. This is the main means of spread of the parasite, but occasionally, long distance spread occurs when birds transport seed on their bodies.

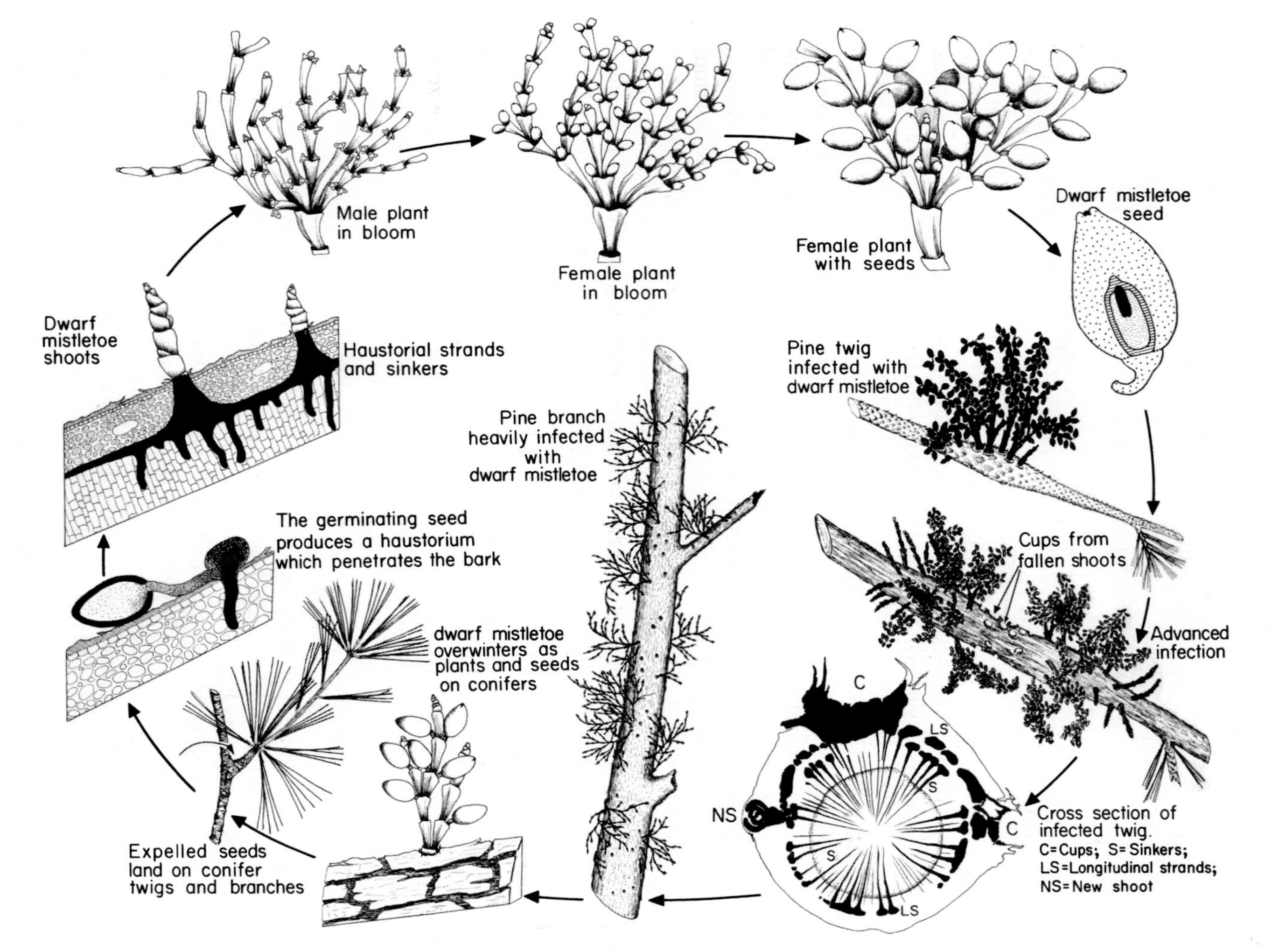

FIGURE 13-6 Disease cycle of dwarf mistletoe *(Arceuthobium sp.)* on conifers.

- *Development of Disease.* When a mistletoe seed lands on and becomes attached to the bark of a twig or a young branch of a susceptible host, it germinates and produces a germ tube or radicle. This grows along the bark surface until it meets a bud or a leaf base, at which point the radicle becomes broad and flattened on the side of the bark. A rootlike haustorium is then produced from the center of the flattened area of the radicle, which penetrates the bark directly and reaches the phloem and the cambium. From this haustorium develops the system of longitudinal strands and radial sinkers, all of which absorb from the host the nutrients needed for the development of the parasite (Figure 13-6). The sinkers that reach the cambium of the host become permanently embedded in the wood as the latter is laid down each year, but they always retain their connections with the strands in the phloem. After the endophytic system is well established and developed in the host, it produces buds from which shoots develop the following year or several years later. The shoots first appear near the original point of infection, but later more shoots emerge in concentric zones of increasing diameter. The center of the infection usually deteriorates and becomes easily attacked by various decay-producing fungi. If witches'-brooms are produced on the affected area, the haustoria pervade all branches and produce mistletoe shoots along the proliferating host branches.

  The parasite removes water, minerals, and photosynthates from the host and so starves and kills the portion of the branch lying beyond the point of infection. It also saps the vitality of the branch and, when sufficiently abundant, of the whole tree. Furthermore, it upsets the balance of hormonal substances of the host in the infected area and causes hypertrophy and hyperplasia of the cells with resulting swellings and deformities of various shapes on the branches. This hormonal imbalance also stimulates the normally dormant lateral buds to excessive formation of shoots, forming a dense growth of abnormal appearance. Heavy dwarf mistletoe infections weaken trees and predispose them to wood-decaying and root pathogens, to beetles, and to windthrow and breakage.
- *Control.* The only means of controlling dwarf mistletoes is by physical removal of the parasite. This is done either by pruning infected branches or by cutting and removing entire infected trees. Uninfected stands can be protected from the dwarf mistletoe infections by maintaining a protective zone free of the parasite beween the diseased stand and the stand to be protected.

• True or Leafy Mistletoes

True or leafy mistletoes occur throughout the world, particularly in warmer climates. They attack primarily hardwood forest and shade trees but also many of the common fruit and plantation trees such as apple, cherry, citrus, rubber, cacao, and coffee, and even some gymnosperms such as juniper and cypress. They cause serious economic losses in some areas, although not nearly as severe as those caused by the dwarf mistletoes.

The symptoms are quite similar to those caused by dwarf mistletoes. Infected areas become swollen and produce witches'-brooms. The mistletoe plants sometimes are so numerous that they make up almost half of the green foliage of the tree, and in the winter, they make deciduous trees appear like evergreens with the normal tree branches appearing as though they have died

FIGURE 13-7 True or leafy mistletoe growing on branch of a hardwood tree.

back. Infected trees may survive for many years, but they show reduced growth and portions of the tree beyond the mistletoe infection often become deformed and die.

The pathogens are *Phoradendron serotinum (flavescens)* in North America and *Viscum album* in California, Europe, and the other continents. These mistletoes are parasitic evergreens that have well-developed leaves and stems less than 1 or 2 cm in diameter. (Figure 13-7). In some species of true mistletoe, however, the stems may be up to 30 cm or more in diameter. The height of plants varies from a few centimeters to a meter or more. The true mistletoes produce typical green leaves that can carry on photosynthesis, usually small, dioecious flowers, and sessile berrylike fruits containing a single seed. They produce haustorial sinkers, rather than roots, however, which grow in branches and stems of trees and absorb water and mineral nutrients.

True mistletoes are spread by birds that eat the seed-containing berries and excrete the sticky seeds in the tops of taller trees on which they like to perch. From that point on, infection, disease development, and control of true mistletoes are almost identical to those of dwarf mistletoes. Control in isolated shade or fruit trees can be obtained by pruning of infected branches or periodic removal of mistletoe stems from the branches or trunks.

SELECTED REFERENCES

Anonymous (1957). Witchweed. *U. S., Agric. Res. Serv., Spec. Rep.* **22–41,** 1–17.

Dawson, J. H., Lee, W. O., and Timmons, F. L. (1969). Controlling dodder in alfalfa. *Farmers' Bull.* **2211,** 1–16.

Eplee, R. E. (1981). *Striga's* status as a plant parasite in the United States. *Plant Dis.* **65,** 951–954.

Garman, H. (1903). Broom-rapes. *Bull.—Ky. Agric. Exp. Stn.* **105:,** 1–32.

Gill, L. S. (1935). *Arceuthobium* in the United States. *Conn. Acad. Arts Sci. Trans.* **32,** 111–245.

Gill, S. L. (1953). Broomrapes, dodders and mistletoes. *In* "Plant Diseases," pp. 73–77. U. S. Dept. of Agriculture, Washington, D. C.

Hansen, A. A. (1921). Dodder *Farmers' Bull.* **1161.**

Hawksworth, F. G., and Wiens, D. (1970). Biology and taxonomy of the dwarf mistletoes. *Annu. Rev. Phytopathol.* **8,** 187–208.

Hoffman, J. T., and Hobbs, E. L. (1985). Lodgepole pine dwarf mistletoe in the intermountain region. *Plant Dis.* **69,** 429–431.

Kuijt, J.(1960). Morphological aspects of parasitism in the dwarf mistletoes *(Arceuthobium). Univ. Calif. Publ. Bot.* **30,** 337–436.

Kuijt, J. (1969). "The Biology of Parasitic Flowering Plants." Univ. of California Press, Berkeley.

Kuijt, J. (1977). Houstoria of phanerogamic parasites. *Annu. Rev. Phytopathol.* **17,** 91–118.

Leonard, O. A. (1965). Translocation relationships in and between mistletoes and their hosts. *Hilgardia* **37,** 115–153.

Musselman, L. J. (1980). The biology of *Striga, Orobanche,* and other root-parasitic weeds. *Annu. Rev. Phytopathol.* **18,** 463–489.

Pennypacker, B. W., Nelson, P. E. and Wilhelm, S. (1979). Anatomic changes resulting from the parasitism of tomato by *Orobanche ramosa. Phytopathology* **69,** 741–748.

Scharpf, R. F., and Hawksworth, F. G. (1974). Mistletoes on hardwoods in the United States. *U.S. For. Serv. For. Pest Leafl.* **147,** 1–17.

Scharpf, R. F., and Parmeter, J. R., Jr. (1967). The biology and pathology of dwarf mistletoe *Arceuthobium campylopodium* f. *abietinum* parasitizing true firs *(Abbies* spp.) in California. *U. S. Dep. Agric., For. Serv., Tech. Bull.* **1362,** 1–42.

Thoday, M. G. (1911). On the histological relations between *Cuscuta* and its host. *Ann. Bot.* **25,** 655–682.

Wilhelm, S. *et al.* (1959). Large-scale fumigation against broomrape. *Phytopathology* **49,** 530.

14 PLANT DISEASES CAUSED BY VIRUSES

Introduction

A **virus** is a nucleoprotein that is too small to be seen with a light microscope, multiplies only in living cells, and has the ability to cause disease. All viruses are parasitic in cells and cause a multitude of diseases to all forms of living organisms, from single-celled microorganisms to large plants and animals. Some viruses attack humans or animals, or both, and cause such diseases as influenza, polio, rabies, smallpox, and warts; others attack plants; and still others attack microorganisms, such as fungi, bacteria, and mycoplasmas. The total number of viruses known to date is about two thousand, and new viruses are described almost every month. About one-fourth of all known viruses attack and cause diseases of plants. One virus may infect one or dozens of different species of plants, and each species of plant is usually attacked by many different kinds of viruses. A plant may also be infected by more than one kind of virus at the same time.

Although viruses are agents of disease and share with other living organisms genetic functions and the ability to reproduce, they also behave as chemical molecules. At their simplest, viruses consist of nucleic acid and protein, with the protein forming a protective coat, called a **capsid,** around the nucleic acid. Although viruses can take any of several forms, they are mostly either rod shaped or polyhedral, or variants of these two basic structures. There is always only RNA or only DNA in each virus and, in most plant viruses, only one kind of protein. Some of the larger viruses, however, may have several different proteins, each probably having a different function.

Viruses do not divide and do not produce any kind of specialized reproductive structures such as spores, but they multiply by inducing host cells to form more virus. Viruses cause disease not by consuming cells or killing them with toxins, but by utilizing cellular substances, taking up space in cells, and by disrupting cellular components and processes, which in turn upset the metabolism of cells and lead to the development by the cell of abnormal substances and conditions injurious to the functions and the life of the cell or the organism.

Characteristics of Plant Viruses

Plant viruses differ greatly from all other plant pathogens not only in size and shape, but also in the simplicity of their chemical constitution and physical structure, methods of infection, multiplication, translocation within the host, dissemination, and the symptoms they produce on the host. Because of their small size and the transparency of their bodies, viruses cannot even be viewed and detected by the methods used for other pathogens. Viruses are not cells, nor do they consist of cells.

Detection

When a plant disease is caused by a virus, individual virus particles cannot be seen with the light microscope, although some virus-containing inclusions or crystals may be seen in virus-infected cells. Examination of sections of cells or of crude sap from virus-infected plants under the electron microscope may or may not reveal viruslike particles. Virus particles are not always easy to find under the electron microscope, and even in the rare cases in which such particles are revealed, proof that the particles are a virus, and that this virus causes the particular disease, requires much additional work and time.

A few plant symptoms, such as oak-leaf patterns on leaves and chlorotic or necrotic ring spots, can be attributed to viruses with some degree of certainty. Most other symptoms caused by viruses resemble those caused by mutations, nutrient deficiencies or toxicities, insect secretions, by other pathogens, and other factors. The determination, therefore, that certain plant symptoms are caused by viruses involves the elimination of every other possible cause of the disease and the transmission of the virus from diseased to healthy plants in a way that would exclude transmission of any other causal agent.

The persent methods for detecting plant viruses involve primarily the transmission of the virus from a diseased to a healthy plant by budding, grafting, or by rubbing with plant sap. Certain other methods of transmission, such as by dodder or insect vectors, are also used to demonstrate the presence of a virus. Most of these methods, however, cannot distinguish whether the pathogen is a virus, a mycoplasma, or a fastidious vascular bacterium; only transmission through plant sap is currently considered as proof of the viral nature of the pathogen. The most definitive proof of the presence of a virus in a plant is provided by its purification, electron microscopy, and most commonly, by serology.

Morphology

Plant viruses come in different shapes and sizes, but they are usually described as elongate (rigid rods or flexuous threads), as rhabdoviruses (bacilluslike), and as spherical (isometric or polyhedral) (Figures 14-1, 14-2, 14-3).

Some elongated viruses like tobacco mosaic virus and barley stripe mosaic virus, have the shape of rigid rods with measurements about 15 × 300 nm and 20 × 130 nm, respectively. Most of the elongated viruses appear as long, thin, flexible threads that are usually 10–13 nm wide and range in length from 480 nm (potato virus X) to 2000 nm (citrus tristeza virus). Many of the elongated viruses seem to occur in particles of differing lengths, and the number given usually represents the length that is more common than any other.

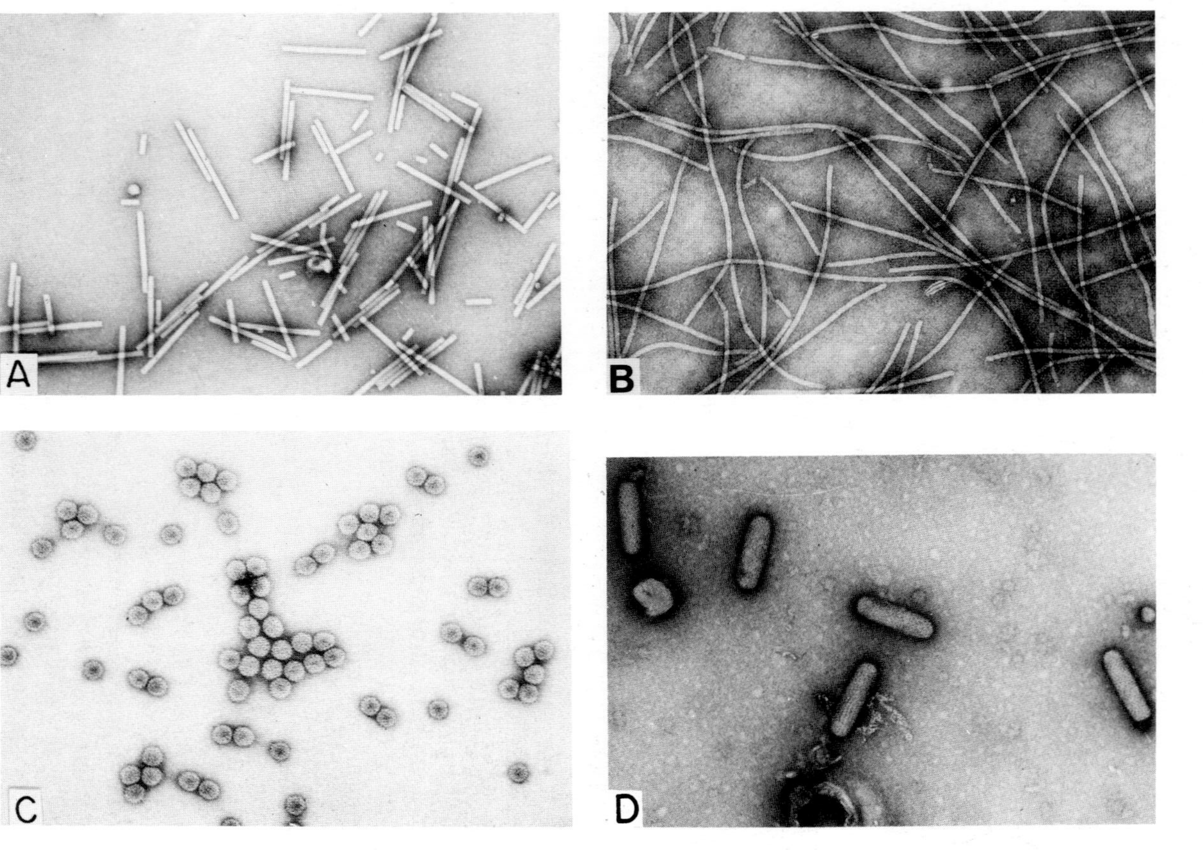

FIGURE 14-1 Electron micrographs of the various shapes of plant viruses. (A) Rod shaped (tobacco mosaic). (B) Flexuous thread (maize dwarf mosaic). (C) Isometric (cowpea chlorotic mottle). (D) Rhabdovirus (broccoli necrotic yellows). (Photo D from Lin and Campbell, *Virology* **48,** 30–40 (1972).)

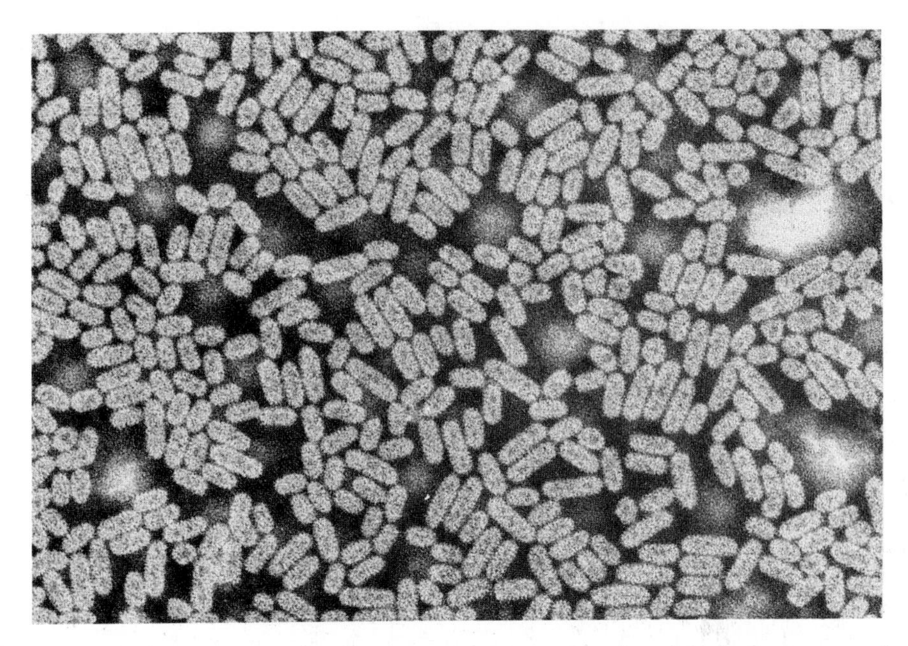

FIGURE 14-2 Electron micrograph of alfalfa mosaic virus showing the various sizes of the five components of this virus. × 168,000. (Photo courtesy E. M. J. Jaspars, University of Leiden, The Netherlands).

The rhabdoviruses are short, bacilluslike rods, approximately three to five times as long as they are wide, as in the cases of potato yellow dwarf virus, which measures 75 × 380 nm, wheat striate mosaic virus (65 × 270 nm), and the lettuce necrotic yellows virus (52 × 300).

Most, and probably all, spherical viruses are actually polyhedral, ranging in diameter from about 17 nm (tobacco necrosis satellite virus) to 60 nm (wound tumor virus). Tomato spotted wilt virus seems to have a flexible, spherical shape 70–80 nm in diameter.

Many plant viruses are split genome viruses, which have two or more distinct nucleic acid strands encapsidated in different-sized particles made of the same protein subunits. Thus, tobacco rattle virus consists of two rods, a long one measuring 195 × 25 nm and a shorter one varying in length from 43 to 110 × 25 nm; alfalfa mosaic virus consists of four components measuring 56 × 18, 43 × 18, 35 × 18, and 30 × 18 nm (Figure 14-2). Also, many isometric viruses have two or three different components of, usually, the same size but different weights as they contain different amounts of nucleic acid. In all the above cases more than one of the components must be present in the plant for the virus to multiply and perform in its usual manner.

The surface of both the elongated and the spherical viruses consists of a definite number of protein subunits, which are spirally arranged in the elongated viruses and packed on the sides of the polyhedral particles of the spherical viruses (Figure 14-3). In cross sections, the elongated viruses appear as hollow tubes with the protein subunits forming the outer coat and the nucleic acid, also spirally arranged, embedded between the inner ends of two successive spirals of the protein subunits. The spherical viruses may or may not be hollow, the visible shell consisting of the protein subunits, with the nucleic acid inside the shell and arranged in an as yet unknown manner.

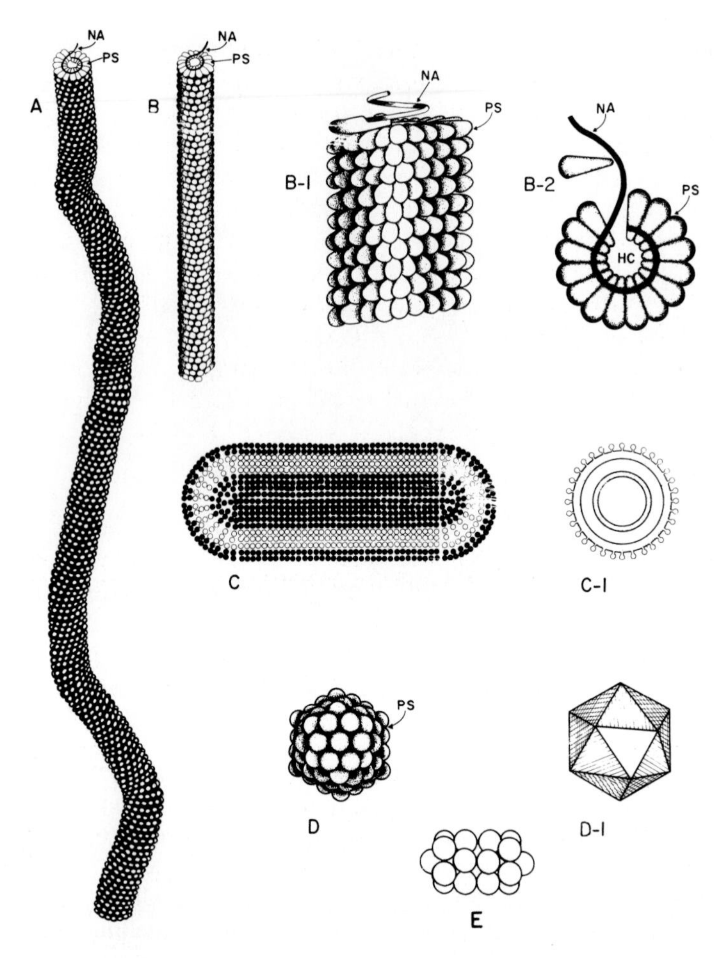

FIGURE 14-3 Relative shapes, sizes, and structures of some representative plant viruses. (A) An elongate virus appearing as a flexous thread. (B) A rigid rod-shaped virus. (B-1) Side arrangement of protein subunits *(PS)* and nucleic acid *(NA)* in viruses A and B. (B-2) Cross section view of the same viruses. *HC* = hollow core. (C) A short, bacilluslike virus. (C-1) Cross-section view of such a virus. (D) A polyhedral virus. (D-1) An icosahedron, representing the 20-sided symmetry of the protein subunits of the polyhedral virus. (E) A geminivirus, consisting of twin particles.

The rhabdoviruses, such as potato yellow dwarf virus, lettuce necrotic yellows, and so on, are provided with an outer envelope or membrane bearing surface projections. Inside the membrane is the nucleocapsid, consisting of helically arranged nucleic acid and associated protein subunits.

Composition and Structure

Each plant virus consists of at least a nucleic acid and a protein. Some viruses consist of more than one size of nucleic acid and proteins, and some of them contain additional chemical compounds, such as polyamines, lipids, or specific enzymes.

The proportions of nucleic acid and protein vary with each virus, nucleic acid making up 5–40 percent of the virus and protein making up the remaining 60–95 percent. The lower nucleic acid and the higher protein percentages are found in the elongated viruses, while the spherical viruses contain higher percentages of nucleic acid and lower percentages of proteins. The total weight of the nucleoprotein of different virus particles varies from 4.6 million molecular weight units (bromegrass mosaic virus) to 39 million (tobacco mosaic virus) to 73 million (tobacco rattle virus). The weight of the nucleic acid alone, however, ranges only between 1 and 3 million ($1-3 \times 10^6$) molecular weight units per virus particle, for most viruses, although some have up to 6×10^6 molecular weight units and the 12-component wound tumor virus nucleic acid weighs approximately 16×10^6 molecular weight units. All of the viral nucleic acid sizes are quite small when compared to 0.5×10^9 for mycoplasmas, 1×10^9 for spiroplasmas, and more than 1.5×10^9 for bacteria.

Composition and Structure of Viral Protein

Viral proteins, like all proteins, consist of amino acids. The sequence of amino acids within a protein is dictated by the genetic material, which in viruses is either deoxyribonucleic acid (DNA) or ribonucleic acid (RNA), and determines the nature of the protein.

The protein components of plant viruses are composed of repeating subunits. The amino acid content and sequence is constant for the identical protein subunits of a virus, but may vary for different viruses, different strains of the same virus, and even for different proteins of the same virus particle. The content and sequences of amino acids are known for the proteins of several viruses. Thus, the protein subunit of tobacco mosaic virus (TMV) consists of 158 amino acids in a constant sequence. Similarly, the protein subunit of turnip yellow mosaic virus (TYMV) has 189 amino acids.

In TMV the protein subunits are arranged in a helix containing $16\frac{1}{3}$ subunits per turn (or 49 subunits per three turns). The central hole of the virus particle down the axis has a diameter of 4 nm, while the maximum diameter of the particle is 18 nm. Each TMV particle consists of approximately 130 helix turns of protein subunits. The nucleic acid is packed tightly between the helices of protein subunits. In the rhabdoviruses the helical nucleoproteins are enveloped in a membrane.

In the polyhedral plant viruses the protein subunits are tightly packed in arrangements that produce 20, or some multiple of 20, facets and form a shell. Within this shell the nucleic acid is folded or otherwise organized.

Composition and Structure of Viral Nucleic Acid

The nucleic acid of most plant viruses consists of RNA, but at least 25 viruses have been shown to contain DNA. Both RNA and DNA are long, chainlike molecules consisting of hundreds or, more often, thousands of units called nucleotides. Each nucleotide consists of a ring compound called the base attached to a 5-carbon sugar (ribose (I) in RNA, deoxyribose (II) in DNA), which in turn is attached to phosphoric acid. The sugar of one nucleotide reacts with the phosphate of another nucleotide, and this is repeated many times, thus forming the RNA or DNA strand. In viral RNA, one of only four bases can be attached to each ribose molecule. These bases are adenine, guanine, cytosine, and uracil. The first two, adenine and guanine, are purines, while cytosine and uracil are pyrimidines. The chemical formulas of the bases and one of their possible relative positions in the RNA chain, are shown in structure (III). DNA is similar to RNA with two small, but very important differences: The oxygen of the sugar hydroxyl is missing, and the base uracil is replaced by the base methyluracil, better known as thymine (IV).

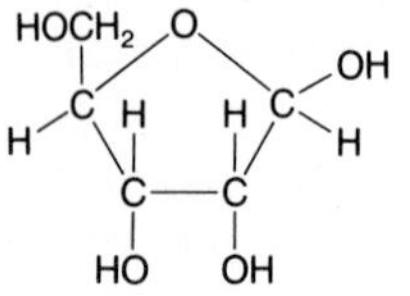

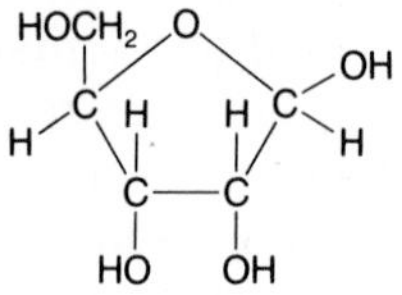

The sequence and the frequency of the bases on the RNA strand vary from one RNA to another, but they are fixed within a given RNA and determine its properties. Healthy cells always contain double-stranded DNA and single-stranded RNA. Most plant viruses (about 400) contain single-stranded RNA, but 10 contain double-stranded RNA, 12 contain double-stranded DNA, and about 15 contain single-stranded DNA.

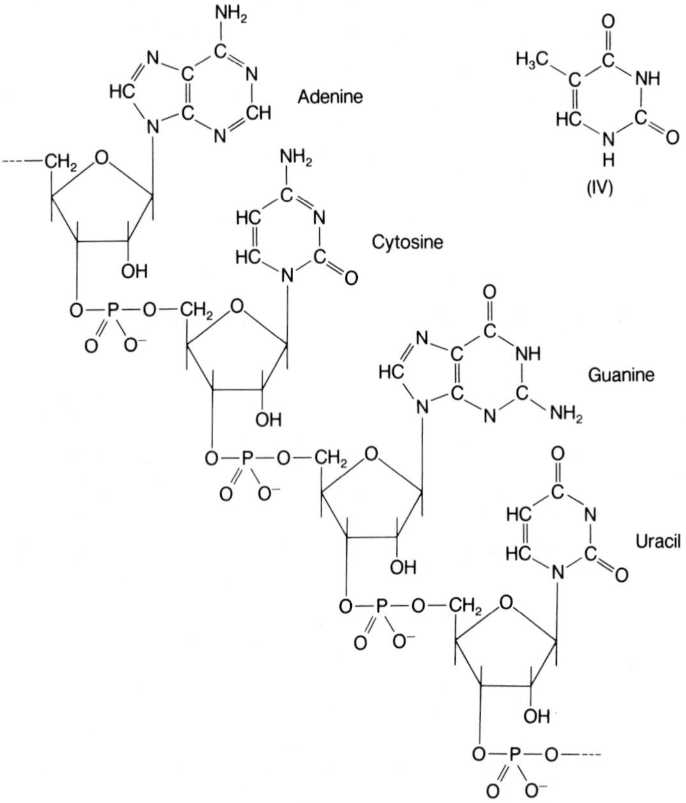

Satellite Viruses, Viroids, Virusoids, and Satellite RNAs

In addition to typical viruses, which consist of one or more rather large strands of nucleic acid contained in a capsid composed of one or more kinds of protein molecules, and which can multiply and cause infection by themselves, at least four other types of viruslike pathogens are associated with plant disease. First, the **satellite viruses,** are viruses associated with certain typical viruses but depend on the latter for multiplication and plant infection and reduce the ability of the typical viruses to multiply and cause disease; that is, satellite viruses act like parasites of the associated typical virus. Second, the **viroids,** are small (250–400 nucleotide), naked, single-stranded, circular RNAs capable of causing disease in plants by themselves (see pages 695–702). Third, the **virusoids,** are viroidlike, small, single-stranded, circular RNAs that are present inside some RNA viruses; virusoids are part of the genetic material of these viruses and therefore form an obligatory association with these viruses so that neither the virus nor the virusoid can multiply and infect a plant in the absence of its partner. Fourth, the **satellite RNAs,** are small, linear RNAs found in virions of certain multicomponent viruses. Satellite RNAs may be related to the RNA of the virus or they may be related to those of the host; satellite RNAs generally attenuate the effects of viral infection and may, possibly, represent a protective response of the host to viral infection.

The Biological Function of Viral Components — Coding

Although apparently each virus produces its own distinct protein coat, the only known function of the protein is to provide a protective sheathing for the nucleic acid of the virus. Protein itself has no infectivity, although its presence generally increases the infectivity of the nucleic acid. In inoculations with intact virus particles (virions), the protein does not seem to assist or to affect the nucleic acid either in its functions or its composition, since inoculations with nucleic acid alone can cause infection and lead to synthesis of new nucleic acid and also of new protein, both being identical with those of the original virus. The synthesis, composition, and structure of the protein, on the other hand, depend entirely on the nucleic acid component, which alone is responsible for the synthesis and assembly of both the RNA and the protein.

The infectivity of viruses in most cases is strictly the property of their nucleic acid, which in most plant viruses is RNA. Some viruses require and carry within them an RNA transcriptase enzyme in order to multiply and infect. The capability, however, of the viral RNA to reproduce both itself and its specific protein, indicates that the RNA carries the genetic determinants of the viral characteristics. The expression of each inherited characteristic depends on the sequence of nucleotides within a certain area (cistron) of the viral RNA, which determines the sequence of amino acids in a particular protein, either structural or enzyme. This is called **coding** and seems to be identical in all living organisms and the viruses.

The code consists of coding units called **codons.** Each codon consists of three adjacent nucleotides and determines the position of a given amino acid.

The amount of RNA, then, contained in each virus indicates the approximate length of, and the number of nucleotides in, the viral RNA. This in turn determines the number of codons in each RNA and, therefore, the number of amino acids that can be coded for. Since the protein subunit of viruses

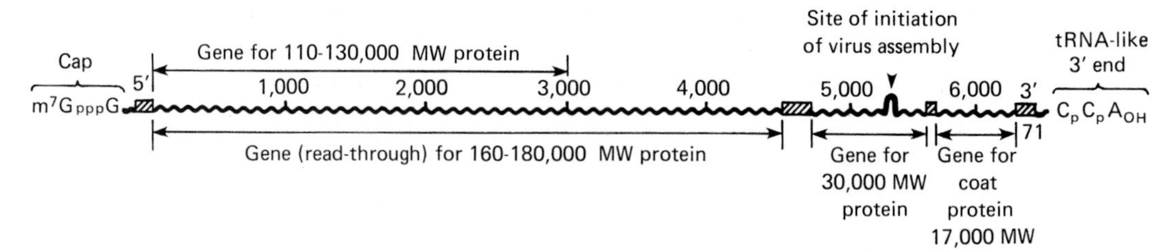

FIGURE 14-4 The 6,400 nucleotide genome of tobacco mosaic virus (TMV). Four genes are translated and produce proteins of 110,000–130,000, 160,000–180,000, 30,000, and 17,000 molecular weight, respectively. The two largest proteins function in the viral replicase(s), the 30,000 protein is thought to facilitate cell-to-cell movement of the virus, and the 17,000 protein makes up the coat protein of the virus. Translation of the viral genome is from left (5′) to right (3′) end. Four short segments of the genome (▨) are not translated and probably include signals for initiation, promotion, and termination of translation. The site of the genome at which assembly with coat proteins takes place to produce complete virus is shown, as are the 5′ cap of the genome and the transfer (tRNA)-like 3′ end. Numbers along the RNA indicate nucleotides.

contains relatively few amino acids (158 in TMV), the number of codons utilized for its synthesis is only a fraction of the total number of codons available (158 out of 2130 in TMV). The remaining codons are presumably involved in the synthesis of other proteins, either structural or enzymes. (Figure 14-4). One of these enzymes is called an RNA-polymerase (RNA-synthetase, or RNA-replicase) and is needed to replicate the RNA of the virus. No specific role is known for most of the other proteins coded for by the viral nucleic acid. So far, it appears that the diseased condition induced in plants by viruses is the result of the interference and disruption of normal metabolic processes caused, in infected parenchyma or specialized cells, by the mere presence and multiplication of the virus and, possibly, by the abnormal or toxic effects of additional virus-induced proteins or their products, although no such substances have been found so far.

Virus Infection and Virus Synthesis

Plant viruses enter cells only through wounds made mechanically or by vectors, or by deposition into an ovule by an infected pollen grain.

The nucleic acid (RNA) of the virus is first freed from the protein coat. It then induces formation by the cell of RNA-polymerase. This enzyme, in the presence of the viral RNA acting as a template and of the nucleotides that compose RNA, produces additional RNA. The first new RNA produced is not the viral RNA but a strand that is a mirror image of that RNA and which, as it is formed, is temporarily connected to the viral strand (Figure 14-5). Thus, the two form a double-stranded RNA that soon separates to produce the original virus RNA and the mirror image (−) strand, the latter then serving as a template for more virus (+strand) RNA synthesis.

The replication of some single-stranded RNA viruses that have parts of their RNA in two or more virus particles, of some rhabdoviruses, and of some double-stranded RNA viruses differs considerably from the above. In viruses in which the different RNA segments are present within two or more virus particles, all the particles must be present in the same cell for the virus to replicate and for infection to develop. In the single-stranded RNA rhabdoviruses the RNA is not infectious because it is the (−) strand. This RNA must be transcribed by a virus-carried enzyme called transcriptase into a (+) strand RNA in the host, and the latter RNA then replicates as above. In the double-stranded RNA isometric viruses, the RNA is segmented within the same virus, is noninfectious, and depends for its replication in the host on a transcriptase enzyme also carried within the virus.

The method of replication of the double-stranded (ds) DNA of plant DNA viruses was at first considered to be identical to the replication of host nuclear DNA. It is now becoming apparent, however, that the replication of dsDNA viruses is a great deal more complex. Briefly, upon infection, the viral dsDNA enters the cell nucleus, where it appears to become twisted and supercoiled and forms a minichromosome. The latter is replicated somewhat and is also transcribed into two single-stranded RNAs: The smaller RNA is transported to the cytoplasm, where it is translated into virus-coded proteins; the larger RNA is also transported to the same location in the cytoplasm, but there it is used as a template for reverse transcription into a complete virion dsDNA, which is promptly encapsidated with protein subunits to form complete virions. The method of replication of the single-stranded (ss) DNA of plant ssDNA viruses has not yet been determined with any degree of certainty. In animal and bacterial ssDNA viruses, however, the ssDNA replicates by forming a rolling circle that produces a multimeric (−) strand, which serves as a template for the production of multimeric (+) strands that are then cleaved to produce unit length (+) strands.

FIGURE 14-5 Hypothetical schematic representation of vital RNA replication.

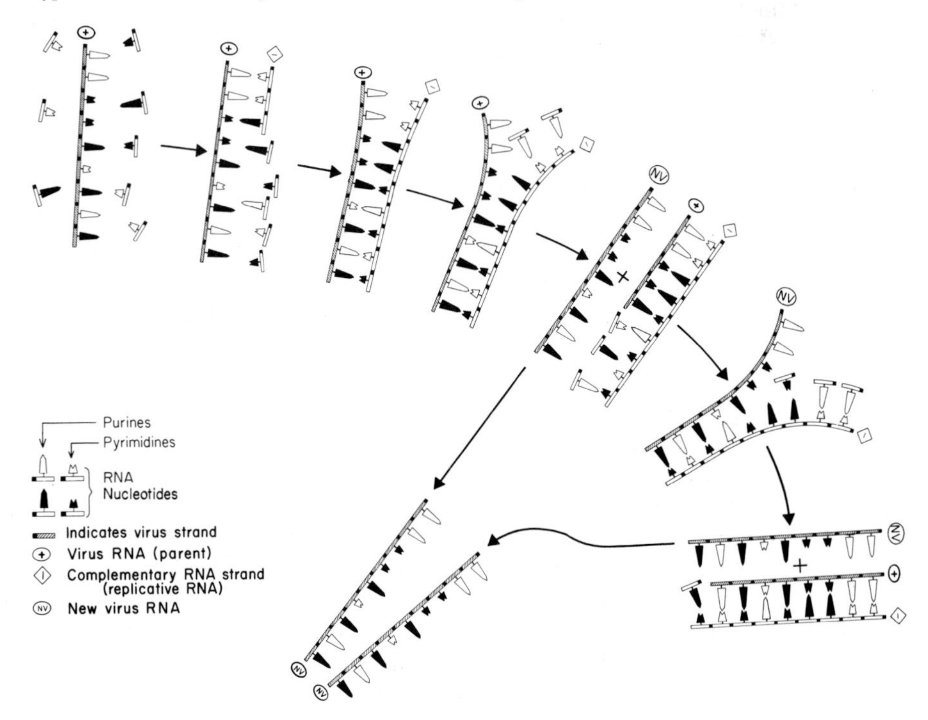

As soon as new viral nucleic acid is produced, some of it is translated, that is, it induces the host cell to produce the protein molecules that will be the protein subunits and that will form the protein coat of the virus. Apparently, only a part of the viral RNA or DNA strand is needed to participate in the formation of the viral protein. Since each amino acid on the protein subunit molecule is "coded" by three nucleotides of the viral RNA, for TMV, whose RNA consists of 6400 nucleotides (Figure 14-4) and its protein of 158 amino acids, only 474 nucleotides are required to code the arrangement of the amino acids in the protein subunit.

Protein synthesis in healthy cells depends on the presence of amino acids and the cooperation of ribosomes, messenger RNA, and transfer RNAs. Each transfer RNA is specific for one amino acid, which it carries toward and along the messenger RNA. Messenger RNA, which is produced in the nucleus and reflects part of the DNA code, determines the kind of protein that will be produced by coding the sequence in which the amino acids will be arranged. The ribosomes seem to travel along the messenger RNA and to provide the energy for the bonding of the prearranged amino acids to form the protein (Figure 14-6)

FIGURE 14-6 Schematic representation of the basic functions in a living cell.

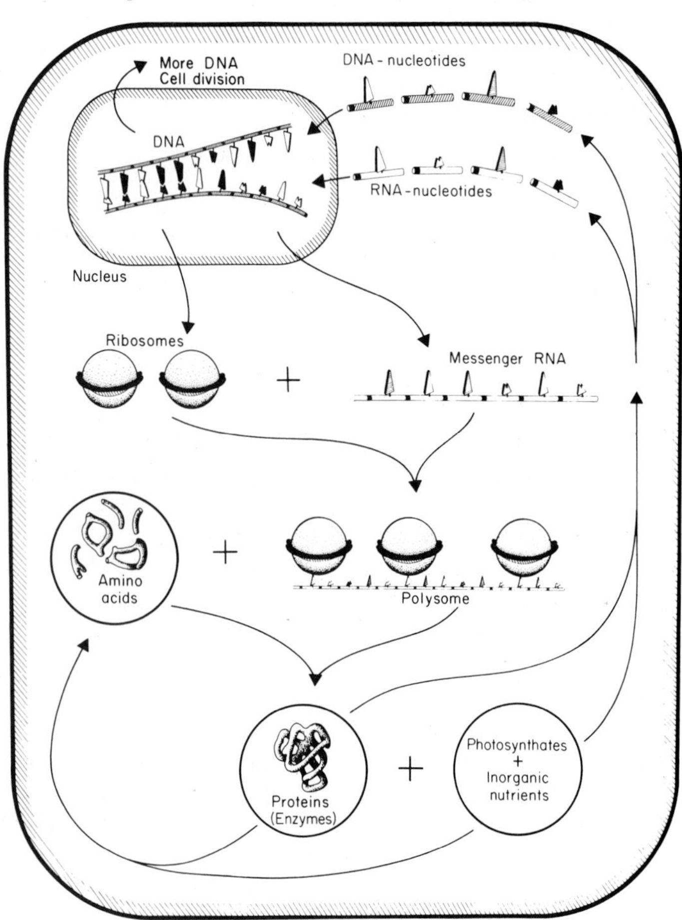

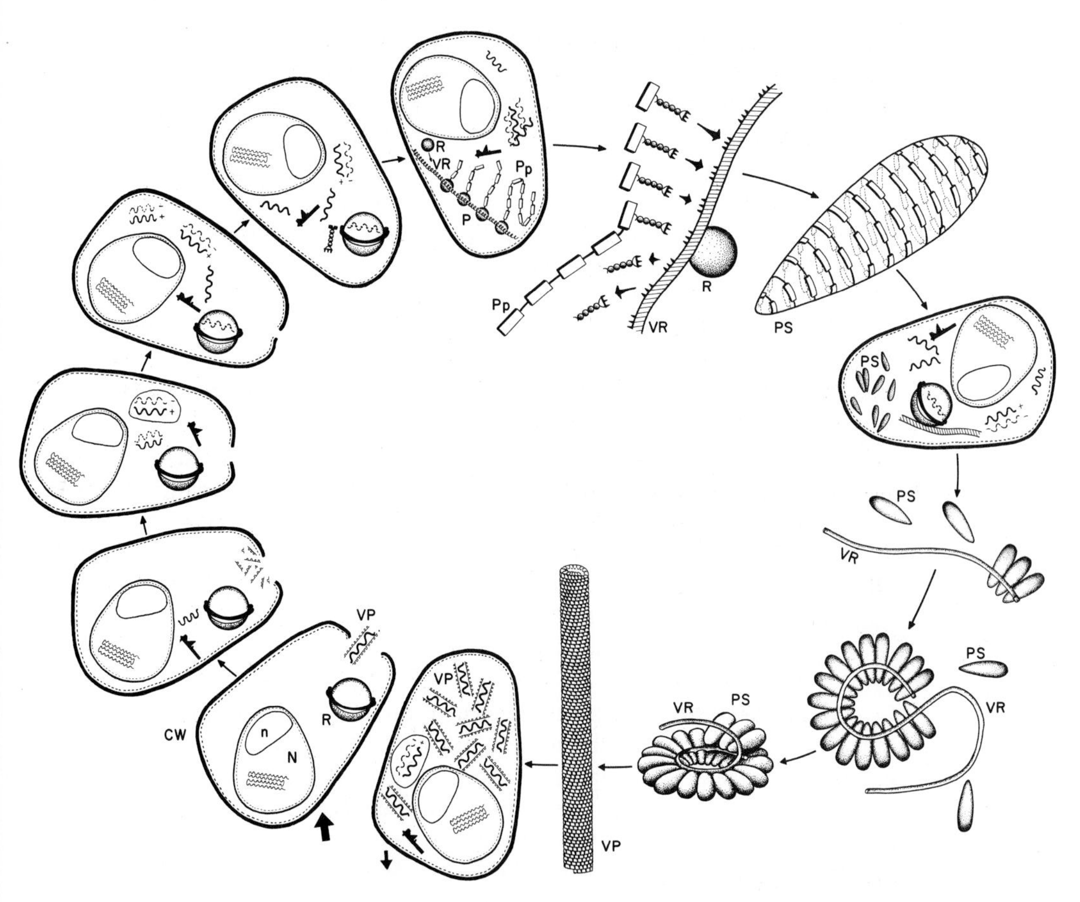

FIGURE 14-7 Sequence of events in virus infection and biosynthesis. CW = cell wall, R = ribosome, N = nucleus, n = nucleolus, P = polyribosome (polysome), Pp = protein subunit, VP = viral particle. ▭ Amino acid, ▬ Viral RNA replicase, ∞∞ Transfer RNA, ∿∿ or VR Viral RNA.

For virus protein synthesis, the part of the viral RNA coding for the viral protein plays the role of messenger RNA. The virus utilizes the amino acids, ribosomes, and transfer RNAs of the host, but it becomes its own blueprint (messenger RNA), and the protein formed is for exclusive use by the virus as a coat (Figure 14-7) or other functions.

During virus synthesis, parts of its nucleic acid also become involved with synthesis of proteins other than the viral coat protein (Figure 14-4). Some of these proteins are enzymes, for example replicases, needed for replication of the viral nucleic acid, but the role of most such proteins is still unknown.

When new virus nucleic acid and virus protein subunits have been produced, the nucleic acid organizes the protein subunits around it, and the two are assembled together to form the complete virus particle, the virion.

The site, or sites, of the cell in which virus nucleic acid and protein are synthesized and in which these two components are assembled to produce the virions varies with the particular group of virus. For most RNA viruses, the virus RNA, after it is freed from the protein coat, replicates itself in the cytoplasm, where it also serves as a messenger RNA and, in cooperation with the ribosomes and transfer RNAs, produces the virus protein subunits. The assembly of virions follows, also in the cytoplasm. In other viruses, for example, those with ssDNA, the synthesis of viral nucleic acid and protein, as well as their assembly into virions, seem to take place in the nucleus, from which the virus particles are then released into the cytoplasm.

The first intact virions appear in plant cells approximately 10 hours after inoculation. The virus particles may exist singly or in groups and may form amorphous or cyrstalline inclusion bodies within the cell areas (cytoplasm, nucleus) in which they happen to be.

Translocation and Distribution of Viruses in Plants

For infection of a plant by a virus to take place, the virus must move from one cell to another and must multiply in most, if not all, cells into which it moves. In their movement from cell to cell, viruses follow the pathways through the plasmodesmata connecting adjacent cells (Figure 14-8). Viruses, however, do not seem to move through parenchyma cells unless they infect the cells and multiply in them, thus resulting in continuous and direct cell-to-cell invasion. In leaf parenchyma cells the virus moves approximately 1 mm, or 8–10 cells, per day.

Although some viruses appear to be more or less restricted to cell-to-cell movement through parenchyma cells, a large number of viruses are known to be rapidly transported over long distances through the phloem. Transport of viruses in the phloem apparently occurs in the sieve tubes, in which they can move fairly rapidly. However, most viruses require 2–5 or more days to move out of an inoculated leaf. Once the virus has entered the phloem, it moves

FIGURE 14-8 Mechanical inoculation and early stages in the systemic distribution of viruses in plants.

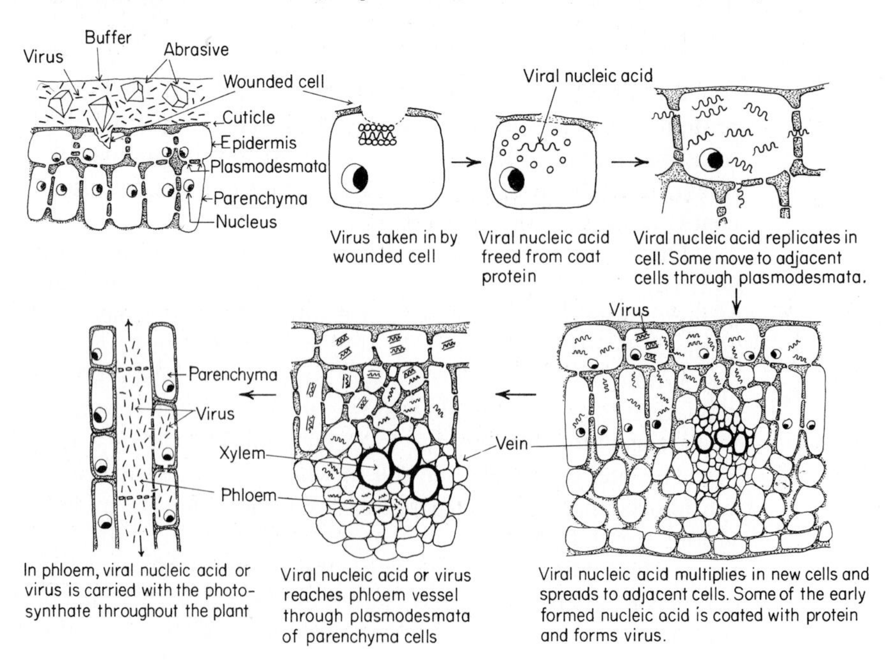

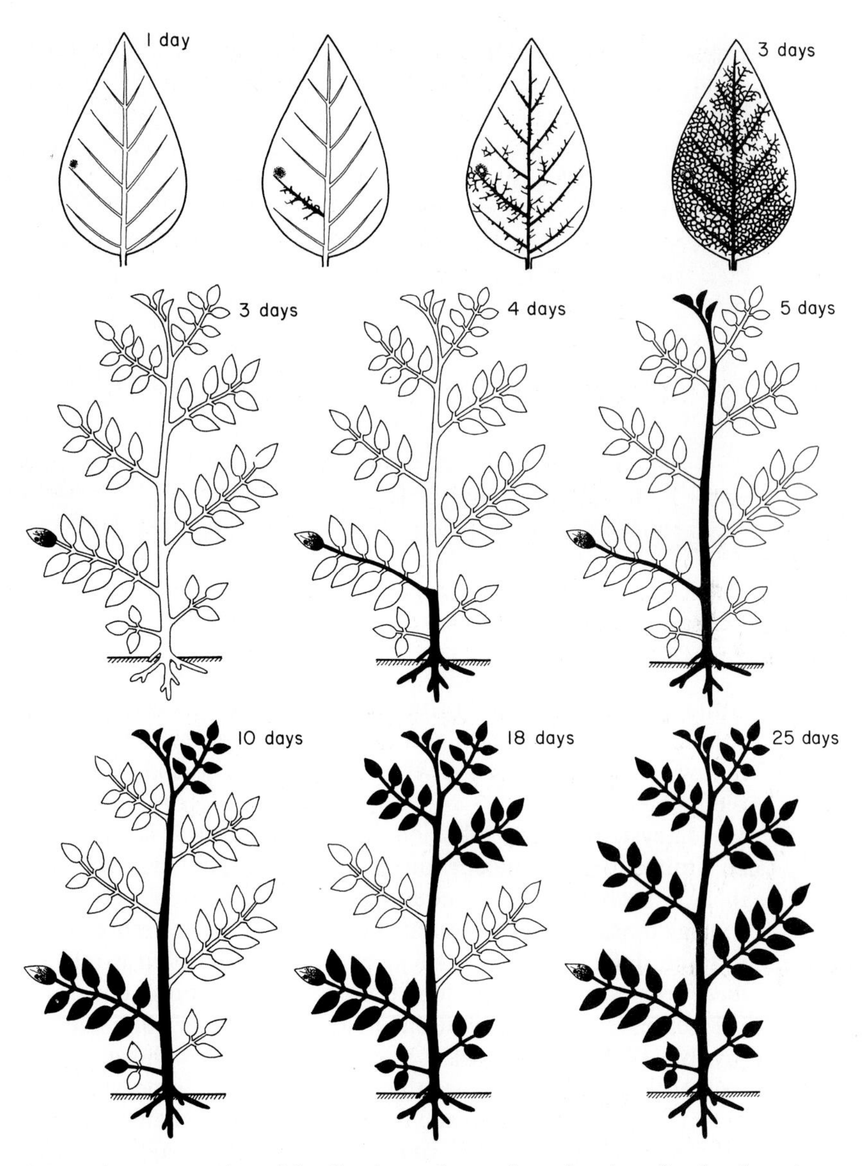

FIGURE 14-9 Schematic representation of the direction and rate of translocation of a virus in a plant. [Adapted from G. Samuel, *Ann. Appl. Biol.* **21,** 90–111 (1934).]

rapidly in the phloem toward growing regions (apical meristems) or other regions of food utilization in the plant, such as tubers and rhizomes (Figure 14-9). For example, when potato virus is introduced into the basal leaves of young potato plants, it moves rapidly up the stem, but when plants already forming tubers are similarly inoculated, the virus does not move upward for more than 30 days while it moves downward into the tubers. Once in the phloem, the virus spreads systemically throughout the plant and reenters the parenchyma cells adjacent to the phloem through plasmodesmata.

The distribution of viruses within plants varies with the virus and the plant. The development of local lesion symptoms has been considered as an indication of the localization of the virus within the lesion area (Figure 14-10).

FIGURE 14-10 Local lesions caused by two strains of a virus (tobacco ring spot) on mechanically inoculated leaves (cowpea).

Although this is probably true in some cases, in several diseases the lesions continue to enlarge and, sometimes, development of systemic symptoms follows, indicating that the virus continued to spread beyond the borders of the lesions.

In systemic virus infections, some phloem-translocated viruses seem to be limited to the phloem and to a few adjacent parenchyma cells. These include such diseases as potato leaf roll and cereal yellow dwarf. Viruses causing mosaic-type diseases are not generally tissue-restricted, although there may be different patterns of localization. Mosaic virus-infected plant cells have been estimated to contain between 100,000 and 10,000,000 virus particles per cell. Systemic distribution of some viruses is quite thorough and may involve all living cells of a plant. Other viruses, however, seem to leave segments or gaps of tissues that are virus free. Some viruses invade newly produced apical meristematic tissues almost immediately, while in other cases growing points of stems or roots of affected plants apparently remain free of virus.

Symptoms Caused by Plant Viruses

The most common and sometimes the only kind of symptom produced by virus infection is reduced growth rate of the plant, resulting in various degrees of dwarfing or stunting of the entire plant. Almost all viral diseases seem to cause some degree of reduction in total yield, and the length of life of virus-infected plants is usually shortened. These effects may be severe and easily noticeable, or they may be very slight and easily overlooked.

The most obvious symptoms of virus-infected plants are usually those appearing on the leaves, but some viruses may cause striking symptoms on the stem, fruit, and roots, with or without symptom development on the leaves (Figure 14-11, A, B). In almost all virus diseases of plants occurring in the field, the virus is present throughout the plant (systemic infection) and the symptoms produced are called systemic symptoms. In many plants inoculated artifically with certain viruses, and probably in some natural infections, the virus causes the formation of small, usually necrotic lesions only at the points of entry (local infections), and the symptoms are called **local lesions.** Many viruses may infect certain hosts without ever causing development of visible symptoms on them. Such viruses are usually called **latent viruses,** and the hosts are called **symptomless carriers.** In other cases, however, plants that usually develop symptoms upon infection with a certain virus may remain temporarily symptomless under certain environmental conditions (for example, high or low temperature), and such symptoms are called **masked.** Finally, plants may show acute or severe symptoms soon after inoculation that may lead to death of the host; if the host survives the initial shock phase, the symptoms tend to become milder (chronic symptoms) in the subsequently developing parts of the plant, leading to partial or even total recovery. On the other hand, symptoms may progressively increase in severity and may result in gradual (slow) or quick decline of the plant.

The most common types of plant symptoms produced by systemic virus infections are **mosaics** and **ring spots.**

Mosaics are characterized by light-green, yellow, or white areas intermingled with the normal green of the leaves or fruit, or of whitish areas intermingled with areas of the normal color of flowers or fruit. Depending on the intensity or pattern of discolorations, mosaic-type symptoms may be described as mottling, streak, ring pattern, line pattern, veinclearing, veinbanding, or chlorotic spotting.

Ring spots are characterized by the appearance of chlorotic or necrotic rings on the leaves and sometimes also on the fruit and stem. In many ring spot diseases the symptoms, but not the virus, tend to disappear after onset and to reappear under certain environmental conditions.

A large number of other less common virus symptoms have been described (Figure 14-11) and include stunt (for example, tomato bushy stunt), dwarf (barley yellow dwarf), leaf roll (potato leaf roll), yellows (beet yellows), streak (tobacco streak), pox (plum pox), enation (pea enation mosaic), tumors (wound tumor), pitting of stem (apple stem pitting), pitting of fruit (pear stony pit), and flattening and distortion of stem (apple flat limb). These symptoms may be accompanied by other symptoms on other parts of the same plant.

Physiology of Virus-Infected Plants

Plant viruses do not contain any enzymes, toxins, or other substances considered to be involved in the pathogenicity of other types of pathogens and yet cause a variety of symptoms on the host. The viral nucleic acid (RNA or DNA) seems to be the only determinant of disease, but the mere presence of viral nucleic acid or complete virus in a plant, even in large quantities, does not seem to be sufficient reason for the disease syndrome, since some plants

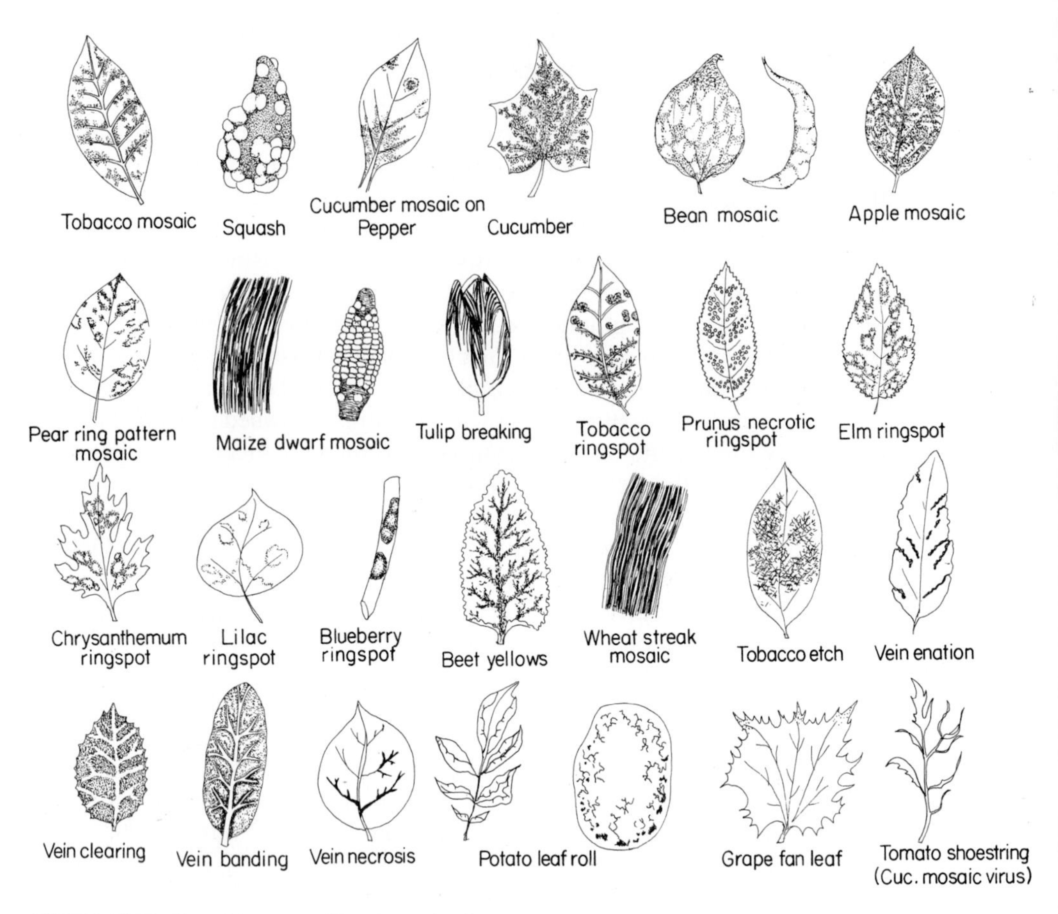

FIGURE 14-11 Symptoms caused by viruses in plants.

containing much higher concentrations of virus than others may show milder symptoms than the latter or they may even be symptomless carriers. This indicates that viral diseases of plants are not due primarily to depletion of nutrients that have been diverted toward synthesis of the virus itself, but to other more indirect effects of the virus on the metabolism of the host. These effects are brought about probably through the virus-induced synthesis of new proteins by the host, some of which are biologically active substances (enzymes, toxins, hormones, and so on) and can interfere with the normal metabolism of the host.

Viruses generally cause a decrease in photosynthesis through a decrease in chlorophyll per leaf, in chlorophyll efficiency, and in leaf area per plant. Viruses usually cause a decrease in the amount of growth-regulating substances (hormones) in the plant, frequently by inducing an increase in growth-inhibiting substances. A decrease in soluble nitrogen during rapid virus synthesis is rather common in virus diseases of plants, and in the mosaic diseases there is a chronic decrease in the levels of carbohydrates in the plant tissues.

Respiration of plants is generally increased immediately after infection with a virus, but after the initial increase the respiration of plants infected with some viruses remains higher, while with other viruses it becomes lower

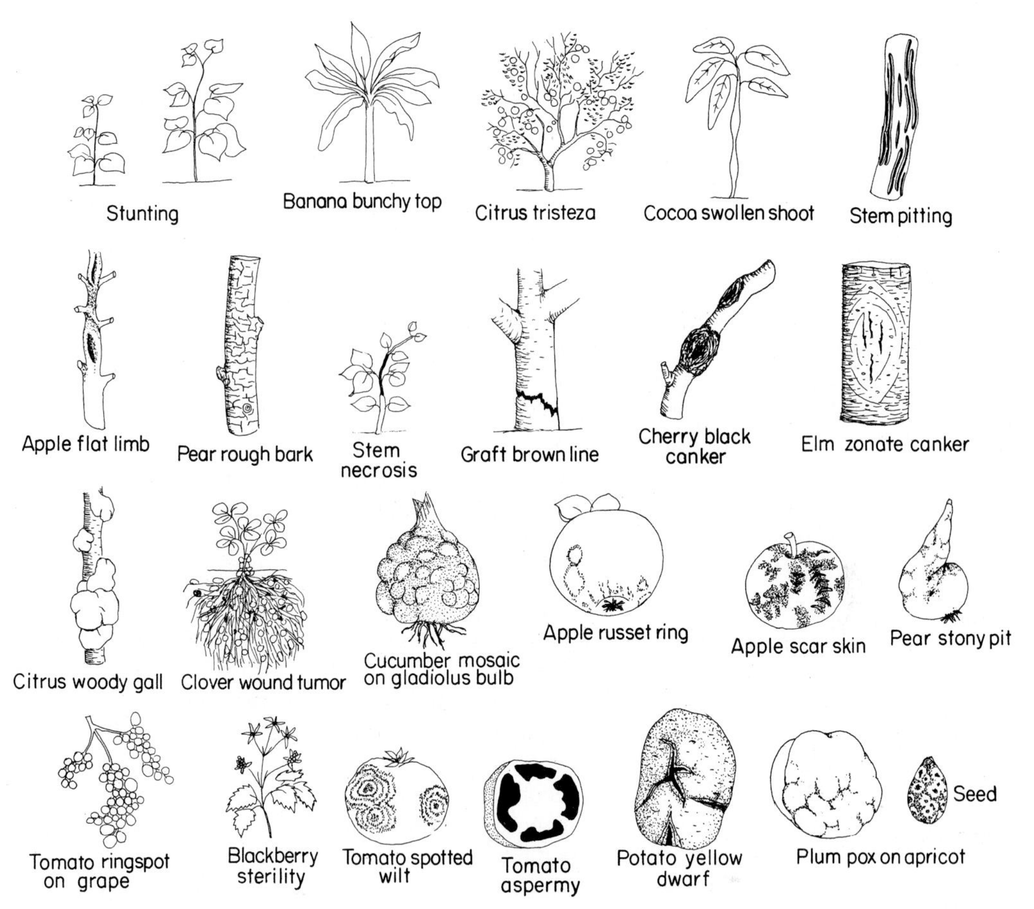

FIGURE 14-11 *(continued)*

than that of healthy plants, and with still other viruses it may return to normal.

The amounts of nonvirus nitrogenous compounds in diseased plants seem to be generally lower than those found in healthy plants, probably because the virus, which in some virus–host systems may account for 33–65 percent of the total nitrogen in the plant, is formed at the expense of the normal levels of nitrogenous compounds in the plant. When the plant, however, is provided with high nitrogen nutrition, the amount of total nitrogen in diseased plants may be higher than that in healthy plants, especially after completion of the phase of rapid virus synthesis.

It appears, therefore, that many of the functional systems of the plant are directly or indirectly affected by virus infection. Certain degrees or types of such metabolic derangements can probably be tolerated by the plant and do not cause any symptoms, while others probably have a deleterious effect on the host and contribute to symptom development. The effects of virus on nitrogenous compounds, on growth regulators, and on phenolics, have often been considered to be the immediate causes of various types of symptoms, since the first two are so profoundly involved in anything concerned with plant growth and differentiation, and since the oxidized products of phenolics may themselves, because of their toxicity, be responsible for the development of certain kinds of necrotic symptoms.

Transmission of Plant Viruses

Plant viruses rarely, if ever, come out of the plant spontaneously. For this reason, viruses are not disseminated as such by wind or water, and even when they are carried in plant sap or debris they generally do not cause infections unless they come in contact with the contents of a wounded living cell. Viruses, however, are transmitted from plant to plant in a number of ways such as vegetative propagation; mechanically through sap; and by seed, pollen, insects, mites, nematodes, dodder, and fungi.

Some of the methods of virus transmission (for example, through vegetative propagation and through seed) are important primarily in the transmission of virus from one plant generation to another but play no role in the spread of virus from diseased to healthy plants of the same plant generation. By themselves, these methods of transmission result only in primary infections of plants and, therefore, only in monocyclic diseases. On the other hand, the other methods of virus transmission, particularly those involving vectors, such as insects, not only bring the virus into a crop (primary infection) but they also result in transmission of the virus from infected to healthy plants within the same plant generation and during the same growth season (secondary infections). The rate of secondary spread of viruses varies with the particular vector, and it increases as the size of the vector population increases and as the weather, as it affects the movement of the vector, remains favorable. Diseases caused by vector-transmitted viruses are polycyclic, the number of disease cycles per season varying from a few (2–5 for nematode-transmitted viruses) to many (10–20 or more, for aphid-transmitted viruses). Of course, when viruses that are transmitted by vegetative propagation or by seed are also transmitted by vectors, the availability of both (large primary inoculum in the crop and effective secondary virus spread by the vectors) often results in early and total infection of the crop plants with subsequent severe losses.

Transmission of Viruses by Vegetative Propagation

Whenever plants are propagated vegetatively by budding or grafting, by cuttings, or by the use of tubers, corms, bulbs, or rhizomes, any viruses present in the mother plant from which these organs are taken will almost always be transmitted to the progeny (Figure 14-12). Considering that almost all fruit and many ornamental trees and shrubs are propagated by budding, grafting, or cuttings, and that many field crops, such as potatoes, and most florist's crops are propagated by tubers, corms, or cuttings, this means of transmission of viruses is the most important for all these types of crop plants. Transmission of viruses by vegetative propagation not only makes the new plants diseased, but in the cases of propagation by budding or grafting, the presence of a virus in the bud or graft may result in appreciable reduction of successful bud or graft unions with the rootstock and, therefore, in poor stands.

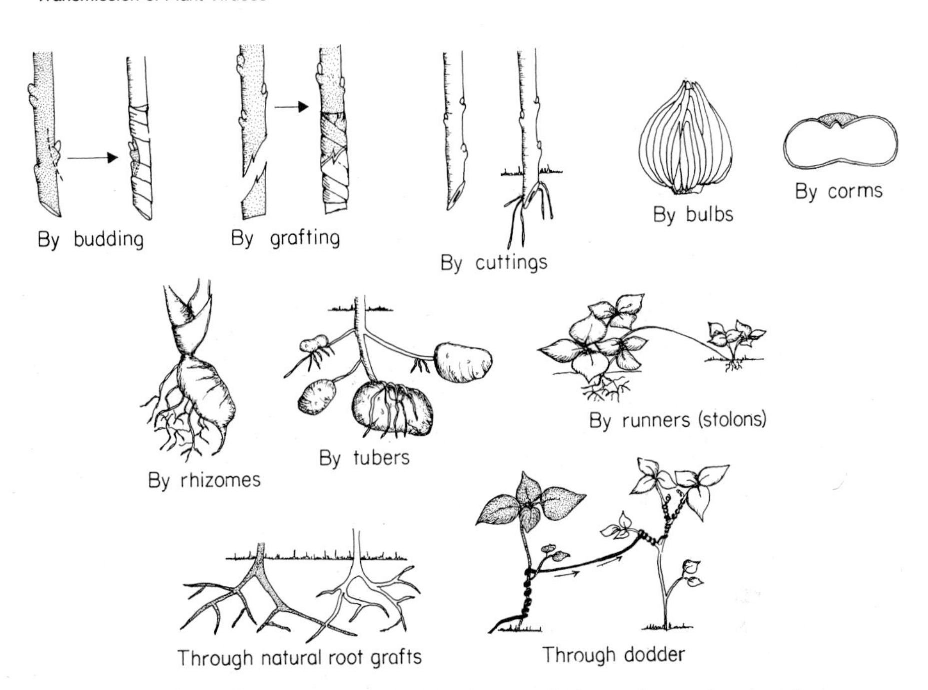

FIGURE 14-12 Transmission of viruses, mycoplasmalike organisms, and other pathogens by vegetative propagation, natural root grafts, and through dodder.

Transmission of viruses may also occur through natural root grafts of adjacent plants, particularly trees, the roots of which are often intermingled and in contact with each other. For several tree viruses, natural root grafts are the only known means of tree-to-tree spread of the virus within established orchards.

Mechanical Transmission of Viruses Through Sap

Mechanical transmission of plant viruses in nature by direct transfer of sap through contact of one plant with another is uncommon and relatively unimportant. Such transmission may take place between closely spaced plants after a strong wind that could cause the leaves of adjacent plants to rub together and, if wounded, to exchange some of their sap, and thus transmit any virus present in the sap (Figures 14-13 and 14-14). Potato virus X (PVX) seems to be one of the viruses most easily transmitted that way. When plants are wounded during cultural practices in the field or greenhouse and some of the virus-infected sap adhering to the tools, hands, or clothes is accidentally transferred to subsequently wounded plants, virus transmission through sap may be rapid and widespread and, as in the case of TMV on tobacco and tomato, may result in serious losses. Virus-infected sap transferred from plant to plant on the mouthparts or body of animals feeding on and moving among the plants may on rare occasions lead to virus transmission.

The greatest importance of mechanical transmission of plant viruses stems from its indispensability in studying almost every facet of the viruses that cause plant diseases, since all investigations of virus outside the host are dependent on the ability to demonstrate and measure the infectiousness of the material.

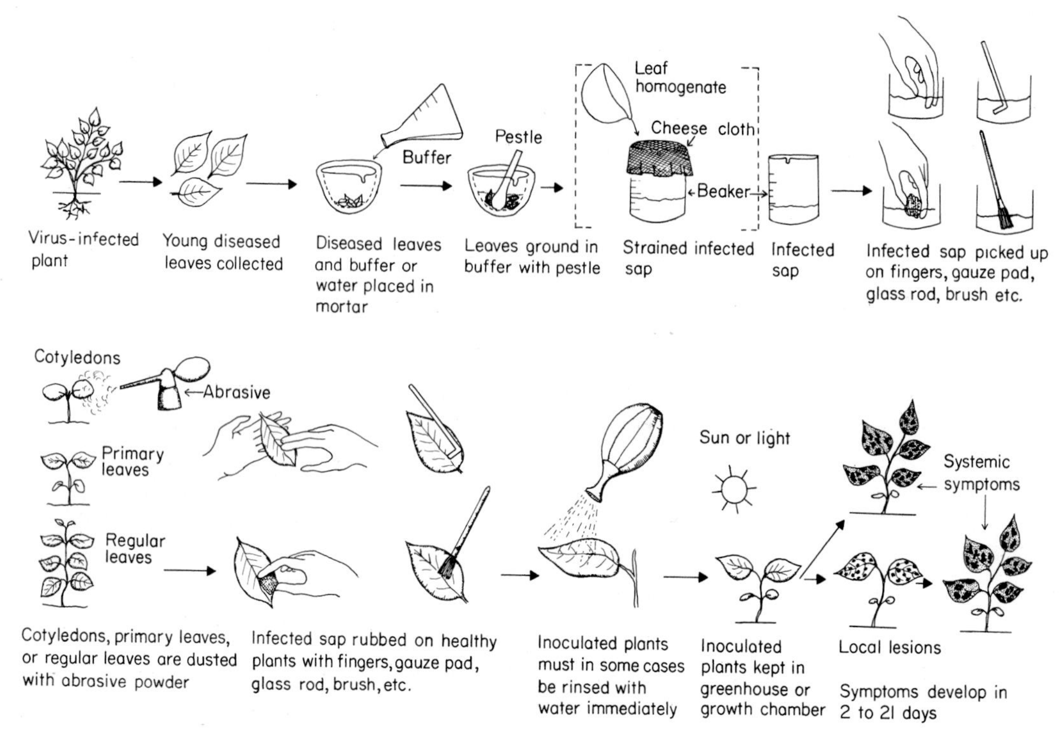

FIGURE 14-13 Typical mechanical or sap transmission of plant viruses.

For mechanical transmission of a virus from one plant to another, tissues of the infected plant believed to contain a high concentration of the virus, that is, young leaves and flower petals, are ground with a mortar and pestle or with some other grinder (Figure 14-13). Breakage of the cells results in release of the virus in the sap. Sometimes a buffer solution, usually phosphate buffer, is added for stabilization of the virus. The expressed sap is then strained through cheesecloth and is centrifuged at low speeds to remove tissue fragments, or at alternate low and high speeds if concentration or purification of the virus is desired. The crude or partially purified sap is then applied to the surface of leaves of young plants that have been previously dusted with an abrasive such as 600-mesh Carborundum added to aid in wounding of the cells. Application of the sap is usually made by gently rubbing the leaves with a cheesecloth or gauze pad dipped in the sap, with the finger, a glass spatula, a painter's brush, or with a small sprayer. In successful inoculations, the virus enters the leaf cells through the wounds made by the abrasive or through broken leaf hairs and initiates new infections. In local-lesion hosts, symptoms usually appear within 3–7 or more days, and the number of local lesions is proportional to the concentration of the virus in the sap. In systemically infected hosts, symptoms usually take 10–14 or more days to develop. Sometimes the same plants may first develop local lesions and then systemic symptoms. In mechanical transmission of viruses, the taxonomic relationship of the donor and receiving (indicator) plants is unimportant, since virus from one kind of plant, whether herbaceous or a tree, may be transmitted to dozens of unrelated herbaceous plants (vegetables, flowers, or weeds).

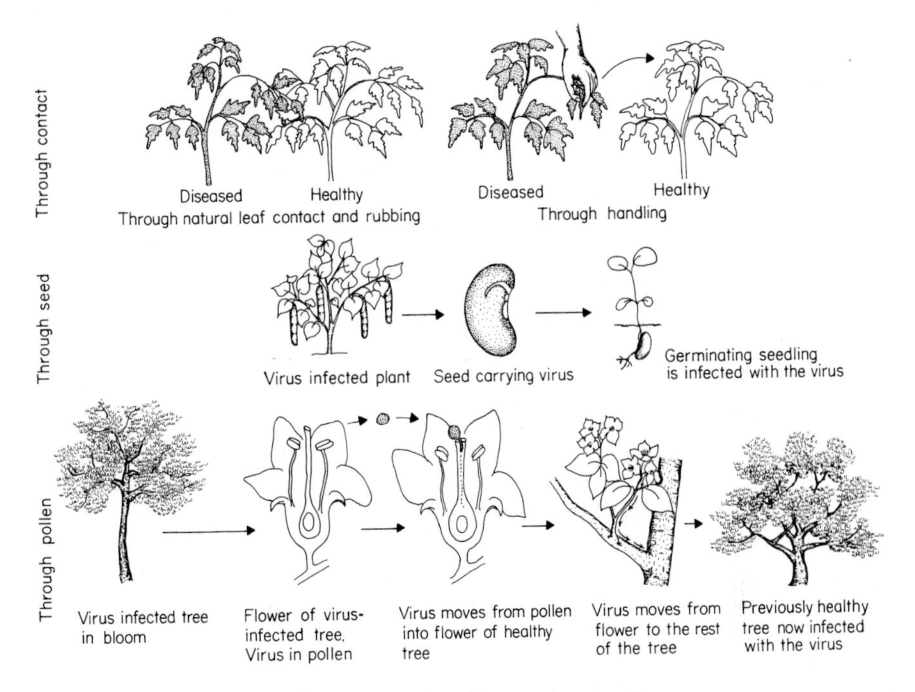

FIGURE 14-14 Virus transmission through direct contact, handling, seed, and pollen.

Although viruses are almost always transmitted by budding or grafting, several viruses, especially of woody plants, have not yet been transmitted mechanically. The possible reasons for this failure seem to be that some viruses are not present in high enough concentration in the donor plant, they are unstable in sap or are quickly inactivated by inhibitory substances released or formed upon grinding of the cells, and also because some viruses, for example, those causing yellows-type diseases, apparently require that they be introduced into specific tissues (phloem) if they are to cause infection.

Seed Transmission

About one hundred viruses have been reported to be transmitted by seed. As a rule, however, only a small portion (1–30 percent) of the seeds derived from virus-infected plants transmit the virus, and the frequency varies with the host-virus combination (Figure 14-14). In a few cases, such as tobacco ringspot virus in soybean, the virus may be transmitted by almost 100 percent of the seeds of infected plants, and in others, seed transmission may be quite high, for example, 28–94 percent in squash mosaic virus in muskmelon and 50–100 percent in barley stripe mosaic virus in barley. Even within a species, however, different varieties or plants inoculated at different stages of their growth may vary in the percentages of their seeds that transmit the virus.

In most seed-transmitted viruses, the virus seems to come primarily from the ovule of infected plants, but several cases are known in which the virus in the seed seems to be just as often derived from the pollen that fertilized the flower.

Pollen Transmission

Virus transmitted by pollen may infect not only the seed and the seedling that will grow from it, but more important, it can spread through the fertilized flower and down into the mother plant, which thus becomes infected with the virus (Figure 14-14). Such plant-to-plant transmission of virus through pollen is known to occur, for example, in stone fruit ring spot virus in sour cherry.

Although pollination of flowers with virus-infected pollen may result in considerably lower fruit set than is produced with virus-free pollen, transmission of pollen-carried virus from plant to plant is apparently quite rare or it occurs with only a few of the viruses.

Insect Transmission

Undoubtedly the most common and economically most important means of transmission of viruses in the field is by insect vectors. Members of relatively few groups of insects, however, can transmit plant viruses (Figure 14-15). The order Homoptera, which includes both aphids (Aphidae) and leafhoppers (Cicadellidae or Jassidae), contains by far the largest number and the most important insect vectors of plant viruses. Certain species of several other families of the same order also transmit plant viruses, but neither their numbers nor their importance compares with the Aphidae and Cicadellidae.

FIGURE 14-15 Insect vectors of plant viruses. Insects in second row also transmit mycoplasmas and fastidious vascular bacteria.

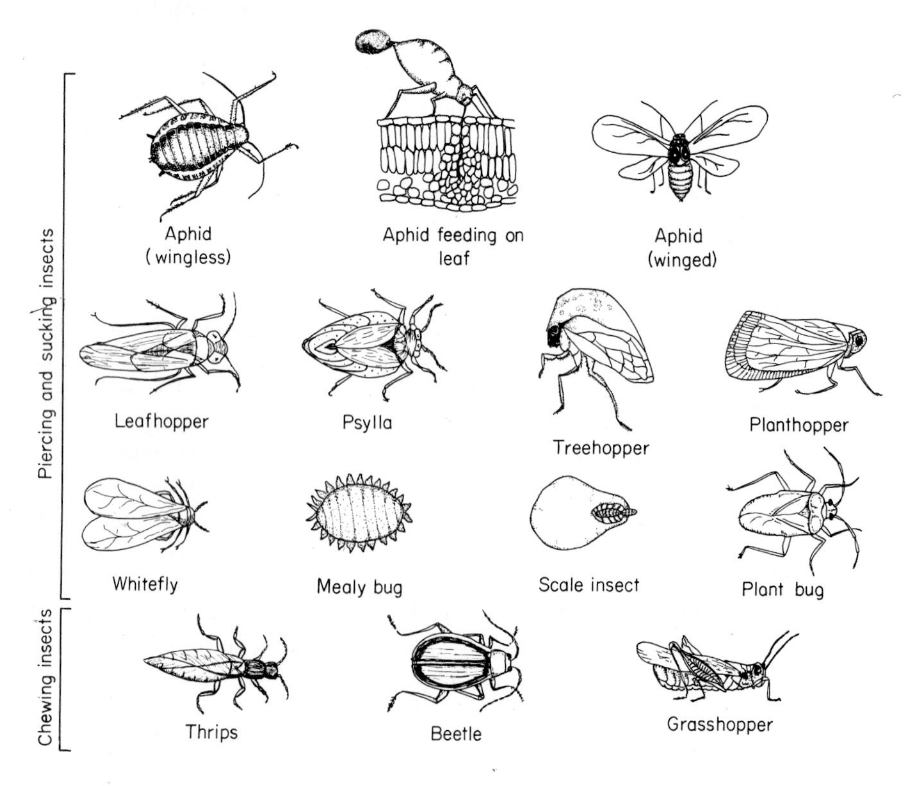

FIGURE 14-16 A winged aphid vector of a plant virus (barley yellow dwarf) sucking up juices, and possibly virus, from an oat stem. (Photo courtesy Department of Plant Pathology, Cornell University.)

Among these families are the white flies (Aleurodidae), the mealy bugs and scale insects (Coccoidae), and the treehoppers (Membracidae). A few insect vectors of plant viruses belong to other orders, such as the true bugs (Hemiptera), the thrips (Thysanoptera), the beetles (Coleoptera), and the grasshoppers (Orthoptera). The most important virus vectors, that is, aphids, leafhoppers, and the other groups of Homoptera, as well as the true bugs, have piercing and sucking mouthparts; all the other groups of insect vectors have chewing mouthparts, and virus transmission by the latter is much less common.

Insects with sucking mouthparts carry plant viruses on their stylets—**style-borne** or **nonpersistent viruses**—or they accumulate the virus internally, and after passage of the virus through the insect tissues, they introduce the virus into plants again through their mouthparts—**circulative** or **persistent viruses.** Some circulative viruses may multiply in their respective vectors and are then called **propagative viruses.** Viruses transmitted by insects with chewing mouthparts may also be circulative or they may be carried on the mouthparts.

Aphids are the most important insect vectors of plant viruses and transmit the great majority (about 170) of all stylet-borne viruses (Figure 14-16). As a rule several aphid species can transmit the same stylet-borne virus and the same aphid species can transmit several viruses, but in many cases the vector–virus relationship is quite specific. Aphids generally acquire the stylet-borne virus after feeding on a diseased plant for only a few seconds (30 seconds or less) and can transmit the virus after transfer to and feeding on a healthy plant for a similarly short time of a few seconds. The length of time aphids remain viruliferous after acquisition of a stylet-born virus varies from a few minutes to several hours, after which they can no longer transmit the virus. Stylet-borne viruses are said to be transmitted in the nonpersistent manner. In the few cases of aphid transmission of circulative viruses, aphids cannot transmit the virus immediately but must wait several hours after the acquisition feeding, but once they start to transmit the virus, they continue to do so for many days after the removal of the insects from the virus source (persistent transmission). In aphids transmitting style-borne viruses, the virus seems to be borne on the tips of the stylets, it is easily lost through the scouring that occurs during probing of host cells, and it does not persist through the molt or egg.

At least 40 plant viruses are transmitted by leafhoppers, including viruses with double-stranded RNA, bacilliform viruses, and small isometric viruses.

Leafhopper-transmitted viruses cause disturbances in plants that arise primarily in the region of the phloem. All leafhopper-transmitted viruses are circulatory, several are known to multiply in the vector (propagative), and some persist through the molt and are transmitted to a greater or lesser degree through the egg stage of the vector. Most leafhopper vectors require a feeding period of one to several days before they become viruliferous, but once they have acquired the virus they may remain viruliferous for the rest of their lives. There is usually an incubation period of 1–2 weeks between the time a leafhopper acquires a virus and the time it can transmit it for the first time.

Mite Transmission

Mites of the family Eriophyidae have been shown to transmit nine viruses, including wheat streak mosaic, peach mosaic, and fig mosaic viruses. These mites have piercing and sucking mouthparts (Figure 14-17). Virus transmission by eriophyid mites seems to be quite specific, since each of these mites has a restricted host range and is the only known vector for the virus or viruses it transmits. Some of the mite-transmitted viruses are stylet borne, while others are circulatory, and of the latter, at least one persists through the molts.

Nematode Transmission

Approximately twenty plant viruses have been shown to be transmitted by one or more species of four genera of soil-inhabiting, ectoparasitic nematodes

FIGURE 14-17 Transmission of plant viruses by nematodes, mites, and fungi.

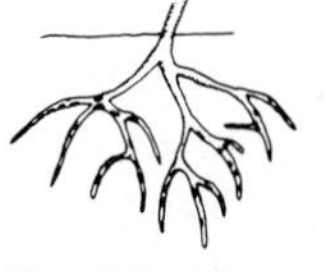

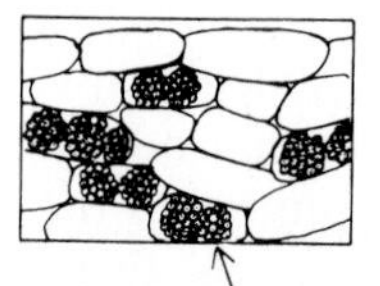

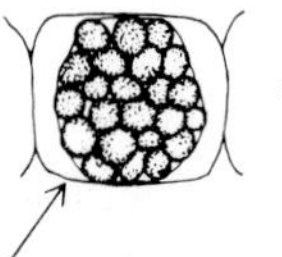

(Figure 14-17). Nematodes of the genera *Longidorus* and *Xiphinema* are vectors of polyhedral-shaped viruses such as tobacco ring spot, tomato ring spot, raspberry ring spot, tomato black ring, cherry leaf roll, brome mosaic, grape fanleaf, and other viruses, while nematodes of the genera *Trichodorus* and *Paratrichodorus* transmit two rod-shaped viruses, tobacco rattle, and pea early browning viruses. Nematode vectors transmit viruses by feeding on roots of infected plants and then moving on to roots of healthy plants. Larvae as well as adult nematodes can acquire and transmit viruses, but the virus is not carried through the larval molts or through the eggs, and after molting, the larvae or the resulting adults must feed on a virus source before they can transmit again.

Fungus Transmission

The root-infecting fungus *Olpidium* transmits at least four plant viruses: tobacco necrosis, cucumber necrosis, lettuce big vein, and tobacco stunt viruses. Another fungus, *Polymyxa,* transmits wheat mosaic virus and beet necrotic yellow vein virus, while a third fungus, *Spongospora,* transmits potato mop top virus. Some of these viruses apparently are borne internally in, and others are carried externally on, the resting spores and the zoospores, which upon infection of new host plants introduce the virus and cause symptoms characteristic of the virus they transmit (Figure 14-17).

Dodder Transmission

Several plant viruses can be transmitted from one plant to another through the bridge formed between the two plants by the twining stems of the parasitic plant dodder (*Cuscuta* sp.) (Figure 14-12). A large number of viruses have been transmitted in this way, frequently between plants belonging to families widely separated taxonomically. The virus is usually transmitted passively in the food stream of the dodder plant, being acquired from the vascular bundles of the infected plant by the haustoria of dodder and, after translocation through the dodder phloem, it is introduced in the next plant by the new dodder haustoria produced in contact with the vascular bundles of the inoculated plant.

Purification of Plant Viruses

Isolation or, as it is usually called, purification of viruses is most commonly obtained by ultracentrifugation of the plant sap. This involves 1–3 cycles of alternate high (40,000–100,000 g or more) and low (3000–10,000 g) speeds. Ultracentrifugation concentrates the virus and separates it from host cell contaminants. Several modifications of the ultracentrifugation technique, particularly density-gradient centrifugation, are currently employed in virus purification with excellent results (Figure 14-18). In all these methods, the virus is finally obtained as a colorless pellet in a test tube and may be used for infections, electron microscopy, and serology.

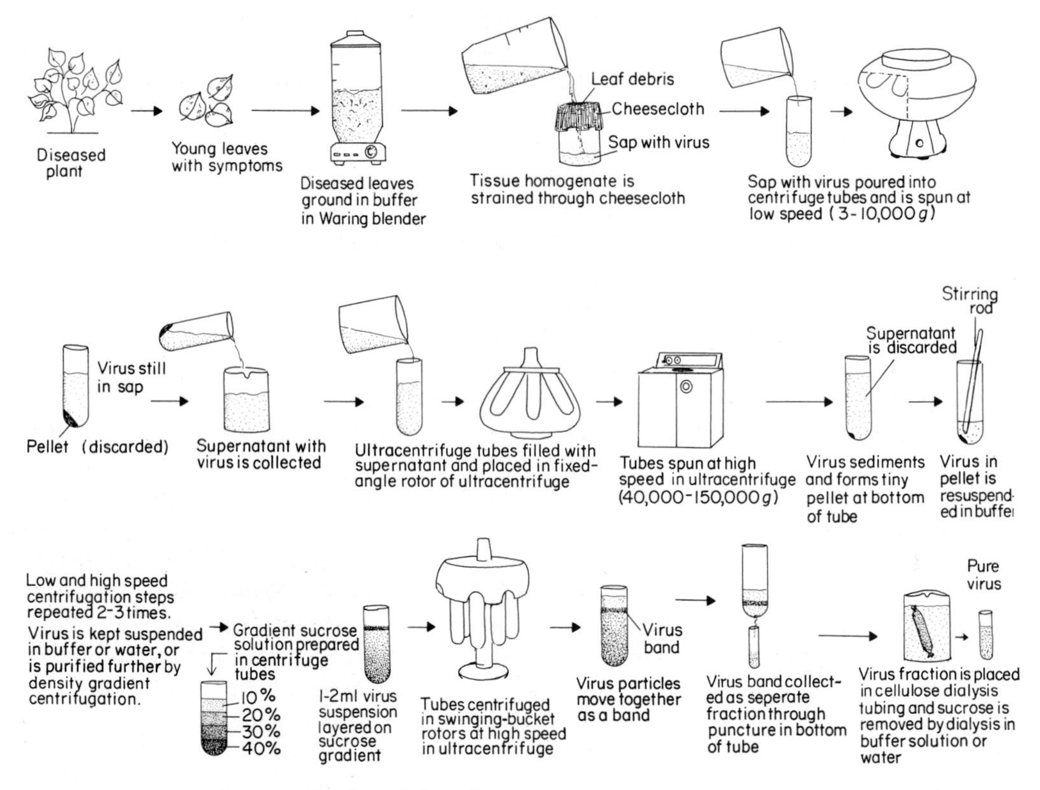

FIGURE 14-18 Steps in the purification of plant viruses.

Serology of Plant Viruses

When an **antigen,** that is, a virus protein or any other foreign protein, is injected into a mammal (rabbit, mouse, horse), or bird (chicken, turkey), it results in the appearance of specific new proteins called **antibodies** in the blood fluid—**serum**—of the animal. The antibodies react specifically with and bind to a small area, known as the **antigenic determinant,** of the antigen injected. Each antigen, such as a virus, has many different antigenic determinants (distinct groups of 6–10 amino acids) at its surface, and since each of them prescribes the production of a different kind of antibody, the **antiserum** (serum containing antibodies) of the animal contains a mixture of many different antibodies. Such mixtures of antibodies are called **polyclonal antibodies,** each antibody reacting with the antigen but at a different surface locality. Recently, it has been possible to produce pure lines (clones) of antibodies that react only with a single antigenic determinant of a protein (or a pathogen), and such antibodies are called **monoclonal antibodies.** Production of monoclonal antibodies is possible because each cell of the immune system, for example, of the spleen, of the animal is capable of producing many copies of only a single kind of antibody. Such cells, unfortunately, do not divide, and therefore their usefulness is limited. If such an antibody-producing cell is fused with a mouse myeloma (cancer) cell, however, it produces a hybrid cell that can grow in culture indefinitely and continues to produce monoclonal antibodies for a long time. Such antibody-producing hybrid cells are called **hybridomas** and can be grown in culture for months or years and produce large quantities of identical, monoclonal antibodies. Monoclonal antibodies can be obtained in high concentration and purity from the liquid of hybridoma cultures and can be used to detect, identify, and measure the antigen that induced their production. Monoclonal antibodies, however, are very specific and may not detect even strains of the same virus that happen to lack the specific antigenic determinant responsible for the monoclonal antibody. For this reason, mixtures of several monoclonal antibodies are often used in virus detection screening tests.

The virus and its antibody are brought together in several ways, the most common being the precipitin reaction. In this, the antibodies and antigens are mixed in solution—**precipitin test**—or they meet at the interface between two solutions containing each separately—**ring interface test**—or they diffuse toward each other through an agar gel and meet in a zone in suitable concentrations—**Ouchterlony gel diffusion test**. Sometimes the antigen is absorbed on the surface of a large particle such as a cell, plastid, or a latex sphere, and these are precipitated by addition of antibodies. This is known as **agglutination reaction.** In all these tests the reaction of antigen and antibody becomes visible either by precipitation of the two on the bottom of the test tube or by formation of a band at the interface where the two meet (Figure 14-19).

FIGURE 14-19 Production of antisera and serological tests for identification of unknown pathogens.

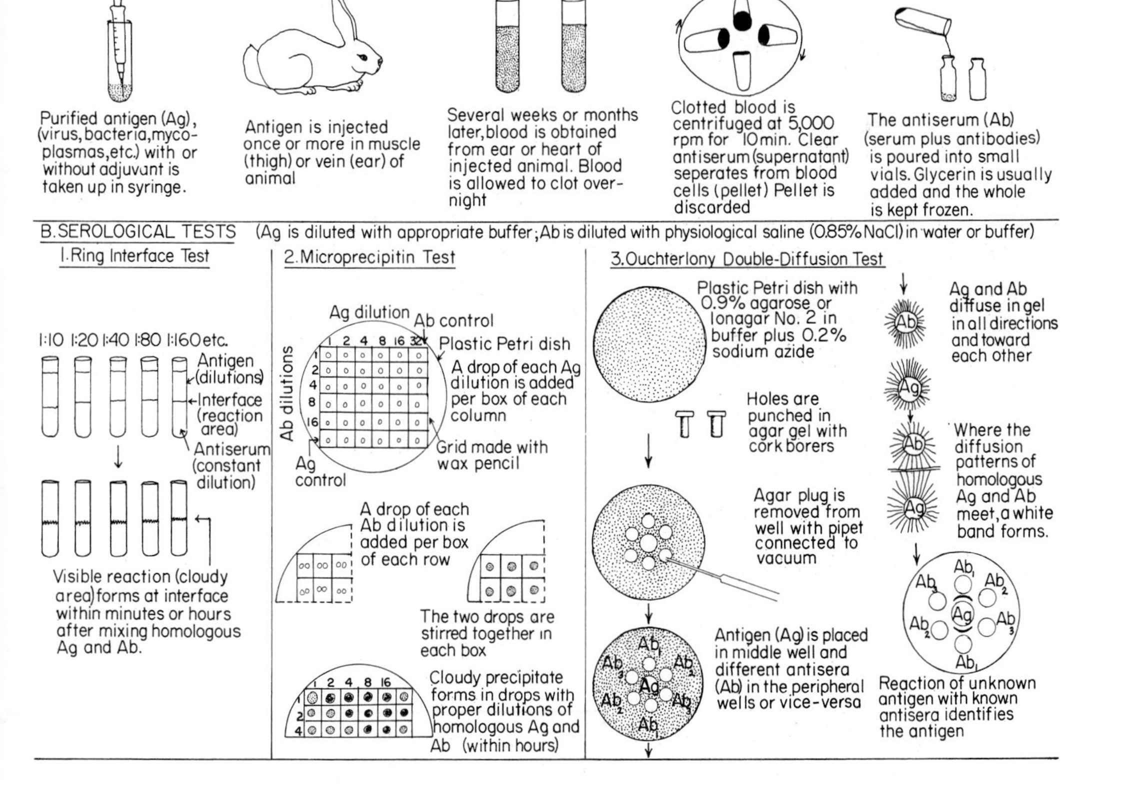

A most useful serological technique called **ELISA** (for Enzyme-Linked Immuno Sorbent Assay), developed in the late 1970s, has been used widely by pathologists of all kinds and has increased tremendously the ability of plant pathologists to detect and study plant viruses and the diseases they cause. Several variations of ELISA are currently in use. In the double antibody sandwitch ELISA, usually referred to as **direct ELISA,** the wells (capacity 0.4 ml) of a polystyrene microtiter plate are first half-filled and then emptied with, sequentially, (a) antibodies to the virus, (b) virus preparation or sap from an infected plant, (c) antibodies to the virus to which molecules of a particular enzyme have been attached, and (d) a substrate for the enzyme, that is, a substance that the enzyme can break down and cause change in its color. The substrate is not emptied but is kept in the well. Within 30 or 60 minutes, the wells are "read" visually or with a colorimeter that measures the amount of color in each well. Presence of color in the well indicates that there was virus in the sample (step *b,* above). The degree of visible coloration or the size of the reading given by the colorimeter is proportional to, and therefore a measure of, the amount of virus present in the sample.

In a variation of the above, in the **indirect ELISA** (Figure 14-20), the sequence of steps *a* and *b* is reversed. Also, in step *c* the antibodies in the antibody–enzyme complex are not those against the virus but rather they are antibodies against the antibody proteins of the animal in which the virus antibody was produced; that is, they are anti-rabbit antibodies produced in still another animal, such as a goat. All other procedures are the same. The advantage of indirect ELISA is that the same goat anti-rabbit antibody–enzyme complex can be used in step *c* of ELISA tests for any virus, as long as the first antibodies, that is, those against the virus used in step *b,* were produced in rabbit.

The advantages of ELISA tests lie in their extreme sensitivity, in the large number of samples that can be tested concurrently, in the small amount of antisera required, in that the results are quantitative, in that the procedure can be semi-automated, and in that ELISA tests can be run regardless of virus morphology and virus concentration. Because of these advantages, ELISA is rapidly replacing all other serodiagnostic techniques.

Two other serological techniques, each with several variations, are also gaining acceptance among plant virologists for finding and identifying virus present in low concentration through the electron microscope and for detecting the virus inside infected cells. In **immunosorbent electron microscopy**

FIGURE 14-20 Schematic presentation of the steps in an indirect ELISA test.

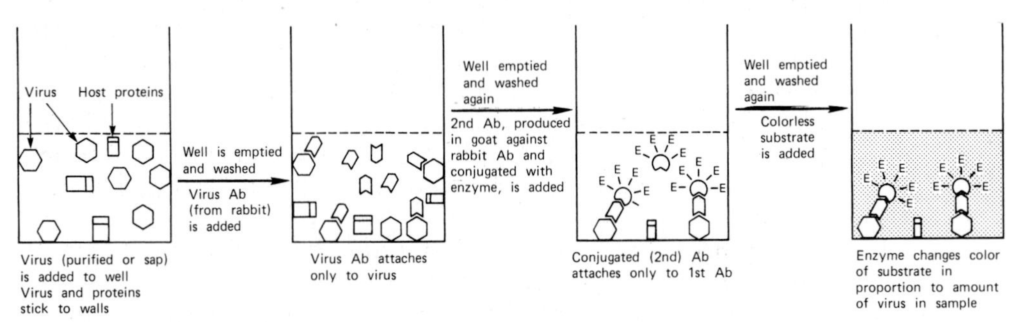

(ISEM), grids, prepared for electron microscopy of a virus present in low concentration or in a mixture with other viruses, are first coated with antibody to that virus. Then the virus sample is placed on the antibody-coated grid, and the antibodies trap the virus from the sample and concentrate it on the grid where it can be easily found with the electron microscope and identified because of its reaction with the antibodies. Identification of the virus is further facilitated by coating the virus particles already on the grid with antibodies (decoration) that make them appear quite distinctive under the electron microscope. In the **immunofluorescent staining** technique, parts of a plant leaf, whole cells, or cell sections are first "fixed," that is, killed with acetone or other organic compounds. The fixed leaf tissues are then treated with antibodies to a virus that had been previously labeled with a compound, such as fluorescent isothiocyanate (FITC), which fluoresces under ultraviolet light. If the treated cells are infected with the virus, the virus traps the antibodies and the attached fluorescent compound. When such cells are viewed with a microscope supplied with ultraviolet light, cells or cell parts that contain virus appear fluorescent while the rest of the cells or cell areas appear dark.

The uses of plant virus serology are numerous. It is used to determine relationships between viruses, to identify a virus causing a plant disease, to detect virus in foundation stocks of plants, and to detect symptomless virus infections. It can also be used to measure virus quantitatively, to locate the virus within a cell or tissue, to detect plant viruses in insects, and to purify a virus.

Nomenclature and Classification of Plant Viruses

Naming of plant viruses has been based on the most conspicuous symptom they cause on the first host they have been studied in. Thus, a virus causing a mosaic on tobacco is called tobacco mosaic virus, while the disease itself is called tobacco mosaic; another virus causing ringspot symptoms on tomato is called tomato ringspot virus, and the disease is called tomato ringspot, and so forth. Considering, however, the variability of symptoms caused by the same virus on the same host plant under different environmental conditions, by different strains of a virus on the same host, or by the same virus on different hosts, it becomes apparent that this system of nomenclature leaves much to be desired.

All viruses belong to the kingdom VIRA. Within the kingdom there are two virus divisions, DNA viruses and RNA viruses, depending on whether the nucleic acid of the virus is DNA or RNA. Viruses within each division are either helical or cubical (polyhedral). Within each subdivision there may be viruses possessing one or two strands of RNA or DNA, possessing or lacking a membrane around the protein coat, containing or lacking certain substances, having certain symmetry of helix in the helical viruses or number of subunits in the cubical (polyhedral) viruses, size of the virus, and finally, any other physical, chemical, or biological properties.

In many plant diseases assumed to be caused by viruses, no virus has yet been observed, and it is possible that some of these diseases will be proved later to be caused by pathogens other than viruses or by as yet uncharacterized

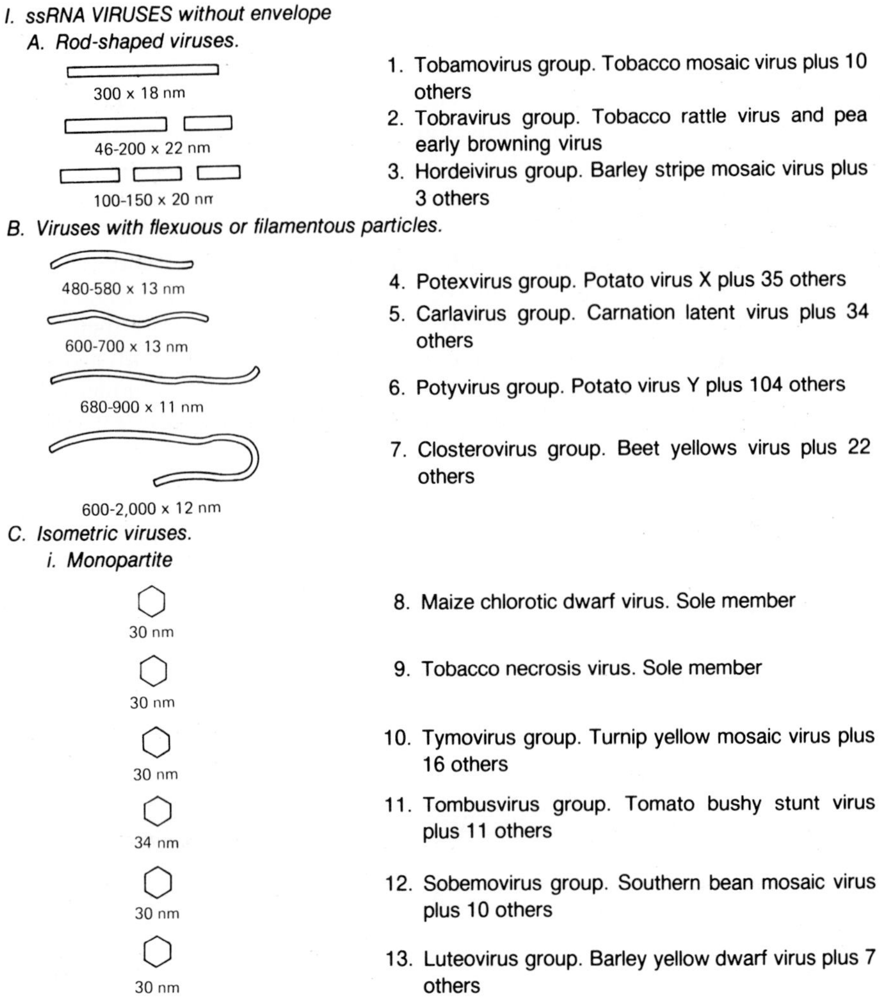

I. ssRNA VIRUSES without envelope

A. Rod-shaped viruses.

300 x 18 nm — 1. Tobamovirus group. Tobacco mosaic virus plus 10 others

46-200 x 22 nm — 2. Tobravirus group. Tobacco rattle virus and pea early browning virus

100-150 x 20 nm — 3. Hordeivirus group. Barley stripe mosaic virus plus 3 others

B. Viruses with flexuous or filamentous particles.

480-580 x 13 nm — 4. Potexvirus group. Potato virus X plus 35 others

600-700 x 13 nm — 5. Carlavirus group. Carnation latent virus plus 34 others

680-900 x 11 nm — 6. Potyvirus group. Potato virus Y plus 104 others

600-2,000 x 12 nm — 7. Closterovirus group. Beet yellows virus plus 22 others

C. Isometric viruses.

i. Monopartite

30 nm — 8. Maize chlorotic dwarf virus. Sole member

30 nm — 9. Tobacco necrosis virus. Sole member

30 nm — 10. Tymovirus group. Turnip yellow mosaic virus plus 16 others

34 nm — 11. Tombusvirus group. Tomato bushy stunt virus plus 11 others

30 nm — 12. Sobemovirus group. Southern bean mosaic virus plus 10 others

30 nm — 13. Luteovirus group. Barley yellow dwarf virus plus 7 others

FIGURE 14-21 Groups of plant viruses.

viruses. For those plant diseases proved to be caused by viruses, however, a system of nomenclature and classification has been proposed (Figure 14-21), in which the viruses are grouped according to the above-listed criteria, according to shape and size and several additional criteria unique to plant viruses. The groups are named after a typical virus in the group.

Detection and Identification of Plant Viruses

Once the cause of a disease has been established as a virus, a series of tests may be necessary to determine its identity. The host range of the virus, that is, the hosts on which the virus induces symptoms and the kinds of symptoms produced, may help to differentiate this virus from several others. Transmission studies should indicate whether the virus is transmitted mechanically

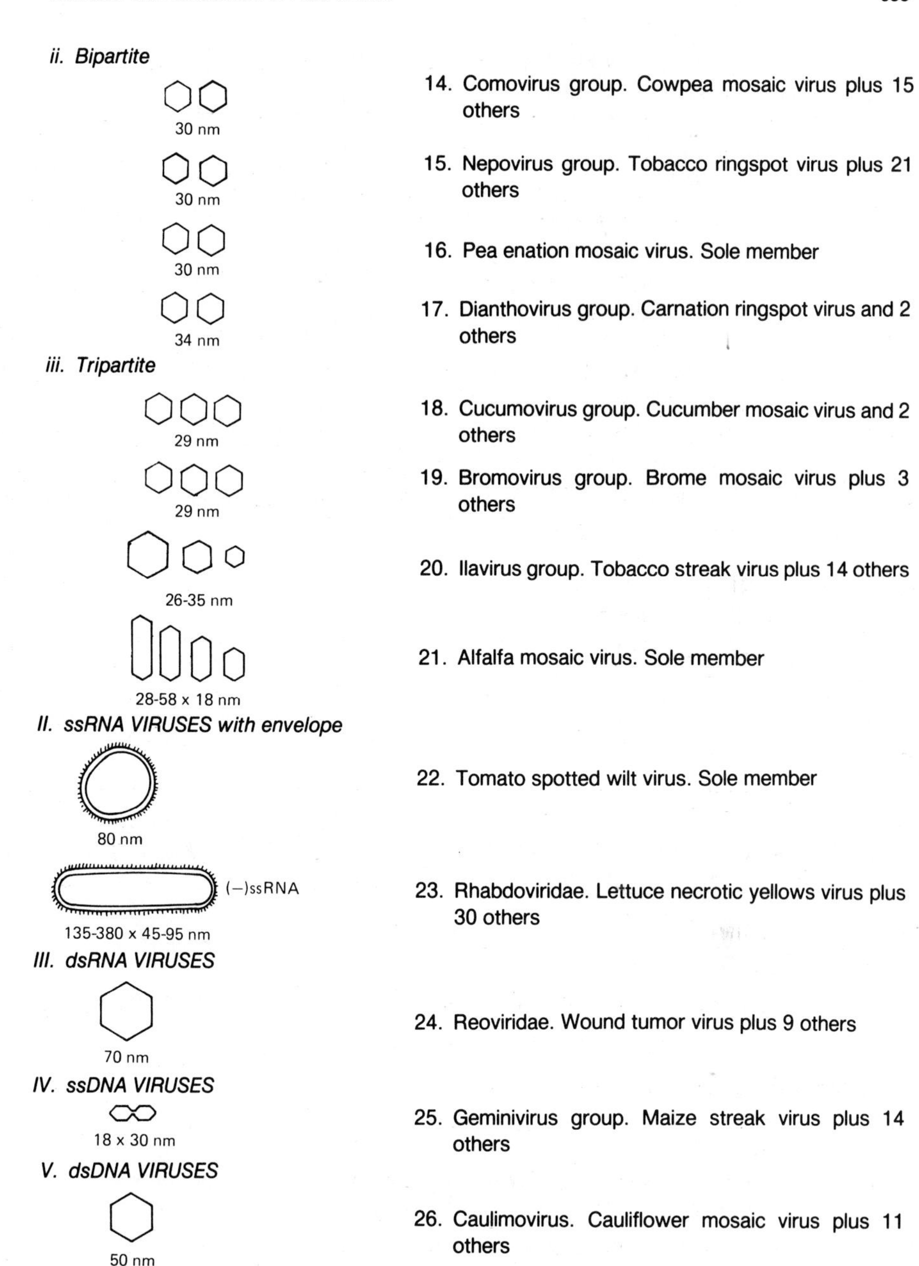

FIGURE 14-21 *(continued)*

and to what hosts, or by insects and which insects, and so on, each new property discovered helping to further characterize the virus. If the virus is transmitted mechanically, certain properties of the virus such as its thermal inactivation point, that is, the temperature required for complete inactivation of the virus in untreated crude juice during a 10-minute exposure, its longevity *in vitro,* its dilution end point, that is, the highest dilution of the juice at which the virus can still cause infection, may be used to narrow the possibilities to just a few viruses. If, at this stage, or at the beginning, the identity of the virus is suspected, serological tests may be used, and if they are positive, a tentative identification may be made. Examination of the virus in the electron microscope, and inoculation of certain plant species, is also usually sufficient for a tentative identification of the virus.

In the viruslike diseases of woody (and other) plants in which no pathogens have been observed so far, identification of the pathogens, which are at present presumed to be viruses, is made strictly by **indexing.** Indexing involves inoculation by grafting (Figure 14-22) of certain plant species or varieties called **indicators;** these are sensitive to specific viruses and upon inoculation with these viruses develop characteristic symptoms and vice versa, that is, development of the characteristic symptoms by an indicator identifies the virus with which the indicator was inoculated.

Since viruses are too small to be detected with the naked eye or seen through the microscope, their presence has been detected primarily by the symptoms exhibited by the host plant; by the symptoms induced in an

FIGURE 14-22 Indexing for viral, mycoplasmal, and fastidious bacterial diseases.

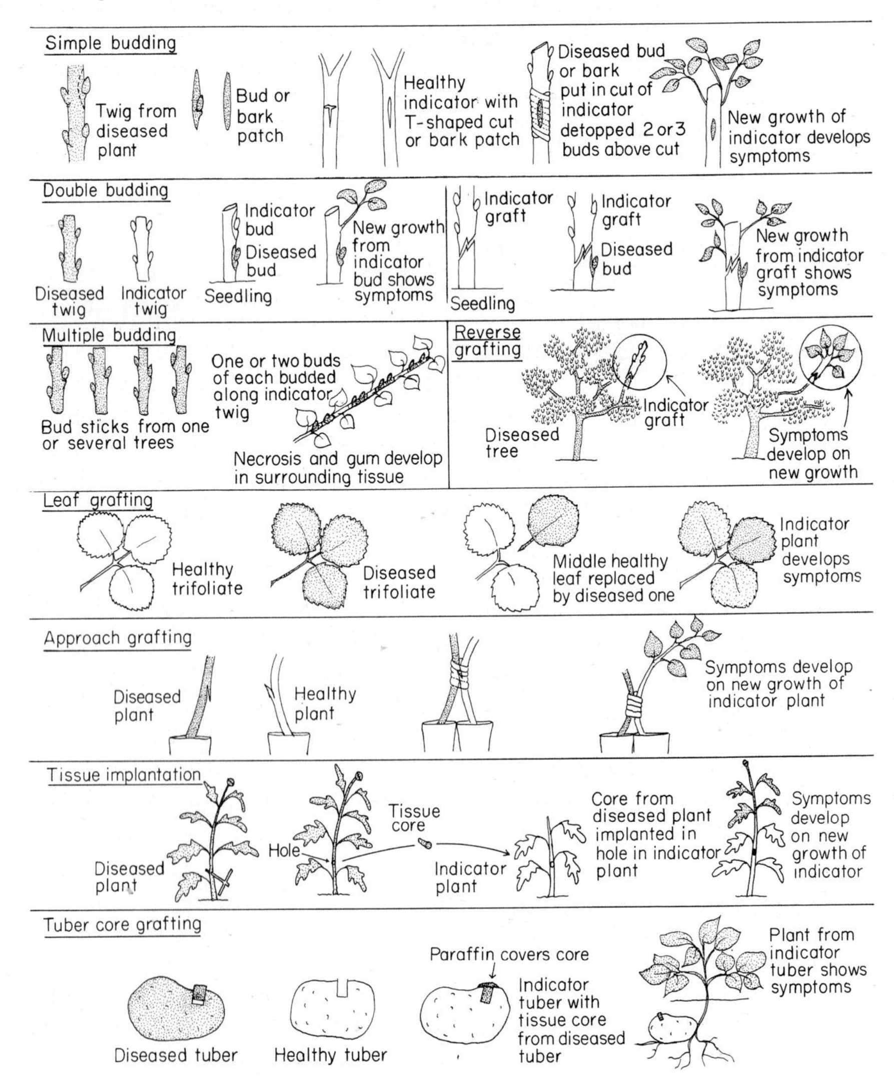

indicator plant after transmission of the virus by grafting, mechanical inoculation, or by one of the vectors; by electron microscopy; and by one of the serological tests such as ELISA.

Recently, it has been possible to detect, and even identify, RNA viruses in plants by isolating, and subsequently analyzing through electrophoresis, the dsRNA of viruses replicating in plants since healthy plants do not produce such dsRNAs. Although laborious, this technique has the advantage that it can be used to detect known as well as unknown viruses for which no antiserum or not much information is available, and it can therefore be used for detection of even woody plant viruses.

Isolated single- or double-stranded RNA can be further used to produce complimentary DNA (cDNA) to the RNA. The cDNA, if produced in the presence of radioactive molecules of at least one of the four DNA nucleotides, can then be used immediately for hybridization experiments with viral RNA. The viral RNA is partially purified from suspected infected plants and is allowed to react with cDNA in test tubes or on nitrocellulose filter supports. The cDNA : RNA hybrids are detected and counted by autoradiography or by liquid scintillation counting. In addition to use for virus detection via formation of cDNA:RNA hybrids, cDNA to the virus RNA can be further converted to dsDNA, which can then be cloned into suitable vectors (for example, *E. coli* bacteria) and produce almost unlimited amounts of dsDNA and, from this, unlimited amounts of cDNA for further hybridization experiments for virus detection.

Economic Importance of Plant Viruses

Viruses attack all forms of plant life, from mycoplasmas, bacteria, fungi, and algae to herbaceous plants and trees.

Plant virus diseases may damage leaves, stems, roots, fruits, seed, or flowers and may cause economic losses by reduction in yield and quality of plant products. Losses may be catastrophic or they may be mild and insignificant. On a nationwide basis, viruses account for a considerable portion of the losses suffered annually from diseases of the various crops.

The severity of individual virus diseases may vary with the locality, the crop variety, and from one season to the next. Some virus diseases have destroyed entire plantings of certain crops in some areas, for example, plum pox, hoja blanca of rice, sugar beet yellows, and citrus tristeza. Most virus diseases, however, occur year after year on crops on which they cause small to moderate but generally unspectacular losses, sometimes without even inducing any visible symptoms. For example, potato virus X, which used to be present in all potato plants grown in the United States, reduces yields by about 10 percent although the potato plants show no obvious symptoms in the field.

Control of Plant Viruses

The best way to control a virus disease is to keep it out of an area through systems of quarantine, inspection, and certification. The existence of symptomless hosts, the incubation period after inoculation, and the absence of obvious symptoms in seeds, tubers, bulbs, and nursery stock make quarantine sometimes ineffective. Eradication of diseased plants to eliminate inoculum from the field may, in some cases, help to control the disease. Plants may be protected against certain viruses by protecting them against the virus vectors. Controlling the insect vectors and removing weeds that serve as hosts may help to control disease. The losses caused by nematode-transmitted viruses can be reduced considerably by soil fumigation to control the nematodes.

The use of virus-free seed, tubers, budwood, and so on is the single most important measure for avoiding virus diseases of many crops, especially those lacking insect vectors. Periodic indexing of the mother plants producing such propagative organs is necessary to ascertain their continuous freedom from viruses. Several types of inspection and certification programs are now in effect in various states producing seeds, tubers, and nursery stock used for propagation. Serological testing of mother plants, seeds, and nursery stock for virus by the ELISA technique has helped greatly in reducing the frequency of viruses in the propagating stock of crop plants.

Although health or vigor of host plants confers no resistance or immunity to virus disease, breeding plants for hereditary resistance to virus is of great importance, and many plant varieties resistant to certain virus diseases have already been produced. In some host–virus combinations, the disease caused by severe strains of the virus can be avoided if the plants are inoculated first with a mild strain of the same virus—**cross protection**—which then protects the plant from infection by the severe strain of the virus.

Once inside a plant, some viruses can be inactivated by heat. Dormant, propagative organs are usually dipped in hot water (35–54°C) for a few minutes or hours, while actively growing plants are usually kept in greenhouses or growth chambers at 35–40°C for several days, weeks, or months, after which the virus in some of them is inactivated and the plants are completely healthy. Plants free of virus may also be produced from virus-infected ones by culture of short (0.1 mm to 1 cm or more) tips of apical and root meristems, especially at elevated (28–30°C) temperatures.

No chemical substances (viricides) are yet available for controlling virus diseases of plants in the field, although some of them, such as ribavirin, applied as a spray or injected into the plant reduce symptoms drastically and, with some viruses, apparently eliminated the virus from the treated host plant. Foliar application of certain growth-regulating substances, such as gibberellic acid, has been effective in stimulating growth of the virus-suppressed axillary buds in sour cherry yellows, resulting in increased fruit production. Similarly, sprays with gibberellic acid can overcome the stunting induced by severe etch virus on tobacco.

SELECTED REFERENCES

Anonymous (1985). Virus diseases: A dilemma for plant breeders. A symposium. *HortScience* **20,** 833–859.

Bock, K. R. (1982). Geminivirus diseases in tropical crops. *Plant Dis* **66,** 266–270.

Cheplick, S. M., and Agrios, G. N. (1983). Effect of injected antiviral compounds on apple mosaic, scar skin, and dapple apple diseases of apple trees. *Plant Dis.* **67,** 1130–1133.

Clark, M. F. (1981). Immunosorbent assays in plant pathology. *Annu. Rev. Phytopathol.* **19,** 83–106.

D'Arcy, C. J., and Nault, L. R. (1982). Insect transmission of plant viruses and mycoplasmalike and rickettsialike organisms. *Plant. Dis.* **66,** 99–104.

Davis, J. E., ed. (1985). "Molecular Plant Virology," 2 vols. CRC Press, Boca Raton, Florida.
Francki, R. I. B., ed. (1985). "The Plant Viruses," Vol. 1. Plenum, New York.
Francki, R. I. B., Milne, R. G., and Hatta, R. (1985). "Atlas of Plant Viruses," 2 vols. CRC Press, Boca Raton, Florida.
Fridlund, P. R. (1980). The IR-2 program for obtaining virus-free fruit trees. *Plant Dis.* **64,** 831–834.
Grogan, R. G. (1980). Control of lettuce mosaic with virus-free seed. *Plant Dis.* **64,** 446–449.
Hansen, A. J. (1984). Effect of ribavirin on green ring mottle causal agent and necrotic ringspot virus in *Prunus* species. *Plant Dis.* **68,** 216–218.
Harris, K. F. (1981). Arthropod and nematode vectors of plant viruses. *Annu. Rev. Phytopathol.* **19,** 391–426.
Harris, K. F., and Maramorosch, K., eds. (1977). "Aphids as Virus Vectors." Academic Press, New York.
Harris, K. F., and Maramorosch, K., eds. (1980). "Vectors of Plant Pathogens." Academic Press, New York.
Harrison, B. D. (1985). Advances in geminivirus research. *Annu. Rev. Phytopathol.* **23,** 55–82.
Hill, S. A. (1984). "Methods in Plant Virology." Blackwell, Oxford.
Hollings, M. (1965). Disease control through virus-free stock. *Annu. Rev. Phytopathol.* **3,** 367–396.
Hollings, M. (1982). Mycoviruses and plant pathology. *Plant Dis.* **66,** 1106–1112.
Kado, C. I., and Agrawa, H. O., eds. (1972). "Principles and Techniques in Plant Virology," Van Nostrand-Reinhold, New York.
Lamberti, F. (1981). Combatting nematode vectors of plant viruses. *Plant Dis.* **65,** 113–117.
Lawson, R. H. (1981). Controlling virus diseases of major international flower and bulb crops. *Plant Dis.* **65,** 780–786.
Maramorosch, K., and Harris, K. F., eds. (1979). "Leafhopper Vectors and Plant Disease Agents." Academic Press, New York.
Maramorosch, K., and Koprowski, H., eds. (1967–1984). "Methods in Virology," 7 vols. Academic Press, New York.
Maramorosch, K., and McKelvey, J. J., Jr., eds. (1985). "Subviral Pathogens of Plants and Animals: Viroids and Prions." Academic Press, Orlando, Florida.
Matthews, R. E. F. (1981). "Plant Virology," 2nd ed. Academic Press, New York.
Milstein, C. (1980). Monoclonal antibodies. *Sci. Am.* 66–74.
Plumb, R. T., and Thresh, J. M., eds. (1983). "Plant Virus Epidemiology: The Spread and Control of Insect-Borne Viruses." Blackwell, Oxford.
Robertson, H. D., Howell, S. H., Zaitlin, M., and Malmberg, R. L., eds. (1983). "Plant Infectious Agents: Viruses, Viroids, Virusoids, and Satellites." Cold Spring Harbor Lab., Cold Spring Harbor, New York.
Smith, K. M. (1972). "A Textbook of Plant Virus Diseases," 3rd ed. Academic Press, New York.
Thresh, J. M. (1982). Cropping practices and virus spread. *Annu. Rev. Phytopathol.* **20,** 193–218.
Van Regenmortel, M. H. V. (1982). "Serology and Immunochemistry of Plant Viruses." Academic Press, New York.
Walkey, D. G. A. (1985). "Applied Plant Virology." Wiley, New York.
Wyss, U. (1982). Virus-transmitting nematodes: Feeding behavior and effect on root cells. *Plant Dis.* **66,** 639–644.

Specific Crop Diseases Caused by Viruses

The great majority of viruses cause various degrees of leaf mottling, mosaic, or yellowing, accompanied by various degrees of stunting, bushiness, and reduced yield. In many of the same diseases, leaves are curled or rolled, although in some, leaf rolling or curling are the most prominent symptoms. In some viral diseases, necrotic areas may develop in leaves or stems, and in some, parts of the plants become malformed.

Since the particle morphology of the virus generally is not related to the symptoms it induces, to the hosts it infects, to the mode of transmission of the virus, or to its epidemiology and control, it is difficult to group plant viruses in a way that would make it easier for us to study them and the diseases they cause in groups rather than as single viruses and diseases. Nevertheless, some grouping must be attempted.

One method of grouping plant viruses is according to the plant families or genera they infect, for example, viruses that infect solanaceous plants (potato, tobacco, tomato, pepper, eggplant), viruses that infect cucurbitaceous plants (cucumber, squash, melons), viruses that infect citrus plants, and so forth. Generally, viruses tend to infect only plants within the same plant family although different viruses do not always cause similar diseases nor do they have similar epidemiologies and controls. There are of course some viruses that infect plants in many different families.

Another method of grouping plant viruses is according to their method of transmission, since these groups (insect-transmitted viruses, soil-borne viruses transmitted by nematodes or by fungi, seed-borne viruses, viruses transmitted through contact, and viruses transmitted only by vegetative propagation) have different epidemiologies and therefore require different controls. In this method of grouping, however, insect-transmitted, particularly aphid-transmitted, viruses make up a huge group—indeed they make up the bulk of viruses of all annual, seed-propagated plants and many of the viruses of some perennial, vegetatively propagated plants such as potatoes and strawberries. On the other hand, almost all viruses of trees are transmitted only by vegetative propagation, although a few are transmitted by insects (for example, citrus tristeza virus, plum pox virus, cacao swollen shoot virus). Furthermore, many of the insect-, nematode-, and vegetatively-transmitted viruses are also transmitted through the seed! Obviously, this grouping, too, leaves much to be desired.

In the following pages, several virus diseases, representing various aspects of the above groupings, are described in some detail. Many more diseases, equally and often more important, are described only briefly. Still many more are barely mentioned, although each of them may be prevalent and severe in a particular area. Moreover, many viruses are more severe and destructive in plants of other genera or families than in those under which they are listed, yet space does not allow mention, much less description, of the effects of these other hosts.

Virus Diseases of Tobacco

Tobacco is affected by more than 20 viruses, of which the most important are tobacco mosaic virus, tobacco etch, tobacco ring spot, tomato spotted wilt, tobacco rattle, tobacco streak, and tobacco leaf curl viruses. Other viruses affecting tobacco include tobacco necrosis, tobacco yellow dwarf, potato virus Y, potato virus X, cucumber mosaic, and alfalfa mosaic.

• Tobacco Mosaic

Tobacco mosaic is worldwide in distribution. It affects more than 150 genera of primarily herbaceous, dicotyledonous plants, including many vegetables, flowers, and weeds. It causes serious losses on tobacco, tomato, and some other crop plants but is almost symptomless on crops like grape and apple.

Tobacco mosaic affects plants by damaging the leaves, flowers, and fruit and by causing stunting of the plant. It almost never kills plants. In tobacco, the disease lowers the quantity and especially the quality of the crop, particularly when plants are infected while young. Thus, plants inoculated at transplanting time, a month later, or at topping time produced yields that were lower than those of healthy plants by 33, 20, and 5 percent, while the quality of the crop, as measured by its market value, was reduced by 50, 42, and 23 percent, respectively. In tomato, also, yield reductions may vary from 5 to more than 25 percent, and the fruit quality is proportionally lower, depending on the age of plants at the time of infection and on environmental conditions.

○ *Symptoms.* The symptoms of tobacco mosaic virus-infected plants consist of various degrees of chlorosis, curling, mottling, dwarfing, distortion, and blistering of the leaves, dwarfing of the entire plant, dwarfing, distortion, and discoloration of flowers, and in some plants, development of necrotic areas on the leaf.

The most common symptom on tobacco is the appearance of mottled dark-green and light-green areas on leaves developing after inoculation (Figure 14-23A). The dark green areas are thicker and appear somewhat elevated in a blisterlike fashion over the thinner, chlorotic, light-green areas. Stunting of young plants is common and is accompanied by a slight downward curling and distortion of the leaves, which may also become narrow and elongated rather than the normal oval shape. Inoculation of plants approaching maturity usually causes no symptoms on the older leaves, but it does affect any new ones that may be produced.

On tomato, mottling of the older leaves and mottling with or without malformation of the leaflets are also produced. Leaflets become long and pointed and, sometimes, shoestringlike. Infections of young plants reduce fruit set and may occasionally cause blemishes and internal browning on the fruit that does form.

○ *The Pathogen: Tobacco Mosaic Virus (TMV).* Tobacco mosaic virus (TMV) is a tobamovirus, that is, it is rod shaped, 300 nm long by 15 nm in diameter (Figure 14-23, B, C). Its protein consists of approximately 2130 protein subunits, and each subunit consists of 158 amino acids. The protein subunits are arranged in a helix. The TMV nucleic acid is single-stranded ribonucleic acid (RNA) and consists of approximately 6400 nucleotides. The RNA strand also forms a helix that is parallel with that of the protein and is located on the protein subunits and approximately 2 nm out from the inner end of the protein subunits. The weight of each virus particle is between 39 and 40 million molecular weight units.

TMV is one of the most thermostable viruses known, the thermal inactivation point of the virus in undiluted plant juice being 93°C. In dried, mosaic-infected leaves, however, the virus retains its infectiousness even when heated to 120°C for 30 minutes. TMV-infected tobacco plants may contain up to 4 g of virus per liter of plant juice, and the virus retains its infectivity even at dilutions of 1 : 1,000,000. The virus is inactivated in 4–6 weeks in ordinary plant sap, but in sterile, bacteria-free sap the virus may survive for five years, and in TMV-infected leaves kept dry in the laboratory the virus remains infectious for more than 50 years. Tobacco mosaic virus is transmitted readily through sap, grafting, and dodder, and, in some hosts such as apple, pear, and grape, through seed. Tobacco mosaic virus is not transmitted by insects, except occasionally by contaminated jaws and feet of insects

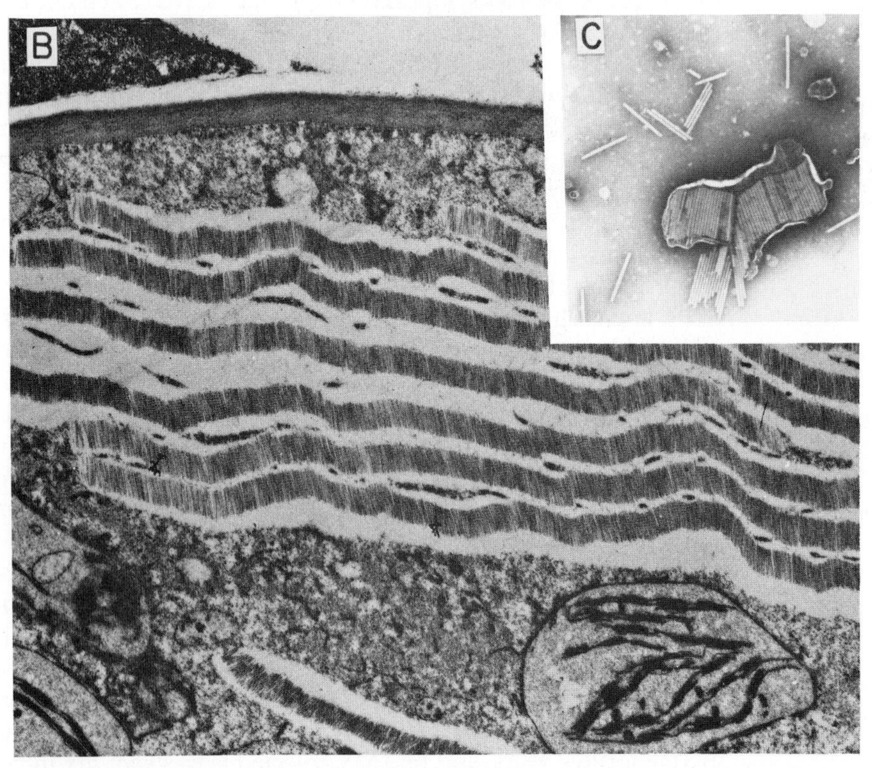

FIGURE 14-23 (A) Symptoms of tobacco mosaic (upper) and healthy tobacco leaf. (B) Layers of tobacco mosaic virus particles in a tobacco epidermal leaf cell. (C) TMV particles in sap from an infected tobacco leaf. The virus was negatively stained with phosphotungstate. (Photos B and C courtesy of H. E. Warmke).

feeding on TMV-infected and healthy plants. The most common means of transmission of TMV in the field and in the greenhouse is through the hands of workers handling infected and healthy plants indiscriminately.

Tobacco mosaic virus exists in numerous strains, which differ from each other in one or more characteristics.

◦ *Development of Disease.* Tobacco mosaic virus overwinters in infected tobacco stalks and leaves in the soil, on the surface of contaminated tobacco seeds, on contaminated seedbed cloth, and in natural leaf and manufactured tobacco, including cigarettes, cigars, and snuff. Contact of the virus with wounded tissues of tobacco seedlings in the seedbed or of transplants in the fields results in initial infections of a few plants. These subsequently serve as a source of inoculum for further spread of the virus to more plants through contaminated hands, tools, or equipment during handling of tobacco plants in the routine cultural practices with that crop. The virus can, of course, be introduced into a field on transplants already infected in the seedbed. The spread of the virus in the field continues throughout the season, the number of plants infected increasing progressively during the season and doubling approximately with each handling or cultivating of the crop.

In almost all host plants, TMV produces systemic infections, invading all parenchymatous cells of the plant. The virus moves from cell to cell and through the phloem.

Within the cell, TMV seems to occur primarily in the cytoplasm as individual particles, as crystalline aggregates (Figure 14-23, B), and as amorphous bodies (X-bodies), ranging in size from submicroscopic to those seen with the light microscope.

The TMV-infected leaves show thin, light-green or yellowing areas intermixed with thicker, dark green areas. In the light-green areas both palisade and spongy parenchyma cells are round rather than their normal elongated shape and, due to reduction of the intercellular spaces among such cells, they are arranged much more compactly than cells in the dark-green areas or in healthy leaves. In the light areas, the number of chloroplasts is reduced appreciably; they seem to contain much less chlorophyll than those of healthy or dark-green areas. Thus, chlorophyll synthesis is impaired while some of the chlorophyll produced is destroyed or its activity is impaired as a consequence of virus infection. This leads to reduced photosynthesis and, therefore, reduced levels of carbohydrates in TMV-infected plants. Tobacco mosaic virus also induces alterations in a number of other physiological processes in infected plants.

◦ *Control.* Sanitation and use of resistant varieties are the two main means of control of TMV in tobacco and tomato fields or greenhouses.

Tobacco should not be grown for at least 2 years in seedbeds or fields where a diseased crop was grown. Removal of infected plants and of certain solanaceous weeds that harbor the virus early in the season helps reduce or eliminate the subsequent spread of the virus to other plants during the various cultural practices. The chewing and smoking of tobacco during cultural practices requiring handling of tobacco and other susceptible plants should be avoided. Workers using tobacco products or involved in removing TMV-infected plants should wash their hands with a phosphate detergent or soap and water before handling healthy tobacco or tomato plants.

Several TMV-resistant varieties of tobacco have been developed but are generally of low quality. Tomato varieties resistant to TMV are also available.

Tomatoes in greenhouses, which in most cases become infected with TMV, are in some countries protected against virulent strains of TMV by infecting young plants with a mild strain. This practice results in an increase in yield of up to 15 percent.

Because infection by TMV is inhibited by milk, some states are now recommending spraying the plants with milk several hours before transplanting or otherwise handling them or dipping the hands in milk during transplanting and handling, since these practices greatly reduce the spread of TMV from plant to plant. Milk treatments should be combined with washing hands with a phosphate detergent.

- **Tobacco Etch**

Tobacco etch occurs in North and South America. It is caused by the tobacco etch virus (TEV), which also infects pepper and tomato. It causes severe losses on all three hosts. Infected tobacco leaves are narrowed and show mottling and necrosis. Pepper leaves show mottling, mosaic, and distortion; pepper fruit are distorted, and the entire plant may be stunted. Tomato plants are also stunted, and the leaves are mottled and distorted. TEV is a potyvirus, that is, it has a flexuous filamentous particle 730 nm long by 12 nm in diameter. The virus is transmitted by more than 10 species of aphids in the nonpersistent manner.

- **Tobacco Ring Spot**

Widespread and common in North America, tobacco ring spot is found less frequently in other parts of the world. It has a wide host range including many annual and perennial crop and ornamental plants. It causes ring spot diseases on leaves, flowers, stems, and fruit of plants such as tobacco, cucumber, Easter lily, iris, and blueberry. It also causes bud blight of soybean. Tobacco ring spot virus (TobRSV) is a nepovirus, that is, a polyhedral virus 28 nm in diameter transmitted by a nematode (*Xiphinema* sp.). TobRSV is also transmitted through the seed, the frequency varying from very low in most hosts to 100 percent in soybean.

- **Tobacco Rattle**

Tobacco rattle occurs in many parts of the world. It affects many hosts and causes symptoms ranging from curly or crumpled leaves, collapse of the veins, and systemic necrosis in tobacco to stem mottle and corky ring spot (spraing) in potato and to ring spot and yellow blotch diseases on pepper, sugar beet, and others. Tobacco rattle virus (TRV) is a tobravirus, that is, it consists of two rod-shaped particles 180–215 and 46–114 nm long by 22 nm in diameter. TRV is transmitted by several species of the nematodes *Trichodorus* and *Paratrichodorus*.

- **Tobacco Streak**

Tobacco streak occurs in North and South America, Europe, New Zealand, and probably elsewhere. It affects many dicots and monocots. Tobacco plants first develop a systemic necrotic disease and then recover and show no symptoms. Many hosts (such as dahlia, strawberry, and cotton) show mottling or no symptoms, while tomato plants show yellow ring spots and malformation, and plants such as pea, potato, and soybean exhibit systemic necrotic symptoms. Tobacco streak virus (TSV) is an ilarvirus, that is, it is a

ntaining virus with particle sizes 25, 28,

s is transmitted primarily by vegetative

igh the seed. No vector of TSV is known.

tropics and subtropics but is also reported

nd Japan. It affects tobacco, tomato, pep-

and twisting of stem, curling, puckering,

leaves and swelling of veins. Yields are

irl virus (TLCV) is a geminivirus, that is, it

ng 18 × 28 nm. TLCV is transmitted only

baci.

f tobacco. *U. S. Dep. Agric., Bull.* **40,** 1–33.

iruses," Specific tobacco viruses. Nos. 12, 14, 17, 44,

n of tobacco mosaic virus to the host cells. *J. Cell Biol.*

saic diseases on yield and quality of tobacco. *J. Agric.*

d Frey, S. (1964). The anatomy of the tobacco mosaic

al symptomless mutant of tobacco mosaic for seedling

ant Pathol. **78,** 110–112.

ses, the most important of which are tomato

per mosaic, tomato ring spot, tomato spotted

ato virus Y. Other viruses affecting tomato

acco etch, tobacco rattle, and tobacco streak,

k ring virus.

- **Tomato Mosaic and Tobacco Mosaic**

Tomato mosaic and tobacco mosaic and the two viruses that cause them are nearly identical and are described under "Tobacco Mosaic" (see page 658). They differ only slightly in some host, serological, and cross protection reactions.

- **Tomato Ring Spot**

Tomato ring spot is widespread in North America and has also been reported from other parts of the world. It is of minor importance to tomato production, but it infects many other hosts and causes particularly severe losses on many perennial hosts. On annual and some perennial hosts, tomato ring spot virus (TomRSV) causes mostly mosaic and ring spot diseases, sometimes accompanied by various degrees of systemic necrosis. On perennial hosts, however, TomRSV usually causes no distinctive symptoms on the foliage, but rather, it affects the base of the plant. In grafted trees, such as stone fruits and pome fruits, TomRSV affects the area above, at, and below the graft union and causes stem pitting, brown line, or union necrosis, which is then followed by decline and gradual death of the tree. Certain rootstocks and varieties of peach, apricot, plum, cherry, and apple are very sensitive to TomRSV, as are grapes and raspberries. The virus is a nepovirus, that is, it is polyhedral, 28 nm in diameter, and is transmitted by the nematode *Xiphinema*. In some hosts, TomRSV is also transmitted through the seed.

• Tomato Spotted Wilt

Tomato spotted wilt occurs in all temperate and subtropical regions of the world and has an extremely wide host range, including tomato, tobacco, dahlia, and pineapple. In tomato leaves, it causes characteristic bronzing and one-sided growth, while in all its hosts tomato spotted wilt virus (TSWV) causes various degrees of chlorotic, necrotic, stunting, and enation symptoms. Losses from tomato spotted wilt are often great. The virus is unique in that its particles are roughly spherical, are surrounded by a membrane, and are quite large, measuring about 85 nm in diameter. TSWV is transmitted by at least four species of thrips insects (*Thrips* and *Frankliniella*).

SELECTED REFERENCES

Broadbent, L. (1976). Epidemiology and control of tomato mosaic virus. *Annu. Rev. Phytopathol.* **14,** 189–210.

"C. M. I./A. A. B. Descriptions of Plant Viruses," Specific tomato viruses, Nos. 38, 39, 69, 79, 151, 156, 290.

Virus Diseases of Potato

Potatoes are affected by about 20 viruses and one viroid (see page 698). The most important viruses of potato are potato leafroll virus, potato viruses Y, X, and S, and the potato spindle tuber viroid. Other viruses affecting potato include potato viruses A, M. and T, potato aucuba mosaic, potato mop-top, tobacco rattle, alfalfa mosaic, potato yellow dwarf, and tomato spotted wilt virus.

• Potato Leafroll

Potato leafroll occurs worldwide. It is caused by the potato leafroll virus (PLRV) and affects only potato. It causes high yield losses and is probably the most important virus of potato. It causes a prominent rolling of the leaves, and the plants are stunted and have a stiff upright growth (Figure 14-24A). In some varieties, phloem becomes necrotic, and carbohydrates accumulate in the leaves. There is phloem necrosis in tubers also (Figure 14-24B). Potato leafroll virus is a luteovirus, 24 nm in diameter, and is confined to the phloem

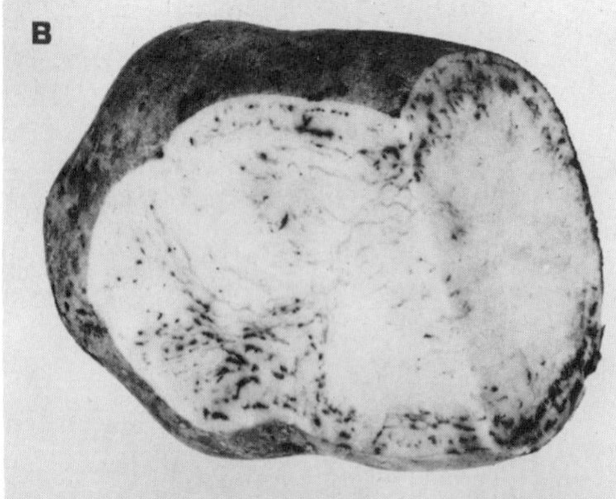

FIGURE 14-24 Rolling of leaves and stunting of young potato plant infected with potato leafroll virus (A). Veins of potato tubers infected with leafroll virus often show necrosis of the phloem (B). (Photo A courtesy Department of Plant Pathology, Cornell University, Photo B courtesy USDA)

tissues. It is transmitted through infected potato seed tubers and, in the field, by more than 10 species of aphids in the persistent manner.

- **Potato Virus Y (PVY)**

Potato virus Y (PVY) occurs worldwide and is of great economic importance. It affects potato, pepper, tomato, and tobacco and causes severe losses on all these hosts. The symptoms it causes vary from a mild to severe mottle on most hosts to a streak or "leaf-drop streak" resulting from long necrotic lesions along the veins on the underside of leaflets of some potato varieties. When present together with potato virus X, PVY causes "rugose mosaic," in which the plants are dwarfed and the tubers reduced in size. Potato virus Y is a potyvirus, that is, it has flexuous threadlike particles 730 nm long by 11 nm in diameter (Figure 14-1B). Several distinct strains of PVY exist in nature. It is transmitted through infected potato seed tubers and by at least 25 species of aphids in the nonpersistent manner.

- **Potato Virus X (PVX)**

Potato virus X (PVX) occurs worldwide. It causes light to moderate losses but is quite important because it is so universally present in potato. It also infects tomato and tobacco. In potato it may be completely latent or it may cause mild mosaic to severe necrotic streaks. In tomato it causes mosaic and mild stunting, while in tobacco it causes mottle or necrotic ring spotting. PVX is a potexvirus, that is, it has filamentous particles 515 nm long by 13 nm in diameter. It is often present with PVY and causes "rugose mosaic." It is transmitted through infected potato seed tubers and by contact between adjacent plants, hands, tools, and so on. It is not transmitted by vectors.

Control of all potato viruses is based on the use of certified virus-free potato seed tubers. Some control of the persistently transmitted PLRV can be obtained by insecticidal sprays to control the aphid vector.

SELECTED REFERENCES

"C. M. I./A. A. B. Descriptions of Plant Viruses," Specific potato viruses, Nos. 4, 35, 54, 60, 87, 98, 138, 187, 200, 203, 206, 242, 248, 291.

de Bokx, J. A., ed. (1981). "Viruses of Potatoes and Seed-Potato Production." Pudoc Wageningen, The Netherlands.

Hooker, W. J., ed. (1981). "Compendium of Potato Diseases." Am. Phytopathol. Soc., St. Paul, Minnesota.

Rich, A. E. (1983). "Potato Diseases." Academic Press, New York.

Virus Diseases of Cucurbits

Cucurbits, that is, cucumber, squash, pumpkin, cantaloupe, and watermelon, are affected by about 20 viruses, the most important of which are cucumber mosaic, squash mosaic, watermelon mosaic (1 and 2), and zucchini yellow mosaic viruses. Other viruses affecting cucurbits include wild cucumber mosaic, cucumber green mottle mosaic, squash leaf curl, lettuce infectious yellows, cucumber necrosis, bean yellow mosaic, and beet curly top.

• Cucumber Mosaic

Cucumber mosaic is worldwide in distribution. The virus causing cucumber mosaic has, perhaps, a wider range of hosts and attacks a greater variety of vegetables, ornamentals, and other plants than any other virus. Among the most important vegetables and ornamentals affected by cucumber mosaic are cucumbers, melons, squash, peppers, spinach, tomatoes, celery, beets, bean, banana, crucifers, delphinium, gladiolus, lilies, petunias, zinnias, and many weeds.

Cucumber mosaic affects plants by causing mottling or discoloration and distortion of leaves, flowers, and fruit. Infected plants may be greatly reduced in size or they may be killed. Crop yields are reduced in quantity and are often lower in quality. Plants are seriously affected in the field as well as the greenhouse. In some localities one-third to one-half of the plants may be destroyed by the disease, and susceptible crops, such as summer squash, may have to be replaced by other crops.

◦ *Symptoms.* Young seedlings are seldom attacked in the field during the first few weeks. Most general field infections occur when the plants are about 6 weeks old and growing vigorously. Four or five days after inoculation, the young developing leaves become mottled, distorted, and wrinkled, and their edges begin to curl downward (Figure 14-25). All subsequent growth is reduced drastically, and the plants appear dwarfed as a result of shorter stem internodes and petioles and of leaves developing to only half their normal size. Such plants produce few runners and also few flowers and fruit. Instead, they have a bunched or bushy appearance, with the leaves forming a rosette-like clump near the ground. The older leaves of infected plants develop at first chlorotic and then necrotic areas along the margins, which later spread over the entire leaf. The killed leaves hang down on the petiole or fall off, leaving part or most of the older vine bare.

Fruit produced on the plant after the infection shows pale green or white areas intermingled with dark green, raised areas; the latter often form rough, wartlike projections and cause distortion of the fruit. Cucumbers produced by

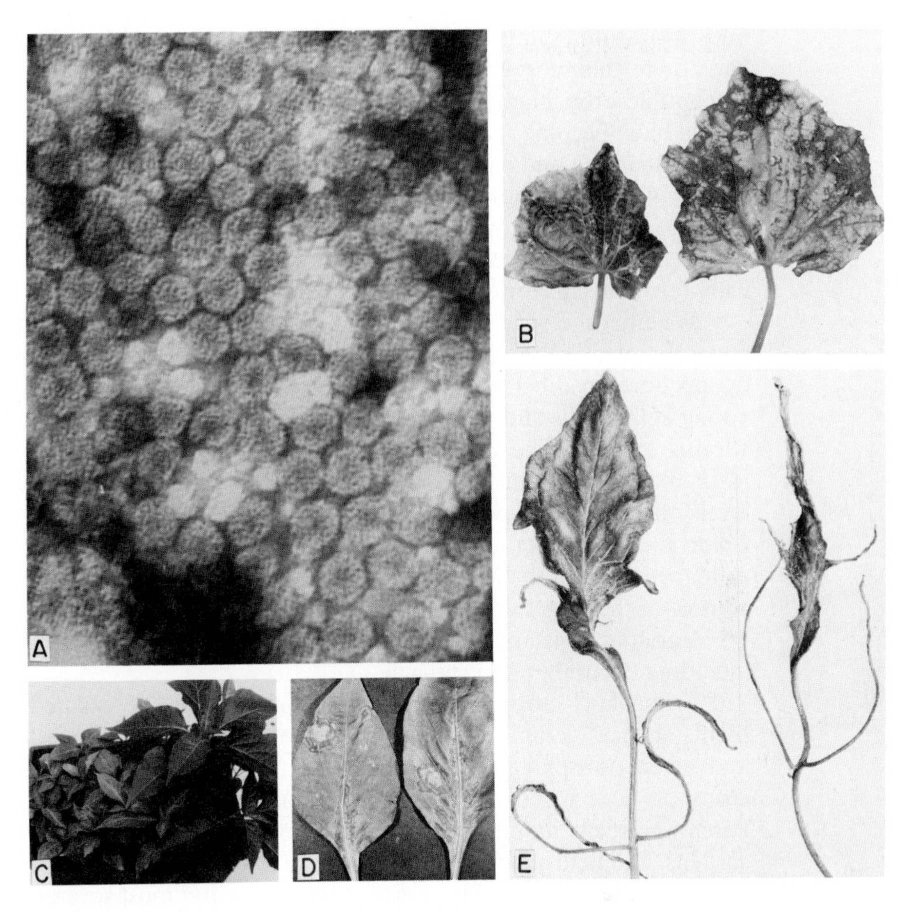

FIGURE 14-25 Cucumber mosaic virus (A) and some of the symptoms it causes. Cucumber mosaic on cucumber leaves (B). Stunting of infected pepper plants is shown at C (left) compared to two healthy plants, and leaf symptoms on pepper (D). CMV-infected tomato leaves often become filiform or shoestringlike (E).

the plants in the later stages of the disease are somewhat misshapen but have smooth gray-white color with some irregular green areas and are often called "white pickle." Cucumbers infected with cucumber mosaic often have a bitter taste and upon pickling become soft and soggy.

◦ *The Pathogen: Cucumber Mosaic Virus (CMV).* Cucumber mosaic virus is a cucumovirus, that is, a tripartite polyhedral virus, with a diameter of 30 nm (Figure 14-25, A). The virus consists of 180 protein subunits, one of three different single-stranded RNAs, and a hollow core. The molecular weight of CMV falls in the range of 5.8–6.7 million, of which 18 percent is RNA and the remaining 82 percent protein. Cucumber mosaic virus exists in numerous strains that differ somewhat in their hosts, in the symptoms they produce, in the ways they are transmitted, and in other properties and characteristics.

The virus is readily transmitted by sap and also by many aphids, such as the common green peach aphid, in the nonpersistent manner.

◦ *Development of Disease.* Cucumber mosaic virus overwinters in many perennial weeds, flowers, and crop plants. Perennial weeds such as white cockle, wild ground cherry, horse nettle, milkweed, ragweed, pokeweed, nightshade, and the various mints harbor the virus in their roots during the winter and carry it to their top growth in the spring from which aphids transmit it to susceptible crop plants. Once a few squash, tomato, pepper, or cucumber plants have become infected with CMV, insect vectors and humans during their cultivating and handling of the plants, especially at picking time, spread the virus to many more healthy plants. Entire fields of cucurbits sometimes begin to turn yellow with mosaic immediately after the first pick has been made, indicating the ease and efficiency of transmission of CMV mechanically through sap carried on the hands and clothes of workers.

Whether the virus is transmitted by insects or through sap, it produces a systemic infection of curcurbit and most other host plants. Older tissues and organs developed before infection are not, as a rule, affected by the virus, but young active cells and tissues developing after infection may be affected with varying severity. The virus concentration in CMV-infected plants continues to increase for several days after inoculation, and then it decreases until it levels off or until the plant dies.

◦ *Control.* Cucumber mosaic in vegetables and flowers can be controlled primarily through the use of resistant varieties, elimination of weed hosts, and control of the insect vectors.

Varieties resistant to CMV have been developed for several host crops, including cucumber and spinach.

Transplant crops kept in greenhouses should be isolated from other plants such as geraniums, lilies, and cucumbers that may harbor the virus, and when transplanted they should not be planted near early, susceptible crops or near woods in which there may be weeds harboring the virus. Perennial weeds should be eradicated from around greenhouses, cold frames, gardens, and fields to eliminate the source of CMV likely to be carried to crop plants by insects or sap. Since most of the early and most severe infections are initiated by virus brought in by aphids from outside the field, any measures, such as vertical sticky traps, border trap crops, or polyethylene reflective mulches, that delay the arrival or reduce the number of aphids in the crop delay the onset and spread of the virus and reduce losses from the disease. Similarly, losses from CMV are reduced in some crops by spraying the plants several times with certain kinds of oils that have been shown to interfere with virus transmission by the aphids. Early sprays with insecticides to control the aphid vectors before they carry the virus into the young, rapidly growing plants have been only marginally helpful.

• Squash Mosaic

Squash mosaic occurs widely in the Western Hemisphere and probably elsewhere. In nature, it infects only cucurbits on which it causes symptoms generally indistinguishable from those caused by cucumber mosaic and watermelon mosaic. Squash mosaic virus (SqMV) is a comovirus, that is, a bipartite virus about 30 nm in diameter. SqMV survives in cucurbit, mostly squash, seed through which it is carried to new plants the following season and then is transmitted from them to other cucurbits by the spotted and striped cucumber beetles *Diabrotica* sp. and *Acalymma* sp.

• Watermelon Mosaic

Watermelon mosaic occurs worldwide. It causes mosaic and mottle diseases on all cucurbits and reduces fruit production and quality. It also infects peas and other leguminous, malvaceous, and chenopodiaceous crop plants, ornamentals, and weeds. Watermelon mosaic virus (WMV-2) is a potyvirus, that is, it has flexuous filamentous particles 760 nm long by 11 nm in diameter. It is transmitted by at least 38 species of aphids in the nonpersistent manner.

• Zucchini Yellow Mosaic

Zucchini yellow mosaic is probably worldwide. It causes economically important diseases in zucchini squash, muskmelon, cucumber, and watermelon. The symptoms consist of severe mosaic, yellowing, shoestringing, stunting, and distortions of fruit and seed. The virus is a potyvirus, 750 × 11 nm. It infects many hosts experimentally, although so far in nature it has not been found in hosts other than cucurbits. It is transmitted by at least four aphid species in the nonpersistent manner.

SELECTED REFERENCES

Agrios, G. N., Walker, M. E., and Ferro, D. N. (1985). Effect of cucumber mosaic virus inoculation at successive weekly intervals on growth and yield of pepper *(Capsicum annuum)* plants. *Plant Dis.* **69,** 52–55.

"C. M. I./A. A. B. Descriptions of Plant Viruses," Specific cucurbit viruses, Nos. 43, 63, 82, 84, 154, 213, 282, 292, 293.

Doolittle, S. P. (1920). The mosaic disease of cucurbits. *U. S. Dep. Agric., Bull.* **879,** 1–69.

Doolittle, S. P., and Walker, M. N. (1925). Further studies on the overwintering and dissemination of cucurbit mosaic. *J. Agric. Res. (Washington, D.C.)* **31,** 1–58.

Lovisolo, O. (1980). Virus and viroid diseases of cucurbits. *Acta Hortic.* **88,** 33–82.

Nameth, S. T., Dodds, J. A., Paulus, A. O., and Laemmlen, F. F. (1986). Cucurbit viruses of California: An ever-changing problem. *Plant Dis.* **70,** 8–11.

Porter, C. A. (1954). Histological and cytological changes induced in plants by cucumber mosaic virus. *Contrib. Boyce Thompson Inst.* **17,** 453–471.

Sherf, A. F. (1965). Cucumber mosaic virus in New York vegetables. *Cornell Ext. Bull.* **1144,** 1–8.

Virus Diseases of Crucifers

Crucifers, that is, cabbage, cauliflower, radish, turnip, and so on, are affected by only about 6–8 viruses, all of them infecting only crucifers in nature. The most important of these are turnip mosaic and cauliflower mosaic viruses. Others are radish mosaic, turnip yellow mosaic, turnip crinkle, turnip rosette, and broccoli necrotic yellows viruses.

• Turnip Mosaic

Turnip mosaic occurs worldwide. It affects all vegetable and ornamental crucifers. It appears as mottling, black necrotic spots and ring spots in cabbage, cauliflower, and Brussels sprouts, while in the other crucifers it causes mosaic, leaf distortions, and stunting. Turnip mosaic virus is a potyvirus, 720 × 11 nm. It is transmitted by about 50 species of aphids in the nonpersistent manner.

• Cauliflower Mosaic

Cauliflower mosaic occurs throughout the temperate regions of the world. It causes mosaic and mottle diseases on most cruciferous crops and ornamentals, in which it is often found in mixed infections with turnip mosaic virus. Cauliflower mosaic virus is a caulimovirus, that is, it is isometric, 50 nm in diameter, and its nucleic acid is double stranded DNA. It is transmitted by about 30 aphid species in the nonpersistent manner.

SELECTED REFERENCES

"C. M. I./A. A. B. Descriptions of Plant Viruses," Specific crucifer viruses, Nos. 2, 8, 109, 121, 125, 214, 230, 243, 295.

Smith, K. M. (1972). "A Textbook of Plant Virus Diseases," 3rd ed. Academic Press, New York.

Virus Diseases of Legumes

The legumes consist of the crops grown for their edible seeds, such as beans, broad beans, soybeans, peas, cowpeas, and peanuts, and of those grown for forage and hay, such as alfalfa and the various clovers. Legumes are affected by at least 40 viruses. The most important viruses of legumes include bean common mosaic and bean yellow mosaic, soybean mosaic, pea mosaic, pea seed-borne mosaic, and pea streak, cowpea mosaic and cowpea aphid-borne mosaic, peanut mottle and peanut stunt, and alfalfa mosaic virus. Most of these viruses occur worldwide, they are common, and the losses they cause vary from moderate to quite severe. Several more of the legume viruses cause severe losses where present, but they are not as common or widespread as the above. Some of these viruses affect only or primarily the crop after which they are named; for example, common bean mosaic virus affects only beans, soybean mosaic virus only soybeans, and pea seed-borne mosaic only peas. Most of the other viruses mentioned, however, as well as many of the rest, such as bean yellow mosaic, pea mosaic, cowpea aphid-borne mosaic, peanut mottle, peanut stunt, and alfalfa mosaic, infect all or most legumes and cause losses of varying severity on all of them. In addition, many other viruses also affect legumes, for example, cucumber mosaic, beet curly top, and tobacco ring spot.

• Bean Common Mosaic and Bean Yellow Mosaic Diseases

Bean common mosaic and bean yellow mosaic diseases occur wherever beans are grown. Bean common mosaic affects primarily the French or snap beans *(Phaseolus vulgaris)* but also other *Phaseolus* sp., while bean yellow mosaic also affects peas, clovers, vetch, black locust, gladiolus, and yellow summer squash. Both diseases are widespread in bean fields, with common mosaic being more widespread than yellow mosaic. Often, 80–100 percent of the plants in some fields are infected. Depending on the growth stage of the plant at the time of infection, the plants may be stunted to a smaller or greater extent, and losses may vary from slight up to 35 percent for common mosaic and up to 100 percent for yellow mosaic. Usually both diseases occur in the same fields and often on the same plants.

○ *Symptoms.* Bean common mosaic causes stunting of the plants and mottling and malformation of the leaves (Figure 14-26, A). The leaves show mild mottling or they have rather large, irregularly shaped light-yellow and light-green areas. Often, leaves are narrower and longer than normal and show considerable puckering, consisting of raised dark-green areas along the main veins, while the leaf margins curl downward. The younger the bean plants at the time of infection the more dwarfed and spindling they remain and the smaller the crop they produce. Pods may be mottled or malformed and the seeds are shriveled and undersized. In some varieties the roots turn dark to almost black and become necrotic. External discoloration may also develop on young stems and petioles, while vascular necrosis is evident in the root, stem, leaves, and pods.

Bean yellow mosaic produces symptoms similar to bean common mosaic, and in the field it is usually impossible to distinguish it from bean common mosaic. Generally, however, bean yellow mosaic produces a much more yellow mottling of the leaves with an intense contrast between the yellow and the green areas. Also, bean plants infected with yellow mosaic are

FIGURE 14-26 Symptoms of bean common mosaic on bean leaves (A), and of bean yellow mosaic on bean leaves (B) and pods (C). (Photo A courtesy U.S.D.A.)

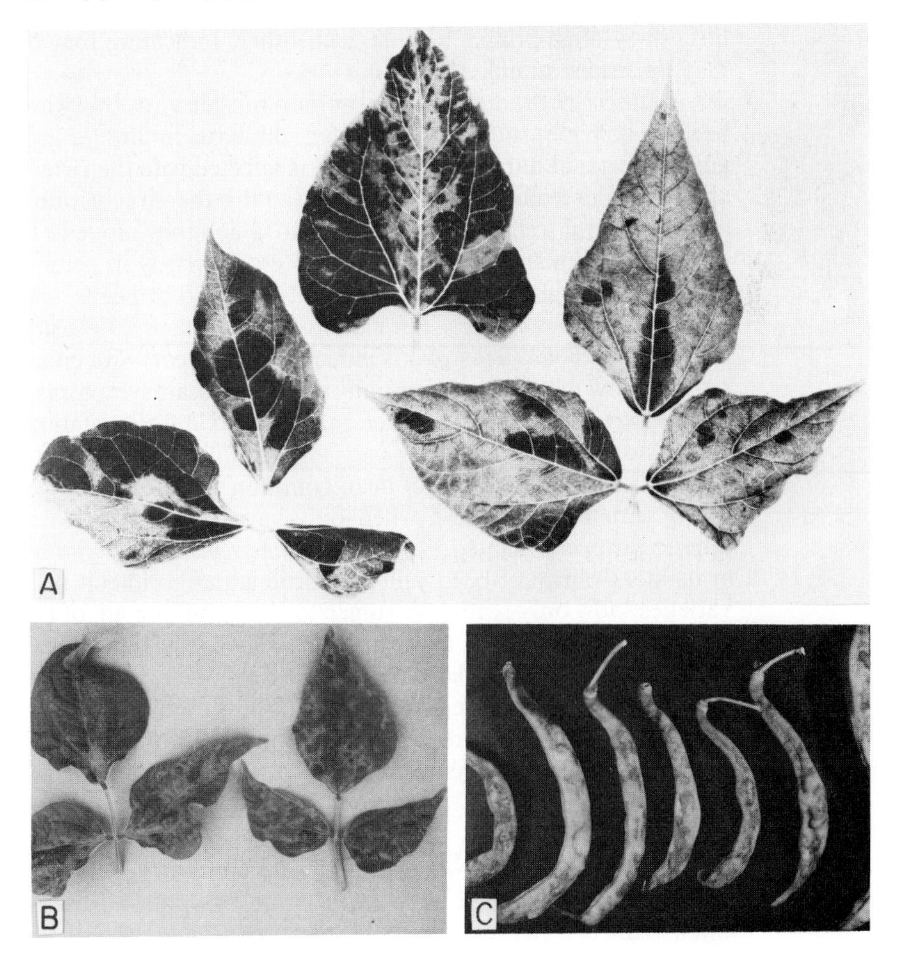

much more dwarfed and bunchy than those infected with common mosaic. Yellow mosaic also produces greater leaf malformation and pod distortion than common mosaic (Figure 14-26, B, C). In both diseases, however, symptoms vary greatly with the variety and with the virus strain prevailing in the area.

◦ *The Pathogens: Bean Common Mosaic Virus and Bean Yellow Mosaic Virus.* Both viruses are potyviruses, that is, they have particles that are filamentous and measure 750 × 15 nm.

Both viruses are transmitted by several species of aphids in the nonpersistent manner, most of the vectors being common to both viruses. Both viruses are also readily transmitted by sap inoculation. Bean common mosaic is, moreover, transmitted readily through bean seeds, especially when the mother plants are infected while young. As much as 83 percent of the seed of diseased plants may produce virus-infected plants. Seed transmission is the most important source of initial crop infection in bean fields. Bean common mosaic virus is also transmitted to new plants through pollen. Bean yellow mosaic virus is not transmitted through the seed in beans but is transmitted in about 3–6 percent of the seeds of several other legumes.

Although bean common mosaic and bean yellow mosaic viruses differ in the kinds of hosts they attack and in seed transmissibility, they resemble each other in numerous characteristics, they are serologically related, and in some hosts they cross-protect against each other, indicating that the two viruses may be distant strains of the same virus.

◦ *Development of Disease.* Bean common mosaic virus overwinters in infected bean seed. When such seed is planted, the virus multiplies in the cells of the growing plant and the produced plant is infected with the virus. Subsequently, aphid vectors feeding on such plants acquire the virus within a few seconds and transmit it to healthy bean plants to which they move to feed.

Bean yellow mosaic virus overwinters primarily in perennial hosts such as clovers and gladiolus, from which it is spread to beans and other annual hosts by its aphid vectors. The same vectors, of course, transmit both viruses from bean to bean. Bean plants inoculated by insects with either the common or the yellow mosaic viruses usually develop mosaic symptoms within 10 days of infection. Symptoms, however, may be mild and almost unnoticeable if the weather remains cool.

◦ *Control.* The best control of bean common mosaic is obtained through the use of virus-free seed and when only varieties resistant to bean common mosaic virus are planted. Several varieties resistant to common mosaic are available. Control of bean yellow mosaic is more difficult because few bean varieties show only partial resistance to some, but not all, of the strains of the virus and because of the overwintering of the virus in perennial hosts such as clovers and gladiolus. Planting beans in fields removed from gladiolus fields and destruction of clovers from around bean fields are recommended, but such control measures are difficult to carry out in practice and, besides, their effectiveness is not always apparent.

• Soybean Mosaic

Soybean mosaic occurs wherever soybeans are grown. Affecting only soybean in nature, it is regarded as one of the most important diseases of soybean, often reducing yields in the field by 25 percent or more. The leaves show mosaic symptoms and exhibit chlorotic areas among darker green rolled, distorted, or puckered areas. The plants remain somewhat stunted and produce fewer pods, which sometimes are malformed and seedless. Symptoms are more severe at cool temperatures (around 18°C), than they are at mild temperatures (24–25°C), and are masked at temperatures above 30°C. Soybean mosaic virus is carried in 30 percent or more of the seeds from diseased plants and can remain viable in the seed for at least 2 years. Infected seeds fail to germinate or produce diseased seedlings. The virus is a potyvirus, 750 × 11 nm. In addition to being transmitted by seed, soybean mosaic virus is also transmitted by at least 20 species of aphids in the nonpersistent manner.

- **Peanut Mottle**

Peanut mottle occurs worldwide. It affects peanuts, bean, soybean, and pea. The symptoms range from a mild systemic mottle to a limited or severe necrosis, depending on host variety and virus strain. Losses are slight to moderate (10–20 percent). Peanut mottle virus is a potyvirus, 750 × 11 nm. It is present in 10–30 percent of the seed from infected plants and is also transmitted by several aphids in the nonpersistent manner.

- **Alfalfa Mosaic**

Alfalfa mosaic occurs worldwide and is common. It affects most legumes, potato, tomato and tobacco, and many herbaceous and woody plants in other families. It is generally severe and reduces yields. Infected plants generally show mottles, mosaics and malformations, stunting, and necrosis. Alfalfa mosaic virus has bacilliform particles of four different lengths from 28 to 58 nm long by 18 nm wide. The virus is transmitted by 10–50 percent of the seed and by at least 14 aphid species in the nonpersistent manner.

SELECTED REFERENCES

"C. M. I./A. A. B. Descriptions of Plant Viruses," Specific legume viruses: Alfalfa and clovers, Nos. 111, 131, 211, 229; Beans, Nos. 20, 29, 40, 81, 101, 108, 192, 223, 231, 246, 274, 286; Soybeans, Nos. 93, 179; Cowpeas, Nos. 49, 134, 140, 197, 209, 212; Peas, Nos. 112, 120, 146, 257, 286; Peanuts, Nos. 92, 141, 150, 235.

Edwardson, J. R. and Christie, R. G., eds. (1986). "Viruses Infecting Forage Legumes." Vols. I–III. I.F.A.S. Univ. of Florida, Gainesville.

Graham, J. H., Stuteville, D. L., Frosheiser, F. I. and Erwin, D. C. (1979). "A Compendium of Alfalfa Diseases." Am. Phytopathol. Soc., St. Paul, Minnesota.

Hagedorn, D. J. (1974). "Virus Diseases of Pea, *Pisum sativum*," Monogr. No. 9. Am. Phytopathol. Soc., St. Paul, Minnesota.

Pierce, W. H. (1934). Viruses of the bean. *Phytopathology* **24,** 87–115.

Porter, D. M., Smith, D. H., and Rodriguez-Kábana, R., eds. (1984). "Compendium of Peanut Diseases." Am. Phytopathol. Soc., St. Paul, Minnesota.

Sinclair, J. B. ed. (1982). "Compendium of Soybean Diseases." Am. Phytopathol. Soc., St. Paul, Minnesota.

Virus Diseases of Beets

Sugar beets and table beets are affected by about 15 viruses, some of which cause huge losses in beet weight and sugar content. The most important viruses affecting beets are beet curly top, beet yellows, beet mosaic, and beet western yellows viruses. Other viruses affecting beets include beet yellow stunt, beet leaf curl, and beet necrotic yellow vein viruses.

• Curly Top of Sugar Beets

Curly top occurs primarily and is most destructive in the western half of North America and in Turkey. Curly top is most destructive on sugar beet, bean, tomato, flax, melons, and spinach. The virus infects more than 150 species of herbaceous plants belonging to more than 50 families.

Curly top damages plants by killing young plants and causing stunting, malformations, reduced yields, and lower quality in older plants. Losses from curly top of sugar beets and tomato have sometimes been so severe that vast areas previously planted to these crops had to be completely abandoned after years of destructive outbreaks of curly top.

◦ *Symptoms.* The first symptoms appear as a clearing and swelling of the veins of the younger leaves, the edges of which begin to roll and curl inward. If the plant is infected at the seedling stage, while the root is still about 1 cm or less in diameter, it makes little further growth; it either dies shortly or remains as a tight ball of stunted leaves for several weeks or months, and finally dies. Infected larger plants remain stunted and produce more but smaller leaves than healthy plants. Most of these leaves become curled and their veins become swollen and give rise to little nipplelike swellings on the lower side of the leaves (Figure 14-27). At times, a sticky brownish fluid may be exuded from the veins, and this collects in droplets along the leaf stalks. Affected leaves generally remain dark green for a time, but they eventually become

FIGURE 14-27 (A) Sugar beet plant infected with curly top virus. (B) Reaction of sugar beet plants of the same variety to three strains of the virus of different degrees of virulence: mild, moderately severe, and severe. (C) Longitudinal sections of CTV-infected (left) and healthy sugar beet roots. The dark lines in the diseased root represent necrotic vascular bundles. (D) Bean plant infected with CTV compared to healthy plants on both sides. (Photos A, B, and C courtesy C. W. Bennett, photo D courtesy U.S.D.A.)

yellow, then brown, and usually die prematurely. Leaves that were mature at the time of infection develop no curling or swelling of the veins, but they soon turn yellow and die. Infection of fully matured plants late in the season usually has little or no effect on the appearance or yield of the plant.

The roots of curly-top-infected sugar beets are affected almost proportionally to the damage caused by the disease on the tops of the plants. The younger the plant at the time of infection, the smaller the root. In many instances the roots of the diseased beets are exceedingly hairy, and the root tissue is woody and tough. In cross sections, infected roots show brownish rings indicating degenerative changes in the vascular tissues. In longitudinal sections, the same tissues appear as a discolored line (Figure 14-27, C).

◦ *The Pathogen: Curly Top Virus (CTV).* Curly top virus is a geminivirus, that is, it has paired particles about 18 × 30 nm. The virus is not sap transmitted. In nature it is transmitted by the leafhopper *Circulifer tenellus.* In beet leaf juice the virus retains its infectivity for about 7 days, but it can remain infectious for 4 months in dried beet leaves, and 6 months in dried insect vectors.

The leafhopper *Circulifer tenellus,* transmits the virus after feeding for short periods or after incubation periods varying from 4 hours to more than 5 days depending on the concentration of the virus in the vector's food. Viruliferous leafhoppers transmit the virus efficiently for 2 or 3 days after the incubation period, but subsequently their efficiency declines steadily. Thus, although the virus appears to be both stylet borne and circulative, it does not multiply in the vector.

In the plant the virus seems to be limited almost entirely to the phloem and adjacent parenchyma cells.

◦ *Development of Disease.* The virus overwinters primarily in infected perennial and biennial weeds such as plantago, pepper grass, Russian thistle, and filaree. It also overwinters in perennial ornamental hosts, in annuals in the greenhouse, and occasionally in the overwintering adults of the insect vector. The insects feed on the infected wild plants in the winter and spring, become viruliferous, and carry the virus to cultivated crops in late spring or summer. The insect feeds by inserting its stylet into the phloem of infected or healthy plants and transmits the virus in the process. Once inside the phloem, the virus moves rather rapidly through it, that is about 2–3 cm per minute, while at the same time it causes destructive changes in the sieve elements of the phloem.

After inoculation, the first symptoms on the plant may appear within 24 hours when the temperature is high, but usually there is an incubation period of 7–14 days under normal temperatures, and even longer during cool weather. The virus, however, spreads throughout the plant quickly, so that a plant may become a source of virus for new leafhoppers within 5 hours of its inoculation.

Curly top virus–infected plants exhibit hypertrophy, hyperplasia, and necrosis of the phloem elements. The hyperplastic sieve elements apparently are not functional and sometimes spread beyond the limits of the phloem, into the cortex and the xylem. Hypertrophy and hyperplasia also occur in parenchyma cells adjacent to the phloem. These cells become closely packed, leave no intercellular spaces, and their chloroplasts are few, small, and pale and result in the appearance of vein clearing. Further hypertrophy and hyperplasia of these cells produce thickening and distortion of the veins, result in the formation of protuberances, and since they occur primarily on the underside of the leaves, cause upward rolling of the leaves. Degeneration and necrosis of phloem also occur in the stem and root. The root is retarded in growth and produces numerous laterals. Although hypertrophy and hyperplasia occur in the phloem and adjacent parenchyma cells, most other cells remain hypoplastic and result in dwarfing and stunting of the whole plant.

◦ *Control.* Insecticide sprays carried out systematically and over a large area simultaneously have been effective in controlling the vector. Statewide programs to eradicate the leafhopper by mapping and spraying the breeding ground of the leafhopper with insecticides have markedly reduced the disease in some areas.

The most effective and most widespread means of curly top control today is through the use of resistant varieties. Several sugar beet varieties resistant to curly top are available. Resistant varieties to curly top have also been developed for tomato, for bean, and for other crops.

• Beet Yellows

Beet yellows occurs in all major sugar beet growing areas of the world. It causes a yellows disease in sugar beets, table beets, and spinach. The outer and middle leaves of infected plants become yellow, thickened, and brittle and may become necrotic. Beet production is reduced drastically, as is sugar content in the beets produced. Beet yellows virus is a closterovirus, 1250 nm long by 10 nm in diameter. It is transmitted by more than 20 aphid species in the semipersistent manner.

• Beet Mosaic

Beet mosaic occurs worldwide. It affects sugar beets and spinach and causes slight to moderate reduction in yield. Leaves show a mottle and are sometimes distorted, and plants are slightly stunted. Beet mosaic virus is a potyvirus, 730 × 11 nm. It is transmitted by at least 30 aphid species in the nonpersistent manner.

• Beet Western Yellows

Beet western yellows probably occurs worldwide. It affects sugar beets, spinach, lettuce, and many crucifers. It causes chlorosis and stunting and moderate reductions in yield. Beet western yellows virus is a luteovirus, that is, it is isometric, 26 nm, causing yellowing symptoms. It is transmitted by eight species of aphids in the persistent (circulative) manner, persisting in the vector for more than 50 days.

SELECTED REFERENCES

Bennett, C. W. (1971). "The Curly Top Disease of Sugarbeet and Other Plants," Monogr. No. 7. Am. Phytopathol. Soc., St. Paul, Minnesota.

"C. M. I./A. A. B. Descriptions of Plant Viruses," Specific beet viruses, Nos. 13, 53, 144, 207, 210, 268.

Mink, G. I., and Thomas, P. E. (1974). Purification of curly top virus. *Phytopathology* **64,** 140–142.

Mumford, D. L. (1974). Purification of curly top virus. *Phytopathology* **64,** 136–139.
Whitney, E. D. and Duffus, J. E. (1986). "Compendium of Beet Diseases and Insects." APS Press, Am. Phytopathol. Soc., St. Paul, Minnesota.

Virus Diseases of Grain Crops and Sugarcane

Wheat, barley, oats, rye, turf and pasture grasses, rice, corn, sorghum, millets, and sugarcane are affected by more than 50 viruses that are almost entirely restricted to crops within the family Gramineae. Many of these viruses infect several or all of the listed crops in nature, while others seem to infect only one or two of these crops. Virus infections in grain crops range from latent to lethal. Most of them cause patterned foliar mosaics consisting of chlorotic mottles, dashes, blotches, or streaks. The same viruses may also cause stunting, yellowing, rosetting, and necrosis. As with other viruses, most of those affecting Gramineae survive from one season to the next in perennial grasses, a few of them in seeds, and some, possibly, in their vectors. The vectors of several viruses affecting Gramineae are aphids, but many of these viruses are transmitted by leafhoppers, eriophyid mites, nematodes, and fungi.

The most important viruses of gramineous crops are barley yellow dwarf, barley stripe mosaic, wheat soil-borne mosaic, wheat streak mosaic, rice hoja blanca, rice tungro, maize dwarf mosaic, and sugarcane mosaic viruses. Among the other viruses, many of them very important in certain countries or continents, are wheat spindle mosaic, wheat striate mosaic, maize streak, maize chlorotic dwarf, rice black-streaked dwarf, rice dwarf, rice transitory yellowing, sugarcane fiji disease, and sugarcane Sereh disease.

• Barley Yellow Dwarf

Barley yellow dwarf occurs throughout the world. The barley yellow dwarf virus attacks a wide variety of gramineous hosts, including barley, oats, wheat, rye, and many lawn, weed, pasture, and range grasses.

Barley yellow dwarf affects plants by causing stunting, reduced tillering, suppressed heading, sterility, and failure to fill the kernels. In some localities, plant damage may be so severe that entire fields are destroyed and the crops are not worth harvesting. Of the three main crops, barley, oats, and wheat, oats is the most severely affected and suffers serious losses annually. In years of barley yellow dwarf outbreaks, some states reported yield losses ranging from 30 to 50 percent of their entire oat crop, while barley and wheat losses ranged between 5 and 30 percent. To the losses in yield of these cereals should be added losses in quality of the grain and losses in forage crops from the resulting failure or reduced productivity of pasture, range, and meadow grasses.

◦ *Symptoms.* The first symptoms on barley yellow dwarf–infected plants appear as yellowish, reddish, or purple areas along the margins, tips, or lamina of the older leaves. The discolored areas soon enlarge and frequently surround still unaffected green areas. The tissues along the midrib usually remain green longer than the rest, but finally they, too, become discolored. In late infections the flag leaf may be the only one that develops the characteristic discoloration. In seedling infections, leaves may emerge distorted, curled,

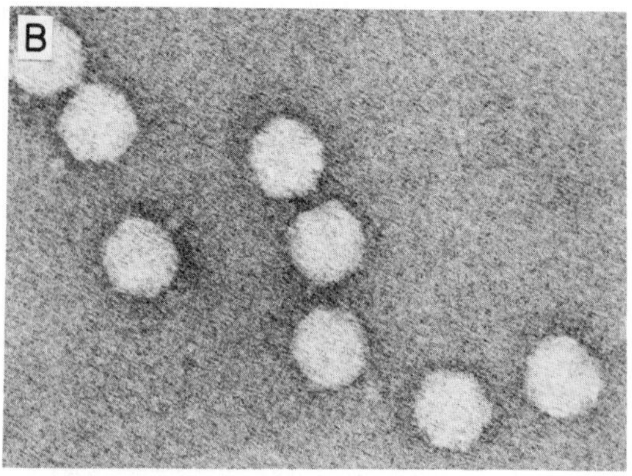

FIGURE 14-28 (A) Symptoms of barley yellow dwarf virus on barley plants infected in early tillering stage (left), and in jointing stage (middle). Healthy plant at right. (B) Particles of barley yellow dwarf virus. (Photo A courtesy U.S.D.A. Photo B taken by H. W. Israel and supplied courtesy W. F. Rochow.)

and with serrations. Leaves developing after the infection are progressively shortened, narrower and stiffer than normal, and grow more upright than normal.

The stem internodes of infected plants are shorter, and sometimes the head fails to emerge (Figure 14-28). Tillering is reduced or completely suppressed in oat and wheat plants, but severely stunted barley plants may show excessive tillering. Inflorescences of diseased plants emerge later and are smaller. Many of the flowers are also sterile, and the number and weight of kernels are reduced. The root systems of diseased plants are drastically reduced in weight but show no characteristic symptoms.

◦ *The Pathogen: Barley Yellow Dwarf Virus (BYDV).* The virus is a luteovirus, that is, it has a polyhedral particle about 22 nm in diameter (Figure 14-28, B) and causes yellowing of infected plants. The virus concentration in infected plants is very low; 1 liter of plant juice contains only about 25–50 micrograms (μg) of virus. Barley yellow dwarf virus is not mechanically transmissible to plants but is readily transmitted by insects. At least 14 species of aphids (Figure 14-16) serve as vectors of BYDV. Most aphids require an acquisition feeding period of about 24 hours and an inoculation feeding period of 4–8 hours or more. Some of the aphid vectors seem to be much more efficient vectors of BYDV than others. BYDV is circulative in its vectors, all viruliferous aphids remaining so for 2–3 weeks. The virus is not passed from adults to their progeny, but it does persist through molting.

Barley yellow dwarf virus consists of numerous strains, which differ in their relative virulence on different host varieties, in the symptoms they produce, and in their transmission by different aphid vectors.

◦ *Development of Disease.* Barley yellow dwarf virus in the northern areas overwinters in perennial grass hosts, while in the south it may overwinter in annual grasses and fall-sown cereals and in viruliferous adult aphids. The spread of the virus depends on the spread of the aphid vectors. In a few areas, such as parts of Oregon and Washington, the climate is favorable or tolerable for some of the aphid vectors throughout the year, and they annually carry the virus from wild grasses to cultivated cereals, causing frequent outbreaks of the disease and subsequent serious losses. In most of the other main cereal-producing areas of the United States and Canada, the winter temperatures are too low to allow overwintering of the adult aphids, but they can overwinter in the egg stage. It appears, however, that the aphid populations resulting from eggs is rather small, and since they must first feed on virus-infected perennial grasses in order to become viruliferous, the rate of BYDV spread by them is rather limited and does not result in severe outbreaks of the disease unless weather conditions become extremely favorable for aphid multiplication and a virus reservoir is plentiful and readily accessible to the aphids. The worst epidemics, however, develop from virus brought into cereal fields in the spring by viruliferous aphids migrating northward from the south. Winter survival of large populations of aphids in north Texas and Oklahoma, followed by properly timed south winds, could move the aphids northward in stages to capitalize on the northward progression of spring and the successive appearance of wheat, oat, and other susceptible grass seedlings on which these aphids thrive. Barley yellow dwarf epidemics generally occur when the spring and early summer weather is cool and moist.

The stage of host development at the time of infection is a crucial factor in disease development. The most severe symptoms result only from infection of the annual cereals in the seedling stage. Infected seedlings may die as in a seedling "blight" or they may survive for a time with the third or fourth leaf emerging distorted. Such plants usually fail to head, and if they do, the inflorescence and entire plant are extremely small. In later stages of infection, in which the virus has progressively less time in which to affect the host, the disease severity is reduced proportionately, and only the last formed leaf may show mild symptoms. In fall-sown cereals, BYDV infections increase winter killing of plants as well as reduce yields, and the effects are much more pronounced in young seedling infections than in infections of more developed plants.

◦ *Control.* Control of barley yellow dwarf through control of the aphid vectors with insecticides has been attempted repeatedly, but the results have been disappointing.

The main hope for control of BYDV is the use of resistant varieties. Most of the commercial varieties of oats, barley, and wheat commonly grown in the United States are susceptible to BYDV, but some are less susceptible than others. A number of varieties have been found or developed that show some resistance to BYDV. An extensive breeding program to develop varieties of the three main cereals that can withstand heavy barley yellow dwarf epidemics is currently being carried out.

• Barley Stripe Mosaic

Barley stripe mosaic occurs worldwide. It affects primarily barley and wheat. Losses may vary from slight to 35 percent, depending on the percentage of plants infected. Symptoms vary from mild yellow-brown stripes, chlorotic spots, or mottle to necrotic areas on leaves. Infected plants show moderate to severe stunting, sterility of some florets, and poorly developed heads and kernels. Barley stripe mosaic virus is a hordeivirus, that is, it is a tripartite rod-shaped virus with particles 100–150 nm long by 20 nm in diameter. It is transmitted only by seed and pollen. No vector is known.

- **Wheat Soil-Borne Mosaic**

Wheat soil-borne mosaic probably occurs worldwide. It affects winter wheat, barley, and rye. Infected plants show a yellow-green mosaic and, depending on the variety, slight to extreme stunting of tops and roots and excessive tillering or rosetting. Wheat soil-borne mosaic virus is considered a possible member of the tobamovirus group, but it has particles of two lengths, 300 × 20 nm and 110–160 × 20 nm. It is transmitted by the soil fungus *Polymyxa graminis.*

- **Wheat Streak Mosaic**

Wheat streak mosaic occurs in North America and Europe. It affects all small-grain crops and grasses and causes severe losses on most of them. It causes mild to severe mosaic, slight to severe stunting and leaf necrosis, death of tillers, and reduced seed set and seed weight. The wheat streak mosaic virus is a potyvirus, 700 × 11 nm. It is transmitted by the mite *Aceria tulipae.*

- **Rice Hoja Blanca (White Leaf)**

Rice hoja blanca occurs in North and South America and in Japan. It affects rice and is sometimes severe on most other small-grain crops. It is generally severe on rice. Leaves of infected plants are striped, mottled, or white. The plants are shorter than normal, the panicles smaller, and the floral parts absent or sterile. Rice hoja blanca virus consists of filaments 7–8 nm in diameter of uncertain length and taxonomy. The virus is transmitted by two planthoppers in the circulative, propagative manner.

- **Rice Tungro (Yellow Orange Leaf)**

Rice tungro occurs in southern Asia, from Pakistan to the Philippines. It is the most severe of rice viruses. Infected plants are stunted and show mottling and yellowing of the leaves. Rice tungro virus is probably a member of the maize chlorotic dwarf virus group, that is, it is isometric and 30 nm in diameter, but it is often present together with a bacilliform virus 150–350 nm long by 30 nm in diameter. They are both transmitted by several leafhoppers in the semipersistent manner.

- **Maize Dwarf Mosaic**

Maize dwarf mosaic occurs in the United States and Australia. It affects corn, sorghum, and several grasses. Symptoms on corn and grasses develop only on plants infected early and consist of a stippled mottle, mosaic or narrow streaks on the younger leaves (Figure 14-29B), and shortening of upper internodes. Older leaves show no mosaic but appear yellowish-green and may have yellowish-red streaks. The corn ears remain small and incompletely filled (Figure 14-29C). Yield in susceptible varieties may be reduced by up to 40 percent. Sorghum plants show mosaic followed by red striping and necrotic areas on the leaves. Maize dwarf mosaic virus is a potyvirus, 750 × 11 nm (Figure 14-29A). It apparently consists of at least two, more or less host-specific strains of the sugarcane mosaic virus that in nature infect and damage

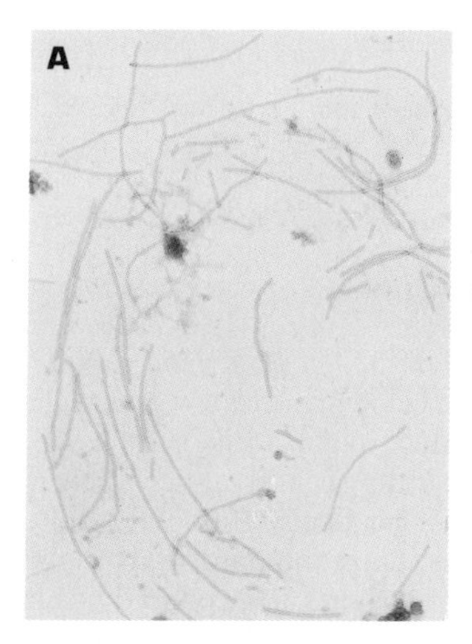

FIGURE 14-29 (A) The threadlike particles of maize dwarf mosaic virus (×38,000). (B) Mosaic symptoms in the form of narrow streaks on the younger leaves of a corn plant infected with MDMV. (C) Smaller and incompletely filled ear of corn produced by MDMV infected plant.

primarily corn. Strain A infects and overwinters in the perennial weed Johnsongrass, while strain B does not infect Johnsongrass. Both strains infect corn, sorghum, and several other annual grain crops and grasses. Both are transmitted by several aphid species in the nonpersistent manner and, rarely, by seed.

• Sugarcane Mosaic

Sugarcane mosaic occurs worldwide, wherever sugarcane is grown. Its many strains also infect corn, sorghum, and the other Gramineae. It can be very severe. The symptoms appear as pale patches or blotches on the leaves, not of uniform width and not confined between the veins. Stems may show mottling or marbling, the affected areas later becoming necrotic. The stems become small and deformed, the shoots remain stunted and produce few twisted or distorted leaves. Cane and sugar yield are reduced severely. Sugarcane mosaic virus is, of course, a potyvirus, 750 × 11 nm. It is transmitted primarily vegetatively during propagation of the crop and also by several aphid species in the nonpersistent manner.

SELECTED REFERENCES

Bruehl, G. W. (1961). "Barley Yellow Dwarf, A Virus Disease of Cereals," Monogr. No. 1. Am. Phytopathol. Soc., St. Paul, Minnesota.

"C. M. I./A. A. B. Descriptions of Plant Virus," Specific Gramineae viruses: Small grains, Nos. 32, 48, 68, 77, 99, 123, 143, 145, 157, 167, 169, 217; Rice, Nos. 67, 100, 102, 135, 149, 172, 248, 269; Maize, sorghum, sugarcane, Nos. 72, 88, 94, 119, 133, 194, 220, 283, 284.

Gordon, D. T., Knoke, J. K., and Scott, G. E., eds. (1981). "Virus and Viruslike Diseases of Maize in the United States," South. Coop. Ser. Bull. No. 247. Ohio Agric. Res. and Develop. Center, Wooster, Ohio.

Mathre, D. E., ed. (1982). "Compendium of Barley Diseases." Am. Phytopathol. Soc., St. Paul, Minnesota.

Shurtleff, M. C., ed. (1980). "Compendium of Corn Diseases," 2nd Ed. Am. Phytopathol. Soc., St. Paul, Minnesota.

Slykhuis, J. T. (1976). Virus and virus-like diseases of cereal crops. *Annu. Rev. Phytopathol.* **14,** 189–210.
Wiese, M. V. (1987). "Compendium of Wheat Diseases." Am. Phytopathol. Soc., St. Paul, Minnesota.

Virus Diseases of Fruit Trees and Grapevines

Most of the fruits produced and consumed in the temperate and subtropical regions of the world are produced by plants belonging to a few genera (families) of plants. Thus, the genera that produce pome fruits such as apples (*Malus* sp.), pears (*Pyrus* sp.), and stone fruits such as apricots, cherries, peaches, and plums (*Prunus* sp.) belong to the family Rosaceae, while all citrus fruits, such as oranges, lemons, limes, tangerines, and grapefruits are produced by plants in the genus *Citrus,* and all grapes are produced by plants in the genus *Vitis.* In order to maintain the identity of the fruit variety in these cross-pollinated crops, and for other reasons, all fruit trees and vines are propagated vegetatively, that is, by budding or grafting the variety on seedling or clonal rootstocks or, in grapevines, by rooting vine cuttings. Vegetative propagation tends to carry and perpetuate viruses from any infected mother plants to all the offspring plants and to spread viruses from scion varieties into rootstocks and vice versa.

Over time, many viruses have developed, or have been collected, in each of these genera and families. Thus, it appears that about 25 viruses are already known to affect apple (for example, apple chlorotic leafspot, apple mosaic, tomato ring spot virus); about 10 viruses affect pear (for example, pear stony pit and pear vein yellows); more than 50 viruses affect plants within the genus *Prunus* (for example, cherry leaf roll, peach mosaic, peach rosette mosaic, plum pox, prune dwarf, prunus necrotic ring spot, and tomato ring spot); about 20 viruses affect citrus (for example, citrus tristeza, citrus psorosis, citrus infectious variegation); and another 20 affect grapevines (for example, grapevine fanleaf, grapevine leaf roll, tomato ring spot). Most of these viruses affect plants only within a single genus (such as *Malus, Prunus, Citrus,* or *Vitis*), but a few viruses affect plants in more than one genus within the same family (*Malus* and *Pyrus,* or *Malus* and *Prunus*). Some viruses, however, particularly tomato ring spot virus and tobacco ring spot virus, infect and cause severe diseases on most of these genera (*Prunus, Malus,* and *Vitis*).

The losses caused by viruses in fruit trees vary from slight to extremely severe. Some viruses, such as apple chlorotic leafspot, are latent in most apple varieties—that is, they cause no visible symptoms. Even latent viruses, however, cause on the average an 8–12 percent reduction in plant growth and yield. Other viruses, such as prunus necrotic ring spot, cause severe symptoms and losses in the first few years after the tree is infected (shock phase of the disease), but the symptoms become milder and the losses less severe in subsequent years. Still other viruses, such as citrus tristeza, plum pox, grapevine fanleaf, and tomato ring spot of pome and stone fruits and grapevines, cause symptoms and losses that become progressively worse and lead to decline and even death of the plant. The losses are the result of reduced tree growth, reduced fruit set, reduced fruit size and quality, and of the need to replace trees after substantial investment of time, space, and capital. Fruit tree viruses transmitted by vectors cause much more severe losses than those that lack vectors. Of the viruses transmitted only by vegetative propagation (the majority of fruit tree viruses), latent viruses cause more severe losses than viruses causing obvious symptoms. This occurs because the presence of latent viruses easily escapes detection, and therefore the viruses often are present in all trees of a variety or genus, while viruses that cause symptoms are easily eliminated by avoiding them during vegetative propagation by obtaining propagative material from only symptomless plants.

The vast majority of viruses of fruit trees lack vectors and are spread only by vegetative propagation. These include all viruses affecting apple and pear, all but two affecting citrus, and most viruses affecting stone fruits and grapes. Therefore, the main and only control measure necessary for such viruses is that the mother plants from which buds, grafts, or cuttings are taken be free of viruses and that these propagative organs be placed on rootstocks derived from similarly virus-free mother plants. There are, however, some viruses that are transmitted by pollen (cherry leaf roll, prune dwarf, and prunus necrotic ring spot) and the same viruses plus pear vein yellows, are also transmitted by seed. Control of such viruses, in addition to using virus-free scion wood and rootstocks, also requires that new orchards be planted at some distance (approximately 200 m) from bearing orchards that contain virus-infected trees, so that the new trees will not become infected with virus carried in the pollen arriving from the infected trees. Moreover, several viruses are transmitted by vectors such as mites (peach mosaic), nematodes (cherry leaf roll, peach rosette mosaic, grapevine fanleaf, tomato ring spot, and tobacco ring spot), or aphids (plum pox, citrus tristeza, and citrus vein enation and woody gall). Control of vector-transmitted viruses of fruit trees is difficult. Control of mites and aphids is attempted by spraying the trees with miticides and aphicides, while control of nematodes is attempted primarily by preplanting fumigation of the orchard field with nematicides. None of these controls is very effective, and therefore resistant rootstocks or scion varieties must be employed whenever possible.

• Prunus Necrotic Ring Spot

Necrotic ring spot occurs worldwide in temperate regions. The disease affects most cultivated stone fruits, including sour cherry, cherry, almond, peach, apricot, and plum, many wild and flowering cherries, peaches, and plums, and also some ornamental species such as rose. Necrotic ring spot virus is present in all trees infected with cherry yellows (prune dwarf virus), and it is frequently associated with other viruses.

Necrotic ring spot is the most widespread virus disease of stone fruit trees. In the fruit-producing areas almost all orchard trees in production are infected. The losses caused by necrotic ring spot vary with the *Prunus* species or variety affected and with the time from inoculation with the virus. Successful commercial budding is lower in combinations in which the bud or the rootstock carry the virus than when both are virus free or virus infected. The growth of virus-infected trees may be reduced by 10–30 percent or more, while the yield of virus-infected trees may be 20–56 percent or more lower than that of healthy trees. Trees affected with necrotic ring spot also show increased susceptibility to winter injury.

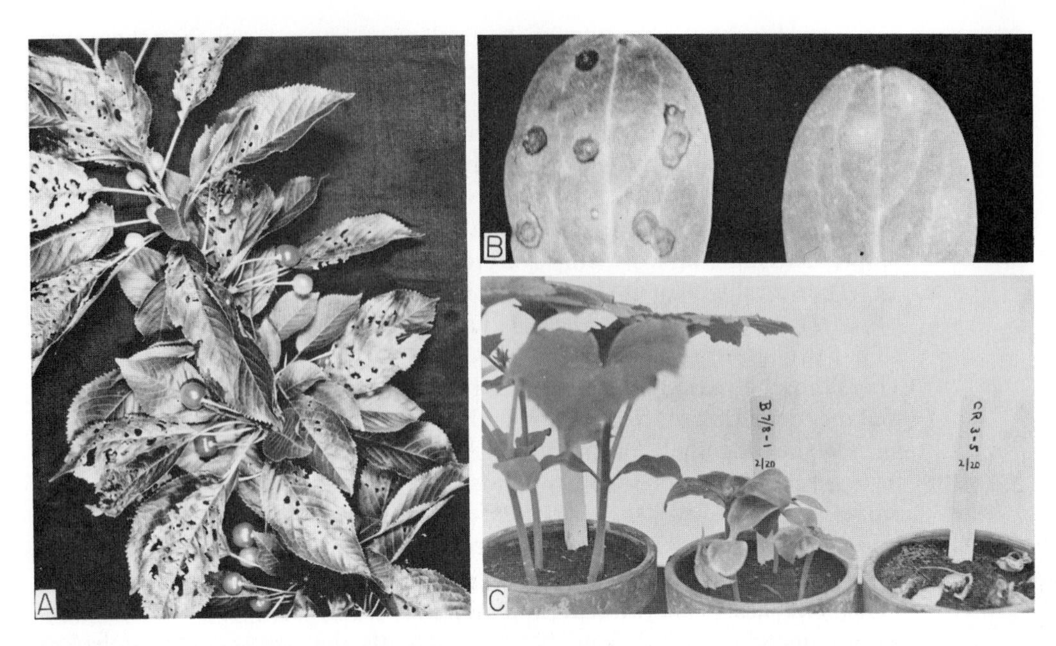

FIGURE 14-30 (A) Necrotic ring spot symptoms ("shock" phase) in sweet cherry. (B) Local lesion reactions of two PNRV strains in watermelon cotyledons about 10 days after inoculation. (C) Cucumber plants inoculated with sap from cherry infected with a mild (middle) and a severe (right) strain of PNRV. Healthy plant at left. (Photos courtesy R. M. Gilmer.)

◦ *Symptoms.* The first symptoms appear as a pronounced delayed foliation of individual branches or entire trees. Leaves on affected branches are small and show light green spots and dark rings 1–5 mm in diameter. In later stages of the disease affected areas may become necrotic, fall out, and give a "shredded leaf" or "tatter-leaf" effect (Figure 14-30, A). Such symptoms, called shock or acute symptoms, are usually limited to the first leaves that unfold, while leaves formed later generally do not show marked symptoms. Affected trees, however, usually have fewer leaves and therefore have a thin appearance.

Blossoms of affected trees have short or no pedicels, the calyx and corolla may be twisted and distorted, and the sepals may develop chlorotic or necrotic rings or arcs. Such severely affected blossoms ordinarily do not set fruit, and occasional fruits also develop small rings similar to those on the leaves.

As a rule, trees severely affected one year show few or no symptoms in subsequent years except for the thinness of foliage. If severe symptoms are present only on a few branches the first year, other branches may show striking symptoms the following year. In many areas, however, trees may continue to show striking ring symptoms and wavy leaf margins for 4–6 years or more.

◦ *The Pathogen: Prunus Necrotic Ring Spot Virus (PNRV).* Necrotic ring spot virus is an ilarvirus, that is, it has a small polyhedral particle about 23 nm in diameter. The virus is very unstable in undiluted plant extracts, beginning to lose infectivity within a few minutes and being completely inactived in a few hours.

PNRV can be transmitted by budding and grafting and by sap from virus-infected tree leaves or petals to leaves of cucumber and of several other herbaceous plants (Figure 14-30, B, C). PNRV is transmitted through seed, the percentage of transmission varying among various species from 5 to 70 percent. The virus is also transmitted through pollen to seeds and to pollinated plants. No vector of PNRV is known.

◦ *Development of Disease.* The virus overwinters in infected stone fruit trees from which it spreads to healthy trees in the spring primarily through infected pollen. PNRV spreads slowly in orchards less than four years old but can spread rapidly in older orchards, probably because older trees have more bloom and therefore are much more subject to infection through pollen than young ones. PNRV can spread over a distance of at least 800 meters, but most infections occur within 15 meters of a known infected tree. Symptoms on trees infected by virus-infected pollen usually develop in the spring one year after inoculation.

Although PNRV becomes systemic and spreads throughout the tree in one or, at most, two seasons, it is not known whether the virus moves from cell to cell, is spread through the phloem, or both. The virus seems to move in the tree at first upward from the point of inoculation, then on parts below this point along a direct path to the base of the tree without affecting side branches. Leaf buds that are just opening at the time the virus reaches them react with development of acute symptoms. Leaves that are fully opened at the time of invasion by the virus develop no symptoms during that growing season. Invasion of the new buds formed at the base of these leaves results in symptoms the following spring when the buds will produce new leaves or flowers. The virus, however, even in systemic infections, does not invade all parts of the tree, but it may leave short gaps along twigs or branches and therefore some buds may be virus free although the entire tree seems to be virus infected.

◦ *Control.* The control of necrotic ring spot of stone fruits is based almost exclusively on starting with virus-free nursery stock and on eliminating PNRV-infected *Prunus* trees from the area where the virus-free trees are grown.

The production of virus-free nursery stock depends on the use of rootstock seedlings derived from virus-free seed and of scion buds derived from virus-free trees of the desired variety. Since no PNRV symptoms are usually present on trees several years after infection, the mother trees providing the seed or the buds are indexed on PNRV-sensitive indicator hosts, which reveal the presence or absence of PNRV in the mother trees. The indicator hosts most commonly used for detection of PNRV are seedlings of peach, *Prunus tomentosa,* or cucumber, and limbs of Shirofugen flowering cherry *(Prunus serrulata).* Peach, *P. tomentosa,* and Shirofugen are inoculated with buds taken from the trees being indexed and inserted into the stem of the seedlings or along the limbs of Shirofugen. Inoculated peach and *P. tomentosa* seedlings produce shock symptoms on the foliage and dieback of terminals, while inoculated limbs of Shirofugen produce a local necrotic reaction around the inserted buds (Figure 14-31). Cucumber seedlings are inoculated at the cotyledon stage with sap obtained from immature leaves of the suspected trees and react by producing local lesions, systemic mottling, or death of the cucumber seedlings (Figure 14-30, B, C). Seed or scion mother trees indexing positive for PNRV are destroyed; those indexing negative for PNRV are used for propagation, but they must be reindexed annually or biannually since they

FIGURE 14-31 Indexing of PNRV on Shirofugen. (A) Buds grafted on twig on left were virus free and are growing. Buds grafted on other twigs were infected with PNRV, caused local necrosis on Shirofugen, and failed to grow. (B) Necrosis of bark around the areas on which PNRV-infected buds have been grafted (right), compared to healthy (left). (C) Localized necrosis and gumming on Shirofugen inoculated with PNRV. (Photos A and B courtesy Department of Plant Pathology, Cornell University. Photo C courtesy R. M. Gilmer).

may become infected with the virus through virus-infected pollen. In recent years, mother trees have been tested for PNRV primarily by serological tests, particularly by ELISA.

After a new orchard has been established with virus-free trees, it is necessary to remove all wild *Prunus* trees from a radius of about 200 meters around the periphery of the orchard to avoid spread of the virus into the orchard. A new orchard should not be planted next to an older one containing infected trees, and any infected trees appearing in the new orchard should be removed immediately to prevent further spread of the virus in the orchard.

• Cherry Leaf Roll

Cherry leaf roll occurs in North America and Europe. It affects cherry, elm, blackberry, raspberry, walnut, dogwood, and other trees and shrubs, and causes severe losses. Infected cherry trees leaf and flower later, and their leaves roll upward in the summer. Leaves may show pale rings and spots. Trees decline in vigor, exude gum through splits in the bark, and usually die within a few years of infection. Cherry leaf roll virus is a nepovirus, that is, a nematode-transmitted polyhedral virus 30 nm in diameter. It is transmitted by budding and grafting, by 0.6 to 100 percent of the seed, by pollen, and by nematodes of the genus *Xiphinema*.

• Plum Pox

Plum pox occurs in Europe and Asia Minor. It affects plum, peach, nectarine, and apricot. It causes devastating losses of fruit quantity and quality and debilitates infected trees. Plum pox, where present, is the most important disease of these trees. Leaves of infected trees show severe mottling, diffuse or bright rings, or vein yellowing and elongated line patterns (Figure 14-32A). Infected plum fruits develop severe pox symptoms (Figure 14-32B) with dark-colored rings or patches on the skin, brown or reddish discoloration on the flesh, and brown spots on the stones (pits). Most of the infected fruits fall prematurely. Peach fruits show mottled rings and distortion (Figure 14-32C), while apricot fruits also show rings but are more deformed, have necrotic rings and bumps (Figure 14-32D), and the stones show striking whitish-yellow rings (Figure 14-32E). Plum pox virus is a potyvirus, 760 × 12 nm. The virus is transmitted by budding and grafting and by several aphid species in the nonpersistent manner.

• Grapevine Fanleaf

Grapevine fanleaf occurs worldwide. It affects only grapes. It occurs in many strains and causes variable symptoms but always severe losses. Depending on virus strain, infected leaves show a green or yellow mosaic, rings, line patterns

FIGURE 14-32 Symptoms caused by plum pox virus. (A) Severe mottle on plum leaves. (B) Pox symptoms on plums. (C) Ring patterns on peaches. (D) Rings and bumps on apricots. (E) Rings on apricot pits.

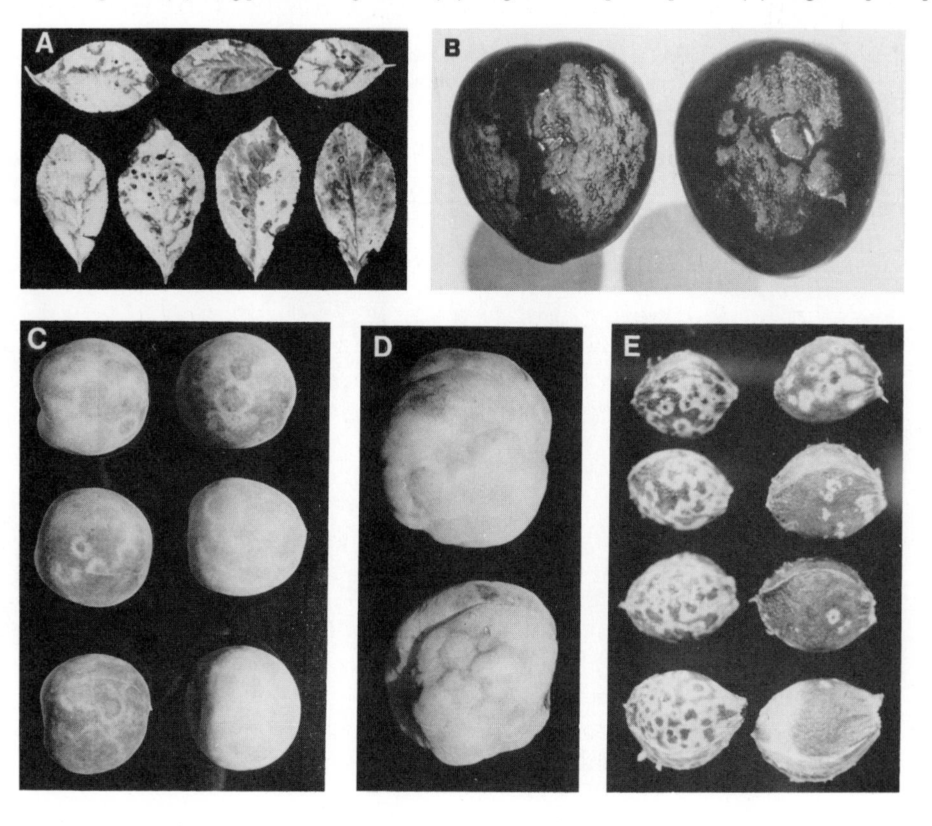

or flecks, with most varieties having smaller, slightly asymmetric leaves, while in others the veins are spread abnormally, giving the leaf a fanlike appearance. Leaves may show a chrome-yellow mottle, the mottled areas later becoming paler, then necrotic, and finally dropping, or leaves may show chrome-yellow areas along main veins of mature leaves. Canes are often deformed, having uneven internode lengths, double nodes, flat leaves and trunks, and bark- and wood-pitting. Fruit production is low. Many flowers shell from clusters, and small, seedless berries develop along with a few normal berries. The vigor and yield of grapevines are reduced progressively, and the vines gradually degenerate and die. Grapevine fanleaf virus is a nepovirus 30 nm in diameter. It is transmitted by budding and grafting, by cuttings, and by nematodes of the genus *Xiphinema*.

- **Prunus Stem Pitting and Decline, Apple Graft Union Necrosis and Decline, and Grapevine Mottle, Yellow Vein and Decline Caused by Tomato Ring Spot Virus**

Many pome fruit and stone fruit varieties and rootstocks, as well as many small fruits, such as grapes, raspberries, and strawberries, are affected by tomato ring spot virus (TomRSV) in North America and suffer severe losses by diseases sketchily described by the names listed above. In apple, the most common symptoms are slight stem pitting on either side of the graft union (Figure 14-33A), followed by gradual necrosis of the graft union (Figure 14-33A & B). This occurs when hypersensitive apple varieties are grafted on

FIGURE 14-33 Stem pitting (small arrow) and graft union brown line (A) and necrosis (B) on apple trunks. They are caused by tomato ringspot virus infection of TomRSV-tolerant rootstocks onto which were grafted TomRSV-sensitive apple varieties.

tomato ring spot–tolerant apple rootstocks such as MM 106, which later become infected with TomRSV via nematode vectors of the virus. Eventually, affected trees show yellowing of foliage, twig dieback, and general decline and death within 3–5 years of the appearance of symptoms at the graft union. In *Prunus* species, there is more extensive and severe pitting of the scion or rootstock, or both, on either side of the graft union, various degrees of necrosis at the union plate, and again foliage yellowing, twig dieback and general decline and death of the trees within 3–5 years. In grapevines and raspberries, the leaves may show mottling, rings, yellow veins, the vines remain stunted, fruit clusters develop poorly or not at all, and berry size may be uneven. All of these diseases are caused by the nematode-transmitted tomato ring spot virus (see page 663) and are among the most important diseases in each of the respective fruit trees or vines.

• Citrus Tristeza

Tristeza occurs in all citrus-growing areas of the world. Tristeza affects practically all kinds of citrus plants but primarily orange, grapefruit, and lime.

Tristeza causes collapse of trees through a more or less sudden wilting and drying of the leaves, followed by death of the tree or by dieback of twigs and partial recovery. Many host plants of tristeza develop stem pitting, followed by poor growth and decline of the tree. Although certain citrus trees can be affected by tristeza even when they are grown on their own roots (seedlings), tristeza causes its most severe damage on trees budded or grafted to rootstocks of certain other species, for example, sweet orange trees growing on sour orange rootstocks. Losses from tristeza vary greatly with the particular scion-rootstock combination grown in a particular area, the strain of the tristeza virus, and the abundance and efficiency of the vectors. In the first 15 years after the discovery of the disease in California, almost 400,000 trees were destroyed or made worthless by tristeza even before the disease had invaded some of the important citrus-growing areas. In the state of Sao Paolo, Brazil, 9,000,000 trees, or about 75 percent of the orange trees of the state, were destroyed by tristeza within 12 years of the appearance of the disease in that state.

◦ *Symptoms.* Tristeza symptoms vary on different hosts and even on the same hosts if they are grown on different rootstocks. The typical tristeza symptoms appear as a quick or chronic decline of trees budded on susceptible rootstocks (Figure 14-34, A–C), but seedling or budded trees may also develop stem pitting (Figure 14-34, D).

The typical tristeza symptoms in older orange trees (Figure 14-34, A–C) appear as suppression of new growth and bronzed to yellow leaves that tend to stand upright. As the disease progresses, the older leaves begin to fall, abscission often taking place between the petiole and the leaf blade, leaving the twigs defoliated or with a few younger leaves. Twigs begin to die back from the tip and later smaller limbs die and only a few weak shoots on the main limbs still have leaves. Twig growth becomes weaker each season until the tree dies, but some trees seem to linger on for many years. In some cases, affected trees collapse quickly after a sudden wilting and drying of the leaves. Tristeza-affected trees also show root symptoms consisting of a marked depletion

FIGURE 14-34 Tristeza disease of citrus. (A) Healthy and diseased sweet orange trees, the latter showing quick decline. (B) Chronic decline of sweet orange tree. (C) Vein clearing of Mexican lime trees caused by tristeza. (D) Effects of three different isolates (A–C) of tristeza of Eureka lemon seedlings. Healthy seedling at right. (E) Stem pitting symptoms on Mexican lime caused by tristeza. (F) Split trunks and roots of sweet orange on sour orange rootstock treated with potassium iodide and showing absence of starch below bud union of tristeza-infected tree and normal starch throughout in healthy tree. (Photos courtesy J. M. Wallace. Photo B taken by L. J. Klotz.)

of starch (Figure 14-34, F), death and decay of the feeder rootlets, and the injury later extends to the larger roots.

Seedlings and budded trees of many different varieties of citrus, especially lime and grapefruit, develop stem pitting as a result of infection with tristeza. Stem pitting consists most commonly of longitudinal grooves or depressions in the stem paralleling the grain of the wood (Figure 14-34, E). Stem pitting symptoms are always associated with decreased vigor of the trees, poor bushy growth, and small and distorted fruit.

Seedlings of certain citrus species, such as sour orange, lemon, grapefruit, and citron, when inoculated with tristeza develop the so-called seedling yellows symptoms. The leaves produced subsequent to inoculation are small and yellow (Figure 14-34, D). Growth ceases after the first few leaves are formed, though a certain amount of restricted growth may occur at a later stage.

The Pathogen: Citrus Tristeza Virus. The citrus tristeza virus is a closterovirus and is one of the longest plant viruses, appearing as a threadlike particle, approximately 2000 nm long by 12 nm in diameter. It is present only in phloem cells. The virus is not transmitted mechanically but is transmitted by grafting and, in nature, by insects. The insect vectors of tristeza are all aphids, the most important vector being the tropical citrus aphid, *Toxoptera citricidus.* The tristeza virus is stylet borne; its vectors become viruliferous after feeding for a few seconds and transmit the virus after equally short feedings.

Tristeza virus exists in nature in numerous strains. At least three distinct strains inducing typical tristeza symptoms are known, and many are known to induce the stem-pitting or yellows syndrome or both. Each strain may infect trees alone or in combination with one or more of the others.

◦ *Development of Disease.* The citrus tristeza virus or virus complex is widely distributed among citrus trees in the citrus-growing areas. In some localities 100 percent of the trees carry the virus. The spread of the virus is accomplished readily through the use of tristeza-infected propagative material, both scion and rootstocks, and through the insect vectors. The tristeza virus is not transmitted through seed.

Infection of citrus plants with tristeza virus apparently occurs only when the virus is introduced into phloem sieve tubes. The virus seems to be limited to a few of the phloem cells in each bundle, and this may account for the ability of a second strain of the virus to infect the same plant by multiplying in some of the remaining phloem cells.

After infection of citrus plants with the tristeza virus, cells adjacent to a sieve tube begin to degenerate and become necrotic. In hosts that develop stem pitting, the degeneration spreads first into the cambium and inhibits the formation of normal xylem and phloem cells. The tissue produced in the lesion is soft and disorganized and usually remains attached to the bark so that when the bark is removed it leaves a pit in the wood. In some feeder roots the cambium is affected in its entirety rather than in localized areas, and therefore no normal xylem or phloem is produced after the infection.

In hosts that develop seedling yellows, the degeneration appears in cells adjacent to phloem sieve tubes, which in leaves and stems are only mildly affected, but most of those in feeder roots become extensively necrotic. Abnormal cambium activity and phloem formation are followed by the eventual deterioration of entire clusters of feeder roots.

In trees budded to rootstocks on which typical tristeza symptoms develop, sieve-tube necrosis appears below the bud union about 7 or 8 months after inoculation with the virus and top symptoms about 10–23 months later. During this time the root tissues utilize and finally exhaust the reserve starch previously stored in them (Figure 14-34, F). Also, during the growing season, new phloem is produced intensively and functions for a time before becoming necrotic. When the reserve starch is finally depleted, the roots rot and decline, or collapse follows.

Tristeza is primarily a disease of citrus trees on sour orange stocks caused by a virus that is transmitted by grafting and by some kinds of vectors. The tristeza symptoms shown by scion trees of certain combinations of scion and rootstock are due not to the susceptibility of the scions but to harmful effects produced by the virus on the phloem cells of the rootstock just below the union.

◦ *Control.* Control of tristeza disease on existing plantings of susceptible scion-stock combinations is difficult or impossible. In tristeza-affected areas, some success has resulted from top working the existing, still healthy sweet orange trees on sour orange rootstocks to tristeza-resistant tops such as lemon, and also from changing existing trees over to resistant rootstocks by inarching. Both practices, however, are expensive and time consuming, and their success is influenced by several factors, particularly, the environment. For these reasons it is generally more satisfactory and economical to remove susceptible trees after they become infected and to replant with a resistant combination.

Avoiding losses in new citrus plantings depends mainly on the use of tolerant scion-stock combinations. The rootstocks most generally recommended are sweet orange, Rough lemon, Cleopatra mandarin, and Troyer citrange. Rootstocks of several other species or varieties are also tolerant to tristeza, but they are objectionable because of their susceptibility to other viruses or to root diseases caused by other pathogens.

In recent years, satisfactory control of tristeza has been obtained by inoculating young citrus trees with selected mild strains of the virus. This practice appears to cross-protect these trees indefinitely from subsequent infection by more severe natural strains of tristeza brought in by its aphid vectors.

SELECTED REFERENCES

Agrios, G. N., and Buchholtz, W. F. (1967). Virus effect on union and growth of peach scions on *Prunus besseyi* and *P. tomentosa* understocks. *Iowa State J. Sci.* **41,** 385–391.

Bennett, C. W., and Coasta, A. S. (1949). Tristeza disease of citrus. *J. Agric. Res. (Washington, D.C.)* **78,** 207–237.

"C. M. I./A. A. B. Descriptions of Plant Viruses," Specific fruit tree viruses: Citrus, Nos. 33, 164, 208; Grapevines, Nos. 28, 103, 186; Pome fruits, Nos. 30, 31, 42, 83; Stone fruits, Nos. 70, 80, 159, 280.

Costa, A. S., and Muller, G. W. (1980). Tresteza control by cross protection: A U.S.-Brazil cooperative success. *Plant Dis.* **64,** 538–541.

Davidson, T. R., and George, J. A. (1965). Effects of necrotic ring spot and sour cherry yellows on the growth and yield of young sour cherry trees. *Can. J. Plant. Sci.* **45,** 525–535.

Frazier, N. W., ed. (1970). "Virus Diseases of Small Fruits and Grapevines." University of California, Div. Agric. Sci., Berkeley.

Klotz, L. J., Calavan, E. C., and Weathers, L. G. (1972). Viruses and viruslike diseases of citrus. *Circ.—Calif. Agric. Exp. Stn.* **559,** 1–42.

Pine, T. S., ed. (1976). "Virus Diseases and Noninfectious Disorders of Stone Fruits in North America." U.S.D.A. Agric. Handbook No. 437.

Posnette, A. F., ed. (1963). "Virus Diseases of Apples and Pears," Tech. Commun. No. 30. Commonw. Agric. Bur., Farnham Royal, Bucks, England.

Wallace, J. M., ed. (1957). "Citrus Virus Diseases." University of California, Div. Agric. Sci., Berkeley.

Virus Diseases of Small Fruits

Small fruits are produced by plants belonging to a few genera (or families) such as *Fragaria* (strawberries) and *Rubus* (raspberries and blackberries) of the family Rosaceae, *Ribes* (gooseberries and currants), and *Vaccinium* (blueberries and cranberries) of the family Ericaceae. Since all small fruits are propagated vegetatively, viruses tend to become transmitted to, and be perpetuated from generation to generation in, the new plants. Furthermore, many of the viruses affecting small fruits are also transmitted by aphids or nematodes, and this further increases their spread, complicates their epidemiology and control, and increases their significance.

Several of the small fruit viruses cause obvious, and severe, symptoms and losses, but several others cause mild symptoms or are latent, and the losses they cause are not nearly as obvious. It has been shown repeatedly, however, that virus-free strawberry and raspberry plants outlive and outproduce even latent virus-infected plants of the same kind by a wide margin. As a result, production and culture of virus-free strawberries and raspberries has been one of the earliest and most successful commercial endeavors of growers to produce virus-free plants.

Production of virus-free plants has generally been attempted by obtaining one or a few virus-free plants of a particular strawberry or raspberry variety from a government agency and multiplying that plant for 2–3 years in a screenhouse under strict vector-free conditions. The new plants obtained from runners or cuttings, respectively, are then planted in fields fumigated for control of nematodes, and are sprayed in the field with aphicides in order to keep low the populations of the vectors and the introduction of viruses by the vectors. More recently, many small fruits and other vegetatively propagated plants have been multiplied on nutrient media in the laboratory by meristem tip culture. This step not only multiplies the initial plant a millionfold more rapidly but under certain conditions also excludes or reduces viruses from new plants, and these plants are then planted in the field.

The number of viruses affecting each of the small fruits is not known with certainty because many of them have yet to be purified and characterized. It appears, however, that strawberries are affected by at least 20 viruses, raspberries and blackberries by about 15 viruses, and each of the others by about 6–10 viruses. The most important viruses of strawberries include strawberry mottle, strawberry vein banding, strawberry mild yellow edge, and strawberry crinkle, all four occurring worldwide and being transmitted by aphids in the semipersistent or persistent manner. Also, strawberry ring spot, present in Europe and transmitted by nematodes. The most important viruses of raspberries (also of blackberries and, some, of strawberries, too) include raspberry mosaic (worldwide), raspberry leaf curl (North America), both transmitted by aphids in the semipersistent manner, and arabis mosaic and raspberry ring spot, both in Europe and both transmitted by 10–100 percent of the seed and by nematodes.

SELECTED REFERENCES

"C. M. I./A. A. B. Descriptions of Plant Viruses." Specific viruses of small fruits, Nos. 16, 106, 126, 163, 165, 174, 198, 204, 219, 267.

Frazier, N. W., ed. (1970). "Virus Diseases of Small Fruits and Grapevines." University of California, Div. Agric. Sci., Berkeley.

Maas, J. E., ed. (1984). "Compendium of Strawberry Diseases." Am. Phytopathol. Soc., St. Paul, Minnesota.

Viruses of Tropical Plants

Some of the most severe viruses of a few of the most common and important tropical crops are discussed in the following sections.

• Banana Bunchy Top

Banana bunchy top occurs in Australia, Asia, the Pacific Islands, and Africa. It causes severe losses. Leaves of infected plants have irregular, dark green streaks along small veins and are bunched together at the top of the plant forming a rosette. Plants are stunted and produce no fruit. Banana bunchy top virus is a luteovirus, that is, it is isometric, 30 nm in diameter. It is transmitted by vegetative propagation and by aphids in the semipersistent manner.

- **Cacao Swollen Shoot**

Cacao swollen shoot occurs in West Africa and in Ceylon. It affects cacao and cola and causes severe losses. Infected plants develop swellings in stems and tap roots, necrosis of side roots, and chlorosis of leaves and pods and produce small, rounded pods that contain fewer, smaller beans. Trees decline and die or they may linger on. Cacao swollen shoot virus is a bacilliform virus, 142 × 27 nm, but has not yet been assigned to a definite taxonomic group. It is transmitted by several species of mealybugs in the semipersistent manner.

- **African Cassava Mosaic**

African cassava mosaic occurs in Africa and Java, and it causes severe losses. The leaves show typical mosaic symptoms, often accompanied by malformations. Entire plants show general stunting. The African cassava mosaic virus is a geminivirus. It is transmitted by whiteflies.

- **Papaya Ring Spot Virus**

Papaya ring spot virus occurs wherever papayas are grown. It infects papaya and also cucurbits. Papaya losses are severe. Infected papaya leaves appear mottled and distorted. Fruits develop rings and spots on their surfaces, while stems and petioles develop streaks. Plants are stunted, and yields are reduced considerably. Papaya ring spot virus is a potyvirus, 780 × 12 nm. It is transmitted by more than 20 species of aphids in the nonpersistent manner.

SELECTED REFERENCES

"C. M. I./A. A. B. Descriptions of Plant Viruses," Specific viruses of tropical fruits, Nos. 10, 11, 56, 90, 173, 292.
Smith, K. M. (1972). "A Textbook of Plant Virus Diseases," 3rd ed. Academic Press, New York.

Virus Disease of Forest Trees

Relatively few virus diseases have been detected and identified in forest trees, and most of them are caused by viruses already described under some of the other crops. Thus, ash ring spot is caused by tobacco ring spot virus, ash line pattern is caused by tobacco mosaic virus, birch line pattern is caused by apple mosaic virus, elm mosaic is caused by cherry leaf roll virus. Maple is infected by peach rosette virus and by tobacco ring spot virus. About the only known virus diseases of forest trees that are caused by distinct viruses are elm mottle, caused by elm mottle virus (an ilarvirus), and poplar mosaic, caused by poplar mosaic virus (a carlavirus). While it is not clear why so few viruses have been detected in forest trees, it is assumed to be due to three reasons: (1)

few virologists have examined forest trees for viruses; (2) forest trees, because of their location and size are more difficult to observe for viruses; (3) forest trees are propagated from seed in nature, which not only is less conducive to transmission of viruses but also it does not involve grafting onto other stocks, which often helps reveal the presence of a virus in fruit trees.

Virus Diseases of Ornamental Plants

Many ornamental plants are affected by some of the same viruses that affect crop plants (for example, tobacco mosaic, tomato ring spot, and cucumber mosaic) and develop symptoms that may be similar to or quite different from those produced by crop plants infected with these viruses. Many ornamental plants, however, are also affected by some viruses that are more or less specific for and cause disease only on that particular genus or family of ornamental plants. Many of these viruses are present worldwide, while others seem to be restricted so far to a single continent, such as North America or Europe. Many of these viruses are transmitted by aphids. Examples are carnation etched ring, carnation vein mottle, cattleya mosaic, dahlia mosaic, iris mild mosaic, narcissus yellow stripe, and tulip breaking. Some are transmitted by nematodes, for example, carnation ring spot; some by whiteflies, for example, abutilon mosaic; and some, such as hydrangea ring spot, by the pruning knife. Several viruses of ornamentals are transmitted only by budding or grafting. Examples are lilac ring mottle and rose mosaic. For several viruses no vector is known, but the viruses are transmitted by contact or sap transfer. Examples are cactus virus X, carnation mottle, cymbidium mosaic, narcissus mosaic, odontoglossum ring spot, pelargonium leaf curl, and peony ring spot. Virus diseases of many ornamentals cause severe losses because they reduce the growth of these plants, the number and size of flowers, and the aesthetic quality of foliage and flowers of infected plants.

SELECTED REFERENCES

"C. M. I./A. A. B. Descriptions of Plant Viruses." (Numerous viruses of ornamental plants).

Horst, R. K. (1983). "Compendium of Rose Diseases." Am. Phytopathol. Soc., St. Paul, Minnesota.

Smith, K. M. (1972). "A Textbook of Plant Virus Diseases," 3rd ed. Academic Press, New York.

Plant Diseases Caused by Viroids

To date, ten plant diseases, including potato spindle tuber, citrus exocortis, chrysanthemum stunt, chrysanthemum chlorotic mottle, hop stunt, tomato bunchy-top, avocado sunblotch, coconut cadang-cadang, and cucumber pale fruit, have been shown to be caused by viroids. So far, no animal or human disease has been shown to be caused by a viroid. It is likely, however, that viroids will be soon implicated as the causes of several "unexplained" diseases in plants, animals, and humans. Almost all the information on viroids up to now has been obtained from studies with the potato spindle tuber viroid and the citrus exocortis viroid.

Viroids are small, low-molecular-weight ribonucleic acids (RNA) that can infect plant cells, replicate themselves and cause disease (Figure 14-35). Viroids differ from viruses in at least two main characteristics: (1) the size of RNA, which has a molecular weight of 110,000 to 130,000 in viroids compared to 1,000,000–10,000,000 for self-replicating viruses, (2) the fact that virus RNA is enclosed in a protein coat while the viroids lack a protein coat and apparently exist as free RNA.

The small size of RNA of viroids indicates that they consist of about 250–400 nucleotides and therefore lack sufficient information to code for even one enzyme (replicase) that may be required to replicate the viroid. The existence of viroids as free RNAs rather than as nucleoproteins necessitates the use of quite different methods of extraction, isolation, and purification than those used for viruses, and makes their visualization with the electron microscope extremely difficult even in purified preparations, while in plant tissues or plant sap their detection with the electron microscope is currently impossible.

Viroids are circular, single-stranded RNA molecules with extensive base-pairing in parts of the RNA strand. The base-pairing results in some sort of hairpin structure with single-stranded and double-stranded regions of the same viroid. Although viroids have many of the properties of single-stranded RNAs, when seen with the electron microscope they appear about 50 nm in length and have the thickness of double-stranded RNA (Figure 14-35).

Viroids seem to be associated with the cell nuclei, particularly the chromatin, and possibly with the endomembrane system of the cell.

How viroids replicate themselves is still not known. Their small size is barely sufficient to code for a very small protein, and such a protein would be considerably smaller than known RNA polymerase (replicase) subunits and

FIGURE 14-35 Electron micrograph of potato spindle tuber viroids (arrows) mixed with a double-stranded DNA of a bacterial virus (T_7) for comparison. (Photo taken by T. Koller and J. M. Sogo, and supplied by courtesy of T. O. Diener.)

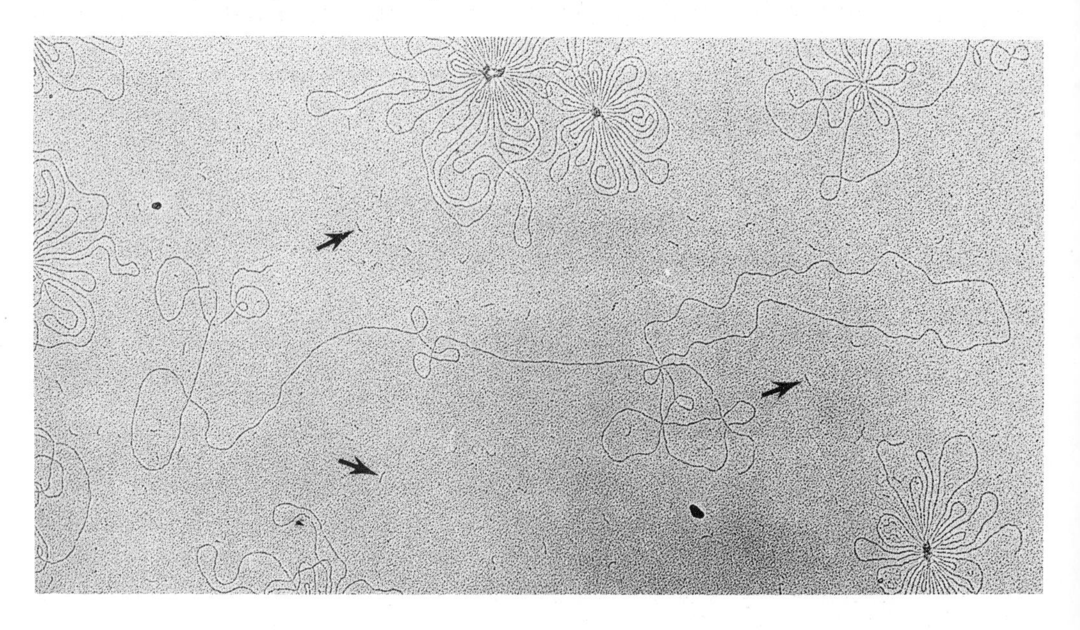

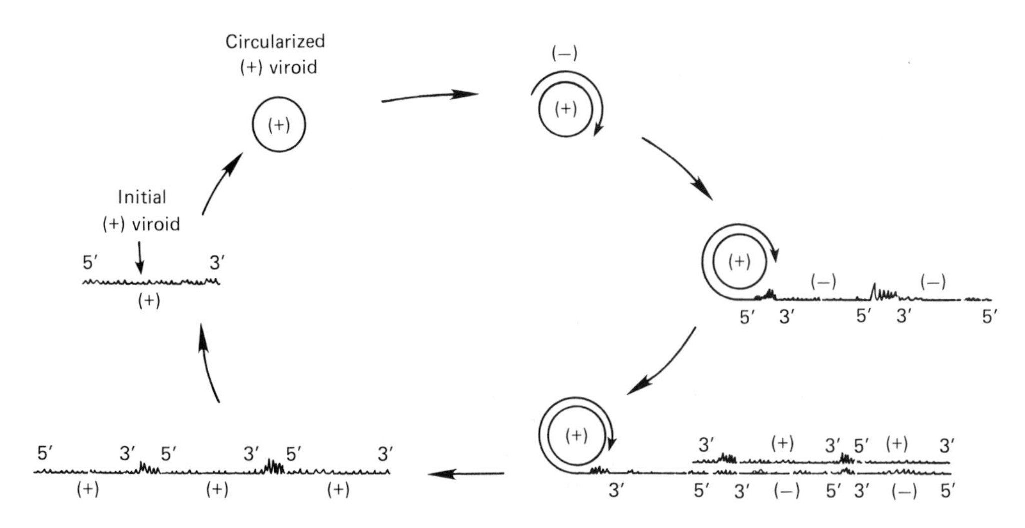

FIGURE 14-36 Schematic representation of presumed viroid replication.

would therefore be unable to carry out the replication of the viroid. Besides, viroids have been shown to be inactive as a messenger RNA in several *in vitro* protein-synthesizing systems and no new proteins could be detected in viroid-infected plants. It has been proposed recently that viroids replicate by direct RNA copying, in which all components required for viroid replication, including the RNA polymerase, are provided by the host. During viroid replication, the circular (+) strand of the viroid is replicated while it acts as a rolling drum producing multimeric linear strands of (−) RNA. The linear (−) strand then serves as a template for replication of multimeric strands of (+) RNA. The (+) RNA is subsequently processed (cleaved) by enzymes that release linear unit-length viroid (+) RNAs and these circularize and produce many copies of the original viroid RNA (Figure 14-36).

How viroids cause disease is also not known. Viroid diseases show a variety of symptoms (Figure 14-37) that resemble those caused by virus infections. The amount of viroids formed in cells seems to be extremely small, and it is therefore unlikely that they cause a shortage of RNA nucleotides in cells. Besides, as with viruses, many infected hosts show no obvious damage, although viroids seem to be replicated in them as much as in the sensitive hosts. So, viroids apparently interfere with the host metabolism in ways resembling those of viruses but which ways are also unknown.

Viroids are spread from diseased to healthy plants primarily by mechanical means, that is, through sap carried on the hands or tools during propagation or cultural practices and, of course, by vegetative propagation. Some viroids, such as potato spindle tuber, chrysanthemum stunt, and chrysanthemum chlorotic mottle viroids, are transmitted through sap quite readily while others, such as citrus excortis, are transmitted through sap with some difficulty. Some viroids, for example, potato spindle tuber, are transmitted through the pollen and seed in rates ranging from 0 to 100 percent. No specific insect or other vectors of viroids are known, although viroids seem to be transmitted on the mouthparts or feet of some insects.

Viroids apparently survive in nature outside the host or in dead plant matter for periods of time varying from a few minutes to a few months.

and bugs. Insect transmission is apparently nonspecific and incidental, that is, on contaminated mouthparts and feet of insects visiting the plants.

- *Development of Disease.* After inoculation of a tuber with PSTV by means of a contaminated knife, or of a growing plant with sap from an infected plant, the viroid replicates itself and spreads systemically throughout the plant. There is no information on the spread of the viroid within the plant nor on the mechanism(s) by which the viroid brings about development of symptoms in infected plants.
- *Control.* Potato spindle tuber can be controlled effectively by planting only PSTV-free potato tubers in fields free of diseased tubers that may have survived from the previous year's crop. The presence of PSTV in tubers can be detected by attaching viroid RNA present in infected sap to a nitrocellulose membrane, hybridizing it with highly radioactive DNA complimentary to viroid RNA, and subjecting the treated membrane to autoradiography. Spots on the membrane that had received sap from infected tubers appear dark after autoradiography, while spots that received healthy sap show little or no change in color.

• Citrus Exocortis

Exocortis is worldwide in distribution and affects trifoliate oranges, citranges, Rangpur and other mandarin and sweet limes, some lemons, and citrons. Orange, lemon, grapefruit, and other citrus trees grafted on exocortis-sensitive rootstocks show slight to great reductions in growth, and yields are reduced by as much as 40 percent.

- *Symptoms.* Infected susceptible plants show vertical splits in the bark and narrow, vertical, thin strips of partially loosened outer bark that give the bark a cracked and scaly appearance (Figure 14-39). Since many of the exocortis-susceptible plants, such as trifoliate orange, are used primarily as rootstocks for other citrus trees, and because the scions make poor growth on such rootstocks, the enlarged, scaly rootstocks have given the disease the name "scale butt." Infected exocortis-susceptible plants may also show yellow blotches on young infected stems, and some citrons show leaf and stem epinasty, and cracking and darkening of leaf veins and petioles. All infected plants usually appear stunted to a smaller or greater extent and have lower yields.
- *The Pathogen: Citrus Exocortis Viroid (CEV).* Citrus exocortis viroid is apparently similar to, but not identical with, the potato spindle tuber viroid. It consists of 371 nucleotides arranged in a circular or linear form. CEV is readily transmitted from diseased to healthy trees by budding knives, pruning shears, or other cutting tools, by hand, and possibly by scratching and gnawing of animals; CEV is also transmitted by dodder and by sap to *Gynura, Petunia,* and other herbaceous plants. On contaminated knife blades, CEV retains its infectivity for at least 8 days and, when partially purified, CEV remains infective at room temperature for several months. The thermal inactivation point of extracted sap is about 80°C for 10 minutes, but partially purified CEV remains infectious even after boiling for 20 minutes. The viroid also survives brief heating of contaminated blades in the flame of a propane torch (blade temperature about 260°C!) and flaming of blades dipped in alcohol. The viroid also survives on contaminated blades treated with almost all common chemical sterilants except sodium hypochlorite solution.

FIGURE 14-39 Exocortis symptoms on the trifoliate rootstock portion of an orange tree. (Photo courtesy L. C. Knorr, Agricultural Research and Education Center, Lake Alfred, Florida.)

- *Development of Disease.* CEV survives in most citrus and many herbaceous hosts and is spread to healthy citrus plants by budding or grafting and by contaminated cutting tools or other cultivating equipment. The viroid apparently enters the phloem elements and spreads in them throughout the plant. The viroid is associated with the nuclei and internal membranes of host cells and results in aberrations of the plasma membranes. Although the viroid apparently lacks the ability to serve as a messenger molecule or as an amino acid acceptor, it brings about several metabolic changes in infected plants. These changes include an increase in oxygen uptake and respiration, and also in sugars and certain enzymes. Marked changes also occur in several amino acids.
- *Control.* Exocortis can be controlled only by propagating exocortis-free nursery trees from certified healthy foundation stock and use of sanitary budding, nursery, and field practices. Tools should be disinfected between cuts into different plants by dipping in a 10–20 percent solution of household bleach (sodium hypochlorite).

SELECTED REFERENCES

Branch, A. D., and Robertson, H. D. (1984). A replication cycle for viroids and other small infectious RNAs. *Science* **223,** 450–455.

"C. M. I./A. A. B. Descriptions of Plant Viruses," Specific viroids, Nos. 66, 226, 254, 287.

Diener, T. O. (1971). Potato spindle tuber "virus." IV. A replicating, low molecular weight RNA. *Virology* **45,** 411–428.

Diener, T. O. (1979). "Viroids and Viroid Diseases." Wiley, New York.

Diener, T. O. (1982). Viroids: Minimal biological systems. *BioScience* **32,** 38–44.

Diener, T. O. (1984). Subviral pathogens: Viroids and prions. *Plant Dis.* **68,** 4.

Diener, T. O., and Lawson, R. H. (1973). Chrysanthemum stunt: A viroid disease. *Virology* **51,** 94–101.

Folsom, D. (1923). Potato spindle tuber. *Maine, Agric. Exp. Stn., Bull.* **321,** 1–4.

Maramorosch, K., and McKelvey, J. J., Jr., eds. (1985). "Subviral Pathogens of Plants and Animals: Viroids and Prions." Academic Press, Orlando, Florida.

Mohamed, N. A., Bautista, R., Buenaflor, G., and Imperial, J. S. (1985). Purification and infectivity of the coconut radang-cadang viroid. *Phytopathology* **75,** 79–83.

Owens, R. A., and Diener, T. O. (1981). Sensitive and rapid diagnosis of potato spindle tuber viroid disease by nucleic acid hybridization. *Science* **213,** 670–671.

Robertson, H. D., Howell, S. H., Zaitlin, M., and Malmberg, R. L., eds. (1983). "Plant Infectious Agents: Viruses, Viroids, Virusoids, and Satellites." Cold Spring Harbor Lab., Cold Spring Harbor, New York.

Van Dorst, H. J. M., and Peters, D. (1974). Some biological observations on pale fruit, a viroid-incited disease of cucumber. *Neth. J. Plant Pathol.* **80,** 85–96.

15 PLANT DISEASES CAUSED BY NEMATODES

Introduction

Nematodes belong to the animal kingdom. Sometimes called eelworms, nematodes are wormlike in appearance but quite distinct taxonomically from the true worms. Most of the several thousand species of nematodes live freely in fresh or salt waters or in the soil and feed on microscopic plants and animals. Numerous species of nematodes attack and parasitize humans and animals, in which they cause various diseases. Several hundred species, however, are known to feed on living plants, causing a variety of plant diseases.

Characteristics of Plant-Pathogenic Nematodes

Morphology

Plant-parasitic nematodes are small, 300–1000 μm, with some up to 4 mm long, by 15–35 μm wide (Figure 15-1). Their small diameter makes them invisible to the naked eye, but they can be easily observed under the microscope. Nematodes are, in general, eel-shaped, and round in cross section, with smooth, unsegmented bodies, without legs or other appendages. The females of some species, however, become swollen at maturity and have pear-shaped or spheroid bodies (Figure 15-2).

Anatomy

The nematode body (Figure 15-1) is more or less transparent. It is covered by a colorless cuticle, which is usually marked by striations or other markings. The cuticle molts when a nematode goes through its successive larval stages. The cuticle is produced by the hypodermis, which consists of living cells and extends into the body cavity as four chords separating four bands of longitudinal muscles. These muscles enable the nematode to move. Additional specialized muscles exist at the mouth and along the digestive tract and the reproductive structures.

The body cavity contains a fluid through which circulation and respiration take place. The digestive system is a hollow tube extending from the mouth through the esophagus, intestine, rectum, and anus. Lips, usually six in number, surround the mouth. All plant-parasitic nematodes have a hollow stylet or spear, which is used to puncture plant cells.

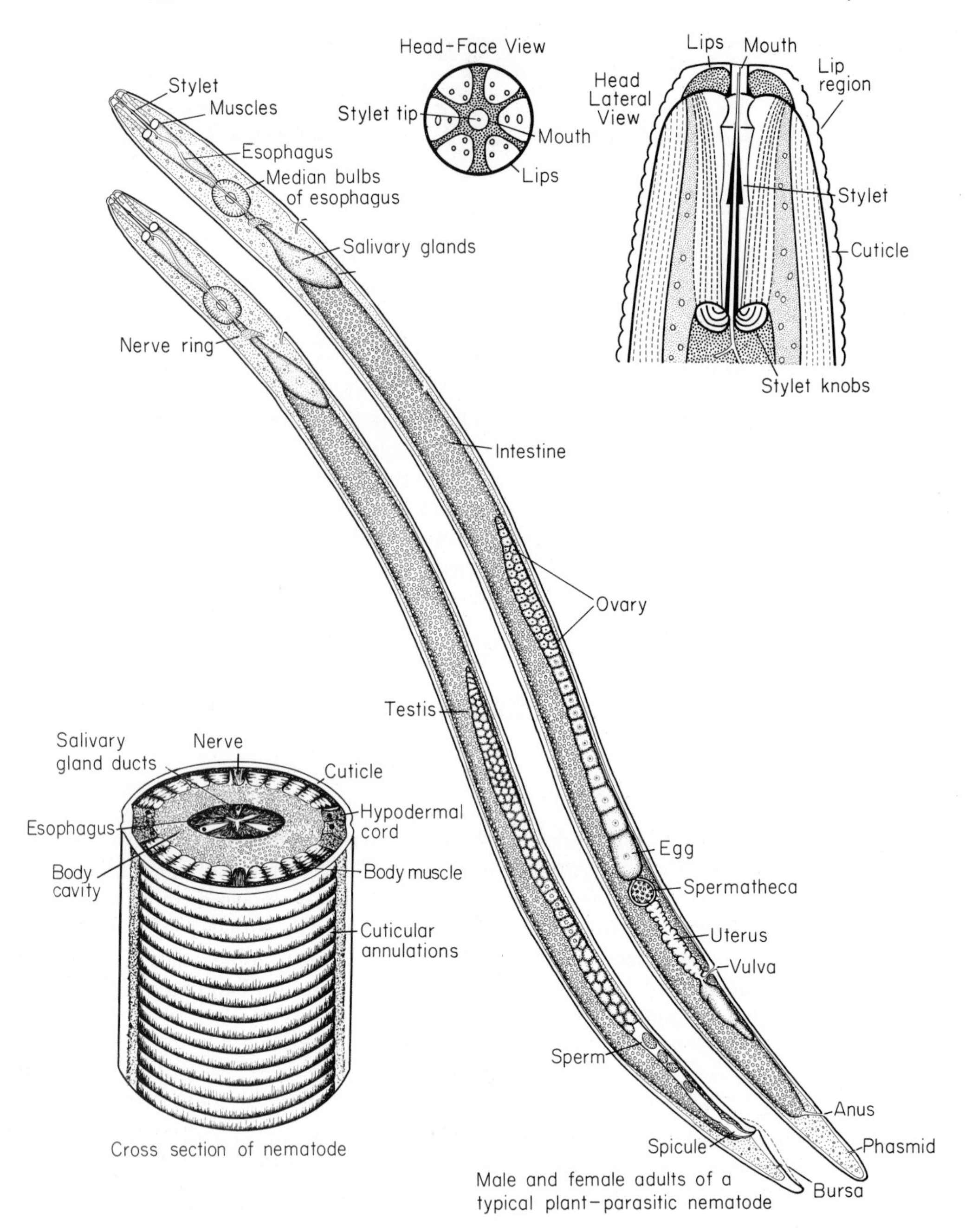

FIGURE 15-1 Morphology and main characteristics of typical male and female plant parasitic nematodes.

The reproductive systems are well developed. Female nematodes have one or two ovaries, followed by an oviduct and uterus terminating in a vulva. The male reproductive structure is similar to the female, but there is a testis, seminal vesicle, and a terminus in a common opening with the intestine. A pair of protrusible, copulatory spicules are also present in the male. Reproduction in nematodes is through eggs and may be sexual, hermaphroditic, or parthenogenetic. Many species lack males.

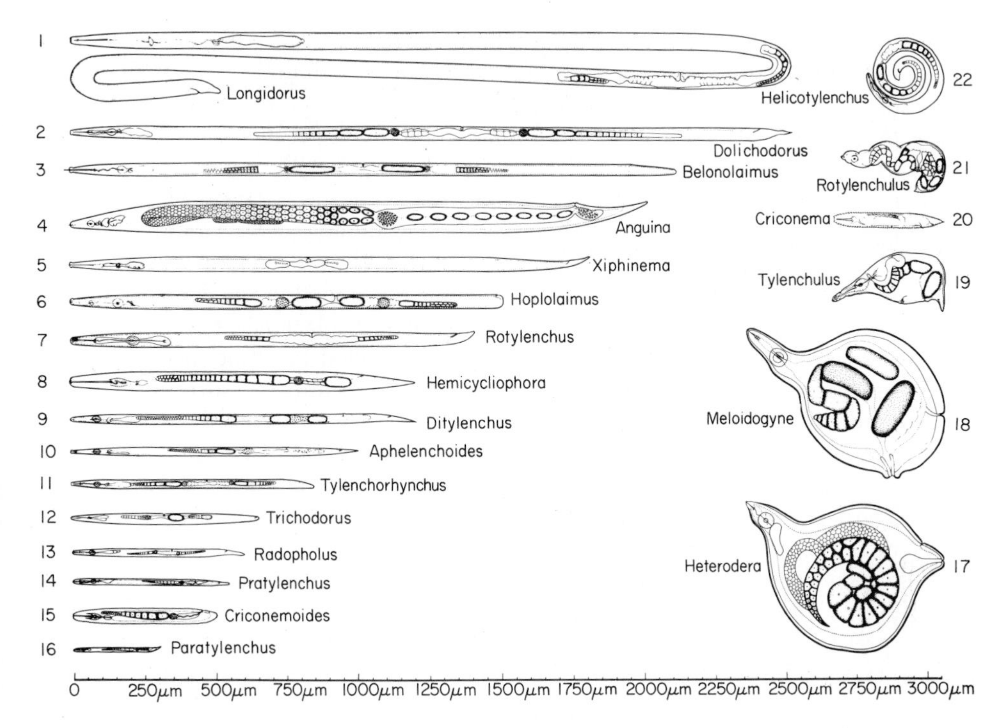

FIGURE 15-2 Morphology and relative size of some of the most important plant-parasitic nematodes.

Life Cycles

The life histories of most plant parasitic nematodes are, in general, quite similar. Eggs hatch into larvae, whose appearance and structure are usually similar to those of the adult nematodes. Larvae grow in size, and each larval stage is terminated by a molt. All nematodes have four larval stages, with the first molt usually occurring in the egg. After the final molt the nematodes differentiate into adult males and females. The female can then produce fertile eggs either after mating with a male or, in the absence of males, parthenogenetically, or can produce sperm herself.

A life cycle from egg to egg may be completed within 3 or 4 weeks under optimum environmental, especially temperature, conditions but will take longer in cooler temperatures. In some species of nematodes the first or second larval stages cannot infect plants and depend for their metabolic functions on the energy stored in the egg. When the infective stages are produced, however, they must feed on a susceptible host or starve to death. Absence of suitable hosts may result in the death of all individuals of certain nematode species within a few months, but in other species the larval stages may dry up and remain quiescent, or the eggs may remain dormant in the soil for years.

Ecology and Spread

Almost all plant-pathogenic nematodes live part of their lives in the soil. Many of these live freely in the soil, feeding superficially on roots and underground stems, but even in the specialized sedentary parasites, the eggs, the preparasitic larval stages, and the males are found in the soil for all or part of their lives. Soil temperature, moisture, and aeration affect survival and movement of nematodes in the soil. Nematodes occur in greatest abundance in a layer of soil from 0 to 15 cm deep, although distribution of nematodes in cultivated soil is irregular and is greatest in or around the roots of susceptible plants, which they follow sometimes to considerable depths (30–150 cm or more). The greater concentration of nematodes in the region of host plant roots is due primarily to their more rapid reproduction on the food supply available and also to attraction of nematodes by substances released into the rhizosphere. To these must be added the so-called hatching factor effect of substances originating from the root that diffuse into the surrounding soil, markedly stimulating the hatching of eggs of certain species. Most nematode eggs, however, hatch freely in water in the absence of any special stimulus.

Nematodes spread through the soil very slowly under their own power. The overall distance traveled by a nematode probably does not exceed a meter per season. Nematodes move faster in the soil when the pores are lined with a thin (a few micrometers) film of water than when the soil is waterlogged. In addition to their own movement, however, nematodes can be easily spread by anything that moves and can carry particles of soil. Farm equipment, irrigation, flood or drainage water, animal feet, and dust storms spread nematodes in local areas, while over long distances nematodes are spread primarily with farm produce and nursery plants. A few nematodes that attack the aboveground parts of plants not only spread through the soil as described above, but they are also splashed to the plants by falling rain or overhead watering, or they ascend wet plant stem or leaf surfaces on their own power. Further spread takes place upon contact of infected plant parts with adjacent healthy plants.

Three genera of the family Aphelenchoididae, that is, *Aphelenchoides* (bud and leaf nematodes), *Bursaphelenchus* (the pine wilt nematode), and *Rhadinaphelenchus* (the coconut red-ring nematode), seldom, if ever, enter the soil. They survive instead in the tissues of the plants they infect and, the last two, in their insect vector.

Classification

All plant-parasitic nematodes (Figure 15-2) belong to the phylum Nematoda. Most of the important parasitic genera belong to the order Tylenchida, but a few belong to the order Dorylaimida.

Phylum: NEMATODA
 Order: Tylenchida.
  Suborder: Tylenchina
   Superfamily: Tylenchoidea
    Family: Tylenchidae
     Genus: *Anguina,* wheat or seed-gall nematode
      Ditylenchus, stem or bulb nematode of alfalfa, onion, narcissus, etc.
    Family: Tylenchorhynchidae
     Genus: *Tylenchorhynchus,* stunt nematode of tobacco, corn, cotton, etc.

Family: Pratylenchidae
Genus: *Pratylenchus,* lesion nematode of almost all crop plants and trees
Radopholus, burrowing nematode of banana, citrus, coffee, sugarcane, etc.
Family: Hoplolaimidae
Genus: *Hoplolaimus,* lance nematode of corn, sugarcane, cotton, alfalfa, etc.
Rotylenchus, spiral nematode of various plants
Helicotylenchus, spiral nematode of various plants
Family: Belonolaimidae
Genus: *Belonolaimus,* sting nematode of cereals, legumes, cucurbits, etc.
Superfamily: Heteroderoidea
Family: Heteroderidae
Genus: *Globodera,* cyst nematode of potato
Heterodera, cyst nematode of tobacco, soybean, sugar beets, cereals
Meloidogyne, root-knot nematode of almost all crop plants
Family: Nacobbidae
Genus: *Nacobbus,* false root-knot nematode
Rotylenchulus, reniform nematode of cotton, papaya, tea, tomato, etc.
Superfamily: Criconematoidea
Family: Criconematidae
Genus: *Criconemella,* ring nematode of woody plants
Hemicycliophora, sheath nematode of various plants
Family: Paratylenchidae
Genus: *Paratylenchus,* pin nematode of various plants
Family: Tylenchulidae
Genus: *Tylenchulus,* citrus nematode of citrus, grapes, olive, lilac, etc.
Suborder: Aphelenchina
Superfamily: Aphelenchoidea
Family: Aphelenchoididae
Genus: *Aphelenchoides,* foliar nematode of chrysanthemum, strawberry, begonia, rice, coconut, etc.
Bursaphelenchus, the pine wood nematode
Rhadinaphelenchus, the coconut red ring nematode
Order: Dorylaimida
Family: Longidoridae
Genus: *Longidorus,* needle nematode of some plants
Xiphinema, dagger nematode of trees, woody vines, and of many annuals
Family: Trichodoridae
Genus: *Paratrichodorus,* stubby root nematode of cereals, vegetables, cranberry, apple
Trichodorus, stubby root nematode of sugar beet, potato, cereals, apple

In terms of habitat, pathogenic nematodes are either *ectoparasites,* that is species that do not normally enter root tissue but feed only on the cells near the root surfaces, or *endoparasites,* that is, species that enter the host and feed from within. Both of these can be either *migratory,* that is, they live freely in the soil and feed on plants without becoming attached, or move around inside the plant, or *sedentary,* that is, species that, once within a root, do not move about. The ectoparasitic nematodes include the ring nematodes (sedentary) and the dagger, stubby root, and sting nematodes (all migratory). The endoparasitic nematodes include the root knot, cyst, and citrus nematodes (all sedentary), and the lesion, stem and bulb, burrowing, leaf, stunt, lance, and spiral nematodes (all somewhat migratory). Of these, the cyst, lance, and spiral nematodes may be somewhat ectoparasitic, at least during part of their lives.

Isolation of Nematodes

Plant-parasitic nematodes are generally isolated from the roots of plants they infect or from the soil surrounding the roots on which they feed (Figure 15-3). A few kinds of nematodes, however, attack aboveground plant parts, for example, chrysanthemum foliar nematode, grass seed gall nematode, and the stem, leaf, and bulb nematode, and these nematodes can be isolated primarily from the plant parts they infect.

Isolation of Nematodes from Soil

From a freshly collected soil sample of about 100–300 cm³, the nematodes in it can be isolated by either the Baermann funnel method or by sieving.

A Baermann funnel consists of a fairly large (12- to 15-cm diameter) glass funnel to which a piece of rubber tubing is attached, with a pinchcock placed on the tubing. The funnel is placed on a stand and filled with water. The soil sample is placed in the funnel on porous, wet-strength paper, sometimes supported by a 5- to 6-cm circular piece of screen, or in a beaker over which a piece of cloth is fastened with a rubber band. The beaker is then inverted in the funnel with the cloth and all the soil below the surface of the water and allowed to stand overnight or for several hours. The live nematodes move actively and migrate through the cloth or porous paper into the water and sink to the bottom of the rubber tubing just above the pinchcock. More than 90 percent of the live nematodes are recovered in the first 5–8 ml of water drawn from the rubber tubing, and this sample is placed in a shallow dish for examination and, if desired, single nematode isolation.

The sieving method is based on the fact that when a small soil sample, such as 300 cm³ is mixed with considerably more water, for example, 2 liters, the nematodes float in the water and can be collected on sieves with pores of certain sizes. Thus, the soil-water mixture is stirred and then allowed to stand for 30 seconds. The supernatant is poured through a 20-mesh sieve (20 holes per square inch), which holds large debris but allows the nematodes to pass into a bucket. The liquid containing the nematodes is then poured through a 60-mesh sieve, which holds the larger nematodes and some debris but lets the smaller ones pass through into another bucket. The latter is then passed through a 200-mesh sieve, which holds the small nematodes and some debris. Both the 60- and the 200-mesh sieves are washed two or three times to remove as much of the debris as possible, and the nematodes are then washed into

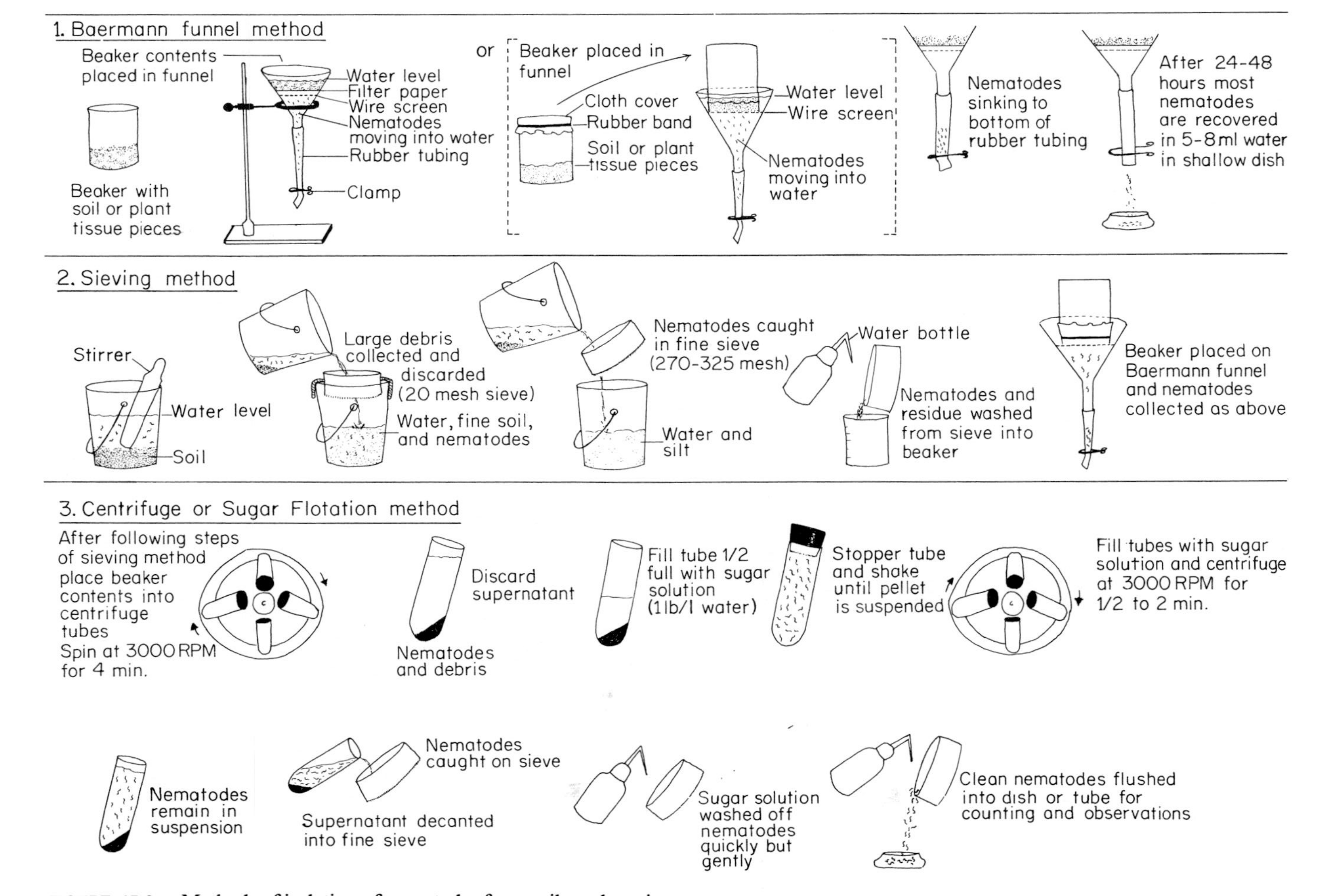

FIGURE 15-3 Methods of isolation of nematodes from soil or plant tissues.

shallow dishes for direct examination and further isolation. A "semiautomatic elutriator" developed recently combines the steps described into one continuous process requiring one person where three people were needed with the old hand method.

Isolation of Nematodes from Plant Material

Regardless of the type of plant material containing nematodes, the material is cut into very small pieces by hand or with the use of a blender for a few seconds, and is then placed in the Baermann funnel as described above. The nematodes leave the tissue and move into the water in the tubing, from which they are collected in a shallow dish.

Symptoms Caused by Nematodes

Nematode infections of plants result in the appearance of symptoms on roots as well as on the aboveground parts of plants (Figure 15-4). Root symptoms may appear as root knots or root galls, root lesions, excessive root branching, injured root tips, and root rots when nematode infections are accompanied by plant-pathogenic or saprophytic bacteria and fungi.

These root symptoms are usually accompanied by noncharacteristic symptoms in the aboveground parts of plants, appearing primarily as reduced growth, symptoms of nutrient deficiencies such as yellowing of foliage, excessive wilting in hot or dry weather, reduced yields, and poor quality of products.

Certain species of nematodes invade the aboveground portions of plants rather than the roots, and on these they cause galls, necrotic lesions and rots, twisting or distortion of leaves and stems, and abnormal development of the floral parts. Certain nematodes attack grains or grasses forming galls full of nematodes in place of seed.

How Nematodes Affect Plants

The direct mechanical injury inflicted by nematodes while feeding causes only slight damage to plants. Most of the damage seems to be caused by a secretion of saliva injected into the plants while the nematodes are feeding. Some nematode species are rapid feeders. They puncture a cell wall, inject saliva into the cell, suck part of the cell contents, and move on within a few seconds. Others feed much more slowly and may remain at the same puncture for several hours or days. These, as well as the females of species that become permanently established in or on roots, inject saliva intermittently as long as they are feeding.

The feeding process causes the affected plant cells to react, resulting in dead or devitalized root tips and buds, lesion formation and tissue breakdown, swellings and galls of various kinds, and crinkled and distorted stems

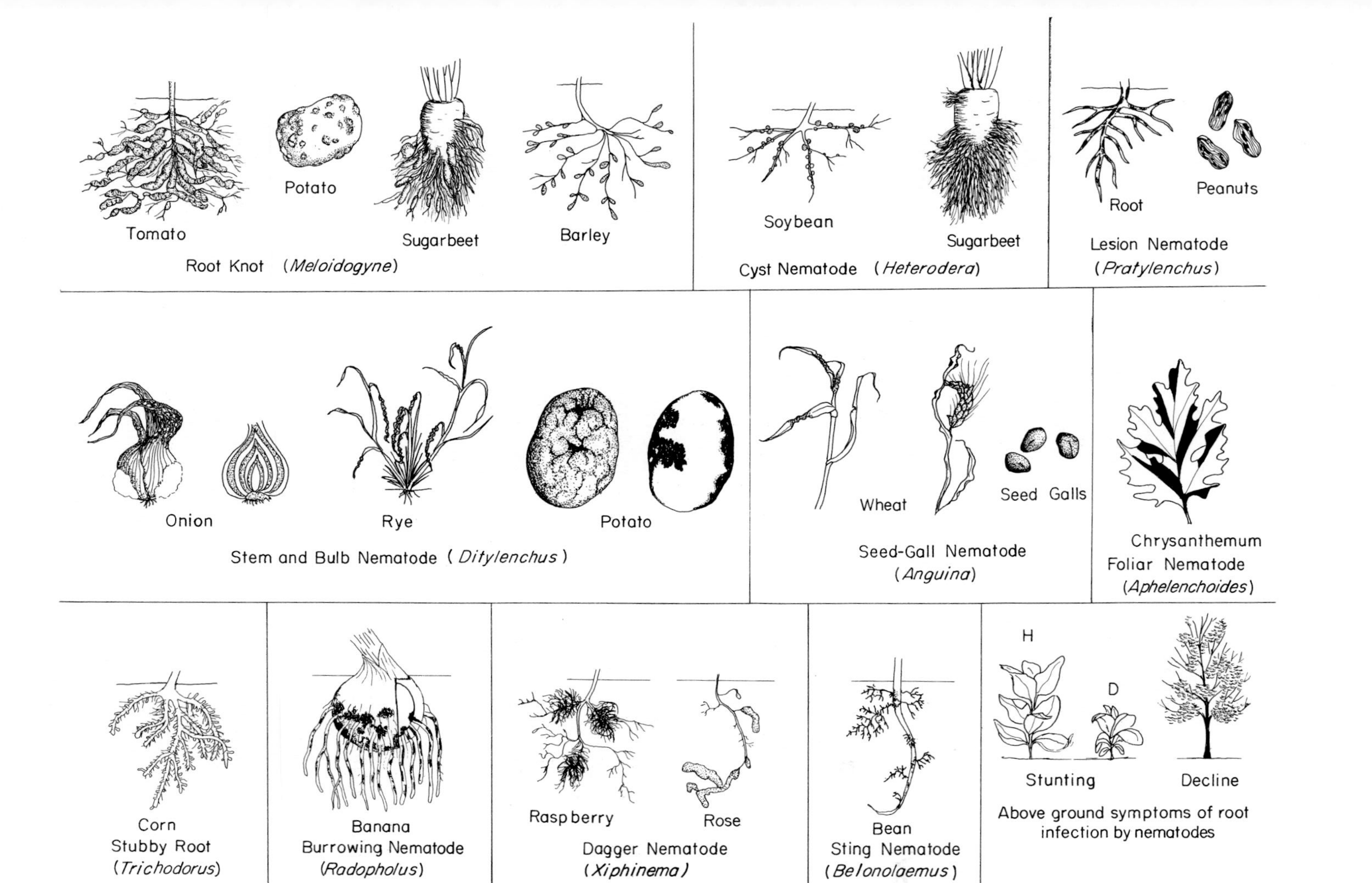

FIGURE 15-4 Types of symptoms caused by some of the most important plant parasitic nematodes.

and foliage. Some of these manifestations are caused by dissolution of infected tissues by nematode enzymes, which, with or without the help of toxic metabolites, cause tissue disintegration and the death of cells. Others are caused by abnormal cell enlargement (hypertrophy), by suppression of cell divisions, or by stimulation of cell division proceeding in a controlled manner and resulting in the formation of galls or of large numbers of lateral roots at or near the points of infection.

Plant disease syndromes caused by nematodes are complex. Root-feeding species often decrease the ability of plants to take up water and nutrients from soil and thus cause symptoms of water and nutrient deficiencies in the aboveground parts of plants. In some cases, however, it is the plant–nematode biochemical interactions that impair the overall physiology of plants and the role nematodes play in providing courts for entry of other pathogens that are primarily responsible for plant injury. The mechanical damage or withdrawal of food from plants by nematodes is generally less significant but may become all-important when nematode populations become very large.

Interrelationships between Nematodes and Other Plant Pathogens

Although nematodes can cause diseases to plants by themselves, most of them live and operate in the soil, where they are constantly surrounded by fungi and bacteria, many of which can also cause plant diseases. In many cases an association develops between nematodes and certain of the other pathogens. Nematodes then become a part of an etiological complex resulting in a combined pathogenic potential far greater than the sum of the damage either of the pathogens can produce individually.

Several nematode–fungus disease complexes are known. *Fusarium* wilt of several plants increases in incidence and severity when the plants are also infected by the root knot, lesion, sting, reniform, burrowing, or stunt nematodes. Similar effects have also been noted in disease complexes involving nematodes and *Verticillium* wilt, *Pythium* damping off, *Rhizoctonia* and *Phytophthora* root rots, and in some other instances. In none of these cases is the fungus transmitted by the nematode. However, plant varieties susceptible to the respective fungi are damaged even more when the plants are infected with the nematodes, the combined damage being considerably greater than the sum of the damage caused by each pathogen acting alone. Also, varieties ordinarily resistant to the fungi apparently become infected by them after previous infection by nematodes. The importance of nematodes in these complexes in indicated by the fact that soil fumigation aimed at eliminating the nematode but not the fungus greatly reduces the incidence and the damage caused by the fungus-induced disease.

Although it seems quite probable that the mechanical wounding caused to plants by nematodes is an important factor in providing avenues of entry for the fungus, the continuation of the effect that nematodes have on host susceptibility in later stages of plant development suggests that the nematodes may also cause some host response that lowers natural resistance to the fungus. It should also be noted that, in at least some such complexes, there is a greater mass of mycelium present in nematode-infected than in nematode-free tissues of the same plant and also that higher populations of nematodes are present in fungus-infected than in fungus-free tissues of a diseased plant.

Relatively few cases of nematode–bacterial disease complexes are known. Thus, root-rot nematode increases the frequency and severity of the bacterial wilt of tobacco caused by *Pseudomonas solanacearum*, of the bacterial wilt of alfalfa caused by *Clavibacter michiganense subsp. insidiosum*, and of the bacterial scab of gladiolus caused by *Pseudomonas marginata*. In most of these the nematode role seems to be that of providing the bacteria with an infection court and to assist bacterial infection by wounding the host. On the other hand, root infection of plum trees with the ring nematode *Criconemella xenoplax* changed the physiology of the trees and resulted in the development of more extensive cankers by the bacterium *Pseudomonas syringae* pv. *syringae* on branches of nematode-infected trees than on nematode-free trees.

Much better known are the interrelationships between nematodes and viruses. Several plant viruses such as grapevine fanleaf, arabis mosaic, tobacco ring spot, tomato ring spot, tomato black ring, raspberry ring spot, tobacco rattle, and pea early browning virus are transmitted through the soil by means of nematode vectors. All these viruses, however, are transmitted by only one or more of the four genera of dagger, needle, and stubby root nematodes: *Xiphinema, Longidorus, Trichodorus,* and *Paratrichodorus*. *Xiphinema* and *Longidorus* transmit only round, that is, polyhedral viruses, which include most of the nematode-transmitted viruses, while *Trichodorus* and *Paratrichodorus* transmit two rod-shaped viruses, tobacco rattle and pea early browning viruses. These nematodes can transmit some of the viruses after feeding on infected plants for as short a time as one hour, but the percentage of transmission increases with longer feedings up to four days. Once they have acquired virus from an infected plant the nematodes remain infective for periods of 2–4 months and sometimes even longer. All stages, larval and adult nematodes, can transmit viruses, but the virus is not carried from one larval stage to another and to adults through molts nor does the virus pass from adults, through eggs, to larvae. Although nematodes can ingest and carry within them several plant viruses, they can only transmit certain of them to healthy plants, which suggests that there is a close biological association between the nematode vectors and the viruses they can transmit.

Control of Nematodes

Several methods of effectively controlling nematodes are available, although certain factors, such as expense and types of crops, limit their applicability in some cases. Four general types of control methods are employed. Control through cultural practices, biological control through resistant varieties and certain other means, control by means of physical agents, such as heat, and control through chemicals. In practice, a combination of several of these methods is usually employed for controlling nematode diseases of plants.

SELECTED REFERENCES

Anonymous (1972 and annually afterwards). "Commonweath Institute of Helminthology Descriptions of Plant-Parasitic Nematodes." Commonw. Agric. Bur. Farnham Royal, Bucks, England.

Barker, K. R. and Imbriani, J. L. (1984). Nematode advisory programs. Status and prospects. *Plant Dis.* **68,** 735–741.

Christie, J. R. (1959). "Plant Nematodes, Their Bionomics and Control." Florida University, Agric. Exp. Stn., Gainesville.

Dropkin, V. H. (1980). "Introduction to Plant Nematology." Wiley, New York.

Endo, B. Y. (1975). Pathogenesis of nematode-infected plants. *Annu. Rev. Phytopathol.* **12,** 213–238.

Ferris, H. (1981). Dynamic action thresholds for diseases induced by nematodes. *Annu. Rev. Phytopathol.* **19,** 427–436.

Giebel, J. (1982). Mechanism of resistance to plant nematodes. *Annu. Rev. Phytopathol.* **20,** 257–279.

Jenkins, W. R., and Taylor, D. P. (1967). "Plant Nematology." Reinhold, New York.

Peachy, J. E., ed. (1969). "Nematodes of Tropical Crops," Tech. Commun., Commonw. Bur. Helminthol., No. 40.

Powell, N. T. (1971). Interactions between nematodes and fungi in disease complexes. *Annu. Rev. Phytopathol.* **9,** 253–294.

Smart, G. G., Jr., and Perry, V. G., eds. (1968). "Tropical Nematology." Univ. of Florida Press, Gainesville.

Thorne, G. (1961). "Principles of Nematology." McGraw-Hill, New York.

Webster, J. M. (1969). The host-parasite relationships of plant-parasitic nematodes. *Adv. Parasitol.* **7,** 1–40.

Webster, J. M., ed. (1972). "Economic Nematology." Academic Press, New York.

Wyss, U. (1982). Virus-transmitting nematodes: Feeding behavior and effect on root cells. *Plant Dis.* **66,** 639–644.

Zuckerman, B. M., Mai, W. F., and Rohde, R. A., eds. (1971–1981). "Plant Parasitic Nematodes," 3 vols. Academic Press, New York.

Zuckerman, B. M., Mai, W. F., and Harrison, M. B., eds. (1985). "Plant Nematology Laboratory Manual." University of Massachusetts, Agric. Exp. Stn., Amherst.

Veech, J. A., and Dickson, D. W., eds. (1987). "Vistas on Nematology." Soc. of Nematologists, Hyattsville, Maryland.

• Root-Knot Nematodes: *Meloidogyne*

Root-knot nematodes occur throughout the world but are found more frequently and in greater numbers in areas with warm or hot climates and short or mild winters. Root-knot nematodes are also found in greenhouses everywhere when nonsterilized soil is used. They attack more than 2000 species of plants, including almost all cultivated plants.

Root-knot nematodes damage plants by devitalizing root tips and either stopping their growth or causing excessive root production, but primarily by causing formation of swellings of the roots, which not only deprive plants of nutrients but also disfigure and reduce the market value of many root crops. When susceptible plants are infected at the seedling stage, losses are heavy and may result in complete destruction of the crop. Infections of older plants may have only slight effects on yield or they may reduce yields considerably.

○ *Symptoms.* The aboveground symptoms are similar to those caused by many other root diseases or environmental factors that result in reduced amounts of water available to the plant. Infected plants show reduced growth and fewer, small, pale green, or yellowish leaves that tend to wilt in warm weather. Blossoms and fruits are either lacking or are dwarfed and of poor quality. Affected plants usually linger through the growing season and are seldom killed prematurely.

The most characteristic symptoms of the disease are those appearing on the underground parts of the plants. Infected roots swell at the point of invasion and develop into the typical root-knot galls which are two or three times as large in diameter as the healthy root (Figure 15-5, A). Several infections take place along the same root, and the developing galls give the root a rough, clubbed appearance. Roots infected by certain species of this nematode develop, in addition to galls several short root branches that rise from the upper part of the gall and result in a dense, bushy root system (Figure 15-5, B). Usually, however, infected roots remain smaller and show various stages of necrosis. Rotting of the roots frequently develops, particularly late in the season. When tubers or other fleshy underground organs are attacked, they produce small swellings over their surface, which become quite prominent at times and may cause distortion of the organs or cracking of their skin.

◦ *The Pathogen: Meloidogyne sp.* The adult male and female root-knot nematodes are easily distinguishable morphologically (Figures 15-5 and 15-6). The males are wormlike and about 1.2–1.5 mm long by 30–36 μm in diameter. The females are pear shaped and about 0.40–1.30 mm long by 0.27–0.75 mm wide. Each female lays approximately 500 eggs in a gelatinous substance

FIGURE 15-5 (A) Galls on the roots of tomato plant caused by the root-knot nematode *Meloidogyne* sp. (B) Healthy and root-knot nematode-infected carrots. (C) Cross section of young tomato root showing part of root-knot nematode (arrow) and giant cells in the stele. (D) Section of tomato root showing a root-knot nematode feeding on the giant cells surrounding its head. (E) Female root-knot nematode feeding on young root and laying its egg mass in a matrix outside the root. (Photos A and B courtesy U.S.D.A. Photos C,D, and E courtesy R. A. Rohde.)

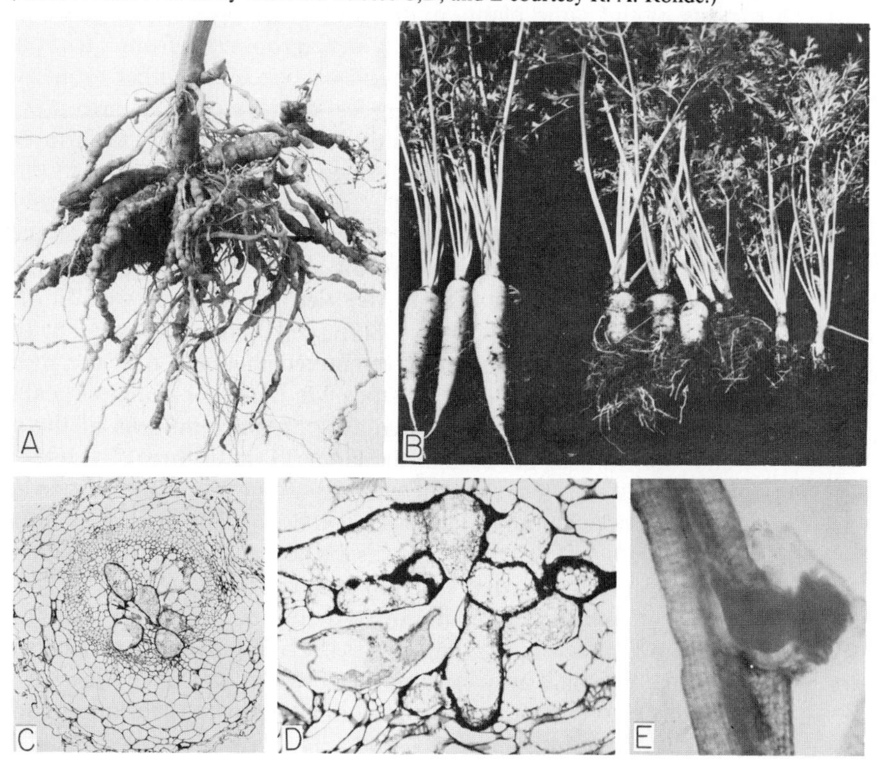

produced by the nematode. The first-stage larva develops inside each egg, and after undergoing the first molt within the egg it becomes second-stage larva. The latter emerges from the egg into the soil, where it moves until it finds a susceptible root. The second-stage larva is wormlike and is the only infective stage of this nematode. If a susceptible host is present in its vicinity, the larva enters the root, becomes sedentary, and grows in thickness, assuming a sausage-shaped form. The nematode feeds on the cells around its head by inserting its stylet and secreting saliva into these cells. The saliva stimulates cell enlargement and also liquefies part of the contents of the cells, which are then sucked by the nematode through its stylet. The nematode undergoes a second molt and gives rise to the third-stage larva, which is similar to the second-stage larva but lacks a stylet and is stouter. The third-stage larva goes through the third molt and gives rise to the fourth-stage larva, which can be distinguished as either male or female. A male fourth-stage larva becomes wormlike and is coiled within the third cuticle. It undergoes the fourth and final molt and emerges from the root as the wormlike adult male, which becomes free living in the soil. The fourth-stage female larva continues to grow in thickness and somewhat in length, undergoes the fourth and final molt, and becomes an adult female, which appears pear shaped. The adult female continues to swell and, with or without fertilization by a male, produces eggs that are laid in a gelatinous protective coat. The eggs may be laid inside or outside the root tissues, depending on the position of the female. Eggs may hatch immediately or they may overwinter and hatch in the spring. A life cycle is completed in 25 days at 27°C, but it takes longer at lower or higher temperatures. When the eggs hatch, the infective second-stage larvae may migrate from within galls to adjacent parts of the root and cause new infections in the same root, or they may emerge from the root and infect other roots of the same plants or roots of other plants. The greatest numbers of root-knot nematodes are usually in the root zone from 5 to 25 cm below the surface, but galls have been found on peach and other roots 2–2.5 m deep. The ability of root-knot nematodes to move on their own power is limited, but they can be spread by water or by soil clinging to farm equipment or otherwise transported into uninfested areas.

◦ *Development of Disease.* Infected second-stage larvae usually enter roots behind the root tip and push their way between or through cells until they reach positions behind the growing point. There they become permanently established with their head in the plerome (Figure 15-6). In older roots the head is usually in the pericycle. Some cell damage occurs along the path of the larva and, if several larvae have entered, the cells near the root tip cease to divide and growth of the root stops. On the other hand, cortical cells near the point of entry begin to enlarge as sometimes do cells of the pericycle and endodermis near the path of the larvae. Two or three days after the larva has become established, some of the cells around its head begin to enlarge. Their nuclei divide, but no cell walls are laid down. The existing walls between some of the cells break down and disappear, and the protoplasmic contents of several coalesce, giving rise to giant cells (Figures 15-5, C, D, and 15-6). Enlargement and coalescing of cells continues for 2–3 weeks, and the giant cells invade the surrounding tissues irregularly. Each gall usually contains 3–6 giant cells, which may form in the cortex as well as in the stele. The enlargement of the cells seems to be brought about by the substances contained in the

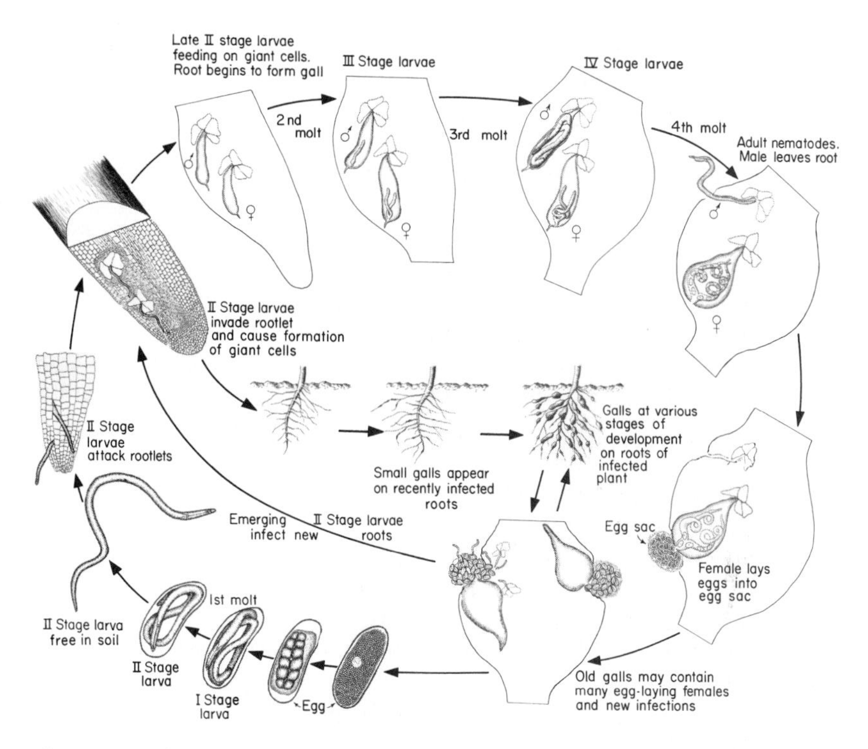

FIGURE 15-6 Disease cycle of root knot caused by nematodes of the genus *Meloidogyne*.

saliva secreted by the nematode in the giant cells during feeding. The giant cells degenerate when nematodes cease to feed or die. When giant cells form in the stele, irregular xylem elements develop or their development may be interrupted. Xylem elements already present may be crushed by the mechanical pressure exerted by the enlarging cells. In the early stages of gall development the cortical cells enlarge in size but, during the later stages, they also divide rapidly. Swelling of the root results also from hypertrophy and hyperplasia of the vascular parenchyma, pericycle, and endodermis cells surrounding the giant cells and from enlargement of the nematode. As the females enlarge and egg sacs are formed, they push outward, split the cortex, and may become exposed on the surface of the root or may remain completely covered, depending on the position of the nematode in relation to the root surface.

In addition to the disturbance caused to plants by the nematode galls themselves, frequently damage to infected plants is increased by certain parasitic fungi, which can easily attack the weakened root tissues and the hypertrophied, undifferentiated cells of the galls. Moreover some fungi, for example, *Pythium, Fusarium,* and *Rhizoctonia,* grow and reproduce much faster in the galls than in other areas of the root, thus inducing an earlier breakdown of the root tissues.

Control. Root knot can be effectively controlled in the greenhouse with steam sterilization of the soil or soil fumigation with nematicides. In the field the best control of root knot is obtained by fumigating the soil with chemicals such as methyl bromide plus chloropicrin, metam sodium, or methyl isothiocyanate. Several newer nematicides such as aldicarb, oxamyl, and fenamiphos are being used effectively. Each treatment usually gives satisfactory control of root knot for one season. In recent years, good experimental control of root knot has been obtained by application to the soil of granular formulations of avermectins, which are antibiotic substances produced by the actinomycete *Streptomyces avermitilis.* Biological control of root knot has also been obtained by treating nematode-infested soil with spores of *Bacillus (Pasteuria) penetrans,* which is an obligate parasite of some plant-parasitic nematodes; by treating transplants or infested soils with spores of the fungus *Dactylella oviparasitica,* which parasitizes the eggs of *Meloidogyne* nematodes; and in some experiments by treating transplants or infested soils with spores of the vesicular-arbuscular mycorrhizal fungi *Gigaspora* and *Glomus.* In several crops, varieties resistant to root-knot nematodes are also available.

SELECTED REFERENCES

Bird, A. F. (1974). Plant response to root-knot nematode. *Annu. Rev. Phytopathol.* **12,** 69–85.

Carter, W. W., and Nieto, S., Jr. (1975). Population development of *Meloidogyne incognita* as influenced by crop rotation and fallow. *Plant Dis. Rep.* **59,** 402–403.

Christie, J. R. (1936). The development of root-knot nematode galls. *Phytopathology* **26,** 1–22.

Dropkin, V. H., and Nelson, P. E. (1960). The histopathology of root-knot nematode infections in soybeans. *Phytopathology* **50,** 442–447.

Sasser, J. N. (1954). Identification and host-parasite relationships of certain root-knot nematodes (*Meloidogyne* sp.). *Md., Agric. Exp. Stn., Bull.* **A-77,** 1–30.

Sasser, J. N. *et al.* (1983). The international *Meloidogyne* project—its goals and accomplishments. *Annu. Rev. Phytopathol.* **21,** 271–288.

Stirling, G. R. (1984). Biological control of *Meloidogyne javanica* with *Bacillus penetrans. Phytopathology* **74,** 55–60.

Cyst Nematodes: *Heterodera* and *Globodera*

Cyst nematodes cause a variety of plant diseases, mostly in the temperate regions of the world. Some species of cyst nematodes attack only a few plant species and are present over limited geographic areas, while others attack a large number of plant species and are widely distributed. A cyst nematode particularly severe on potato but also on tomato and eggplant is *Globodera rostochiensis* (Figure 15-7, A). Other common cyst nematodes and their most important hosts are *Heterodera avenae* on cereals, *H. glycines* on soybeans, *H. schachtii* on sugar beets (Figure 15-7, B), crucifers, and spinach, *H. tabacum* on tobacco, and *H. trifolii* on clover. The diagnostic feature of cyst nematode infections is the presence of cysts on the roots and usually the proliferation of roots and production of shallow, bushy root systems.

• Soybean Cyst Nematode: *Heterodera glycines*

The soybean cyst nematode has been found in northeastern Asia, Japan, and in the United States in an area from Virginia to Florida to Arkansas to Missouri, Iowa, Minnesota, and Illinois. It continues to spread slowly to new areas in spite of the strict quarantine measures imposed on the currently infested areas. The most severely affected host is soybean, but several other legumes, such as common bean, vetch, lespedeza, lupine, and a few nonleguminous plants are also attacked by this nematode. Depending on the degree of infestation, it can cause losses varying from slight to complete destruction of the crop. Usually, however, in heavily infested fields yield is reduced from 30 to 75 percent.

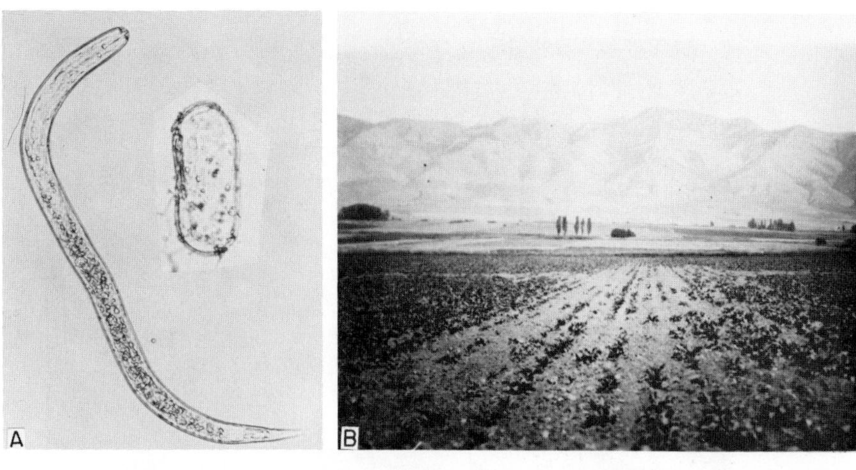

FIGURE 15-7 (A) Larva and egg of the golden nematode of potato *(Globodera rostochiensis).* (B) Bare spots in sugar beet field caused by the sugar beet nematode *Heterodera schachtii.* (Photos courtesy U.S.D.A.)

◦ *Symptoms.* Infected soybean plants appear stunted and have an unthrifty appearance. The foliage turns yellow prematurely and falls off early. The plants bear only a few flowers and a few small seeds. Infected plants growing on sandy soil usually die. Infected plants growing on fertile soils with plenty of moisture may show only slight chlorosis of the older leaves, little or no stunting, and may produce a nearly normal yield for a year or two. In subsequent years, however, due to the tremendous buildup of nematodes in the soil, plants in these areas also become severely chlorotic and dwarfed.

The root system of infected plants appears smaller than that of healthy plants, but no macroscopic lesions, galls, or other type of abnormalities are evident on infected roots. Roots of infected plants bind only a few nodule-inducing *Rhizobia* and therefore usually have considerably fewer bacterial nodules than those of healthy plants. The most characteristic symptom of this disease is the presence of female nematodes in varying stages of development and of cysts attached on the soybean roots (Figure 15-8). Young females are small, white, and partly buried in the root, with only part of them protruding on the surface. Older females are larger, almost completely on the surface of the root, and appear yellowish or brown, depending on maturity. Dead, brown cysts are also present on the roots.

◦ *The Pathogen: Heterodera glycines.* The soybean cyst nematode overwinters as a brown cyst in the upper 90–100 cm of soil. The cysts are the leathery skins of the females and are filled with eggs. The eggs contain fully developed second-stage larvae (Figure 15-9). When temperature and moisture become favorable in the spring, the larvae emerge from the cysts and infect roots of host plants. At least four races of the pathogen are known.

At 4–6 days after penetrating the roots, the larvae molt and produce the third-stage larvae. The third-stage larvae are much stouter than the second-stage larvae, and 5–6 days later fourth-stage larvae begin to appear. The female fourth-stage larva loses its somewhat slender appearance and develops the typical flask shape, measuring approximately 0.40 mm in length by 0.12–0.17 mm in width. By day 12–15, adult males and females appear.

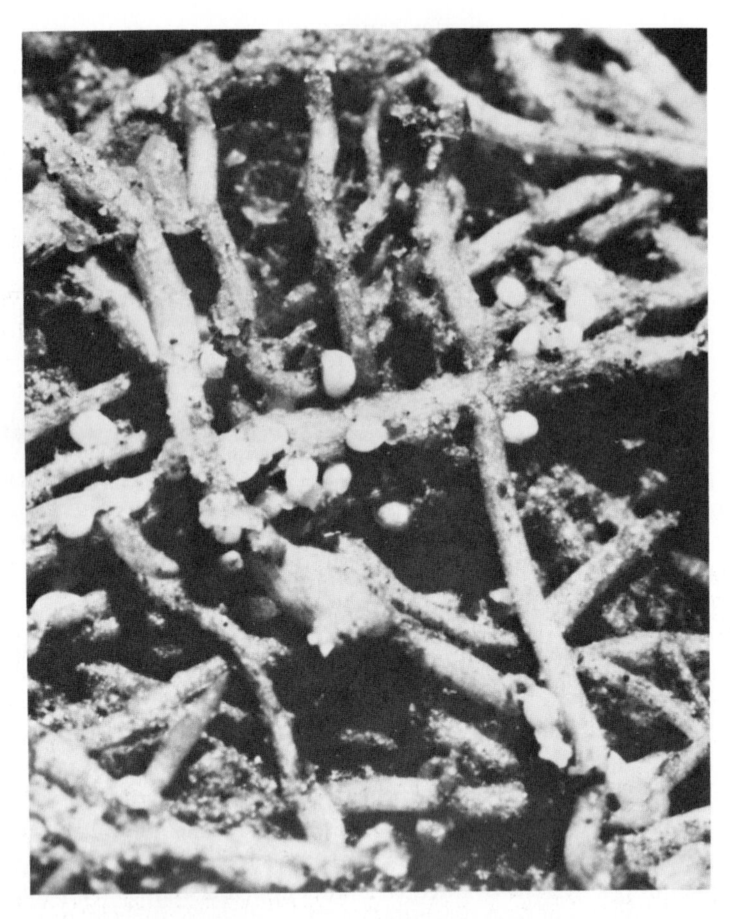

FIGURE 15-8 Lemon-shaped encysted female nematodes attached to soybean roots. (Photo courtesy U.S.D.A.)

The adult male is wormlike, about 1.3 mm long by 30–40 μm in diameter. The males remain in the root for a few days, during which they may or may not fertilize the females, then move into the soil and soon die.

The adult females when fully developed are lemon shaped, measuring 0.6–0.8 mm in length and 0.3–0.5 mm in diameter. They are white to pale yellow at first, becoming yellowish-brown as they mature. The body cavity of the female is almost completely filled by the ovaries, and as the ova gradually develop into fully formed eggs, the body cavity of the female becomes completely filled with eggs. As the female body distends during egg production, it crushes cortical cells, splits the root surface, and protrudes until it is almost entirely exposed through the root surface. A gelatinous mass, usually mixed with dirt and debris, surrounds the posterior end of the females, and the nematodes deposit some of their eggs in it. Each female produces 300–600 eggs, most of which remain inside her body when the female dies. Eggs in the gelatinous matrix may hatch immediately, and the emerging second-stage larvae may cause new infections. Finally, the old body wall, darkening to brown, becomes the cyst. Approximately 21–24 days are required for the completion of a life cycle of this nematode. The cyst consists of the female cuticle transformed through the secretions of the nematode into a tough, brown sac that persists in the soil for many years and protects the eggs, which have been formed within the body.

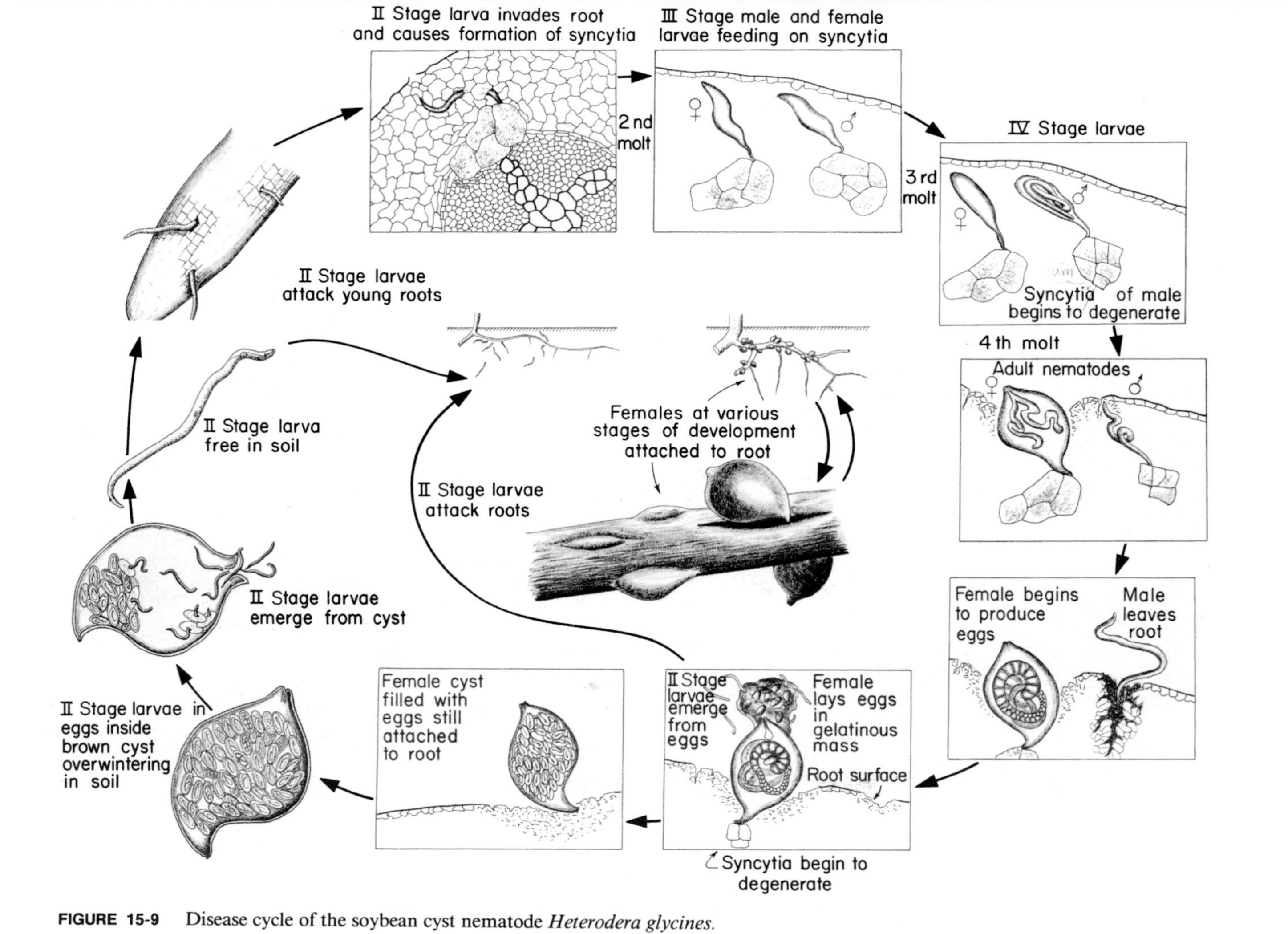

FIGURE 15-9 Disease cycle of the soybean cyst nematode *Heterodera glycines.*

◦ *Development of Disease.* The infective second-stage larvae penetrate young primary roots or apical meristems of secondary roots directly (Figure 15-9). The advance into the cortex is mostly intracellular and results in distortion and death of invaded cells. The larvae often pass through the cortex and pierce their stylets into cells of the endodermis or the pericycle. Within 2 days of penetration larvae come to rest and feed on cells of the cortex and stele tissues, causing the enlargement of these cells. Such groups of enlarged cells, called **syncytia,** are surrounded by a single layer of small, hyperplastic cells, the walls of which undergo further dissolution and allow enlargement of the syncytia. During the development of the third larval stage cortical cells surrounding the nematode are crushed by the expanding nematode body, especially by developing females. Syncytial development is either restricted largely to pericyclic tissue or occurs in tissues of the phloem and secondary cambium. Syncytia in contact with developing third- or fourth-stage males begin to show signs of degeneration, indicating cessation of feeding. Syncytia in contact with females remain active up to and beyond the stage of egg deposition. Degeneration of syncytia is accompanied by reduction of syncytial volume and results in the receding and collapse of the syncytial wall. The resulting space is only partly occupied by surrounding parenchymatous tissue.

When soybean varieties resistant to the soybean cyst nematode are attacked, there is no apparent inhibition of penetration of the organism into the host tissues. Syncytia are formed within 2–3 days of inoculation, but by day 5 many of them degenerate and most second-stage larvae associated with them are dead. A few nematodes advance to the third stage, but no adult males or females are produced. Development of syncytia and subsequent degeneration and necrosis is restricted to the periphery of the stele and to regions in the cortex that are invaded and stimulated by infective larvae. The root regions vacated by degenerate syncytia are quickly filled by adjacent rejuvenated parenchyma cells.

Syncytial development into the secondary cambial region of susceptible varieties results in inhibition of secondary growth of both phloem and xylem. Since a short portion of a root may be attacked by many larvae, the large number of syncytia that develop may cause widespread reduction of the conductive elements, resulting in the restricted growth and yield of soybean plants, especially under stresses of moisture.

◦ *Control.* Soil fumigation of soybean cyst nematode-infested fields with ethylene dibromide (EDB) (the use of which was recently suspended in the United States), or soil treatment with nonfumigant nematicides such as aldicarb and carbofuran, temporarily increases plant growth and soybean yield. Nematode cysts and larvae, however, are almost never eradicated from a field completely by fumigation, and a small nematode population left over after fumigation can build up rapidly on the vigorous soybean grown in newly fumigated soil. In addition, the cost of fumigation per acre makes its use impractical.

The most practical method of control of the soybean cyst nematode is through the use of resistant varieties and through a 1- to 2-year crop rotation with nonhost crops, since some legumes are the only other cultivated crops that are hosts of this nematode. The effectiveness of crop rotation is increased by planting the more resistant soybean varieties, which do not allow a quick and excessive buildup of nematode populations.

Cysts and eggs of soybean and other cyst nematodes are often found infected with one of several fungi such as *Fusarium, Verticillium, Nema-*

tophthora, and *Catenona.* In the near future it may be possible to use one of these or another parasite as biological control of the cyst nematodes.

Quarantine regulations are currently enforced to prevent the parasite from spreading into nematode-free areas by means of contaminated soil, products, machinery, or other articles.

• Sugar Beet Nematode: Heterodera Schachtii

The sugar beet nematode occurs wherever sugar beets are grown in North America, Europe, the Middle East, and Australia and is the most important nematode pest of sugar beet production. It also affects spinach and crucifers. The sugar beet nematode causes yield losses of 25–50 percent or more, especially in warmer climates or late-planted crops. The losses on sugar beet are mostly the result of reduced root weight, but in warm climates the sugar content is also reduced and, generally, the nematode aggravates losses caused by other pathogens such as *Cercospora, Rhizoctonia,* and beet viruses. In fields infested with the sugar beet nematode, small to large patches of wilting or dead young plants or stunted older sugar beets appear (Figure 15-7, B). The latter have an excessive number of hairlike roots. Small white or brownish cysts of female nematodes can be seen clinging to the roots. The morphology, biology, and spread of the sugar beet nematode is similar to that of the soybean cyst nematode. Control of the sugar beet nematode in red table beets is based on early sowing so that plants can grow as much as possible at temperatures at which the nematodes are more or less inactive, on crop rotations with alfalfa, cereals, or potatoes, which are not hosts of this nematode, and soil fumigation with DCP. No sugar beet varieties resistant to this nematode are commercially available yet.

SELECTED REFERENCES

Anonymous (1961). Soybean cyst nematoe. *U. S. Agric. Res. Serv. ARS, Spec. Rep.* **22-72,** 1-20.

Brown, R. H. (1984). Cereal cyst nematode and its chemical control in Australia. *Plant Dis.* **68,** 922-928.

Endo, B. V. (1964). Penetration and development of *Heterodera glycines* in soybean roots and related anatomical changes. *Phytopathology* **54,** 79-88.

Endo, B. V. (1965). Histological responses of resistant and susceptible soybean varieties and backcross progeny to entry and development of *Heterodera glycines. Phytopathology* **55,** 375-381.

Franklin, M. T. (1972). *Heterodera schachtii.* C. I. H. Descriptions of Plant-Parasitic Nematodes, Set 1, No. 1, pp. 1-4.

Gipson, I., Kim, K. S., and Riggs, R. D. (1971). An ultrastructural study of syncytium development in soybean roots infected with *Heterodera glycines. Phytopathology* **40,** 135-152.

Hartwig, E. E. (1981). Breeding productive soybean cultivars resistant to the soybean cyst nematode for the southern United States. *Plant Dis.* **65,** 303-307.

Huang, J. S., Barker, K. R., and Van Dyke, C. G. (1984). Suppression of binding between *Rhizobia* and soybean roots by *Heterodera glycines. Phytopathology* **74,** 1381-1384.

Kerry, B. (1981). Fungal parasites: A weapon against cyst nematodes. *Plant Dis.* **65,** 390-394.

Raski, D. J. (1950). The life history and morphology of the sugar beet nematode *Heterodera schachtii. Phytopathology* **40,** 135-152.

Stone, A. R. (1973). *Heterodera rostochiensis.* C. I. H. Descriptions of Plant-Parasitic Nematodes, Set 2, No. 16, pp. 1-4.

Wrather, J. A., Anand, S. C., and Dropkin, V. H. (1984). Soybean cyst nematode control. *Plant Dis.* **68,** 829-833.

• The Citrus Nematode: *Tylenchulus semipenetrans*

The citrus nematode is present wherever citrus trees are grown. In some regions, in addition to citrus the nematode also attacks grapevines, olive, lilac, and other plants. Infected trees show a slow decline, that is, they grow poorly, their leaves turn yellowish and drop early, their twigs die back, and fruit production is gradually reduced to unprofitable levels.

The pathogen, *Tylenchulus semipenetrans,* is a semiendoparasitic sedentary nematode. The larvae and males are wormlike, but the female body is swollen irregularly behind the neck. The nematodes measure about 0.4 mm long by 18–80 μm in diameter, the larger diameters found only in the maturing and mature females. The females, whose front end of the body is buried in the root tissue while the rear end remains outside (Figure 15-10), lay eggs in a gelatinous substance. The life cycle of *T. semipenetrans* is completed within 6–14 weeks at 24°C. The eggs hatch and second-stage larvae emerge. The male larvae and adults do not feed and apparently do not play a role either in the disease or the reproduction of the nematode. The second-stage female larva is the only infective stage of the nematode and cannot develop without feeding, but it can survive for several years. In the soil, the citrus nematode occurs as deep as 4 meters.

The female second-stage larvae usually attack the 4- to 5-week-old feeder roots and feed on the surface cells of the roots. There they undergo the three additional molts and produce adult females. The young females then penetrate deeper into the cortex and may reach as deep as the pericycle. The head of the nematode develops a tiny cavity around it and feeds on the surrounding 3–4 layers of parenchyma cells known as "nurse cells." Later on, the cells around the feeding site become disorganized and break down. After invasion by secondary fungi and bacteria, the affected areas turn into dark necrotic lesions, which may be so numerous that they give the root a darkened appearance. In severe infections, one hundred or more females may be feeding per centimeter of root. The females, along with soil particles that cling to

FIGURE 15-10 *Tylenchulus semipenetrans* females feeding on citrus roots with their heads embedded in individual cells. (Photo courtesy U.S.D.A.)

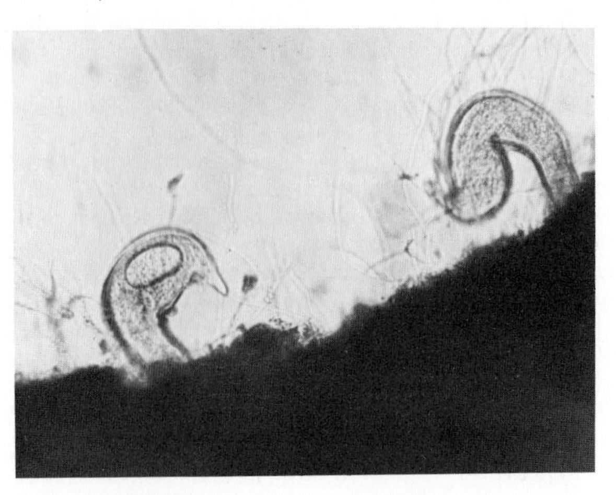

the gelatinous substance of the egg mass, result in dark, bumpy, and often decayed young roots.

The spread of the nematode through the soil is slow, the rate being approximately 1.5 cm per month when the roots of adjacent citrus plants are in contact. The nematode, however, is spread over long distances by movement of nematode-infested soil on equipment, animals, and by irrigation water, and to even longer distances by transfer of infested citrus nursery plants. The nematodes reach high populations in infected trees, which begin to show decline 3–5 years after the initial infection. When the trees show advanced stages of decline, the nematode populations also decline in numbers.

Control of the citrus nematode is based on preventing its introduction into new areas by growing nursery stock in mematode-free fields and by treating nursery stock with hot water at 45°C for 25 minutes or with fensulfothion. Because of the great depth at which the citrus nematode can survive, soil fumigation is not always effective. Satisfactory control has been obtained by preplant fumigation with DD, or methyl bromide; also by postplant treatment with aldicarb applied by chisel injection, or as granules that are incorporated into the soil immediately. Some citrus clones are resistant to the nematode populations of some regions but not to those of others.

SELECTED REFERENCES

Cohn, E. (1972). Nematodes diseases of citrus. *In* "Economic Nematology," (J. M. Webster, ed.), pp. 215–244. Academic Press, New York.

Siddiqi, M. R. (1974). *Tylenchulus semipenetrans.* C.I.H. Descriptions of Plant-Parasitic Nematodes, Set 3, No. 34, pp. 1–4.

Van Gundy, S. D. (1985). The life history of the citrus nematode *Tylenchulus semipenetrans. Nematologica* **3,** 283–294.

• Lesion Nematodes: *Pratylenchus*

Lesion or meadow nematodes occur in all parts of the world, where they attack the roots of all kinds of plants, such as field crops, cereal crops, vegetable crops, fruit trees, and many ornamentals.

The severity of damage caused by lesion nematodes varies with the crop attacked and consists in root reduction or inhibition by formation of local lesions on young roots, which may be followed by root rotting due to secondary fungi, bacteria, and so on. As a result of the root damage, affected plants grow poorly, produce low yields, and may finally die.

○ *Symptoms.* Susceptible herbaceous host plants affected by lesion nematodes appear stunted and chlorotic as though they are suffering from mineral deficiencies or lack of water. Usually several plants are affected in one area, producing patches of plants with reduced growth and a yellowish-green color, that can be seen from a distance. As the season progresses, stunting becomes more pronounced, the foliage wilts during hot summer days, and the color of the leaves becomes yellowish brown. Such plants can be easily pulled from the soil because of the extensive destruction of the root system. Yields of affected plants are reduced in varying degrees, and in severe infections the plants are killed.

When shrubs or trees are attacked by lesion nematodes, damage is usually slow to appear; it is less obvious than that on herbaceous hosts, and it rarely kills the plants. The symptoms usually consist of isolated trees or patches of trees gradually becoming unthrifty and producing poor crops. The leaves are smaller in size, their color being a dull green or yellow. Terminal branches may lose their leaves prematurely and die back. The whole appearance of affected trees indicates that the trees are weakened and are in a condition of decline. The patches of affected trees may slowly increase in size, although this happens over a rather long period.

The root symptoms of affected plants consist of lesions that appear as tiny, elongate, water-soaked, or cloudy yellow spots but that soon turn brown to almost black. The lesions appear mainly on the young feeder roots, and they are most concentrated in the area of the root hairs, but they may appear anywhere along the roots. The lesions enlarge mostly longitudinally following the root axis, and they may coalesce with other lesions, but at the same time they slowly expand laterally until they finally girdle the entire root, which they kill. As the lesions enlarge, the affected cells in the cortex collapse, and the discolored area appears constricted. Secondary fungi and bacteria usually accompany nematode infections in the soil and contribute to further discoloration and rotting of the affected root areas, which may slough off. Moderately affected plants exhibit varying degrees of root survival, and in some hosts production of adventitious roots may be stimulated by the infection; but generally the individual roots are discolored and stubby, and the whole root system is severely reduced by the root pruning that results from the formation of lesions (Figure 15-11, A).

FIGURE 15-11 (A) Roots of bean plants showing symptoms caused by lesion nematodes. Healthy plant on left. (B) Lesion nematodes (*Pratylenchus* sp.) penetrating young carrot root. (C) Lesion nematode feeding withinn a young carrot root. (D) Cross section of young carrot showing six lesion nematodes in the destroyed cortex and one in the stele. (Photo A courtesy Department of Plant Pathology, Cornell University. Photos B, C, and D courtesy R. A. Rohde.)

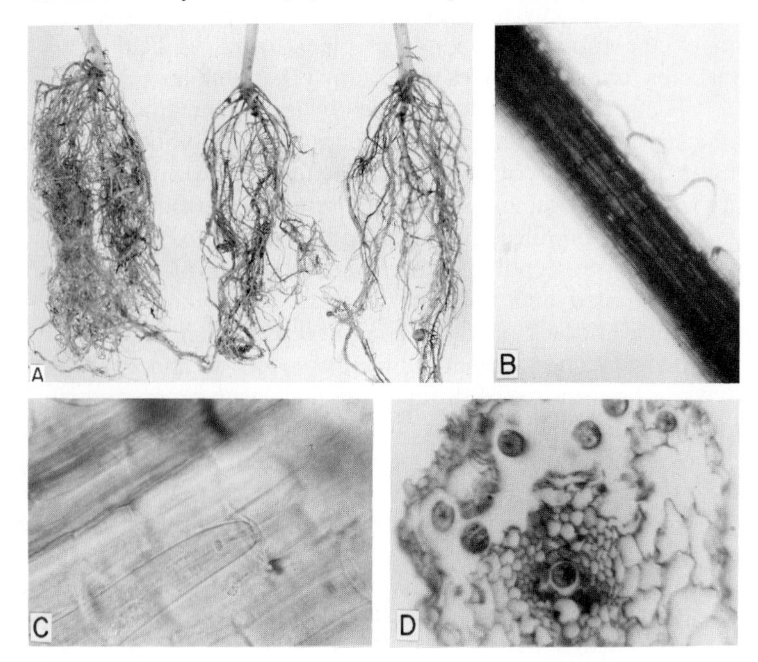

- *The Pathogen: Pratylenchus sp.* The nematodes are approximately 0.4–0.7 mm long and 20–25 μm in diameter. They appear as stout, cylindroid nematodes with a blunt head, strong, stout spear, and bluntly rounded tail (Figure 15-11, B). They are migratory, endoparasitic nematodes affecting the roots of many kinds of plants. Development and reproduction of *Pratylenchus* nematodes is rather slow, the life cycle of the various species being completed within 45–65 days. These nematodes overwinter in infected roots or in soil as eggs, larvae, or adults, except for the egg-producing females, which seem to be unable to survive the winter. Adults and larvae of various ages can enter and leave roots of susceptible hosts. The females, with or without fertilization, lay their eggs singly or in small groups inside infected roots. The eggs remain in the roots and hatch there, or when the root tissues break down, they are released into the soil. The first larval stage and the first molt occur in the egg. The emerging second-stage larva moves about in the soil or enters the root, in either case developing into the subsequent larval stages and the adults. When in the soil the nematodes are susceptible to drying, and during periods of drought they lie quiescent until the moisture increases and the plants resume growth.
- *Development of Disease.* Larvae and adult *Pratylenchus* nematodes enter roots usually in a radial direction anywhere along the roots (Figure 15-11, B). Intracellular penetration is accomplished by a persistent thrusting of the stylet and head, which seems to soften and break the cell wall. The cell walls and the adherent cytoplasm usually turn light brown in color and appear as small,

FIGURE 15-12 Disease cycle of the lesion nematode *Pratylenchus* sp.

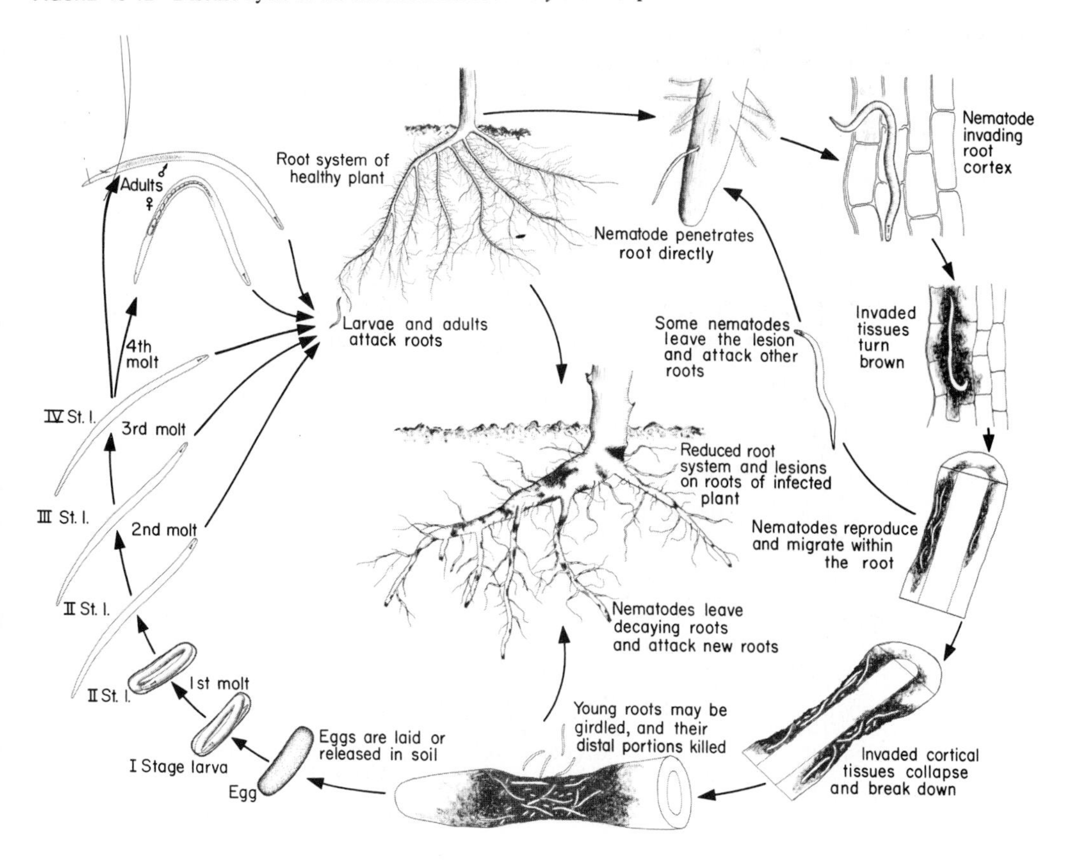

discolored spots within a few hours after inoculation. The nematodes move into the cortex, where they feed and reproduce (Figure 15-11, C, D). The endodermis is not attacked even when the nematodes completely fill the area between the endodermis and the epidermis (Figure 15-12). Necrosis of cortical cells follows the path of nematodes, but discoloration of the affected and adjoining cells varies with the host plant. Sometimes only 1 or 2 cells on each side of the nematode tunnels are affected, but at other times the lesion involves more than half the circumference of the root. The part of the endodermal layer adjacent to the nematode also takes on a deep brown color, which extends into rather large groups of cells. As the feeding of the nematode on cortical cells continues, cell walls break down, and cavities appear in the cortex with their walls sometimes lined with brown deposits.

Each lesion is usually inhabited by more than one nematode, and sometimes single host cells are simultaneously transversed by four or more nematodes. The females lay their eggs in the cortex, and frequently eggs, larvae, and a few adults form "nests"; these nests occur in great numbers in the cortex. Upon hatching of the eggs, the nematodes feed on the parenchyma cells and move mostly lengthwise within the cortex, thus enlarging the lesion (Figure 15-12). Some of the nematodes leave the lesion, emerge from the root, and travel to other points of the root or other roots where they cause new infections. Necrotic cortical tissues of large lesions are sloughed off or are invaded by secondary fungi and bacteria with resulting rotting and breakdown of the root tissues around the point of infection and subsequent death of the distal part of the root beyond the point of infection. Thus, the number of functioning roots of the plant is drastically reduced, absorption of water and nutrients becomes insufficient, and the aboveground parts of the plant become stunted and chlorotic, showing symptoms of water and nutrient deficiencies.

◦ *Control.* Root-lesion nematodes can best be controlled by overall or row treatment of the soil with nematicides before the crop is planted. DD and Brozone give good control of these nematodes, but they usually fail to eradicate them completely. Oxamyl, applied as a soil drench or foliar spray, has also given good control of lesion nematodes in some cases.

In hot and dry climates fairly good control of lesion nematodes can be achieved by summer fallow, which reduces nematode populations by exposing them to heat and drying and by eliminating host plants. Control through crop rotation is at present rather unsuccessful due to the wide host ranges of the lesion nematodes and to the lack of information on their precise host preferences.

SELECTED REFERENCES

Dickerson, O. J., Darling, H. M., and Griffin, G. D. (1964). Pathogenicity and population trends of *Pratylenchus penetrans* on potato and corn. *Phytopathology* **54,** 317–322.

Good, J. M., Boyle, L. W., and Hammons, R. O. (1958). Studies of *Pratylenchus brachyurus* on peanuts. *Phytopathology* **48,** 530–535.

Mountain, W. B., and Patrick, Z. A. (1959). The peach replant problem. VII. The pathogenicity of *Pratylenchus penetrans*. *Can. J. Bot.* **37,** 459–470.

Parker, K. G., and Mai, W. F. (1974). Root diseases of fruit trees in New York State. VI. Damage caused by *Pratylenchus penetrans* to apple trees in the orchard growing on different rootstocks. *Plant Dis. Rep.* **58,** 1007–1011.

Rebois, R. V., and Golden, R. M. (1985). Pathogenicity and reproduction of *Pratylenchus agilis* in field microplots of soybeans, corn, tomato, or corn-soybean cropping systems. *Plant Dis.* **69,** 927–929.

• The Burrowing Nematode: *Radopholus similis*

The burrowing nematode occurs widely in tropical and subtropical regions of the world and in greenhouses in Europe. It is the most important banana root pathogen in most banana-growing areas, where it causes the so-called root rot, blackhead, toppling disease, or decline of banana. It also causes the spreading decline disease of citrus in Florida, a decline of avocados in Florida and of tea in Ceylon, and the yellows disease of black pepper in Indonesia. Furthermore, it attacks coffee and other fruit, ornamental and forest trees, sugarcane, corn, vegetables, grasses, and weeds.

Infected banana plants appear to be growing poorly, have fewer and smaller leaves, show premature defoliation and reduced weight of fruits. Often entire banana plants topple over. Banana roots at first show browning and cavities in the cortex, followed by deep cracks with raised margins on the root surface (Figure 15-13, D). The nematodes, along with fungi and bacteria that invade the cracked roots, cause the roots to rot and eventually only a few short root stubs remain, which cannot anchor the plant sufficiently and the latter topples over. Diffuse black, rotten areas also develop in the rhizome cortex surrounding infected roots (Figure 15-13, E). As a result of this disease the profitable life of a banana plantation in many areas is decreased from indefinite to as little as one year, and the costs of annual replanting and losses in production are tremendous.

In the spreading decline of citrus, blocks of affected trees also have fewer and smaller leaves and fruits, and many of the twigs and branches die back (Figure 15-13, A). Yields of infected trees are reduced by 40–70 percent. Even during periods of mild moisture stress, infected trees wilt readily, but they generally do not die and often recover temporarily after rainy periods. The symptoms of decline spread steadily to more trees each year, the diameter of the decline area increasing approximately 10–20 meters per year. The symptoms on the aboveground parts follow infection of the roots by about a year. Infected feeder roots have numerous lesions that appear puffy and cracked and are often invaded by primary and secondary fungal parasites that result in the rotting and destruction of the feeder roots. Feeder roots seem to be attacked and destroyed most at depths of 50 cm or more, leaving less than half the feeder roots functional.

The pathogen, *Radopholus similis,* usually known as the burrowing nematode, is wormlike, measuring about 0.65 mm long by 25 μm wide (Figure 15-13, B, C). It spends its life and reproduces inside cavities in the root cortex, where it completes a life cycle in about 20 days. All larvae and the adults can infect roots, and although they can emerge from the roots and spread through the soil, most of the spread of the nematode from plant to plant is through root contact or near contact. Long-distance spread of the nematode is primarily with infected plant material, such as infected banana sets. Although the nematodes infecting banana and citrus are morphologically identical, the "banana race" can attack banana but not citrus while the "citrus race" can attack citrus as well as banana and several other hosts. The citrus race, however, is so far known to occur only in Florida. Other races probably exist in other parts of the world.

The burrowing nematode enters feeder roots and moves intercellularly in the cortical parenchyma, feeding on nearby cells, destroying them, and causing the formation of cavities (Figure 15-14). As the nematodes continue to

A

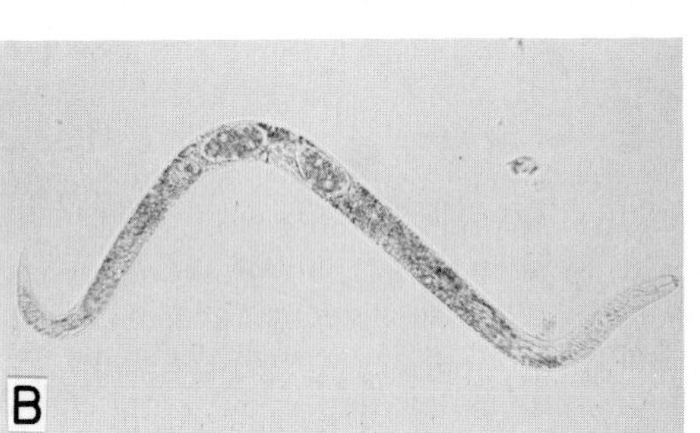
B

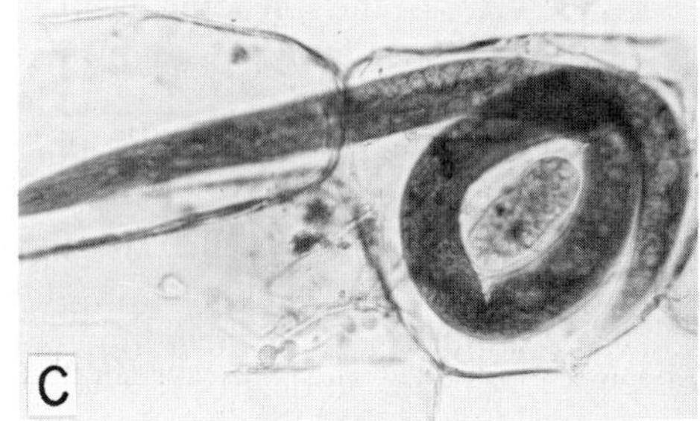
C

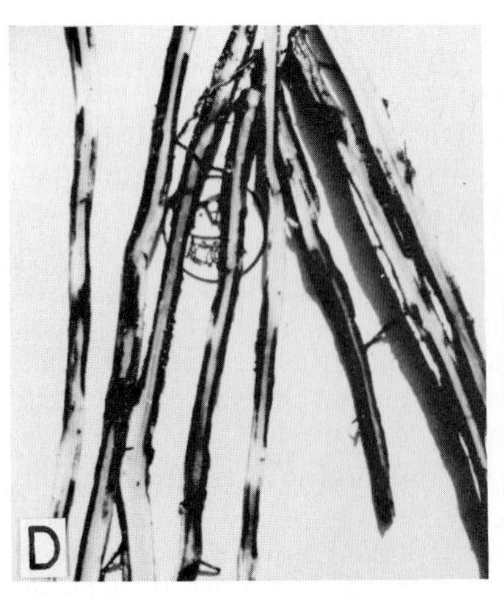
D

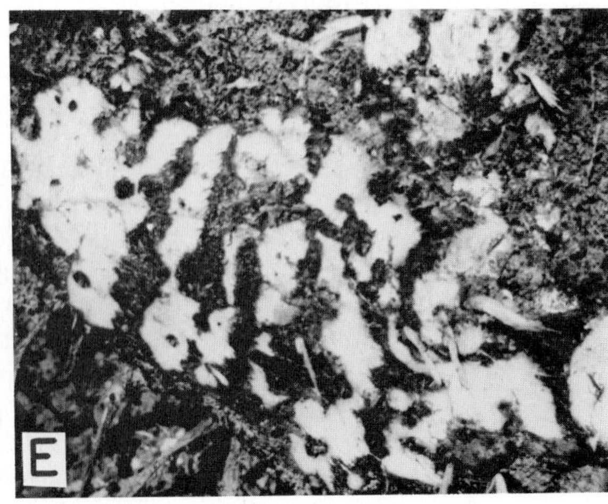
E

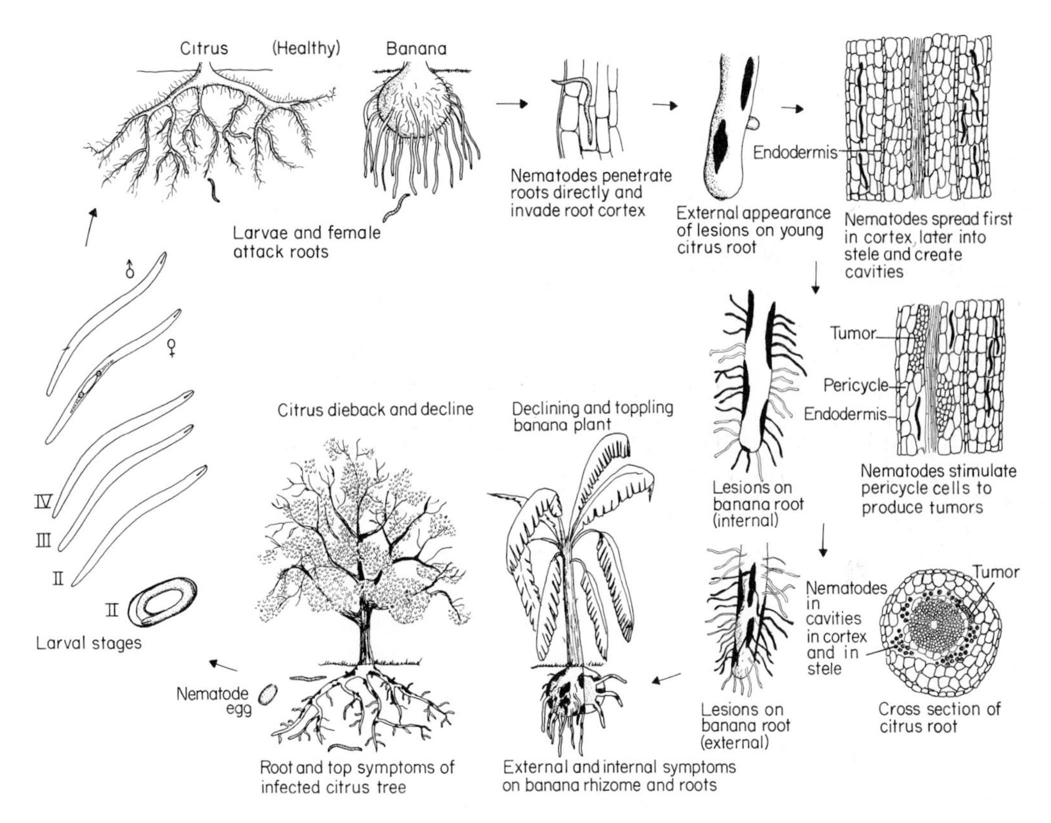

FIGURE 15-14 Disease cycle of the burrowing nematode *Radopholus similis* in citrus, banana, and so on.

feed, the cavities enlarge and coalesce with others, forming long and lateral tunnels. In banana, the tunnels are limited to the cortex between the epidermis and the endodermis. From the feeding roots the nematodes also move into the rhizome. In citrus, however, the nematodes not only form cavities in the cortex but they also enter the stele through endodermal passage cells. There they accumulate in the phloem and cambium, which they destroy in time and form nematode-filled cavities. At the same time, gum is deposited in the cortex, and cells of the pericycle divide excessively and produce groups of tumorlike cells. The cavities within the roots appear externally as brown-reddish lesions throughout the cortex, and three to four weeks from infection the lesions develop one or more deep cracks. Each female lays one or a few eggs per day for many days and as they hatch, develop, and reproduce, nematode populations increase rapidly. As many as 800 nematodes may be present in a single lesion and a single tree may be supporting hundreds of thousands of

FIGURE 15-13 (A) Healthy and infected orange trees at the margin of a spreading decline area caused by *Radopholus similis.* (B) *R. similis* female containing two eggs. (C) *R. similis* and egg in banana cells. (D) Lesions on banana roots caused by *Radopholus similis.* (E) Lesions on banana rhizome caused by *R. similis.* (Photo A courtesy Agricultural Research and Educational Center, Lake Alfred, Florida. Photos B–E courtesy R. H. Stover.)

burrowing nematodes. Fungi such as *Fusarium* and *Sclerotium* invade nematode-infected roots much more readily and further increase their rotting and destruction.

Control of the burrowing nematode in banana can be obtained by removing discolored tissues from banana sets by paring and then dipping the sets in hot water at 55°C for 20 minutes; by flooding the field for 5–6 months where possible; and by soil fumigation with nematicides.

Control of spreading decline of citrus is much more difficult and depends primarily on: (1) preventive regulatory measures that inhibit the spread and establishment of the nematode in new areas by treating nursery trees with hot water at 50°C for 10 minutes or dipping them in nematicides such as ethoprop and fensulfothion; disinfesting equipment; and fumigating every 6 months with DD 5-meter-wide strips of land around the area of decline and keeping it free of weeds with cultivation or herbicides. (2) Fumigation of decline areas with heavy doses of DD after removal of all declining trees and at least two rows around them. (3) Use of tolerant or resistant rootstocks.

SELECTED REFERENCES

Blake, C. D. (1972). Nematode diseases in banana plantations. *In* "Economic Nematology," (J. M. Webster, ed., pp. 245–267. Academic Press, New York.

DuCharme, E. P. (1959). Morphogenesis and histopathology of lesions induced on citrus roots by *Radopholus similis*. *Phytopathology* **49,** 388–395.

Poucher, C. *et al.* (1967). Burrowing nematode in citrus. *Fla., Dep. Agric. Bull.* **7,** 1–63.

Williams, K. J. O., and Siddiqi, M. R. (1973). *Radopholus similis*. C.I.H. Descriptions of Plant-Parasitic Nematodes, Set 2, No. 27, pp. 1–4.

• Stem and Bulb Nematode: *Ditylenchus*

Of the several species of *Ditylenchus* that cause diseases in plants, *D. dipsaci* is the most common and most serious of all and is the one generally referred to as the stem and bulb nematode, while *D. destructor* is known as the potato rot nematode.

The stem and bulb nematode is worldwide in distribution but is particularly prevalent and destructive in areas with temperate climate. It is one of the most destructive plant-parasitic nematodes. It attacks a large number of host plants including onions, narcissus, hyacinth, tulip, oats, rye, alfalfa (Figure 15-15, A), red clover, strawberry, and phlox. Although all these hosts are attacked by nematodes belonging to one species, different populations or biotypes of the stem and bulb nematode have certain host preferences and, when transferred to them, will attack one or a few of these hosts but not the others. On most crops it causes heavy losses by killing seedlings, dwarfing plants, destroying bulbs or making them unfit for propagation or consumption, by causing the development of distorted, swollen and twisted stems and foliage, and, generally, reducing yields greatly.

◦ *Symptoms.* In fields infested with stem and bulb nematodes, emergence of seedlings such as onion is retarded and stands are reduced considerably. Half or more of the emerging seedlings may be diseased, appearing pale, twisted, arched, and with enlarged areas along the cotyledon. Cotyledons are usually puffy with the epidermis cracked in a lacelike fashion. Most infected seedlings die within 3 weeks of planting and the remainder usually die later.

When bulbs are planted in infested soil, the developing plants within about 3 weeks show stunting, light yellow spots, swellings ("spikkles"), and

FIGURE 15-15 (A) Alfalfa plant with shortened and thickened stems and small crowded leaves as a result of infection with the stem nematode *Ditylenchus dipsaci*. (B) Infected onion plants showing the prostrate condition, thin stand, stunting, and many dead outer leaves. (C) Infected narcissus bulbs showing complete or incomplete brown rings. (Photo A courtesy U.S.D.A. Photos B and C courtesy Department of Plant Pathology, Cornell University.)

open lesions on the foliage. Young plants and sprouts develop swellings on the stem and foreshortening and curling of the leaves. Many outer leaves often become flaccid, their leaf tips die back, and the leaves are so weakened that they cannot maintain their erect growth and fall to the ground (Figure 15-15, B). The stem and the neck of the bulb become softened, and the softening gradually proceeds downward into individual scales, which become soft, loose, and pale gray in color. Affected scales appear as discolored rings in cross sections of infected bulbs (Figure 15-15, C) and as discolored, unequal lines in longitudinal sections. In more advanced cases large areas or the whole bulb may be affected. Infected bulbs may also split and become malformed or may produce sprouts and double bulbs. The outer scales may be loosened and detached by applying a little oblique pressure with the thumb on the upper half of the bulb and reveal the soft mealy, frosty-looking tissue beneath. In dry weather the bulbs become desiccated, odorless, and very light in weight. In wet seasons a soft rot due to secondary invaders sets in, destroying the bulb and giving off a foul odor. Infected bulbs are sometimes superficially healthy, but they continue to decay in storage, where the outer scale often sloughs off exposing the lower puffy, soft scales with the characteristic frosty mealiness.

○ *The Pathogen: Ditylenchus dipsaci.* The nematode is 1.0–1.3 mm long and about 30 μm in diameter (Figure 15-16). Each female lays 200–500 eggs. Second-stage larvae emerge from the egg and quickly undergo the second and third molt and produce the preadult or infective larva. The latter can withstand adverse conditions of freezing and of extreme drying for long periods in fragments of plant tissue, stems, leaves, bulbs, seeds, or in the soil. Under favorable moisture and temperature conditions the preadult larvae become

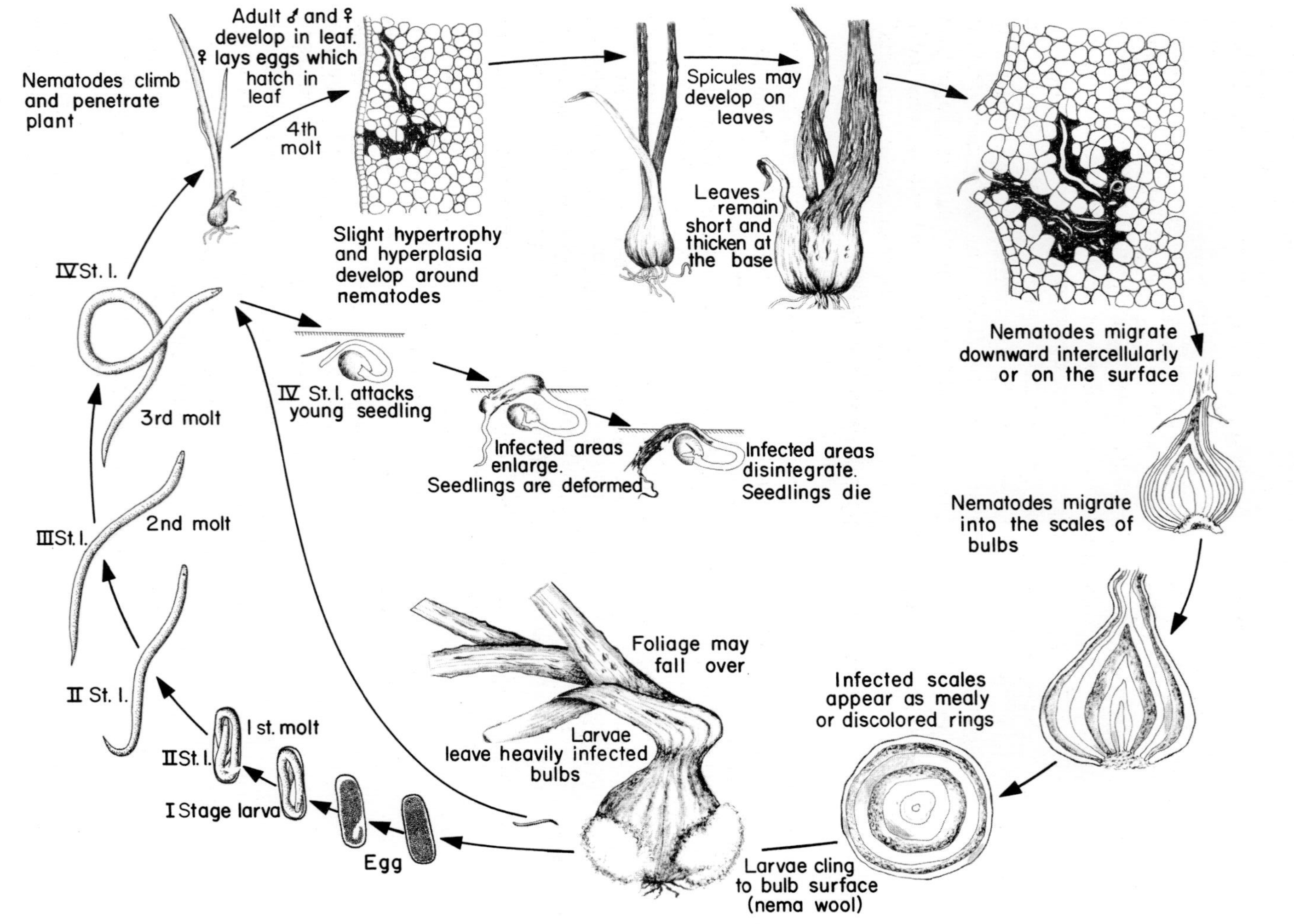

FIGURE 15-16 Disease cycle of the stem and bulb nematode *Ditylenchus dipsaci*.

active, enter the host, pass through the fourth molt, and become males and females. The females then lay eggs, mostly after fertilization by the males. A complete cycle from egg to egg usually lasts about 19–25 days. Reproduction takes place in succulent, rapidly growing tissues or in storage organs and continues throughout the year, although it is retarded or stopped by low temperatures. *Ditylenchus dipsaci* is an internal parasite of bulbs, stems, and leaves and passes generation after generation in these tissues, escaping to the soil only when living conditions in the plant tissues become unfavorable. When heavily infected bulbs decay, preadult larvae pass out and sometimes accumulate about the basal plates of dried bulbs as grayish-white, cottony masses, called nematode "wool," where they can remain alive for years.

◦ *Development of Disease.* When nematodes attack a germinating seed or young seedlings, they enter near the root cap or at points still within the seed. The nematodes remain mostly intercellular, feeding on the parenchymatous cells of the cortex. Cells near the heads of the nematodes lose all or a portion of their contents, while cells surrounding these divide and enlarge, resulting in development of swellings on the seedlings. The seedling may become malformed. Splitting of the epidermis often follows the enlargement and opens the way to secondary invaders, such as bacteria and fungi.

In older seedlings or young plants, nematodes enter the leaves through stomata or penetrate directly through the epidermis in leaf bases (Figure 15-16). Cell enlargement, disappearance of chloroplasts, and an increase of intercellular spaces in parenchyma tissue follow penetration. The nematodes usually remain and reproduce in the intercellular spaces, feeding on the nearby parenchyma cells whose contents they consume without causing appreciable discoloration on the cell remains. As the bulbs enlarge, the nematodes migrate down from the leaves either intercellularly or by traveling on the surface of the leaves and entering again at the outer sheaths of the stem or neck, through which they infect the outer scales of the bulbs. Heavily infected stems become soft and puffy due to the formation of large cavities through the breakdown of the middle lamella and of the cells the nematodes feed on. Such stems can no longer remain rigid under the weight of the foliage and they frequently collapse. The nematodes continue their progress intercellularly through the outer scales of the bulbs by breaking down the parenchymatous tissue. Parenchyma cells are separated from each other and from the vessels, the latter giving a lacy appearance to the scale. The macerated parenchyma cells have a white mealy texture at first, but secondary invaders usually set in and cause them to turn brown. In early stages of infection the nematodes remain within individual scales, and in sections the infection appears as complete or incomplete rings of frosty white or brownish tissue. In later stages of infection the nematodes pass from one scale to the next, and thus more scales may be involved in one ring. The spread of the infection within a bulb continues in the field and in storage until, usually, the entire bulb becomes affected.

◦ *Control.* Populations of *Ditylenchus dipsaci* parasitic on certain crops can be reduced by long (2–3 years at least) rotations with resistant crops, such as spinach, carrots, potatoes, and lettuce. Since this nematode also overwinters in infected bulbs and seeds, the use of nematode-free sets or seeds is extremely important. Infested seeds or bulbs can be disinfested by treating them in hot water for an hour at 46°C. Onion seed can also be freed from nematodes by enclosing it with methyl bromide gas for 24 hours at 24°C in a gas-tight container, and flower bulbs can be successfully disinfested by placing them in 0.5 percent formaldehyde solution at 43°C for 4 hours.

Control of this nematode in large fields, although often too expensive, can be achieved by fall fumigation of the soil, by preplant row treatment, and by treatment at or soon after planting with appropriate nematicides.

SELECTED REFERENCES

Chitwood, B. G., Newhall, A. G., and Clement, R. L. (1940). Onion bloat or eelworm rot, a disease caused by the bulb or stem nematode, *Ditylenchus dipsaci* (Kuhn) Filipjev. *Proc. Helminthol. Soc. Wash.* **7**, 44–51.

Darling, H. M., Adams, J., and Norgren, R. L. (1983). Field eradication of the potato rot nematode, *Ditylenchus destructor:* A 29-year history. *Plant Dis.* **67**, 422–423.

Krusberg, L. R. (1961). Studies on the culturing and parasitism of plant-parasitic nematodes, in particular *Ditylenchus dipsaci* and *Aphelenchus ritzemabosi* on alfalfa tissues. *Nematologica* **6**, 81–200.

Newhall, A. G. (1943). Pathogenesis of *Ditylenchus dipsaci* in seedlings of *Allium cepa. Phytopathology* **33**, 61–69.

Sayre, R. M., and Mountain, W. B. (1962). The bulb and stem nematode *(Ditylenchus dipsaci)* on onion in Southwestern Ontario. *Phytopathology* **52**, 510–516.

• Seed-Gall Nematodes: *Anguina*

Seed-gall nematodes were the first recorded plant-parasitic nematodes, discovered in 1743, when an infected wheat seed (seed gall) was crushed in a drop of water under a microscope. Several species of *Anguina* are known, and all of them cause formations of galls on seeds, leaves, and other aboveground parts of plants. Of the seed-gall nematodes, *A. agrostis* is probably the most widespread in Europe and North American and causes severe injury to bentgrasses (*Agrostis* spp.), but *A. tritici* has been important on wheat and rye in the past and in some areas still is. The wheat seed gall, which is described below, is present wherever wheat is grown, but in most countries it is quite rare due to the use of fresh and cleaned seed. The wheat seed-gall nematode is still common, however, in some Mediterranean countries, eastern Europe, and Asia.

The symptoms appear on plants in all growth stages. Infected seedlings are more or less severely stunted and show characteristic rolling, twisting, curling, or wrinkling of the leaves (Figure 15-17). A rolled leaf often traps the next emerging leaf or the inflorescence within it and causes it to become looped or bent and badly distorted. Stems are often enlarged near the base, frequently bent, and generally stunted. Diseased heads are shorter and thicker than healthy ones and the glumes are spread further apart by the nematode-filled seed galls. A diseased head may have one, a few, or all of its kernels turned into nematode galls. The galls are shiny green at first, but turn brown or black as the head matures. Diseased heads remain green longer than healthy ones, and galls are shed off the heads more readily than kernels. Mature galls are hard, dark, rounded, and shorter than normal wheat kernels and often resemble cockle seeds, smutted grains, or ergot sclerotia.

The pathogen, *Anguina tritici,* is a large nematode about 3.2 mm long by 120 μm in diameter. The nematode lays its eggs and produces all its larval stages and the adults in seed galls.

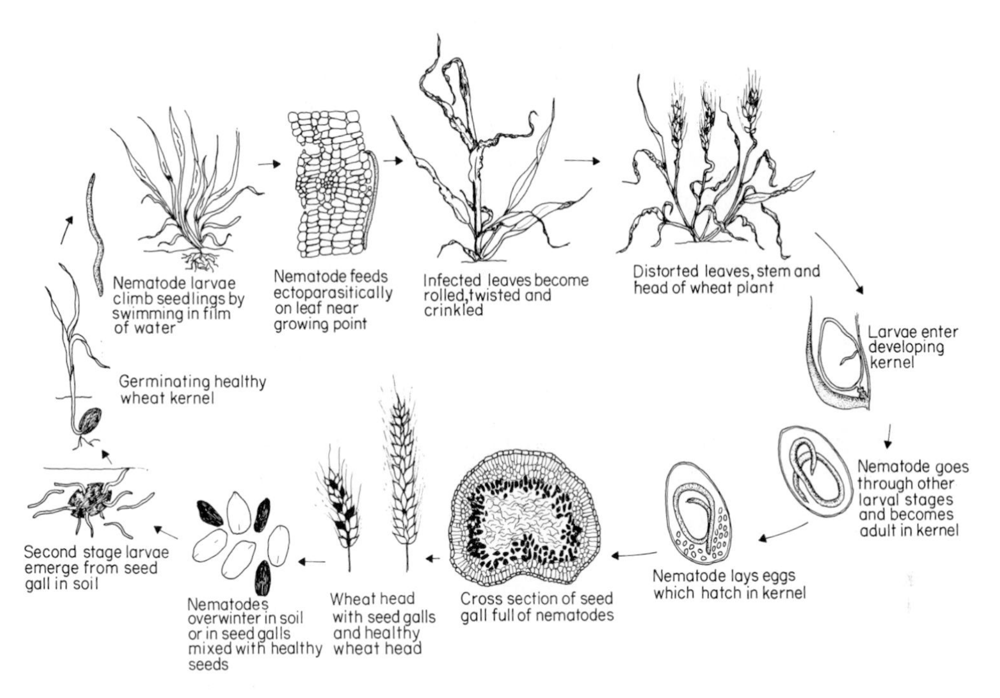

FIGURE 15-17 Disease cycle of wheat seed gall caused by *Anguina tritici.*

The seed-gall nematode overwinters as second-stage larvae in seed galls or in plants infected in the fall. Galls fallen to the ground or sown with the seed soften during warm moist weather and release infective second-stage larvae. When a film of water is present on the surface of the plants the larvae swim upward and feed ectoparasitically on the tightly compacted leaves near the growing point, causing the leaves and stem to become malformed. When the inflorescence begins to form, the larvae enter the floral primordia and produce the third- and fourth-stage larvae and the adults. Each infected floral primordium becomes a seed gall and may contain 80 or more adults of both sexes. Each of the females then lays up to 2000 eggs over several weeks within the freshly formed gall so that each gall contains 10,000–30,000 eggs. The adults die soon after the eggs are laid. The eggs then hatch and the first-stage larvae emerge, but these soon molt and by harvest produce the second-stage larvae, which are very resistant to desiccation and can survive in the galls for up to 30 years (Figure 15-18). The seed-gall nematode produces only one generation per year. The nematode is spread in infected seed.

Control of the seed gall nematode depends on the use of clean seed free of nematode-containing galls. Fields infested with seed-gall nematodes should not be planted to wheat or rye for at least a year. In moist weather the seed galls release the second-stage larvae and, if no susceptible hosts are present, they die before they can infect and reproduce. In dry weather, however, nematodes can survive in the seed galls for many years.

SELECTED REFERENCES

Leukel, R. W. (1929). The nematode disease of wheat and rye. *Farmers' Bull.* **1607,** 1–12.

Southey, J. F. (1972). *Anguina tritici.* C.I.H. Descriptions of Plant-Parasitic Nematodes, Set 1, No. 13, pp. 1–4.

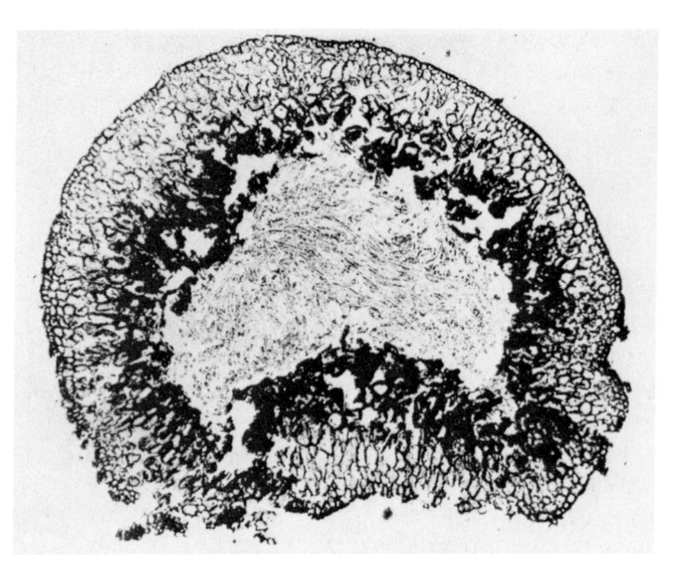

FIGURE 15-18 Cross section of a wheat seed gall showing the white mass of dormant nematodes in the center. (Photo courtesy U.S.D.A.)

• Foliar Nematodes: *Aphelenchoides*

Several species of *Aphelenchoides* feed ectoparasitically and endoparasitically on aboveground plant parts. Some of the most important species are: *A. ritzemabosi,* the chrysanthemum foliar nematode; *A. fragariae,* the spring crimp or spring dwarf nematode of strawberry and which also attacks many ornamentals; and *A. besseyi,* the nematode causing summer dwarf or crimp of strawberry and white tip of rice.

The foliar nematode of chrysanthemums, also known as chrysanthemum eelworm, is known to be widespread in the United States and Europe. It results in fairly severe losses. In addition to chrysanthemum, the foliar nematode, or closely related species, attacks several other plants, including aster, dahlia, delphinium, phlox, zinnia, and sometimes tobacco and strawberry.

○ *Symptoms.* Affected buds or growing points of stems produce short and often abnormally bushy-looking plants with short internodes. The growing point may be so damaged that the shoot does not grow and turns brown. The leaves produced from infested buds are small and distorted. The stem and petioles show brown scars caused by nematodes feeding externally on these tissues while still in the bud. Shoots so infested seldom develop into normal plants. Uninfested shoots may exist on the same stool with infested ones, and they develop into normal new plants. As the season progresses, however, nematodes climb up the stem and attack first the lower and then the upper leaves on which they cause small yellowish spots, later turning brownish black. These spots soon coalesce and form large blotches, which at first are contained between the larger leaf veins (Figure 15-19). Eventually the entire leaf is covered with spots or blotches, and it soon shrinks, becomes brittle, and falls to the ground. Defoliation, like infection, progresses from the lower to the upper leaves. The nematodes also infest the ray flowers and prevent their development. Severely infected plants die without producing much normal foliage or marketable flowers.

- *The Pathogen: Aphelenchoides Ritzemabosi.Aphelenchoides ritzemabosi* is a long, slender nematode measuring about 1 mm long by 20 μm in diameter. It may live its entire life inside leaves or at the surface of other plant organs. The female adult lays its eggs in the intercellular spaces of leaves. The eggs hatch and produce the four larval stages, and finally adults, all inside the leaf. The life cycle is completed in about 2 weeks. These nematodes do not have to spend any part of their lives in the soil but are frequently found there carried by infected, dead leaves fallen to the ground, or washed down by rain or irrigation water when they happen to be on the surface of plant tissues. The foliar nematodes overwinter as adults in dead leaves or between the scales of buds of infected tissues.
- *Development of Disease.* Nematodes overwintering between the bud scales or the growing point of shoots become activated in the spring and feed ectoparasitically by inserting their stylets into the epidermal cells of the organs in their vicinity. Thus, stem areas near infested buds and the petioles and leaves derived from such buds show brown scars consisting of groups of cells killed by the nematodes. In addition to direct killing of cells, the nematodes, through their secretions, cause shortening of the internodes, which results in a bushy appearance of the plant; browning and failure of the shoot to grow (blindness), production of low, premature side-shoots; and development of distorted leaves.

  Nematodes infest new, healthy plants by swimming up the stem when it is covered with a film of water during rainy or humid weather. When they

FIGURE 15-19 Discolored areas on chrysanthemum leaves caused by the foliar nematode *Aphelenchoides ritzemabosi.* (Photo courtesy Department of Plant Pathology, Cornell University.)

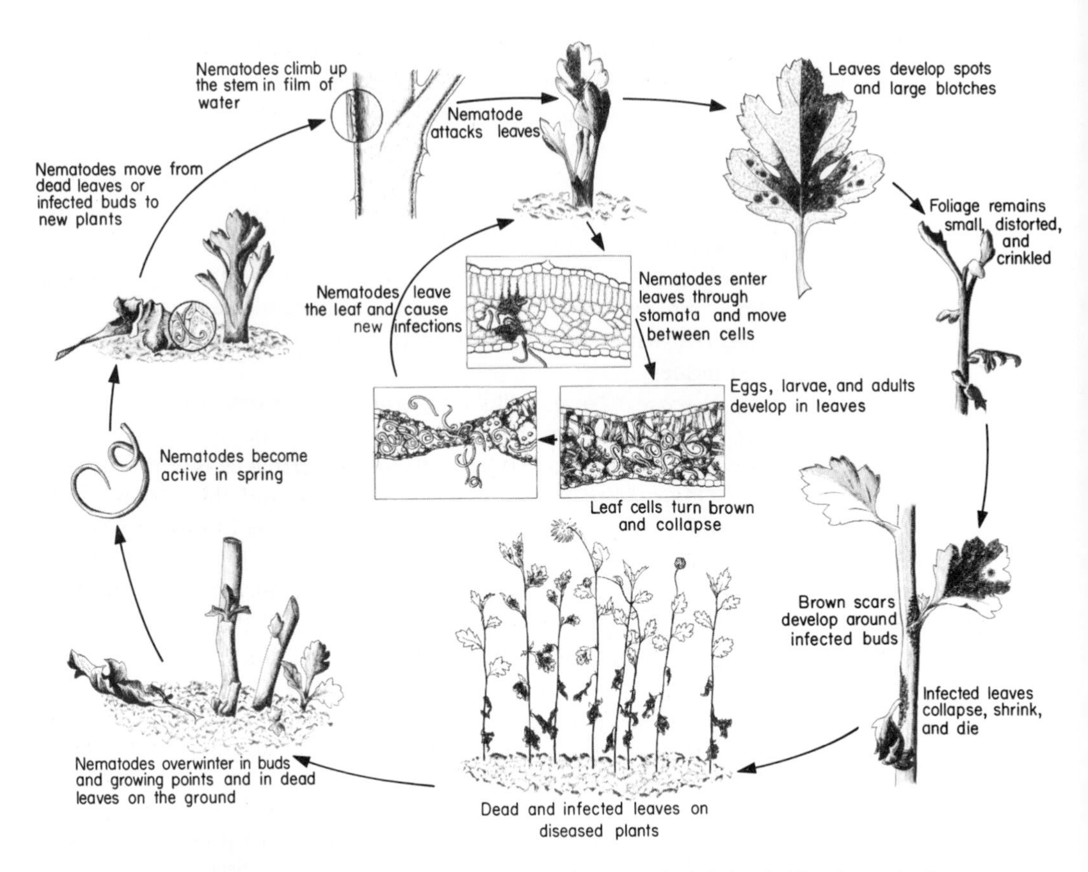

FIGURE 15-20 Disease cycle of the foliar (chrysanthemum) nematode *Aphelenchoides ritzemabosi.*

reach the leaves, the nematodes enter through the stomata (Figure 15-20). The presence of nematodes between the leaf cells causes browning in cells. Cells in the mesophyll begin to break down, creating large cavities in the mesophyll. In the early stages of infection the cells of the vein sheath, which in large veins extend from the upper to the lower surface, do not allow penetration of the nematodes through their intercellular spaces and thus block extension of leaf necrosis across the veins. In advanced stages of infection, even these cells break down and the nematodes and leaf necrosis spread across the veins over the entire leaf. In heavily infected leaves the disintegrated cells have a thick layer of brown substance on their walls, the epidermis is broken down in places, and the leaves shrink and, after clinging to the stem for some time, fall to the ground.

◦ *Control.* Several sanitary practices are quite important in controlling the foliar nematode. The leaves and stem should be kept dry, especially indoors, to prevent movement and spreading of the nematodes. Cuttings should be taken only from the tops of long, vigorous branches, not from shoots near the base of the plant. Early in the spring the soil surface around chrysanthemums should be mulched to cover old infested leaves and prevent nematodes overwintering in them from reaching the lower leaves. Suspected dormant cuttings or stools may be disinfested by dipping in hot water (50°C) for 5 minutes or at 44°C for 30 minutes. Excellent control of this nematode can be obtained by spraying plants with parathion or malathion from July to early September and by applying thionazin as a drench twice with a 2-week interval.

SELECTED REFERENCES

Bryden, J. W., and Hodson, W. E. H. (1957). Control of chrysanthemum eelworm by parathion. *Plant Pathol.* **6,** 20–24.

French, N., and Barraglough, R. M. (1964). Observations on eelworm on chrysanthemum stools. *Plant Pathol.* **13,** 32–37.

Hesling, J. J., and Wallace, H. R. (1961). Observations on the biology of chrysanthemum eelworm. *Aphelenchoides ritzemabosi* (Schwartz) Steiner in florist's chrysanthemum. I. Spread of eelworm infestation. *Ann. Appl. Biol.* **49,** 195–203, 204–209.

• Pine Wood Nematode: *Bursaphelenchus xylophilus*

The pine wood nematode has been known to occur in Japan for more than thirty years, in France, and since 1979, in the United States, where it was first found in Missouri and then in most states east of the Rocky Mountains and in California. It affects, with different severity, more than 28 species of pine and several other conifers, being most severe on Scotch *(Pinus sylvestris),* Austrian *(P. nigra),* Monterey *(P. radiata)* and shortleaf *(P. echinata)* pines. It is the causal agent of pine wilt, a lethal disease of trees of many host species. The disease spread rapidly and caused severe losses of pines in several localities in Japan. The disease seems to have been spreading quite rapidly since 1979 in planted conifers at numerous localities in the United States also, but it has not yet become a significant problem in the vast expanses of native forests. The potential threat of this nematode lies in both its ability to kill whole or parts of trees and in the fact that it is transmitted from dead to live pines by certain insects.

◦ *Symptoms.* Infected branches or whole trees suddenly develop light grayish-green foliage. Resin, which normally flows from the wounds of healthy trees, has ceased to flow from wounds of infected trees. One or two weeks later, the foliage appears generally yellowish green, while random needles turn brown. Soon the foliage turns yellowish brown and only random needles remain light green. At this point, the wood of affected trees shows obvious blue stain symptoms. Within 4–6 weeks from the appearance of symptoms, the tree or branch has totally brown foliage, appears wilted, or the needles are retained without obvious droop. In many affected tres blue stain in wood is heavy. Infected trees invariably die.

◦ *The Pathogen: Bursaphelenchus Xylophilus.* The nematodes are about 800 μm long by 22 μm in diameter. They develop and reproduce rapidly, completing a life cycle within 4 days during the summer. Each female lays about 80 eggs, which hatch, and the four larval stages develop through the four molts and produce adults in about 4 days. While the tree is still living, the nematodes feed on plant cells, but after its death they feed on fungi that invade the dying or dead tree. In late stages of infection, a different form of third-stage larvae, called the dispersal stage, appear, characterized by large amounts of nutritional reserves and a thick cuticle. These molt to the fourth-stage dispersal larvae, called **dauerlarvae,** which are especially adapted to survive in the respiratory system of certain cerambycid beetles, by which they are transmitted to healthy trees. *Bursaphelenchus xylophilus* is mycophagous, that is, it feeds and can complete its life cycle feeding on many kinds of fungi, for example, *Botrytis cinerea* and the blue stain fungi (*Ceratocystis* spp.).

◦ *Development of Disease.* The pine wood nematode overwinters in the wood of infected dead trees, which also contain instars (larvae) of one or more cerambycid beetles such as *Monochamus alternatus* in Japan and *M. carolinensis* in the United States. In early spring, the instars, while still in the wood, excavate small chambers in which they pupate. At the same time, third-stage nematode larvae aggregate around the pupal chambers and molt to produce fourth-stage dauerlarvae. As the adult beetles emerge from the pupae in mid to late spring, large numbers of dauerlarvae enter the beetles through the metathoracic spiracles and more or less fill many of the tracheae of the insect's respiratory system. The emerging adult beetles bore their way out of the wood, each carrying with it an average of 15,000–20,000 dauerlarvae, although as many as 230,000 dauerlarvae have been obtained from single insects. The emerging beetles fly to succulent branch tips of healthy trees and feed for about 30 days by stripping the bark to reach cambial tissue. As the beetles feed, the dauerlarvae emerge from the insect through the spiracles and enter the pine tree through the wound. Once in the plant, the dauerlarvae undergo the final molt and produce adult nematodes, which reproduce. The nematodes migrate to the resin canals, where they feed on the epithelial cells lining the canals and cause their death as well as the death of the surrounding parenchyma cells. The nematodes move quickly through resin canals in both the xylem and the cortex, reproduce rapidly, and within a few weeks, build up enormous populations in the host. For example, seedlings of susceptible Scotch pines, inoculated with 2,000 nematodes each, had approximately 50,000 nematodes within 30 days from inoculation and died shortly afterwards.

The destruction of the resin canals leads to the early cessation of resin flow from artificial wounds of branches and trunks within about 10 days of inoculation. This is followed by reduced, and later complete cessation of, transpiration by the foliage within about 30 days of inoculation, followed by sudden wilting and loss of foliage coloration. Rapid nematode population increases take place primarily after cessation of resin flow, they continue after foliar discoloration and wilting and after the death of the tree, and reach a maximum level about one month after inoculation. From 1,000 to 10,000 nematodes are often present per gram of dry wood of infected trees. In later stages of the disease, and as the condition of the tree deteriorates, nematode populations decline. At the same time, there is a gradual increase in the proportion of the dispersal third-stage larvae in relation to the total population of the nematode in the wood. The third-stage dispersal larvae make up the resting stage of the nematode.

In the meantime, the adult *Monochamus* sp. beetles, that is, the vectors of *Bursaphelenchus xylophilus,* after they have fed on tender pine twigs for about one month, look for and deposit their eggs under the bark of stressed and dead pine trees, including trees showing symptoms or dying from infection by the pine wood nematode. The first two instars of the insect feed under the bark, but the third penetrates the wood, where, after a molt, it produces the fourth instar, which overwinters in the wood. In early spring, the fourth instar excavates a chamber in the wood, in which it pupates, and attracts numerous third-stage nematode larvae all around it. The latter molt to produce dauerlarvae, which infect the adult insect as soon as it emerges from the pupa—and the cycle is completed.

In some temperate regions, primarily pine trees stressed by various diseases and insects are attacked by the pine wood nematode, and typical wilt symptoms are not usually produced.

◦ *Control.* Two types of control measures, that is, insecticide treatment to control the beetles and early removal and burning of dead and dying pine trees to eliminate the breeding habitat of the nematode and of the beetle, are only moderately effective and practical only in restricted localities. Neither of these controls is possible in large forests. Affected susceptible pine species planted as shade trees should be replaced with more resistant pine species or with other types of trees.

SELECTED REFERENCES

Dropkin, V. H., Foudin, A., Kondo, E., Linit, M., Smith, M., and Robbins, K. (1981). Pinewood nematode: A threat to U. S. forests. *Plant Dis.* **65,** 1022–1027.

Kondo, E. *et al.* (1982). Pine wilt diseases—nematological, entomological, and biochemical investigations. *Univ. Mo., Columbia, Agric. Exp. Stn., Bull.* **SR282,** 1–56.

Malek, R. B., and Appleby, J. E. (1984). Epidemiology of pine wilt in Illinois. *Plant Dis.* **68,** 180–186.

Mamiya, Y. (1983). Pathology of the pine wilt disease caused by *Bursaphelenchus xylophilus. Annu. Rev. Phytopathol.* **21,** 201–220.

Wingfield, M. J. (1987). "Pathogenicity of the Pine Wood Nematode." Symposium Series. APS Press, St. Paul.

• Stubby-Root Nematodes: *Trichodorus*

Stubby-root nematodes occur all over the world. They attack a wide variety of plants, including oats, cabbage, tomato, clover, corn, bean, grape, and peach. They affect plants by devitalizing root tips and stopping their growth, resulting in reduction of the root system of plants. This results in severe stunting and chlorosis of the whole plant, reduced yields, and poor quality of produce. Infected plants, however, are seldom, if ever, killed by these nematodes.

◦ *Symptoms.* Infected plants appear stunted within 2–3 weeks of inoculation. They have fewer and smaller leaves and branches than healthy plants, although at first their color appears normal. As the growing season progresses, the difference in size between healthy and stubby-root-infected plants increases, healthy plants being 3–4 times as large as infected ones. The latter also begin to show changes in color, appearing chlorotic instead of normal green.

The root symptoms of infected plants appear as an abnormal growth of lateral roots and proliferation of branch roots. Parasitized root tips show no necrosis or other injury although they are usually darker than normal in color. In such root tips meristematic activity and root growth stops, but cells already formed may enlarge abnormally and cause swelling of the root tip (Figure 15-21). Frequently, affected roots produce numerous lateral roots, which are in turn attacked by nematodes. Repeated infections of lateral roots and their branches produce a smaller root system, devoid of feeder roots and characterized instead by short, stubby, swollen root branches, the growth of which was stopped by the nematode infections (Figures 15-21 and 15-22).

◦ *The Pathogen: Trichodorus christiei. Trichodorus christiei* is a small nematode about 0.65 mm long by 40 μm wide. It lives in the upper 30 cm of the soil. It is an ectoparasite, feeding on the epidermal cells at or near the root-tip

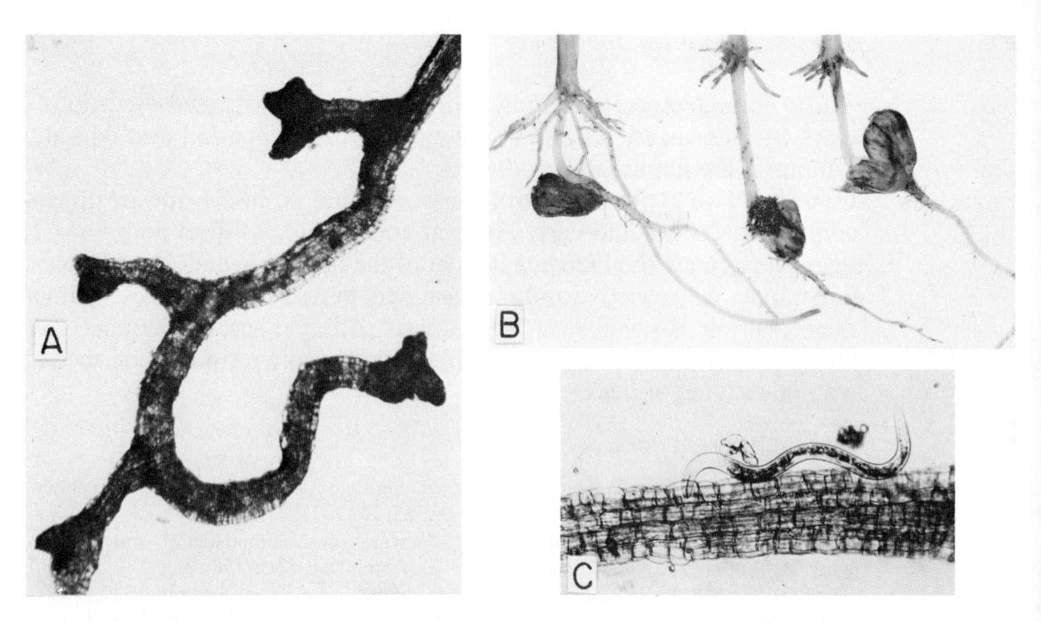

FIGURE 15-21 (A) Blackberry root showing stubby-root symptoms caused by *Trichodorus christiei.* (B) Roots of young corn plants injured by the stubby-root nematode. (C) *Trichodorus christiei* nematode feeding externally on blueberry root. (Photos A and C courtesy B. M. Zuckerman. Photo B courtesy U.S.D.A.)

FIGURE 15-22 Disease cycle of the stubby-root nematode *Trichodorus christiei.*

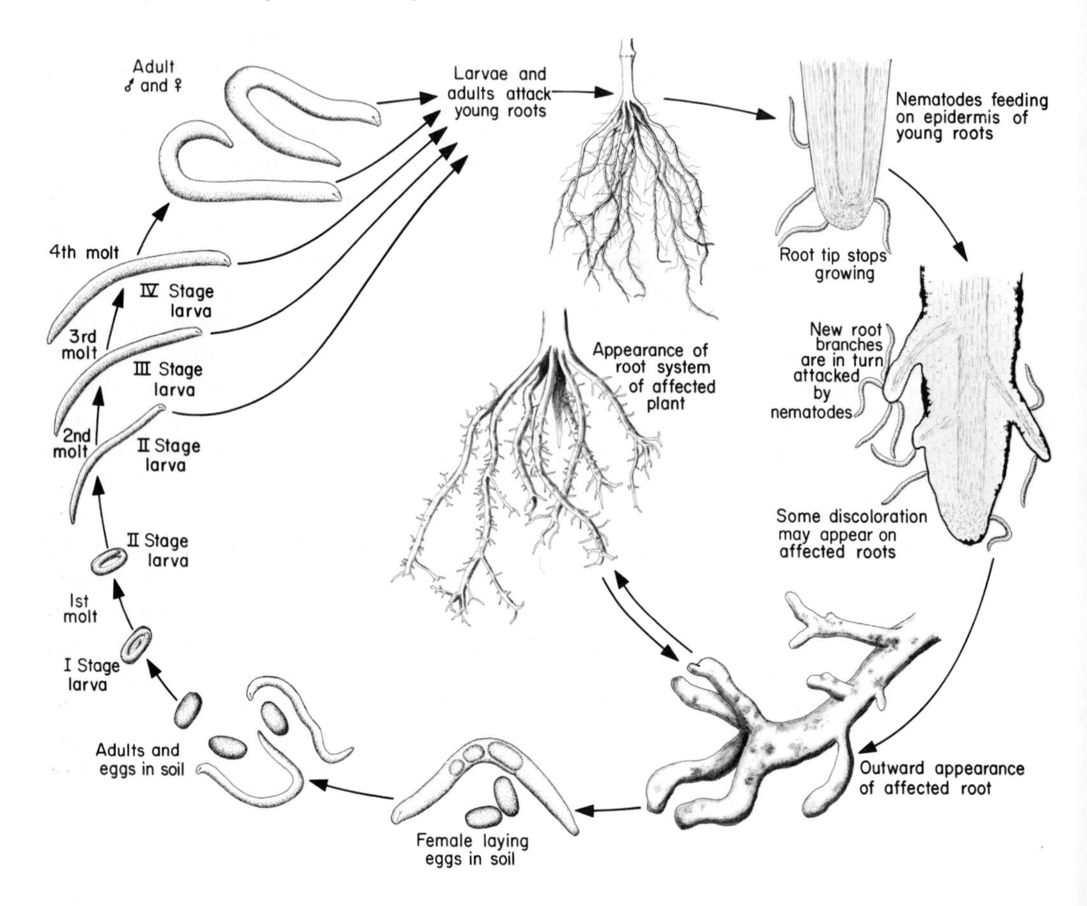

region, never entering the root tissue (Figure 15-21, C). It lays eggs in the soil, which hatch to produce larvae and then adults. The life cycle of this nematode is completed within about 20 days (Figure 15-22). Populations of *T.christiei* build up quickly around susceptible hosts but decline when host plants become old and do not produce new root tips or when good host plants are absent. Eggs, larvae, and adults are usually found in the soil throughout the year, although preadults and eggs seem to be the stages found mostly during winter.

Several species of *Trichodorus, T. christiei* included, are capable of transmitting plant viruses from one plant to another. Tobacco-rattle virus and pea early browning virus, both rod-shaped viruses, are the only ones known to be transmitted by *Trichodorus* nematodes. Several other plant viruses, all polyhedral shaped, are transmitted by the only other nematode vectors: *Longidorus* and *Xiphinema.*

◦ *Development of Disease.* When susceptible host plants are growing in soil infested with *Trichodorus,* as soon as the nematode comes in contact with young roots or root tips, it bends its head at approximately a right angle to the root surface, places its lip region against the cell wall, and punctures the wall with direct thrusts of the stylet. Once the stylet is inside the cell, a viscous substance is released through it into the cell, causing the cytoplasm of the plant cell to aggregate around the stylet tip. Part of the cytoplasm is then ingested by the nematode, after which it moves on to another cell within seconds or perhaps a few minutes from the beginning of feeding. Although a small opening up to 0.5 μm in diameter can be seen on the cell wall for many hours after the nematode has left the cell, no cell contents appear to be lost through the hole, and gradually the aggregated cytoplasm is dispersed and the cell returns to normal.

All free larval stages and the adults can attack plants and feed on them. Feeding is restricted to the epidermal cells at or near the root tip on older roots, and the whole length of young succulent roots (Figure 15-22).

Although one root tip may be attacked by many nematodes simultaneously or over a period of time, the mechanical damage caused by *Trichodorus* in feeding is slight and does not account for the gross changes on roots nor for the symptoms of the aboveground part of the plant. Parasitized roots show loss of meristematic activity at the root tip, have no definite root cap or region of elongation, and their region of mitosis is much smaller than that of healthy roots. Branch roots are more abundant and closer together in infected than in healthy roots. These effects seem to be the result of inhibitory or stimulatory action of substances secreted by nematodes into cells rather than of direct mechanical injury.

◦ *Control.* Stubby-root nematodes can be controlled through application of nematicides over the entire field. Methyl bromide–chloropicrin mixtures, Telone or DD give good, but temporary, control of this nematode. Six to eight weeks after treatment, stubby-root nematodes begin to reappear in the field, and if susceptible hosts are present, nematode populations build up rapidly. Slow-acting nematicides, such as ethoprop, retard or prevent the rapid buildup of nematodes, thus increasing the effectiveness of the treatment. Fallow or fallow and dry cultivation give fairly effective control of *Trichodorus.*

SELECTED REFERENCES

Allen, M. W. (1957). A review of the nematode genus *Trichodorus* with descriptions of ten new species. *Nematologica* **2,** 32–62.

Rohde, R. A., and Jenkins, W. R. (1957). Host range of a species of *Trichodorus* and its host–parasite relationships on tomato. *Phytopathology* **4,** 295–298.

Russell, C. C. and Perry, V. G. (1966). Parasitic habit of *Trichodorus christiei* on wheat. *Phytopathology* **56,** 357–358.

Zuckerman, B. M. (1962). Parasitism and pathogenesis of the cultivated highbush blueberry by the stubby root nematode. *Phytopathology* **52,** 1017–1019.

16 PLANT DISEASES CAUSED BY FLAGELLATE PROTOZOA

Certain trypanosomatid flagellates, that is, protozoa of the class Mastigophora, order Kinetoplastida, family Trypanosomatidae, have been known to parasitize plants for almost 80 years. That flagellates may be pathogenic to their host plants was suggested several times by the investigators of these parasites, and rather good evidence was presented that some plant diseases are caused by flagellates. However, because these parasites could not be isolated in pure culture and could not be inoculated into healthy plants so that they could reproduce the disease, as Koch's rules dictate, flagellates have not yet been fully accepted as plant pathogens. Yet, the pathogenicity of mycoplasmas and of some fastidious vascular bacteria in plants is almost universally accepted, although the same Koch's rules are equally unfulfilled with these organisms as they are with flagellates. Since the evidence supporting the pathogenicity of flagellates is no less than that available for mycoplasmas and fastidious vascular bacteria, it is reasonable to assume that at least some flagellates are considered capable of causing disease in plants, and it is apparent that the role of flagellates, as well as the role of other protozoa, in plant pathology deserves more attention than it has received in the past.

The protozoa are mostly one-celled, microscopic animals, generally motile, and have typical nuclei. They may live alone or in colonies, may be free living, symbiotic, or parasitic. Some protozoa subsist on other organisms such as bacteria, yeasts, algae, and other protozoa, some saprophytically on dissolved substances in the surroundings, and some by photosynthesis as in plants. Protozoa move by flagella (class Mastigophora, the flagellates), by pseudopodia (class Rhizopoda, the amoebae), by cilia (class Ciliata, the ciliates), or by movements of the cell itself (class Sporozoa, a diverse group of parasitic protozoa).

Of the protozoa, apparently only the flagellates have been reported as associated with plant diseases so far, but there are no good reasons why the other classes might not be found in the future also to be parasitic on plants.

The Mastigophora, or flagellates, are characterized by one or more long slender flagella at some or all stages of their life cycle (Figures 1-2, 1-3, 16-1, and 16-2). The flagella are used for locomotion and food capture, and perhaps as sense organs. The body of the flagellates usually has a definite long, oval, or spherical form that is maintained by a thin, flexible covering membrane or, in some groups, it may be armored. Flagellates generally reproduce by longitudinal fission (Figures 1-3, 16-1, C, 16-2, B). Although many flagellates are saprophytic and some contain plastids with colored pigments, including

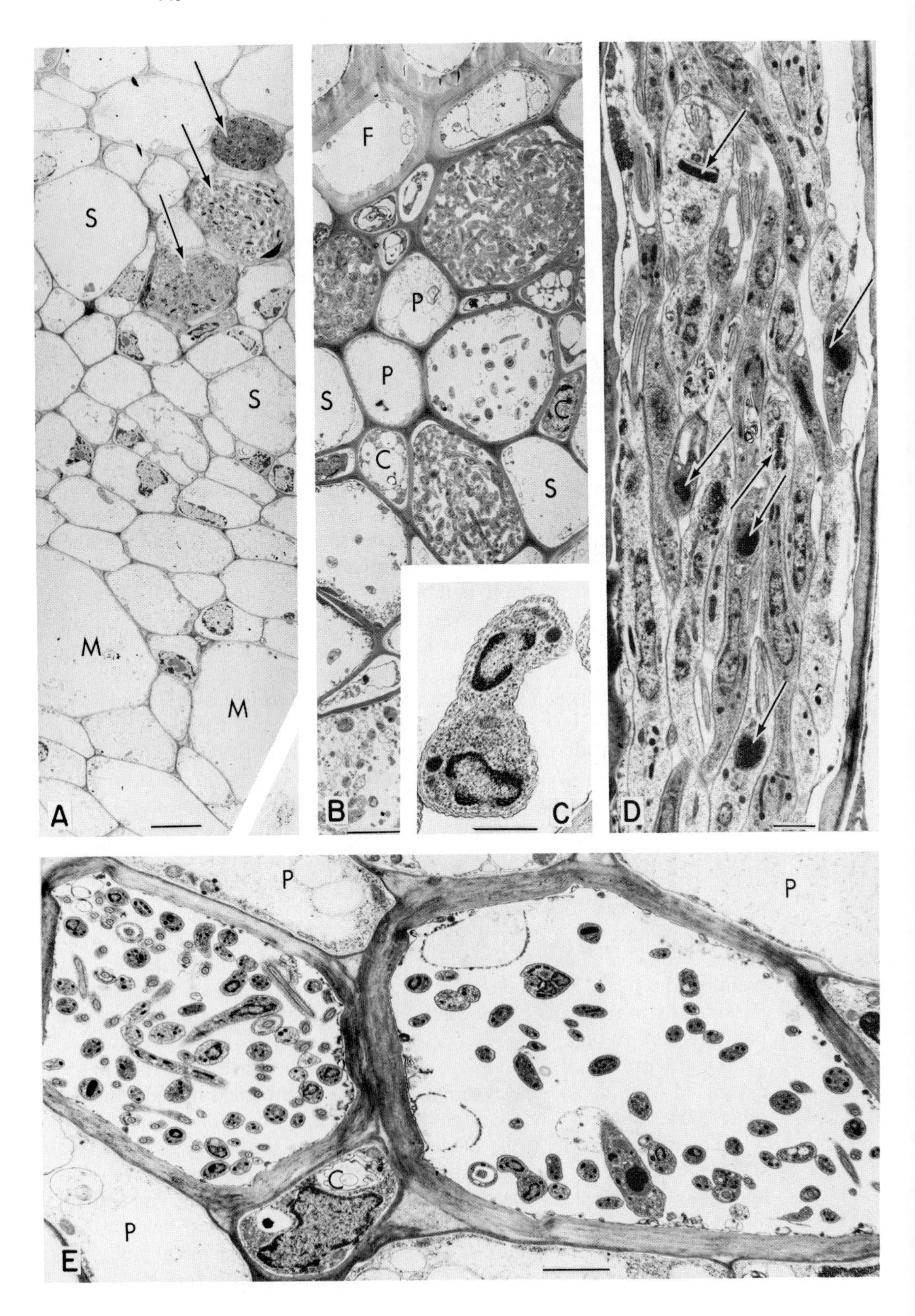
S
S
M
M
A
F
P
P
S
C
C
S
B
C
D
P
P
C
P
E

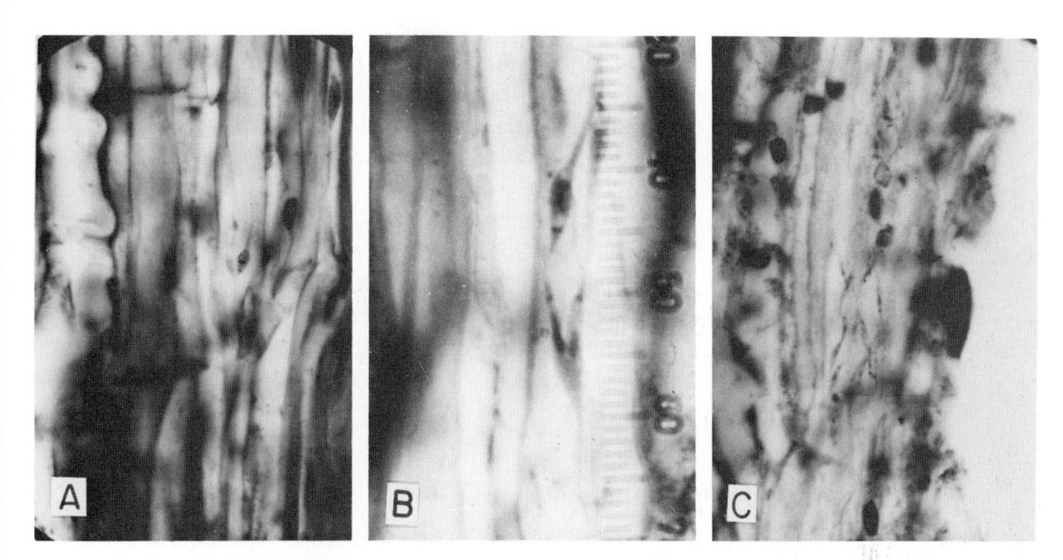

FIGURE 16-2 Flagellates associated with the coffee wilt disease. (A) Single protozoon in vascular vessel of diseased *Coffea liberica.* (B) Flagellates in vessel of *C. liberica,* one of them in the process of division. (C) Long and thin flagellates in vessels of coffee tree showing advanced symptoms of the disease. (Photos courtesy J. H. van Emden.)

functional chlorophyll, others are parasites of humans and various animals and some cause serious diseases. The best known flagellate pathogenic to humans is the blood parasite, *Trypanosoma,* the cause of sleeping sickness in Africa, which is transmitted by tsetse flies.

Flagellates were first found to be associated with plants in 1909, when Lafont reported that *Phytomonas davidi* parasitizes the latex-bearing cells—the laticifers—of the laticiferous plant *Euphorbia* (Euphorbiaceae). Since then several other species of *Phytomonas* have been reported from plants belonging to the families Asclepiadaceae (for example, *P. elmassiani* on milkweed), Moraceae (for example, *P. bancrofti* on a ficus species), Rubiaceae (for example, *P. leptovasorum* on coffee), and species of unknown identity on coconut palm and on oil palm. All plant flagellates belong to the order Kinetoplastida, family Trypanosomatidae. The plant-infecting Phytomonads are apparently transmitted by insects, but so far insect vectors are known only for *P. elmassiani.* Many of the investigators who studied the flagellates in

FIGURE 16-1 Electron micrographs of the trypanosomatid flagellate *Phytomonas* in the phloem of young inflorescences of coconut palms affected with hartrot. (A) Cross section of a differentiating vascular bundle in a palm that had early symptoms of the disease. Recently matured sieve elements are filled with flagellates; M, immature metaxylem; S, immature sieve elements (scale bar 10 μm). (B) Cross section of the phloem in a palm that had advanced symptoms of the disease, showing C, companion cell; F, fiber; P, phloem parenchyma cell; S, sieve elements free of flagellates (scale bar 5 μm). (C) Cross section of a flagellate undergoing longitudinal fission (scale bar 0.5 μm). (D) Longitudinal section of a sieve element filled with the flagellates. Arrows point to the DNA portion of kinetoplasts (scale bar 1 μm). (E) Similar to B but at a higher magnification; C, companion cell; P, parenchyma cell (scale bar 2 μm). (Photo courtesy M. V. Parthasarathy, from *Science* **192,** 1346–1348. Copyright © 1976 by the American Association for the Advancement of Science.)

laticiferous plants feel that although the flagellates parasitize the plants—since they live off their latex—the plants do not become diseased, and therefore, the flagellates are not pathogenic to these plants. According to some reports, however, symptoms apparently do develop in some flagellate-infected laticiferous plants, which would indicate that the flagellates are pathogenic to their hosts.

The nonlaticiferous hosts, coffee, coconut palm, and oil palm, are apparently infected by pathogenic *Phytomonas* species and develop characteristic internal and external symptoms and severe and economically important diseases. Flagellates apparently cause the phloem necrosis disease of coffee, the hartrot disease of coconut palm, and the marchitez sorpresiva (sudden wilt) disease of oil palms. All three diseases are so far known to occur only in South America.

• Phloem Necrosis of Coffee

Phloem necrosis of coffee occurs in Surinam, British Guiana, and probably Brazil, San Salvador, and Colombia. It affects trees of *Coffea liberica, C. arabica,* and of other coffee species. Infected trees show sparse yellowing and dropping of leaves, and as these increase gradually, only the young top leaves remain on the otherwise bare branches. As the roots begin to die back, the condition of the tree worsens and the tree dies (Figure 16-3, A). Sometimes, in the beginning of the dry season, trees wilt and die within 3–6 weeks (Figure 16-3, B). Internally, the roots and trunk of trees show multiple division of cambial cells and production of a zone of smaller and shorter phloem vessels of disorderly structure right next to the wood cylinder (Figure 16-3, C, D). At this stage the bark in the roots and the trunk is firmly attached to the wood and cannot be separated from it.

The pathogen, *Phytomonas leptovasorum,* is a trypanosomatid flagellate. When symptoms first appear there are only a few, big ($14-18 \times 1.0-1.2\ \mu m$), spindle-shaped flagellates in the phloem (Figure 16-2, A, B). As multiple division of cambial cells and abnormal phloem production become apparent and many leaves turn yellow and fall, the flagellates are numerous, slender, and spindle shaped, $4-14 \times 0.3-1.0\ \mu m$ (Figure 16-2, C). A few shorter ($2.0-3.0\ \mu m$) forms of the flagellate, called "leishmania forms," also appear in the oldest sieve tubes. When the multiple division of cambial cells results in a multilayered sheath around the wood cylinder that extends from the roots up to 2 meters above the ground line and the tree is almost dead, there is a great abundance of small ($3-4 \times 0.1-0.2\ \mu m$), "spaghetti" flagellates only in the living tissues of the stem, while previously occupied cells are evacuated.

The flagellates can be traced from the roots upward into the trunk, where they seem to migrate vertically in the phloem and laterally through the sieve plates into healthy sieve tubes. They also seem to move downward into unaffected roots. Flagellates could not be found in the tree outside areas that show multiple division.

The disease can be transmitted through root grafts but not through green branch or leaf grafts. After grafting of healthy trees with roots infected with flagellates, the flagellates can be observed in the previously healthy roots within a few weeks, the tree begins to develop external symptoms 4–5 months later, and it then dies shortly afterward. The disease spreads in the field from

FIGURE 16-3 Coffee wilt of *Coffea liberica* caused by the flagellate protozoon *Phytomonas leptovasorum.* (A) Affected tree during rainy season. Note loss of leaves and yellowing but no acute wilting. (B) Affected tree at the onset of the dry season. Note sudden wilting. (C) Cross section of abnormal phloem tissue from flagellate-affected and wilting coffee tree. (D) Cross section of healthy phloem tissue of coffee tree. (Photos courtesy J. H. van Emden.)

one tree to another, and healthy trees often become infected when transplanted in areas from which a diseased tree had been removed. No vector of the disease is known.

• Hartrot of Coconut Palms

Hartrot has been known in Surinam since 1906, sometimes under the names lethal yellowing, bronze-leaf wilt, Coronie wilt, and "unknown disease." The disease also occurs in Colombia and Equador and, under the local name

FIGURE 16-4 "Hartrot" of coconut palms caused by flagellate protozoa. (A) Malayan Dwarf palms, 3 years old, suffering from "hartrot" disease. Note broken leaves and the collapsed spear. (B) Unopened inflorescence of Mayalan dwarf showing necrotic spike tops. It is inflorescence No. 1, and the first symptom of the disease. (C) Inflorescence showing necrotic spike tops. (D) Ceylonese dwarf-yellow, 4 years old, suffering from "hartrot" disease. Note nuts on the ground, one of the first symptoms. (Photos courtesy W. G. van Slobbe.)

Cedros wilt, in Trinidad. Many of the symptoms of hartrot (Figure 16-4, A–D) are similar to those caused by the lethal yellowing disease of coconut palms in the Caribbean, West Africa, and Florida, but the causes of the two diseases seem to be unrelated. Trees affected with lethal yellowing contain only mycoplasmas and no flagellates, while trees affected with hartrot contain only flagellates and no mycoplasmas. Trees infected with hartrot die within a few months of the appearance of external symptoms.

Flagellates of the genus *Phytomonas* occur in mature sieve elements of young leaves and inflorescences of hartrot-affected coconut palms (Figure 16-1). In advanced stages of the disease, 10–100 percent of the mature sieve elements contain flagellates, and many of them are plugged with flagellates, which are usually oriented longitudinally within the phloem. The flagellates measure $12-18 \times 1.0-2.5$ μm. The number and spread of the flagellates in sieve tubes increase proportionally with the development of the disease.

Hartrot spreads very rapidly. For example, about 15,000 coconut trees died in 3 years in the Cedros region of Trinidad. So far, however, no vector and no means of transmission of the disease are known.

SELECTED REFERENCES

Dollet, M. (1984). Plant diseases caused by flagellated protozoa *(Phytomonas). Annu. Rev. Phytopathol.* **22,** 115–132.

Harvey, R. D., and Lee, S. B. (1943). Flagellates of laticiferous plants. *Plant Physiol.* **18,** 633–655.

Holmes, F. O. (1924). Herpetomonad flagellates in the latex of milkweed in Maryland. *Phytopathology* **14,** 146–151.

Holmes, F. O. (1925). The relationship of *Herpetomonas elmassiani* to its plant and insect hosts. *Biol. Bull. (Woods Hole, Mass.)* **49,** 323–327.

Lafont, A. (1909). Sur la presence d'un parasite de la classe des flagellés dans le latex de *l'Euphorbia pilulifera. C. R. Soc. Biol.* **66,** 1011–1013.

McCoy, R. E., and Martinez-Lopez, G. (1982). *Phytomonas staheli* associated with coconut and oil palm diseases in Colombia. *Plant Dis.* **66,** 675–677.

McGhee, R. B., and Hanson, W. L. (1964). Comparison of the life cycle of *Leptomonas oncopelti* and *Phytomonas elmassiani. J. Protozool.* **11,** 555–562.

McGhee, R. B., and McGhee, A. H. (1971). The relation of migration of *Oncopeltus fasciatus* to distribution of *Phytomonas elmassiani* in the eastern United States. *J. Protozool.* **18,** 344–352.

Parthasarathy, M. V., vanSlobbe, W. G., and Soudant, C. (1976). Trypanosomatid flagellate in the phloem of diseased coconut palms. *Science* **192,** 1346–1348.

Stahel, G. (1933). Zur Kenntnis der Siebröhren-krankheit (Phloëmnekrose) des Kaffeebaumes in Surinam. III. *Phytopathol. Z.* **6,** 335–357.

Thomas, D. L., McCoy, R. E., Norris, R. C., and Espinoza, A. S. (1979). Electron microscopy of flagelated protozoa associated with marchitez sopresiva disease of African oil palm in Equador. *Phytopathology* **69,** 222–226.

van Emden, J. H. (1962). On flagellates associated with a wilt of *Coffea liberica. Meded. Landbouwhogesch. Opzoekingsstn. Staat Gent* **27,** 776–784.

Vermeulen, H. (1963). A wilt of *Coffee liberica* in Surinam and its association with a flagellate, *Phytomonas leptovasorum. J. Protozool.* **10,** 216–222.

Vermeulen, H. (1968). Investigations into the cause of the phloem necrosis disease of *Coffea liberica* in Surinam, South America. Neth. J. Plant Pathol. **74,** 202–218.

Waters, H. (1978). A wilt disease of coconuts from Trinidad, associated with *Phytomonas* sp., a sieve tube-restricted protozoan flagellate. *Ann. Appl. Biol.* **90,** 293–302.

III

Biotechnology and Plant Pathology

17 APPLICATIONS OF BIOTECHNOLOGY IN PLANT PATHOLOGY

In modern terms, **biotechnology** is defined as the manipulation, genetic modification, and multiplication of living organisms through novel technologies, such as tissue culture and genetic engineering, resulting in the production of improved or new organisms and products that can be used in a variety of ways. For example, excess ova are removed from high-milk-producing cows, are fertilized *in vitro* with sperm from prized bulls, and are implanted in local low-breed cows, which then give birth to top-quality calves expected to be high-milk producers. Or the gene for human insulin is inserted into a nonpathogenic bacterium, which then multiplies and produces unlimited amounts of insulin that can be used by people suffering from diabetes. Or anthers of a plant are plated on a nutrient medium and produce haploid plants that can be further manipulated to produce homozygous dihaploid plants, and through tissue culture, millions of identical plants can be obtained in just a few months.

Plant biotechnology is based on a thorough understanding of (plant) molecular biology, on the use of a variety of plant tissue culture techniques, and on our ability to identify, isolate, and transfer specific genes from one kind of organism (plant) to another plant or other organism. Plant biotechnology makes possible the rapid clonal propagation of plants, it accelerates and expands the limits of plant breeding, and makes possible the production of specialized industrial plant products under tissue culture conditions.

Plant biotechnology impinges upon plant pathology in several ways. The most obvious of these ways are discussed in the following paragraphs.

(1) Increased production of plants through rapid clonal propagation is likely to result in a greater need to obtain pathogen-free mother plants and subsequently protect the daughter plants from pathogens. The latter will be necessary because the increased crowding of these plants in the early stages of tissue culture cultivation, the increased genetic uniformity of the plants, and the prolonged exposure of the plants to marginal nutritional and environmental conditions before and after they are set out in the field are likely to make them susceptible to catastrophic sudden outbreaks of pathogen infections.

(2) New plant varieties to which genes have been added through genetic engineering are likely to exhibit greater or unexpected instability toward certain unpredictable sets of environmental conditions and toward the pathogenic microflora of their habitats.

(3) So far, and possibly in the future, the main vehicles for moving genes from donor plants or other organisms to recipient plants are plant pathogens,

particularly the bacterium *Agrobacterium tumefaciens* and the cauliflower mosaic virus, while several more viruses are being further developed as vectors.

(4) The study of plant genes for resistance to disease and of pathogen genes for virulence to pathogens is already aided considerably by genetic engineering and is expected to be greatly advanced by it in the future.

(5) Control of many plant diseases is likely to come about either by inserting resistance genes into plants by genetic engineering techniques, or by genetically engineering microorganisms that can effectively antagonize or compete with particular pathogens.

Tissue Culture Techniques of Importance to Plant Pathology

Almost all tissue culture techniques used by plant scientists are of importance to plant pathology. Some of them, for example, plant micropropagation, carry with them the danger of disseminating pathogens, or conversely, they are used to produce pathogen-free plants. More importantly, however, many of them can be used to study, locate, and isolate the genes of resistance to certain pathogens, and others are used for the modification and transfer of such genes to susceptible plants. The most important tissue culture techniques and their importance to plant pathology are described briefly.

Rapid Clonal Propagation in Culture

Sections (explants) of shoot or root tips, shoot nodes, germinated seedlings, and in some plants, of leaves are trimmed, surface-sterilized, washed repeatedly in sterile distilled water, and then placed in test tubes containing suitable culture media in either liquid or, more often, semi-solid form (a small amount of agar added). The explants, depending on the kinds and relative amounts of growth regulators included in the culture medium, produce numerous shoots, roots, or both. These explants usually are further subcultured at frequent intervals by subdividing single mother cultures into several daughter cultures. Complete plants are eventually set out in the greenhouse or the field.

Explants used for rapid clonal propagation (micropropagation) in culture are usually one to several millimeters long. In general, the smaller the explants, the more difficult it is to regenerate plants in culture. Large explants, however, are likely to carry with them virus, viroid, mycoplasma, and other pathogens, unless they are taken from known pathogen-free mother plants or are placed at high temperatures and treated with antiviral or antibiotic substances that inhibit the respective pathogens.

Although difficult to manipulate and to regenerate into plants, apical meristems 0.4–0.8 mm long are sometimes used to obtain pathogen-free plants from mother plants of unknown health condition. The rate of plants obtained free of pathogens increases if the meristem tips are placed in culture media containing substances inhibitory to the pathogen but not to the plant tissue or if they are exposed to high temperatures that destroy the pathogen but not the plant tissue.

Callus and Single-Cell Culture

Callus, that is, an unorganized mass of undifferentiated dividing and enlarging cells, is produced when explants are placed on or in nutrient media containing auxins. Repeated subdivision and subculturing of callus in liquid nutrient media under continuous shaking produces numerous single cells and smaller or larger clumps of cell aggregates. If the calli, or the single cells and cell aggregates are placed on or in nutrient media containing appropriate combinations of auxins and cytokinins, the subsequently produced cells become organized and differentiate into embryoids, then into shoots and roots, and eventually develop into complete plants. From one to dozens of new plants may be produced from a single callus, and with some kinds of plants, thousands of plantlets may be produced in a single flask. These plants can then be subcultured on nutrient media and later potted in soil and transplanted in the field. In this way, thousands and even millions of nearly identical plants can be produced in culture in a much shorter time than would have been possible otherwise.

Cultures of callus, single cells, and the plantlets regenerated from them are beginning to be used extensively in plant pathology. Most commonly they are used to study the behavior of pathogens, particularly obligate parasites, such as some fungi (that is, the rusts, powdery mildews, and downy mildews) many viruses, and nematodes, but also nonobligate parasitic fungi and bacteria. They are also used to study the effects of and possible roles in pathogenicity played by substances, particularly toxins, secreted by pathogens and to study the structural barriers and biochemical substances (for example, phytoalexins) produced as defense mechanisms by cells and tissues of resistant varieties, compared to the reactions of susceptible varieties. Similarly, infected calli and single cells are used for *in vitro* testing of chemicals, such as fungicides, viricides, or antibotics used, respectively, for control of fungi, viruses, and mycoplasmas. Although cultured callus and cells often react to a pathogen in a way that parallels that of intact plants, in many cases their reactions to pathogens differ drastically from those of intact plants. Therefore, conclusions drawn from studies with infected cultured callus and cells may not always apply to events in infected intact plants.

Plant Protoplasts

Plant protoplasts are plant cells from which the cell wall has been removed by digestion with cell wall–degrading enzymes. Protoplasts are obtained most frequently from leaves from which the lower epidermis has been removed but can also be obtained from narrow strips of intact leaves, tender shoots and roots, from callus, and from cultured single cells. The strips of the organ from which protoplasts are to be obtained are placed in a solution containing about 0.5 percent pectinase, 2 percent cellulase, and 8–13 percent mannitol. Pectinase degrades the middle lamella and cell wall pectins, resulting in the release of single cells with still rigid cell walls. At the same time, the cellulase degrades the cell wall cellulose, until eventually no cellulose is left and the naked, membrane-bound protoplast is released. Cells and protoplasts, because of the salts and sugars dissolved in them, have high osmotic pressure, and if placed in water they tend to absorb more and more water. In cells, the cell wall limits the extent of cell expansion and the amount of water that can be absorbed. In the absence of the cell wall, however, the protoplast would continue to absorb water until the membrane burst. This is avoided with the addition of 8–13 percent mannitol, which is a sugar that cells do not metabolize to a great extent but which brings the osmotic pressure of the liquid to the same level as, and in equilibrium with, that of the cell contents, thereby stabilizing the size and viability of the protoplasts. After the protoplasts are released from the cells or tissues, they are washed several times with appropriate mannitol or nutrient solutions, or both, to remove the enzymes and cellular debris and are ready for a variety of subsequent uses.

The most common uses of protoplasts in plant pathology are the following:

1. *Protoplast inoculation with viruses and study of virus replication and physiology.* It has been possible to inoculate the protoplasts of many plants with one or more of the viruses that infect the plant. The inoculation procedure involves mixing protoplasts with a small amount of purified virus to which has been added a fusion-promoting agent, called a **fusagen,** such as poly-L-ornithine or polyethylene glycol. The protoplast–fusagen–virus mixture is incubated with slight stirring at room temperature for 10–20 minutes, and the protoplasts are then washed with mannitol or nutrient solution or both to remove excess virus and the fusagen. The proportion of protoplasts that become synchronously infected with the virus varies with the virus–host combination but can be as high as 70–95 percent. Virus replication in protoplasts is usually completed 24–36 hours from inoculation. The rate of virus replication can be monitored by using a portion of the inoculated protoplasts for electron microscopy, for bioassay on a local lesion host, or for testing serologically (usually by ELISA) at certain time intervals after inoculation. Individual infected protoplasts can be detected and counted after treatment with virus antibodies labeled with a fluorescent compound such as FITC and observation of the treated protoplasts under a microscope provided with an ultraviolet light source. Such synchronously infected protoplasts are ideal for studying the various stages and compounds that appear during virus replication and the changes in the cell physiology as a result of virus infection.

2. *Protoplast inoculation with genetically engineered vectors of genetic material.* So far, two types of vectors have been used to successfully introduce foreign genetic material into plant cells: the bacterium *Agrobacterium tumefaciens* or its modified Ti-plasmid and the double-stranded DNA virus cauliflower mosaic virus. Other viruses, such as the single-stranded DNA geminiviruses, multipartite viruses, and satellite viruses, viroids, transposable elements, chromosomes, and even other bacteria (such as protoplasts of transformed *E. coli*) are at various stages of development as vectors of foreign genetic material into plant protoplasts. Genetic material (nuclear, plasmid, or viral DNA, or RNA) may also be introduced into protoplasts, either by incubating with protoplasts in the presence of a fusagen or by encapsulating the genetic material in small artificial lipid vesicles, called **liposomes.** The latter, upon incubation with protoplasts are taken in by the protoplasts and bring into them the genetic material they contain.

Protoplast inoculation with genetic material has been aided greatly by plant pathology through its contribution to the knowledge of the plant pathogens (*Agrobacterium tumefaciens,* plant viruses, viroids, and so on) already used or being developed for use as vectors of genetic material. Protoplast inoculation with genetic material also appears at present to be the most likely method for introducing genes controlling disease resistance into important agronomic plants, once these genes have been identified and isolated.

3. *Selection of protoplasts and of protoplast-derived plants resistant to pathogen infection, to pathogen toxins, and to other toxic substances.* One of the most surprising discoveries of plant tissue culture has been the observation that plants, regenerated from callus, single cells, or protoplasts derived from a single plant, often exhibit one or more different characteristics than the other plants or the parent. This is called **somaclonal variation.** Many such plants differ from the parent and from each other in the degree of resistance they exhibit to a particular pathogen. Once protoplasts have been obtained, they may be inoculated with pathogens, such as viruses, or they may be placed in nutrient media to which have been added varying concentrations of a pathogen toxin or of an antibiotic, fungicide, or viricide. The protoplasts and protoplast-derived plants that survive the pathogen or one of the other treatments are apparently resistant to the particular pathogen, toxin, or chemical and are therefore selected for further propagation, study, and incorporation into the breeding program.

4. *Evaluation of antiviral compounds through treatment of virus-infected protoplasts.* Antiviral compounds may be tested quickly if various concentrations of a particular compound are added to the medium into which protoplasts are placed immediately after inoculation with a virus. An antiviral compound that would inhibit or greatly reduce virus replication in infected protoplasts (as determined by bioassay or ELISA) without affecting the survival and regeneration ability of the protoplasts would obviously deserve more extensive and more expensive testing, while ineffective or highly toxic compounds would be eliminated from further testing and further expense.

5. *Protoplast fusion for transmission of resistance genes into sexually incompatible hosts.* When freshly obtained protoplasts of even unrelated plant species are mixed together in the presence of a chemical fusagen or short pulses of direct electric current, many of them fuse with other protoplasts of the same species or of other species. This is called **somatic hybridization.** Somatic hybrids often display a wide variety of genome combinations containing nuclear and cytoplasmic (especially mitochondrial) DNA from both partners, although for some reason, only one species of chloroplast DNA seems to remain in such hybrids. Somatic hybrids are likely to be more viable and regenerate into fertile plants when they are produced from species within the same genus or between closely related genera. Many somatic hybrids do not grow or are sterile, and many become unstable and continue to lose more and more of the DNA of one of the two partners. Several somatic hybrids, however, survive and develop normally, containing the complete genome of one partner plus limited parts of the genome of the other partner. Such somatic hybrids, in which the additional genome contains effective genes of resistance against one or more pathogens, are of great importance to plant pathology. Furthermore, somatic hybrids are also possible from haploid protoplasts of sexually compatible or incompatible species. This further makes possible either the production of stable homozygous or heterozygous amphiploids or, after loss of most of the genome of one of the partners and subsequently induced diploitization of the surviving partner, the production

of homozygous dihaploids. When haploid protoplasts are used for fusion, any somatic hybrids that can be selected for desirable disease resistance genes are likely to be diploid, rather than tetra- or polyploid.

Haploids

Haploid plants of many species can be produced by placing anthers of a certain stage of development on specialized nutrient media. The microspores in the anther at first produce haploid callus, which subsequently grows and may produce complete haploid plants. In a few species, haploid plants may also be obtained by direct culture of pollen microspores. Since microspores are the products of meiosis, the genetic make-up of each microspore and of each derived callus and haploid plant is different from the genetic make-up of the other microspores and haploid calli and plants. Since each haploid callus and plant can be propagated and multiplied by subdivision and culture, clones of nearly identical calli and plants can be produced from each microspore-derived callus or plant. Haploid plants, or protoplasts obtained from haploid callus, plants, or directly from pollen, may be used for various studies in plant pathology. The studies may include inoculation with viruses and other pathogens and subsequent selection for resistance, diploidization, and inoculation of the dihaploid tissue, followed by selection for resistance, and fusion with other haploid protoplasts of the same or other (related) plants, followed by inoculation and selection for resistance.

Because cells of haploid tissues and plants contain haploid (1N) genones, each gene can express itself, and therefore it may be possible to detect, locate, and isolate even minor genes for resistance to a particular pathogen. Since cells of haploid tissues often diploidize spontaneously, or can be induced to diploidize by treatment with colchicine and other chemicals, it is easy to obtain dihaploid tissues and plants homozygous for all genes. The homozygous dihaploid stage is also useful for studying the reaction of such plants to certain pathogens or pathogen races, determining the number, kind, and location of the genes on the chromosomes, and selecting the resistant individuals or clones for further propagation or study.

Genetic Engineering Techniques of Importance to Plant Pathology

Probably most, if not all, techniques used in molecular biology, and plant molecular biology in particular, are used in the genetic engineering of plants or their pathogens in relation to disease development and control. A brief outline of plant molecular biology and of some of the most important genetic engineering techniques relevant to plant pathology follows.

Plant Molecular Biology

Plant cells contain three separate but interacting double-stranded DNA genomes: nuclear, mitochondrial, and chloroplast (plastid) genomes. The **nuclear genome** is divided into chromosomes and is the largest of the three, ranging in size from 0.2×10^9 to more than 40×10^9 base pairs among the various plants. The **mitochondrial genome** usually consists of a circular, but sometimes linear, DNA molecule; it may exist as more than one molecule and may even contain small, plasmidlike DNA molecules. Its total size varies among plants from 0.2 to 2.5×10^6 base pairs. The **chloroplast genome** consists of a circular DNA molecule composed of $0.12-0.19 \times 10^6$ base pairs. Each chloroplast contains 30–200 copies of its genome DNA.

The DNA of each genome does two basic things: (1) it **replicates,** that is, it produces exact copies of either certain parts of each DNA molecule or of the entire genome, and the copies then either remain in the same cell or are distributed in each of the daughter cells; and (2) it is **transcribed** into RNA, which can be ribosomal (rRNA), transfer (tRNA), or messenger RNA (mRNA). Each segment of DNA that is transcribed into a particular functional RNA is known as a **gene.** Messenger RNAs are the only RNAs that can be **translated** into proteins. The coding sequences of many mRNAs are interrupted by intervening sequences (introns) that are removed before each mRNA is translated to produce a protein. Almost all RNAs are **processed,** that is, they are modified by the cell before they **mature** and become functional in the cell.

It should be noted that, while the nuclear genetic system (DNA, mRNA, and ribosomes) is typical of eukaryotic organisms, the genetic system in the mitochondria is partly eukaryotic and partly prokaryotic in nature. On the other hand, the genetic system within the chloroplasts is essentially prokaryotic in nature, that is, its mRNA has regulatory control sequences similar to those of bacterial mRNA, it lacks a terminal polyadenylated sequence, it lacks introns, and its expression is inhibited by the same antibiotics that inhibit expression of bacterial mRNA. Furthermore, chloroplast ribosomes are smaller than those of the nuclear genetic system in the cytoplasm and are equal in size to those of bacteria, while mitochondrial ribosomes are somewhat intermediate but closer in size to the cytoplasmic ribosomes. Moreover, mitochondria use a genetic code slightly altered from the "universal" genetic code in that two nucleotide triplets in the mitochondrial mRNA code for two different amino acids than those coded by the same triplets in the "universal" genetic code.

A gene is said to be **expressed** when it is transcribed into mRNA. The mRNA is then **translated** into a protein, which is subsequently translocated to the proper locus in the cell where it carries out its normal function. The timing, duration, and rate of expression of a gene are regulated by numerous internal and external control mechanisms. Such mechanisms may involve other regulatory genes, promoter regions in front of the actual gene, initiation and termination signals on the DNA and the mRNA, growth regulators present at certain developmental stages of cells and plants, environmental conditions such as light and nutrition, and several others.

Molecular Biology of Plant Pathogens

Plant pathogens are either eukaryotes (fungi, nematodes, parasitic higher plants, flagellated protozoa) or prokaryotes (bacteria, mycoplasmalike organisms), or viruses (including viroids). Certain similarities exist among the genetic systems of all pathogens, but considerable differences also exist. For example, the genetic systems of some eukaryotic pathogens (fungi and parasitic plants) resemble those of plants, and the genetic systems of others (nematodes, flagellated protozoa) resemble those of animals. The genetic systems of plant pathogenic bacteria and mycoplasmas resemble those of all bacteria, and the genetic systems of plant pathogenic viruses are, of course, different from those of either eukaryotes or prokaryotes but depend absolutely on the interaction of the viral DNA or RNA with the cell genomes of their hosts.

The molecular biology of several plant pathogens has been receiving tremendous attention lately, the interest begun in the last ten years or so and increasing rapidly with each passing year. The increased interest is due to the realization that the genetic systems of at least some pathogens (such as *Agrobacterium tumefaciens* and cauliflower mosaic virus) can be used as vehicles of foreign genetic material into plant genomes and, thereby, to genetically modify the plants. Interest increased even more when it was realized that the genetic systems of the pathogens themselves and of their antagonists may be modified in ways that may be used to biologically control the plant pathogens.

Gene Cloning

Gene cloning is the isolation and multiplication of an individual gene or gene sequence by inserting it into a bacterium or yeast cell where it can be replicated. Gene cloning allows the production of sufficiently large quantities of a gene or its mRNA, which then can be used to study the structure of the gene and the sequences that regulate its expression, to resynthesize or modify the gene, and to move the gene into other organisms that can then produce the product coded by the gene. The techniques used in gene cloning are complex and make up the core of genetic engineering of organisms. A sketchy outline of the steps in gene cloning follows.

Cloning Complimentary DNA from mRNA

The mRNA of cells is extracted at a cell stage when a particular gene is most active (for example, the gene of an enzyme or a toxin produced by an attacking pathogen, or the gene of a phytoalexin produced by a host plant during infection by a pathogen). The mRNAs are exposed to the enzyme reverse transcriptase, which synthesizes single-stranded DNAs from, and complementary to, the mRNAs. The complementary single-stranded DNAs —cDNAs—are then exposed to another enzyme (DNA polymerase), which makes them double-stranded cDNAs. The double-stranded cDNAs are then inserted into special bacterial plasmids, each plasmid usually accepting one cDNA. The plasmids used carry genes for resistance to two antibiotics, for example, ampicillin and tetracycline, and the cDNA gene is inserted within the ampicillin resistance gene, thereby destroying (eliminating) resistance to ampicillin. These **recombinant plasmids** are then mixed with bacteria such as *E. coli,* which take up the plasmids, that is, the bacteria become **transformed.** When placed in nutrient media containing tetracycline, only those bacteria that have taken up the plasmid with the tetracycline resistance gene survive.

The surviving bacteria are then plated out so that each can produce a separate colony. Each bacterium multiplies rapidly, making billions of copies of itself and of the plasmid and the gene it took up.

Detecting which colony of bacteria contains the gene of interest requires another series of complicated steps. These may involve either detection by immunological or enzymological techniques of the specific protein coded by the gene or, more frequently, detection of the gene DNA itself by hybridizing it with a **radioactive probe,** that is, radioisotope-labeled smaller piece of the DNA of interest or a labeled related DNA or RNA. When a bacterial colony that contains the gene of interest is found, the bacteria are allowed to multiply and to produce sufficient amounts of the gene DNA, or its protein product. That bacterial colony is then used for further study, including gene isolation , manipulation, and transfer.

Cloning Genes from Genome DNA

Mature mRNAs often lack not only any introns that may have been spliced out but also the sequences surrounding the gene DNA in the genome. Therefore, it is necessary to also make **genomic clones,** that is clones of the various parts of the genome. Genomic clones allow study of the structure and arrangement of the genes in the genome as well as of the DNA sequences flanking, and often regulating the expression of, the genes. To make genomic clones, nuclei, chloroplasts, and mitochondria are isolated separately, and the DNA of each is purified. Each DNA is partially digested with a **restriction enzyme** to give DNA pieces of about 20,000 base pairs. The DNA pieces are then mixed with appropriate bacterial plasmids or, more frequently, with an especially modified genome of the bacteriophage lambda. The DNA pieces are taken up by the plasmids or the bacteriophage genome. These are used to insert the DNA into *E. coli* bacteria, which subsequently increase it manyfold. Detection of genes of interest is the same as above, and since cDNA cloning usually precedes genome cloning, specific probes for many genes are often available and are used to detect and isolate genes during genomic cloning.

Expression of Cloned Genes

Detection, cloning, and isolation of genes provides a great deal of useful information; however, the final goal of most genetic engineering studies is to enable a gene to be expressed in a new or different genetic environment. Expression of plant (eukaryotic) genes in bacterial (prokaryotic) genomes is not always possible because of the different promotor genes involved, the inability of bacterial genomes to remove introns from mRNA, or the inability to properly modify the primary proteins to produce stable, functional proteins. Nevertheless, several plant nuclear, chloroplast, and even mitochondrial genes have been cloned and have been expressed in bacteria by using cDNA-derived clones of these genes. Additional new strategies and plasmid or virus vectors have been developed that make possible the expression of many plant genes in bacteria. Cloning of some plant and fungal pathogen genes in eukaryotic hosts, such as yeasts, has already been obtained and is expected to prove useful for future work since yeasts can carry out modification of primary translational products and produce functional mature proteins.

More recently, cloned plant genes have been transferred, incorporated, and expressed in plants of species other than that from which they were obtained, opening whole new vistas for genetic engineering experimentation with genes that code for a variety of proteins or functions, including resistance to disease.

Vectors Used for Gene Cloning in Plants

Vectors are organisms or agents that can transfer genetic material from one organism, called the donor, to another organism, called the recipient, in a way that the genetic material will survive and be expressed in the recipient cell. Just as plasmids, viruses (such as bacteriophage for bacteria and SV40 for animal cells), and cosmids (specially engineered plasmids derived from bacteriophage lambda) are used as vectors of genetic material into bacteria, yeasts, and animals, plasmids of the plant-infecting bacterium *Agrobacterium tumefaciens* and the plant-infecting double-stranded DNA cauliflower mosaic virus are used as vectors of genetic material into plants. In addition, the single-stranded DNA geminiviruses, the single-stranded RNA tobacco mosaic virus, some other virus systems, and transposon elements are being developed as plant gene vectors.

Bacterial Plasmids as Vectors

Ti-plasmid of Agrobacterium tumefaciens When virulent *A. tumefaciens* bacteria infect plants, they induce formation of tumors, called crown galls, by means of a plasmid they contain, called Ti-plasmid (for tumor-inducing) (See pages 559–564). The Ti-plasmid is a circular double-stranded DNA molecule containing up to 200,000 base pairs organized into several genes. When the bacterium comes in contact with a wounded plant cell, the Ti-plasmid is transferred from the bacterium into the cell. A specific region of the plasmid, the T-DNA, is transferred from the plasmid to the nucleus of the plant cell, becomes integrated into the plant nuclear genome, and is transcribed.

The Ti-plasmid consists of many genes, some of which have been indentified (Figure 17-1). The T-DNA region contains several genes, including: (1) the gene for opine synthase, that is, the enzyme that synthesizes **opines,** which are substances produced only by transformed (tumorous) plant cells and can be used as carbon and nitrogen source only by bacteria that contain the Ti-plasmid and the specific gene for catabolism of the particular opines; (2) the gene or genes that control cytokinin biosynthesis, the inactivation of which results in production of rooty tumors; (3) the genes that control auxin biosynthesis, inactivation of which results in production of shooty tumors; and (4) the right and left border 25 base-pairs that seem to be needed for transfer of the T-DNA to the plant genome, since removal of the right 25 base-pair T-DNA border from a Ti-plasmid abolishes transfer of T-DNA into the plant cell genome and, therefore, virulence. Both these events, however, that is, transfer of T-DNA into plant cell genomes and virulence, are also controlled by genes in the virulence region, quite far from the T-DNA region. Some of the other genes located on the Ti-plasmid include those controlling (1) the conjugative transfer of the Ti-plasmid from virulent to nonvirulent bacteria; (2) catabolism of the opine synthesized by the plant cell after inte-

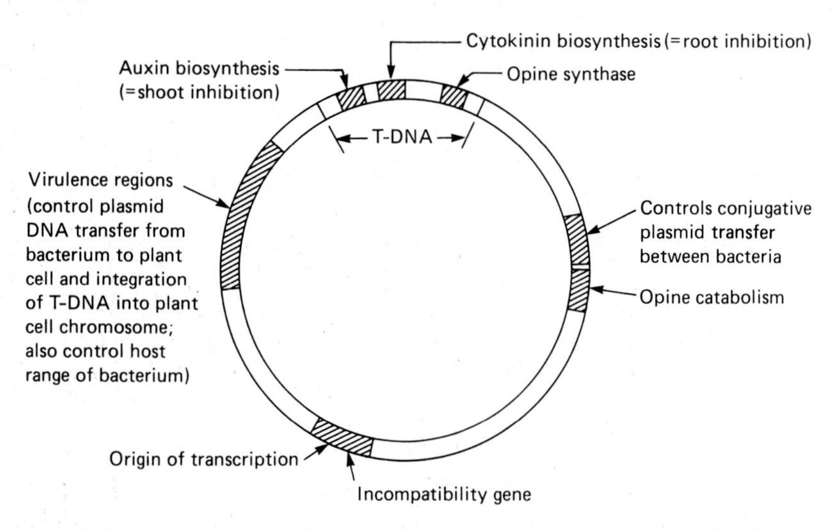

FIGURE 17-1 The Ti-plasmid and location of some of the genes known to control certain functions.

gration of the bacterial gene for opine synthase into the plant genome; (3) the incompatibility properties of the plasmid toward certain strains of the bacterium; and (4) the origin of replication of the plasmid.

Although Ti-plasmids are, apparently, efficient natural vectors of bacterial DNA into plant cells, their use as vectors of other foreign genetic material was difficult because the plasmid is too large and difficult to manipulate genetically, and because infected cells are converted into tumor cells and therefore are difficult to regenerate into whole plants. It was discovered, however, that none of the genes inside the T-DNA region, other than the T-DNA border regions, is necessary for the transfer and integration of the T-DNA. Moreover, genetic material from animals, bacteria, and unrelated plants inserted in the T-DNA was transferred to and was integrated into the plant nuclear genome. Such genes, however, were not expressed in the plant. It was only after the promoter region of the opine synthase gene was added to the foreign gene, thereby producing a so-called **chimaeric gene,** that some of these genes could be recognized by the host enzymes and could be expressed.

It has now been possible to remove the **oncogenic** (tumor-inducing) region, consisting of the genes controlling auxin and cytokinin biosynthesis, from the T-DNA of the Ti-plasmid without removing the border sequences that allow transfer of T-DNA. Plant cells transformed by such plasmids have lost their independence from plant growth regulators, that is, they are transformed with the new genetic material but are no longer tumor cells and, therefore, can be regenerated into whole transformed plants! The positions of the auxin and cytokinin genes on the T-DNA are, in some engineered plasmids, taken by chimaeric genes for resistance to the highly toxic antibiotics kanamycin and methotrexate, which are then used to select for transformed cells or plants.

The problem of the large size of the Ti-plasmid has been solved by using a disarmed, that is, nontumorous, Ti-plasmid in which the border sequences of T-DNA are preserved but the genes inside the border regions are replaced by a small well-understood *E. coli* plasmid, for example, the one known as pBR322. The desired foreign gene DNA is also cloned into another pBR322 plasmid (Figure 17-2).

By allowing *Agrobacterium* bacteria to take up both plasmids. homologous recombination between the pBR322 plasmids takes place within some bacteria and results in cointegration of the foreign gene in pBR322 into the T-DNA region of the Ti-plasmid. When bacterial cells containing such plasmids infect plants, they transfer to the plant chromosomes all the DNA between the border sequences of T-DNA, which includes the foreign gene. Such plasmids have already been used successfully to transfer foreign genes into cells that subsequently regenerated into normal, fertile plants and transmitted the introduced gene through meiosis.

Another strategy to bypass the large size of the Ti-plasmids involves cloning the T-DNA border regions, and any foreign genes between them, in one small plasmid and cloning the virulence (vir) region in another small plasmid. Alone, neither plasmid causes infection. When the two plasmids co-infect *Agrobacterium* bacteria, however, and the bacteria are allowed to infect plant cells, the mini-Ti is virulent and the T-DNA and the foreign gene or genes it contains become integrated into the plant chromosome.

Our knowledge of the Ti-plasmid and its uses as a vector increases rapidly, and many new advances are made and techniques are developed that will make its use dependable and commonplace. Transfer of plant, bacterial, or animal genes into plants is now possible, but little is known about the regulatory controls of expression of such genes. *Agrobacterium* and the Ti-plasmid infect only dicotyledonous plants, while most major food crops are monocotyledons. The use of protoplasts for direct inoculation with Ti-plasmids or with bacterial protoplasts, called **sphaeroplasts,** makes possible inoculation of even monocots with Ti-plasmid vectors, but regeneration of monocot plants from protoplasts has not yet been reported. Knowledge about the Ti-plasmid not only increases our knowledge in the pathology of this plant pathogen, but it also increases our potential to selectively move genes for resistance from one plant to other, perhaps unrelated plants without introducing any undersirable genes or losing desirable ones and without the need for time-consuming crosses and backcrosses between plants. The main obsta-

FIGURE 17-2 Transfer of a foreign gene into a Ti-plasmid by using an additional smaller plasmid.

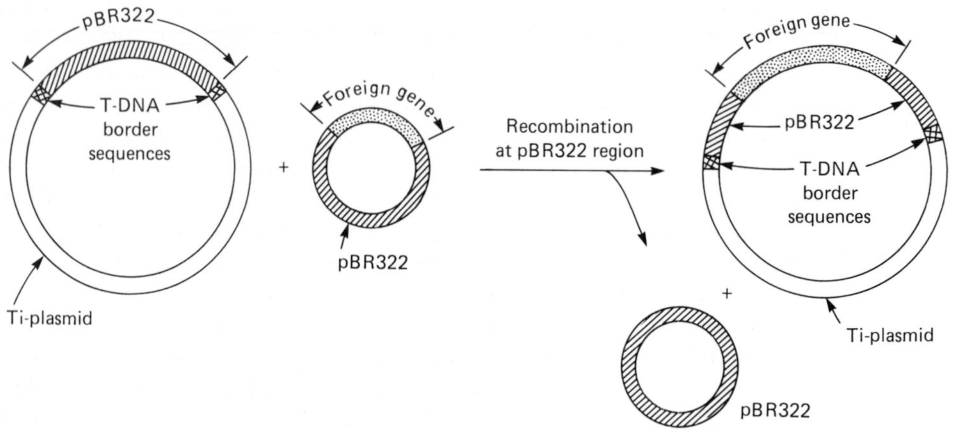

cle to using the Ti-plasmid in plant pathology is the lack of knowledge concerning where in the genome of cells of resistant plants are the genes that control disease resistance and how they may be expressed in the new plants.

Plant Viruses as Vectors

Some of the most effective vectors of genetic material in bacteria and in animals are viruses. So far, the best, if not the only vector of plant genes has been the Ti-plasmid. With the rapidly accelerating studies of plant viruses as potential plant gene vectors, it is believed that one or more virus vector systems will soon become available for effective transfer of genes between plants.

It should be pointed out that plant viral vectors are not likely to be integration-type vectors such as the Ti-plasmid. Rather, they are likely to deliver a gene into a plant cell, where it would replicate about a millionfold with the virus, and would also spread the gene systemically throughout the plant. Introduction of desired genes by viruses into perennial or annual plants propagated vegetatively, or into plants in which the virus is transmitted readily through the seed, would amount to a perpetual existence of the gene in the plant. Or, the gene may be introduced into annual plants by mechanical inoculation or, particularly, into existing perennials (such as trees) by grafting. In these cases, the gene could serve to provide cross protection or some type of systemic resistance against a pathogen that had recently become important and threatening to the crop.

It is assumed, of course, that plant viruses that will be used as gene vectors will be so selected or engineered that they will be able to infect plant cells and replicate themselves and the foreign gene or genes they carry without causing symptoms and loss of yield in the plant. All these objectives are likely to be realized before the end of this century.

Caulimoviruses Caulimoviruses are isometric viruses about 50 nm in diameter and contain circular double-stranded DNA of about 8,000 base pairs. Each DNA strand has one or two discontinuities, consisting of short 6–18 base-pair overlaps at specific sites.

The caulimovirus best studied as a gene vector is cauliflower mosaic virus (CaMV) (see pages 653 and 670 and Figure 17-3). The virus is transmitted by mechanical inoculation of leaves and by aphids to plants in the family Cruciferae. Both the virus and its isolated DNA are infectious, causing systemic infections of the host and producing about half a million virus particles per cell. Although the virus DNA is transcribed in the plant nucleus, the transcript is a mRNA, which is transported to the cytoplasm. There, the mRNA either undergoes reverse transcription and produces the (−) strand DNA, from which the (+) strand DNA and the double-stranded DNA are produced, or it is translated into several (6–8) proteins, including viral coat protein. The viral DNA does not seem to be integrated into the plant genome and is not even transmitted through the seed.

Although the DNA of CaMV has been cloned in plasmids in *E. coli* bacteria, and was shown to be infectious after reisolation from the bacteria, it has not yet been possible to use CaMV as a gene vector. One of the problems is that most of the viral genome is packed with coding regions needed by the

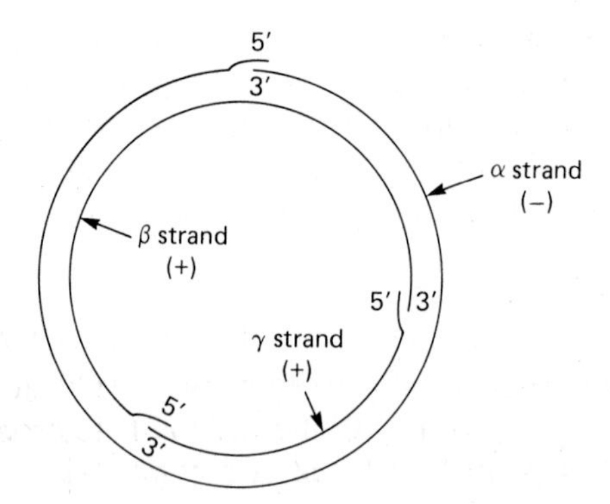

FIGURE 17-3 DNA of cauliflower mosaic virus.

virus, since most deletions of any significant size destroy the infectivity of the virus. In one coding region of the virus, however, inserts of up to 400 base-pairs long are tolerated. The size limitation problem can possibly be overcome by using a helper virus system, in which a substantial portion of the viral genome in some of the virus is deleted and replaced with foreign DNA, while the loss of function could be complemented by co-infection of the same plant cells with a normal viral DNA or with viral DNA from which DNA coding for different functions has been deleted. These steps, however, have yet to be carried out successfully enough to allow the practical use of caulimoviruses as plant gene vectors.

Geminiviruses Geminiviruses are paired (gemini) isometric viruses 18–20 nm by 30 nm in size. Each virus pair contains a single circular single-stranded DNA of about 2,500 bases, but several geminiviruses have a divided genome consisting of two DNA molecules of similar size but coding for different things. Each geminivirus seems to consist of two populations of paired particles that have identical coat proteins but DNAs of different nucleotide sequence. Geminiviruses enter and replicate in the plant nucleus. In nature they are transmitted by leafhoppers or white flies in the persistent manner and with difficulty, if at all, mechanically.

Little is known about the number, kind, and location of the genes in geminiviruses and about their promotor sequences. Geminiviruses, although containing single-stranded DNA, produce a double-stranded DNA intermediate stage in the nucleus. Double-stranded DNA has been able to infect plant protoplasts and could be used for *in vivo* manipulation in bacterial plasmids. Viruses of this group infect both dicot and monocot plants. In spite of the potential problems presented by the small size of the geminivirus genome, and of the difficulty of transmission of geminiviruses through sap, a great deal of effort is being made in the development and future use of geminiviruses as plant gene vectors.

RNA Viruses RNA viruses may become important in the future, particularly the multicomponent viruses and the satellite viruses and RNAs. In some multicomponent viruses, such as brome mosaic virus, the smallest of the three components contains the coat protein RNA, which could be totally replaced by foreign genetic material (RNA or cDNA) without affecting the infectivity of the virus. In satellite viruses and RNAs, whose sizes range from 270 to 1,500 base pairs, the RNAs are not needed by the virus and could, therefore, be replaced totally or in part by foreign genetic material (RNA or cDNA) that would presumably be introduced into the plant upon infection with the virus. Their construction and use as plant gene cloning vectors has yet to be achieved, however.

Viroids Viroids are small, circular, single-stranded, naked RNAs, 300–400 bases long. They are mechanically transmissible, they replicate in the host nucleus, and they infect plants systemically. Several of these properties make them attractive candidates as vectors of plant genetic material, but so far no such vectors have been reported.

Plant Genomic Components as Vectors

Several chromosomal and extrachromosomal elements seem to be capable of introducing and integrating genetic material into plant cell genomes and may have potential as vectors.

Transposable Elements Transposable elements are segments of DNA probably present in the genome of all types of organisms. They have diverse properties but all share the property that, although they spend most of their time integrated into the genome of a cell, they can move around, that is, transpose, in the genome and integrate at different sites. If such a transposable element transposes into a particular gene and disrupts its expression, it gives rise to a mutant. Transposable elements vary in size from 400 to 20,000 base pairs. They all seem to have inverted terminal repeats of about 11 base pairs. It is believed that after a transposable element is isolated, a foreign plant gene could be inserted into it and the hybrid transposable element then could be introduced into plant cells or protoplasts where it could become integrated in the cell genome. Such a transfer and integration of foreign genes by transposable elements has already been carried out in the fruit fly but has not yet been reported with plants. Transposable elements, of course, could be used with monocot as well as dicot plants, while the Ti system only infects dicots.

Extrachromosomal Elements Extrachromosomal elements are autonomously replicating circular or linear segments of DNA present in the cytoplasm. They vary in size from 1,000 to 6,000 base pairs. At least some of them have inverted terminal repeats and appear to be capable of transposition within the mitochondrial genome. Their potential as plant gene vectors is still unknown, but their properties make them inviting targets for study and possible development as vectors.

Chromosomal Elements Efforts are being made to develop chromosomal elements as vectors that will allow integration of additional genetic material into a host's genome with or without replacement of the homologous host genes. Although this has not yet been accomplished with plants, it has been successfully carried out in yeast fungi. Moreover, in yeast, successful vectors have been obtained by using portions of the centromeric fragments of the chromosome, which contain autonomously replicating sequences of DNA representing origins of DNA replication. Also in yeast, great progress has been made toward construction of an artificial chromosome. Whether or not the success with yeast genomes can be transferred to plant genomes is not known, but several scientists are working toward the solution of the problem already. Successful yeast transformation vectors are expected to soon find use in the study of the genes for virulence and avirulence in the plant pathogenic fungi and of fungi antagonistic to the pathogenic ones.

Chemical and Mechanical Delivery Systems of DNA into Cells

Natural vectors, whether bacterial, viral, or plant genomic components, have the advantage that they introduce active genomic material into plant cells that is not normally degraded by the cell's enzymes. For some hosts, however, no vectors are available, and some vectors, especially the viral and bacterial (Ti) ones, also infect the cell with unwanted genetic material. In these cases use of other delivery systems is necessary or desirable. Such systems include:

1. Incorporation of pieces of DNA on chromosome fragments or on plasmids, and of whole chromosomes, into protoplasts by mixing nucleic acid and protoplasts in the presence of cations, such as poly-L-ornithine, or of polyethylene glycol.
2. Electroporation or electrotransformation, that is, treating a mixture of plant protoplasts and foreign DNA with short pulses of electricity, which temporarily puncture the cell membrane and allow the foreign DNA to enter. The foreign DNA can then be expressed and the protoplasts can be regenerated into plants. Both dicots and monocots can be transformed by this technique.
3. Direct microinjection of nucleic acid into plant cells.
4. Enclosure of DNA or RNA in liposomes (small, artificial lipid vesicles) and fusion of the liposomes with plant protoplasts.
5. Direct fusion of plant protoplasts with transformed bacteria or bacterial sphaeroplasts, that is, bacteria that contain foreign DNA-carrying plasmids.

There is little doubt that more and better vectors and nonvector delivery systems will be developed in the next few years that will simplify many of the existing ones and will certainly utilize many new concepts.

Biotechnology and Plant Pathology: The Beginning

First came the Ti-plasmid and T-DNA, then cauliflower mosaic virus and the other virus and viruslike systems. Now numerous laboratories throughout the world are cloning, mapping, and studying the genes of several plant pathogenic bacteria. Similar studies are revealing the number, kinds, and regulation of the genes of several plant viruses. Work has just begun to clone genes of plant pathogenic fungi and to study their physical structures, metabolic functions, and mechanisms of regulation. Once the genes have been identified and isolated, it will be possible to manipulate them, modify them, transfer them, study their communication with the host genes, and eventually inhibit them or neutralize them. Once known and available, genes for virulence or inhibition may be transferred to microorganisms antagonistic to the pathogens, which then may be used on plant surfaces to protect plants from the pathogens.

Beyond the study of the virulence genes of pathogens, knowledge of the nature and expression of host genes for disease resistance is likely to produce the greatest dividends of biotechnology to plant disease control. It is hard to imagine how far this will lead, but expectations are high.

The first tangible effects of genetic engineering and biotechnology are beginning to trickle in. A gene for resistance to the crown gall bacterium has been transfered to tobacco and none of the more than one thousand genetically altered plants developed crown gall after exposure to the pathogen! The gene for coat protein of tobacco mosaic virus has been transferred to tobacco plants, making the plants resistant to TMV infection, presumably through some sort of cross protection phenomenon! Resistance to potato leaf roll virus has been transfered by protoplast fusion from a wild, resistant but nontuber-forming potato species into the cultivated and susceptible potato. A tripartite plant RNA virus (brome mosaic virus) has already been used to transfer foreign DNA into plant (barley) protoplasts and the DNA was expressed in the protoplasts, making this a powerful vehicle for transferring genes into monocots as well as dicot plants. Plants can be genetically engineered to be resistant to antibiotics effective against certain pathogens and are therefore not injured when treated to control the disease. Genetically engineered bacteria or viruses (bacteriophage) can be sprayed on plants to prevent frost injury. Monoclonal antibodies have been produced against several virus and bacterial plant pathogens and diagnostic kits for these pathogens are available commercially! And, nucleic acid hybridization tests and test kits can be used for the detection and diagnosis of viroids.

We all know this is just the beginning. We can hardly wait for what tomorrow will bring!

SELECTED REFERENCES

Abelson, P. H., ed. (1984). "Biotechnology and Biological Frontiers." Am. Assoc. Adv. Sci. Washington, D. C.

Davis, J. W., ed. (1985). "Molecular Plant Virology," Vol. 2. CRC Press, Boca Raton, Florida.

Dixon, R. A., ed. (1985). "Plant Cell Culture: A Practical Approach." IRL Press, Oxford.

Callow, J. A., ed. (1983). "Biochemical Plant Pathology." Wiley, New York.

Evans, D. A., Sharp, W. R., Ammirato, P. V., and Yamada, Y., eds. (1983). "Handbook of Plant Cell Culture." Vol. 1. Macmillan, New York.

Fowke, L. C., and Constabel, F., eds. (1985). "Plant Protoplasts." CRC Press, Boca Raton, Florida.

Gardner, R. C. (1983). Plant viral vectors: CaMV as an experimental tool. *In* "Genetic Engineering of Plants" (T. Kosuge, C. P. Meredith, and A. Hollaender, eds.), pp. 121–142. Plenum, New York.

Glover, D. M., ed. (1985). "DNA Cloning," Vol. 2, IRL Press, Oxford.

Gustafson, J. P., ed. (1984). "Gene Manipulation in Plant Improvement." Plenum, New York.

Helgeson, J. P., and Deverall, B. J., eds. (1983). "Use of Tissue Culture and Protoplasts in Plant Pathology." Academic Press, New York.

Ingram, D. S., and Helgeson, J. P., eds. (1980). "Tissue Culture Methods for Plant Pathologists." Blackwell Oxford.

Kahl, G., and Schell, J. S. (eds. (1982). "Molecular Biology of Plant Tumors." Academic Press, New York.

Kosuge, T., and Nester, E. W., eds. (1984). "Plant-Microbe Interactions: Molecular and Genetic Perspectives," Vol. 1. Macmillan, New York.

Kosuge, T., Meredith, C. P., and Hollaender, A., eds. (1983). "Genetic Engineering of Plants: An Agricultural Perspective." Plenum, New York.

Mantell, S. H., Matthews, J. A., and McKee, R. A. (1985). "Principles of Plant Biotechnology: An Introduction to Genetic Engineering in Plants." Blackwell, Oxford.

Olson, S. (1986). "Biotechnology: An Industry Comes of Age." Natl. Acad. Press, Washington, D. C.

Owen, L. D., ed. (1983). "Genetic Engineering: Applications to Agriculture." Beltsville Symp. Agric. Res., Vol. 7. Wiley, New York.

Scandalios, J. G., ed. (1984). "Molecular Genetics of Plants." Academic Press, Orlando, Florida.

Science (1983). Biotechnology issue, Vol. 219, No. 4585. *Science* (1985). Biotechnology issue, Vol. 229, No. 4719.

Vasil, I. K., ed. (1984). "Cell Culture and Somatic Cell Genetics of Plants," Vol. 1. Academic Press, Orlando, Florida.

Vasil, I. K., Scowcroft, W. R., and Frey, K. J., eds. (1982). "Plant Improvement and Somatic Cell Genetics." Academic Press, New York.

Watson, J. D., Tooze, J., and Kurtz, D. T. (1983). "Recombinant DNA: A Short Course." Freeman, New York.

Yoder, O. C. (1983). Use of pathogen-produced toxins in genetic engineering of plants and pathogens. *In* "Genetic Engineering of Plants" (T. Kosuge, C. P. Meredith, and A. Hollaender, eds.), pp. 335–353. Plenum, New York.

Zaitlin, M., Day, P., and Hollaender, A., eds. (1985). "Biotechnology in Plant Science: Relevance to Agriculture in the Eighties." Academic Press, Orlando, Florida.

GLOSSARY

Acervulus A subepidermal, saucer-shaped, asexual fruiting body producing conidia on short conidiophores.

Actinomycetes A group of bacteria forming branching filaments.

Aeciospore A dikaryotic rust spore produced in an aecium.

Aecium A cup-shaped fruiting body of the rust fungi which produces aeciospores.

Aerobic A microorganism that lives or a process that occurs in the presence of molecular oxygen.

Agar A gelatinlike material obtained from seaweed and used to prepare culture media on which microorganisms are grown and studied.

Agglutination A serological test in which viruses or bacteria suspended in a liquid collect into clumps whenever the suspension is treated with antiserum containing antibodies specific against these viruses or bacteria.

Allele One of two or more alternate forms of a gene occupying the same locus on a chromosome.

Alternate host One of two kinds of plants on which a parasitic fungus (e.g., rust) must develop to complete its life cycle.

Anaerobic Relating to a microorganism that lives or a process that occurs in the absence of molecular oxygen.

*NOTE:
To make the plural of Latin words ending in:
-us (e.g., acervulus), change *us* to *i* (e.g., acervuli);
-um (e.g., aecium), change *um* to *a* (e.g., aecia);
-a (e.g., hypha), change *a* to *ae* (e.g., hyphae);
-is (e.g., tylosis), change *is* to *es* (e.g., tyloses).

Anastomosis The union of a hypha or vessel with another resulting in intercommunication of their contents.

Antheridium The male sexual organ found in some fungi.

Anthracnose A disease that appears as black, sunken leaf, stem, or fruit lesions and caused by fungi that produce their asexual spores in an acervulus.

Antibiotic A chemical compound produced by one microorganism which inhibits or kills other microorganisms.

Antibody A protein produced in a warm-blooded animal in reaction to an injected foreign antigen and capable of reacting specifically with that antigen.

Antigen Foreign proteins, and occasionally complex lipids, carbohydrates, and some nucleic acids, which upon injection into a warm-blooded animal induce the production of antibodies.

Antiserum The blood serum of a warm-blooded animal that contains antibodies.

Apothecium An open cup- or saucer-shaped ascocarp of some ascomycetes.

Appressorium The swollen tip of a hypha or germ tube that facilitates attachment and penetration of the host by the fungus.

Ascocarp The fruiting body of ascomycetes bearing or containing asci.

Ascogenous hypha Hyphae arising from the fertilized ascogonium and producing the asci.

Ascogonium The female gametangium or sexual organ of ascomycetes.

Ascomycetes A group of fungi producing their sexual spores, ascospores, within asci.

Ascospore A sexually produced spore borne in an ascus.

Ascus A saclike cell of a hypha in which meiosis occurs and which contains the ascospores (usually eight).

Asexual reproduction Any type of reproduction not involving the union of gametes or meiosis.

Autoecious fungus A parasitic fungus that can complete its entire life cycle on the same host.

Avirulent Lacking virulence.

Bacillus A rod-shaped bacterium.

Bactericide A chemical compound that kills bacteria.

Bacteriocins Bactericidal substances produced by certain strains of bacteria and active against some other strains of the same or closely related species.

Bacteriophage A virus that infects bacteria and usually kills them.

Bacteriostatic A chemical or physical agent that prevents multiplication of bacteria without killing them.

Base An alkaline, usually nitrogenous organic compound; used particularly for the purine and pyrimidine moieties of the nucleic acids of cells and viruses.

Basidiomycetes A group of fungi producing their sexual spores, basidiospores, on basidia.

Basidiospore A sexually produced spore borne on a basidium.

Basidium A club-shaped structure on which basidiospores are borne.

Bioassay The use of a test organism to measure the relative infectivity of a pathogen or toxicity of a substance.

Biological control Total or partial destruction of pathogen populations by other organisms.

Biotechnology The use of genetically modified organisms and/or modern techniques and processes with biological systems for industrial production.

Biotroph An organism that can live and multiply only on another living organism.

Biotype A subgroup within a species usually characterized by the possession of a single or a few characters in common.

Blight A disease characterized by general and rapid killing of leaves, flowers, and stems.

Blotch A disease characterized by large, and irregular in shape, spots or blots on leaves, shoots, and stems

Budding A method of vegetative propagation of plants by implantation of buds from the mother plant onto a rootstock.

Callus A mass of thin-walled undifferentiated cells, developed as the result of wounding or culture on nutrient media.

Canker A necrotic, often sunken lesion on a stem, branch, or twig of a plant.

Capsid The protein coat of viruses forming the closed shell or tube that contains the nucleic acid.

Capsule A relatively thick layer of mucopolysaccharides that surrounds some kinds of bacteria.

Carbohydrates Foodstuffs composed of carbon, hydrogen, and oxygen (CH_2O) with the last two in a 2 to 1 ratio, as in water, H_2O.

Cellulase An enzyme that breaks down cellulose.

Cellulose A polysaccharide composed of hundreds of glucose molecules linked in a chain and found in the plant cell walls.

Chemotherapy Control of a plant disease with chemicals (chemotherapeutants) that are absorbed and are translocated internally.

Chlamydospore A thick-walled asexual spore formed by the modification of a cell of a fungus hypha.

Chlorosis Yellowing of normally green tissue due to chlorophyll destruction or failure of chlorophyll formation.

Chronic symptoms Symptoms that appear over a long period of time.

Circulative viruses Viruses that are acquired by their vectors through their mouthparts, accumulate internally, then are passed through their tissues and introduced into plants again via the mouthparts of the vectors.

Cistron The sequence of nucleotides within a certain area of a nucleic acid (DNA or RNA) that codes for a particular protein.

Cleistothecium An entirely closed ascocarp.

Clone The group of genetically identical individuals produced asexually from one individual.

Coding The process by which the sequence of nucleotides within a certain area of RNA determines the sequence of amino acids in the synthesis of the particular protein.

Codon The coding unit, consisting of three adjacent nucleotides, which codes for a specific amino acid.

Complementary DNA (cDNA) DNA synthesized by reverse transcriptase from an RNA template.

Conjugation A process of sexual reproduction involving the fusion of gametes morphologically similar.

Conidiophore A specialized hypha on which one or more conidia are produced.

Conidium An asexual fungus spore formed from the end of a conidiophore.

Coremium An asexual fruiting body consisting of a cluster of erect hyphae bearing conidia.

Cork An external, secondary tissue impermeable to water and gases. It is often formed in response to wounding or infection.

Cosmid An artificial plasmid that combines the transduction and replication properties of a phage and a plasmid and is used as a cloning vehicle for genetic engineering.

Cross protection The phenomenon in which plant tissues infected with one strain of a virus are protected from infection by other strains of the same virus.

Culture To artificially grow microorganisms or plant tissue on a prepared food material; a colony of microorganisms or plant cells artificially maintained on such food material.

Culture medium The prepared food material on which microorganisms or plant cells are cultured.

Cuticle A thin, waxy layer on the outer wall of epidermal cells consisting primarily of wax and cutin.

Cutin A waxy substance comprising the inner layer of the cuticle.

Cyst An encysted zoospore (fungi); in nematodes, the carcass of dead adult females of the genus *Heterodera* or *Globodera* which may contain eggs.

Cytokinins A group of plant growth-regulating substances that regulate cell division.

Damping-off Destruction of seedlings near the soil line, resulting in the seedlings falling over on the ground.

Denatured protein Protein whose properties have been altered by treatment with physical or chemical agents.

Density-gradient centrifugation A method of centrifugation in which particles are separated in layers according to their density.

Dieback Progressive death of shoots, branches, and roots generally starting at the tip.

Dikaryotic Mycelium or spores containing two sexually compatible nuclei per cell. Common in the basidiomycetes.

Disease Any malfunctioning of host cells and tissues that results from continuous irritation by a pathogenic agent or environmental factor and leads to development of symptoms.

Disease cycle The chain of events involved in disease development, including the stages of development of the pathogen and the effect of the disease on the host.

Disinfectant A physical or chemical agent that frees a plant, organ, or tissue from infection.

Disinfestant An agent that kills or inactivates pathogens in the environment or on the surface of a plant or plant organ before infection takes place.

Downy mildew A plant disease in which the sporangiophores and spores of the fungus appear as a downy growth on the lower surface of leaves and stems, fruit, etc., caused by fungi in the family Peronosporaceae.

Ectoparasite A parasite feeding on a host from the exterior.

Egg A female gamete. In nematodes, the first stage of the life cycle containing a zygote or a larva.

ELISA A serological test in which one antibody carries with it an enzyme that releases a colored compound.

Elicitors Molecules produced by the host (or pathogen) that induce a response by the pathogen (or host).

Enation Tissue malformation or overgrowth induced by certain virus infections.

Endoparasite A parasite which enters a host and feeds from within.

Enzyme A protein produced by living cells that can catalyze a specific organic reaction.

Epidemic A widespread and severe outbreak of a disease. A disease increase in a population.

Epidemic rate The amount of increase of disease per unit of time in a plant population.

Epidemiology The study of factors affecting the outbreak and spread of infectious diseases.

Epidermis The superficial layer of cells occurring on all plant parts.

Epiphytically Existing on the surface of a plant or plant organ without causing infection.

Epiphytotic A widespread and destructive outbreak of a disease of plants. Epidemic.

Eradicant A chemical substance that destroys a pathogen at its source.

Eradication Control of plant disease by eliminating the pathogen after it is established or by eliminating the plants that carry the pathogen.

Facultative parasite Having the ability to be a parasite.

Fermentation Oxidation of certain organic substances in the absence of molecular oxygen.

Fertilization The sexual union of two nuclei resulting in doubling of chromosome numbers.

Filamentous Threadlike; filiform.

Fission Transverse splitting in two of bacterial cells, asexual reproduction.

Flagellum A whiplike structure projecting from a bacterium or zoospore and functioning as an organ of locomotion. Also called a cilium.

Forma specialis (f. sp.) A group of biotypes of a pathogen species that can infect only plants within a certain host genus or species.

Free-living Of a microorganism that lives freely, unattached, or a pathogen living in the soil, outside its host.

Fructification Production of spores by fungi. Also, a fruiting body.

Fruiting body A complex fungal structure containing spores.

Fumigant A toxic gas or volatile substance that is used to disinfest certain areas from various pests.

Fumigation The application of a fumigant for disinfestation of an area.

Fungicide A compound toxic to fungi.

Fungistatic A compound that prevents fungus growth without killing the fungus.

Gall A swelling or overgrowth produced on a plant as a result of infection by certain pathogens.

Gametangium A cell containing gametes or nuclei that act as gametes.

Gamete A male or female reproductive cell or the nuclei within a gametangium.

Gene A linear portion of the chromosome which determines or conditions one or more hereditary characters. The smallest functioning unit of the genetic material.

Gene cloning The isolation and multiplication of an individual gene sequence by its insertion into a bacterium where it can multiply.

Genetic engineering The alteration of the genetic composition of a cell by various procedures (transformation, protoplast fusion, etc.) in tissue culture.

Genotype The genetic constitution of an organism.

Germ tube The early growth of mycelium produced by a germinating fungus spore.

Gibberellins A group of plant growth-regulating substances with a variety of functions.

Grafting A method of plant propagation by transplantation of a bud or a scion of a plant on another plant. Also, the joining of cut surfaces of two plants so as to form a living union.

Growth regulator A natural substance that regulates the enlargement, division, or activation of plant cells.

Gum Complex polysaccharidal substances formed by cells in reaction to wounding or infection.

Gummosis Production of gum by or in a plant tissue.

Guttation Exudation of water from plants, particularly along the leaf margin.

Habitat The natural place of occurrence of an organism.

Haploid A cell or an organism whose nuclei have a single complete set of chromosomes.

Haustorium A projection of hyphae into host cells which acts as an absorbing organ.

Hectare An area of land equal to 2.5 acres.

Herbaceous plant A higher plant that does not develop woody tissues.

Hermaphrodite An individual bearing both functional male and female reproductive organs.

Heteroecious Requiring two different kinds of hosts to complete its life cycle. Pertaining particularly to rust fungi.

Heterokaryosis The condition in which a mycelium contains two genetically different nuclei per cell.

Heteroploid A cell, tissue, or organism that contains more or fewer chromosomes per nucleus than the normal 1N or 2N for that organism.

Heterothallic fungi Fungi producing compatible male and female gametes on physiologically distinct mycelia.

Homothallic fungus A fungus producing compatible male and female gametes on the same mycelium.

Hormone A growth regulator. Frequently referring particularly to auxins.

Horizontal resistance Partial resistance equally effective against all races of a pathogen.

Host A plant that is invaded by a parasite and from which the parasite obtains its nutrients.

Host range The various kinds of host plants that may be attacked by a parasite.

Hyaline Colorless, transparent.

Hybrid The offspring of two individuals differing in one or more heritable characteristics.

Hybridization The crossing of two individuals differing in one or more heritable characteristics.

Hybridoma A hybrid animal cell produced by the fusion of a spleen cell and a cancer cell and able to multiply and to produce monoclonal antibodies.

Hydathodes Structures with one or more openings that discharge water from the interior of the leaf to its surface.

Hydrolysis The enzymatic breakdown of a compound through the addition of water.

Hyperparasite A parasite parasitic on another parasite.

Hyperplasia A plant overgrowth due to increased cell division.

Hypersensitivity Excessive sensitivity of plant tissues to certain pathogens. Affected cells are killed quickly, blocking the advance of obligate parasites.

Hypertrophy A plant overgrowth due to abnormal cell enlargement.

Hypha A single branch of a mycelium.

Hypovirulence Reduced virulence of a pathogen strain as a result of the presence of transmissible double-stranded RNA.

Immune Cannot be infected by a given pathogen.

Immunity The state of being immune.

Imperfect fungus A fungus that is not known to produce sexual spores.

Imperfect stage The part of the life cycle of a fungus in which no sexual spores are produced. The anamorph stage.

Incubation period The period of time between penetration of a host by a pathogen and the first appearance of symptoms on the host.

Indexing A procedure to determine whether a given plant is infected by a virus. It involves the transfer of a bud, scion, sap, etc. from one plant to one or more kinds of (indicator) plants that are sensitive to the virus.

Indicator A plant that reacts to certain viruses or environmental factors with production of specific symptoms and is used for detection and identification of these factors.

Infection The establishment of a parasite within a host plant.

Infectious disease A disease that is caused by a pathogen which can spread from a diseased to a healthy plant.

Infested Containing great numbers of insects, mites, nematodes, etc. as applied to an area or field. Also applied to a plant surface or soil contaminated with bacteria, fungi, etc.

Injury Damage of a plant by an animal, physical, or chemical agent.

Inoculate To bring a pathogen into contact with a host plant or plant organ.

Inoculation The arrival or transfer of a pathogen onto a host.

Inoculum The pathogen or its parts that can cause infection. That portion of individual pathogens that are brought into contact with the host.

Integrated control An approach that attempts to use all available methods of control of a disease or of all the diseases and pests of a crop plant for best control results but with the least cost and the least damage to the environment.

Intercalary Formed along and within the mycelium—not at the hyphal tips.

Intercellular Between cells.

Intracellular Within or through the cells.

Invasion The spread of a pathogen into the host.

In vitro In culture. Outside the host.

In vivo In the host.

Isolate A single spore or culture and the subcultures derived from it. Also used to indicate collections of a pathogen made at different times.

Isolation The separation of a pathogen from its host and its culture on a nutrient medium.

L-form bacteria Bacteria that have, temporarily or permanently, lost the ability to produce a cell wall as a result of growth in the presence of antibiotics inhibiting cell wall synthesis.

Larva The life stage of a nematode between the embryo and the adult; an immature nematode.

Latent infection The state in which a host is infected with a pathogen but does not show any symptoms.

Latent virus A virus that does not induce symptom development in its host.

Leaf spot A self-limiting lesion on a leaf.

Lectins A group of plant proteins that bind to specific carbohydrates.

Lesion A localized area of discolored, diseased tissue.

Life cycle The stage or successive stages in the growth and development of an organism that occur between the appearance and reappearance of the same stage (e.g., spore) of the organism.

Lipids Substances whose molecules consist of glycerin and fatty acids and sometimes certain additional types of compounds.

Local lesion A localized spot produced on a leaf upon mechanical inoculation with a virus.

mm (millimeter) A unit of length equal to 1/10 of a centimeter (cm) or 0.03937 of an inch.

μm (micrometer) A unit of length equal to 1/1000 of a millimeter.

nm (nanometer) A unit of length equal to 1/1000 of a micrometer.

Macroscopic Visible to the naked eye without the aid of a magnifying lens or a microscope.

Malignant Used of a cell or tissue that divides and enlarges autonomously, i.e., its growth can no longer be controlled by the organism on which it is growing.

Masked symptoms Virus-induced plant symptoms that are absent under certain environmental conditions but appear when the host is exposed to certain conditions of light and temperature.

Mechanical inoculation Inoculation of a plant with a virus through transfer of sap from a virus-infected plant to a healthy plant.

Messenger RNA A chain of ribonucleotides that codes for a specific protein.

Metabolism The process by which cells or organisms utilize nutritive material to build living matter and structural components or to break down cellular material into simple substances to perform special functions.

Microscopic Very small; can be seen only with the aid of a microscope.

Middle lamella The cementing layer between adjacent cell walls; it generally consists of pectinaceous materials, except in woody tissues, where pectin is replaced by lignin.

Migratory Migrating from plant to plant.

Mildew A fungal disease of plants in which the mycelium and spores of the fungus are seen as a whitish growth on the host surface.

Mold Any profuse or woolly fungus growth on damp or decaying matter or on surfaces of plant tissue.

Molt The shedding or casting off of the cuticle.

Monoclonal antibodies Identical antibodies produced by a single clone of lymphocytes.

Monocyclic Having one cycle per season.

Mosaic Symptom of certain viral diseases of plants characterized by intermingled patches of normal and light green or yellowish color.

Mottle An irregular pattern of indistinct light and dark areas.

Mummy A dried, shriveled fruit.

Mutant An individual possessing a new, heritable characteristic as a result of a mutation.

Mutation An abrupt appearance of a new characteristic in an individual as the result of an accidental change in genes or chromosomes.

Mycelium The hypha or mass of hyphae that make up the body of a fungus.

Mycoplasmalike organisms Microorganisms found in the phloem and phloem parenchyma of diseased plants and assumed to be the causes of the disease; they resemble mycoplasmas in all respects except that they cannot yet be grown on artificial nutrient media.

Mycoplasmas Pleomorphic prokaryotic microorganisms that lack a cell wall.

Mycorrhiza A symbiotic association of a fungus with the roots of a plant.

Mycotoxicoses Diseases of animals and humans caused by consumption of feed and foods invaded by fungi that produce mycotoxins.

Mycotoxins Toxic substances produced by several fungi in infected seeds, feeds, or foods and capable of causing illnesses of varying severity and death to animals and humans that consume such substances.

Necrotic Dead and discolored.

Nectarthode An opening at the base of a flower from which nectar exudes.

Nematicide A chemical compound or physical agent that kills or inhibits nematodes.

Nematode Generally microscopic, wormlike animals that live saprophytically in water or soil, or as parasites of plants and animals.

Noninfectious disease A disease that is caused by an environmental factor, not by a pathogen.

Nucleic acid An acidic substance containing pentose, phosphorus, and pyrimidine and purine bases. Nucleic acids determine the genetic properties of organisms.

Nucleoprotein Referring to viruses: consisting of nucleic acid and protein.

Nucleoside The combination of a sugar and a base molecule in a nucleic acid.

Nucleotide The phosphoric ester of a nucleoside. Nucleotides are the building blocks of DNA and RNA.

Obligate parasite A parasite that in nature can grow and multiply only on or in living organisms.

Oogonium The female gametangium of Oomycetes containing one or more gametes.

Oomycete A fungus that produces oospores. A water mold.

Oospore A sexual spore produced by the union of two morphologically different gametangia (oogonium and antheridium).

Operon A cluster of functionally related genes regulated and transcribed as a unit.

Osmosis The diffusion of a solvent through a differentially permeable membrane from its higher concentration to its lower concentration.

Ostiole A porelike opening in perithecia and pycnidia through which the spores escape from the fruiting body.

Ovary The female reproductive structure that produces or contains the egg.

Oxidative phosphorylation The utilization of energy released by the oxidative reactions of respiration to form high-energy ATP bonds.

Ozone (0_3) A highly reactive form of oxygen that in relatively high concentrations may injure plants.

Papillate Bearing a papilla, i.e., a hump or swelling.

Paraphysis A sterile hypha present in some fruiting bodies of fungi.

Parasexualism A mechanism whereby recombination of hereditary properties occurs within fungal heterokaryons.

Parasite An organism living on or in another living organism (host) and obtaining its food from the latter.

Parenchyma A tissue composed of thin-walled cells which usually leave intercellular spaces between them.

Pathogen An entity that can incite disease.

Pathogenicity The capability of a pathogen to cause disease.

Pathovar In bacteria, a subspecies or group of strains that can infect only plants within a certain genus or species.

Pectin A methylated polymer of galacturonic acid found in the middle lamella and the primary cell wall.

Pectinase An enzyme that breaks down pectin.

Penetration The initial invasion of a host by a pathogen.

Perfect stage The sexual stage in the life cycle of a fungus. The teleomorph.

Perithecium The globular or flask-shaped ascocarp of the Pyrenomycetes, having an opening or pore (ostiole).

Phage A virus that attachs bacteria; also called bacteriophage.

Phenolic Applied to a compound that contains one or more phenolic rings.

Phenotype The external visible appearance of an organism.

Phloem Food-conducting tissue, consisting of

sieve tubes, companion cells, phloem parenchyma, and fibers.

Phytoalexin A substance which inhibits the development of a fungus on hypersensitive tissue, formed when host plant cells come in contact with the parasite.

Phytopathogenic Term applicable to a microorganism that can incite disease in plants.

Phytotoxic Toxic to plants.

Plasmalemma The cytoplasmic membrane found on the outside of the protoplast adjacent to the cell wall.

Plasmid A self-replicating, extrachromosomal, hereditary circular DNA found in certain bacteria and fungi, generally not required for survival of the organism.

Plasmodesma (Plural = plasmodesmata) A fine protoplasmic thread connecting two protoplasts and passing through the wall which separates the two protoplasts.

Plasmodium A naked, slimy mass of protoplasm containing numerous nuclei.

Plasmolysis The shrinking and separation of the cytoplasm from cell wall due to exosmosis of water from the protoplast.

Plerome The plant tissues inside the cortex.

Polycyclic Completes many (life or disease) cycles in one year.

Polyetic Requires many years to complete one life or disease cycle.

Polygenic A character controlled by many genes.

Polyhedron A spheroidal particle or crystal with many plane faces.

Polymerase An enzyme that joins single small molecules into chains of such molecules (e.g., DNA, RNA).

Polysaccharide A large organic molecule consisting of many units of a simple sugar.

Polysome (or polyribosome) A cluster of ribosomes associated with a messenger RNA.

Precipitin The reaction in which an antibody causes visible precipitation of antigens.

Primary infection The first infection of a plant by the overwintering or oversummering pathogen.

Primary inoculum The overwintering or oversummering pathogen, or its spores that cause primary infection.

Probe A radioactive nucleic acid used to detect the presence of a complementary strand by hybridization.

Promoter A region on a DNA or RNA which is recognized by RNA polymerase in order to initiate transcription.

Promycelium The short hypha produced by the teliospore; the basidium.

Propagative virus A virus that multiplies in its insect vector.

Propagule The part of an organism that may be disseminated and reproduce the organism.

Protectant A substance that protects an organism against infection by a pathogen.

Protein subunit A small protein molecule that is the structural and chemical unit of the protein coat of a virus.

Protoplast A plant cell from which the cell wall has been removed. The organized living unit of a single cell; the cytoplasmic membrane and the cytoplasm, nucleus, and other organelles inside it.

Purification The isolation and concentration of virus particles in a pure form, free from cell components.

Pustule Small blisterlike elevation of epidermis created as spores form underneath and push outward.

Pycnium Also called a spermagonium.

Pycnidium An asexual, spherical, or flask-shaped fruiting body lined inside with conidiophores and producing conidia.

Pycniospore Also called a spermatium. A spore produced in a pycnium.

Quarantine Control of import and export of plants to prevent spread of diseases and pests.

Race A genetically and often geographically distinct mating group within a species; also a group of pathogens that infect a given set of plant varieties.

Recognition factors Specific molecules or structures on the host (or pathogen) that can be recognized by the pathogen (or host).

Resistance The ability of an organism to exclude or overcome, completely or in some degree, the effect of a pathogen or other damaging factor.

Resistant Possessing qualities that hinder the development of a given pathogen. Infected little or not at all.

Resting spore A sexual or other thick-walled spore of a fungus that is resistant to extremes in temperature and moisture and which often germinates only after a period of time from its formation.

Restriction enzymes A group of enzymes from bacteria which break internal bonds of DNA at highly specific points.

Reverse transcription Copying of an RNA into DNA.

Rhizoid A short, thin hypha growing in a rootlike fashion toward the substrate.

Rhizosphere The soil near a living root.

Ribosome A subcellular particle involved in protein synthesis.

Rickettsiae Microorganisms similar to bacteria in most respects but generally capable of multiplying only inside living host cells; parasitic or symbiotic.

Ringspot A circular area of chlorosis with a green center; a symptom of many virus diseases.

RNA (Ribonucleic acid) A nucleic acid involved in protein synthesis; also, the most common nucleic acid (genetic material) of plant viruses.

RNase (Ribonuclease) An enzyme that breaks down RNA.

Rosette Short, bunchy habit of plant growth.

Rot The softening, discoloration, and often disintegration of a succulent plant tissue as a result of fungal or bacterial infection.

Russet Brownish roughened areas on skin of fruit as a result of cork formation.

Rust A disease giving a "rusty" appearance to a plant and caused by one of the Uredinales (rust fungi).

Sanitation The removal and burning of infected plant parts, decontamination of tools, equipment, hands, etc.

Saprophyte An organism that uses dead organic material for food.

Scab A roughened, crustlike diseased area on the surface of a plant organ. A disease in which such areas form.

Scion A piece of twig or shoot inserted on another in grafting.

Sclerotium A compact mass of hyphae with or without host tissue, usually with a darkened rind, and capable of surviving under unfavorable environmental conditions.

Scorch "Burning" of leaf margins as a result of infection or unfavorable environmental conditions.

Secondary infection Any infection caused by inoculum produced as a result of a primary or a subsequent infection; an infection caused by secondary inoculum.

Secondary inoculum Inoculum produced by infections that took place during the same growing season.

Sedentary Staying in one place; stationary.

Septate Having cross walls.

Septum A cross wall (in a hypha or spore).

Serology A method using the specificity of the antigen-antibody reaction for the detection and identification of antigenic substances and the organisms that carry them.

Serum The clear, watery portion of the blood remaining after coagulation.

Sexual Participating in or produced as a result of a union of nuclei in which meiosis takes place.

Shock symptoms The severe, often necrotic symptoms produced on the first new growth following infection with some viruses; also called acute symptoms.

Shot-hole A symptom in which small diseased fragments of leaves fall off and leave small holes in their place.

Sieve plate Perforated wall area between two phloem sieve cells through which they are connected.

Sieve tube A series of phloem cells forming a long cellular tube through which food materials are transported.

Sign The pathogen or its parts or products seen on a host plant.

Slime molds Fungi of the class Myxomycetes; also, superficial diseases caused by these fungi on lowlying plants.

Smut A disease caused by the smut fungi (Ustilaginales); it is characterized by masses of dark, powdery spores.

Soil inhabitants Microorganisms able to survive in the soil indefinitely as saprophytes.

Soil transients Parasitic microorganisms that can live in the soil for short periods.

Somaclonal variation Variability in clones generated from a single mother plant, leaf, etc., by tissue culture.

Somatic hybridization Production of hybrid cells by fusion of two protoplasts with different genetic makeup.

Sooty mold A sooty coating on foliage and fruit formed by the dark hyphae of fungi that live in the honeydew secreted by insects such as aphids, mealybugs, scales, and whiteflies.

Sorus A compact mass of spores or fruiting structure found especially in the rusts and smuts.

Spermatium (formerly pycniospore) The male gamete or gametangium of the rust fungi.

Spermogonium (formerly pycnium) A fruiting body of rust fungi in which the gametes or gametangia are produced.

Spiroplasmas Pleomorphic, wall-less microorganisms that are present in the phloem of diseased plants. They are often helical in culture and are thought to be a kind of mycoplasma.

Sporangiophore A specialized hypha bearing one or more sporangia.

Sporangiospore Nonmotile, asexual spore borne in a sporangium.

Sporangium A container or case of asexual spores. In some cases it functions as a single spore.

Spore The reproductive unit of fungi consisting of one or more cells; it is analogous to the seed of green plants.

Sporidium The basidiospore of the smut fungi.

Sporodochium A fruiting structure consisting of a cluster of conidiophores woven together on a mass of hyphae.

Sporophore A hypha or fruiting structure bearing spores.

Sporulate To produce spores.

Stem-pitting A symptom of some viral diseases characterized by depressions on the stem of the plant.

Sterigma A slender protruberance on a basidium that supports the basidiospore.

Sterile fungi A group of fungi that are not known to produce any kind of spores.

Sterilization The elimination of pathogens and other living organisms from soil, containers, etc., by means of heat or chemicals.

Strain The descendants of a single isolation in pure culture; an isolate. Also a group of similar isolates; a race. In plant viruses, a group of virus isolates having most of their antigens in common.

Stroma A compact mycelial structure on or in which fructifications are usually formed.

Stylet A long, slender, hollow feeding structure of nematodes and some insects.

Stylet-borne A virus borne on the stylet of its vector; a noncirculative virus.

Substrate The material or substance on which a microorganism feeds and develops. Also, a substance acted upon by an enzyme.

Suppressive soils Soils in which certain diseases are suppressed because of the presence in the soil of microorganisms antagonistic to the pathogen.

Suscept Any plant that can be attacked by a given pathogen; a host plant.

Susceptible Lacking the inherent ability to resist disease or attack by a given pathogen; nonimmune.

Susceptibility The inability of a plant to resist the effect of a pathogen or other damaging factor.

Symbiosis A mutually beneficial association of two or more different kinds of organisms.

Symptom The external and internal reactions or alterations of a plant as a result of a disease.

Symptomless carrier A plant which, although infected with a pathogen (usually a virus), produces no obvious symptoms.

Syncytium A multinucleate mass of protoplasm surrounded by a common cell wall.

Synergism The concurrent parasitism of a host by two pathogens in which the symptoms or other effects produced are of greater magnitude than the sum of the effects of each pathogen acting alone.

Systemic Spreading internally throughout the plant body; said of a pathogen or a chemical.

Teliospore The sexual, thick-walled resting spore of the rust and smut fungi.

Telium The fruiting structure in which rust teliospores are produced.

Tissue A group of cells of similar structure which performs a special function.

Tolerance The ability of a plant to sustain the effects of a disease without dying or suffering serious injury or crop loss. Also, the amount of toxic residue allowable in or on edible plant parts under the law.

Toxicity The capacity of a compound to produce injury.

Toxin A compound produced by a microorganism and being toxic to a plant or animal.

Transcription Copying of a gene into RNA. Also, copying of a viral RNA into a complementary RNA.

Transduction The transfer of genetic material from one bacterium to another by means of a bacteriophage.

Transfer RNA (tRNA) The RNA that moves amino acids to the ribosome to be placed in the order prescribed by the messenger RNA.

Transformation The change of a cell through uptake and expression of additional genetic material.

Translation Copying of mRNA into protein.

Translocation Transfer of nutrients or virus through the plant.

Transmission The transfer or spread of a virus or other pathogen from one plant to another.

Transpiration The loss of water vapor from the surface of leaves and other aboveground parts of plants.

Transposable element A segment of chromosomal DNA that can move around (transpose) in the genome and integrate at different sites on the chromosomes.

Tumor An uncontrolled overgrowth of tissue or tissues.

Tylosis An overgrowth of the protoplast of a parenchyma cell into an adjacent xylem vessel or tracheid.

Uredium The fruiting structure of the rust fungi in which uredospores are produced.

Uredospore A dikaryotic, repeating spore of the rust fungi.

Variability The property or ability of an organism to change its characteristics from one generation to the other.

Vascular Term applied to a plant tissue or region consisting of conductive tissue; also, to a pathogen that grows primarily in the conductive tissues of a plant.

Vector An animal able to transmit a pathogen. In genetic engineering, *vector (or cloning vehicle)*—A self-replicating DNA molecule, such as a plasmid or virus, used to introduce a fragment of foreign DNA into a host cell.

Vegetative Asexual; somatic.

Vertical resistance Complete resistance to some races of a pathogen but not to others.

Vesicle A bubblelike structure produced by a zoosporangium and in which the zoospores are released or are differentiated.

Vessel A xylem element or series of such elements whose function is to conduct water and mineral nutrients.

Virescent A normally white or colored tissue that develops chloroplasts and becomes green.

Virion A virus particle.

Viroids Small, low-molecular-weight ribonucleic acids (RNA) that can infect plant cells, replicate themselves, and cause disease.

Virulence The degree of pathogenicity of a given pathogen.

Virulent Capable of causing a severe disease; strongly pathogenic.

Viruliferous Said of a vector containing a virus and capable of transmitting it.

Virus A submicroscopic obligate parasite consisting of nucleic acid and protein.

Virusoid The extra-small circular RNA component of some isometric RNA viruses.

Xylem A plant tissue consisting of tracheids, vessels, parenchyma cells, and fibers; wood.

Wilt Loss of rigidity and drooping of plant parts generally caused by insufficient water in the plant.

Witches' broom Broomlike growth or massed proliferation caused by the dense clustering of branches of woody plants.

Yellows A plant disease characterized by yellowing and stunting of the host plant.

Zoosporangium A sporangium which contains or produces zoospores.

Zoospore A spore bearing flagella and capable of moving in water.

Zygospore The sexual or resting spore of zygomycetes produced by the fusion of two morphologically similar gametangia.

Zygote A diploid cell resulting from the union of two gametes.

INDEX